WELCOME to 2nd Semester Calculus

2nd semester calculus is a continuation of the 1st semester and follows a similar structure of Concepts, Applications and Techniques. The major topics this semester are

> Applications of integrals
> Methods of integration
> The calculus of functions given in Polar Coordinates and as Parametric Equations
> Infinite Sequences and Series
> Vectors in 3 Dimensions

So welcome back for more calculus.

Succeeding at calculus.

Probably more than in your previous mathematics classes you need
* to think about the concepts as well as the techniques,
* to think about the patterns as well as the individual steps,
* to think about the meaning of the concepts and techniques in the context of particular applications,
* to think about how the ideas and techniques apply to functions given by graphs and tables as well as by formulas, and
* to spend enough time (1 to 2 hours each day) doing problems to sort out the concepts and to master the techniques and to get better and more efficient with the algebra skills that are vital to success.

Sometimes all this mental stretching can seem overwhelming, but stick with it (and do lots of problems). It can even become fun.

This text book

A free, color PDF version is available online at
http://scidiv.bellevuecollege.edu/dh/Calculus_all/Calculus_all.html

This text is licensed under a Creative Commons Attribution-Share Alike 3.0 United States License.

You are **free**:
> **to Share** – to copy, distribute, display and perform the work
> **to Remix** – to make derivative works

Under the following conditions:
> **Attribution:** You must attribute the work in the manner specified by the author (but not in any way that suggests that they endorse you or your work)
> **Share Alike:** If you alter, transform or build upon this work, you must distribute the resulting work only under the same, similar or compatible license.

CONTEMPORARY CALCULUS II: Contents

Note:

Each section contains Practice Problems throughout the section. The solutions to these Practice Problems are at the end of that section, after the Problem Set for the section.

Each section also contains a Problem Set. The solutions to the odd problems of each Problem Set are at the end of each chapter.

These materials are also available free and in color on the web at:
http://scidiv.bellevuecollege.edu/dh/Calculus_all/Calculus_all.html

Chapter 10: Infinite Series & Power Series

Chapter 11: Vectors in 3D

Calculus Reference Facts

4.7   FIRST APPLICATIONS OF DEFINITE INTEGRALS

The development of calculus by Newton and Leibniz was a vital step in the advancement of pure mathematics, but Newton also advanced the applied sciences and mathematics. Not only did he discover theoretical results, but he immediately used those results to answer important applied questions about gravity and motion. The success of these applications of mathematics to the physical sciences helped establish what we now take for granted: mathematics can and should be used to answer questions about the world.

Newton applied mathematics to the outstanding problems of his day, problems primarily in the field of physics. In the intervening 300 years, thousands of people have continued these theoretical and applied traditions and have used mathematics to help develop our understanding of all of the physical and biological sciences as well as the behavioral sciences and business. Mathematics is still used to answer new questions in physics and engineering, but it is also important for modeling ecological processes, for understanding the behavior of DNA, for determining how the brain works, and even for devising strategies for voting effectively. The mathematics you are learning now can help you become part of this tradition, and you might even use it to add to our understanding of different areas of life.

It is important to understand the successful applications of integration in case you need to use those particular applications. It is also important that you understand the **process** of building models with integrals so you can apply it to new problems. Conceptually, converting an applied problem to a Riemann sum (Fig. 1) is the most valuable step. Typically, it is also the most difficult.

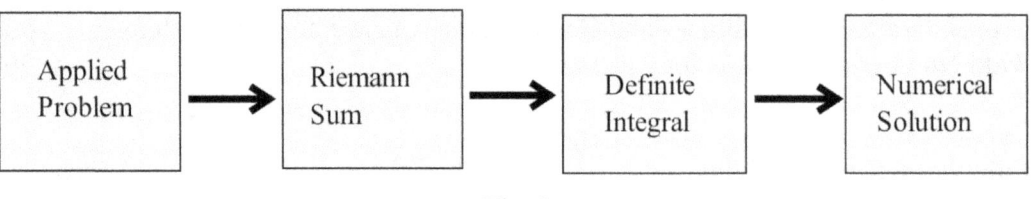

$$\boxed{\begin{array}{c}\text{Applied}\\\text{Problem}\end{array}} \longrightarrow \boxed{\begin{array}{c}\text{Riemann}\\\text{Sum}\end{array}} \longrightarrow \boxed{\begin{array}{c}\text{Definite}\\\text{Integral}\end{array}} \longrightarrow \boxed{\begin{array}{c}\text{Numerical}\\\text{Solution}\end{array}}$$

Fig. 1

Area between f and g

We have already used integrals to find the area between the graph of a function and the horizontal axis. Integrals can also be used to find the area between two graphs.

If $f(x) \geq g(x)$ for all x in [a,b], then we can approximate the area between f and g by partitioning the interval [a,b] and forming a Riemann sum (Fig. 2). The height of each rectangle is $f(c_i) - g(c_i)$ so the area of the ith rectangle is (height)·(base) = $\{f(c_i) - g(c_i)\}\cdot\Delta x_i$. This approximation of the total area is

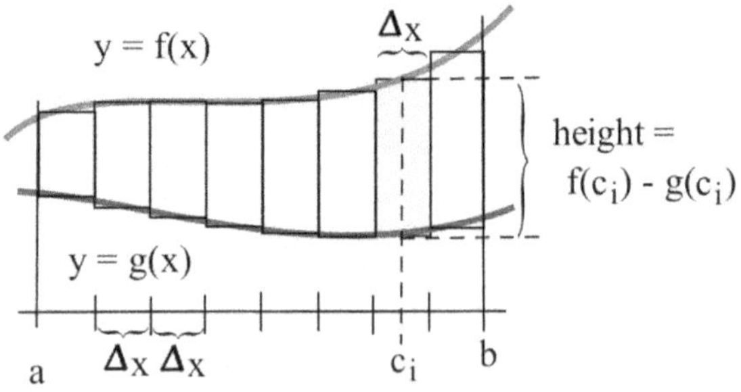

$$\text{area} \approx \sum_{i=1}^{n} \{f(c_i) - g(c_i)\}\cdot\Delta x_i \ , \ \text{a Riemann sum.}$$

Fig. 2

The limit of this Riemann sum, as the mesh of the partitions approaches 0, is the definite integral

$$\int_a^b \{ f(x) - g(x) \}\, dx \ .$$

We will sometimes use an arrow to indicate "the limit of the Riemann sum as the mesh of the partitions approaches zero," and will write

$$\sum_{i=1}^{n} \{ f(c_i) - g(c_i) \} \cdot \Delta x_i \longrightarrow \int_a^b \{ f(x) - g(x) \}\, dx \ .$$

If   **$f(x) \geq g(x)$** on the interval $[a,b]$,

then $\left\{ \begin{array}{l} \text{area bounded by the graphs} \\ \text{of f and g and vertical} \\ \text{lines at } x = a \text{ and } x = b \end{array} \right\} = \int_a^b \{ f(x) - g(x) \}\, dx \ .$

Example 1:   Find the area bounded between the graphs of $f(x) = x$ and $g(x) = 3$ for $1 \leq x \leq 4$. (Fig. 3)

Solution: It is clear from the figure that the area between f and g is 2.5 square inches. Using the theorem,

area between f and g for $1 \leq x \leq 3$ is

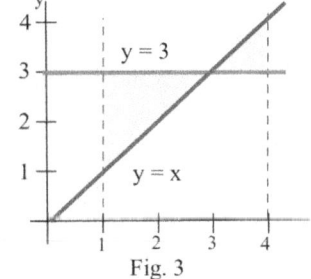

$$\int_1^3 \{ 3 - x \}\, dx = 3x - \frac{x^2}{2} \Big|_1^3 = \left(\frac{9}{2}\right) - \left(\frac{5}{2}\right) = 2 \text{ , and}$$

area between f and g for $3 \leq x \leq 4$ is

$$\int_3^4 \{ x - 3 \}\, dx = \frac{x^2}{2} - 3x \Big|_3^4 = \left(\frac{-8}{2}\right) - \left(\frac{-9}{2}\right) = \frac{1}{2} \ .$$

Fig. 3

The two integrals also tell us that the total area between f and g is 2.5

square inches.

The single integral $\int_1^4 \{ 3 - x \}\, dx = 1.5$ which is not the **area** we want in this problem. The value

of the **integral is 1.5**, and the value of the **area is 2.5** .

Practice 1:   Use integrals and the graphs of $f(x) = 1 + x$ and $g(x) = 3 - x$

to determine the area between the graphs of f and g for $0 \leq x \leq 3$.

Example 2:   Two objects start from the same location and travel along the same

path with velocities $v_A(t) = t + 3$ and

$v_B(t) = t^2 - 4t + 3$ meters per second (Fig. 4).

How far ahead is A after 3 seconds? After 5 seconds?

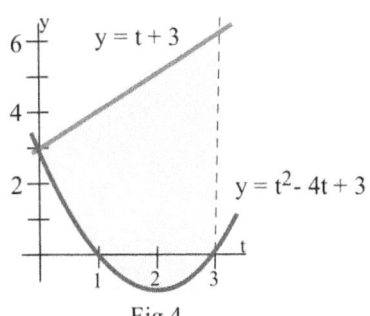

Fig.4

Solution: Since $v_A(t) \geq v_B(t)$, the "area" between the graphs of v_A and v_B represents the distance

between the objects.

After 3 seconds, the distance apart $= \int_0^3 v_A(t) - v_b(t)\ dt\ = \int_0^3 (t+3) - (t^2-4t+3)\ dt$

$$= \int_0^3 5t - t^2\ dt\ = \frac{5}{2}\ t^2 - \frac{t^3}{3}\ \Big|_0^3\ = (\frac{5}{2} \cdot 9 - \frac{27}{3}) - (\frac{5}{2} \cdot 0 - \frac{0}{3}) = 13\frac{1}{2}\ \text{meters.}$$

After 5 seconds, the distance apart $= \int_0^5 v_A(t) - v_b(t)\ dt\ = \frac{5}{2}\ t^2 - \frac{t^3}{3}\ \Big|_0^5\ = 20\frac{5}{6}\ \text{meters.}$

If $f(x) \geq g(x)$, we can use the simpler argument that the area of region A is $\int_a^b f(x)\ dx$ and the area of

region B is $\int_a^b g(x)\ dx$, so the area of region C, the area between f and g, is

$$\text{area of C} = (\text{area of A}) - (\text{area of B}) = \int_a^b f(x)\ dx\ - \int_a^b g(x)\ dx\ = \int_a^b f(x) - g(x)\ dx\ .$$

If the same function is not always greater, then we need to be very careful and find the intervals where

$f \geq g$ and the intervals where $g \geq f$.

Example 3:   Find the **area** of the shaded region in Fig. 5.

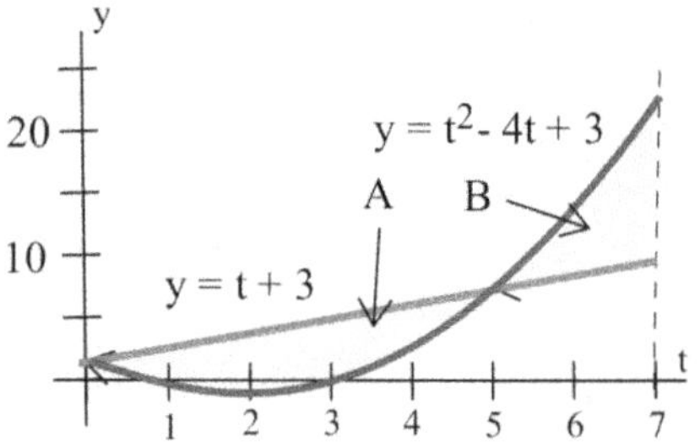

Solution: For $0 \leq x \leq 5, f(x) \geq g(x)$ so the **area** of A is

Fig. 5

$$\int_0^5 f(x) - g(x)\ dx\ = \int_0^5 (x+3) - (x^2-4x+3)\ dx$$

$$= \int_0^5 5x-x^2\ dx\ = \frac{5}{2}\ x^2 - \frac{x^3}{3}\ \Big|_0^5\ = 20\frac{5}{6}\ .$$

For $5 \leq x \leq 7,\ g(x) \geq f(x)$ so the **area** of B is

$$\int_5^7 g(x) - f(x)dx\ = \int_5^7 (x^2-4x+3) - (x+3)\ dx\ = \int_5^7 x^2 - 5x\ dx\ = \frac{x^3}{3} - \frac{5}{2}\ x^2\ \Big|_5^7\ = 12\frac{4}{6}\ .$$

Altogether, the total **area** between f and g for $0 \leq x \leq 7$ is

$$\int_0^5 f(x) - g(x)\ dx\ + \int_5^7 g(x) - f(x)\ dx\ = 20\frac{5}{6}\ + 12\frac{4}{6}\ = 33\frac{1}{2}\ .$$

Average Value of a Function

We know the average of n numbers, $a_1, a_2, \ldots, a_n$, is their sum divided by n: average (mean) $= \dfrac{1}{n} \sum_{k=1}^{n} a_k$.

Finding the average of a function on an interval, an infinite number of values, requires an integral.

To find a Riemann sum approximation of the average value of f on the interval [a,b], we can partition [a,b] into n equally long subintervals of length $\Delta x = (b-a)/n$, pick a value c_i of x in each subinterval, and find the average of the numbers $f(c_i)$. Then

$$\text{average of } f \approx \frac{f(c_1)+f(c_2)+\ldots+f(c_n)}{n} = \frac{1}{n}\sum_{k=1}^{n} f(c_i) = \sum_{k=1}^{n} f(c_i)\cdot\frac{1}{n}$$

This last sum is not a Riemann sum since it does not have the form $\sum f(c_i)\cdot\Delta x_i$, but it can be manipulated into one:

$$\sum_{i=1}^{n} f(c_i)\cdot\frac{1}{n} = \sum_{i=1}^{n} f(c_i)\cdot\frac{b-a}{n}\cdot\frac{1}{b-a} = \frac{1}{b-a}\sum_{i=1}^{n} f(c_i)\cdot\frac{b-a}{n} = \frac{1}{b-a}\sum_{i=1}^{n} f(c_i)\cdot\Delta x.$$

Then $\{\text{average of } f\} \approx \dfrac{1}{b-a}\sum_{k=1}^{n} f(c_i)\cdot\Delta x \longrightarrow \dfrac{1}{b-a}\int_{a}^{b} f(x)\,dx = \{\text{average of f}\}$

as the number of points n gets larger and the mesh , (b–a)/n , approaches 0.

Definition: **Average (Mean) Value of a Function**

For an integrable function f on the interval [a,b],

the average value of f on [a,b] is $\dfrac{1}{b-a}\int_{a}^{b} f(x)\,dx$

The average value of a positive f has a nice geometric interpretation. Imagine that the area under f (Fig. 6a) is a liquid that can "leak" through the graph to form a rectangle with the same area (Fig. 6b). If the height of the rectangle is H, then the area of the rectangle is H·(b–a). We know the area of the rectangle is the same as the area under f so

$$H\cdot(b-a) = \int_{a}^{b} f(x)\,dx \text{ . Then}$$

$$H = \frac{1}{b-a}\int_{a}^{b} f(x)\,dx \text{ , the average value of f on [a,b].}$$

The average value of positive f is the height H of the rectangle whose area is the same as the area under f.

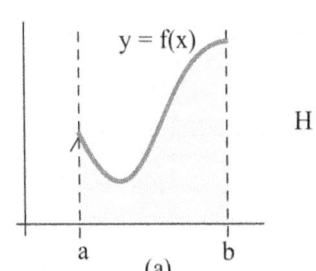 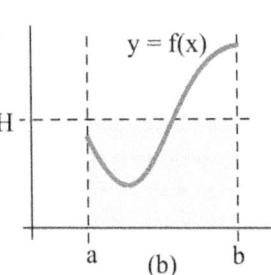

Fig. 6

Example 4:   Find the average value of $f(x) = \sin(x)$ on the

interval $[0, \pi]$. (Fig. 7)

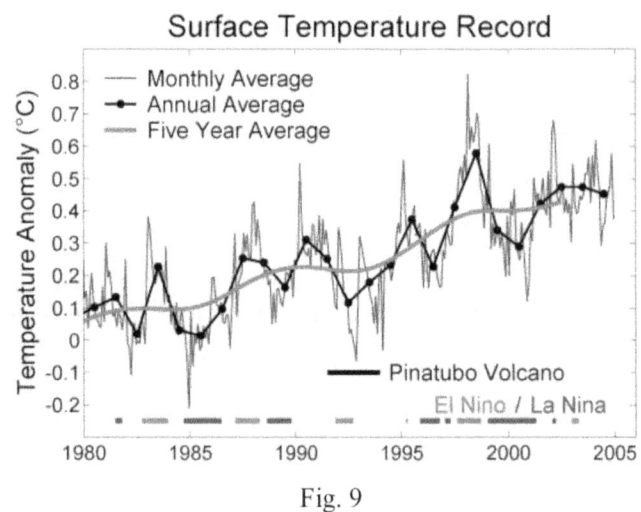

$y = \sin(x)$

Fig. 7

Solution: Average value $= \dfrac{1}{\pi - 0} \displaystyle\int_0^{\pi} \sin(x)\, dx$

$= \dfrac{1}{\pi}\left(-\cos(x)\right)\Big|_0^{\pi} \; = \dfrac{1}{\pi}\{ -(-1) - (-1) \} = \dfrac{2}{\pi} \approx 0.6366$.

A rectangle with height $2/\pi \approx 0.64$ on the interval $[0, \pi]$ encloses the same area as one arch of the

sine curve. The average value of $\sin(x)$ on the interval $[0, 2\pi]$ is 0 since $\dfrac{1}{2\pi} \displaystyle\int_0^{2\pi} \sin(x)dx \; = 0.$

Practice 2:   During a 9 hour work day, the production rate at time t hours was $r(t) = 5 + \sqrt{t}$ cars

per hour. Find the average hourly production rate.

Function averages, involving means and more complicated averages, are used to "smooth" data so that
underlying patterns are more obvious and to remove high frequency "noise" from signals. In these
situations, the original function f is replaced by some "average of f." If f is rather jagged time data, then

the ten year average of f is the integral $g(x) = \dfrac{1}{10} \displaystyle\int_{x-5}^{x+5} f(t)\, dt$ .an average of f over 5 units on each side

of x. For example, Fig. 9 shows the graphs of a

Monthly Average (rather "noisy" data) of surface

temperature data, an Annual Average (still rather

"jagged"), and a Five Year Average (a much

smoother function). Typically the average function

reveals the pattern much more clearly than the

original data. This use of a "moving average"

value of "noisy" data (weather information, stock

prices) is a very common.

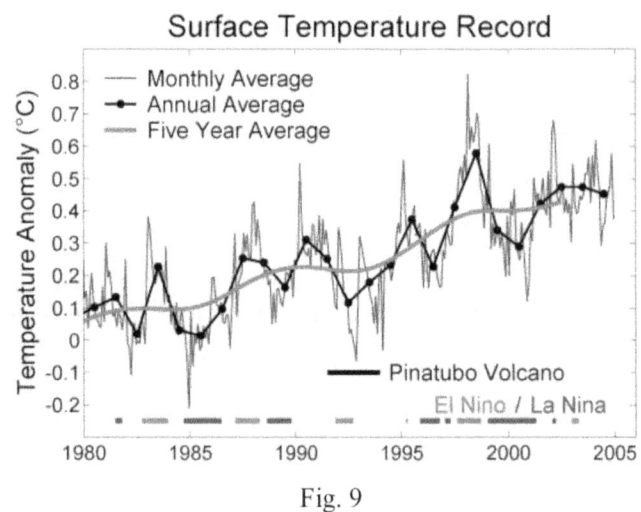

Surface Temperature Record

Fig. 9

Work

The amount of work done on an object is the force

applied to the object times the distance the object is moved while the force is applied (Fig. 10) or, more

succinctly, **work = (force)·(distance)** .

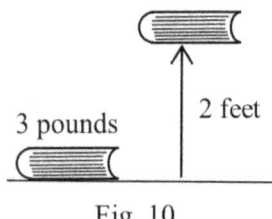

3 pounds

2 feet

Fig. 10

If you lift a 3 pound book 2 feet, then the force is 3 pounds, the weight of the book,

and the distance moved is 2 feet, so you have done (3 pounds)·(2 feet) = 6 foot–

pounds of work. When the applied force and the distance are both constants, then

calculating work is simply a matter of multiplying.

Practice 3:   How much work is done lifting a 10 pound object from the ground to the top of a 30 foot building (assume the cable is weightless).

If either the force or the distance is variable, then integration is needed.

Example 5:   How much work is done lifting a 10 pound object from the ground to the top of a 30 foot building if the cable weighs 2 pounds per foot. (Fig. 11)

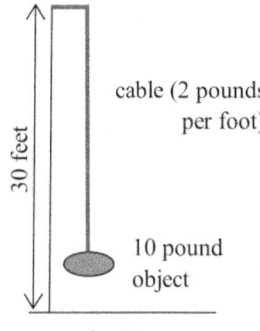

cable (2 pounds per foot)

30 feet

10 pound object

Fig. 11

Solution: This is more difficult than the Practice problem. When the object is at ground level, a force of 70 pounds (10 pounds plus the weight of 30 feet of cable) must be applied, but when the object is 29 feet above the ground, only 12 pounds of force are needed. In general, if the object is x feet above the ground (Fig. 12), then $30 - x$ feet of cable, weighing $2(30-x)$ pounds, is used so the required force is

$$f(x) = 10 + 2(30-x) = 70 - 2x \text{ pounds.}$$

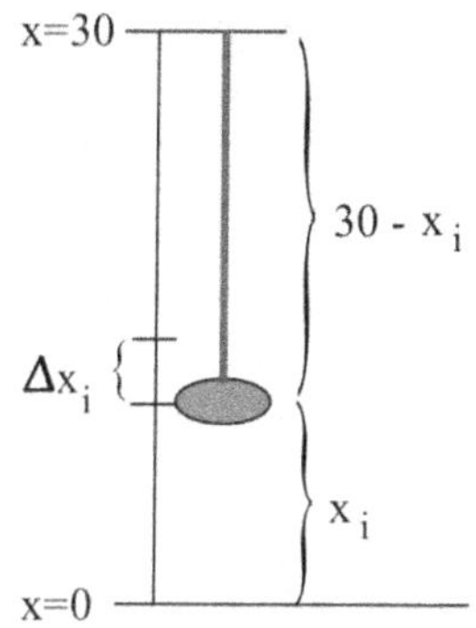

x=30

$30 - x_i$

Δx_i

x=0

x_i

Fig. 12

Let's partition the height of the building into small increments so the force needed in each subinterval does not change much. The force in the ith subinterval will be approximately $f(c_i)$ for some c_i in the subinterval, and the distance moved will be the length of the subinterval, Δx_i . The work done to move the object through the subinterval will be $f(c_i) \cdot \Delta x_i$, and the total work will be the sum of the work on each subinterval:

$$\text{work} \approx \sum \text{(subinterval work)} = \sum f(c_i) \, \Delta x_i = \sum \{70 - 2c_i\} \, \Delta x_i \quad \text{(a Riemann sum)}$$

$$\longrightarrow \int_0^{30} \{70 - 2x\} \, dx \quad \text{as the mesh approaches 0.}$$

We have approximated the solution to an applied problem by a Riemann sum, and obtained an exact solution by taking the limit of the Riemann sum to get a definite integral. Now we just need to evaluate the definite integral:

$$\int_0^{30} (70 - 2x) \, dx = 70x - x^2 \Big|_0^{30} = \{70 \cdot 30 - (30)^2\} - \{70 \cdot 0 - (0)^2\} = (2100 - 900) - (0) = 1200 \text{ foot-pounds.}$$

Practice 4:   Suppose the building in Example 5 is 50 feet tall and the cable weighs 3 pounds per foot.

(a)   How much force is needed when the 10 pound object is x feet above the ground?

(b)   Write an integral for the work done raising the object from the ground to a height of 10 feet. From a height of 10 feet to a height of 20 feet.

In the previous Example and Practice problem, the force was variable and the distance was Δx. In later sections we will examine situations where the force is constant and the distance changes.

Summary

These area, average and work applications simply introduce a few of the applications of definite integrals and illustrate the pattern of going from an applied problem to a Riemann sum, on to a definite integral and, finally, to a number. More applications will be explored in Chapter 5. The rest of this chapter will examine additional ways of finding antiderivatives and of finding the values of definite integrals when an antiderivative can not be found.

x	f(x)	g(x)	h(x)
0	5	2	5
1	6	1	6
2	6	2	8
3	4	2	6
4	3	3	5
5	2	4	4
6	2	5	2

Table 1

PROBLEMS

In problems 1 – 4, use the values in Table 1 to estimate the areas.

1. Estimate the **area** between f and g for $1 \le x \le 4$.

2. Estimate the **area** between f and g for $1 \le x \le 6$.

3. Estimate the **area** between f and h for $0 \le x \le 4$.

4. Estimate the **area** between g and h for $0 \le x \le 6$.

5. Estimate the area of the island in Fig. 13.

6. Estimate the area of the island in Fig. 13 if the distances between the lines is 50 feet instead of 40 feet,.

In problems 7 – 18, sketch the graph of each function and find the **area** between the graphs of f and g for x in the given interval.

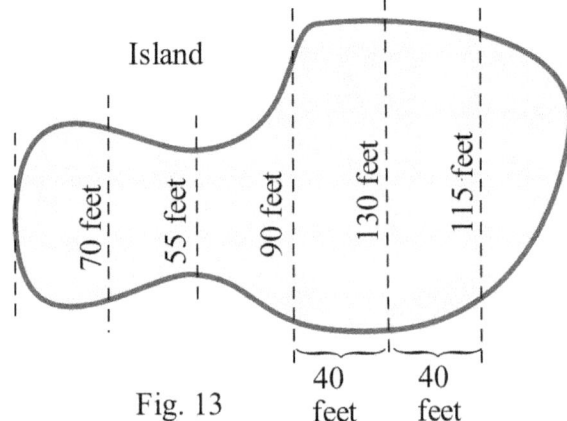

Island

Fig. 13

70 feet   55 feet   90 feet   130 feet   115 feet

40 feet   40 feet

7. $f(x) = x^2 + 3$, $g(x) = 1$ and $-1 \le x \le 2$.

8. $f(x) = x^2 + 3$, $g(x) = 1 + x$ and $0 \le x \le 3$.

9. $f(x) = x^2$, $g(x) = x$ and $0 \le x \le 2$.

10. $f(x) = 4 - x^2$, $g(x) = x + 2$ and $0 \le x \le 2$.

11. $f(x) = \dfrac{1}{x}$, $g(x) = x$ and $1 \le x \le e$.

12. $f(x) = \sqrt{x}$, $g(x) = x$ and $0 \le x \le 4$.

13. $f(x) = x + 1$, $g(x) = \cos(x)$ and $0 \le x \le \pi/4$.

14. $f(x) = (x - 1)^2$, $g(x) = x + 1$ and $0 \le x \le 3$.

15. $f(x) = e^x$, $g(x) = x$ and $0 \le x \le 2$.

16. $f(x) = \cos(x)$, $g(x) = \sin(x)$ and $0 \le x \le \pi/4$.

17. $f(x) = 3$, $g(x) = \sqrt{1 - x^2}$ and $0 \le x \le 1$.   18. $f(x) = 2$, $g(x) = \sqrt{4 - x^2}$ and $-2 \le x \le 2$.

In problems 19 – 22, use the values in Table 1 to estimate the average values.

19. Estimate the average value of f on the interval [0.5, 4.5].

20. Estimate the average value of f on the interval [0.5, 6.5].

21. Estimate the average value of f on the interval [1.5, 3.5].

22. Estimate the average value of f on the interval [3.5, 6.5].

In problems 23 – 32, find the **average value** of f on the given interval.

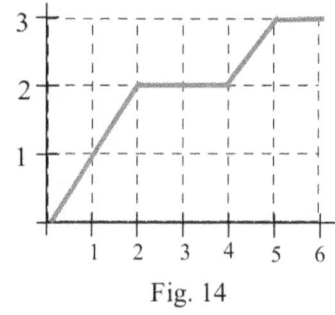
Fig. 14

23. $f(x)$ in Fig. 14 for $0 \le x \le 2$.   24. $f(x)$ in Fig. 14 for $0 \le x \le 4$.

25. $f(x)$ in Fig. 14 for $1 \le x \le 6$.   26. $f(x)$ in Fig. 14 for $4 \le x \le 6$.

27. $f(x) = 2x + 1$ for $0 \le x \le 4$.   28. $f(x) = x^2$ for $0 \le x \le 2$.

29. $f(x) = x^2$ for $1 \le x \le 3$.   30. $f(x) = \sqrt{x}$ for $0 \le x \le 4$.

31. $f(x) = \sin(x)$ for $0 \le x \le \pi$.   32. $f(x) = \cos(x)$ for $0 \le x \le \pi$.

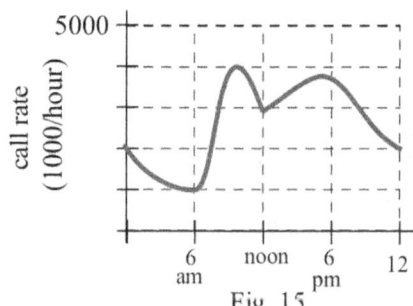
Fig. 15

33. Calculate the average value of $f(x) = \sqrt{x}$ on the interval $[0, C]$ for $C = 1, 9, 81, 100$. What is the pattern?

34. Calculate the average value of $f(x) = x$ on the interval $[0, C]$ for $C - 1, 10, 80, 100$. What is the pattern?

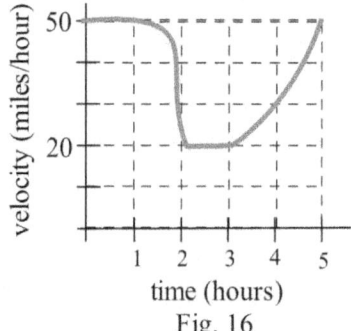
Fig. 16

35. Fig. 15 shows the number of telephone calls per minute at a large company. (a) Estimate the average number of calls per minute from 8 am to 5 pm. (b) From 9 am to 1 pm.

36. Fig. 16 shows the velocity of a car during a 5 hour trip.
    (a) Estimate how far the car traveled during the 5 hours.
    (b) At what **constant** velocity should you drive in order to travel the same distance in 5 hours?

37. (a) How much work is done lifting a 20 pound bucket from the ground to the top of a 30 foot building with a cable which weighs 3 pounds per foot?
    (b) How much work is done lifting the same bucket from the ground to a height of 15 feet with the same cable?

38. (a) How much work is done lifting a 60 pound chair from the ground to the top of a 20 foot building with a cable which weighs 1 pound per foot?

    (b) How much work is done lifting the same chair from the ground to a height of 5 feet with the same cable?

39. (a) How much work is done lifting a 10 pound calculus book from the ground to the top of a 30 foot building with a cable which weighs 2 pounds per foot?

    (b) From the ground to a height of 10 feet?   (c) From a height of 10 feet to a height of 20 feet?

40. How much work is done lifting an 80 pound injured child to the top of a 20 foot hole using a stretcher weighing 14 pounds and a cable which weighs 1 pound per foot?

41. How much work is done lifting an 60 pound injured child to the top of a 15 foot hole using a stretcher weighing 10 pounds and a cable which weighs 2 pound per foot?

42. How much work is done lifting an 120 pound injured adult to the top of a 30 foot hole using a stretcher weighing 10 pounds and a cable which weighs 2 pound per foot?

Section 4.7   **PRACTICE Answers**

Practice 1:   Using geometry (Fig. 17): $A = \frac{1}{2}(2)(1) = 1$ and $B = \frac{1}{2}(4)(2) = 4$ so total area $= A + B = \mathbf{5}$.

Using integrals:

$$A = \int_0^1 (3-x) - (1+x)\, dx = \int_0^1 (2-2x)\, dx = 2x - x^2 \Big|_0^1 = (2-1) - (0) = 1.$$

$$B = \int_1^3 (1+x) - (3-x)\, dx = \int_1^3 (2x-2)\, dx = x^2 - 2x \Big|_1^3 = (9-6) - (1-2) = 4.$$

The single integral $\int_0^3 (1+x) - (3-x)\, dx$ is **not** correct: $\int_0^3 (1+x) - (3-x)\, dx = 3$.

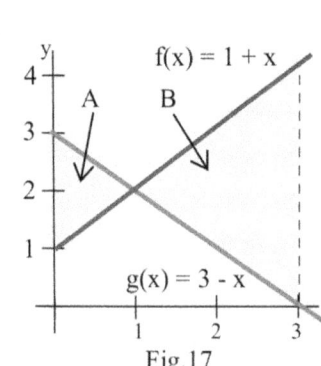

Fig.17

Practice 2:   Average value $= \dfrac{1}{b-a}\displaystyle\int_a^b f(x)\,dx \;=\dfrac{1}{9-0}\displaystyle\int_0^9 5+\sqrt{t}\;dt$

$$= \dfrac{1}{9}\int_0^9 5+t^{1/2}\;dt = \dfrac{1}{9}\left(5t+\dfrac{2}{3}t^{3/2}\right)\Big|_0^9$$

$$= \dfrac{1}{9}(45+\dfrac{2}{3}9^{3/2}) \;-\; \dfrac{1}{9}(0) = \dfrac{1}{9}(45+18) \;=\textbf{7 cars per hour.}$$

Practice 3:   Work = (force)·(distance) = (10 pounds)·(30 feet) = **300 foot·pounds.**

Practice 4:   (a) force = (force for cable) + (force for object)

= (length of cable)(density of cable) + 10 pounds

= $(50-x$ feet)(3 pounds/foot) + 10 pounds = **$160 - 3x$ pounds**

(b) "from the ground to a height of 10 feet:"

Work $\approx \sum$ (subinterval work) $= \sum f(c_i)\,\Delta x_i \;=\sum\{160-3c_i\}\,\Delta x_i$ (a Riemann sum)

$$\xrightarrow{\hspace{2cm}} \int_0^{10}\{160-3x\}\,dx \quad\text{as the mesh approaches 0.}$$

"From a height of 10 feet to a height of 20 feet:"   work $= \displaystyle\int_{10}^{20}\{160-3x\}\,dx$.

4.8   USING TABLES TO FIND ANTIDERIVATIVES

The inside covers of this book show the patterns for many antiderivatives, and there are reference books which contain many more than the ones here. The table of integrals is to help you now while you are learning calculus and to serve as a reference later when you are using calculus. Think of the tables as a dictionary — something to use when you need to spell a difficult word or need the meaning of a new word. It would be difficult to write a letter if you had to look up the spelling of every word, and it will be difficult to learn and do calculus if you have to look up every antiderivative. Tables of antiderivatives are limited and often take longer to use than finding an antiderivative from scratch, but they can also be very valuable and useful. This section shows how to transform some integrals into forms in the tables and how to use the "recursive" formulas in the tables.

The first examples and practice problems illustrate some of the techniques used to change an integral into a standard form. **These techniques are useful whether that standard form resides in a table or in your head.**

Example 1:   Use the table to find $\int \frac{1}{9 + x^2} \, dx$.

Solution: Table entry number 35, $\int \frac{1}{a^2 + x^2} \, dx = \frac{1}{a} \arctan\left(\frac{x}{a}\right) + C$

is the integral pattern of our problem for $a = 3$, so by replacing the a with 3 we have

$$\int \frac{1}{9 + x^2} \, dx = \int \frac{1}{3^2 + x^2} \, dx = \frac{1}{3} \arctan\left(\frac{x}{3}\right) + C.$$

Practice 1:   Use the table to find $\int \frac{1}{25 - x^2} \, dx$. Notice that a small change in the form of the integrand

(from + in the example to – in the practice problem) can lead to a very different looking result. $\left\{ \text{Table} \right.$

entry 37: $\int \frac{1}{a^2 - x^2} \, dx = \frac{1}{2a} \ln\left| \frac{a+x}{a-x} \right| + C \left. \right\}$

Sometimes the choice for a constant in the antiderivative pattern appears unusual.

Example 2:   Use the table to find $\int \frac{1}{5 + x^2} \, dx$.

Solution: The table entry 35, $\int \frac{1}{a^2 + x^2} \, dx = \frac{1}{a} \arctan\left(\frac{x}{a}\right) + C$, for the first example still works if

a^2 is 5. Then $a = \sqrt{5}$,

$$\int \frac{1}{5+x^2} \, dx = \int \frac{1}{(\sqrt{5})^2 + x^2} \, dx = \frac{1}{\sqrt{5}} \arctan\left(\frac{x}{\sqrt{5}}\right) + C.$$

Practice 2:   Use the table to find $\int \frac{1}{7-x^2} \, dx$.

Sometimes algebraic manipulations are needed to change the integrand we have into one that exactly matches the pattern in the table.

Example 3:   Use the table to find $\int \frac{1}{9+4x^2} \, dx$.

Solution: The pattern of this problem is basically an arctangent, but we need to **exactly** match the pattern in the table. For this problem we can use algebra or change of variables.

Algebra: We need x^2 rather than $4x^2$, and we can get it by factoring a 4 from each piece of the denominator: $\frac{1}{9+4x^2} = \frac{1}{4} \frac{1}{(9/4) + x^2}$. Then

$$\int \frac{1}{9+4x^2} \, dx = \frac{1}{4} \int \frac{1}{(9/4)+x^2} \, dx = \frac{1}{4} \int \frac{1}{(3/2)^2 + x^2} \, dx \qquad \text{now put } a = \frac{3}{2}$$

$$= \frac{1}{4} \frac{1}{(3/2)} \arctan\left(\frac{x}{(3/2)}\right) + C = \frac{1}{6} \arctan\left(\frac{2x}{3}\right) + C.$$

Change of variable: Put $u = 2x$. Then $du = 2 \, dx$, $dx = \frac{1}{2} \, du$ and $u^2 = 4x^2$ so

$$\int \frac{1}{9+4x^2} \, dx = \frac{1}{2} \int \frac{1}{9+u^2} \, du = \frac{1}{2} \cdot \frac{1}{3} \arctan\left(\frac{u}{3}\right) + C = \frac{1}{6} \arctan\left(\frac{2x}{3}\right) + C.$$

Practice 3:   Use the table to find $\int \frac{1}{25-9x^2} \, dx$.

Sometimes we have to change the variable.

Example 4:   Use the table to find $\int \frac{e^x}{9+e^{2x}} \, dx$.

Solution: Putting $u = e^x$, then $du = e^x \, dx$ so our integrand is transformed:

$$\int \frac{e^x}{9+e^{2x}} \, dx \quad = \int \frac{1}{9+(e^x)^2} \, (e^x \, dx)$$

$$= \int \frac{1}{3^2 + u^2} \, du \;=\; \tfrac{1}{3} \arctan\left(\tfrac{u}{3}\right) + C \;=\; \tfrac{1}{3} \arctan\left(\tfrac{e^x}{3}\right) + C.$$

Practice 4:   Use the table to find $\displaystyle\int \frac{\cos(x)}{25 - \sin^2(x)} \, dx$.

How should you recognize whether algebra or a change of variable is needed? Experience and practice, practice, practice.

Using "Recursive" Formulas

A recursive formula in the table is one that gives one antiderivative in terms of another antiderivative. Usually the new antiderivative is simpler than the original one. For example, table entry #19 gives the pattern for the antiderivative of $\sin^n(ax)$ in terms of the antiderivative of $\sin^{n-2}(ax)$. If we start with the integral of $\sin^5(x)$, then we can use the table to get an answer in terms of the integral of $\sin^3(x)$. Using the recursion formula again, we can get the integral of $\sin^3(x)$ in terms of the integral of $\sin(x)$, which is easy to integrate.

Table entry 19: $\displaystyle\int \sin^n(ax) \, dx \;=\; \frac{-\sin^{n-1}(ax)\cdot\cos(ax)}{na} + \frac{n-1}{n} \int \sin^{n-2}(ax) \, dx$

Example 5:   Use the given recursive formula to evaluate $\displaystyle\int \sin^3(5x) \, dx$.

Solution: In this example $n = 3$ and $a = 5$. Then

$$\int \sin^3(5x) \, dx \;=\; \frac{-\sin^2(5x)\cdot\cos(5x)}{3\cdot 5} + \frac{2}{3} \int \sin(5x) \, dx$$

$$=\; \frac{-\sin^2(5x)\cdot\cos(5x)}{15} - \frac{2}{3}\frac{1}{5} \cos(5x) + C.$$

Practice 5:   Use the recursion formula in the table to evaluate $\displaystyle\int \cos^3(7x) \, dx$.

PROBLEMS

Use the integral table for the following problems.

1. $\displaystyle\int \frac{1}{4 + x^2} \, dx$ 2. $\displaystyle\int \frac{5}{4 + x^2} \, dx$ 3. $\displaystyle\int 2x + \frac{2}{25 + x^2} \, dx$

4. $\displaystyle\int \frac{1}{4-x^2}\ dx$

5. $\displaystyle\int \frac{2}{9-x^2}\ dx$

6. $\displaystyle\int \cos(x) + \frac{3}{25-x^2}\ dx$

7. $\displaystyle\int \frac{1}{3+x^2}\ dx$

8. $\displaystyle\int \frac{5}{7+x^2}\ dx$

9. $\displaystyle\int e^x + \frac{7}{2+x^2}\ dx$

10. $\displaystyle\int \frac{1}{\sqrt{4-x^2}}\ dx$

11. $\displaystyle\int \frac{3}{\sqrt{5-x^2}}\ dx$

12. $\displaystyle\int \frac{3}{\sqrt{4-x^2}}\ dx$

13. $\displaystyle\int \frac{1}{4+25x^2}\ dx$

14. $\displaystyle\int \frac{2}{\sqrt{9-16x^2}}\ dx$

15. $\displaystyle\int \frac{5}{\sqrt{1-4x^2}}\ dx$

16. $\displaystyle\int \sec(x+5)\ dx$

17. $\displaystyle\int \frac{2}{\sqrt{1+9x^2}}\ dx$

18. $\displaystyle\int x \cdot \sec(2x^2+7)\ dx$

19. $\displaystyle\int \ln(x+1)\ dx$

20. $\displaystyle\int \ln(3x-1)\ dx$

21. $\displaystyle\int 3x \cdot \ln(5x^2+7)\ dx$

22. $\displaystyle\int e^x\ \ln(e^x-3)\ dx$

23. $\displaystyle\int \cos(x)\ \ln(\sin(x))\ dx$

24. $\displaystyle\int \frac{2}{\sqrt{x^2-9}}\ dx$

25. $\displaystyle\int \sqrt{4+x^2}\ dx$

26. $\displaystyle\int \sqrt{9+x^2}\ dx$

27. . $\displaystyle\int \sqrt{16+x^2}\ dx$

28. $\displaystyle\int_0^1 \frac{1}{4+x^2}\ dx$

29. $\displaystyle\int_1^3 2x + \frac{2}{25+x^2}\ dx$

30. $\displaystyle\int_0^2 \frac{2}{9-x^2}\ dx$

31. $\displaystyle\int_{-1}^1 \frac{1}{3+x^2}\ dx$

32. $\displaystyle\int_0^1 e^x + \frac{7}{2+x^2}\ dx$

33. $\displaystyle\int_1^2 \frac{3}{\sqrt{5-x^2}}\ dx$

34. $\displaystyle\int_0^1 \frac{1}{4+25x^2}\ dx$

35. $\displaystyle\int_0^{0.1} \frac{5}{\sqrt{1-4x^2}}\ dx$

36. $\displaystyle\int_0^1 \frac{1}{\sqrt{9-4x^2}}\ dx$

37. $\displaystyle\int_0^6 \ln(x+1)\ dx$

38. $\displaystyle\int_0^3 3x \cdot \ln(5x^2+7)\ dx$

39. $\displaystyle\int_0^{\pi/2} \cos(x)\ \ln(2+\sin(x))\ dx$

40. $\displaystyle\int_0^2 \sqrt{4+x^2}\ dx$

41. $\displaystyle\int_{-3}^3 \sqrt{9+x^2}\ dx$

42. $\displaystyle\int_0^1 \sqrt{16+x^2}\ dx$

In problems 43 – 48 , use the recursion formulas in the table.

43. $\int \sin^3(x)\ dx$ 44. $\int \cos^3(x)\ dx$ 45. $\int \cos^5(x)\ dx$

46. $\int \tan^3(x)\ dx$ 47. $\int x^2 \cos(x)\ dx$ 48. $\int x^2 \sin(x)\ dx$

49. Before doing any calculations, predict which do you expect to be larger; the average value of $\sin(x)$ or of $\sin^2(x)$ on the interval $[0, \pi]$? Then calculate each average to see if your prediction was correct.

50. Find the area between $f(x) = \ln(x)$ and the x–axis for $1 \le x \le C$ when $C = e, 10, 100, 200$.

51. Find the average value of $f(x) = \ln(x)$ on the interval $1 \le x \le C$ when $C = e, 10, 100, 200$.

52. Before doing any calculations, predict which of the following integrals you expect to be the largest?

 (a) $\int_0^1 e^x\ dx$ (b) $\int_0^1 x\,e^x\ dx$ (c) $\int_0^1 x^2 e^x\ dx$. Then calculate the value of each integral.

53. Before doing any calculations, predict which of the following integrals you expect to be the largest?

 (a) $\int_1^2 e^x\ dx$ (b) $\int_1^2 x\,e^x\ dx$ (c) $\int_1^2 x^2 e^x\ dx$. Then calculate the value of each integral.

54. Before doing any calculations, predict which of the following integrals you expect to be the largest?

 (a) $\int_0^\pi \sin(x)\ dx$ (b) $\int_0^\pi x \sin(x)\ dx$ (c) $\int_0^\pi x^2 \sin(x)\ dx$. Then calculate the value of each integral.

55. Evaluate $\int_0^C \dfrac{2}{1+x^2}\ dx$ for $C = 1, 10, 20,$ and 30. Before doing the calculation, estimate the value of the integral when $C = 40$.

Section 4.8   **Practice Answers**

Practice 1:   The integral $\int \dfrac{1}{25 - x^2}\ dx$ matches table entry 37: $\int \dfrac{1}{a^2 - x^2}\ dx = \dfrac{1}{2a}\ \ln\left|\dfrac{a+x}{a-x}\right| + C$

  when we put $a = 5$, so $\int \dfrac{1}{25 - x^2}\ dx = \dfrac{1}{2\cdot 5}\ \ln\left|\dfrac{5+x}{5-x}\right| + C$.

Practice 2:   The integral $\int \dfrac{1}{7 - x^2}\ dx$ matches table entry 37: $\int \dfrac{1}{a^2 - x^2}\ dx = \dfrac{1}{2a}\ \ln\left|\dfrac{a+x}{a-x}\right| + C$

  when $a = \sqrt{7}$, so $\int \dfrac{1}{7 - x^2}\ dx = \dfrac{1}{2\sqrt{7}}\ \ln\left|\dfrac{\sqrt{7}+x}{\sqrt{7}-x}\right| + C$.

Practice 3: $\int \dfrac{1}{25 - 9x^2}\ dx = \dfrac{1}{9}\int \dfrac{1}{(25/9) - x^2}\ dx = \dfrac{1}{9}\int \dfrac{1}{(5/3)^2 - x^2}\ dx$

so use #37 with $a = 5/3$. Then

$$\int \dfrac{1}{25 - 9x^2}\ dx = \dfrac{1}{9}\int \dfrac{1}{(5/3)^2 - x^2}\ dx = \dfrac{1}{9}\dfrac{1}{2(5/3)}\ \ln\left| \dfrac{5/3 + x}{5/3 - x} \right| + C$$

$$= \dfrac{1}{30}\ \ln\left| \dfrac{5/3 + x}{5/3 - x} \right| + C \ \ \text{or} \ \ \dfrac{1}{30}\ \ln\left| \dfrac{5 + 3x}{5 - 3x} \right| + C$$

Practice 4: $\int \dfrac{\cos(x)}{25 - \sin^2(x)}\ dx$. Put $u = \sin(x)$. Then $du = \cos(x)\ dx$ so

$$\int \dfrac{\cos(x)}{25 - \sin^2(x)}\ dx = \int \dfrac{1}{25 - u^2}\ du = \dfrac{1}{2\cdot 5}\ \ln\left| \dfrac{5 + u}{5 - u} \right| + C = \dfrac{1}{10}\ \ln\left| \dfrac{5 + \sin(x)}{5 - \sin(x)} \right| + C$$

Practice 5: We can evaluate $\int \cos^3(7x)\ dx$ by using the recursion formula from the Table with $n = 3$:

20. $\int \cos^n(ax)\ dx = \dfrac{\cos^{n-1}(ax)\sin(ax)}{na} + \dfrac{n-1}{n}\int \cos^{n-2}(ax)\ dx$.   Then

$$\int \cos^3(7x)\ dx = \dfrac{\cos^2(7x)\cdot\sin(7x)}{7\cdot 3} + \dfrac{2}{3}\int \cos(7x)\ dx$$

$$= \dfrac{\cos^2(7x)\cdot\sin(7x)}{21} + \dfrac{2}{3}\cdot\dfrac{1}{7}\cdot\sin(7x) + C$$

4.9   APPROXIMATING DEFINITE INTEGRALS

The Fundamental Theorem of Calculus tells how to calculate the exact value of a definite integral IF the integrand function is continuous and IF we can find an antiderivative of the integrand. In practice, however, we may need the definite integral of a function defined by a table of measurements or a graph, or of a function which does not have an elementary antiderivative. This section includes several techniques for getting approximate numerical values for definite integrals without using antiderivatives. Mathematically, exact answers are preferable and satisfying, but for most applications, a numerical answer with several digits of accuracy is just as useful.

The ideas behind the approximation methods are geometrical and rather simple, but using the methods to get good approximations typically requires lots of arithmetic, something calculators are very good and quick at doing. All of these approximate methods can be easily programmed, and program listings for two of these methods are included after the Practice Answers.

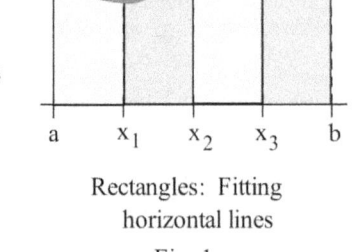

y = f(x)

a x_1 x_2 x_3 b

Rectangles: Fitting
horizontal lines
Fig. 1

The General Approach

The methods in this section approximate the definite integral of a function f by building "easy" functions close to f and then exactly evaluating the definite integrals of the "easy" functions. If the "easy" functions are close enough to f, then the sum of the definite integrals of the "easy" functions will be close to the definite integral of f. The Left, Right and Midpoint approximations fit horizontal lines to f , the "easy" functions are constant functions, and the approximating regions are rectangles (Fig. 1). The Trapezoidal Rule fits slanted lines to f , the "easy" functions are linear, and the approximating regions are trapezoids (Fig. 2). Finally, Simpson's Rule fits parabolas to f, and the "easy" functions are quadratics (Fig. 3).

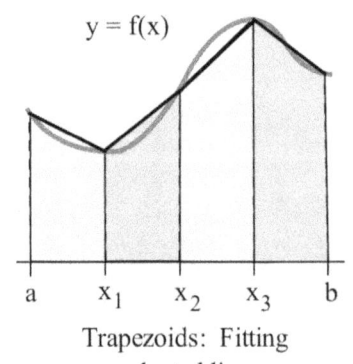

y = f(x)

a x_1 x_2 x_3 b

Trapezoids: Fitting
slanted lines
Fig. 2

The Left and Right approximation rules are simply Riemann sums with the point c_i in each subinterval chosen to be the left or right endpoint of that subinterval.

They typically require a large number of computations to get even mediocre

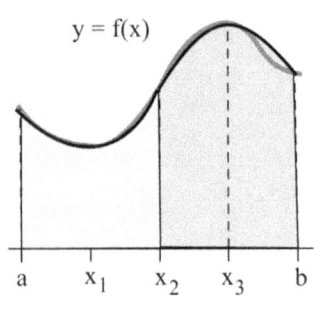

y = f(x)

a x_1 x_2 x_3 b

Parabolas: Fitting
parabolas
Fig. 3

approximations and are seldom used in practice. They and the Midpoint rule are discussed at the end of the problem set.

All of the methods divide the interval [a,b]   into **n equally–long** subintervals. Each subinterval has length $h = \Delta x_i = \frac{b-a}{n}$, and the points of the partition are

$x_0 = a$, $x_1 = a + h$, $x_2 = a + 2 \cdot h$, . . . , $\mathbf{x_i = a + i \cdot h}$, . . . , $x_n = a + n \cdot h = a + n(\frac{b-a}{n}) = b$.

Approximating A Definite Integral Using Trapezoids

If the graph of f is curved, then slanted lines typically come closer to the graph of f than horizontal ones do, and the slanted lines lead to trapezoidal regions (Fig. 2).

The area of a trapezoid is (base)·(average height) so the area of the first trapezoid in Fig. 4 is

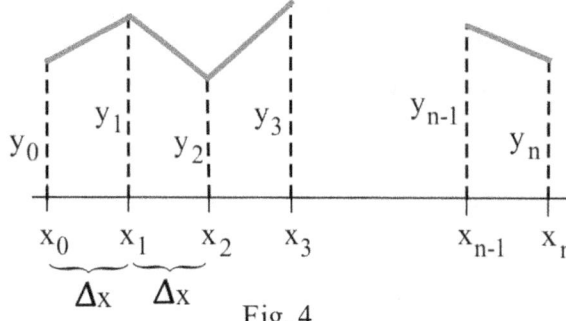

Fig. 4

$$(\Delta x)\cdot\frac{y_0+y_1}{2} \; = \frac{\Delta x}{2}(y_0+y_1) \; .$$

Similarly, the areas of the other trapezoids are

$$\frac{\Delta x}{2}(y_1+y_2)\, , \quad \frac{\Delta x}{2}(y_2+y_3)\, , \quad \dots \, , \quad \frac{\Delta x}{2}(y_{n-1}+y_n) \quad .$$

The sum of the trapezoidal areas is

$$T_n \;\; = \frac{\Delta x}{2}(y_0+y_1) + \frac{\Delta x}{2}(y_1+y_2) + \frac{\Delta x}{2}(y_2+y_3) + \dots + \frac{\Delta x}{2}(y_{n-1}+y_n)$$

$$= \frac{\Delta x}{2}\{y_0 + 2y_1 + 2y_2 + \dots + 2y_{n-1} + y_n\} \;\; \text{or, equivalently,}$$

$$\frac{\Delta x}{2}\left\{ f(x_0) + 2f(x_1) + 2f(x_2) + \dots + 2f(x_{n-1}) + f(x_n) \right\}.$$

Each $f(x_i)$ value, except the first (i = 0) and the last (i = n), is the right–endpoint height of one trapezoid and the left–endpoint height of the next trapezoid so it shows up in the calculation for two trapezoids and is multiplied by two in the formula for the trapezoidal approximation.

Trapezoidal Approximation Rule

If   f is integrable on [a,b], and [a,b] is partitioned into n subintervals of length $h = \dfrac{b-a}{n}$,

then the Trapezoidal approximation of $\displaystyle\int_a^b f(x)\,dx$ is

$$T_n \;\; = \frac{h}{2}\left\{ f(x_0) + 2f(x_1) + 2f(x_2) + \dots + 2f(x_{n-1}) + f(x_n) \right\}$$

---

x	f(x)
1.0	4.2
1.5	3.4
2.0	2.8
2.5	3.6
3.0	3.2

Table 1

Example 1: Calculate T_4 , the Trapezoidal approximation of $\displaystyle\int_1^3 f(x)\,dx$, for the function values in Table 1.

Solution: The step size is $h = (b–a)/n = (3–1)/4 = 1/2$. Then

$$T_4 = \frac{h}{2} \left\{ f(x_0) + 2f(x_1) + 2f(x_2) + 2f(x_3) + f(x_4) \right\}$$

$$= \frac{.5}{2} \left\{ 4.2 + 2(3.4) + 2(2.8) + 2(3.6) + (3.2) \right\} = (.25)(27) = 6.75 .$$

Let's see how well the trapezoidal rule approximates an integral whose exact value we know, $\int_{1}^{3} x^2 \, dx = 8\frac{2}{3}$.

Example 2:   Calculate T_4 , the Trapezoidal approximation of $\int_{1}^{3} x^2 \, dx$ for $n = 4$.

Solution: As in Example 1, $h = .5$ and $x_0 = 1, x_1 = 1.5, x_2 = 2, x_3 = 2.5$, and $x_4 = 3$. Then

$$T_4 = \frac{h}{2} \left\{ f(x_0) + 2f(x_1) + 2f(x_2) + 2f(x_3) + f(x_4) \right\} = \frac{.5}{2} \left\{ f(1) + 2f(1.5) + 2f(2) + 2f(2.5) + f(3) \right\}$$

$$= (.25) \left\{ 1 + 2(2.25) + 2(4) + 2(6.25) + 9 \right\} = 8.75 .$$

Larger values for n give better approximations: $T_{20} = 8.67$ and $T_{100} = 8.6668$.

Practice 1:   On a summer day, the level of the pond in Fig. 5 went down 0.1 feet because of evaporation. Use the trapezoidal rule to approximate the surface area of the pond and then calculate how much water evaporated.

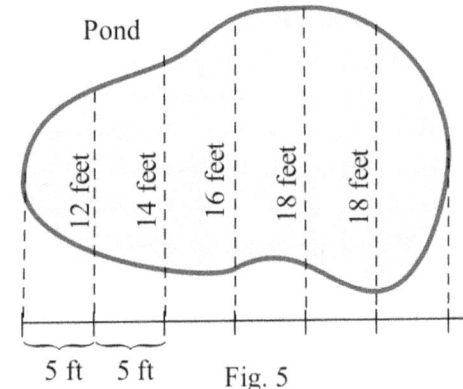

Pond

12 feet   14 feet   16 feet   18 feet   18 feet

5 ft   5 ft   Fig. 5

Approximating A Definite Integral Using Parabolas

If the graph of f is curved, even the slanted lines may not fit the graph of f as closely as we would like, and a large number of subintervals may still be needed with the Trapezoidal rule to get a good approximation of the definite integral. Curves typically fit the graph of f better than straight lines, and the easiest nonlinear curves are parabolas.

Three points $(x_0, y_0), (x_1, y_1), (x_2, y_2)$ are needed to determine the equation of a parabola, and the area under a parabolic region with evenly spaced x_i values (Fig. 6) is

$$\text{Area} = (2\Delta x)\left\{ \frac{y_0 + 4y_1 + y_2}{6} \right\}$$

parabola

y_1

y_2

y_0

x_0 x_1 x_2

$2\Delta x$

Fig/ 6

$$(2\Delta x)\cdot\left\{ \frac{y_0 + 4y_1 + y_2}{6} \right\} = \frac{\Delta x}{3} \cdot\{ y_0 + 4y_1 + y_2 \}.$$

(The steps to verify this formula for parabolas are outlined in problem 32.)

Taking the subintervals in pairs, the areas of the other parabolic regions are

$$\frac{\Delta x}{3} \cdot \{ y_2 + 4y_3 + y_4 \}, \quad \frac{\Delta x}{3} \cdot \{ y_4 + 4y_5 + y_6 \}, \ldots,$$

$$\frac{\Delta x}{3} \cdot \{ y_{n-2} + 4y_{n-1} + y_n \}$$

so the sum of the parabolic areas (Fig. 7) is

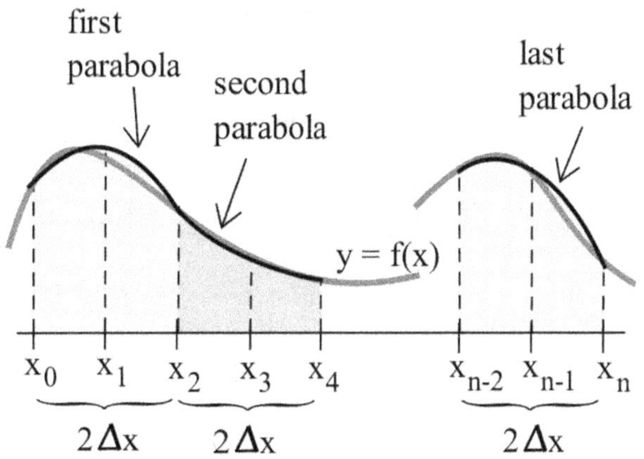

$$S_n = \frac{\Delta x}{3} \cdot \{ y_0 + 4y_1 + y_2 \}$$

$$+ \; \frac{\Delta x}{3} \cdot \{ y_2 + 4y_3 + y_4 \}$$

$$+ \; \ldots \; + \; \frac{\Delta x}{3} \cdot \{ y_{n-2} + 4y_{n-1} + y_n \}$$

$$= \frac{\Delta x}{3} \cdot \{ y_0 + 4y_1 + 2y_2 + 4y_3 + 2y_4 + \ldots + 2y_{n-2} + 4y_{n-1} + y_n \} \quad \text{or, equivalently,}$$

$$\frac{\Delta x}{3} \left\{ f(x_0) + 4f(x_1) + 2f(x_2) + 4f(x_3) + 2f(x_4) + \ldots + 2f(x_{n-2}) + 4f(x_{n-1}) + f(x_n) \right\}.$$

first parabola

second parabola

last parabola

$y = f(x)$

$$x_0 \quad x_1 \quad x_2 \quad x_3 \quad x_4 \qquad\qquad x_{n-2} \; x_{n-1} \; x_n$$

$$2\Delta x \qquad 2\Delta x \qquad\qquad 2\Delta x$$

Fig. 7

In order to use **pairs** of subintervals, the number n of subintervals must be **even**. The coefficient pattern for a single parabola is 1–4–1, but when we put several parabolas next to each other, they share some edges and the pattern becomes 1–4–**2**–4–**2**– ... –**2**–4–1 with the shared edges getting counted twice.

Parabolic Approximation Rule (Simpson's Rule)

If f is integrable on [a,b], and [a,b] is partitioned into an **even number** n of subintervals of

length $h = \frac{b-a}{n}$, then the Parabolic approximation of $\displaystyle\int_a^b f(x)\, dx$ is

$$S_n = \frac{h}{3} \cdot \left\{ f(x_0) + 4f(x_1) + 2f(x_2) + 4f(x_3) + 2f(x_4) + \ldots + 4f(x_{n-1}) + f(x_n) \right\}$$

Example 3:   Calculate S_4 , Simpson's parabolic approximation of $\displaystyle\int_1^3 f(x)\, dx$, for the function in Table 1.

Solution: The step size is h = (b–a)/n = (3–1)/4 = 1/2 . Then

$$S_4 = \frac{h}{3} \left\{ f(x_0) + 4f(x_1) + 2f(x_2) + 4f(x_3) + f(x_4) \right\}$$

$$= \frac{1/2}{3} \left\{ 4.2 + 4(3.4) + 2(2.8) + 4(3.6) + (3.2) \right\} = \frac{1}{6}(41) \approx 6.833 .$$

Example 4:   Calculate S_4 , Simpson's parabolic approximation of $\int_{1}^{3} 2^x\, dx$ for n = 4.

Solution: As in the previous Examples, h = (b–a)/n = .5 and x_0 = 1, x_1 = 1.5, x_2 = 2, x_3 = 2.5, and x_4 = 3.

$$S_4 = \frac{h}{3}\left\{ f(x_0) + 4f(x_1) + 2f(x_2) + 4f(x_3) + f(x_4)\right\} = \frac{.5}{3}\left\{ f(1) + 4f(1.5) + 2f(2) + 4f(2.5) + f(3)\right\}$$

$$= (\frac{1}{6})\left\{ 2 + 4(2.828427) + 2(4) + 4(5.656854) + 8\right\} = \frac{1}{6}(51.941124) = 8.656854 .$$

Larger values for n give better approximations: S_{20} = 8.656171 and S_{100} = 8.656170 .

Practice 2:   Use Simpson's Rule to estimate the area of the pond in Fig. 5.

Which Method Is Best?

The hardest and slowest part of these approximations, whether by hand or by computer, is the evaluation of the function at the x_i values. For n subintervals, all of the methods require about the same number of function evaluations. Table 2 illustrates how closely each method approximates the definite integral of 1/x using several values of n. The values in Table 2 also show how quickly the actual error shrinks as the values of n increase: just doubling n from 4 to 8 cut the actual error of the parabolic approximation of this definite integral by a factor of 9 — a good reward for our extra work. The rest of this section discusses "error bounds" of the approximations so we can know how close our approximation is to the exact value of the integral even if we don't know the exact value.

Table 2: Approximating $\displaystyle\int_{1}^{5} \frac{1}{x}\,dx$ = ln 5 = 1.609437912

Using n=4 (h= (5–1)/4 = 1)

method	approximation	M	error bound	actual error
T_4	**1.6833333**	**2**	**.6666666**	**.07389542**
S_4	**1.6222222**	**24**	**.5333333**	**.01278431**
L_4	2.083333333	1	2	.47389542
R_4	1.283333333	1	2	.32610458
M_4	1.574603175	2	.33333333	.03483474

Using n=8 (h= (5–1)/8 = 1/2)

method	approximation	M	error bound	actual error
T_8	**1.628968254**	**2**	**.1666666**	**.01953034**
S_8	**1.610846561**	**24**	**.0333333**	**.00140865**
L_8	1.828968254	1	1	.21953034
R_8	1.428968254	1	1	.18046966
M_8	1.599844394	2	.08333333	.00959352

Using n=20 (h= (5–1)/20 = 1/5)

method	approximation	M	error bound	actual error
T_{20}	**1.612624844**	**2**	**.0266667**	**.00318693**
S_{20}	**1.609486789**	**24**	**.0008533**	**.00004888**
L_{20}	1.692624844	1	.4	.08318693
R_{20}	1.532624844	1	.4	.07681307
M_{20}	1.607849324	2	.01333333	.00158859

How Good Are the Approximations?

The approximation rules are valuable by themselves, but they are particularly useful because there are "error bound" formulas that guarantee how close the approximations are to the exact values of the integrals. It is useful to know that an integral is "about 3.7," but we can have more confidence if we know that the integral is "within .0001 of 3.7 ." Then we can decide if our approximation is good enough for the job at hand or if we need to improve it. The formulas for the error bounds can also be solved to determine how many subintervals are needed to guarantee that our approximation is within some specified distance of the exact answer. There is no reason to use 1000 subintervals if 18 will give the needed accuracy. Unfortunately, the formulas for the error bounds require information about the derivatives of the integrands, so we can not use these formulas to determine error bounds for the approximations of integrals of functions defined by tables of values.

Error Bound for Trapezoidal Approximation

If   the second derivative of f is continuous on [a,b] and $M_2 \geq$ { maximum of $|f''(x)|$ on [a,b] },

then   the "error" of the T_n approximation is $\left| \int_a^b f(x)\, dx - T_n \right| \leq \frac{(b-a)^3}{12\, n^2} \cdot M_2 =$ "error bound."

The "error bound" formula $\frac{(b-a)^3}{12\, n^2} \cdot M_2$ for the Trapezoidal approximation is a "guarantee:" the actual

error is guaranteed to be no larger than the error bound. In fact, the actual error is usually much smaller than

the error bound. The word "error" does not indicate a mistake, it means the deviation or distance from the

exact answer.

Example 5:   We can be certain that the T_{10} approximation of $\int_0^2 \sin(x^2)\, dx$ is **within** what distance

of the exact value of the integral?

Solution: $b - a = 2, n = 10$, $f(x) = \sin(x^2)$, and $f''(x) = -4x^2 \cdot \sin(x^2) + 2 \cdot \cos(x^2)$ is continuous

on $[0, 2]$. The graph of $f''(x)$ is given in Fig. 8 . Even though we may not know the exact

maximum value M_2 of $|f''(x)|$ on $[0, 2]$, it is

clear from the graph that $M_2 \leq 11$. Then

"actual error" $\leq$ "error bound" $= \frac{(b-a)^3}{12\, n^2} \cdot M_2$

$= \frac{(2)^3}{12\,(10)^2} \cdot (11) = \frac{88}{1200} < 0.074$

so we can be certain that our T_{10} approximation of the definite integral is within .074 of the exact

value:

$$T_{10} - 0.074 \leq \int_0^2 \sin(x^2)\, dx \leq T_{10} + 0.074 \ .$$

$T_{10} = 0.7959247$, so we can be certain that the value of

the integral is between 0.722 and 0.870 .

Practice 3:   Find an error bound for the T_{12} approximation

of $\int_2^5 \frac{1}{x}\, dx$.

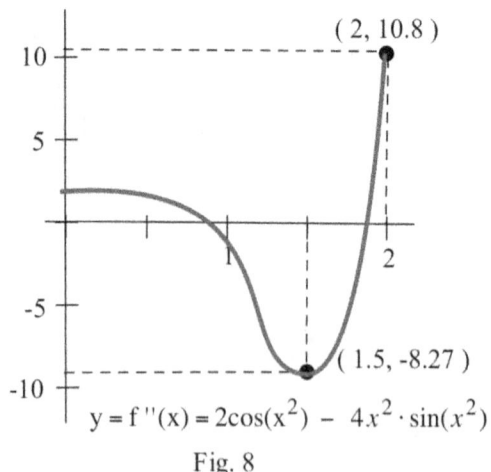

Fig. 8

Example 6:   How large must n be to be certain that T_n is within 0.001 of $\int_0^2 \sin(x^2)\,dx$?

Solution: The "allowable error" of 0.001 is given, and we are asked to find n. From Example 5 we

know that $M_2 \le 11$, so we want the error bound to be less than the allowable error of 0.001. Then

$.001 \ge$ "error bound" $= \dfrac{(2)^3}{12\,n^2} \cdot (11) = \dfrac{88}{12}\,\dfrac{1}{n^2} = \dfrac{22}{3n^2}$. Solving for n, we have $n^2 \ge \dfrac{22}{0.003} > 7334$

so $n \ge \sqrt{7334} \approx 85.6$. Since n must be an integer, we can be certain that T_{86} is within

0.001 of $\int_0^2 \sin(x^2)\,dx$. $T_{86} \approx 0.80465$, so we can be certain that the exact value of the integral is

between 0.80365 and 0.80565 . As is usually the case, T_{86} is even closer than 0.001 to the exact

value, $|\,T_{86} - \text{exact value}\,| \approx 0.00012$.

Practice 4:   How large must n be to be certain that T_n is within 0.001 of $\int_2^5 \dfrac{1}{x}\,dx$?

Error Bound for Simpson's Parabolic Approximation

If   the fourth derivative of f is continuous on [a,b] , and $M_4 \ge \{\text{maximum of } |\,f^{(4)}(x)\,| \text{ on } [a,b]\}$,

then   the "error" of the S_n approximation is $\left|\,\int_a^b f(x)\,dx - S_n\,\right| \le \dfrac{(b-a)^5}{180\,n^4} \cdot M_4 = $ "error bound."

Example 7:   Find an error bound for the S_{10}

approximation of $\int_0^2 \sin(x^2)\,dx$.

Solution: $b - a = 2, n = 10, f(x) = \sin(x^2)$, and

$f^{(4)}(x) = (16x^4 - 12)\sin(x^2) - 48x^2\cos(x^2)$ is

continuous on $[0,2]$. From Fig. 9, the graph of

$f^{(4)}(x)$ on $[0,2]$, we know that $M_4 \le 165$. Then

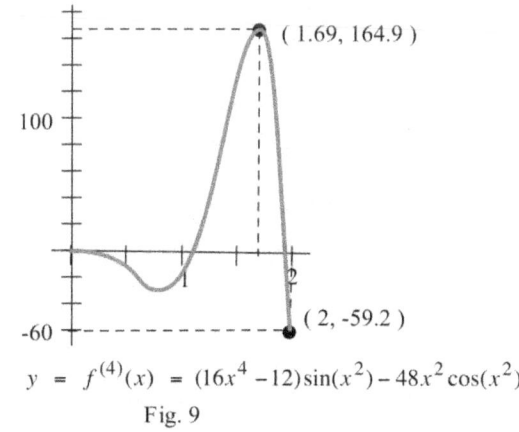

$y = f^{(4)}(x) = (16x^4 - 12)\sin(x^2) - 48x^2\cos(x^2)$

Fig. 9

$\text{"error"} \le \dfrac{(b-a)^5}{180\,n^4} \cdot M_4 = \dfrac{(2)^5}{180\,(10)^4}\,(165) = \dfrac{5280}{1800000} < 0.003$

so we can be certain that our S_{10} approximation of the definite integral is within 0.003 of the exact value:

$$S_{10} - 0.003 \le \int_0^2 \sin(x^2)\,dx \le S_{10} + 0.003 .$$

$S_{10} = 0.80537615$, so we are certain that the exact value of the integral is between 0.80237615 and 0.80837615. Notice that we got a much narrower guarantee using S_{10} compared to using T_{10} to approximate the integral.

Example 8:   How large must n be to be certain that S_n is within 0.001 of $\int_0^2 \sin(x^2)\, dx$?

Solution: We are given an "allowable error" of 0.001 and are asked to find n. From Fig. 9 we know that $M_4 \leq 165$, so we want the error bound to be less than the allowable error of 0.001. Then

$$0.001 \geq \text{"error bound"} = \frac{(2)^5}{180\, n^4}(165) = \frac{5280}{180\, n^4}\ . \text{ Solving for n, we have}$$

$n^4 \geq \frac{5280}{(.001)180} \geq 29{,}333.34$ so $n \geq \sqrt[4]{29333.34} \approx 13.08$. Since n must be an **even** integer, we can take $n = 14$ and be certain that S_{14} is **within** 0.001 of $\int_0^2 \sin(x^2)\, dx$. In fact,

$S_{14} = 0.8049239$ is even closer than 0.001 to the exact value, $|S_{14} - \text{exact value}| \approx 0.00015$.

A variety of other methods for approximating definite integrals can be found in most books on Numerical Analysis. Definite integrals occur often in applied problems, and these approximation methods can get us the numerical answers we need even if we can't find an antiderivative of the integrand. If you have a programmable calculator, program Simpson's rule. It will be useful in Chapter 5.

PROBLEMS

For problems 1 and 2, use the values given in Table 3 to approximate the value of $\int_2^6 f(x)\, dx$.

1.   Calculate T_4 and S_4 .

2.   Calculate T_8 and S_8 .

For problems 3 and 4, use the values given in Table 4 to approximate the value of $\int_{-3}^1 g(x)\, dx$.

3.   Calculate T_8 and S_8 .

4.   Calculate T_4 and S_4 .

x	f(x)
2.0	2.1
2.5	2.7
3.0	3.8
3.5	2.3
4.0	0.3
4.5	−1.8
5.0	−0.9
5.5	0.5
6.0	2.2

Table 3

x	g(x)
−3.0	4.2
−2.5	1.8
−2.0	0.7
−1.5	1.5
−1.0	3.4
−0.5	4.3
0	3.5
0.5	−0.3
1.0	−4.6

Table 4

For problems 5 – 10, calculate (a) T_4 , (b) S_4 , and (c) the exact value of the integral.

5. $\int_1^3 x\ dx$

6. $\int_0^2 (1-x)\ dx$

7. $\int_{-1}^1 x^2\ dx$

8. $\int_2^6 \frac{1}{x}\ dx$

9. $\int_0^\pi \sin(x)\ dx$

10. $\int_0^1 \sqrt{x}\ dx$

For problems 11 – 16, calculate (a) T_6 and (b) S_6 .

11. $\int_0^2 \frac{1}{1+x^3}\ dx$

12. $\int_1^2 2^x\ dx$

13. $\int_{-1}^1 \sqrt{4-x^2}\ dx$

14. $\int_0^1 e^{-x^2}\ dx$

15. $\int_1^4 \frac{\sin(x)}{x}\ dx$

16. $\int_0^1 \sqrt{1+\sin(x)}\ dx$

For problems 17 – 23, calculate (a) the error bound for T_4 , (b) the error bound for S_4 , (c) the value of n so the error bound for T_n is less than 0.001 , and (d) the value of n so the error bound for S_n is less than 0.001 .

17. $\int_1^3 x\ dx$

18. $\int_0^2 (1-x)\ dx$

19. $\int_{-1}^1 x^3\ dx$

20. $\int_2^6 \frac{1}{x}\ dx$

21. $\int_0^\pi \sin(x)\ dx$

22. $\int_1^4 \sqrt{x}\ dx$

23. A friend has asked you to help calculate the area of a piece of land located between a river and a road (Fig. 10). Estimate the area.

24. Estimate the area of the island in Fig. 11.

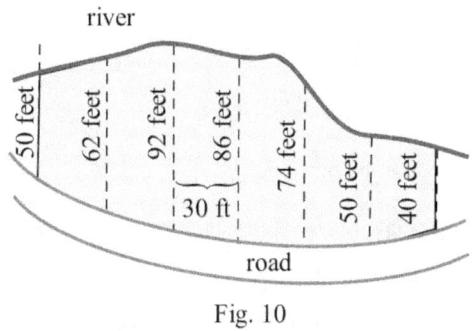

Fig. 10

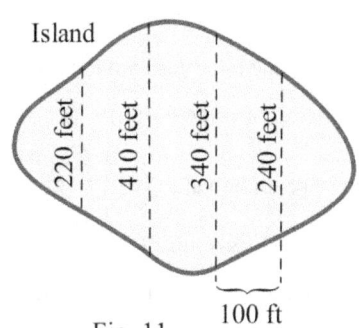

Fig. 11

25. The average depth of the reservoir in Fig. 12 is 22 feet. Estimate the amount of water in the reservoir.

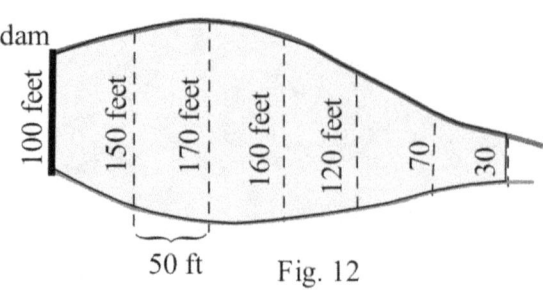

dam

100 feet 150 feet 170 feet 160 feet 120 feet 70 30

50 ft   Fig. 12

26. Table 5 shows the speedometer readings for a car at one minute intervals. Estimate how far the car traveled (a) during the first 5 minutes of the trip and (b) during the first 10 minutes of the trip.

Table 5: Time (minutes) and Velocity (feet/minute) for a car.

Time	0	1	2	3	4	5	6	7	8	9	10
Velocity	0	2000	3000	5000	5000	6000	5200	4400	3000	2000	1200

27. Table 6 shows the speed of a jogger at one minute intervals. Estimate how far the jogger ran during the workout.

Table 6: Time (minutes) and Velocity (feet/minute) for a jogger.

Time	0	1	2	3	4	5	6	7	8	9	10
Velocity	0	420	540	300	500	580	520	440	360	260	180

28. Use the error formula for Simpson's rule to show that the parabolic approximation is the exact value of the integral if the integrand is a polynomial of degree 3 or less, $ax^3 + bx^2 + cx + d$.

29. A trapezoidal region (Fig. 13) with base b and heights h_1 and h_2 (assume $h_1 \leq h_2$) can be cut into a rectangle with base b and height h_1 and a triangle with base b and height $h_1 - h_2$. Show that the sum of the area of the rectangle and the area of the triangle is $b \cdot \left\{ \dfrac{h_1 + h_2}{2} \right\}$.

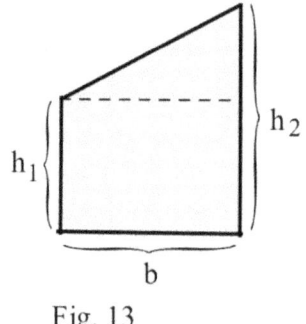

h_2

h_1

b

Fig. 13

30. Let f(m) be the minimum value of f on the interval $[x_0 , x_1]$. Let f(M) be the maximum value of f on $[x_0 , x_1]$. And let $h = x_1 - x_0$. Show that the trapezoidal value, $h \cdot \left\{ \dfrac{f(x_0) + f(x_1)}{2} \right\}$, is between $h \cdot f(m)$ and $h \cdot f(M)$. From this result, it can be shown that the trapezoidal approximation is between the lower and upper Riemann sums for f. Since the limit (as h approaches 0) of these Riemann sums is the definite integral of f , we can conclude that the limit of the trapezoidal sums is the value of the definite integral.

31. Let f(m) be the minimum value of f on the interval $[x_0 , x_2]$. Let f(M) be the maximum of f on $[x_0 , x_2]$. And let $h = x_1 - x_0 = x_2 - x_1$. Show that the parabolic value,

$2h \cdot \{ \dfrac{f(x_0) + 4f(x_1) + f(x_2)}{6} \}$ is between $2h \cdot f(m)$ and $2h \cdot f(M)$. From this result, it can be shown that the parabolic approximation is between the lower and upper Riemann sums for f. Since the limit (as h approaches 0) of these Riemann sums is the definite integral of f , we can conclude that the limit of the parabolic sums is the value of the definite integral.

32. This problem leads you through the steps to show that the area under a parabolic region with evenly spaced x_i values ($x_0 = m-h, x_1 = m, x_2 = m+h$) as in Fig. 14 is

$$\frac{h}{3} \cdot \{ f(x_0) + 4f(x_1) + f(x_2) \} = \frac{h}{3} \cdot \{ y_0 + 4y_1 + y_2 \}$$

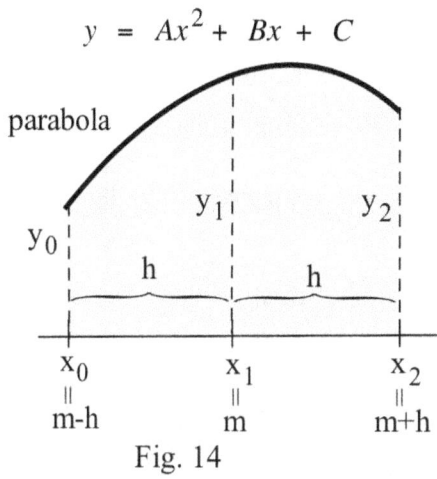

Fig. 14

(a) For $f(x) = Ax^2 + Bx + C$, a parabola, verify that

$$\int_{m-h}^{m+h} f(x) \, dx = \frac{A}{3} x^3 + \frac{B}{2} x^2 + Cx \Big|_{x=m-h}^{x=m+h}$$

$$= 2Am^2 h + \frac{2}{3} Ah^3 + 2Bmh + 2Ch.$$

(b) Expand $y_0 = f(m-h) = A(m-h)^2 + B(m-h) + C$,

$y_1 = f(m) = Am^2 + Bm + C$, and

$y_2 = f(m+h) = A(m+h)^2 + B(m+h) + C$.   Then verify that

$$\frac{h}{3} \cdot \{ y_0 + 4y_1 + y_2 \} = 2h \left\{ \frac{f(m-h) + 4f(m) + f(m+h)}{6} \right\}$$

$$= 2Am^2 h + \frac{2}{3} Ah^3 + 2Bmh + 2Ch .$$

(c) Compare the results of parts (a) and (b) to conclude that for any parabola $f(x) = Ax^2 + Bx + C$,

$$\int_{m-h}^{m+h} f(x) \, dx = 2h \left\{ \frac{f(m-h) + 4f(m) + f(m+h)}{6} \right\} = \frac{h}{3} \cdot \{ y_0 + 4y_1 + y_2 \}.$$

Rectangular Approximations: Left Endpoint, Right Endpoint, and Midpoint Rules

The rectangular approximation methods fit horizontal lines to the integrand. The approximating regions are rectangles, and the sum of the areas of the rectangular regions is a Riemann sum. The Left and Right

Endpoint Rules are easy to understand and use, but they typically require a very large number of subintervals to provide good approximations of a definite integral. The Midpoint Rule uses the value of the function at the midpoint of each subinterval. If these midpoint values of f are available, for example when f is given by a formula, then the Midpoint Rule is often more efficient than the Trapezoidal rule. The rectangular approximation rules and their error bounds are given below.

Left endpoint: L_n = $h \cdot \{ f(x_0) + f(x_1) + f(x_2) + \ldots + f(x_{n-1}) \}$

Right endpoint: R_n = $h \cdot \{ f(x_1) + f(x_2) + f(x_3) + \ldots + f(x_n) \}$

Midpoint Rule: M_n = $h \cdot \{ f(A) + f(A + h) + f(A + 2h) + \ldots + f(A + (n-1)h) \}$ where $A = x_0 + \dfrac{h}{2}$.

(The points $A, A + h, A + 2h, \ldots$ are the **midpoints** of the subintervals.)

The "error bound" for L_n and R_n is $\dfrac{(b-a)^2}{2\,n}$ $\cdot M_1$ where $M_1 \geq \{$maximum of $| f '(x) |$ on [a,b]$\}$.

The "error bound" for M_n is $\dfrac{(b-a)^3}{24\,n^2}$ $\cdot M_2$ where $M_2 \geq \{$maximum of $| f ''(x) |$ on [a,b]$\}$. This is half the error bound of T_n , the trapezoidal approximation.

For problems 33 – 38, calculate (a) L_4 , (b) R_4 , (c) M_4 , and (d) the exact value of the integral.

33. $\displaystyle\int_1^3 x \, dx$

34. $\displaystyle\int_0^2 (1 - x) \, dx$

35. $\displaystyle\int_{-1}^1 x^2 \, dx$

36. $\displaystyle\int_2^6 \frac{1}{x} \, dx$

37. $\displaystyle\int_0^\pi \sin(x) \, dx$

38. $\displaystyle\int_0^1 \sqrt{x} \, dx$

39. Show that the trapezoidal approximation is the average of the left and right endpoint approximations: $T_n = (L_n + R_n)/2$.

40. Which endpoint rule will give a better approximation of $\displaystyle\int_a^b f(x) \, dx$ if f is concave up on [a, b]?

Calculator Problems

The following definite integrals arise in applications, but they do not have easy antiderivatives. Use Simpson's Rule with n = 10 and n = 40 to approximate their values. (Is S_{40} very different from S_{10} ?)

41. $\displaystyle\int_{-1}^2 \sqrt{1 + 4 \cdot x^2} \, dx$. This is the length of the curve $y = x^2$ from (–1, 1) to (2, 4).

42. $\displaystyle\int_0^\pi \sqrt{1 + \cos^2(x)}\ \ dx$.   This is the length of one arch of the curve $y = \sin(x)$.

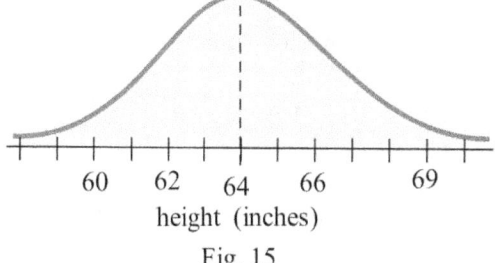

43. $\displaystyle\int_0^{2\pi} \sqrt{16 \cdot \sin^2(x) + 9 \cdot \cos^2(x)}\ \ dx$.

  This is the length of the ellipse $\dfrac{x^2}{16} + \dfrac{y^2}{9} = 1$.

44. $\dfrac{100}{\sqrt{2\pi}} \displaystyle\int_{60}^{69} \text{EXP}\left(-\{(x - 64)/2.5\}^2/2 \right)\ dx$. $\text{EXP}(x) = e^x$.

height (inches)

Fig. 15

  This is the percentage of adult females who are between 60

  and 69 inches tall (Fig. 15). Approximate the value of this integral.

45. Approximate the percentage of adult females who are between 61 and 64 inches tall.

Section 4.9   **Practice Answers**

Practice 1:   Using the Trapezoidal rule to approximate the surface area of the pond in Fig. 5,

$$T \approx \frac{5 \text{ feet}}{2} \cdot \{ 0 + 2 \cdot 12 + 2 \cdot 14 + 2 \cdot 16 + 2 \cdot 18 + 2 \cdot 18 + 0 \text{ feet} \} = 390 \text{ ft}^2.$$

  Then volume = (surface area)(depth) $\approx (390 \text{ ft}^2)(0.1 \text{ ft}) = 39 \text{ ft}^3$.

Practice 2:   Using Simpson's Rule to estimate the area of the pond in Fig. 5,

$$S \approx \frac{5 \text{ feet}}{3} \{ 0 + 4 \cdot 12 + 2 \cdot 14 + 4 \cdot 16 + 2 \cdot 18 + 4 \cdot 18 + 0 \text{ feet}\} = 413.3 \text{ ft}^2.$$

Practice 3: $f(x) = \dfrac{1}{x}$, $b - a = 3$, $n = 12$, $f''(x) = \dfrac{2}{x^3}$. On the interval $[2,5]$, $|f''(x)| = |\dfrac{2}{x^3}| \le \dfrac{2}{2^3} = \dfrac{1}{4}$

  so we can take $M_2 = \dfrac{1}{4}$. Then $|\text{error}| \le \dfrac{(b-a)^3}{12 n^2} \cdot M_2 = \dfrac{3^3}{12(12)^2} \dfrac{1}{4} = \dfrac{27}{6912} \approx \mathbf{0.004}$.

Practice 4:   "error bound" = $\dfrac{(b-a)^3}{12 n^2} \cdot M_2 = \dfrac{(3)^3}{12 n^2} \cdot \dfrac{1}{4} = \dfrac{27}{48 n^2}$. Setting this "error bound" equal to 0.001

  and solving for n, we get $n = \sqrt{\dfrac{27}{48 \cdot (0.001)}} = \sqrt{562.5} \approx 23.7$. Put $n = 24$.

  We can be certain that T_{24} is within 0.001 of the exact value of the integral. (We can not

  guarentee that T_{23} is within 0.001 of the exact value of the integral, but it probably is.)

PROBLEM ANSWERS   Chapter Four

Section 4.7

1.   between 11 (using left endpoints of intervals) and 6 (using right endpoints)

3.   between 4 (using left endpoints of intervals) and 6 (using right endpoints)

5.   Using left endpoint widths: $(0)(40)+(70)(40)+(55)(40)+(90)(40)+(130)(40)+(115)(40) = 18{,}400 \text{ ft}^2$.

   Right endpoint widths $(70, 55, ...)$ and average widths $(70/2, 125/2, ...)$ give the same result, $18{,}400 \text{ ft}^2$.

   All of these are reasonable methods for estimating the area of the island.

7.   9   9.   1   11. $\frac{1}{2} \cdot e^2 - \frac{3}{2}$ 13. $\frac{1}{32} \pi^2 + \frac{1}{4} \pi - \frac{\sqrt{2}}{2}$

15. $e^2 - 3$ 17. $3 - \frac{\pi}{4}$

19. Estimate using midpoints of unit intervals: $\frac{1}{4}\{f(1)(1)+f(2)(1)+f(3)(1)+f(4)(1)\} = \frac{19}{4}$. About $\frac{19}{4}$.

21. Estimate using midpoints of unit intervals: $\frac{1}{2}\{f(2)(1)+f(3)(1)\} = 5$. About 5 .

23. average ≈ 1 25. average $\approx \frac{11}{5}$ 27. average $= 5$ 29. average $= \frac{13}{3}$ 31. average $= \frac{2}{\pi}$

33. (a) C = 1: average $= \frac{2}{3}$ (b) C = 9: average $= 2$ (c) C = 81: average $= 6$ (d) C = 100: average $= \frac{20}{3}$

   In general, average $= \frac{2}{3}\sqrt{C}$.

35. (a) Graphically, average $\approx 3000 \cdot 1000 \frac{\text{calls}}{\text{hour}} = \frac{3000000}{60} \frac{\text{calls}}{\text{min}} \approx 50{,}000 \frac{\text{calls}}{\text{min}}$. (b) About $58{,}333 \frac{\text{calls}}{\text{min}}$.

37. (a) Similar to Example 5: work = 1,950 foot–pounds   (b) work = 1,312.5 foot–pounds

39. (a) work = 1,200 foot–pounds   (b) work = 600 foot–pounds   (c) work = 400 foot–pounds

41. work = 1,275 foot–pounds

Section 4.8

1.   Table #35, a = 2: $\frac{1}{2} \arctan(\frac{x}{2}) + C$ 3.   Table #35, a = 5: $x^2 + \frac{2}{5} \arctan(\frac{x}{5}) + C$

5.   Table #37, a = 3: $(\frac{1}{3}) \ln| \frac{x+3}{x-3} | + C$ 7.   Table #35, a = $\sqrt{3}$: $\frac{1}{\sqrt{3}} \arctan(\frac{x}{\sqrt{3}}) + C$

9.   Table #35, a = $\sqrt{2}$: $e^x + \frac{7}{\sqrt{2}} \arctan(\frac{x}{\sqrt{2}}) + C$

11. Table #34, $a = \sqrt{5}$: $3 \cdot \arcsin(\frac{x}{\sqrt{5}}) + C$ 13. Table #35, $a = \frac{2}{5}$: $\frac{1}{10} \arctan(\frac{5}{2} x) + C$

15. First substitute $u = 2x$, $du = 2\,dx$. Then use Table #34 with $a = 1$: $\frac{5}{2} \cdot \arcsin(u) + C = \frac{5}{2} \cdot \arcsin(2x) + C$

17. Table #43 and substitution $u = 3x$: $\frac{2}{3} \ln|3x + \sqrt{1 + 9x^2}| + C$

19. Table #38 and substitution $u = x+1$: $(x+1) \cdot \ln|x+1| - (x+1) + C$ or $(x+1) \cdot \ln|x+1| - x + C_2$

21. Table #38 and substitution $u = 5x^2+7$: $\frac{3}{10} \{ (5x^2+7) \cdot \ln|5x^2+7| - (5x^2+7) \} + C$

23. Table #38 and substitution $u = \sin(x)$: $\sin(x) \cdot \ln|\sin(x)| - \sin(x) + C$

25. Table #44, $a = 2$: $\frac{x}{2}\sqrt{x^2 + 4} + \frac{1}{2}(4) \cdot \ln|x + \sqrt{x^2 + 4}| + C$

27. Table #44, $a = 4$: $\frac{x}{2}\sqrt{x^2 + 16} + \frac{1}{2}(16) \cdot \ln|x + \sqrt{x^2 + 16}| + C$

29. Table #35, $a = 5$: $x^2 + \frac{2}{5} \arctan(\frac{x}{5}) \Big|_1^3 = 8 + \frac{2}{5} \{ \arctan(\frac{3}{5}) - \mathrm{acrtan}(\frac{1}{5}) \}$

31. Table #35, $a = \sqrt{3}$: $\frac{1}{\sqrt{3}} \mathrm{acrtan}(\frac{x}{\sqrt{3}}) \Big|_{-1}^1 = \frac{1}{\sqrt{3}} \{ \arctan(\frac{1}{\sqrt{3}}) - \arctan(\frac{-1}{\sqrt{3}}) \}$

33. Table #34, $a = \sqrt{5}$: $3 \arcsin(\frac{x}{\sqrt{5}}) \Big|_1^2 = 3 \cdot \{ \arcsin(\frac{2}{\sqrt{5}}) - \arcsin(\frac{1}{\sqrt{5}}) \}$

35. Table #34, $a = 1/2$: $\frac{5}{2} \arcsin(\frac{x}{1/2}) \Big|_0^{0.1} = \frac{5}{2} \arcsin(0.2)$

37. $7 \cdot \ln(7) - 6$ 39. $3 \cdot \ln(3) - 2 \cdot \ln(2) - 1$ 41. $3\sqrt{18} + \frac{9}{2} \cdot \ln(\frac{3 + \sqrt{18}}{-3 + \sqrt{18}})$

43. Table #19a: $\frac{-\sin^2(x) \cdot \cos(x)}{3} - \frac{2}{3} \cos(x) + C$ 45. Table #20: $\frac{\cos^4(x) \cdot \sin(x)}{5} + \frac{4}{5} \{ \text{answer to number 44} \}$

47. Table #29: $x^2 \cdot \sin(x) + 2x \cdot \cos(x) - 2 \cdot \sin(x) + C$

49. Average of $\sin(x)$ on $[0,\pi]$ is $\frac{2}{\pi}$. Average of $\sin^2(x)$ on $[0,\pi]$ is $\frac{1}{2}$. $\frac{2}{\pi} > \frac{1}{2}$.

51. Using results from #50: (a) $\dfrac{1}{e-1}$ (b) $\dfrac{1}{9}\{10{\cdot}\ln(10) - 9\} \approx \dfrac{14.03}{9} \approx 1.56$

     (c) $\dfrac{1}{99}\{100{\cdot}\ln(100) - 99\} \approx 361.52/99 \approx 3.65$ (d) $\dfrac{1}{199}\{200{\cdot}\ln(200) - 199\} \approx 860.66/199 = 4.32$

53. (c) is largest.

55. approximately   (a) 1.57   (b) 2.94   (c) 3.04   (d) 3.07   (e) 3.09

Section 4.9

1. $h = \dfrac{(6-2)}{4} = 1$. $T_4 = \dfrac{1}{2}\{2.1 + 2(3.8) + 2(0.3) + 2(-0.9) + 2.2\} = 5.35$

    $S_4 = \dfrac{1}{3}\{2.1 + 4(3.8) + 2(0.3) + 4(-0.9) + 2.2\} = 5.5$

3. $h = \dfrac{1-(-3)}{8} = 0.5$.

    $T_8 = \dfrac{0.5}{2}\{4.2 + 2(1.8) + 2(0.7) + 2(1.5) + 2(3.4) + 2(4.3) + 2(3.5) + 2(-0.3) + -4.6\} = 7.35$

    $S_8 = \dfrac{0.5}{3}\{4.2 + 4(1.8) + 2(0.7) + 4(1.5) + 2(3.4) + 4(4.3) + 2(3.5) + 4(-0.3) + -4.6\} = 22/3 = 7.3333$

5. $T_4 = 4$ 7. $T_4 = 0.75$ 9. $T_4 = 1.896118898$
    $S_4 = 4$ $S_4 = 2/3 = 0.666$ $S_4 = 2.004559755$
    exact = 4   exact = 2/3   exact = 2

11. $T_6 = 1.088534906$ 13. $T_6 = 3.815780054$
     $S_6 = 1.090560447$ $S_6 = 3.826350295$

15. $T_6 = 0.8159928163$
     $S_6 = 0.8120491229$

17. (a) $M_2 = 0$ so the error bound = 0 (the trapezoidal approximation is exact)
     (b) $M_4 = 0$ so the error bound = 0 (the Simpson's rule approximation is exact)
     (c) n = 1   (d) n = 2 (must be an even integer)

19. $M_2 = \{$ max. of 6x on $[-1,1]\} = 6$.   f ''''(x) = 0 so $M_4 = 0$.

     (a) $|\text{ error }| \le \dfrac{2^3}{(12)(4^2)}(6) = 0.25$

     (b) $M_4 = 0$ so the error bound = 0 (the Simpson's rule approximation is exact)

     (c) Set $0.001 = \dfrac{2^3}{(12)(n^2)}(6)$ and solve for $n^2 = \dfrac{48}{(12)(0.001)} = 4{,}000$ so n = 63.25.

          Take n = 64 (to be certain that we have enough subintervals, always round UP).
     (d) n = 2 (must be an even integer)

21. $M_2 = \{$ max. of $|-\sin(x)|$ on $[0,\pi]\} = 1$. $M_4 = \{$ max. of $|\sin(x)|$ on $[0,\pi]\} = 1$.

   (a) $|$ error $| \le \dfrac{\pi^3}{(12)(4^2)}(1) = 0.1614910244$ (b) $|$ error $| \le \dfrac{\pi^5}{(180)(4^4)}(1) = 0.0066410522$

   (c) Set $0.001 = \dfrac{\pi^3}{(12)(n^2)}(1)$ so $n^2 = \dfrac{\pi^3}{(12)(0.001)} = 2{,}583.9$ and $n = 50.83$. **n = 51**.

   (d) Set $0.001 = \dfrac{\pi^5}{(180)(n^4)}(1)$ so $n^4 = \dfrac{\pi^5}{(180)(0.001)} = 1{,}700.1$ and $n = 6.42$. **n = 8**.

23. $h = 30$. $S_6 = \dfrac{30}{3}\{50 + 4(62) + 2(92) + 4(86) + 2(74) + 4(50) + 40\} = 12{,}140$ square feet.
   ($T_6 = 12{,}270$ square feet.)

25. $h = 50$.
   $S_6 = \dfrac{50}{3}\{100 + 4(150) + 2(170) + 4(160) + 2(120) + 4(70) + 30\} = 37{,}166.7$ square feet area.
      volume = (area)(depth) $= (37{,}166.7 \text{ ft}^2)(22 \text{ ft}) = 817{,}667 \text{ ft}^3$.
   $T_6 = \dfrac{50}{2}\{100 + 2(150) + 2(170) + 2(160) + 2(120) + 2(70) + 30\} = 36{,}750$ square feet area.
      volume = (area)(depth) $= (36{,}750 \text{ ft}^2)(22 \text{ ft}) = 808{,}500 \text{ ft}^3$.

27. Distance traveled by jogger is area under the graph of velocity ($f(x)$) vs. time (x) .
   area $\approx T_{10} = \dfrac{1}{2}\{0+840+1080+600+1000+1160+1040+880+720+520+180\} = 4{,}010$ ft.

29. – 32. on your own

33. (a) $L_4 = 3.5$ (b) $R_4 = 4.5$ (c) $M_4 = 4$ (d) exact $= 4$

35. (a) $L_4 = 0.75$ (b) $R_4 = 0.75$ (c) $M_4 = 0.625$ (d) exact ≈ 0.6667

37. (a) $L_4 = 1.8961$ (b) $R_4 = 1.8961$ (c) $M_4 = 2.0523$ (d) exact $= 2$

5.0   Applications of Definite Integrals

The last chapter introduced the idea of a definite integral as an "area" and a limit of Riemann sums, showed some of the properties of integrals, showed some ways to calculate values of definite integrals, and started to examine a few of their uses. This chapter focuses on several common applications of definite integrals. An obvious goal of the chapter is to enable you to use integration when you encounter these particular applications later in mathematics or in other fields. A deeper goal is to illustrate the process of going from a problem to an integral, a process much broader than these particular applications. If you understand the process, then you can understand the use of integrals in many other fields and can even develop the integrals needed to solve problems in new areas. A final goal is to give you additional practice evaluating definite integrals.

Each section in this chapter follows the same basic format. First a problem is described and some background information presented. Then the solution to the basic problem is approximated using a Riemann sum. An exact answer comes from taking a limit of the Riemann sum, and we get a definite integral. After looking at several examples of the same basic application, we will examine some variations of it.

5.1 VOLUMES OF SOLIDS

The last chapter emphasized a geometric interpretation of definite integrals as "areas" in two dimensions. This section emphasizes another geometrical use of integration, calculating volumes of solid three–

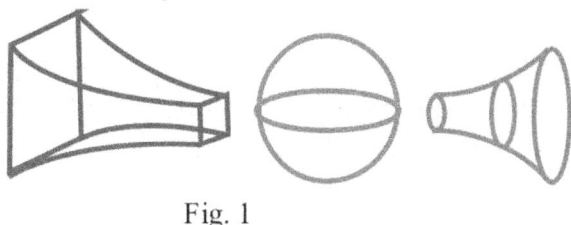

dimensional objects such as those shown in Fig. 1. Our basic approach is to cut the whole solid into thin "slices" whose volumes can be approximated, add the volumes of these "slices" together (a Riemann sum), and finally obtain an exact answer by taking a limit of the sums to get a definite integral.

Fig. 1

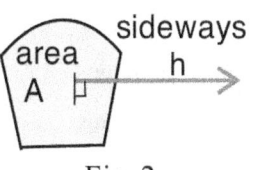

Fig. 2

The Building Blocks: Right Solids

A **right solid** is a three–dimensional shape swept out by moving a planar region A some distance h along a line perpendicular to the plane of A (Fig. 2). The region A is called a **face** of the solid, and the word "right" is used to indicate that the movement is along a line perpendicular, at

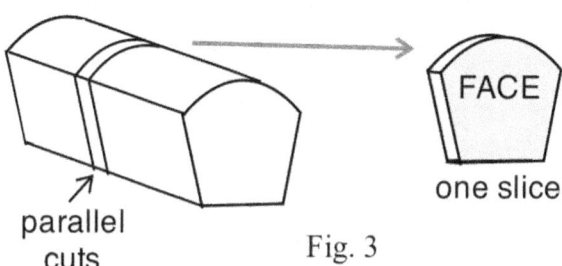

Fig. 3

a right angle, to the plane of A. Two parallel **cuts** produce one **slice** with two **faces** (Fig. 3): a slice has volume, and a face has area.

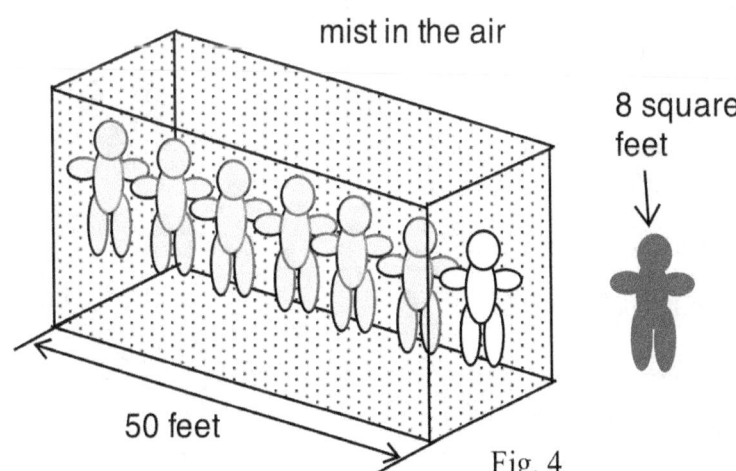

Fig. 4

Example 1: Suppose there is a fine, uniform mist in the air, and every cubic foot of mist contains 0.02 ounces of water droplets. If you run 50 feet in a straight line through this mist, how wet do you get? Assume that the front (or a cross section) of your body has an area of 8 square feet.

Solution: As you run, the front of your body sweeps out a "tunnel" through the mist (Fig. 4). The volume of the tunnel is the area of the front of your body multiplied by the length of the tunnel: volume = $(8 \text{ ft}^2)(50 \text{ ft}) = 400 \text{ ft}^3$. Since each cubic foot of mist held 0.02 ounces of water which is now on you, you swept out a total of $(400 \text{ ft}^3)\cdot(0.02 \text{ oz/ft}^3) = 8$ ounces of water. If the water was truly suspended and not falling, would it matter how fast you ran?

If A is a rectangle (Fig. 5), then the "right solid" formed by moving A along

the line is a 3–dimensional solid box B. The volume of B is

(area of A)·(distance along the line) = (base)·(height)·(width).

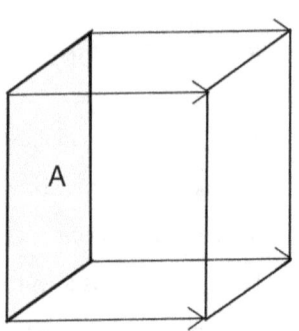

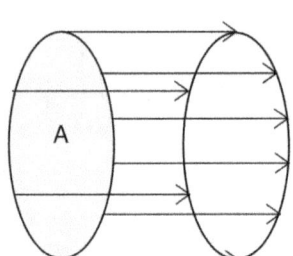

If A is a circle with radius r meters (Fig. 6), then the "right

solid" formed by moving A along the line h meters is a

right circular cylinder with volume equal to

Fig. 5: Solid box

$$\{\text{area of A}\}\cdot\{\text{distance along the line}\}$$
$$= \{\,\pi\,(r\,\text{ft})^2\,\}\cdot\{h\ \text{ft}\} = \{\,\pi{\cdot}r^2\,\text{ft}^2\,\}\cdot\{\ \text{h ft}\ \} = \pi\,r^2h\ \text{ft}^3\ .$$

Fig. 6: Solid cylinder

If we cut a right solid perpendicular to its axis (like cutting a loaf of bread), then each face (cross section) has

the same two–dimensional shape and area. In general, if a 3–dimensional right solid B is formed by moving

a 2–dimensional shape A along a line perpendicular to A, then the volume of B is *defined* to be

volume of B = (area of A)·(distance moved along the line perpendicular to A).

The volume of each right solid in Fig. 7 is (area of the base)·(height).

Example 2: Calculate the volumes of the right solids in Fig. 7.

Solution: (a) This cylinder is formed by moving the circular base

(area = $\pi r^2 = 9\pi$ in^2) along a line perpendicular to the base for 4 inches,

so the volume is (9π in^2)·(4 in) = 36π in^3 .

(b)   volume = (base area)·(distance along the line) = (8 m^2)·(3 m) = 24 m^3.

(c)   This shape is composed to two easy right solids with volumes

$V_1 = (\pi 3^2)\cdot(2) = 18\pi$ cm^3 and $V_2 = (6)(1)\cdot(2) = 12$ cm^3 , so the

total volume is $(18\pi + 12)$ cm^3 or approximately 68.5 cm^3 .

Practice 1: Calculate the volumes of the right solids in Fig. 8.

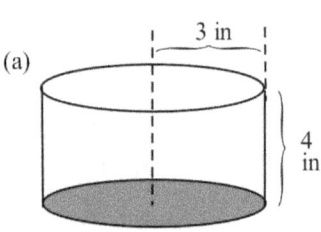

(a)

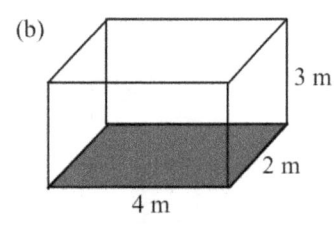

(b)

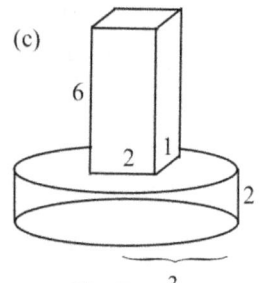

(c)

Fig. 7

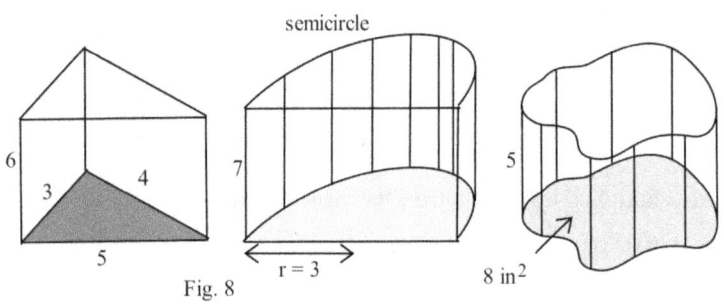

Fig. 8

Volumes of General Solids

A general solid can be cut into slices which are almost right solids. An individual slice may not be exactly a
right solid since its cross sections may have different areas. However, if the cuts are close together, then the
cross sectional areas will not change much within a single slice. Each slice will be almost a right solid, and
its volume will be almost the volume of a right solid.

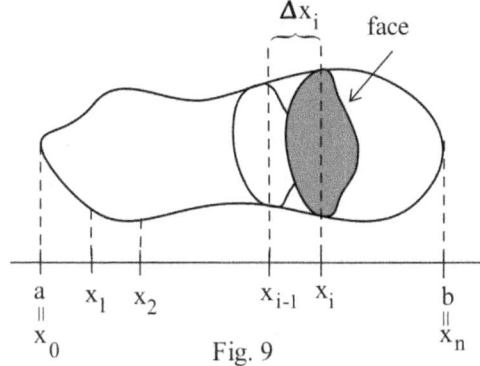

Fig. 9

Suppose an x–axis is positioned below the solid shape (Fig. 9), and let

$A(x)$ be the **area of the face** formed when the solid is cut at x

perpendicular to the x–axis. If $P = \{ x_0 = a, x_1, x_2, \ldots, x_n = b \}$ is a

partition of [a,b], and the solid is cut at each x_i , then each slice of the

solid is almost a right solid, and the volume of each slice is

approximately

(area of a face of the slice)·(thickness of the slice) $\approx A(x_i)\, \Delta x_i$.

The total volume V of the solid is approximately the sum of the volumes of the slices:

$$V = \sum \{\text{volume of each slice}\} \approx \sum A(x_i)\Delta x_i \quad \text{which is a Riemann sum.}$$

The limit, as the mesh of the partition approaches 0 (taking thinner and thinner slices), of the Riemann sum
is the definite integral of $A(x)$:

$$V \approx \sum A(x_i)\Delta x_i \longrightarrow \int_a^b A(x)\, dx \ .$$

Volume By Slices Formula

If   S is a solid and $A(x)$ is the area of the face formed by a cut

at x and perpendicular to the x–axis ,

then   the volume V of the part of S above the interval [a,b] is $V = \int_a^b A(x)\, dx$.

If S is a solid (Fig. 10), and $A(y)$ is the area of a face formed by a cut at
y perpendicular to the **y–axis**, then the volume of a slice with thickness Δy_i is
approximately $A(y_i)\cdot\Delta y_i$. The volume of the part of S between cuts at c and d
on the y–axis is

$$V = \int_c^d A(y)\ dy \ .$$

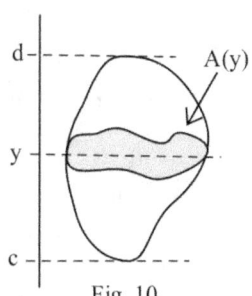

Fig. 10

Example 3:   For the solid in Fig. 11, the face formed by a cut at x is a rectangle with a base of 2 inches

and a height of cos(x) inches. (a) Write a formula for the approximate volume of the slice between x_{i-1}

and x_i . (b) Write and evaluate an integral for the volume of the solid for x

between 0 and $\pi/2$.

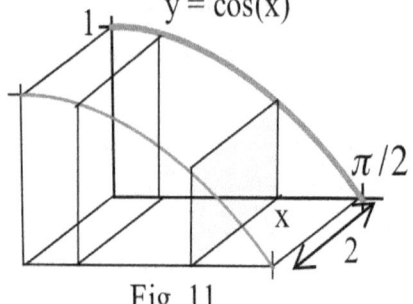

Fig. 11

Solution:

(a)   The volume of the slice $\approx$ (area of the face)·(thickness)

$$= (\text{base})·(\text{height})·(\text{thickness})$$
$$= (2 \text{ in})·(\cos(x_i) \text{ in})·(\Delta x_i \text{ in})$$
$$= 2\cos(x_i) \Delta x_i \text{ in}^3 .$$

(b)   Volume $= \displaystyle\int_a^b A(x)\, dx = \int_0^{\pi/2} 2\cos(x)\, dx = 2\sin(x)\,\Big|_0^{\pi/2}$

$$= 2\sin(\pi/2) - 2\sin(0) = 2 \text{ in}^3 .$$

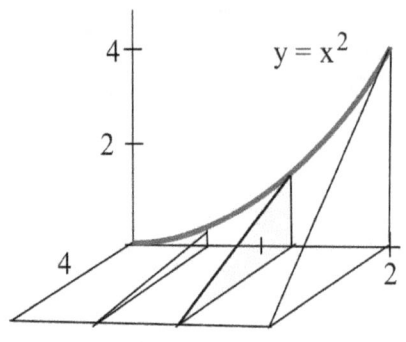

Fig. 12

Practice 2:   For the solid in Fig. 12, the face formed by a cut at x is a

triangle with a base of 4 inches and a height of x^2 inches.

(a)   Write a formula for the approximate volume of the slice between x_{i-1} and x_i . (b) Write and

evaluate an integral for the volume of the solid for x between 1 and 2.

Example 4:   For the solid in Fig. 13, each face formed by a cut at x is a circle with <u>diameter</u> $\sqrt{x}$.

(a)   Write a formula for the approximate volume of the slice

between x_{i-1} and x_i .

(b)   Write and evaluate an integral for the volume of the solid for

x between 1 and 4.

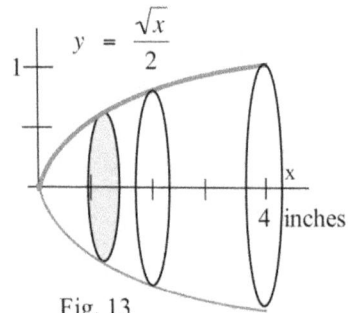

Fig. 13

Solution: (a) Each face is a circle with diameter $\sqrt{x_i}$, and the area of the circle

is

$$A(x_i) = \pi·(\text{radius})^2 = \pi(1/2 \text{ diameter})^2 = \pi(1/2 \sqrt{x_i})^2 = \pi x_i/4 .$$

The volume of the slice $\approx$ (area of the face)·(thickness) $= (\pi x_i/4)·(\Delta x_i)$

(b)   Volume $= \displaystyle\int_a^b A(x)\, dx = \int_1^4 \frac{\pi x}{4}\, dx = \frac{\pi}{4}·\frac{x^2}{2}\,\Big|_1^4 = \frac{\pi}{4}\frac{16}{2} - \frac{\pi}{4}\frac{1}{2} = \frac{15\pi}{8} \approx$

5.89 in^3.

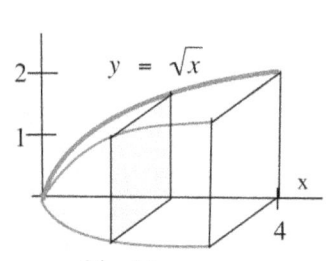

Fig. 14

Practice 3:   For the solid in Fig. 14, each face formed by a cut at x is a

square with height $\sqrt{x}$.

(a)   Write a formula for the approximate volume of the slice between x_{i-1} and

x_i .

(b)   Write and evaluate an integral for the volume of the solid for x between 1 and 4.

Example 5: Find the volume of the square–based pyramid in Fig. 15.

Solution: Each cut perpendicular to the y–axis yields a square face, but in order to find the area of each square we need a formula for the length of one side s of the square as a function of y, the location of the cut. Using similar triangles (Fig. 16), we know that

$$\frac{s}{10-y} = \frac{6}{10} \quad \text{so} \quad s = \frac{6}{10}(10-y) \;.$$

The rest of the solution is straightforward.

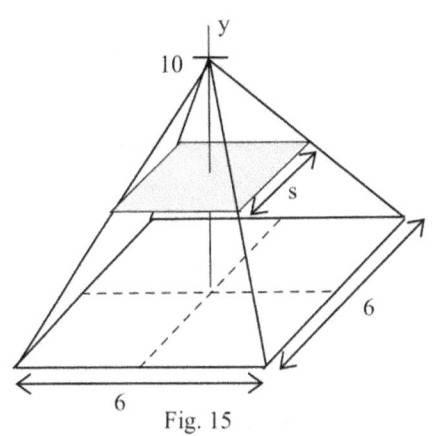

Fig. 15

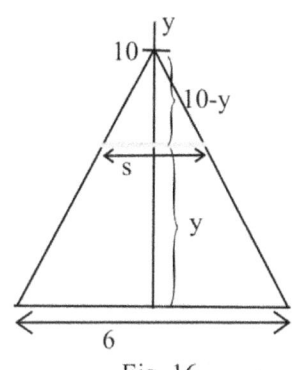

Fig. 16

$$A(y) = (\text{side})^2 = \{ \tfrac{3}{5}(10-y) \}^2 = \tfrac{9}{25}(100 - 20y + y^2) \quad \text{and}$$

$$V = \int_0^{10} A(y)\,dy \; = \; \int_0^{10} \tfrac{9}{25}(100 - 20y + y^2)\,dy \; = \tfrac{9}{25}(100y - 10y^2 + \tfrac{y^3}{3})\;\Big|_0^{10}$$

$$= \tfrac{9}{25}(1000 - 1000 + \tfrac{1000}{3}) \; - (0) = \tfrac{9}{25}\tfrac{1000}{3} = 120\,\text{ft}^3 \;.$$

Example 6:   A solid is built between the graphs of $f(x) = x+1$ and $g(x) = x^2$ by building squares with heights (sides) equal to the vertical distance between the graphs of f and g (Fig. 17). Find the volume of this solid for $0 \le x \le 2$.

Solution: The area of a square face is $A(x) = (\text{side})^2$, and the length of a side is either $f(x)-g(x)$ or $g(x)-f(x)$, depending on which function is higher at x. Fortunately, the side is squared in the area formula so it does not matter which is taller, and $A(x) = \{ f(x) - g(x) \}^2$. Then

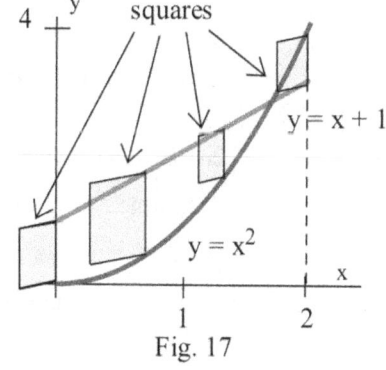

Fig. 17

$$V = \int_a^b A(x)\,dx \; = \int_0^2 \{ f(x) - g(x) \}^2\,dx \; = \int_0^2 \{ (x+1) - x^2 \}^2\,dx \; = \int_0^2 \{ (x+1) - x^2 \}^2\,dx$$

$$= \int_0^2 (1 + 2x - x^2 - 2x^3 + x^4)\,dx \; = x + x^2 - \tfrac{x^3}{3} - \tfrac{x^4}{2} + \tfrac{x^5}{5}\;\Big|_0^2 \; = \tfrac{26}{15} = 1\tfrac{11}{15} \;.$$

We saw earlier that areas can have nongeometric interpretations such as distance and total accumulation. Similarly, volumes can have nongeometric interpretations. If x represents an age in years, and $f(x)$ is the number of females in a population with age exactly equal to x, then the "area," $\int_{a}^{b} f(x)\, dx$, is the total number of females with ages between a and b (Fig. 18). If the birth rate for females of age x is $r(x)$, with units "births per female per year," (Fig. 19) then the "volume" of the solid in Fig. 20 is

$C = \int_{a}^{b} r(x){\cdot}f(x)\, dx$. C is the number of births during a year to females between the ages a and b , and

the units of C will be "births."

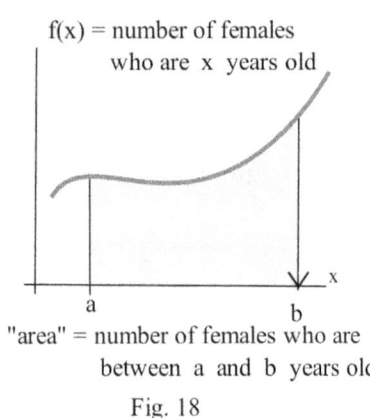

f(x) = number of females
who are x years old

"area" = number of females who are
between a and b years old

Fig. 18

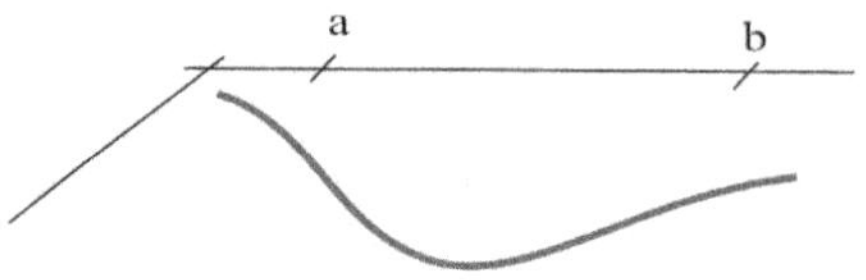

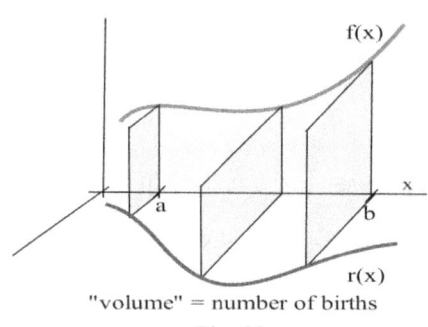

r(x) = reporductive rate of
females at age x

Fig. 19

"volume" = number of births

Fig. 20

Volumes of Revolved Regions

When a region is revolved around a line (Fig. 21) a right solid is formed. The face of each slice of the revolved region is a circle so the formula for the area of the face is easy: $A(x)$ = area of a circle = $\pi(\text{radius})^{2}$ where the radius is often a function of the location x. Finding a formula for the changing radius requires care.

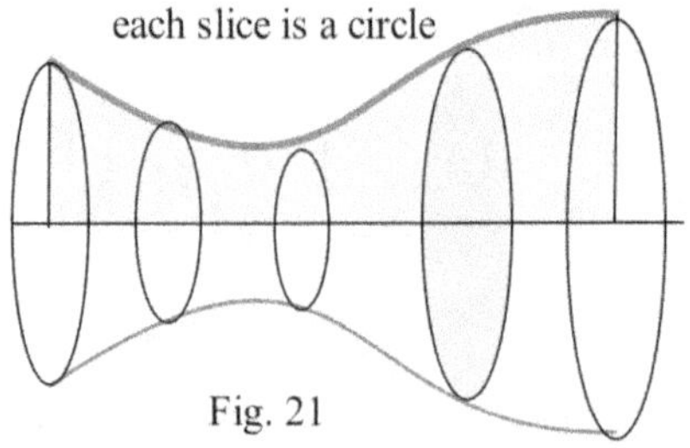

Example 7:   For $0 \le x \le 2$, the area between the graph of $f(x) = x^{2}$ and the horizontal line $y = 1$ is revolved about the horizontal line y=1 to form a solid (Fig. 22). Calculate the volume of the solid.

each slice is a circle

Fig. 21

Solution: The radius function is shown in the figure for several values of x. If $0 \le x \le 1$, then $r(x) = 1 - x^{2}$, and if $1 \le x \le 2$ then $r(x) = x^{2} - 1$. Fortunately, however, $A(x) = \pi \left\{ r(x) \right\}^{2}$ always uses the square of $r(x)$ and the squares of $1 - x^{2}$ and $x^{2} - 1$ are equal.

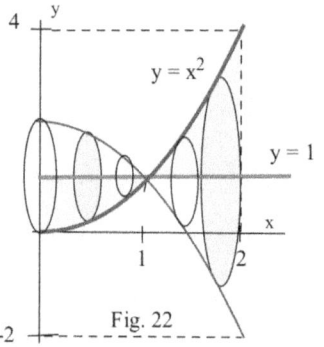

$y = x^2$

$y = 1$

Fig. 22

$A(x) = \pi \{ r(x) \}^2 = \pi \{ x^2 - 1 \}^2 = \pi \{ x^4 - 2x^2 + 1 \}$, and

$$V = \int_0^2 \pi\{ x^4 - 2x^2 + 1 \} \, dx = \pi\left(\frac{x^5}{5} - \frac{2}{3}x^3 + x\right)\Big|_0^2 = \frac{46}{15}\pi \approx 9.63 \ .$$

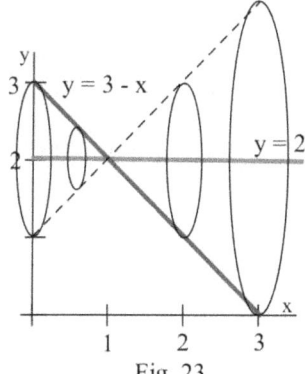

Practice 4:   A solid of revolution is formed when the region between

$f(x) = 3 - x$ and the horizontal line $y = 2$ is revolved about the

line $y=2$ for $0 \le x \le 3$ (Fig. 23). Find the volume of the solid.

Fig. 23

Volumes of Revolved Regions ("Disks")

If   the region formed between f, the horizontal line $y = L$, and the interval [a, b]

is revolved about the horizontal line $y = L$, (Fig. 24)

then   the volume is $V = \int_a^b A(x) \, dx = \int_a^b \pi \cdot (\text{radius})^2 \, dx = \int_a^b \pi\{ f(x) - L \}^2 \, dx$.

This is called the "disk" method because the shape of each thin

slice is a circular disk. If the region between f and the x–axis

(L=0) is revolved about the x–axis, then the previous formula reduces to

$$V = \int_a^b \pi \{ f(x) \}^2 \, dx \ .$$

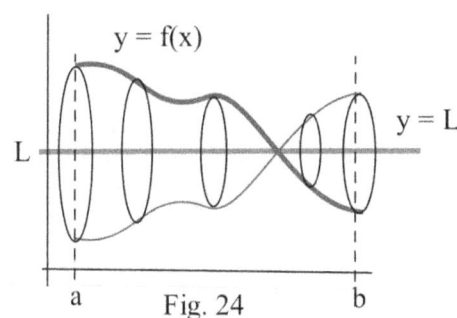

Fig. 24

Example 8:   Find the volume generated when the region between one arch of the sine curve $(0 \le x \le \pi)$

and the x–axis is revolved about (a) the x–axis and (b) the line y=1/2.

Solution: (a) $V = \int_a^b \pi \cdot (\text{radius})^2 \, dx = \int_0^\pi \pi\{ \sin(x) \}^2 \, dx = \pi \int_0^\pi \sin^2(x) \, dx = \frac{\pi}{2}\int_0^\pi 1 - \cos(2x) \, dx$

$= \frac{\pi}{2} \{ x - \frac{\sin(2x)}{2} \}\Big|_0^\pi = \frac{\pi}{2} \{ \pi - 0 \} = \frac{\pi^2}{2} \approx 4.93$.

(b) $V = \int_a^b \pi \cdot (\text{radius})^2 \, dx = \int_0^\pi \pi\{ \sin(x) - \frac{1}{2} \}^2 \, dx = \pi\int_0^\pi \{ \sin^2(x) - \sin(x) + \frac{1}{4} \} \, dx$

$= \pi \{ \frac{\pi}{2} - 2 + \frac{\pi}{4} \} \approx 1.12$.

Practice 5:   Find the volumes swept out when

(a) the region between $f(x) = x^2$ and the x–axis, for $0 \le x \le 2$, is revolved about the x–axis, and

(b) the region between $f(x) = x^2$ and the line y=2 , for $0 \le x \le 2$, is revolved about the line y=2.

Example 9:   Given that $\int_1^5 f(x)\, dx = 4$ and $\int_1^5 \{\, f(x)\, \}^2\, dx = 7$. Represent the volumes of the solids

(a), (b) and (c) in Fig. 25 as definite integrals and evaluate the integrals.

Solution: (a) $V = \int_1^5 \pi\cdot(\text{ radius })^2\, dx = \int_1^5 \pi\cdot\{\, f(x)\, \}^2\, dx = \pi \int_1^5 f^2(x)\, dx = 7\pi$.

(b) $V = \int_1^5 \pi\cdot(\text{ radius })^2\, dx = \int_1^5 \pi\cdot\{\, f(x) - (-1)\, \}^2\, dx = \pi \int_1^5 \{\, f^2(x) + 2f(x) + 1\, \}\, dx$

$$= \pi \left\{ \int_1^5 f^2(x)\, dx + 2 \int_1^5 f(x)\, dx + \int_1^5 1\, dx \right\} = \pi\{\, 7 + 2\cdot 4 + 4\, \} = 19\pi .$$

(c) $V = \int_1^5 \pi\cdot(\text{ radius })^2\, dx = \int_1^5 \pi\cdot\{\, f(x)/2\, \}^2\, dx = \frac{\pi}{4} \int_1^5 f^2(x)\, dx = \frac{7\pi}{4}$.

Practice 6:   Set up and evaluate the integral for the volume of (d) in Fig. 25.

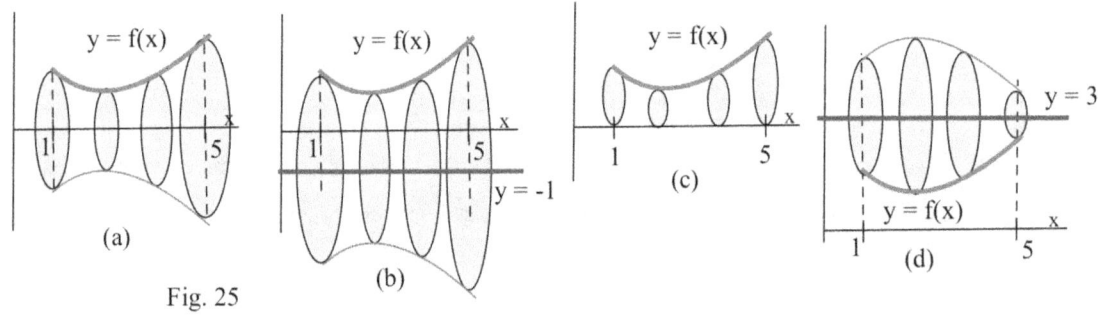

Fig. 25

Solids With Holes

The previous ideas and techniques can also be used to find the volumes of solids with holes in them. If A(x)

is the area of the face formed by a cut at x, then it is still true that the volume is $V = \int_a^b A(x)\, dx$. However, if

the solid has holes, then some of the faces will also have holes and a formula for A(x) may be more

complicated.

Sometimes it is easier to work with two integrals and then subtract: (i) calculate the volume S of the solid without the hole, (ii) calculate the volume H of the hole, and (iii) subtract H from S.

Example 10: Calculate the volume of the solid in Fig. 26.

Solution: The face for a slice at x , has area

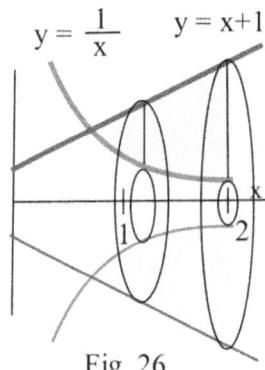

$y = \dfrac{1}{x}$ $y = x+1$

$$A(x) \quad = \{\text{area of large circle}\} - \{\text{area of small circle}\}$$

$$= \pi\{\text{large radius}\}^2 - \pi\{\text{small radius}\}^2$$

$$= \pi\{ x + 1 \}^2 - \pi\{1/x \}^2 \ = \ \pi(x^2 + 2x + 1 - 1/x^2). \quad \text{Then}$$

$$\text{Volume} = \int_a^b A(x)\,dx \ = \ \int_1^2 \pi(x^2 + 2x + 1 - 1/x^2)\,dx$$

Fig. 26

$$= \pi\{ \tfrac{1}{3} x^3 + x^2 + x + 1/x \}\Big|_1^2 \ \approx 18.33 \ .$$

Alternately, the volume of the solid with the large circular faces is $\displaystyle\int_1^2 \pi(x^2 + 2x + 1)\,dx = \dfrac{19\pi}{3} \approx 19.90$,

and the volume of the hole is $\displaystyle\int_1^2 \pi(1/x^2)\,dx \ = \dfrac{\pi}{2} \approx 1.57$ so the

volume we want is 19.90 − 1.57 = 18.33 .

Practice 7:   Calculate the volume of the solid in Fig. 27 .

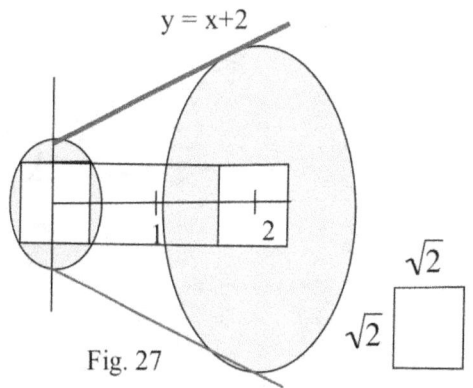

$y = x+2$

$\sqrt{2}$

$\sqrt{2}$ [square] $\sqrt{2}$

Fig. 27

WRAP UP

At first, all of these volumes may seem overwhelming — there are so many possible solids and formulas and different cases. If you concentrate on the differences, it is very complicated. Instead, focus on the pattern of **cutting, finding areas of faces, volumes of slices, and adding**. With that pattern firmly in mind, you can reason your way to the definite integral. Try to make cuts so the resulting faces have regular shapes (rectangles, triangles, circles) whose areas you can calculate. Try not to let the complexity of the whole solid confuse you. Sketch the shape of **one** face and label its dimensions. If you can find the area of **one** face in the middle of the solid, you can usually find the pattern for all of the faces and then you can easily set up the integral.

PROBLEMS

In problems 1 – 6, use the values given in the tables to calculate the volumes of the solids. (Fig. 28 – 33)

Table 1: (Fig. 28)	box	base	height	thickness
	1	5	6	1
	2	4	4	2
	3	3	3	1

Table 2: (Fig. 29)	box	base	height	thickness
	1	5	6	2
	2	5	4	1
	3	3	3	1
	4	2	2	1

Table 3: (Fig. 30)	disk	radius	thickness
	1	4	0.5
	2	3	1
	3	1	2

Table 4: (Fig. 31)	disk	height	thickness
	1	8	0.5
	2	6	1
	3	2	2

Table 5: (Fig. 32)	slice	face area	thickness
	1	9	0.2
	2	6	0.2
	3	2	0.2

Table 6: (Fig. 33)	slice	rock area	min. area	thickness
	1	4	1	0.6
	2	12	1	0.6
	3	20	4	0.6
	4	10	3	0.6
	5	8	2	0.6

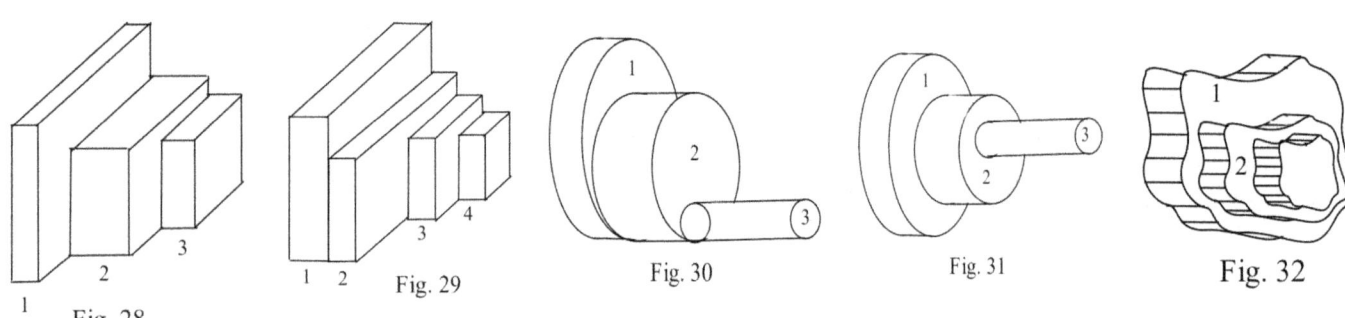

Fig. 28   Fig. 29   Fig. 30   Fig. 31   Fig. 32

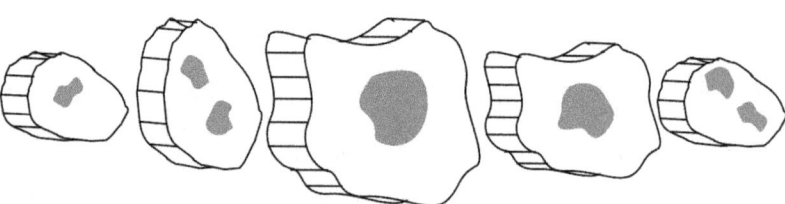

Fig. 33

In problems 7 – 12, represent each volume as an integral and evaluate the integral.

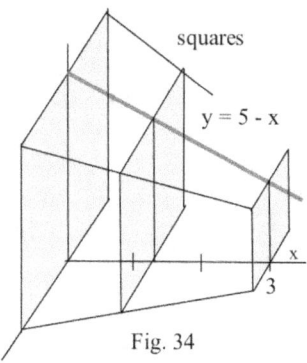

squares

$y = 5 - x$

Fig. 34

7. Fig. 34. For $0 \leq x \leq 3$, each face is a rectangle with base 2 inches and height $5-x$ inches.

8. Fig. 35. For $0 \leq x \leq 3$, each face is a rectangle with base x inches and height x^2 inches.

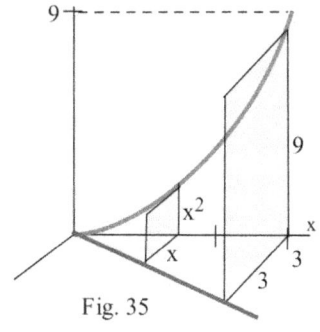

Fig. 35

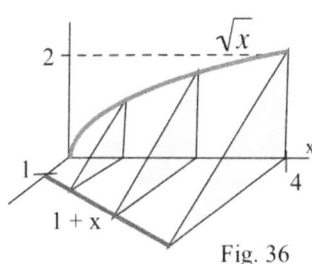

$\sqrt{x}$

$1 + x$

Fig. 36

9. Fig. 36. For $1 \leq x \leq 4$, each face is a triangle with base $x + 1$ meters and height $\sqrt{x}$ meters.

10. Fig. 37. For $0 \leq x \leq 3$, each face is a circle with height (diameter) $4 - x$ meters.

11. Fig. 38. For $2 \leq x \leq 4$, each face is a circle with height (diameter) $4 - x$ meters.

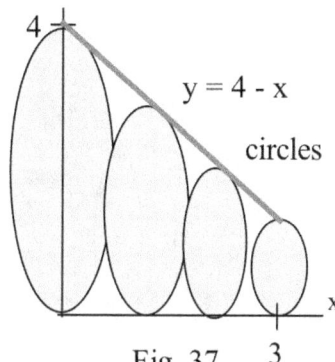

$y = 4 - x$

circles

Fig. 37   3

12. Fig. 39. For $0 \leq x \leq 2$, each face is a square with a side extending from $y = 1$ to $y = x + 2$.

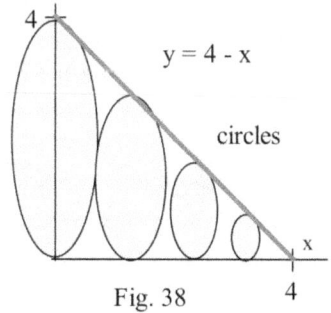

$y = 4 - x$

circles

Fig. 38   4

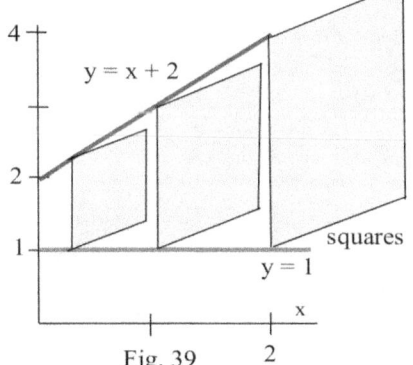

$y = x + 2$

squares

$y = 1$

Fig. 39   2

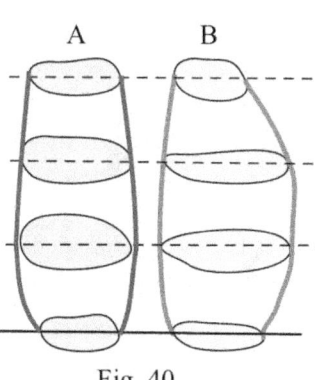

A   B

Fig. 40

13. Suppose A and B are solids (Fig. 40) so that every horizontal cut produces faces of A and B that have equal areas. What can we conclude about the volumes of A and B? Justify your answer.

In problems 14 – 22, represent each volume as an integral and evaluate the integral.

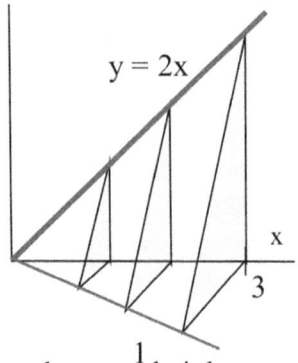

$y = 2x$

base $= \dfrac{1}{2}$ height

Problem 14

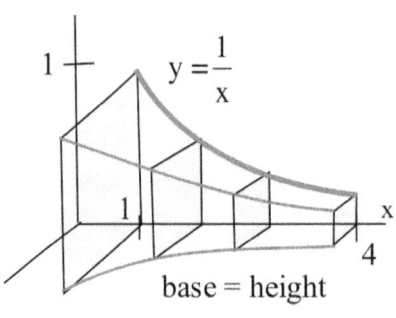

$y = \dfrac{1}{x}$

base = height

Problem 15

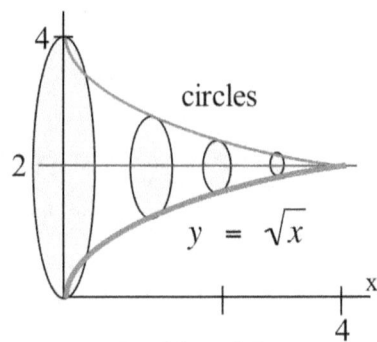

circles

$y = \sqrt{x}$

Problem 16

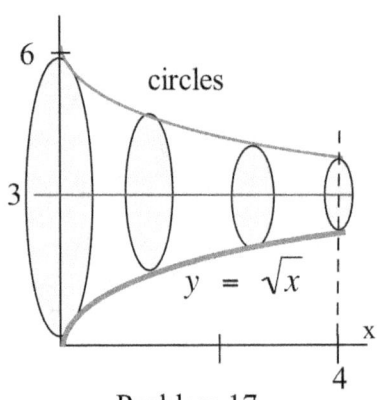

circles

$y = \sqrt{x}$

Problem 17

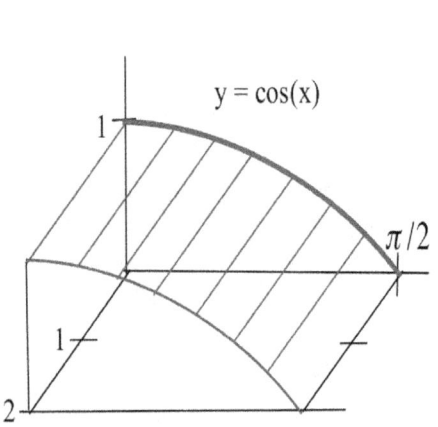

$y = \sin(x)$ circles

Problem 18

$y = \cos(x)$

Problem 20

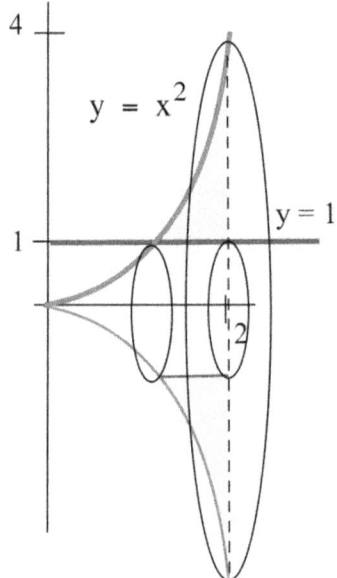

$y = x^2$

$y = 1$

Problem 21

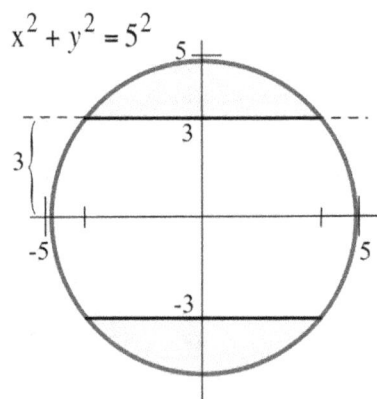

height $= x^2$

base $= x$

Problem 19

$x^2 + y^2 = 5^2$

A sphere with radius 5 has a hole of
radius 3 drilled through its center

Problem 22

23. Calculate the volume of a sphere of radius 2. (A sphere is formed when the region bounded by the
     x–axis and the top half of the circle $x^2 + y^2 = 2^2$ is revolved about the x–axis.)

24. Determine the volume of a sphere of radius r. (A sphere is swept out when the region bounded by the
     x–axis and the top half of the circle $x^2 + y^2 = r^2$ is revolved about the x–axis.)

25. Calculate the volume swept out when the top half of the

     elliptical region bounded by $\dfrac{x^2}{5^2} + \dfrac{y^2}{3^2} = 1$ is revolved

     around the x–axis (Fig. 41). $(y = +3 \sqrt{1 - (x^2/25)} \;)$

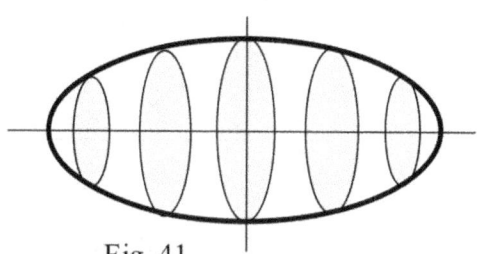

Fig. 41

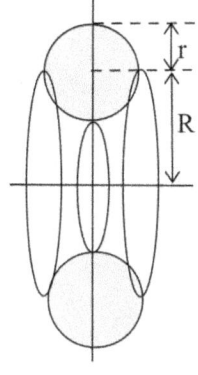

torus ("doughnut")
   Fig. 42

26. Calculate the volume swept out when the top half of the elliptical region bounded by

     $\dfrac{x^2}{a^2} + \dfrac{y^2}{b^2} = 1$ is revolved around the x–axis. $(y = +b \sqrt{1 - (x^2/a^2)} \;)$

27. Determine the volume of the "doughnut" in Fig. 42. (The top half of the circle is

     given by $f(x) = R + \sqrt{r^2 - x^2}$ and the bottom half is given by

     $g(x) = R - \sqrt{r^2 - x^2}$. (It is easier to use a single integral for this problem.)

28. (a) Find the **area** between $f(x) = 1/x$ and the x–axis for $1 \le x \le 10, 1 \le x \le 100,$ and
          $1 \le x \le A.$ What is the limit of the area for $1 \le x \le A$ as $A \to \infty$?

     (b) Find the **volume** swept out when the region in part (a) is revolved about the
          x–axis for $1 \le x \le 10, 1 \le x \le 100,$ and $1 \le x \le A.$ What is the limit of the volumes for $1 \le x \le A$
          as $A \to \infty$?

29. Personal Calculus:" Describe a **practical** way to determine the volume of your hand and arm up to the
     elbow.

30. Personal Calculus:" Most people have a body density between .95 and 1.05 times the density of water
     which is 62.5 pounds per cubic foot. Use your weight to estimate the volume of your body. (If you
     float in fresh water, your body density is less than 1.)

Volumes of "right cones"

31. Calculate (a) the volume of the right solid in Fig. 43a, (b) the volume of the "right cone" in Fig. 43b,

and (c) the ratio of the "right cone" volume to the right solid volume. (square cross sections)

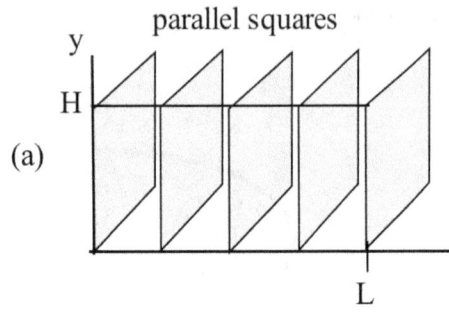

parallel squares

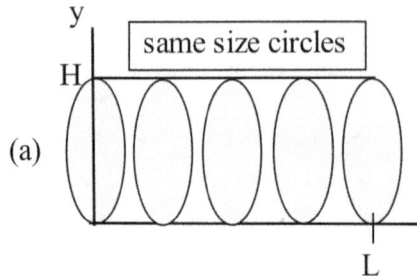

same size circles

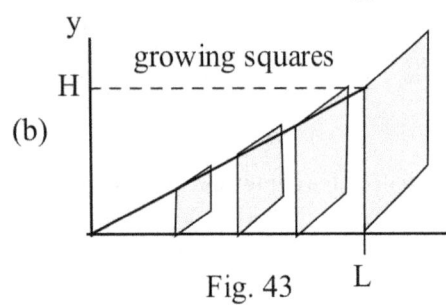

growing squares

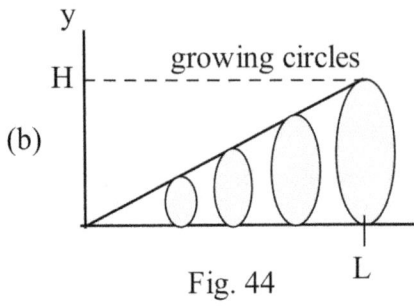

growing circles

Fig. 43   Fig. 44

32. Calculate (a) the volume of the right solid in Fig. 44a, (b) the volume of the "right cone" in Fig. 44b,

and (c) the ratio of the "right cone" volume to the right solid volume. (circular cross sections)

33. The "blob" in Fig. 45 has area B.

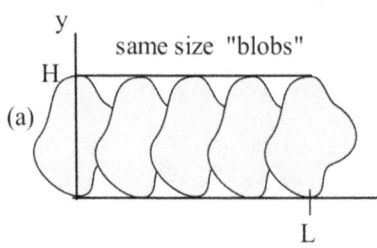

same size "blobs"

(a) Calculate the volume of the right solid in Fig. 45a.

(b) If a "right cone" is formed (Fig. 45b), then the cross

section area at x is $A(x) = (B/L^2)x^2$.

Find the volume of the "right cone".

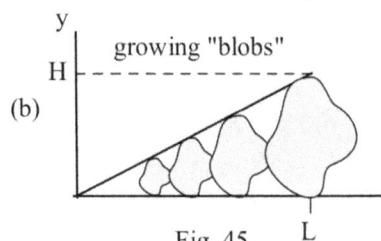

growing "blobs"

(c) Find the ratio of the "right cone" volume to the right solid

volume.

Fig. 45

34. Represent each volume as a definite integral.

A.

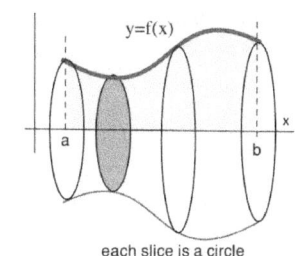

each slice is a circle

B.

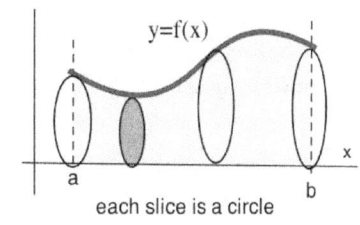

each slice is a circle

C.

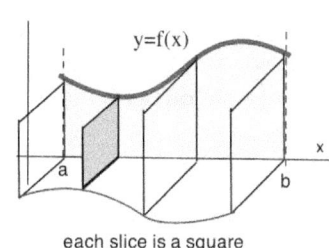

each slice is a square

D.

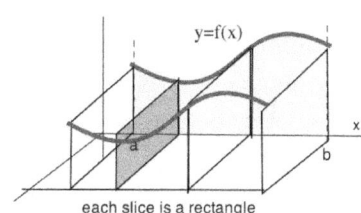

each slice is a rectangle

E.

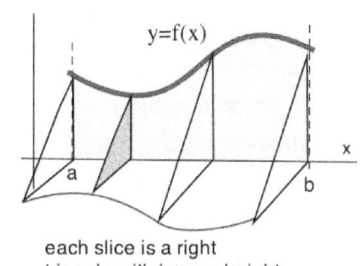

each slice is a right
triangle with base = height

F.

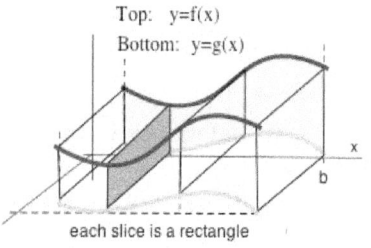

each slice is a rectangle

G.

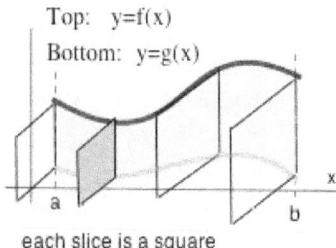

each slice is a square

H.

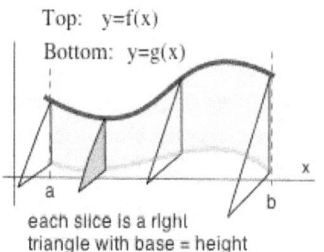

each slice is a right
triangle with base = height

I.

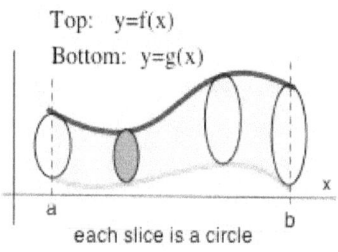

each slice is a circle

J.

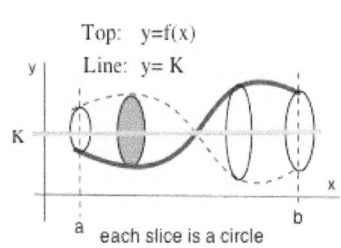

each slice is a circle

K.

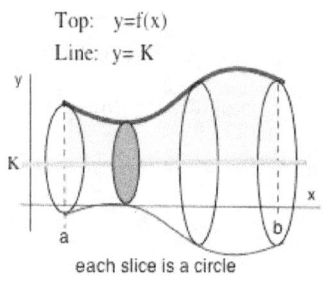

each slice is a circle

L.

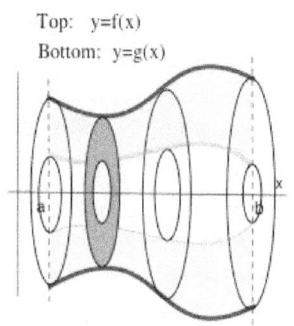

Section 5.1   **PRACTICE Answers**

Practice 1:   (a)   Triangular base: v = (base area)·(height) = ($\frac{1}{2}$ ·3·4)·(6) = 36.

(b)   Semicircular base: v = (base area)·(height) = $\frac{1}{2}$ ·(π ·3^2)·(7) ≈ 98.96 .

(c)   Strangely shaped base: v = (base area)·(height) = (8 in^2)·(5 in) = 40 in^3 .

Practice 2:   (a) $v_i \approx$ (area of face)(thickness) $\approx (\ \frac{1}{2} \cdot 4 \cdot x_i^2\ ) \cdot (\ \Delta x_i\ ) = 2x_i^2\ \Delta x_i$.

(b) $v = \int_1^2 2x^2\ dx = \frac{2}{3}\ x^3\ \Big|_1^2 = \frac{16}{3} - \frac{2}{3} = \frac{14}{3}$ **cubic inches.**

Practice 3:   (a) $v_i \approx$ (area of face)(thickness) $\approx (\ \sqrt{x_i}\ )^2 (\ \Delta x_i\ ) = x_i\ \Delta x_i$.

(b) $v = \int_1^4 x\ dx = \frac{1}{2}\ x^2\ \Big|_1^4 = 8 - \frac{1}{2} = 7.5$.

Practice 4: $v_i \approx$ (area of face)(thickness) $\approx (\ \pi \cdot r^2\ )$(thickness)

$$= (\ \pi \cdot (\ (3-x_i) - 2\ )^2 (\ \Delta x_i\ ) = \pi(1 - 2x_i + x_i^2)\ \Delta x_i$$.

Then volume $= \int_0^3 \pi(1 - 2x + x^2)\ dx = \pi(x - x^2 + \frac{1}{3}\ x^3\ )\ \Big|_0^3 = 3\pi \approx$ **9.42** .

Practice 5:   (a) $v = \int_a^b \pi(\ \text{radius}\ )^2\ dx = \int_0^2 \pi(\ x^2\ )^2\ dx = \frac{\pi}{5}\ x^5\ \Big|_0^2 = \frac{32\pi}{5} = 20.1$.

(b) $v = \int_a^b \pi(\ \text{radius}\ )^2\ dx = \int_0^2 \pi(\ 2 - x^2\ )^2\ dx = \int_0^2 \pi(\ 4 - 4x^2 + x^4\ )\ dx$

$$= \pi(\ 4x - \frac{4}{3}\ x^3 + \frac{1}{5}\ x^5\ )\ \Big|_0^2 = \frac{56\pi}{15} \approx 11.73$$.

Practice 6:   (d) $v = \int_a^b \pi(\ \text{radius}\ )^2\ dx = \int_1^5 \pi(\ 3 - f(x)\ )^2\ dx = \pi \int_1^5 9 - 6f(x) + f^2(x)\ dx$

$$= \pi \int_1^5 (\ 9 - 6f(x) + f^2(x)\ )\ dx \qquad = \pi \int_1^5 9\ dx - 6\pi \int_1^5 f(x)\ dx + \pi \int_1^5 f^2(x)\ dx$$

$$= \pi(36) - 6\pi(4) + \pi(7) = 19\pi \approx \textbf{59.69}$$.

(The values "7" and "4" are given in Example 9) .

Practice 7:   The volume we want can be obtained by subtracting the volume of the "box" from the

volume of the truncated cone generated by the rotated line segment.

volume of truncated cone $= \int_a^b \pi(\ \text{radius}\ )^2\ dx = \int_0^2 \pi(\ x + 2\ )^2\ dx$

$$= \pi \int_0^2 x^2 + 4x + 4\ dx = \pi\{\ \frac{1}{3}\ x^3 + 2x^2 + 4x\ \}\ \Big|_0^2 = \frac{56}{3}\ \pi \approx 58.64$$.

volume of "box" $= $ (length)(width)(height) $= 2\ (\ \sqrt{2}\ )(\ \sqrt{2}\ ) = 4.$

The volume we want is $\frac{56}{3}\ \pi - 4 \approx 54.64$.

5.2   LENGTHS OF CURVES & AREAS OF SURFACES OF REVOLUTION

This section introduces two additional geometric applications of integration: finding the length of a curve and finding the area of a surface generated when a curve is revolved about a line. The general strategy is the same as before: partition the problem into small pieces, approximate the solution on each small piece, add the small solutions together in the form of a Riemann sum, and finally, take the limit of the Riemann sum to get a definite integral.

ARC LENGTH: How Long Is A Curve?

In order to understand an object or an animal, we often need to know how it moves about its environment and how far it travels. We need to know the length of the path it moves along. If we know the object's location at successive times, then it is straightforward to calculate the distances between those locations and add them together to get a total distance.

Example 1:   In order to study the movement of whales, a scientist attached

a small radio transmitter to the fin of a whale and tracked the location of

the whale at 1 hour time intervals over a period of several weeks. The

data for a 5 hour period is shown in Fig. 1. How far did the whale swim

during the first 3 hours?

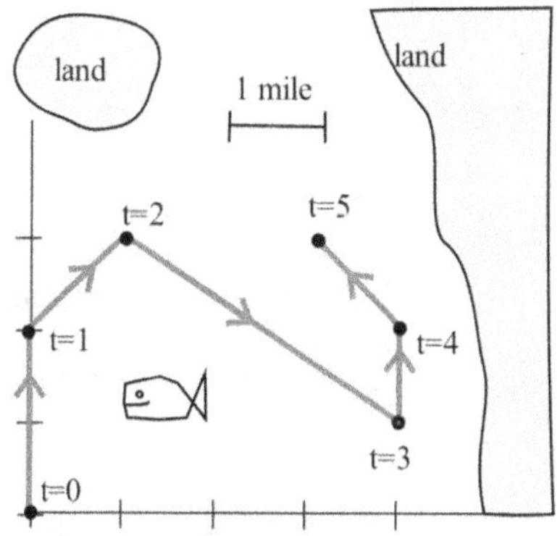

Fig. 1

Solution: In moving from the point (0,0) to the point (0,2), the

whale traveled at least 2 miles. Similarly, the whale traveled at

least $\sqrt{(1-0)^2 + (3-2)^2} \approx 1.4$ miles during the second hour and at least $\sqrt{(4-1)^2 + (1-3)^2} \approx 3.6$

miles during the third hour. The scientist concluded that the whale swam **at least** $2 + 1.4 + 3.6 = 7$

miles during the 3 hours.

Practice 1:   How far did the whale swim during the entire 5 hour period?

The scientist noted that the whale did not swim in a straight line from location to location so its actual swimming distance was more than 7 miles for the first 3 hours. The scientist hoped to get better distance estimates in the future by determining the whale's position over shorter, five–minute time intervals.

Our strategy for finding the length of a curve will be similar to the one the scientist used, and if the locations are given by a formula, then we can calculate the successive locations over very short intervals and get very good approximations of the total path length. In fact, we can get the exact length of the path by evaluating a definite integral.

Suppose C is a curve, and we pick some points (x_i, y_i) along C (Fig. 2) and connect the points with straight line segments. Then the sum of the lengths of the line segments will approximate the length of C. We can think of this as pinning a string to the curve at the selected points, and then measuring the length of the string as an approximation of the length of the curve. Of course, if we only pick a few points as in Fig. 2, then the total length approximation will probably be rather poor, so eventually we want lots of points (x_i, y_i), close together all along C.

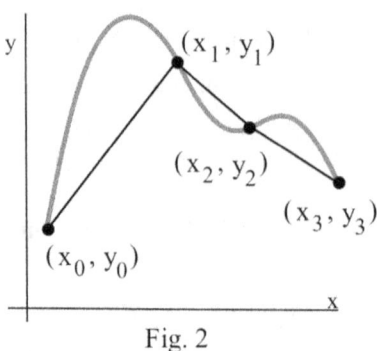

Fig. 2

Suppose the points are labeled so (x_0, y_0) is one endpoint of C and (x_n, y_n) is the other endpoint and that the subscripts increase as we move along C. Then the distance between the successive points (x_{i-1}, y_{i-1}) and (x_i, y_i) is $\sqrt{(\Delta x_i)^2 + (\Delta y_i)^2}$, and the total length of the line segments is simply the sum of the successive lengths. This is an important approximation of the length of C, and all of the integral representations for the length of C come from it.

The length of the curve C is approximately $\sum \sqrt{(\Delta x_i)^2 + (\Delta y_i)^2}$.

Example 2:   Use the points $(0,0), (1,1)$, and $(3,9)$ to approximate the length of $y = x^2$ for $0 \le x \le 3$.

Solution: The lengths of the two linear pieces in Fig. 3 are
$$\sqrt{1^2 + 1^2} = \sqrt{2} \approx 1.41 \text{ and}$$
$$\sqrt{2^2 + 8^2} = \sqrt{68} \approx 8.25 \text{ so the length of the curve}$$
is approximately $1.41 + 8.25 = 9.66$.

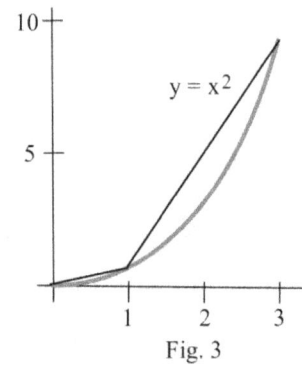

Fig. 3

Practice 2:   Get a better approximation of the length of $y = x^2$ for $0 \le x \le 3$ by using the points $(0,0), (1,1), (2,4)$, and $(3,9)$. Is your approximation longer or shorter than the actual length?

The summation does not have the form $\sum f(c_i) \cdot \Delta x_i$ so it is not a Riemann sum. It is, however,

algebraically equivalent to several Riemann sums, and each one leads to a definite integral representation

for the length of C.

(a) **y = f(x)**: When y is a function of x, we can factor $(\Delta x_i)^2$ from inside the radical and simplify.

$$\text{Length of } C \approx \sum \sqrt{(\Delta x_i)^2 + (\Delta y_i)^2} = \sum \sqrt{(\Delta x_i)^2 \cdot \left\{ 1 + \left(\frac{\Delta y_i}{\Delta x_i}\right)^2 \right\}}$$

$$= \sum \Delta x_i \sqrt{1 + \left(\frac{\Delta y_i}{\Delta x_i}\right)^2} \quad \text{(Riemann sum)} \longrightarrow \boxed{\int_{x=a}^{x=b} \sqrt{1 + \left(\frac{dy}{dx}\right)^2} \ dx}$$

(b) **x = g(y)**: When x is a function of y, we can factor $(\Delta y_i)^2$ from inside the radical and simplify.

$$\text{Length of } C \approx \sum \sqrt{(\Delta x_i)^2 + (\Delta y_i)^2} = \sum \sqrt{(\Delta y_i)^2 \cdot \left\{ \left(\frac{\Delta x_i}{\Delta y_i}\right)^2 + 1 \right\}}$$

$$= \sum \Delta y_i \sqrt{\left(\frac{\Delta x_i}{\Delta y_i}\right)^2 + 1} \quad \text{(Riemann sum)} \longrightarrow \boxed{\int_{y=c}^{y=d} \sqrt{\left(\frac{dx}{dy}\right)^2 + 1} \ dy}$$

(c) **Parametric equations:** When x and y are functions of t, x = x(t) and y = y(t), for $\alpha \le t \le \beta$

we can factor $(\Delta t_i)^2$ from inside the radical and simplify.

$$\text{Length of } C \approx \sum \sqrt{(\Delta x_i)^2 + (\Delta y_i)^2} = \sum \sqrt{(\Delta t_i)^2 \cdot \left\{ \left(\frac{\Delta x_i}{\Delta t_i}\right)^2 + \left(\frac{\Delta y_i}{\Delta t_i}\right)^2 \right\}}$$

$$= \sum \Delta t_i \sqrt{\left(\frac{\Delta x_i}{\Delta t_i}\right)^2 + \left(\frac{\Delta y_i}{\Delta t_i}\right)^2} \quad \text{(Riemann sum)} \longrightarrow \boxed{\int_{t=\alpha}^{t=\beta} \sqrt{\left(\frac{dx}{dt}\right)^2 + \left(\frac{dy}{dt}\right)^2} \ dt}$$

The integrals in (a), (b), and (c) each represent the length of C, and we can use whichever one is more

convenient. Unfortunately, for most functions these integrands do not have easy antiderivatives, and most arc

length integrals must be approximated using one of our approximate integration methods or a calculator.

Example 3: Represent the length of each curve as a definite integral.

(a) The length of $y = x^2$ between (1,1) and (4,16).

(b) The length of $x = \sqrt{y}$ between (1,1) and (4,16).

(c) The length of the parametric curve x(t) = cos(t) and y(t) = sin(t) for $0 \le t \le 2\pi$.

Solution: (a) Length $= \int\limits_{x=1}^{x=4} \sqrt{1 + (dy/dx)^2} \ dx = \int\limits_{x=1}^{x=4} \sqrt{1 + 4x^2} \ dx \approx 15.34$.

(b) Length $= \int\limits_{y=1}^{y=16} \sqrt{(dx/dy)^2 + 1} \ dy = \int\limits_{y=1}^{y=16} \sqrt{\frac{1}{4y} + 1} \ dy \approx 15.34$.

The values of the integrals in (a) and (b) were approximated using Simpson's rule with n = 10. It is not an accident that the lengths in (a) and (b) are equal. Why not?

(c) Length $= \int\limits_{t=0}^{t=2\pi} \sqrt{(-\sin(t))^2 + (\cos(t))^2} \ dt = \int\limits_{t=0}^{t=2\pi} \sqrt{\sin^2(t) + \cos^2(t)} \ dt = \int\limits_{t=0}^{t=2\pi} 1 \ dt = 2\pi$.

The graph of $(x(t), y(t))$ for $0 \le t \le 2\pi$ is a circle of radius 1 so we know that its length is exactly 2π .

Practice 3:   Represent the length of each curve as a definite integral.

(a) The length of one period of y = sin(x).

(b) The length of the parametric path $x(t) = 1 + 3t$ and $y(t) = 4t$ for $1 \le t \le 3$.

AREAS OF SURFACES OF REVOLUTION

Rotated Line Segments

Just as all of the integral formulas for arc length came from the simple distance formula, all of the integral formulas for the area of a revolved surface come from the formula for revolving a single straight line segment. If a line segment of length L, parallel to a line P, (Fig. 4) is revolved about the line P, then the resulting surface can be unrolled and laid flat. The flattened surface is a rectangle with area $A = 2\pi \cdot r \cdot L$.

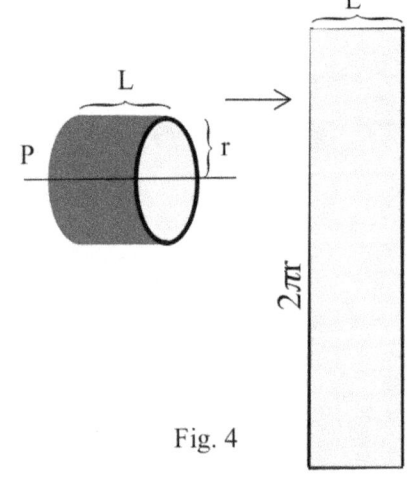

Fig. 4

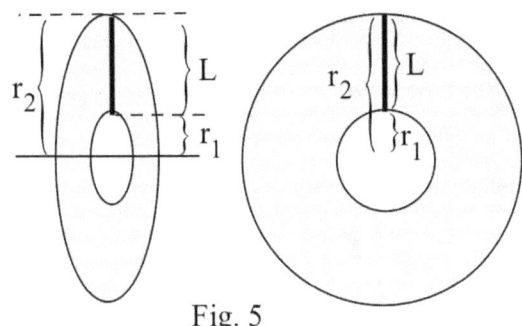

Fig. 5

If a line segment of length L, perpendicular to a line P and not intersecting P, (Fig. 5) is revolved about the line P, then the resulting surface is the region between two concentric circles and its area is

A = (area of large cirle) – (area of small circle) $= \pi(r_2)^2 - \pi(r_1)^2$

$$= \pi\{ (r_2)^2 - (r_1)^2 \} = 2\pi \{ \frac{r_2 + r_1}{2} \} (r_2 - r_1) = 2\pi \cdot \left(\frac{r_1 + r_2}{2} \right) \cdot L$$

In general, if a line segment of length L which does not
intersect a line P (Fig. 6) is revolved about P, then the
resulting surface has area

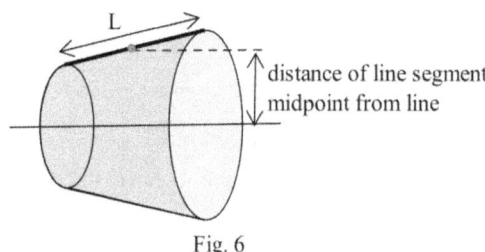

Fig. 6

A   = surface area

    = (distance traveled by midpoint of the line segment)·(length of the line segment)

    = **2π·{ distance of segment midpoint from the line P }· L**

There are several integral formulas for the surface area of a curve rotated about a line, but all of the formulas
come rather easily and quickly from this one fundamental formula for a surface area of a rotated line segment.

Example 4:   Find the surface area generated when each line segment in Fig. 7
                 is rotated about the x–axis and the y–axis.

Solution: Line segment B has length L=2 and its midpoint is at (2,1).

     When B is rotated about the x–axis, the surface area is

          2π·(distance of midpoint from x–axis)·2 = $2\pi(1)2 = 4\pi$.

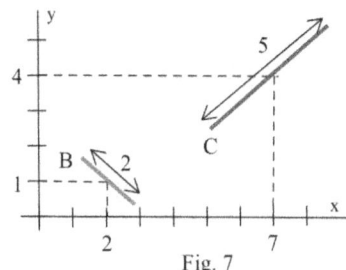

Fig. 7

     When B is rotated about the y–axis, the surface area is

          2π·(distance of midpoint from y–axis)·2 = 8π.

     Line segment C has length 5 and its midpoint is at (7,4). When C is rotated
     about the x–axis the resulting surface area is

          2π·(distance of midpoint from x–axis)·5 = $2\pi(4)5 = 40\pi$.

     When C is rotated about the y–axis, the surface area is

          2π·(distance of midpoint from y–axis)·5 = 70π.

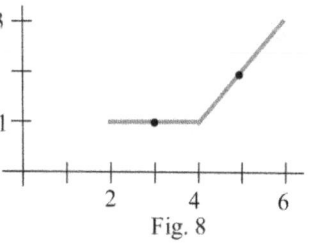

Fig. 8

Practice 4:   Find the surface areas generated when the graph in Fig. 8 is rotated about each axis.

Rotated Curves

When a curve is rotated about a line P, we can use our old strategy again (Fig. 9). Select some points

(x_i , y_i) along the curve, connect the points with line segments, calculate the surface area of each rotated line segment, and add together the surface areas of the rotated line segments. This final sum can be converted to a Riemann sum, and the limit of the Riemann sum is a definite integral for the surface area of the rotated curve.

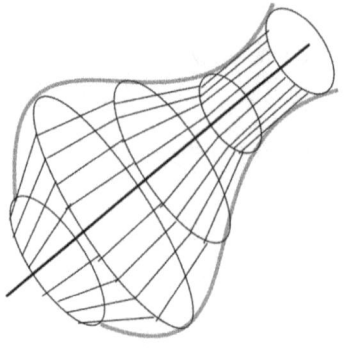

Fig. 9

Suppose we select points (x_i, y_i) along C and number them so (x_0, y_0) is one endpoint of C, (x_n, y_n) is the other endpoint, and the subscripts increase as we move along C. Then each pair of successive points $(x_{i-1} , y_{i-1}), (x_i , y_i)$ are the endpoints of a line segment with

$$\text{length} = \sqrt{(\Delta x_i)^2 + (\Delta y_i)^2} \qquad \text{and} \qquad \text{midpoint} = \left(\frac{x_{i-1}+x_i}{2}, \frac{y_{i-1}+y_i}{2}\right).$$

The midpoint is $(y_{i-1}+y_i)/2$ units from the x–axis and $(x_{i-1}+x_i)/2$ units from the y–axis.

About the x–axis: If C is rotated about the x–axis, then each segment generates an area equal to

$$2\pi \cdot (\text{distance of midpoint from the x–axis}) \cdot (\text{length of the segment}) = 2\pi \left\{ \frac{y_{i-1}+y_i}{2} \right\} \sqrt{(\Delta x_i)^2 + (\Delta y_i)^2} .$$

The sum of these surface areas is

$$\sum 2\pi \left\{ \frac{y_{i-1}+y_i}{2} \right\} \sqrt{(\Delta x_i)^2 + (\Delta y_i)^2} = \sum 2\pi \left\{ \frac{y_{i-1}+y_i}{2} \right\} \sqrt{(\Delta x_i/\Delta x_i)^2 + (\Delta y_i/\Delta x_i)^2} \ \Delta x_i$$

$$\longrightarrow \boxed{\int_a^b 2\pi y \sqrt{1 + (dy/dx)^2} \ dx = \text{area of the surface of revolution of curve C}}$$

About the y–axis: If C is rotated about the y–axis, then each segment generates an area equal to

$$2\pi \cdot (\text{distance of midpoint from the y–axis}) \cdot (\text{length of the segment}) = 2\pi \left\{ \frac{x_{i-1}+x_i}{2} \right\} \sqrt{(\Delta x_i)^2 + (\Delta y_i)^2} .$$

The sum of these surface areas is

$$\sum 2\pi \left\{ \frac{x_{i-1}+x_i}{2} \right\} \sqrt{(\Delta x_i)^2 + (\Delta y_i)^2} = \sum 2\pi \left\{ \frac{x_{i-1}+x_i}{2} \right\} \sqrt{(\Delta x_i/\Delta x_i)^2 + (\Delta y_i/\Delta x_i)^2} \ \Delta x_i$$

$$\longrightarrow \boxed{\int_a^b 2\pi x \sqrt{1 + (dy/dx)^2} \ dx = \text{area of the surface of revolution of curve C}}$$

Example 5:   Use definite integrals to represent the areas of the surfaces generated when the curve

$y = 2 + x^2$, $0 \le x \le 3$, is rotated about each axis.

Solution: Surface area about the x–axis is $\int_{a}^{b} 2\pi \cdot f(x)\sqrt{1+(dy/dx)^2} \, dx = \int_{0}^{3} 2\pi(2 + x^2)\sqrt{1+4x^2} \, dx \approx 383.8$.

Surface area about the y–axis is $\int_{a}^{b} 2\pi \cdot x \sqrt{1+(dy/dx)^2} \, dx = \int_{0}^{3} 2\pi \cdot x \sqrt{1+4x^2} \, dx \approx 117.32$.

Parametric Form For Surface Area of Revolution

If the curve C is described by parametric equations, $x = x(t)$ and $y = y(t)$, then the forms of the surface area integrals are somewhat different, but they still follow from the fundamental surface area formula for the surface area of a line segment rotated about a line P:

Surface area **$= 2\pi \cdot$\{ distance of midpoint from the line P \}$\cdot$ L** .

About x–axis: Starting with the previous equation for the area of a segment rotated about the **x–axis**,

$$2\pi \cdot \left\{ \frac{y_{i-1}+y_i}{2} \right\} \sqrt{(\Delta x_i)^2+(\Delta y_i)^2} \quad ,$$

we can factor Δt_i from the radical, sum the pieces and take the limit, as the mesh apporaches 0, to get

$$\int_{t=\alpha}^{t=\beta} 2\pi \, y \sqrt{(dx/dt)^2 + (dy/dt)^2} \quad dt = \text{area of the surface of revolution of curve C} .$$

About y–axis: Starting with the previous equation for the area of a segment rotated about the **y–axis**,

$$2\pi \left\{ \frac{x_{i-1}+x_i}{2} \right\} \sqrt{(\Delta x_i)^2+(\Delta y_i)^2}$$

we can factor Δt_i from the radical, sum the pieces and take the limit, as the mesh apporaches 0, to get

$$\int_{t=\alpha}^{t=\beta} 2\pi \, x \sqrt{(dx/dt)^2 + (dy/dt)^2} \quad dt = \text{area of the surface of revolution of curve C} .$$

WRAP UP

One purpose of this section was to obtain a variety of integral formulas for two geometric quantities, the length of a curve and the area of the surface generated when a curve is rotated about a line. The integral formulas are useful, but a more basic and fundamental point was to illustate again how relatively simple approximation formulas can lead us, via Riemann sums, to integral formulas. We will see it again.

PROBLEMS

For arc length

1. A squirrel was spotted at the backyard locations in Table 1 at the given times. The squirrel traveled at least how far during the first 15 minutes?

time (min)	location relative to oak tree north	east (feet)
0	10	7
5	25	27
10	1	45
15	13	33
20	24	40
25	10	23
30	0	14

Table 1. Locations of Squirrel

2. The squirrel in Problem 1 traveled at least how far during the first 30 minutes?

3. Use the partition $\{0, 1, 2\}$ to estimate the length of $y = 2^x$ between the points $(0, 1)$ and $(2, 4)$.

4. Use the partition $\{1, 2, 3, 4\}$ to estimate the length of $y = 1/x$ between the points $(1, 1)$ and $(4, 1/4)$.

The graphs of the functions in problems 5 – 8 are straight lines. Calculate each length (a) using the distance formula between 2 points and (b) by setting up and evaluating the arc length integrals.

5. $y = 1 + 2x$ for $0 \le x \le 2$.

6. $y = 5 - x$ for $1 \le x \le 4$.

7. $x = 2 + t$, $y = 1 - 2t$ for $0 \le t \le 3$.

8. $x = -1 - 4t$, $y = 2 + t$ for $1 \le t \le 4$.

9 Calculate the length of $y = \frac{2}{3} x^{3/2}$ for $0 \le x \le 4$.

10. Calculate the length of $y = 4x^{3/2}$ for $1 \le x \le 9$.

Very few functions $y = f(x)$ lead to integrands of the form $\sqrt{1 + (dy/dx)^2}$ which have elementary antiderivatives. In problems 11 – 14, $1 + (dy/dx)^2$ is a perfect square and the resulting arc length integrals can be evaluated using antiderivatives. Do so.

11. $y = \dfrac{x^3}{3} + \dfrac{1}{4x}$ for $1 \le x \le 5$.

12. $y = \dfrac{x^4}{4} + \dfrac{1}{8x^2}$ for $1 \le x \le 9$.

13. $y = \dfrac{x^5}{5} + \dfrac{1}{12x^3}$ for $1 \le x \le 5$.

14. $y = \dfrac{x^6}{6} + \dfrac{1}{16x^4}$ for $4 \le x \le 25$.

In problems 15 – 23, (a) represent each length as a definite integral, and (b) evaluate the integral using your calculator's integral command.

15. The length of $y = x^2$ from (0,0) to (1,1).

16. The length of $y = x^3$ from (0,0) to (1,1).

17. The length of $y = \sqrt{x}$ from (1,1) to (9,3).

18. The length of $y = \ln(x)$ from (1,0) to (e,1).

19. The length of $y = \sin(x)$ from (0,0) to $(\pi/4, \sqrt{2}/2)$ and from $(\pi/4, \sqrt{2}/2)$ to $(\pi/2, 1)$.

20. The length of the ellipse $x(t) = 3\cos(t), y(t) = 4\sin(t)$ for $0 \le t \le 2\pi$.

21. The length of the ellipse $x(t) = 5\cos(t), y(t) = 2\sin(t)$ for $0 \le t \le 2\pi$.

22. A robot was programmed to follow a spiral path and be at location $x(t) = t \cos(t), y(t) = t \sin(t)$ at
     time t . How far did the robot travel between $t = 0$ and $t = 2\pi$?

23. A robot was programmed to follow a spiral path and be at location $x(t) = t \cos(t), y(t) = t \sin(t)$ at
     time t . How far did the robot travel between $t = 10$ and $t = 20$?

24. As a tire of radius R rolls, a small stone stuck in the tread will travel a "cycloid" path,
     $x(t) = R{\cdot}(t - \sin(t))$, $y(t) = R{\cdot}(1 - \cos(t))$. As t goes from 0 to 2π, the tire makes one complete
     revolution and travels forward $2\pi R$ units. How far does the small stone travel?

25. As a tire with a 1 foot radius rolls forward 1 mile, how far does a pebble stuck in the tire tread travel?
     ($x(t) = 1(t - \sin(t))$ and $y(t) = 1(1 - \cos(t))$)

26. (Calculator) Graph $y = x^n$ for $n = 1, 3, 10,$ and 20. As the value of n gets large, what happens to
     the graph of $y = x^n$? Estimate the value of $\displaystyle\lim_{n \to \infty} \left(\int_{x=0}^{x=1} \sqrt{1 + (n \cdot x^{n-1})^2} \ dx \right)$.

27. (Calculator) Find the point on the curve segment $f(x) = x^2$ for $0 \le x \le 4$ which will divide the
     segment into two equally long pieces. Find the points which will divide the segment into 3 equally
     long pieces.

28. Find the pattern for the functions in problems 11 – 14. If $y = \dfrac{x^n}{n} + \dfrac{1}{Ax^B}$, then how are A and B
     related to n? (A = 4(n–2) and B = n–2)

29. Use the formulas for A and B from the previous problem with
     n = 3/2 and write a new function

     $y = \dfrac{2}{3} x^{3/2} + \dfrac{1}{Ax^B}$ so that $1 + (dy/dx)^2$ is a perfect square.

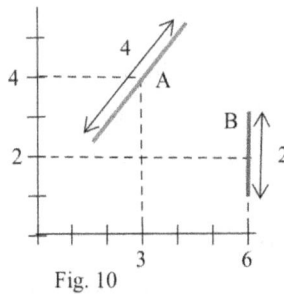

Fig. 10

For surface area of revolution

30. Find the surface area when each line segment in Fig. 10 is rotated about
     the (a) x–axis and (b) y–axis.

31. Find the surface area when each line segment in Fig. 11 is rotated about
     the (a) x–axis and (b) y–axis.

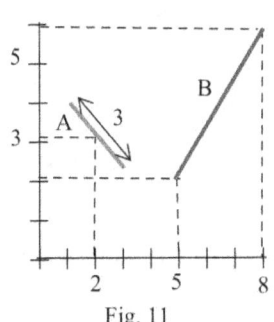

Fig. 11

32. Find the surface area when each line segment in Fig. 10 is rotated about the lines (a) $y = 1$ and (b) $x = -2$.

33. Find the surface area when each line segment in Fig. 11 is rotated about the lines (a) $y = 1$ and (b) $x = -2$.

34. A line segment of length 2 has its center at the point (2,5) and makes an angle of θ with horizontal. What value of θ will result in the largest surface area when the line segment is rotated about the y–axis? Explain your reasoning.

35. A line segment of length 2 has its one end at the point (2,5) and makes an angle of θ with horizontal. What value of θ will result in the largest surface area when the line segment is rotated about the x–axis? Explain your reasoning.

In problems 36 – 44, (a) represent each surface area as a definite integral, and (b) evaluate the integral using your calculator's integral command.

36. Find the area of the surface when the graph of $y = x^3$ for $0 \le x \le 2$ is rotated about the y–axis.

37. Find the area of the surface when the graph of $y = 2x^3$ for $0 \le x \le 1$ is rotated about the y–axis.

38. Find the area of the surface when the graph of $y = x^2$ for $0 \le x \le 2$ is rotated about the x–axis.

39. Find the area of the surface when the graph of $y = 2x^2$ for $0 \le x \le 1$ is rotated about the x–axis.

40. Find the area of the surface when the graph of $y = \sin(x)$ for $0 \le x \le \pi$ is rotated about the x–axis.

41. Find the area of the surface when the graph of $y = x^3$ for $0 \le x \le 2$ is rotated about the x–axis.

42. Find the area of the surface when the graph of $y = \sin(x)$ for $0 \le x \le \pi/2$ is rotated about the y–axis.

43. Find the area of the surface when the graph of $y = x^2$ for $0 \le x \le 2$ is rotated about the y–axis.

44. Find the area of the surface when the graph of $y = \sqrt{4 - x^2}$ is rotated about the x–axis
    (a) for $0 \le x \le 1$, (b) for $1 \le x \le 2$, and (c) for $2 \le x \le 3$.

45. (a) Show that if a thin hollow sphere is sliced into pieces by equally–spaced parallel cuts (Fig. 12), then each piece has the same weight. (Show that each piece has the same surface area).

    (b) What does the result of part (a) mean for an orange cut into slices with equally–spaced parallel cuts?

    (c) Suppose a hemispherical cake with uniformly thick layer of frosting is sliced with equally–spaced parallel cuts. Does everyone get the same amount of cake? Does everyone get the same amount of frosting?

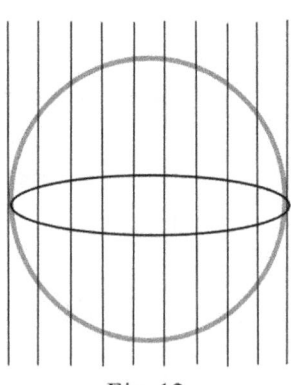

Fig. 12

3–D Arc Length Problems (Optional)

The parametric equation form of arc length extends very nicely to 3 dimensions. If a curve C in 3 dimensions (Fig. 13) is given parametrically by $x = x(t), y = y(t)$, and $z = z(t)$ for $a \leq t \leq b$, then the distance between the successive points $(x_{i-1}, y_{i-1}, z_{i-1})$ and (x_i, y_i, z_i) is

$$\sqrt{(x_{i-1} - x_i)^2 + (y_{i-1} - y_i)^2 + (z_{i-1} - z_i)^2}$$

$$= \sqrt{(\Delta x_i)^2 + (\Delta y_i)^2 + (\Delta z_i)^2} \ .$$

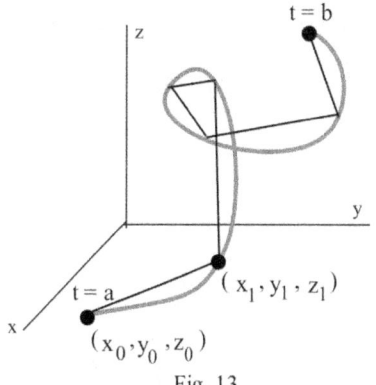

Fig. 13

We can, as before, factor $(\Delta t_i)^2$ from each term under the radical, sum the pieces to get a Riemann sum, and take a limit of the Riemann sum to get a definite integral representing the length of the curve C.

$$\sum \sqrt{(\Delta x_i / \Delta t_i)^2 + (\Delta y_i / \Delta t_i)^2 + (\Delta z_i / \Delta t_i)^2} \ \Delta t_i \longrightarrow \int_{t=a}^{t=b} \sqrt{(dx/dt)^2 + (dy/dt)^2 + (dz/dt)^2} \ dt \ .$$

The length of the curve C is $\displaystyle\int_{t=a}^{t=b} \sqrt{(dx/dt)^2 + (dy/dt)^2 + (dz/dt)^2} \ dt$.

In problems 46 – 50, (a) represent each length as a definite integral, and (b) evaluate the integral using your calculator's integral command.

46. Find the length of the helix $x = \cos(t)$, $y = \sin(t), z = t$ for $0 \leq t \leq 4\pi$. (Fig. 14)

47. Find the length of the straight line segment $x = t, y = t, z = t$ for $0 \leq t \leq 1$.

48. Find the length of the curve $x = t, y = t^2, z = t^3$ for $0 \leq t \leq 1$.

49. Find the length of the "stretched helix" $x = \cos(t)$, $y = \sin(t)$, $z = t^2$ for $0 \leq t \leq 2\pi$.

50. Find the length of the curve $x = 3\cos(t), y = 2\sin(t), z = \sin(7t)$ for $0 \leq t \leq 2\pi$.

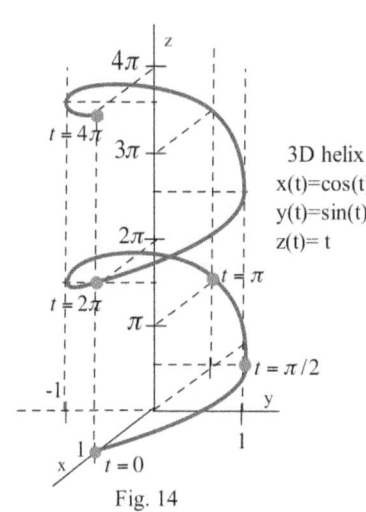

3D helix
x(t)=cos(t)
y(t)=sin(t)
z(t)= t

Fig. 14

Section 5.2   **PRACTICE Answers**

Practice 1:   At least $2 + \sqrt{2} + \sqrt{13} + 1 + \sqrt{2} \approx 9.43$ miles.

Practice 2:   Using the points $(0,0), (1,1), (2,4)$, and $(3,9)$,

$L \approx \sqrt{2} + \sqrt{10} + \sqrt{26} \approx 9.68 <$ actual length.

Practice 3:   (a) Using $\displaystyle\int_{x=a}^{x=b} \sqrt{1 + (\frac{dy}{dx})^2}\ dx$ with $y = \sin(x)$, $L = \displaystyle\int_{x=0}^{x=2\pi} \sqrt{1 + \cos^2(x)}\ dx$.

(b) Using $\displaystyle\int_{t=a}^{t=b} \sqrt{(\frac{dx}{dt})^2 + (\frac{dy}{dt})^2}\ dt$ with $x(t) = 1 + 3t$ and $y(t) = 4t$ for $1 \le t \le 3$,

$L = \displaystyle\int_{t=1}^{t=3} \sqrt{3^2 + 4^2}\ dt = \int_{t=1}^{t=3} 5\ dt = 10$.

Practice 4:   surface area of revolved segment = $2\pi \cdot \{$ distance of segment midpoint from the line P $\} \cdot$ L

surface area of horizontal segment revolved about **x–axis** = $2\pi(1)(2) = 4\pi \approx 12.57$.
surface area of other segment revolved about **x–axis** = $2\pi(2)(\sqrt{8}) \approx 35.54$.
The total surface area about the x–axis is approximately $12.57 + 35.54 = 48.11$ square units.

surface area of horizontal segment revolved about **y–axis** = $2\pi(3)(2) = 12\pi \approx 37.70$.
surface area of other segment revolved about **y–axis** = $2\pi(5)(\sqrt{8}) \approx 88.86$.
The total surface area about the y–axis is approximately $37.70 + 88.86 = 126.56$ square units.

5.3 MORE WORK APPLICATIONS

In Section 4.7 we introduced the problem of calculating the **work** done in lifting an object using a cable which had weight. This section continues that introduction and extends the process to handle situations in which the applied force or the distance or both may be variables. The method we used before is used again here. The first step is to divide the problem into small "slices" so that the force and distance vary only slightly on each slice. Then the work for each slice is calculated, the total work is approximated by adding together (a Riemann sum) the work for each slice, and , finally, a limit is taken to get a definite integral representing the total work. **There are so many possible variations in work problems that it is vital that you understand the process.**

The work done on an object by a constant force is the magnitude of the force applied to the object multiplied by the distance over which the force is applied: **work = (force)·(distance)**.

Example 1:   A 10 pound object is lifted 40 feet from the ground to the top of a building using a cable which weighs 1/2 pound per foot (Fig. 1). How much work is done?

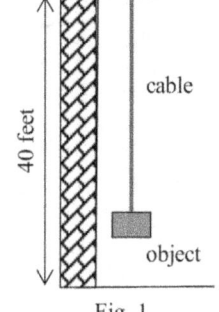

Fig. 1

Solution: This type of problem appeared in section 4.7, but it is a good example of the process of dividing the problem into pieces and analyzing each piece. We can partition the height of the building (Fig. 2). Then the work done to lift the object from the height x_i to the height x_{i+1} is the force applied times the distance moved:

$$\text{force} = (\text{weight of the object}) + (\text{weight of the cable})$$
$$= (10 \text{ pounds}) + (0.5 (\text{length of hanging cable}))$$
$$= 10 + 0.5(40 - x_i) \text{ pounds } = 30 - 0.5x_i \text{ pounds}$$
$$\text{distance} = x_i - x_{i-1} \text{ feet} = \Delta x_i \text{ feet}$$
$$\text{work} = (\text{force})(\text{distance}) = \{ 30 - 0.5x_i \} \Delta x_i \text{ foot–pounds.}$$

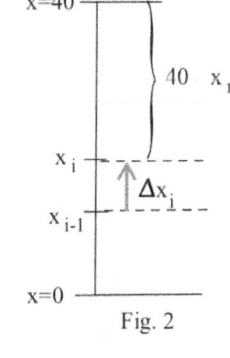

Fig. 2

$$\text{Total work} \approx \sum_{i=1}^{n} \{ \text{work on } i^{\text{th}} \text{ slice} \} = \sum_{i=1}^{n} \{ 30 - 0.5x_i \} \Delta x_i \text{ foot–pounds}$$

$$\longrightarrow \int_{0}^{40} \{ 30 - 0.5x \} dx = (30x - 0.25x^2) \Big|_{0}^{40} = 800 \text{ foot–pounds.}$$

Practice 1:   How much work is done lifting a 130 pound injured person to the top of a 30 foot cliff using a stretcher weighing 10 pounds and a cable weighing 2 pounds per foot?

In the previous Example and Practice problem the distance moved on each part of the partition was always Δx, and the force was more complicated. In some of the following examples, the Δx is part of the force calculation. Analyze each problem.

Example 2:   A cola glass in Fig. 3 has the dimensions given in Table 1. Approximately how much work do you do when you drink a cola glass full of water

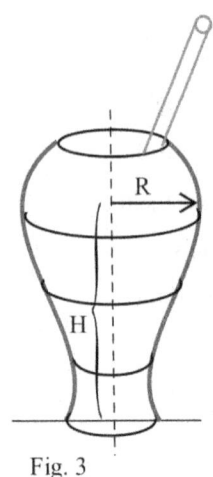

Fig. 3

(weight density = 62.5 pounds/ft^3 = 0.5787 ounces/in^3)
by sucking it through a straw to a point 3 inches
above the top edge of the glass?

Solution: The Table naturally partitions the water into 1 inch
thick "slices" (Fig. 4). The work to move
each slice is approximately the weight of the slice
times the distance it is moved. We can use the radius
at the bottom of each slice to approximate the volume and
then the weight of the slice, and a point half way up each
slice to calculate the distance the slice is moved.

Height above bottom of the glass (inches)	Inside radius (inches)
4	1.4
3	1.6
2	1.5
1	1.0
0	1.1

Table 1: Inside radius of a cola glass

top slice: force = weight = (volume)(density) $\approx \pi(1.6 \text{ in})^2(1 \text{ in})(0.5787 \text{ oz/in}^3) \approx 4.7$ oz.

distance $\approx$ (distance from middle of slice to lips) = 3.5 in.

work $\approx$ (force)(distance) = (4.7 oz)(3.5 in) = 16.4 oz–in.

next slice: force = weight = (volume)(density) $\approx \pi(1.5 \text{ in})^2(1 \text{ in})(0.5787 \text{ oz/in}^3) \approx 4.1$ oz.

distance $\approx$ (distance from middle of slice to lips) = 4.5 in

work $\approx$ (force)(distance) = (4.1 oz)(4.5 in) = 18.4 oz–in.

The work for the last two slices is (1.8 oz)(5.5 in) = 9.9 oz–in and (2.2 oz)(6.5 in) = 14.3 oz–in.

The total work is the sum of the work needed to raise each slice of water:

Total work $\approx$ (16.4 oz–in) + (18.4 oz–in) + (9.9 oz–in) + (14.3 oz–in) = 59 oz–in.

Practice 2:   Approximate the total work needed to raise the water in Example 2 by using the **top** radius
of the slice to approximate the weight and the midpoint of each slice to approximate the distance the
slice is raised.

If we knew the radius of the glass at every
height, then we could improve our
approximation by taking thinner slices. In fact,
if we knew the radius at every height we could
have formed a Riemann sum, taken the limit of
the Riemann sum as the thickness of the slices
approached 0, and obtained a definite integral.
In the next Example we do know the radius of
the container at every height.

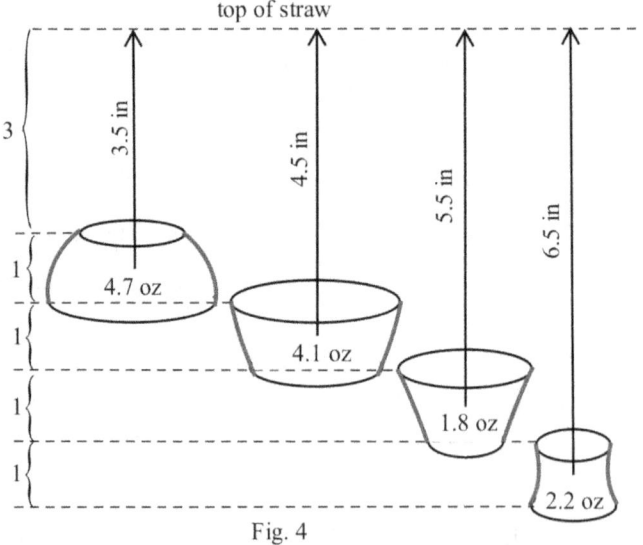

Fig. 4

Example 3:   Find the work needed to raise the water in the cone in Fig. 5a to the top of the straw.

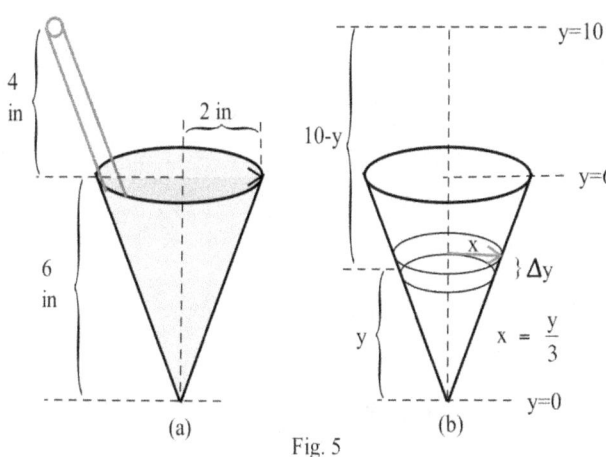

Solution: We can label the cone (Fig. 5b), and partition the height of the cone to get slices of water. The work done raising the ith slice is the distance the slice is raised times the force needed to move it, the weight of the slice. For any c_i in the subinterval $[y_{i-1}, y_i]$, the slice is raised a distance of approximately $(10 - c_i)$ inches.

Each slice is approximately a right circular cylinder so its volume is $\pi(\text{radius})^2 \Delta y$. At the height y, the radius of the cylinder is $x = y/3$ so at the height c_i the radius is $c_i/3$. Then the force is

(a)   Fig. 5   (b)

$$\text{force} = (\text{volume})(\text{density}) \approx \pi(\text{radius})^2(\Delta y_i)(0.5787 \text{ oz/in}^3) = \pi(c_i/3)^2(\Delta y_i)(0.5787) \text{ ounces.}$$

The work to raise the ith slice $\approx \pi(c_i/3)^2(\Delta y_i)(0.5787)(10 - c_i)$ ounce–inches, and the total work is

approximately $\sum_{i=1}^{n} \pi(c_i/3)^2(\Delta y_i)(0.5787)(10 - c_i)$. As the mesh of the partition approaches 0, the

Riemann sum approaches the definite integral:

$$\text{total work} \approx \sum_{i=1}^{n} \pi(c_i/3)^2(\Delta y_i)(0.5787)(10 - c_i) \longrightarrow \int_{0}^{6} \pi(y/3)^2(0.5787)(10 - y)\, dy \ .$$

$$\text{Total work} = \int_{0}^{6} \pi(y/3)^2(0.5787)(10 - y)\, dy = \frac{0.5787\pi}{9} \int_{0}^{6} 10\, y^2 - y^3\, dy$$

$$= \frac{0.5787\pi}{9} \left\{ \frac{10}{3}(6)^3 - \frac{1}{4}(6)^4 \right\} = 79.99 \text{ oz–in}$$

.

In this example, both the force and the distance were variables and both depended on the height of the slice above the bottom of the cone.

Practice 3: How much work is done in drinking just the top 3 inches of the water in Example 3?

Example 4:   The trough in Fig. 6 is filled with a liquid weighing 70 pounds per cubic foot. How much work is done pumping the liquid over the wall next to the trough?

Solution: As before, we can partition the height of the trough to get slices of liquid. In order to form a

Riemann sum for the total work, we need the weight of a typical slice (Fig. 7) and the distance it is raised.

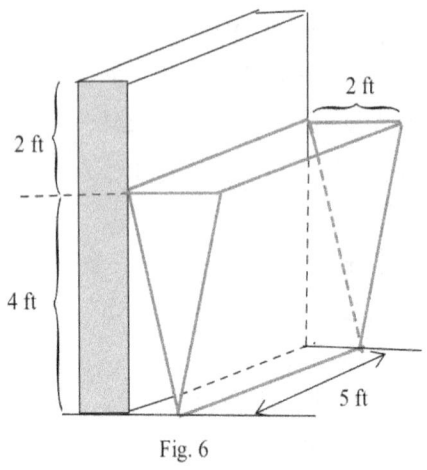

slice at height y_i :   weight   = (volume)(density)

$\qquad$ = (length)(width)(height)(70 pounds/ft^3)

$\qquad$ = (5 ft)($\frac{1}{2}$ y_i ft)(Δy_i ft)(70 pounds/ft^3)

$\qquad$ = 175 y_i Δy_i pounds.

distance raised   = $6 - y_i$ ft

work   = (175 y_i Δy_i pounds)($6 - y_i$ feet)

$\qquad$ = 175 y_i $(6 - y_i)$ Δy_i foot–pounds.

The rest of the solution is straightforward and follows the pattern of the

previous problems:

$$\text{Total work}   \approx   \sum_{i=1}^{n} \{ \text{ work to raise } i^{th} \text{ slice } \}$$

Fig. 6

$$=   \sum_{i=1}^{n} \{ 175 \, y_i \, (6 - y_i) \, \Delta y_i \text{ foot–pounds } \}$$

$$\longrightarrow   \int_{0}^{4} 175 \, y(6 - y) \, dy$$

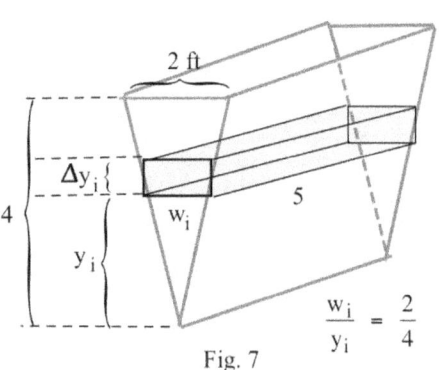

$$=   175(3y^2 - \frac{y^3}{3} \,) \Big|_{0}^{4}   = 4666.7 \text{ foot–pounds.}$$

Fig. 7

$$\frac{w_i}{y_i} = \frac{2}{4}$$

"Raise the liquid" problems can be handled by partitioning the height of the container and then focusing your

attention on **one typical slice**. If you can calculate the **weight** of that slice and the **distance** it is raised, the rest

of the steps are straightforward: form a Riemann sum, form a definite integral, and evaluate the integral to get

the total work.

Work Moving An Object In A Straight Line

Suppose we are pushing a box along a flat surface (Fig. 8a) which is smooth in places and rough in other

places so at some places we only have to push lightly and in other places we have to push hard. If f(x) is

the amount of force we need to use at

location x, and we want to push the box

along a straight line from x=a to x=b, then

we can partition the interval [a, b] into

pieces, $[a,x_1], [x_1,x_2], \ldots, [x_{n-1},b]$ (Fig.

8b). The work to move the box along the ith

piece from x_{i-1} to x_i is approximately

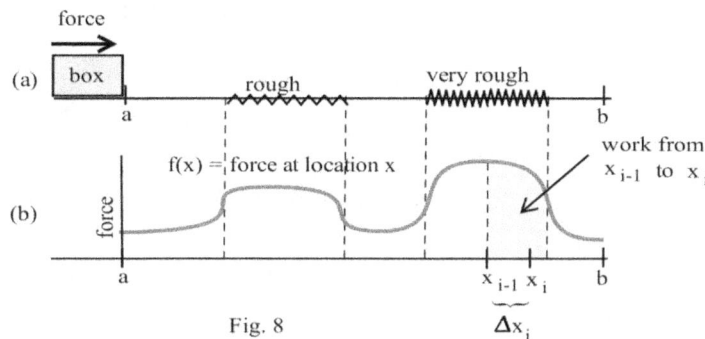

Fig. 8

(force)•(distance) $\approx$ f(c_i)•($x_i - x_{i-1}$) = f(c_i) Δx_i for any c_i in the subinterval [x_{i-1} , x_i].

The total work is the sum of the work along each piece, $\sum\limits_{k=1}^{n}$ f(c_i) Δx_i , a Riemann sum. As we take

smaller and smaller subintervals (as the mesh of the partition approaches 0), the Riemann sum approaches the

definite integral:

$$\sum\limits_{k=1}^{n} f(c_i)\, \Delta x_i \quad \longrightarrow \quad \int\limits_{a}^{b} f(x)\, dx \; = \; \text{total work} .$$

> If   an object starts at $x = a$ and is moved in a straight line to
>
> the location $x = b > a$ by applying a force of f(x) at every
>
> location x between a and b,
>
> then   the total work done on the object is $\int\limits_{a}^{b} f(x)\, dx$.

This has a simple geometric interpretation. If f(x) is the force

applied at the position x (Fig. 9), then the work done to move

from position $x = a$ to position $x = b$ is the area under the graph

of f from $x = a$ to $x = b$.

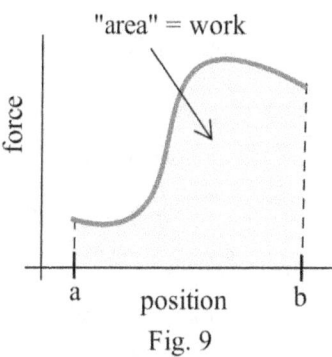

Fig. 9

Example 5:   Suppose a force of $7x$ pounds is required to stretch a
spring (Fig. 10) x inches past its natural length. How much work
will be done stretching the spring from its natural length
($x = 0$) to 5 inches beyond its natural length ($x = 5$ inches)?

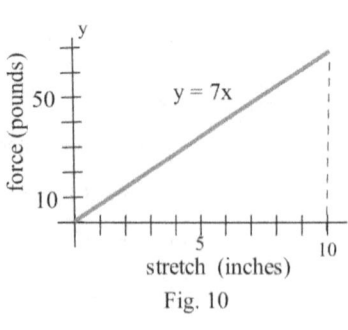

Fig. 10

Solution: Work $= \int_a^b f(x)\, dx = \int_0^5 7x\, dx = \left.\frac{7x^2}{2}\right|_0^5 = 87.5$ inch–pounds.

This can also be don graphically (Fig. 11).

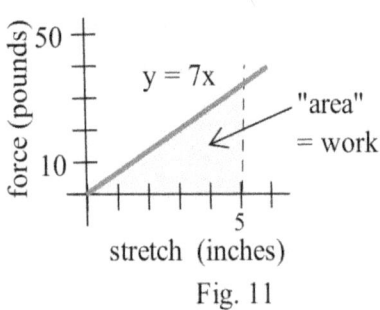

Fig. 11

Practice 4: How much work is done to stretch the spring in the previous example
from 5 inches past its natural length to 10 inches past its natural length?

The spring example is an application of a physical principle discovered by the
English physicist Robert Hooke (1635–1703), a contemporary of Newton.

Hooke's Law: The force needed to stretch or compress a spring x units from its natural length is
proportional to the distance x: force $f(x) = kx$ for some constant k. (Fig. 12)

The "k" in Hooke's Law is called the "spring constant". It varies from spring
to spring (depending on the materials and dimensions of the spring and even
on the temperature), but is constant for each spring as long as the spring is not
overextended or overcompressed. In fact, Hooke's law holds for most solid
objects, at least for limited ranges of force:

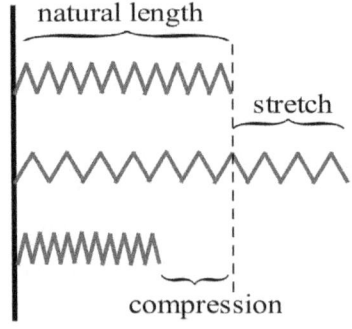

Fig. 12

"Nor is it observable in those bodies only, but in all other springy bodies
whatsoever, whether metal, wood, stones, baked earth, hair, horns, silk,
bones, sinews, glass and the like." (Hooke).

Most bathroom scales use springs and are based on Hooke's Law for compressing a spring.

Example 6:   A spring has a natural length of 43 centimeters, and a weight of 4 grams stretches it to a
total length of 75 centimeters. How much work is done stretching the spring from a total
length of 63 cm to a total length of 93 cm?

Solution: First we need to use the given information to find the value of k, the spring constant. A force
of 4 g produces a stretch of 32 cm (total length of 75 cm minus the rest length of 43 cm). Substituting
$x = 32$ cm and $f(x) = 4$ g into Hooke's Law, $f(x) = kx$, we have 4 g $= k$ (32 cm) so
$k = \frac{4\,g}{32\,cm} = \frac{1\,g}{8\,cm} = .125$ g/cm .

The total length of 63 cm represents a stretch of 20 cm beyond the natural length, and the total length of 93 cm represents a 50 cm stretch. Then the work done is

$$\text{work} = \int_a^b f(x)\,dx = \int_{20}^{50} (.125)\,x\,dx = (.125) \cdot \left.\frac{x^2}{2}\right|_{20}^{50} = 131.25 \text{ g-cm. .}$$

Practice 5:   A spring has a natural length of 3 inches, and a force of 2 pounds stretches it to a total length of 8 inches. How much work is done stretching the spring from a total length of 5 inches to a total length of 10 inches?

Lifting a Payload: The problem of finding the work done lifting a payload from the surface of a moon (or any body with no atmosphere) is very similar. Suppose the moon has a radius of R miles and the payload weighs P pounds at the surface of the moon (at a distance of R miles from the center of the moon). When the payload is x miles from the center of the moon $(x \geq R)$, the gravitational attraction between the moon and the payload is proportional to the reciprocal of the square of the distance x between the centers of the moon and the payload:

$$\text{required force} = f(x) = \frac{R^2 P}{x^2} \text{ pounds (Fig. 13).}$$

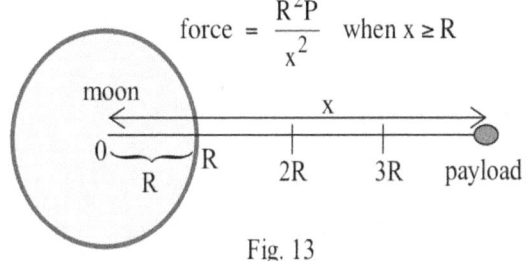

$$\text{force} = \frac{R^2 P}{x^2} \text{ when } x \geq R$$

Fig. 13

The total amount of work done raising the payload from the surface (altitude is 0, so $x = R$) to an altitude of R $(x = R+R = 2R)$ is

$$\text{work} = \int_a^b f(x)\,dx = \int_R^{2R} \frac{R^2 P}{x^2}\,dx = R^2 P \left(-\frac{1}{x}\right)\bigg|_R^{2R} = R^2 P \left(-\frac{1}{2R}\right) - R^2 P \left(-\frac{1}{R}\right) = \frac{RP}{2} \text{ mile-pounds.}$$

Practice 6:   How much work will be needed to raise the payload from the altitude R above the surface $(x = 2R)$ to an altitude of 2R?

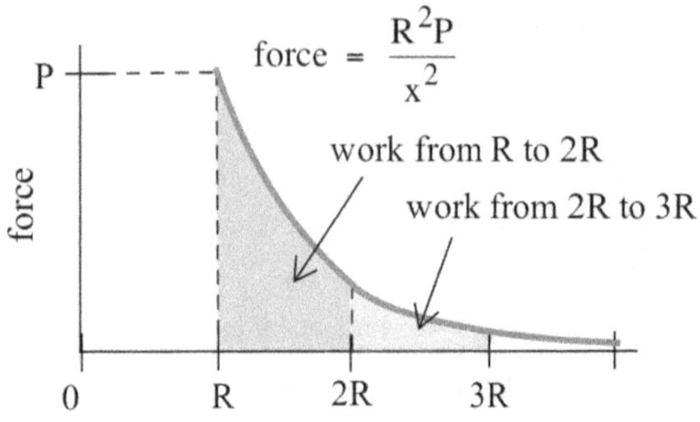

$$\text{force} = \frac{R^2 P}{x^2}$$

work from R to 2R

work from 2R to 3R

force

P

0   R   2R   3R

Fig. 14

The appropriate areas under the force graph (Fig. 14) illustrate why the work to raise the payload from x = R to x = 2R was so much larger than the work to raise it from x = 2R to x = 3R. In fact, the work to raise the payload from x = 2R to x = 100R is 0.49RP which is still less that the 0.5RP needed to raise the payload from x = R to x = 2R.

The real problem of lifting a payload is much more difficult because the rocket doing the lifting must also lift itself (more work) and the mass of the rocket will keep changing as it burns up fuel. Lifting a payload from a body with an atmosphere is even harder: there is friction from the atmosphere, and the frictional force depends on the density of the atmosphere (which varies with height), the speed of the rocket and the shape of the rocket. Life can get complicated.

PROBLEMS

1.   A tank 4 feet long, 3 feet wide and 7 feet tall (Fig. 15) is filled with water which weighs 62.5 pounds per cubic foot. How much work is done pumping the water out over the top of the tank?

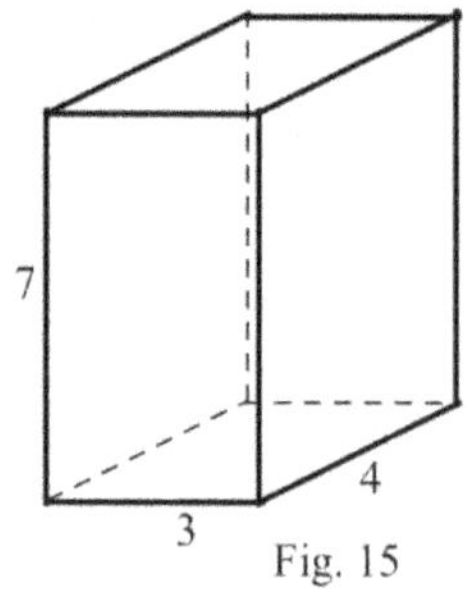

2.   A tank 4 feet long, 3 feet wide and 6 feet tall is filled with a oil which weighs 60 pounds per cubic foot.

   (a) How much work is done pumping the oil over the top edge of the tank?

   (b) How much work is done pumping the 3 feet of oil of the top edge of the tank?

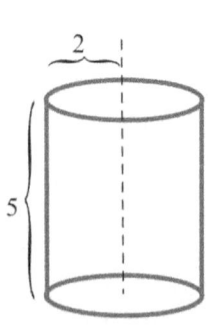

7

3   4

Fig. 15

3.   A tank 5 feet long, 2 feet wide and 4 feet tall is filled with a oil which weighs 60 pounds per cubic foot.

   (a) How much work is done pumping all of the oil out over the top edge of the tank?

   (b) How much work is done pumping the top 36 cubic feet of oil out over the top edge of the tank?

   (c) How long does a 1 horsepower pump take to empty the tank over the top edge of the tank?
        (A 1 horsepower pump works at a rate of 33,000 foot–pounds per minute.) A 1/2 horsepower pump? Which pump does more work?

4.   A cylindrical aquarium with radius 2 feet and height 5 feet (Fig. 16) is filled with salt water (65 pounds/ft^3).

   (a) How much work is done pumping all of the water over the top edge of the tank?

   (b) How much work is done pumping the water to a point 3 feet above the top edge of the tank?

   (c) How long does a 1 horsepower pump take to empty the tank over the top edge of the tank? A 1/2 horsepower pump? Which pump does more work?

2

5

Fig. 16

5. A cylindrical barrel with a radius of 1 foot and a height of 6 feet is filled with oil (60 pounds/ft^3).

   (a) How much work is done pumping all of the oil over the top edge of the barrel?

   (b) How much work is done pumping the top 1 foot of oil to a point 2 feet above the top of the barrel?

   (c) How long will it take a 1 horsepower pump to empty the top 3 feet of oil over the top edge of the barrel?

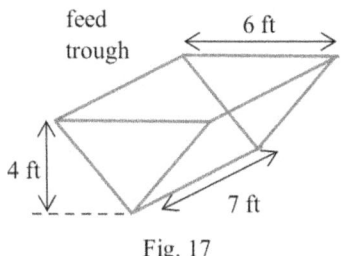

feed trough

6 ft

4 ft

7 ft

Fig. 17

6. An animal feed trough (Fig. 17) is filled with food weighing 80 pounds/ft^3. How much work is done lifting all of the food over the top of the trough?

7. How much work is done lifting the top 1 foot of food over the top of the trough in Problem 6?

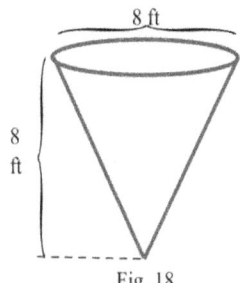

8 ft

8 ft

Fig. 18

8. The conical container in Fig. 18 is filled with loose grain which weighs 40 pounds/ft^3.

   (a) How much work is done lifting all of the grain over the top of the cone?

   (b) lifting the top 2 feet of grain over the top of the cone?

9. If you and a friend share the work equally in emptying the conical container in Problem 8, what depth of grain should the first person leave for the second person to empty?

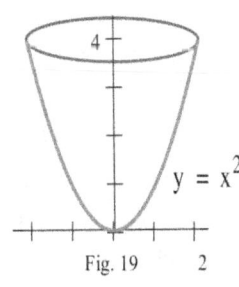

$y = x^2$

Fig. 19   2

10. The parabolic container in Fig. 19 is filled with water. (a) How much work is done pumping the water over the top of the tank? (b) to a point 3 feet above the top of the tank?

11. The parabolic container in Fig. 20 is filled with water.

   (a) How much work is done pumping the water over the top of the tank?

   (b) to a point 3 feet above the top of the tank?

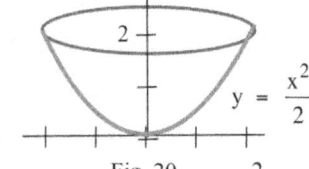

2

$y = \dfrac{x^2}{2}$

Fig. 20   2

12. The spherical tank in Fig. 21 is full of water. How much work is done lifting the water to the top of the tank?

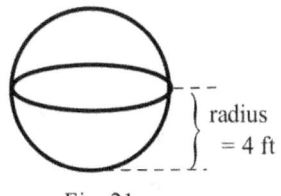

radius = 4 ft

Fig. 21

13. There are two feet of water in the bottom of spherical tank in Fig. 21. How much work is done lifting the water to the top of the tank?

14. The student said, "I've got a shortcut for these tank problems, but it
doesn't always work. I figure the weight of the liquid and multiply that
by the distance I have to move the middle point in the water. It worked for
the first 5 problems and then it didn't."

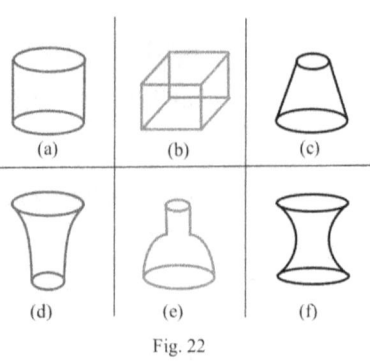

Fig. 22

(a) Does it really give the right answer for the first 5 problems?

(b) How are the containers in the first 5 problems different from the
others?

(c) For which of the containers in Fig. 22 will the "shortcut" work?

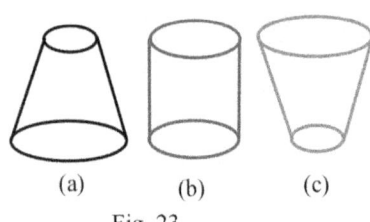

(a)   (b)   (c)

Fig. 23

15. All of the containers in Fig. 23 have the same height and hold the same
volume of water. Which requires

(a) the most work to empty?

(b) the least work to empty?

Explain how you reached your

conclusions.

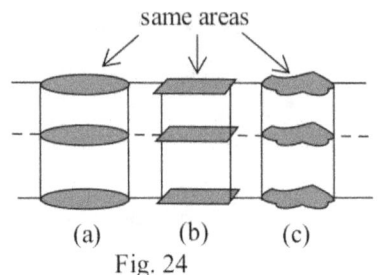

same areas

(a)   (b)   (c)

Fig. 24

16. All of the containers in Fig. 24 have the same height and at each
height x they all have the same cross sectional area. Which requires
(a) the most work to empty? (b) the least work to empty?
Justify your conclusions.

17. Fig. 25 shows the force required to move a box along a rough surface. How much work is done
pushing the box (a) from x = 0 to x = 5 feet? (b) from x = 3 to x = 5 feet?

18. How much work is done pushing the box in Fig. 25
(a) from x = 3 to x = 7 feet? (b) from x = 0 to x = 7 feet?

19. A spring requires a force of 6x ounces to stretch it x inches past
its natural length. How much work is done stretching the spring
(a) from its natural length (x = 0) to 3 inches beyond its natural length?
(b) from its natural length to 6 inches beyond its natural length?

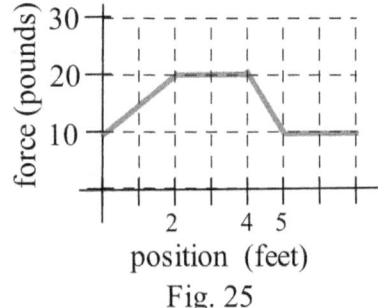

Fig. 25

20. A spring requires a force of 5x grams to compress it x cm. How much work is done
compressing the spring (a) 7 cm from its natural length? (b) 10 cm from its natural length?

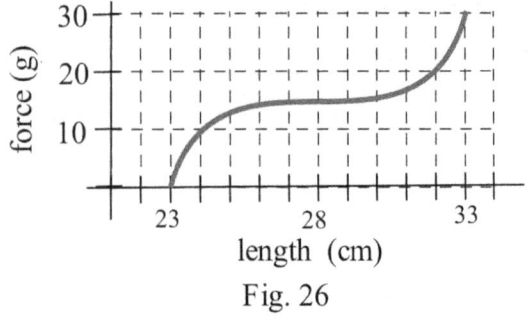

length (cm)

Fig. 26

21. Fig. 26 shows the force needed to stretch a material that
does not obey Hooke's Law. Approximately how much
work is done stretching it
(a) from a total length of 23 cm to 33 cm?
(b) from a total length of 28 cm to 33 cm?

22. Approximately how much work is done stretching the defective spring in the previous problem

(a) from a total length of 23 cm to 26 cm? (b) from a total length of 30 cm to 35 cm?

23. A 3 pound object stretches a spring 5 inches. How much work is done stretching it 4 more inches?

24. A 2 pound fish stretches a spring 3 inches. How much work is done stretching it 3 more inches?

25. A payload weighs 100 pounds at the surface of an asteroid which has a radius of 300 miles. How much

work is done lifting the payload from the asteroid's surface to an altitude of (a) 100 miles?

(b) 200 miles? (c) 300 miles?

26. Calculate the amount of work required to lift **you** from the surface of the moon where your weight is

approximately 1/6 what it is on earth to an altitude of 200 miles? (The moon's radius is approximately

1,080 miles.)

27. Calculate the amount of work required to lift **you** from the surface of the earth to an altitude (a) of 200

miles? (b) of 400 miles? (c) of 1,000,000 miles? (The earth's radius is approximately 4,000 miles.)

28. An object located at the origin repels **you** with a force inversely proportional to your distance from the

object ($f(x) = -\dfrac{1}{kx}$). When you are 10 feet from the origin the repelling force is 0.1 pound. How

much work is done as you move (a) from x = 20 to x = 10? (b) from x = 10 to x = 1? (c) from x =

1 to x = 0.1 ?

29. An object located at the origin repels you with a force inversely proportional to the square of your

distance from the object ($f(x) = -\dfrac{1}{kx^2}$). When you are 10 feet from the origin the repelling force is

0.1 pound. How much work is done as you move (a) from x = 20 to x = 10? (b) from x = 10 to x =

1? (c) from x = 1 to x = 0.1?

30. The student said "I've got a 'work in a line' shortcut that always seems to work. I figure the average

force and then multiply by the total distance. Will it always work?" (a) Will it? Justify your answer.

(Hint: What is the formula for 'average force'?) (b) Is it a shortcut?

Work Along A Curved Path

Suppose the location of a moving object is defined parametrically as $x = x(t)$ and $y = y(t)$ for $a \leq t \leq b$, and the force is a function of $t, f = f(t)$. Then we can represent the work done moving along the path as a definite integral. Partition the time interval $[a, b]$ into short subintervals. For the interval $[t_{i-1}, t_i]$:

force $\approx f(c_i)$ for any c_i in $[t_{i-1}, t_i]$

distance moved $\approx \sqrt{(\Delta x_i)^2 + (\Delta y_i)^2} = \sqrt{(\Delta x_i/\Delta t_i)^2 + (\Delta y_i/\Delta t_i)^2} \ \Delta t_i$

work $\approx f(c_i) \sqrt{(\Delta x_i/\Delta t_i)^2 + (\Delta y_i/\Delta t_i)^2} \ \Delta t_i$

Total work $\approx \sum \{$work along each subinterval $\} \approx \sum f(c_i) \sqrt{(\Delta x/\Delta t)^2 + (\Delta y/\Delta t)^2} \ \Delta t$

$$\longrightarrow \boxed{\int_{t=a}^{t=b} f(t) \sqrt{(dx/dt)^2 + (dy/dt)^2} \ dt = \text{total work along the path } (x(t), y(t)).}$$

In problems 31 – 35, find the total work along the given parametric path. If necessary, approximate the value of the integral using your calculator. f is in pounds, x and y are in feet, t is in minutes.

31. $f(t) = t$. $x(t) = \cos(t), y(t) = \sin(t), 0 \leq t \leq 2\pi$.   32. $f(t) = t$. $x(t) = t, y(t) = t^2, 0 \leq t \leq 1$.

33. $f(t) = t$. $x(t) = t^2, y(t) = t, 0 \leq t \leq 1$.   34. $f(t) = \sin(t)$. $x(t) = 2t, y(t) = 3t, 0 \leq t \leq \pi$. (Fig. 27)

35. $f(t) = t$. $x(t) = \cos(t), y(t) = \sin(t), 0 \leq t \leq 2\pi$ (Fig. 28). (Can you find a geometric way to calculate the shaded area?)

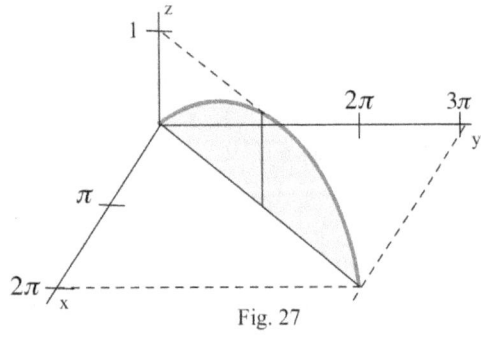

Fig. 27

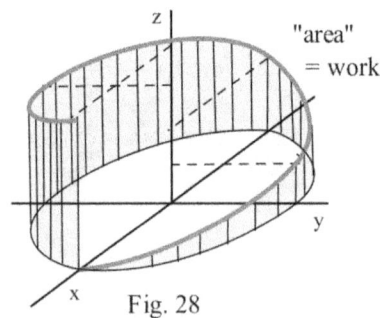

Fig. 28

Section 5.3   **PRACTICE Answers**

Practice 1: x_i = distance to top of cliff, Δx_i = distance lifted (length of "slice")

$w_i = (130 + 10 + 2x_i)\,\Delta x_i = (140 + 2x_i)\,\Delta x_i$.

$$\text{Total work} = \int_{x=0}^{x=30} (140 + 2x)\ dx = 140x + x^2 \Big|_{0}^{30}$$

$$= 4{,}200 + 900 = \mathbf{5{,}100\ \ foot\text{·}pounds}.$$

Practice 2:   Total work $\approx \{ \pi(\mathbf{1.4})^2(3.5) + \pi(\mathbf{1.6})^2(4.5) + \pi(\mathbf{1.5})^2(5.5) + \pi(\mathbf{1.0})^2(6.5) \}\text{·}(0.5787)$

$\approx \mathbf{67.73\ \ inch\text{·}ounces}.$

Practice 3:   Total work to drink top 3 inches in Example 3 is

$$\int_{y=3}^{y=6} \pi(y/3)^2(0.5787)(10 - y)\ dy = \frac{0.5787\pi}{9} \left\{ \frac{10}{3} x^3 - \frac{1}{4} x^4 \right\}\Big|_{3}^{6}$$

$$= \frac{0.5787\pi}{9} \{ 396 - 69.75 \} \approx \mathbf{65.904\ inch\text{·}ounces}.$$

Practice 4:   From Example 5 we know $f(x) = 7x$ so the total work "5 inches past its natural length to 10 inches past its natural length" is

$$\int_{a}^{b} f(x)\ dx = \int_{5}^{10} 7x\ dx = \frac{7x^2}{2}\ \Big|_{5}^{10} = \mathbf{262.5\ inch\text{–}pounds}.$$

Graphically, this total work is the area of the trapezoidal region bounded by $y = 7x$, the x–axis, and vertical lines at $x = 5$ and $x = 10$.

Practice 5:   Be careful to do all calculations based on the amount of **stretch**, not just on the length.

$f(x) = kx$ and we are told that $f(5) = 2$ so $2 = 5k$ and $k = 2/5$. Then the total work is

$$\int_{a}^{b} f(x)\ dx = \int_{2}^{7} \frac{2}{5}\, x\ dx = \frac{x^2}{5}\ \Big|_{2}^{7} = \frac{45}{5} = \mathbf{9\ inch\text{–}pounds}.$$

Practice 6: $\text{work} = \int_{a}^{b} f(x)\ dx = \int_{2R}^{3R} \frac{R^2 P}{x^2}\ dx = R^2 P \left(-\frac{1}{x} \right)\Big|_{2R}^{3R}$

$$= R^2 P \left(-\frac{1}{3R} \right) - R^2 P \left(-\frac{1}{2R} \right) = \frac{\mathbf{RP}}{\mathbf{6}}\ \mathbf{mile\text{–}pounds}.$$

5.4   MOMENTS & CENTERS OF MASS

This section develops a method for finding the center of mass of a thin,

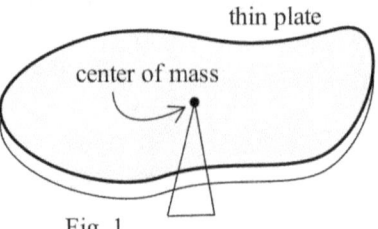

Fig. 1

flat shape — the point at which the shape will balance without tilting (Fig. 1). Centers of mass are important because in many applied

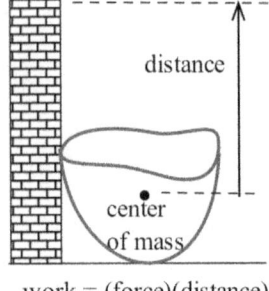

work = (force)(distance)

Fig. 2

situations an object behaves as though its entire mass is located at its center of mass. For example, the work done to pump the water in a tank to a higher point is the same as the work to move the center of mass of the water to the higher point (Fig. 2), a much easier problem, if we know the mass and the center of mass of the water. Also, volumes and surface areas of solids of revolution can be easy to calculate, if we know the center of mass of the region being revolved.

POINT MASSES

Before looking for the centers of mass of complicated regions, we consider point masses and systems of point masses, first in one dimension and then in two dimensions.

Point Masses Along A Line

Two people with different masses can position themselves on a seesaw so that the seesaw balances (Fig. 3). The person on the right causes the

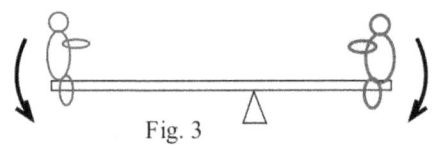

Fig. 3

seesaw to "want to turn" clockwise about the fulcrum, and the person on the left causes it to "want to turn" counterclockwise. If these two "tendencies" are equal, the seesaw will balance. A measure of this tendency to turn about the fulcrum is called the **moment** about the fulcrum of the system, and its magnitude is the mass multiplied by the distance from fulcrum.

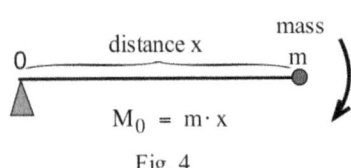

$M_0 = m \cdot x$

Fig. 4

In general, the **moment about the origin, M_0** , produced by a mass m at a location x is $m \cdot x$, the product of the mass and the "signed distance" of the mass from the origin (Fig. 4). For a system of masses $m_1, m_2, \ldots, m_n$ at locations $x_1, x_2, \ldots, x_n$, respectively,

$$M = \text{total mass of the system} \ = \ \sum_{i=1}^{n} m_i \ , \text{ and}$$

$$M_0 = \text{moment about the origin} = x_1 \cdot m_1 + x_2 \cdot m_2 + x_3 \cdot m_3 + \ldots + x_n \cdot m_n \ = \ \sum_{i=1}^{n} x_i \cdot m_i \ .$$

If the moment about the origin is positive then the system tends to rotate clockwise about the origin. If the moment about the origin is negative then the system tends to rotate counterclockwise about the origin. If the moment about the origin is zero, then the system does not tend to rotate in either direction about the origin; it balances on a fulcrum at the origin.

The **moment about the point p, M_p** , produced by a mass m at the location x is the signed distance of x from p times the mass m: $(x-p) \cdot m$. The moment about the point p produced by masses $m_1, m_2, \ldots, m_n$ at locations $x_1, x_2, \ldots, x_n$, respectively, is

$$M_p = \text{moment about the point p} = (x_1-p) \cdot m_1 + (x_2-p) \cdot m_2 + \ldots + (x_n-p) \cdot m_n = \sum_{i=1}^{n} (x_i-p) \cdot m_i .$$

The point at which the system balances is called the **center of mass** of the system and is written $\overline{x}$ (pronounced "x bar") . Since the system balances at $\overline{x}$, the moment about $\overline{x}$ must be zero. Using this fact and properties of summation, we can find a formula for $\overline{x}$.

$$0 = M_{\overline{x}} = \text{moment about } \overline{x} = \sum_{i=1}^{n} (x_i - \overline{x}) \cdot m_i = \sum_{i=1}^{n} (x_i m_i - \overline{x} m_i)$$

$$= \sum_{i=1}^{n} x_i m_i - \sum_{i=1}^{n} \overline{x} \, m_i = \left(\sum_{i=1}^{n} x_i m_i \right) - \overline{x} \left(\sum_{i=1}^{n} m_i \right) , \text{ since } \overline{x} \text{ is a constant.}$$

So $\left(\sum_{i=1}^{n} x_i m_i \right) = \overline{x} \left(\sum_{i=1}^{n} m_i \right)$, and $\overline{x} = \dfrac{\sum_{i=1}^{n} x_i m_i}{\sum_{i=1}^{n} m_i} = \dfrac{M_0}{M} = \dfrac{\text{moment about the origin}}{\text{total mass}}$.

The **center of mass** of a system of masses $m_1, m_2, \ldots, m_n$ at locations $x_1, x_2, \ldots, x_n$ is the point $\overline{x}$ at which the system balances. The moment of the system about $\overline{x}$ is zero.

$$\overline{x} = (\text{moment about the origin}) / (\text{total mass}) = M_0 / M = \sum_{i=1}^{n} x_i \cdot m_i \bigg/ \sum_{i=1}^{n} m_i .$$

The single point mass with mass M located at $\overline{x}$, the center of mass of the system, produces the same moment about any point on the line as the whole system. For many purposes, the mass of the entire system can be thought of as "concentrated at $\overline{x}$."

Example 1:   Find the center of mass of the first three point–masses given in Table 1.

Solution: $M = 2 + 3 + 1 = 6$. $M_0 = (2)(-3) + (3)(4) + (1)(6) = 12$.

$\overline{x} = M_0 / M = 12/6 = 2$.

i	m_i	x_i
1	2	−3
2	3	4
3	1	6
4	5	−2
5	3	4

Table 1

The first three point–masses will balance on a fulcrum located at 2.

Practice 1:   Find the center of mass of the last three point–masses given in Table 1.

Point Masses In The Plane

The ideas of moments and centers of mass extend nicely from one dimension to a system of masses located at points in the plane. For a "knife edge" fulcrum located along the y–axis (Fig. 5), the moment of a mass m at the point (x, y) is the mass times the signed distance of the mass from the y–axis:

(mass)(signed distance from the y–axis) = m·x. This "tendency to rotate about the y–axis" is called the **moment about the y–axis**, written M_y : **$M_y = m·x$**. Similarly, the mass M at the point (x, y) has a **moment about the x–axis** (Fig. 6): **$M_x = m·y$** . For a system of masses m_i located at the points (x_i , y_i) ,

$$M = \text{total mass} = \sum_{i=1}^{n} m_i \qquad\qquad M_y = \text{moment about y–axis} = \sum_{i=1}^{n} m_i·x_i$$

$$M_x = \text{moment about x–axis} = \sum_{i=1}^{n} m_i·y_i$$

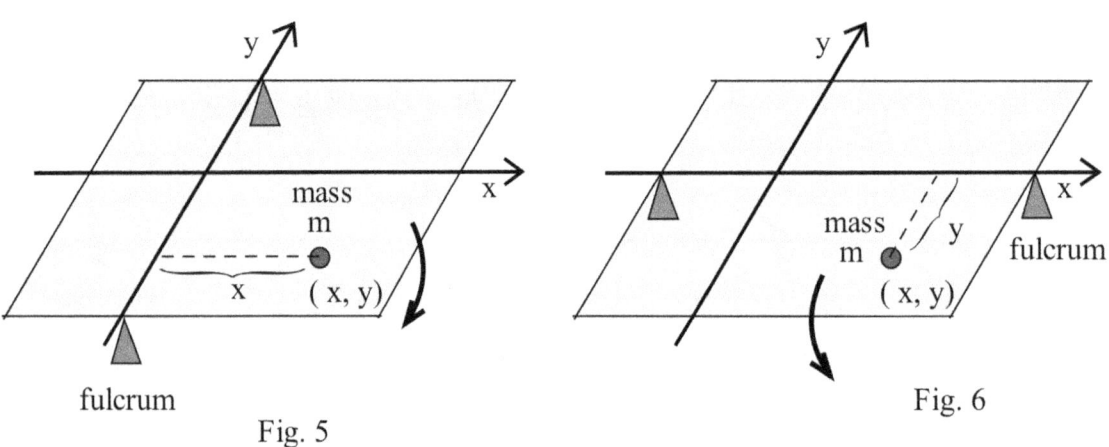

Fig. 5   Fig. 6

The total mass M of the system, located at the center of mass $(\bar{x} , \bar{y})$, has the same moment about any line as the entire system has about that line. For the moment about the y–axis, $M·\bar{x} = M_y$ so $\bar{x} = M_y / M$. Similarly, for the moment about the x–axis, $M·\bar{y} = M_x$ so $\bar{y} = M_x/M$.

Point Masses:	Along a Line	In the Plane
masses:	$m_1, m_2, \ldots, m_n$	$m_1, m_2, \ldots, m_n$
locations:	$x_1, x_2, \ldots, x_n$	$(x_1, y_1), (x_2, y_2), \ldots, (x_n, y_n)$
total mass:	$M = \displaystyle\sum_{i=1}^{n} m_i$	$M = \displaystyle\sum_{i=1}^{n} m_i$
moments:	$M_0 = \displaystyle\sum_{i=1}^{n} m_i \cdot x_i$	$M_y = \displaystyle\sum_{i=1}^{n} m_i \cdot x_i \quad, \quad M_x = \displaystyle\sum_{i=1}^{n} m_i \cdot y_i$
center of mass:	$\bar{x} = M_0 / M$	$\bar{x} = M_y / M \quad, \quad \bar{y} = M_x / M$

Example 2:   Find the center of mass of the first three point–masses in Table 2.

Solution: $M = 2 + 3 + 1 = 6$. $M_y = (2)(-3) + (3)(4) + (1)(6) = 12$.

$M_x = (2)(4) + (3)(-7) + (1)(-2) = -15$.

$\bar{x} = M_y / M = 12/6 = 2$. $\bar{y} = M_x / M = -15/6 = -2.5$.

The first three point–masses will balance at the point $(2, -2.5)$.

Practice 2:   Find the center of mass of all five point–masses in Table 2.

i	m_i	x_i	y_i
1	2	–3	4
2	3	4	–7
3	1	6	–2
4	5	–2	1
5	3	4	–6

Table 2

CENTER OF MASS OF A REGION

When we move from discrete point masses to whole, continuous regions in the plane, we move from finite sums and arithmetic to limits of Riemann sums, definite integrals, and calculus. The following material extends the ideas and calculations from point masses to uniformly thin, flat plates that have a constant density given as mass per area such as "grams/cm^2" (Fig. 7). The center of mass of one of these plates is the point $(\bar{x}, \bar{y})$ at which the plate balances without tilting. It turns out that the center of mass $(\bar{x}, \bar{y})$ of such a plate depends only on the region of the plane covered by the plate and not on its density.

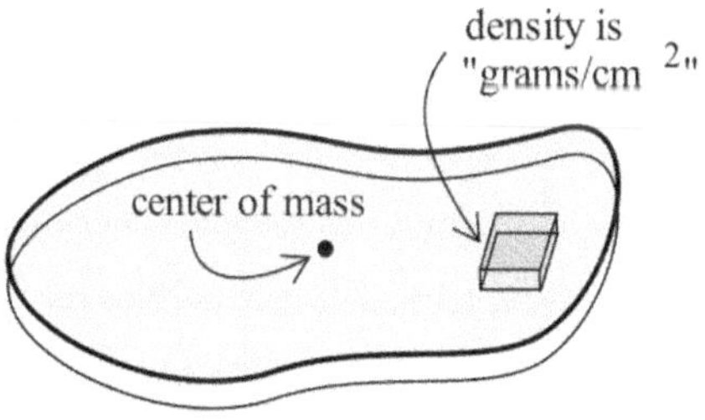

density is "grams/cm^2"

center of mass

Fig. 7

The point $(\bar{x}, \bar{y})$ is also called the **centroid of the region.** (Fig. 8)

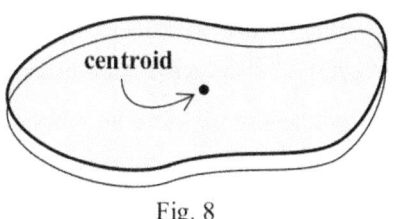

centroid

Fig. 8

In the following discussion, you should notice that each finite sum that appeared in the discussion of point masses has a counterpart for these thin plates in terms of integrals.

Rectangles

The rectangle is the basic shape used to extend the point mass ideas to regions.
The total mass of a rectangular plate is the area of the plate multiplied by the
density constant: mass M = {area}{density}.

We assume that the center of mass of a thin, rectangular plate is located half
way up and half way across the rectangle, at the point where the diagonals of
the rectangle cross (Fig. 9). Then the moments of the rectangle can be found
by treating the rectangle as a point with mass M located at the center of mass
of the rectangle.

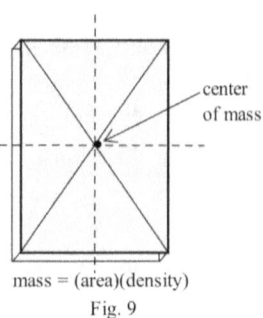

mass = (area)(density)

Fig. 9

Example 3:   Find the moments about the x–axis, y–axis, and the line x = 5

of the rectangular plate in Fig. 10.

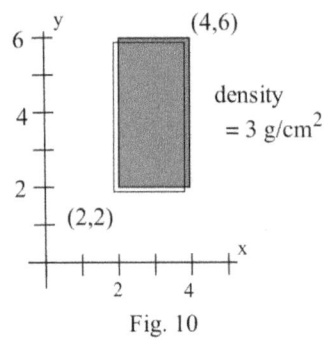

Fig. 10

Solution: The density of the plate is 3 g/cm^2 and the area of the plate is

(2 cm)(4 cm) = 8 cm^2 so the total mass is M = (8 cm^2)(3 g/cm^2) = 24 g.

The center of mass of the rectangular plate is $(\overline{x}, \overline{y}) = (3, 4)$.

The moment about the x–axis is the mass multiplied by the signed distance

of the mass from the x–axis:

M_x = (24 g)•{signed distance of (3,4) from the x–axis} = (24 g)(4 cm) = 96 g–cm.

Similarly,

M_y = (24 g)•{signed distance of (3,4) from the y–axis} = (24 g)(3 cm) = 72 g–cm.

The moment about the line x = 5 is

(24 g)•{signed distance of (3,4) from the line x = 5} = (24 g)(2 cm) = 48 g–cm.

To find the moments and center of mass of a plate made up of several rectangular
regions, just treat each of the rectangular pieces as a point mass concentrated at its
center of mass. Then the plate is treated as a system of discrete point masses..
The plate in Fig. 11 can be divided into two rectangular plates, one with mass 24 g
and center of mass (1,4), and one with mass 12 g and center of mass (3,3). The
total mass of the pair is M = 36 g, and the moments about the axes are

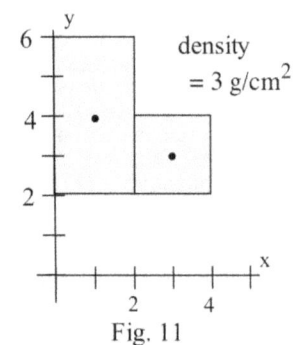

Fig. 11

M_x = (24 g)(4 cm) + (12 g)(3 cm) = 132 g–cm, and

M_y = (24 g)(1 cm) + (12 g)(3 cm) = 60 g–cm.

Then $\overline{x}$ = M_y/M = (60 g–cm)/(36 g) = 5/3 cm and $\overline{y}$ = M_x/M = (132 g–cm)/(36 g) = 11/3 cm so the center
of mass of the plate is at $(\overline{x}, \overline{y}) = (5/3, 11/3)$.

Practice 3:   Find the center of mass of the region in Fig. 12 .

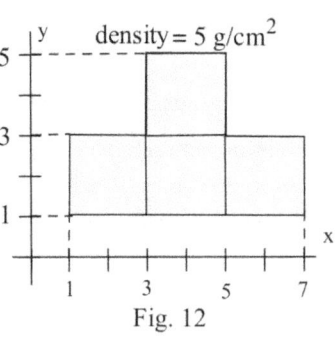

To find the center of mass of a thin plate, we will "slice" the plate into

narrow rectangular plates and treat the collection of rectangular plates as a

system of point masses located at the centers of mass of the rectangles.

The total mass and moments about the axes for the system of point masses

will be Riemann sums. Then, by taking limits as the widths of the

rectangles approach 0, we will obtain exact values for the mass and

moments as definite integrals

$\bar{x}$ For A Region

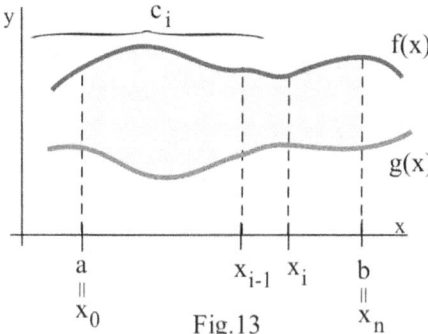

Suppose $f(x) \geq g(x)$ on [a,b] and R is a plate on the region

between the graphs of f and g for $a \leq x \leq b$ (Fig. 13). If the

interval [a, b] is partitioned into subintervals $[x_{i-1} , x_i]$ and the

point c_i is the midpoint of each subinterval, then the slice between

vertical cuts at x_{i-1} and x_i is approximately rectangular and has

mass approximately equal to

$$(\text{area})(\text{density}) = (\text{height})(\text{width})(\text{density}) \approx \{ f(c_i) - g(c_i) \}(x_{i-1} - x_i) k = \{ f(c_i) - g(c_i) \} (\Delta x_i) k .$$

The mass of the whole plate is approximately

$$M = \sum \{f(c_i) - g(c_i)\} (\Delta x_i) k \quad \longrightarrow \quad k \int_a^b \{ f(x) - g(x) \} dx \; = k \cdot \{ \text{ area of the region between f and g } \}.$$

The moment about the y–axis of each rectangular piece is

$$(\text{distance from the y–axis to the center of mass of the piece}) \cdot (\text{mass}) = c_i \cdot \{f(c_i) - g(c_i)\} (\Delta x_i) k$$

so $$M_y = \sum c_i \{ f(c_i) - g(c_i) \} (\Delta x_i) k \quad \longrightarrow \quad k \int_a^b x \cdot \{ f(x) - g(x) \} dx .$$

Then the center of mass of the plate is $\bar{x} \; = \dfrac{M_y}{M} = \dfrac{\displaystyle\int_a^b x \{ f(x) - g(x) \} dx}{\displaystyle\int_a^b f(x) - g(x) \, dx}$.

The density constant k is a factor of M_y and of M, so it has no effect on the value of $\bar{x}$. The value of

$\bar{x}$ depends only on the shape and location of the region, and $\bar{x}$ is the x coordinate of the centroid of the

region between the graphs of $y = f(x)$ and $y = g(x)$ for $a \le x \le b$.

If the bottom curve is the x–axis, then $g(x) = 0$, and the previous results simplify to

$$M = k \int_a^b f(x)\, dx , \quad M_y = k \int_a^b x \cdot f(x)\, dx , \text{ and } \bar{x} = \frac{M_y}{M} .$$

Practice 4:   Find the x–coordinate of the center of mass of the region between $f(x) = x^2$ and the x–axis

for $0 \le x \le 2$. (In this case, $g(x) = 0$.)

$\bar{y}$ For A Region

To find $\bar{y}$, the y–coordinate of the center of mass of a plate R, we
need to find M_x , the moment of the plate about the x–axis. When

R is partitioned vertically (Fig. 14), the moment of each (very narrow) strip
about the x–axis, M_x , is

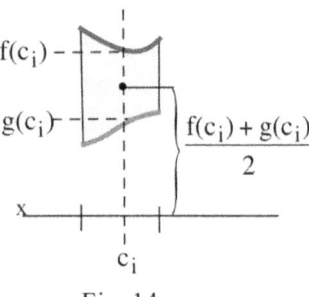

Fig. 14

(the signed distance from the x–axis to the center of mass of the strip)•(mass of strip).

Since each thin strip is approximately rectangular, the y–coordinate of the center of mass of each strip is
approximately **half way** up the strip: $\bar{y}_i \approx \{ f(c_i)+g(c_i) \}/2$. Then

M_x for the strip   = (signed distance from the x–axis to the center of mass of the strip)•(mass of strip)

= (signed distance from x–axis)•(height of strip)•(width of strip)•(density constant)

$$= \frac{f(c_i)+g(c_i)}{2} \cdot (f(c_i) - g(c_i))\cdot(\Delta x_i)\cdot k .$$

The total moment about the x–axis is $M_x = \displaystyle\sum_{i=1}^{n} \frac{f(c_i)+g(c_i)}{2} \cdot \{ f(c_i) - g(c_i) \} (\Delta x_i) k$

$$\longrightarrow \quad k \int_a^b \frac{f(x)+g(x)}{2} \cdot\{ f(x) - g(x)\}\, dx = \frac{k}{2} \int_a^b \{ f^2(x) - g^2(x)\}\, dx .$$

If the lower curve is the x–axis, then $g(x) = 0$ and the formulas simplify to

$$M = k \int_a^b f(x)\, dx , \qquad M_x = k \int_a^b \frac{f(x)}{2} \cdot f(x)\, dx = \frac{k}{2} \int_a^b f^2(x)\, dx , \text{ and } \bar{y} = \frac{M_x}{M} .$$

Example 4:   Find the y–coordinate of the centroid of the region between

the x–axis and the top half of a circle of radius r (Fig. 15).

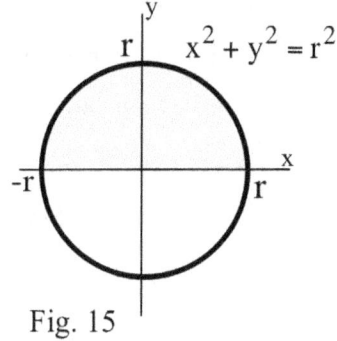

Solution: The equation of the circle is $x^2 + y^2 = r^2$ so $f(x) = y = \sqrt{r^2 - x^2}$.

$$M = k \int_a^b f(x)\,dx = k \int_{-r}^r \sqrt{r^2 - x^2}\,dx$$

Fig. 15

$$= k\,\frac{1}{2}\,(\text{area of the circle of radius r}) = k\,\frac{1}{2}\,(\,\pi r^2\,) = \frac{1}{2}\,k\pi r^2 \ .$$

The region is symmetric about the y–axis so $\bar{x} = 0$. The moment of the region about the x–axis is

$$M_x = \frac{k}{2} \int_a^b f^2(x)\,dx = \frac{k}{2} \int_{-r}^r \left(\sqrt{r^2 - x^2}\right)^2 dx = \frac{k}{2} \int_{-r}^r r^2 - x^2\,dx = \frac{k}{2}\left\{\, r^2 x - \frac{x^3}{3}\,\right\}\Big|_{-r}^r = \frac{2}{3}\,k\,r^3 \ .$$

Finally, $\bar{y} = \dfrac{M_x}{M} = \dfrac{\frac{2}{3}\,k\,r^3}{\frac{1}{2}\,k\pi r^2} = \dfrac{4}{3}\,\dfrac{r}{\pi} \approx 0.4244\,r$.

Practice 5:   Show that the centroid of a triangular region with vertices $(0,0), (0,h)$ and $(b,0)$ is

$(\bar{x}, \bar{y}) = (\,b/3\,, h/3\,)$.

The integral formulas for moments are given below in a form useful for actually calculating moments of regions between the graphs of two functions, but it is also important that you understand the process used to derive the formulas.

Point masses in the plane	Region between f and g for $a \leq x \leq b$ ($f \geq g$)
total mass: $M = \sum_{i=1}^{n} m_i$	$M = \int_a^b \{area\}\{density\} = k \cdot \int_a^b f(x) - g(x)\ dx$
moments: $M_y = \sum_{i=1}^{n} x_i \cdot m_i$	$M_y = \int_a^b \{dist.\ of\ c.m.\ of\ slice\ to\ y\text{–}axis\}\{mass\}$
	$= k \cdot \int_a^b x \cdot \{ f(x) - g(x) \}\ dx$
$M_x = \sum_{i=1}^{n} y_i \cdot m_i$	$M_x = \int_a^b \{dist.\ of\ c.m.\ of\ slice\ to\ x\text{–}axis\}\{mass\}$
	$= \dfrac{k}{2} \cdot \int_a^b \{ f^2(x) - g^2(x) \}\ dx$
center of mass: $\bar{x} = \dfrac{M_y}{M}$, $\bar{y} = \dfrac{M_x}{M}$	$\bar{x} = \dfrac{M_y}{M}$, $\bar{y} = \dfrac{M_x}{M}$

Example 5:   Find the centroid of the region bounded between the graphs of $y = x$ and $y = x^2$ for $0 \leq x \leq 1$.

Solution: $M = k \cdot \int_0^1 (x - x^2)\ dx\ = \dfrac{k}{6}$, $M_y = k \cdot \int_0^1 x(x - x^2)\ dx\ = \dfrac{k}{12}$ and

$M_x = \dfrac{k}{2} \cdot \int_0^1 (x)^2 - (x^2)^2\ dx\ = \dfrac{k}{15}$. Then $\bar{x} = M_y / M = 1/2$ and $\bar{y} = \dfrac{M_x}{M} = 2/5$.

Symmetry

Symmetry is a very powerful geometric concept that can simplify many mathematical and physical problems, including the task of finding centroids of regions. For some regions, we can use symmetry alone to determine the centroid.

Geometrically, a region R is **symmetric about a line L** if, when R is folded along the line L, each point of R on one side of the fold matches up with one point of R on the other side of the fold (Fig. 16).

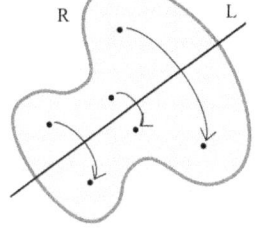

Fig. 16

Example 6:   Sketch two lines of symmetry for each region in Fig. 17.

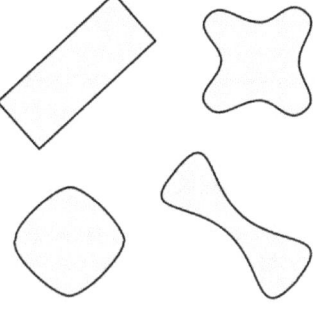

Fig. 17

Solution: The lines of symmetry are shown in Fig. 18. Every line through the center of the circular region is a line of symmetry.

A very useful fact about symmetric regions is that the centroid $(\bar{x}, \bar{y})$ of a symmetric region must lie on **every** line of symmetry of the region. If a region has two different lines of symmetry, then the centroid must lie on each of them, so the centroid is located at the point where the lines of symmetry intersect.

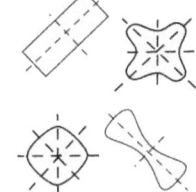

Fig. 18

Practice 6:   Locate the centroid of each region in Fig. 17.

WRAP UP

The purpose of this section is to illustrate again the process of going from an applied problem, to an approximate solution as a Riemann sum, to an exact solution represented as a definite integral. The emphasis is on using calculus to solve an applied problem. However, centroids and centers of mass can themselves be used to solve other applied problems. If we know the center of mass of a region, then some work, volume of revolution, and surface area of revolution problems become simple. These applications of centroids and centers of mass are discussed very briefly in the "optional" section of the problems, and a physical method for determining the centroid of a region is described. Centers of mass of regions with variable density are discussed in a later chapter.

PROBLEMS

1. (a) Find the total mass and the center of mass for a system consisting of the 3 masses

    in Table 3.

m	x
2	4
5	2
5	6

Table 3

(b) Where should you locate a new object with mass 8 so the new system has its center

of mass at x = 5?

(c) How much mass should be located at x = 10 so the original system plus the new mass

at x = 10 has is center of mass at x = 6?

2. (a) Find the total mass and the center of mass for a system consisting of the 4 masses in Table 4.

m	x
5	1
3	7
2	5
6	2

Table 4

(b) Where should you locate a new object with mass 10 so the new system has its

center of

mass at x = 6?

(c) How much mass should be located at x = 14 so the original system

plus the new mass at   x = 14 has is center of mass at x = 6?

3. (a) Find the total mass and the center of mass for a system consisting of the 3 masses

in Table 5.

m	x	y
2	4	3
5	2	4
5	6	2

Table 5

(b) Where should you locate a new object with mass 10 so the new system has its

center of mass at (5, 2)?

m	x	y
1	5	4
2	2	7
3	1	0
5	3	8

Table 6

4. (a) Find the total mass and the center of mass for a system consisting of the 4

masses in Table 6.

(b) Where should you locate a new object with mass 12 so the new system has its

center of mass at (3, 5)?

In problems 5 – 10, divide the flat plate in each Figure into rectangles and semicircles, calculate the mass,

moments and centers of mass of each piece, and use those results to find the center of mass of the plate.

Assume that the density of the plate is 1 g/cm^2 . Plot the location of the center of mass for each shape.

(See Example 4 for the centroid of a semicircular region.)

5.   Fig. 19.   6.   Fig. 20.   7.   Fig. 21.   8.   Fig. 22.   9.   Fig. 23.   10. Fig. 24.

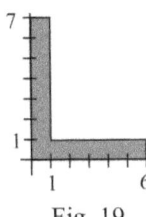

Fig. 19

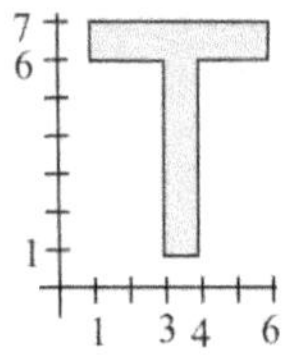

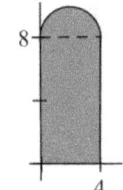

Fig. 20

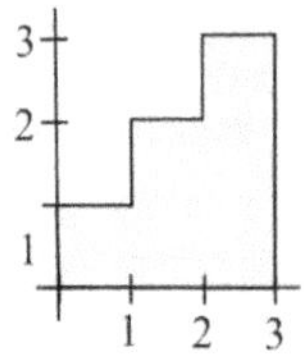

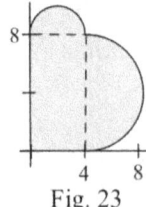

Fig. 21

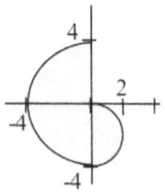

Fig. 22

Fig. 23

Fig. 24

In problems 11 – 26, sketch the region bounded between the given functions on the interval and calculate the centroid of each region (use Simpson's rule with n = 20 if necessary). Plot the location of the centroid on your sketch of the region.

11. $y = x$ and the x–axis for $0 \le x \le 3$.

12. $y = x^2$ and the x–axis for $-2 \le x \le 2$.

13. $y = x^2$ and the line $y = 4$ for $-2 \le x \le 2$.

14. $y = \sin(x)$ and the x–axis for $0 \le x \le \pi$.

15. $y = 4 - x^2$ and the x–axis for $-2 \le x \le 2$.   16. $y = x^2$ and $y = x$ for $0 \le x \le 1$.

17. $y = 9 - x$ and $y = 3$ for $0 \le x \le 3$.   18. $y = \sqrt{1 - x^2}$ and the x–axis for $0 \le x \le 1$.

19. $y = \sqrt{x}$ and the x–axis for $0 \le x \le 9$.   20. $y = \ln(x)$ and the x–axis for $1 \le x \le e$.

21. $y = e^x$ and the line $y = e$ for $0 \le x \le 1$.   22. $y = x^2$ and the line $y = 2x$ for $0 \le x \le 2$.

23. An empty one foot square tin box (Fig. 25) weighs 10 pounds and its center of mass is 6 inches above the bottom of the box. When the box is **full** with 60 pounds of liquid, the center of mass of the box–liquid system is again 6 inches above the bottom. (a) Write the height of the center of mass of the box–liquid system as a function of the x, the height of liquid in the box.
(b) What height of liquid in the bottom of the box results in the box–liquid system having the lowest center of mass (and the greatest stability)?

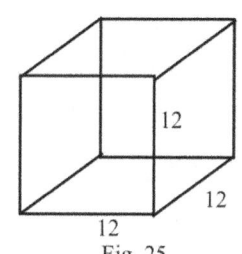
Fig. 25

24. The empty can in Fig. 26 weighs 1 ounce when empty and 13 ounces when full. Write the height of the center of mass of the can–liquid system as a function of the height of the liquid in the can.

25. The empty glass in Fig. 27 weighs 4 ounces when empty and 20 ounces when full. Write the height of the center of mass of the glass–liquid system as a function of the height of the liquid in the glass.

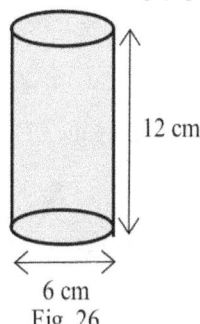
12 cm
6 cm
Fig. 26

26. Give a **practical** set of directions someone could actually use to find the **height** of the center of gravity of their body with their arms at their sides (Fig. 28). How will the height of the center of gravity change if they lift their arms? (In a uniform gravitational field such as at the surface of the earth, the center of mass and center of gravity are at the same point.)

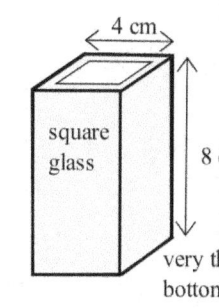
4 cm
square glass
8 c
very th
bottom
Fig. 27

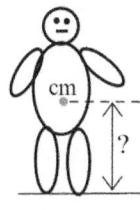
cm
?
Fig. 28

27. Try the following experiment. Stand straight with your back and heels against a wall. Slowly raise one leg, keeping it straight, in front of you. What happened? Why?

28. Why can't two dancers stand in the position shown in Fig. 29?

29. Determine the centroid of your state.

30. (a) Sketch regions with exactly 2 lines of symmetry, exactly 3 lines of symmetry, and
         exactly 4 lines of symmetry.

Fig. 29

   (b) If a shape has exactly two lines of symmetry, the lines can meet at right angles.
         Do they have to meet at right angles?

Work

In a uniform gravitational field, the center of gravity of an object is at the same point as the center of mass
of the object, and the work done to lift an object is the weight of the object multiplied by the distance
that the center of gravity of the object is raised:

   work = (total weight of object)·(distance the center of gravity of the object is raised).

In the high jump, this explains the effectiveness of the "Fosbury Flop", a technique in
which the jumper assumes an inverted U position while going over the bar (Fig. 30).
In this way, the jumper's body goes over the bar while the jumper's center of gravity
goes under it, so a given amount of upward thrust produces a higher bar cleared.

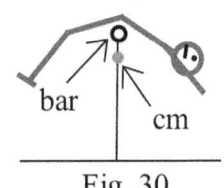

Fig. 30

If the center of gravity of an object is known, some work problems become easy.

31. A rectangular box is filled to a depth of 4 feet with 300 pounds of liquid. How
     much work is done pumping the liquid to a point 10 feet high? (How high is the center of gravity of
     the liquid, and how much must it be raised?)

32. A cylinder is filled to a depth of 2 feet with 40 pounds of liquid. How much work is done pumping the
     liquid to a point 7 feet high? (How high is the center of gravity of the liquid, and how much must it be
     raised?)

33. A sphere of radius 1 foot is filled with 250 pounds of liquid. How much work is done pumping the
     liquid to a point 3 feet above the top of the sphere? (Draw a picture.)

34. A sphere of radius 2 foot is filled with 2000 pounds of liquid. How much work is done pumping the
     liquid to a point 5 feet above the top of the sphere?

If the amount of work is already known, it can be used to find the height of the center of gravity.

Theorems of Pappus

When location of the center of mass of an object is known, the theorems of Pappus make some volume and surface area calculations very easy.

Volume of Revolution:

If   a plane region with **area** A and centroid $(\bar{x}, \bar{y})$

is revolved around a line in the plane which does not go through the region (touching the boundary is alright),

then   the **volume** swept out by one revolution is the area of the region times the distance traveled by the centroid (Fig. 31):

Volume about line L = A $\cdot 2\pi \cdot$ { distance of $(\bar{x}, \bar{y})$ from the line L }.

Volume about x–axis = A$\cdot 2\pi \cdot \bar{y}$.   Volume about y–axis = A$\cdot 2\pi \cdot \bar{x}$.

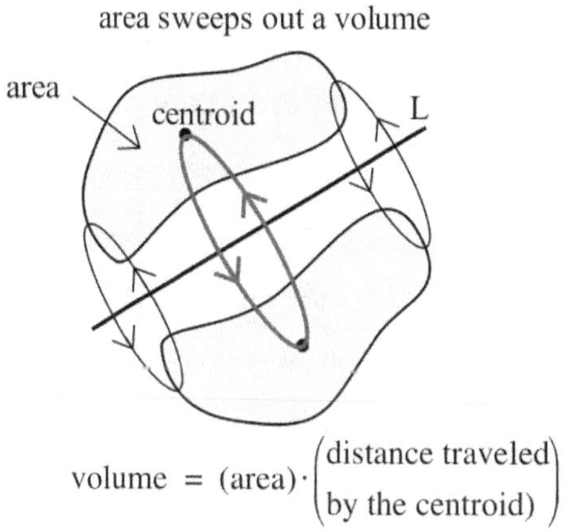

area sweeps out a volume

$$\text{volume} = (\text{area}) \cdot \begin{pmatrix} \text{distance traveled} \\ \text{by the centroid} \end{pmatrix}$$

Fig. 31

Surface Area of Revolution

If   a plane region with **perimeter** P and centroid of the
edge $(\bar{x} , \bar{y})$ is revolved around a line in the plane
which does not go through the region (touching the
boundary is alright),

then   the **surface area** swept out by one revolution
      is the perimeter of the region times the distance
      traveled by the centroid (Fig. 32):

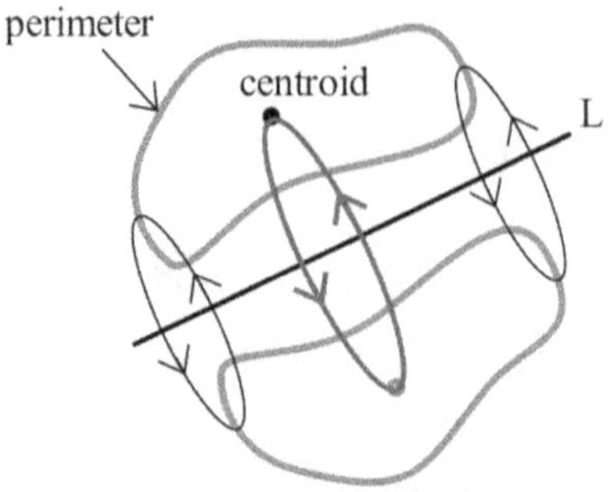

perimeter sweeps out an area (surface area)

perimeter

centroid   L

$$\text{surface area} = (\text{perimeter}) \cdot \left(\begin{array}{l} \text{distance traveled} \\ \text{by the centroid} \end{array} \right)$$

Fig. 32

Surface area about line L = P·2π·{ distance of $(\bar{x} , \bar{y})$ from the line L }.
Surface area about x–axis = P· 2π· $\bar{y}$.   Surface area about y–axis = P· 2π· $\bar{x}$.

35. The center of a square region with 2 foot sides is at the point (3,4). Use the Theorems of Pappus to
     find the volume and surface area swept out when the square is rotated   (a) about the x–axis,
     (b) about the y–axis, and (c) about the horizontal line y = 6.

36. The lower left corner of a rectangular region with an 8 inch base and a 4 inch height is at the point
     (3,5). Use the Theorems of Pappus to find the volume and surface area swept out
     when the rectangle is rotated (a) about the x–axis, (b) about the y–axis, and (c)
     about the line y = x + 5.

37. The center of a circle with radius 2 feet is at the point (3,5). Use the Theorems of
     Pappus to find the volume and surface area swept out when the circular region is
     rotated   (a) about the x–axis, (b) about the y–axis, and
     (c) about the vertical line x = 6.

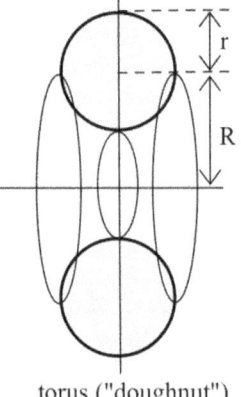

torus ("doughnut")
Fig. 33

38. The center of a circle (Fig. 33) with radius r is at the point (0, R). Use the
     Theorems of Pappus to find the volume and surface area swept out when the
     circular region is rotated about the x–axis.

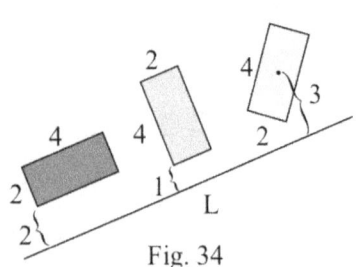

Fig. 34

39. Find the volumes and surface areas swept out when the rectangles in
     Fig. 34 are rotated about the line L. (Measurements are in feet.)

Physically Approximating Centoids of Regions

The centroid of a region can be approximated experimentally, even if the region, such as a state or country, is not described by a formula.

Cut the shape out of a piece of some uniformly thick material such as paper and pin an edge to a wall. The shape will pivot about the pin until its center of mass is directly below the pin (Fig. 35) so the center of mass of the shape must lie directly below the pin, on the line connecting the pin with the center of mass of the earth. Repeat the process using a different point near the edge of the shape and a different line can be found. The center of mass also lies on the new line, so we can conclude that the centroid of the shape is located where the two lines intersect, the only point located on both lines (Fig. 36). It is a good idea to pick a third point near the edge and plot a third line. This line should also pass through the point of intersection of the other two lines.

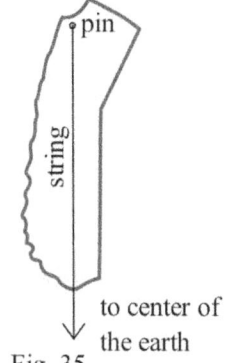

Fig. 35

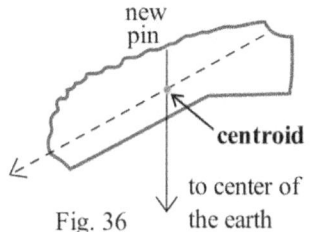

new pin

centroid

to center of the earth

Fig. 36

The "population center" of a region can be physically approximated by attaching weights proportional to the populations of the different areas and then repeating the "pin" process with this weighted model. The point on the new model where the lines intersect is the approximate "population center" of the region.

Section 5.4   **PRACTICE Answers**

Practice 1: $M = 1 + 5 + 3 = 9$. $M_0 = (1)(6) + (5)(-2) + (3)(4) = 8$.

$\bar{x} = M_0 / M = 8/9$

The last three point–masses will balance on a fulcrum located at $x = 8/9$.

i	m_i	x_i
1	2	-3
2	3	4
3	1	6
4	5	-2
5	3	4

Table 1

Practice 2: $M = 2 + 3 + 1 + 5 + 3 = 14$.

$M_y = (2)(-3) + (3)(4) + (1)(6) + (5)(-2) + (3)(4) = 14$.

$M_x = (2)(4) + (3)(-7) + (1)(-2) + (5)(1) + (3)(-6) = -28$.

$\bar{x} = M_y / M = 14/14 = 1$. $\bar{y} = M_x / M = -28/14 = -2$.

The five point–masses balance at the point $(1, -2)$.

i	m_i	x_i	y_i
1	2	-3	4
2	3	4	-7
3	1	6	-2
4	5	-2	1
5	3	4	-6

Table 2

Practice 3:   There are several ways to break the region in Fig. 12 into "easy"

pieces — one way is to consider the four 2–by–2 cm squares.

The cm of each square is located at the center of the square (at $(2,2)$, $(4,2)$, $(6,2)$, and $(4,4)$),

and each square has mass $(4 \text{ cm}^2)(5 \text{ g/cm}^2) = 20$ g.

$M = 4(20 \text{ g}) = 80$ g.

$M_y = 2(20) + 4(20) + 6(20) + 4(20) = 320$ g·cm

$M_x = 2(20) + 2(20) + 2(20) + 4(20) = 200$ g·cm

Then $\bar{x} = M_y / M = \dfrac{320 \text{ g·cm}}{80 \text{ g}} = \mathbf{4\ cm}$ and

$\bar{y} = M_x / M = \dfrac{200 \text{ g·cm}}{80 \text{ g}} = \mathbf{2.5\ cm}$.

The center of mass is $(4, 2.5)$.

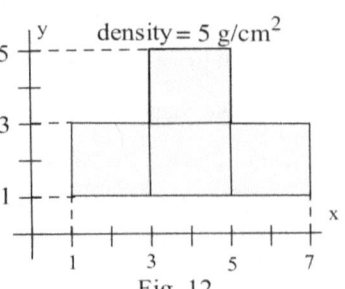

Fig. 12

Practice 4: $M = k \displaystyle\int_a^b f(x)\, dx == k \int_0^2 x^2\, dx = \dfrac{8}{3} k$, $M_y = k \displaystyle\int_a^b x \cdot f(x)\, dx = k \int_0^2 x \cdot x^2\, dx = 4k.$

Then $\bar{x} = M_y / M = \dfrac{4k}{\frac{8}{3} k} = \mathbf{1.5}$.

Practice 5:   The triangular region is shown in Fig. 36 : $f(x) = h - \dfrac{h}{b} x$ for $0 \le x \le b$.

$M = k \displaystyle\int_a^b f(x)\, dx = k \int_0^b \left(h - \dfrac{h}{b} x \right)\, dx = k \left\{ hx - \dfrac{h}{b} \dfrac{1}{2} x^2 \right\}\Big|_0^b$

$= k\left\{ hb - \dfrac{h}{b} \dfrac{b^2}{2} \right\} = \dfrac{k}{2} hb$.

$M_y = k \displaystyle\int_a^b x \cdot f(x)\, dx = k \int_0^b x \cdot \left(h - \dfrac{h}{b} x \right)\, dx = k \left\{ h \dfrac{x^2}{2} - \dfrac{h}{b} \dfrac{x^3}{3} \right\}\Big|_0^b$

$= k\left\{ h \dfrac{b^2}{2} - \dfrac{h}{b} \dfrac{b^3}{3} \right\} = \dfrac{k}{6} hb^2$.

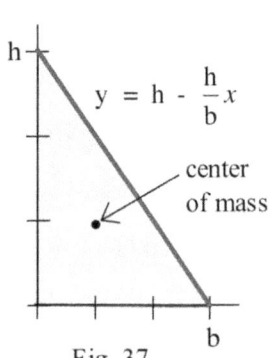

$y = h - \dfrac{h}{b} x$

center of mass

Fig. 37

$M_x = \dfrac{k}{2} \displaystyle\int_a^b f^2(x)\, dx = \dfrac{k}{2} \int_0^b \left(h - \dfrac{h}{b} x \right)^2 dx = \dfrac{kh^2}{2} \left\{ x - \dfrac{2}{b} \dfrac{x^2}{2} + \dfrac{1}{b^2} \dfrac{x^3}{3} \right\}\Big|_0^b$

$= \dfrac{kh^2}{2} \left\{ b - b + \dfrac{b}{3} \right\} = \dfrac{k}{6} h^2 b^2$.

Fig. 18

Finally, $\bar{x} = \dfrac{M_y}{M} = \dfrac{\frac{k}{6} hb^2}{\frac{k}{2} hb} = \dfrac{b}{3}$ and $\bar{y} = \dfrac{M_x}{M} = \dfrac{\frac{k}{6} h^2 b^2}{\frac{k}{2} hb} = \dfrac{h}{3}$ so $\mathbf{cm = (b/3, h/3)}$.

Practice 6:   The centroid of each region in Fig. 18 is located at the point where the lines of symmetry intersect.

5.5   ADDITIONAL APPLICATIONS

This section **introduces** three additional applications of integrals and once again illustrates the process of going from a problem to a Riemann sum and on to a definite integral. A fourth application is included which does not follow this process. It uses the idea of "area" to try to model an election and to qualitatively understand why certain election outcomes occur.

The main point of this section is to show the power of definite integrals to solve a wide variety of applied problems. Each of these new applications is treated more briefly than those in the previous sections.

1.   Liquid Pressures and Forces

The hydrostatic pressure on an immersed object is the density of the fluid times the depth of the object: **pressure = (density)(depth)**. The total hydrostatic force against an immersed object is the sum of the hydrostatic forces against each part of the object. If our entire object is at the same depth, we can determine the total hydrostatic force simply by multiplying the density times the depth times the area. If the unit of density is "pounds per cubic foot" and the depth is given in "feet," then the unit of pressure is "pounds per square foot," a measure of **force per area**. If a pressure, with the units "pounds per square foot," is multiplied by an area with the units "square feet," the result is a force, "pounds."

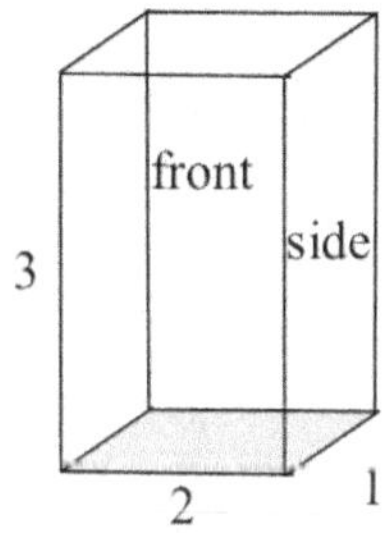

Example 1:   Find the total hydrostatic force against the bottom of the aquarium
shown in Fig. 1.

Solution: The density of water is 62.5 pounds/ft^3, so

total hydrostatic force = (density)•(depth)•(area)

= (62.5 pounds/ft^3)•(3 feet)•(2 square feet) = 375 pounds.

Fig. 1

Finding the total hydrostatic force against the front of the aquarium is a very different problem because different parts of the front are at different depths and are subject to different pressures. To find the force against the front of the aquarium, we can partition it into thin horizontal slices (Fig. 2) and focus on one of them. Since the slice is very thin, every part of the ith slice is at almost the same depth so every part of the slice is subject to almost the same pressure. We can approximate the total hydrostatic force against the slice at the depth x_i as

(density)•(depth)•(area) = (62.5 pounds/ft^3)(x_i feet)(2 feet)(Δx_i feet)

= $125 x_i \cdot \Delta x_i$ pounds.

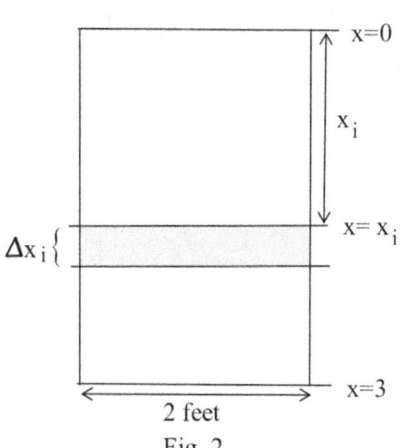

Fig. 2

The total hydrostatic force against the front is the sum of the forces against each slice,

total hydrostatic force $\approx \sum 125x_i \, \Delta x_i$, a Riemann sum.

The limit of the Riemann sum as the slices get thinner $(\Delta x \to 0)$ is a definite integral:

total hydrostatic force $\approx \sum 125x_i \, \Delta x_i \longrightarrow \int_{x=0}^{3} 125x \, dx = 62.5 \, x^2 \Big|_{x=0}^{3} = 562.5$ pounds.

Practice 1:   Find the total hydrostatic force against one side of the aquarium.

Hydrostatic Force

If   the width of a slice of a horizontal object at depth x is w(x) (Fig. 3)

then   the total hydrostatic force against the object between

           depths a and b is

           total hydrostatic force $\approx \sum$ (density)·(depth)·w(x_i) Δx_i

$$\longrightarrow \int_{x=a}^{b} (\text{density}) \cdot x \cdot w(x) \, dx \ .$$

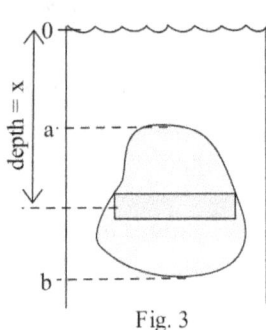

Fig. 3

Example 2:   Find the total hydrostatic force against windows A and B in Fig. 4.

Solution: Window A: Using similar triangles,

$\dfrac{w}{6-x} = \dfrac{3}{2}$ so $w(x) = \dfrac{3}{2}(6-x)$.

Then total hydrostatic force

$$= \int_{x=4}^{6} (\text{density}) \cdot x \cdot w(x) \, dx$$

$$= \int_{4}^{6} (60) \cdot x \cdot \frac{3}{2}(6-x) \, dx = 90\left(3x^2 - \frac{x^3}{3} \right)\Big|_{4}^{6} = 840$$

pounds.

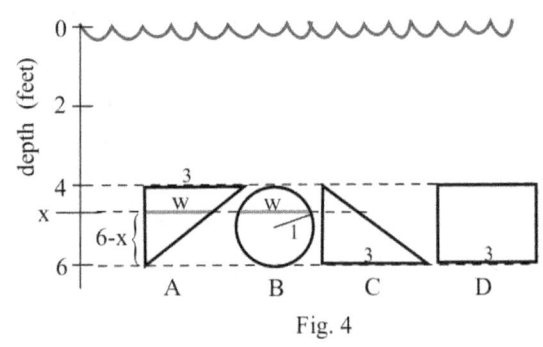

Fig. 4

Window B: The equation of the circle is $(5-x)^2 + (w/2)^2 = 1$ so $w(x) = 2\sqrt{1 - (5-x)^2}$. Then

total hydrostatic force $= \displaystyle\int_{4}^{6} (60) \cdot x \cdot 2\sqrt{1 - (5-x)^2} \ dx \ \approx 938.1$ pounds (using a calculator).

Practice 2:   Find the total hydrostatic force against windows C and D in Fig. 4.

Because the total force at even moderate depths is so large, the underwater windows at aquariums are made of thick glass or plastic and strongly secured to their frames. Similarly, the bottom of a dam is much thicker than the top in order to withstand the greater force against the bottom.

2.   Kinetic Energy of a Rotating Object

The **Kinetic Energy** (the energy of motion) of an object is defined to be half the mass of the object multiplied by the square of the velocity of the object: $KE = \frac{1}{2} m \cdot v^2$.

The larger an object is or the faster it is moving, the greater its kinetic energy. If every part of the object has the same velocity, then it is easy to compute its kinetic energy. Sometimes, however, different parts of the object move with different velocities. For example, if an ice skater is spinning with an angular velocity of 2 revolutions per second, then her arms travel further in one second (have a greater linear velocity) when they are extended than when they are drawn in close to her body (Fig. 5). So the ice skater, spinning at 2 revolutions per second, has greater kinetic energy when her arms are extended. Similarly, the tip of a rotating propeller or of a swinging baseball bat has a greater has a greater linear velocity than other parts of the propeller or bat. If the units of mass are "grams" and the units of velocity are "centimeters per second," then the units of kinetic energy $KE = \frac{1}{2} m \cdot v^2$ are "ergs."

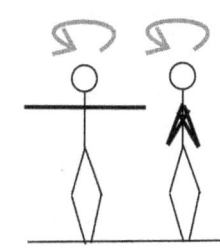

Fig. 5

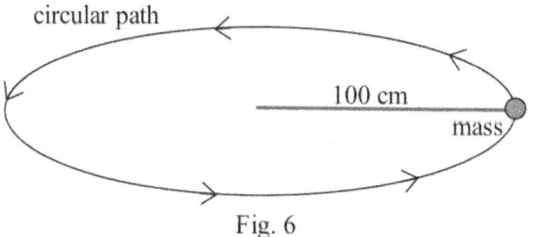

circular path

100 cm

mass

Fig. 6

Example 3:   A point–mass of 1 gram at the end of a (massless) 100 centimeter long string is rotated at a rate of 2 revolutions per second (Fig. 6). (a) Find the kinetic energy of the point–mass. (b) Find its kinetic energy if the string is 200 centimeters long.

Solution: (a) In one second, the mass travels twice around a circle with radius 100 centimeters so it travels 2·(2π·radius) = 400π centimeters. The velocity is v = 400π cm/s , and

$$KE = \frac{1}{2} m \cdot v^2 = \frac{1}{2}(1 \ g) \cdot (400\pi \ cm/s)^2 = 80{,}000 \ \pi^2 \ ergs.$$

(b) If the string is 200 centimeters long, then the velocity is 2·(2π·radius)/second = 800π cm/s ,

and $KE = \frac{1}{2} m \cdot v^2 = \frac{1}{2}(1 \ g) \cdot (800\pi \ cm/sec)^2 = 320{,}000 \ \pi^2 \ ergs$.

When the length of the string doubles, the velocity is twice as large and the kinetic energy is 4 times as large.

Practice 3:   A 1 gram point–mass at the end of a 2 meter (massless) string is rotated at a rate of 4
revolutions per second. Find the kinetic energy of the point mass.

If different parts of a rotating object are different distances from the axis of rotation, then those parts have
different linear velocities, and it is more difficult to calculate the total kinetic energy of the object. By now the
method should seem very familiar: partition the object into small pieces, approximate the kinetic energy of
each piece, and add the kinetic energies of the small pieces (a Riemann sum) to approximate the total kinetic
energy of the object. The limit of the Riemann sum as the pieces get smaller is a definite integral.

Example 4:   The density of a narrow bar (Fig. 7) is 5 grams per meter of

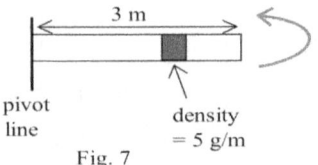

length. Find the kinetic energy of the 3 meter long bar when it is

rotated at a rate of 2 revolutions per second.

Fig. 7

Solution: If the length of the bar is partitioned (Fig. 8), the mass of the ith piece is

$$m_i \approx (length) \cdot (density) = (\Delta x_i \text{ meters})(5 \text{ grams/meter}) = 5 \Delta x_i \text{ grams.}$$

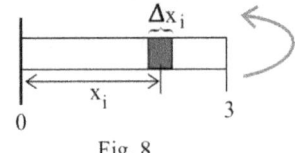

In one second the ith piece will make two revolutions and will travel approximately

$$2 \cdot (2\pi \text{ radius}) = 400\pi \cdot x_i \text{ centimeters so } v_i \approx 400\pi \, x_i \text{ cm/sec.}$$

Fig. 8

The kinetic energy of the ith piece is

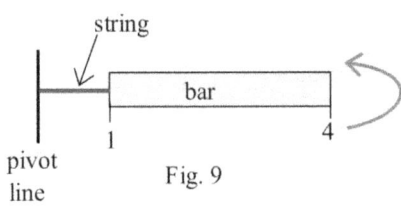

$$ke_i = \frac{1}{2} \, m_i \cdot v_i^2 \quad = \frac{1}{2} (5 \, \Delta x_i \text{ grams}) \cdot (400\pi \, x_i \text{ cm/sec})^2$$

$$= 400{,}000\pi^2 \, (x_i)^2 \, \Delta x_i \text{ ergs,}$$

Fig. 9

and the total kinetic energy of the rotating bar is

$$KE = \sum ke_i \; = \sum 400{,}000\pi^2 \, (x_i)^2 \, \Delta x_i \;\longrightarrow\; \int_0^3 400{,}000\pi^2 \cdot x^2 \; dx$$

$$= 400{,}000 \, \pi^2 \frac{x^3}{3} \Big|_0^3 \; = 3{,}600{,}000 \, \pi^2 \text{ ergs.}$$

Practice 4:   Find the kinetic energy of the bar in the previous example if it
is rotated at 2 revolutions per second at the end of a 100 centimeter (massless) string (Fig. 9).

density = 5 g/cm^2

Example 5:   Find the kinetic energy of the object in Fig. 10 when

it rotates at 2 revolutions per second.

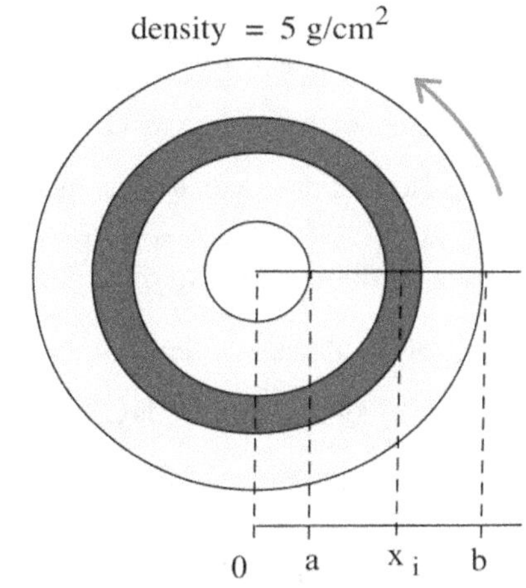

Solution: We can partition along one radial line, and form circular

"slices." Then the "slice" between x_i and $x_i + \Delta x$ is a thin circular

band with mass $m_i = $ (volume)·(density) $\approx 2\pi x_i \Delta x_i \cdot h \cdot d$ Newtons.

Each part of the "slice" is approximately x_i centimeters from the

axis of rotation so each part has approximately the same velocity:

$v_i \approx$ (2 rev/sec)·($2\pi x_i$ centimeters/rev) = $4\pi \cdot x_i$

centimeters/sec.

Fig. 10

The kinetic energy of the ith piece is

$$ke_i \approx = \frac{1}{2} m_i \cdot v_i^2 = \frac{1}{2}(2\pi \cdot x_i \Delta x_i \cdot h \cdot d) \cdot (4\pi \cdot x_i)^2 = 16\pi^3 \cdot h \cdot d \cdot (x_i)^3 \Delta x_i ,$$

so $KE = \sum ke_i \quad = \sum 16\pi^3 \cdot h \cdot d \cdot (x_i)^3 \Delta x$

$$\longrightarrow \int_a^b 16\pi^3 \cdot h \cdot d \cdot x^3 \, dx = 16\pi^3 hd \int_a^b x^3 \, dx = 4\pi^3 hd \cdot (b^4 - a^4) .$$

Since b is raised to the 4th power, a small increase in the value of b leads to a large increase in the

kinetic energy.

3.   Volumes of Revolution Using "Tubes" (Shells)

In Section 5.2 the "disk" method was used for finding the volume swept out when a region is revolved about

a line (Fig. 11). To find the volume swept out when a region was revolved about the x–axis, we made cuts

perpendicular to the x–axis so each slice was a "disk" with volume $\pi(\text{radius})^2 \cdot (\text{thickness})$.

After adding the volumes of the slices together (a Riemann sum) and taking a limit as the thicknesses

approached 0, we obtained a definite integral representation for the exact volume:

{ volume of revolution about the x–axis }

$$= \int_a^b \pi f^2(x) \, dx .$$

However, the disk method can be cumbersome if we

want the volume when the region in

the figure is revolved about

the y–axis or some other

vertical line. To revolve the region about the y–axis, the

disk method requires that we represent the original

equation $y = f(x)$ as a function of y: $x = g(y)$.

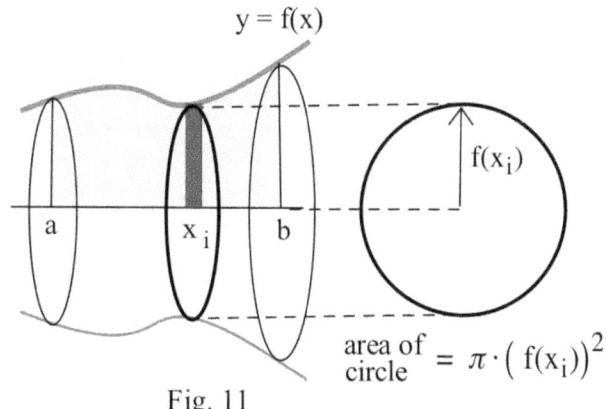

$y = f(x)$

$f(x_i)$

area of circle $= \pi \cdot \left(f(x_i) \right)^2$

Fig. 11

Sometimes that is easy: if $y = 3x$ then $x = y/3$. But sometimes it is not easy at all: if $y = x + e^x$, then we can not solve for y as an elementary function of x. The "tube" method lets us use the original equation $y = f(x)$ to find the volume when the region is revolved about a vertical line.

We partition the x–axis to cut the region into thin, almost rectangular "slices." When the thin "slice" at x_i is revolved about a vertical line (Fig. 12a), the volume of the resulting "tube" can be approximated by cutting the wall of the tube and laying it out flat (Fig. 12b) to get a thin, solid rectangular box. The volume of the tube is approximately the same as the volume of the solid box:

$$V_{tube} \approx V_{box} = (length) \cdot (height) \cdot (thickness) = \mathbf{(2\pi \cdot radius)} \cdot \mathbf{(height)} \cdot (\Delta x_i) = (2\pi\, x_i) \cdot (f(x_i))\, \Delta x_i .$$

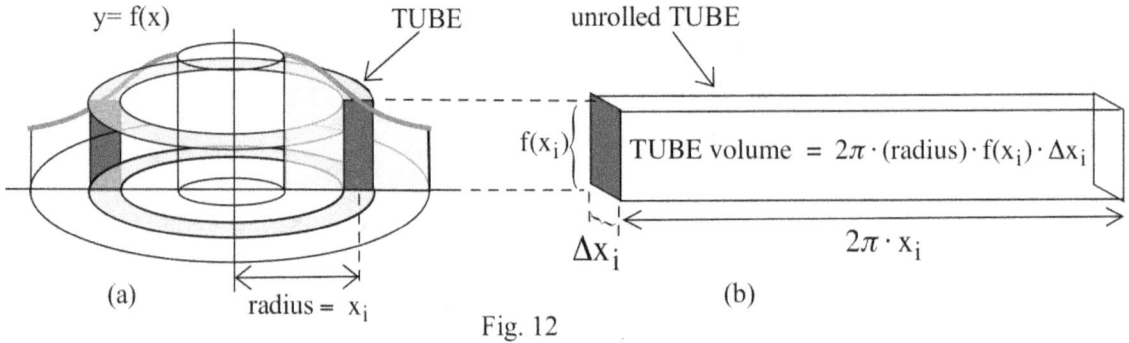

Fig. 12

The volume swept out when the whole region is revolved is the sum of the volumes of these "tubes", a Riemann sum. The limit of the Riemann sum is

$$\{ \text{volume of rotation about a vertical line} \} = \int_{x=a}^{b} 2\pi \cdot (radius) \cdot (height)\ dx .$$

The "tube" pattern for the volume of a region defined by a single function extends easily to regions between two functions.

Volume of Revolution Using "Tubes"   (Shells)

If   region R is bounded between the functions $f(x) \geq g(x)$ for $0 \leq a \leq b$ (Fig. 13),

then   { volume obtained when R is revolved about a vertical line } = $\displaystyle\int_{x=a}^{b} 2\pi \cdot (\text{radius}) \cdot \{ f(x) - g(x) \} \ dx$.

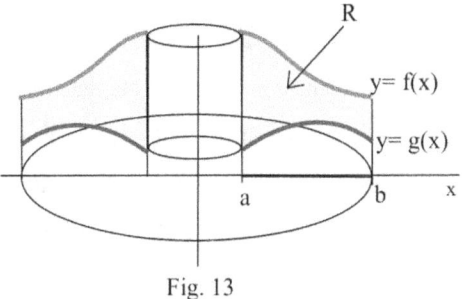

Fig. 13

Example 6:   Find the volume when the region R in

Fig. 14 is revolved about a vertical line.

Solution: We can partition the interval $[2, 4]$ on the x–axis to get

thin slices of R. When the slice at x_i is

revolved around the y–axis, a tube is swept out,

and the volume if this i^{th} tube is

$$v_i \ \approx (2\pi \cdot \text{radius}) \cdot (\text{height}) \cdot (\text{thickness}) \approx 2\pi (x_i)(x_i^2 - x_i)(\Delta x_i) = 2\pi (x_i^3 - x_i^2) \Delta x_i$$

The total volume is the sum of the volumes of the tubes:

$$V = \sum v_i \ = \sum 2\pi (x_i^3 - x_i^2) \Delta x_i \quad \text{(a Riemann sum)}$$

$$\longrightarrow \quad \int_2^4 2\pi (x^3 - x^2) \ dx = 2\pi (\frac{x^4}{4} - \frac{x^3}{3}) \Big|_2^4 = \frac{248 \pi}{3} \approx 259.7 .$$

Example 7:   Write a definite integral to represent the volume swept out when

the region in Fig. 15 is revolved about the vertical line $x = 4$.

Solution: $V = \displaystyle\int_a^b 2\pi \cdot (\textbf{radius}) \cdot (\textbf{height}) \ dx$

$$= \int_0^{\pi} 2\pi \cdot (4 - x) \cdot \sin(x) \ dx$$

≈ 30.526 (using calculator integrate command))

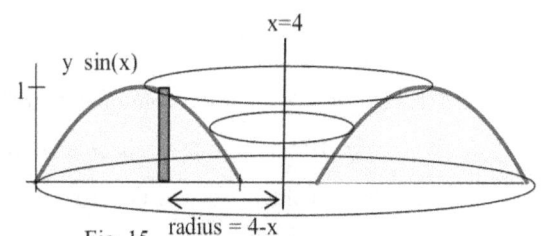

Fig. 15   radius = 4-x

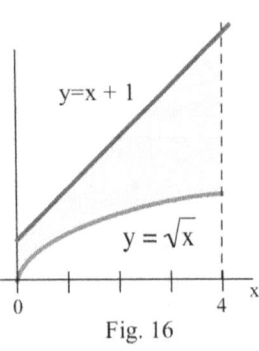

y=x + 1

y = √x

0   4   x

Fig. 16

Practice 5:   Represent the volume when the region in Fig. 16 is revolved about

the y–axis.

In theory, each method works for each volume of revolution problem. In

practice, however, for any particular problem one of the methods is often

easier to use than the other.

4.   Areas & Elections

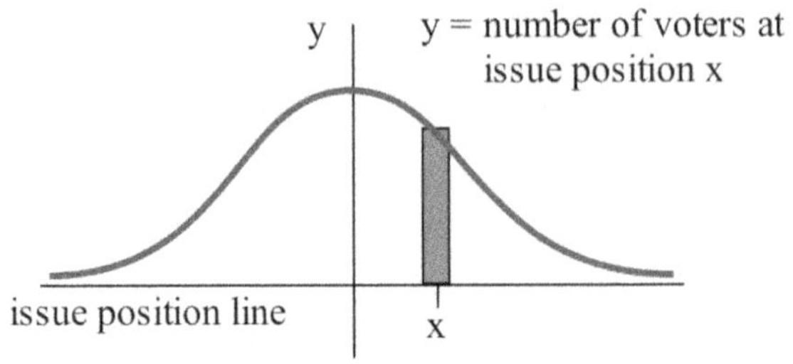

The previous applications used definite integrals to determine

areas, volumes, pressures, and energies **exactly**. But

exactness and numerical precision are not the same as

"understanding," and sometimes we can gain insight and

understanding simply by determining which of two areas or

integrals is larger. One situation of this type involves

elections.

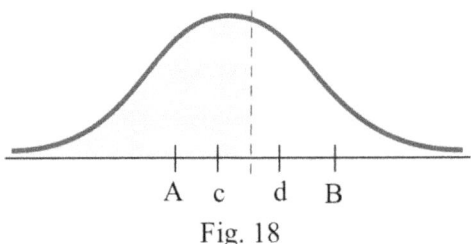

y   y = number of voters at
        issue position x

issue position line   x

Fig. 17

Suppose the voters of a state have been surveyed about their

positions on a **single** issue, and the distribution of voters who

place themselves at each position on this issue is shown in Fig.

17. Also suppose that each voter votes for the candidate closest to

that voter. If two candidates have taken the positions labeled A

and B in Fig. 18, then a voter at position c votes for the

candidate at A since A is closer to c than B is to c.

A   c   d   B

Fig. 18

Similarly, a voter at position d votes for the candidate at B. The total votes for the

candidate at A in this election is the shaded area under the curve, and the

candidate with the larger number of votes, the larger

area, is the winner. In this illustration, the candidate at A wins.

A   B

Fig. 19

Example 8:   The distribution of voters on an issue is shown in

Fig. 19. If these voters decide between candidates on the basis

of that single issue, which candidate will win the election?

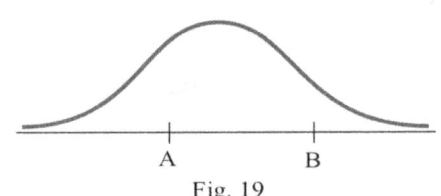

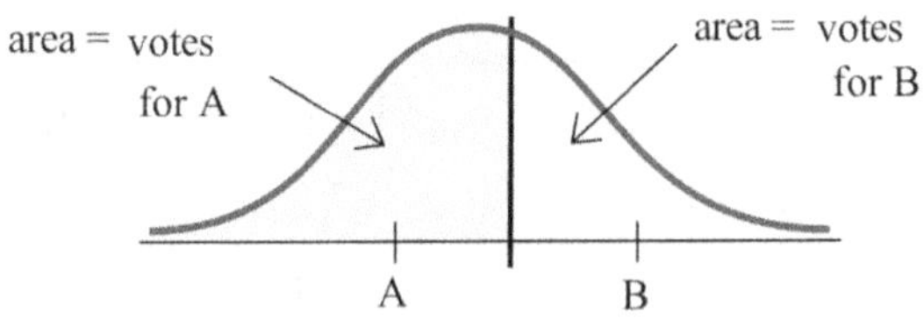

area = votes
for A

area = votes
for B

A   B

Fig. 20

Solution: Fig. 20 illustrates that A has a larger area (more

votes) than B. A will win.

Practice 6:   In an election between candidates with

positions A and B in Fig. 21, who will win?

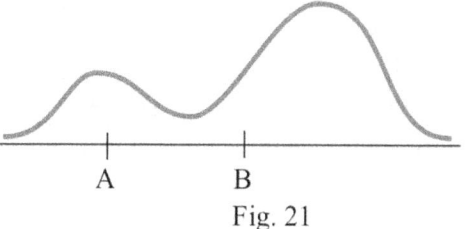

Fig. 21

If voters behave as described and if the election is between **2 candidates**, then we can give the candidates some advice.

The best position for a candidate is at the "median point," the location that divides the voters into two equal sized (equal area) groups so half of the voters are on one side of the median point and half are on the other side (Fig. 22). A candidate at the median point gets more votes than a candidate at any other point. (Why?)

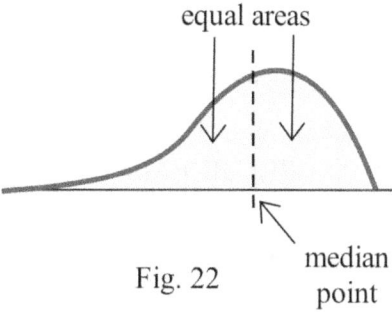

equal areas

Fig. 22   median point

If two candidates have positions on opposite sides of the median point

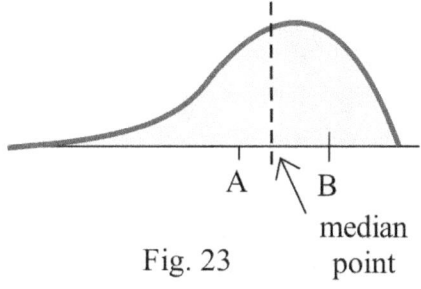

A   B

median point

Fig. 23

(Fig. 23), then a candidate can get more votes by moving a bit toward the median point. This "move toward the middle ground commonly occurs in elections as candidates try to sell themselves as "moderates" and their opponents as "extremists."

If there are more than two candidates running in an election, then the situation changes dramatically, and a candidate at the median position, the unbeatable place in a 2–candidate election, can even get the fewest votes. If Fig. 24 is the distribution of voters on the single issue in the

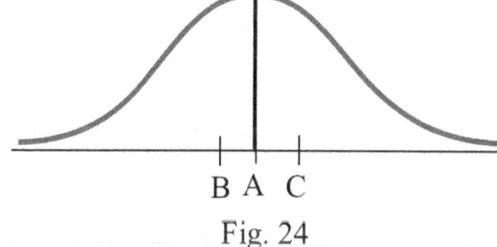

B A C

Fig. 24

election, then candidate A would beat B in an election just between A and B (Fig. 25a); and A would beat C in an election just between A and C (Fig. 25b). But in an election among all 3 candidates, A would get the fewest votes of the 3 candidates (Fig. 25c). This type of situation really does occur. It leads to the political saying about a primary election with lots of candidates and a general election between the final nominees of the two parties:

"extremists can win primaries, but moderates are elected to office."

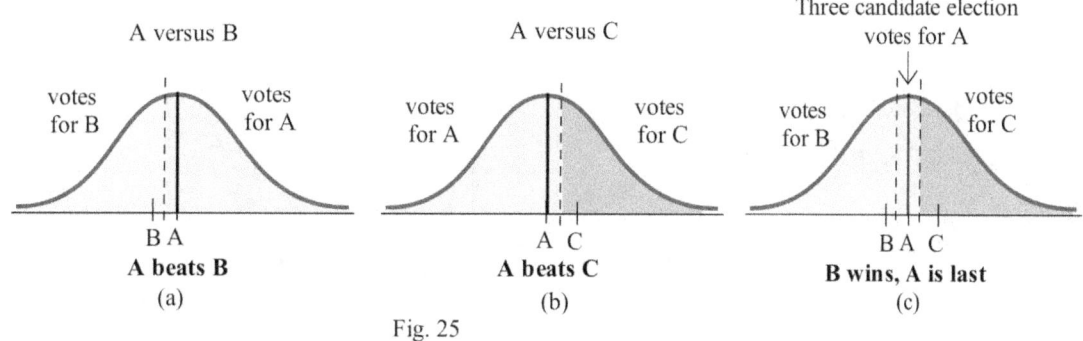

Fig. 25

The previous discussion of elections and areas is greatly oversimplified. Most elections involve several issues of different importance to different voters, and the views of the voters are seldom completely known before the election. Many candidates take "fuzzy" positions on issues. And it is not even certain that real voters vote for the "closest" candidate: perhaps they don't vote at all unless some candidate is "close enough" to their position. But the very simple model of elections can still help us understand how and why some things happen in elections. It is also a starting place for building more sophisticated models to help understand more complicated election situations and to test assumptions about how voters really do make voting decisions.

PROBLEMS

Liquid Pressure

For problems 1–5, assume that the liquid has density d.

1. Calculate the total force against windows A and B in Fig. 26

2. Calculate the total force against windows C and D in Fig. 26.

3. Calculate the total force against each end of the tank in Fig. 27. How does the total force against the ends of the tank change if the length of the tank is doubled?

4. Calculate the total force against the ends of the tank in Fig. 28.

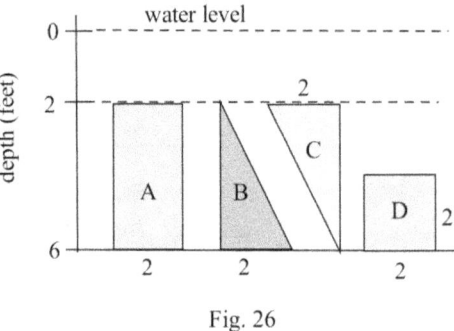
Fig. 26

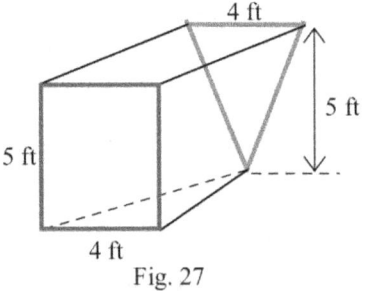
Fig. 27

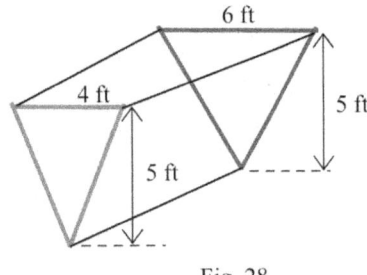
Fig. 28

5. Calculate the total force against the end of the tank in Fig. 29.

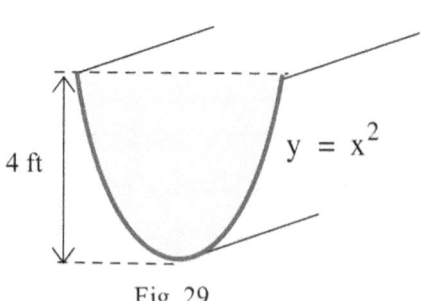
Fig. 29

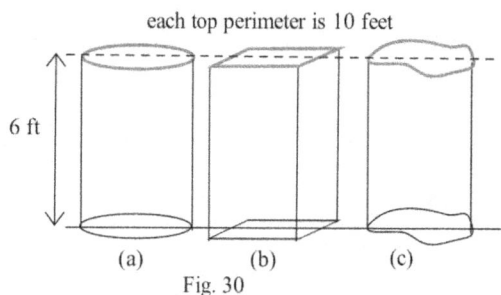

Fig. 30

6. The 3 tanks in Fig. 30 are all 6 feet tall and the top perimeter of each tank is 10 feet. Which tank has the greatest total force against its sides?

7. The 3 tanks in Fig. 31 are all 6 feet tall and the cross sectional area of each tank is 16 square feet. Which tank has the greatest total force against its sides?

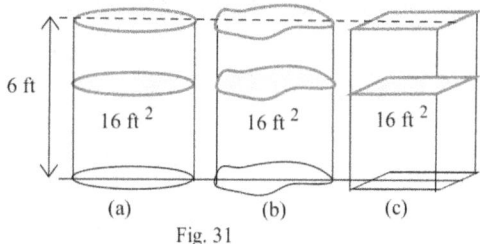

Fig. 31

8. Calculate the total force against the bottom 2 feet of the sides of a 40 foot by 40 foot tank which is filled (a) to a depth of 30 feet with water and (b) to a depth of 35 feet.

9. Calculate the total force against the bottom 2 feet of the sides of a cylindrical tank with a radius of 20 feet which is filled (a) to a depth of 30 feet with water and (b) to a depth of 35 feet.

10. Calculate the total force against the sides and bottom of a 12 ounce can of cola (height = 12 cm, radius = 3 cm). (The density of cola is approximately the same as water.) Why is the total force more than 12 ounces?

Kinetic energy

11. Calculate the kinetic energy of a 20 gram object rotating at 3 revolutions per second at the end of (a) a 15 cm (massless) string, and (b) a 20 cm string.

12. One centimeter of a metal bar weighs 3 grams. Calculate the kinetic energy of a 50 centimeter bar which is rotating at a rate of 2 revolutions per second about one end.

13. One centimeter of a metal bar weighs 3 grams. Calculate the kinetic energy of a 50 centimeter bar which is rotating at a rate of 2 revolutions per second at the end of a 10 cm piece of string.

14. Calculate the kinetic energy of a 20 gram meter stick which is rotating at a rate of 1 revolution per second about one end.

15. Calculate the kinetic energy of a 20 gram meter stick if it is rotating at a rate of 1 revolution per second about its middle point.

16. A flat, circular plate is made from material which weighs 2 grams per cubic centimeter. The plate is 5 centimeters thick, has a radius of 30 centimeters and is rotating about its center at a rate of 2 revolutions per second. (a) Calculate its kinetic energy. (b) Find the radius of the plate that would have twice the kinetic energy of this plate? (Assume the density, thickness, and rotation rate are the same.)

17. Each "washer" in Fig. 32 is made from material weighing 1 gram per cubic centimeter, and each is rotating about its center at a rate of 3 revolutions per second. Calculate the kinetic energy of each washer.

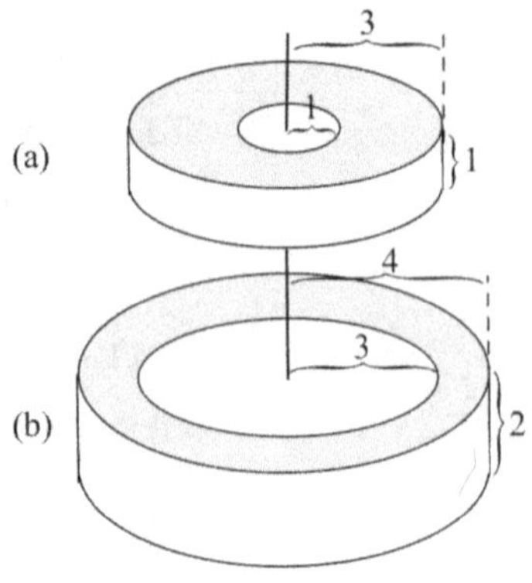

18. The rectangular plate is 1 cm thick, 10 cm long and 6 cm wide and is made of material which weighs 3 grams per cubic centimeter. Calculate the kinetic energy of of the plate if it is rotated at a rate of 2 revolutions per second

    (a) about its 10 cm side (Fig. 33) and

    (b) about its 6 cm side.

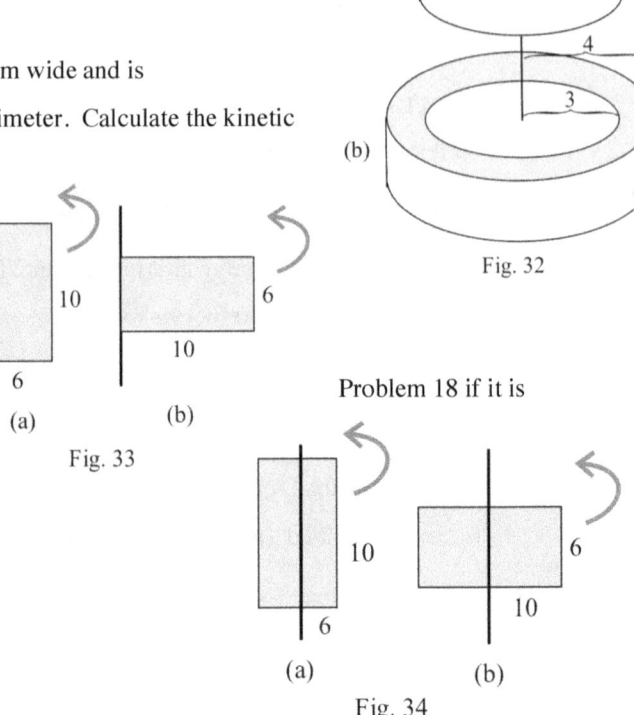

(a)

(b)

Fig. 32

19. Calculate the kinetic energy of of the plate in Problem 18 if it is rotated at a rate of 2 revolutions per second about a vertical line

    (a) through the center of the plate (Fig. 34a) and

    (b) through the center of the plate (Fig. 34b).

Fig. 33

Fig. 34

Volumes by "Tubes"

In problems 20 – 25, sketch each region and calculate the volume swept out when the region is revolved about the specified vertical line.

20. The region between $y = 2x - x^2$ and the x–axis for $0 \le x \le 2$ is rotated about the y-axis.

21. The region between $y = \sqrt{1 - x^2}$ and the x–axis for $0 \le x \le 1$ is rotated about the y-axis.

22. The region between $y = \dfrac{1}{1 + x^2}$ and the x–axis of $0 \le x \le 3$ is rotated about the y-axis.

23. The region between $y = 2x$ and $y = x^2$ for $0 \le x \le 3$ is rotated about the x = 4 line.

24. The region between $y = x$ and $y = 2x$ for $1 \le x \le 3$ is rotated about the x = 1 line.

25. The region between $y = 1/x$ and $y = 1/3$ for $1 \le x \le 3$ is rotated about the x = 5 line.

In problems 26 – 30, write a definite integral representing the volume swept out when the region is revolved about the y–axis, and use a calculator to evaluate the integral.

26. The region between $y = e^x$ and $y = x$ for $0 \le x \le 2$.

27. The region between $y = \ln(x)$ and $y = x$ for $1 \le x \le 4$.

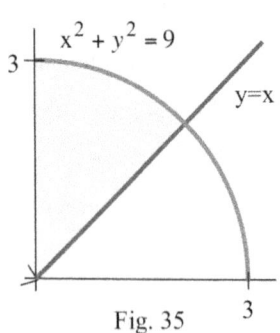

Fig. 35

28. The region between $y = x^2$ and $y = 6 - x$ for $1 \le x \le 4$.

29. The shaded region in Fig. 35.

30. The shaded region in Fig. 36.

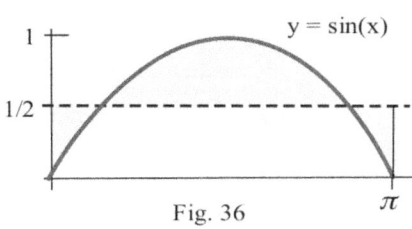

Fig. 36

Areas & Elections

31. For the voter distribution in Fig. 37, which candidates would the
    voters at positions a , b and c vote for?

32. For the voter distribution in Fig. 38, which candidates would the
    voters at positions a , b and c vote for?

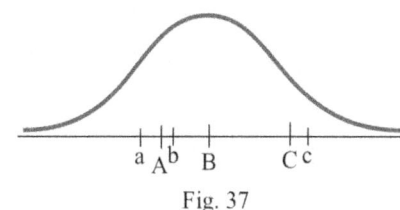

Fig. 37

33. Shade the region representing votes for candidate A
    in Fig. 39. Which candidate wins?

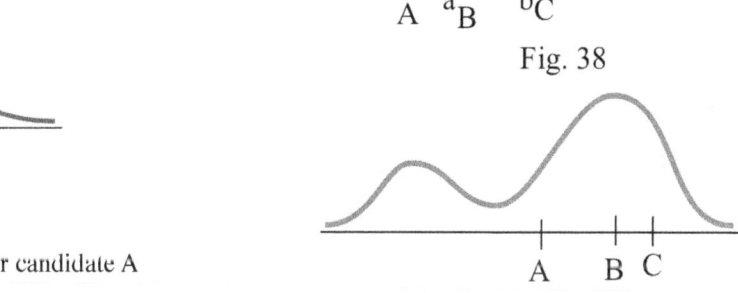

Fig. 38

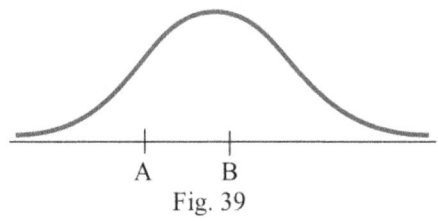

Fig. 39

34. Shade the region representing votes for candidate A
    in Fig. 40. Which candidate wins?

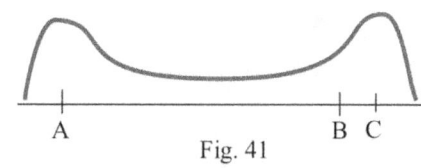

Fig. 40

35. In Fig. 41, (a) which candidate wins?

    (b) If candidate B withdraws before the
        election then which candidate will win?

    (c) If candidate B stays in the election, but C withdraws, then
        who will win?

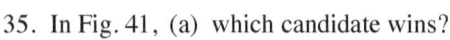

Fig. 41

36. In Fig. 42, (a) which candidate wins?

    (b) If candidate A withdraws before the election, which candidate wins?

    (c) If candidate B stays in the election, but C withdraws, then who wins?

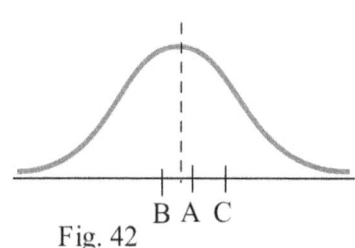

Fig. 42

37. In Fig. 43,

    (a) if the election was only between A and B, who would win?

    (b) If the election was only between A and C, who would win?

    (c) If the election was among A, B, and C, who would win?

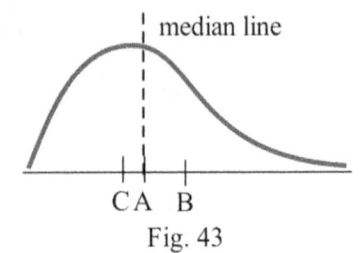

Fig. 43

38. In Fig. 44,

    (a) if the election was only between A and B, who would win?

    (b) If the election was only between A and C, who would win?

    (c) If the election was among A, B, and C, who would win?

39. Sketch what a voter distribution might look like for a 2 issue election.

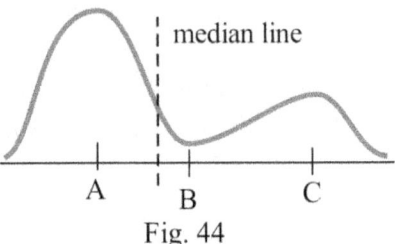

Fig. 44

Section 5.5   Practice Answers

Practice 1:   The reasoning is the same as in Example 1 except that the width of the front is 2 feet but the width of the side is 1 foot. Then

$$(\text{density}) \cdot (\text{depth}) \cdot (\text{area}) = (62.5 \text{ pounds/ft}^3)(x_i \text{ feet})(1 \text{ feet})(\Delta x_i \text{ feet}) = 62.5 \ x_i \cdot \Delta x_i \text{ pounds}$$

and

$$\text{hydrostatic force} \approx \sum 125 x_i \ \Delta x_i \longrightarrow \int_{x=0}^{3} 62.5 x \ dx = 31.25 \ x^2 \Big|_{x=0}^{3} = 281.25 \text{ pounds.}$$

Practice 2:   C: $\dfrac{w}{x-4} = \dfrac{3}{2}$ in Fig. 45 so $w = \dfrac{3}{2}(x-4)$. Then

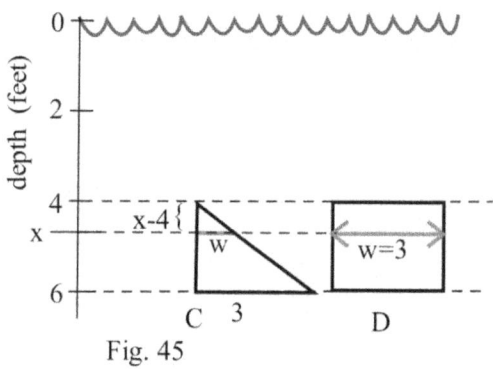

Fig. 45

$$\text{hydrostatic force} = \int_{x=a}^{b} (\text{density}) \cdot (\text{depth}) \cdot w(x) \ dx$$

$$= \int_{x=4}^{6} (60) \cdot (x) \cdot \frac{3}{2}(x-4) \ dx$$

$$= 90 \int_{x=4}^{6} (x^2 - 4x) \ dx = 90 \left(\frac{x^3}{3} - 2x^2 \right) \Big|_{4}^{6}$$

$$= \textbf{960 pounds.}$$

D: w = 3 at all depths so

$$\text{hydrostatic force} = \int_{x=a}^{b} (\text{density}) \cdot (\text{depth}) \cdot w(x) \ dx = \int_{x=4}^{6} (60) \cdot (x) \cdot (3) \ dx$$

$$= 180 \int_{x=4}^{6} x \ dx = 90 \ x^2 \Big|_{4}^{6} = \textbf{1800 pounds.}$$

Windows A and C (with a flip) fit together to form window D, and it is encouraging that the sum of the total hydrostatic forces on A and C is the same as the total hydrostatic force on D.

Practice 3:   The object travels $2\pi(\text{radius}) = 2\pi(2 \text{ meters}) = 4\pi$ meters in one revolution so in the 1 second it takes to make 4 revolutions it travels 16π meters: v = 1600π cm/second.

$$KE = \frac{1}{2} \ m \cdot v^2 = \frac{1}{2}(1 \ g) \cdot (1600\pi \text{ cm/s})^2 = 1{,}280{,}000\pi^2 \text{ ergs} \approx 12{,}633{,}094 \text{ ergs.}$$

Practice 4:   Since the bar and the number of revolutions per second are the same as in Example 4,

$$ke_i = \frac{1}{2}\, m_i \cdot v_i^2 = \frac{1}{2}(5\,\Delta x_i \text{ grams})\cdot(400\pi\, x_i \text{ cm/sec})^2 = 400{,}000\pi^2\, (x_i)^2\, \Delta x_i \text{ ergs.}$$

Then, since the bar is at the end of a 1 meter string, we integrate from x = 1 to x = 1+3 = 4:

$$KE = \sum ke_i\ = \sum 400{,}000\pi^2\, (x_i)^2\, \Delta x_i \text{ æææÆ} \qquad \int_1^4 400{,}000\pi^2 \cdot x^2 \ dx$$

$$= 400{,}000\pi^2\, \frac{x^3}{3}\ \Big|_1^4 \ = 8{,}400{,}000\ \pi^2 \text{ ergs.}$$

Practice 5: $\{$ volume obtained when R is revolved about the y–axis $\} = \displaystyle\int_{x=a}^{b} 2\pi \cdot x \cdot \{\ f(x) - g(x)\ \}\ dx$

so volume $= \displaystyle\int_{x=0}^{4} 2\pi \cdot x \cdot \{\ (x+1) - \sqrt{x}\ \}\ dx\ = 2\pi \int_{x=0}^{4} (\ x^2 + x - x^{3/2}\ )\ dx$

$$= 2\pi\,(\ \frac{x^3}{3}\ +\ \frac{x^2}{2}\ -\ \frac{2}{5}\, x^{5/2}\ \}\Big|_0^4 \ \approx 2\pi(\ 16.53\ ) \approx 103.9\ .$$

Practice 6:   The shaded regions in Fig. 46 show the total votes for each candidate: **B wins**.

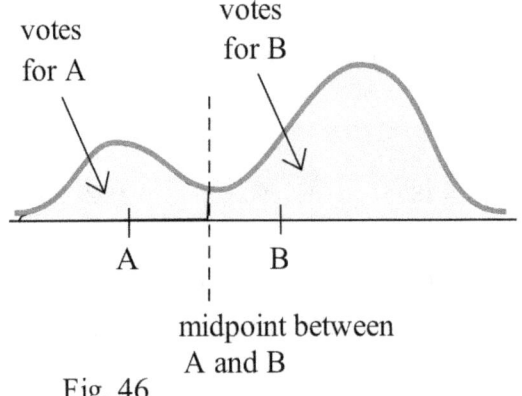

Fig. 46

PROBLEM ANSWERS Chapter Five

Section 5.1

1.   volume $= (5 \cdot 6 \cdot 1) + (4 \cdot 4 \cdot 2) + (3 \cdot 3 \cdot 1) = 71$ 2. $v = 60 + 20 + 9 + 4 = 93$

3. $v = (\pi \cdot 4^2 \cdot 0.5) + (\pi \cdot 3^2 \cdot 1) + (\pi \cdot 1^2 \cdot 2) = 19\pi$ 5.   volume $= (9 \cdot 0.2) + (6 \cdot 0.2) + (2 \cdot 0.2) = 3.4$

7. $\displaystyle\int_0^3 (\text{base})(\text{height})(\text{thickness}) = \int_0^3 (2)(5-x)\ dx = 10x - x^2\Big|_0^3 = (30-9) - (0) = 21\ \text{in}^3$

8. $\displaystyle\int_0^3 (\text{base})(\text{height})(\text{thickness}) = \int_0^3 (x)(x^2)\ dx = \int_0^3 x^3\ dx = \frac{1}{4}x^4\Big|_0^3 = \frac{81}{4}\ \text{in}^3 = 20.25\ \text{in}^3$

9. $\displaystyle\int_1^4 \frac{1}{2}(\text{base})(\text{height})(\text{thickness}) = \int_1^4 \frac{1}{2}(x+1)(x^{1/2})\ dx = \frac{1}{2}\int_1^4 x^{3/2} + x^{1/2}\ dx$

$$= \frac{1}{2}\left\{\frac{2}{5}x^{5/2} + \frac{2}{3}x^{3/2}\right\}\Big|_1^4 = \frac{1}{2}\left\{\frac{64}{5} + \frac{16}{3}\right\} - \frac{1}{2}\left\{\frac{2}{5} + \frac{2}{3}\right\} = \frac{128}{15}\ \text{m}^3$$

11. $\displaystyle\int_2^4 \pi(\text{radius})^2 (\text{thickness}) = \int_2^4 \pi(\tfrac{1}{2}(4-x))^2\ dx = \frac{\pi}{4}\int_2^4 16 - 8x + x^2\ dx$

$$= \frac{\pi}{4}(16x - 4x^2 + \tfrac{1}{3}x^3)\Big|_2^4 = \frac{\pi}{4}\left\{\frac{64}{3}\right\} - \frac{\pi}{4}\left\{\frac{56}{3}\right\} = \frac{2\pi}{3}\ \text{m}^3$$

13. Solids A and B have equal volumes. Why?

14. $\displaystyle\int_0^3 \frac{1}{2}(\text{base})(\text{height})(\text{thickness}) = \int_0^3 \frac{1}{2}(x)(2x)\ dx = \int_0^3 x^2\ dx = (\tfrac{1}{3}x^3)\Big|_0^3 = 9\ \text{cubic units}.$

15. $\displaystyle\int_1^4 (\text{side})(\text{side})(\text{thickness}) = \int_1^4 (\tfrac{1}{x})(\tfrac{1}{x})\ dx = \int_1^4 x^{-2}\ dx = -x^{-1}\Big|_1^4 = (\tfrac{-1}{4}) - (\tfrac{-1}{1}) = \frac{3}{4}\ \text{cubic units}.$

16. $\displaystyle\int_0^4 \pi(\text{radius})^2(\text{thickness}) = \int_0^4 \pi(2 - \sqrt{x})^2\ dx = \pi(4x - 4(\tfrac{2}{3})x^{3/2} + \tfrac{1}{2}x^2)\Big|_0^4 = \frac{8}{3}\pi \approx 8.38.$

17. $\displaystyle\int_0^4 \pi(\text{radius})^2(\text{thickness}) = \int_0^4 \pi(3 - \sqrt{x})^2\ dx = \pi\int_0^4 (9 - 6\sqrt{x} + x)\ dx$

$$= \pi(9x - 6(\tfrac{2}{3})x^{3/2} + \tfrac{1}{2}x^2)\Big|_0^4 = \pi(36 - 32 + 8) - \pi(0) = 12\pi \approx 37.7\ \text{cubic units.}$$

18. $\displaystyle\int_0^\pi \pi(\text{radius})^2(\text{thickness}) = \int_0^\pi \pi(\tfrac{1}{2} - \sin(x))^2\ dx = \pi\int_0^\pi (\tfrac{1}{4} - \sin(x) + \sin^2(x))\ dx$

$$= \pi\left\{\tfrac{1}{4}x + \cos(x) + \tfrac{1}{2}x - \tfrac{1}{4}\sin(2x)\right\}\Big|_0^\pi$$

$$= \pi \left\{ \frac{\pi}{4} + (-1) + \frac{\pi}{2} - 0 \right\} - \pi \left\{ 0 + 1 + 0 - 0 \right\} = \frac{3}{4} \pi^2 - 2\pi \approx 1.12 \text{ cubic units.}$$

19. $\displaystyle\int_0^2 \frac{1}{2}(\text{base})(\text{height})(\text{thickness}) = \int_0^2 \frac{1}{2}(x)(x^2)\, dx = \frac{1}{2}\int_0^2 x^3\, dx = \frac{1}{2}\left(\frac{1}{4}x^4\right)\Big|_0^2 = \frac{1}{8}(16) - \frac{1}{8}(0) = 2 \text{ cubic units }.$

20. $\displaystyle\int_0^{\pi/2} 2\cdot\cos(x)\, dx = 2 \text{ cubic units }.$

21. $\displaystyle\int_1^2 \pi(\text{radius})^2 (\text{thickness}) = \int_1^2 \pi(x^2)^2\, dx - \int_1^2 \pi(1)^2\, dx = \pi\int_1^2 x^4 - 1\, dx = \pi\left(\frac{1}{5}x^5 - x\right)\Big|_1^2$

$$= \pi\left(\frac{32}{5} - 2\right) - \pi\left(\frac{1}{5} - 1\right) = \frac{26}{5}\pi \approx 16.34 \text{ cubic units }.$$

22. $\dfrac{256}{3}\pi \approx 268.08$ cubic units .   23. $\dfrac{32}{3}\pi$ 25. 60π 27. $2Rr^2\pi^2$

29. on your own   31. (a) $H^2 L$ (b) $\dfrac{1}{3}H^2 L$ (c) ratio $= \dfrac{1}{3}$

33. (a) BL   (b) Volume $= \dfrac{1}{3}$ BL   (c) ratio $= \dfrac{1}{3}$

Section 5.2

1. $\sqrt{15^2 + 20^2} + \sqrt{(-24)^2 + 18^2} + \sqrt{12^2 + (-12)^2} \approx 71.97$ feet.

2. $71.97 + \sqrt{11^2 + 7^2} + \sqrt{(-14)^2 + (-17)^2} + \sqrt{(-10)^2 + (-9)^2} \approx 120.48$ feet.

3. $\sqrt{1^2 + 1^2} + \sqrt{1^2 + 2^2} \approx 3.65$

4. $\sqrt{1^2 + (1/2)^2} + \sqrt{1^2 + (1/6)^2} + \sqrt{1^2 + (1/12)^2} \approx 3.135$

5. (a) (0,1) to (2,5): $\sqrt{2^2 + 4^2} = \sqrt{20} = 2\sqrt{5}$

    (b) $y' = 2$: $L = \displaystyle\int_0^2 \sqrt{1 + (2)^2}\, dx = \sqrt{5}\int_0^2 1\, dx = \sqrt{5}\, x\Big|_0^2 = 2\sqrt{5}$ (same as in part (a))

7. (a) (2,1) to (5, –5): $\sqrt{3^2 + 6^2} = \sqrt{45} = 3\sqrt{5}$

    (b) $x' = 1, y' = -2$: $L = \displaystyle\int_0^3 \sqrt{1^2 + (-2)^2}\, dt = \sqrt{5}\int_0^3 1\, dt = \sqrt{5}\, t\Big|_0^3 = 3\sqrt{5}$

9. $y' = \sqrt{x} = x^{1/2}$: $L = \displaystyle\int_0^4 \sqrt{1 + (y')^2}\, dx = \int_0^4 \sqrt{1 + (\sqrt{x})^2}\, dx = \int_0^4 \sqrt{1 + x}\, dx$

$$= \frac{2}{3}(1 + x)^{3/2}\Big|_0^4 = \frac{2}{3}(5)^{3/2} - \frac{2}{3}(1)^{3/2} \approx 6.787$$

11. $y' = x^2 - \dfrac{1}{4x^2}$: $1 + (y')^2 = 1 + \left(x^4 - \dfrac{1}{2} + \dfrac{1}{16x^4}\right) = x^4 + \dfrac{1}{2} + \dfrac{1}{16x^4} = \left(x^2 + \dfrac{1}{4x^2}\right)^2$

$$L = \int_1^5 \sqrt{1 + (y')^2} \ dx \ = \int_1^5 \sqrt{x^2 + \frac{1}{4x^2}} \ dx \ = \int_1^5 \sqrt{x^2 + \frac{1}{4} x^{-2}} \ dx$$

$$= \frac{1}{3} x^3 - \frac{1}{4} x^{-1} \Big|_1^5 \ = \left(\frac{1}{3}(5)^3 - \frac{1}{4(5)} \right) - \left(\frac{1}{3}(1)^3 - \frac{1}{4(1)} \right) \approx 41.53$$

13. $y' = x^4 - \frac{1}{4} x^{-4}$: $1 + (y')^2 = 1 + (x^8 - \frac{1}{2} + \frac{1}{16x^8}) = x^8 + \frac{1}{2} + \frac{1}{16x^8} = (x^4 + \frac{1}{4x^4})^2$

$$L = \int_1^5 \sqrt{1 + (y')^2} \ dx \ = \int_1^5 \sqrt{x^4 + \frac{1}{4} x^{-4}} \ dx \ = \frac{1}{5} x^5 - \frac{1}{12} x^{-3} \Big|_1^5 \quad \backslash$$

$$= \left(\frac{1}{5}(5)^5 - \frac{1}{12(5)^3} \right) - \left(\frac{1}{5}(1)^5 - \frac{1}{12(1)^3} \right) \approx 624.88$$

15. $y' = 2x$: $L = \int_0^1 \sqrt{1 + (y')^2} \ dx \ = \int_0^1 \sqrt{1 + (2x)^2} \ dx$

$$= \int_0^1 \sqrt{1 + 4x^2} \ dx \ \approx 1.479 \ \text{(using calculator)}$$

16. $y' = 3x^2$: $L = \int_0^1 \sqrt{1 + (y')^2} \ dx \ = \int_0^1 \sqrt{1 + 9x^4} \ dx \ \approx 1.548 \ \text{(using calculator)}$

17. $y' = \frac{1}{2} x^{-1/2}$: $L = \int_1^9 \sqrt{1 + (y')^2} \ dx \ = \int_1^9 \sqrt{1 + \frac{1}{4x}} \ dx \ \approx 8.268 \ \text{(using calculator)}$

18. $y' = \frac{1}{x}$: $L = \int_1^e \sqrt{1 + (y')^2} \ dx \ = \int_1^e \sqrt{1 + \frac{1}{x^2}} \ dx \ \approx 2.003 \ \text{(using calculator)}$

19. $y' = \cos(x)$: (a) $L = \int_0^{\pi/1} \sqrt{1 + \cos^2(x)} \ dx \ \approx 1.058 \ \text{(using calculator)}$

(b) $L = \int_{\pi/4}^{\pi/2} \sqrt{1 + \cos^2(x)} \ dx \ \approx 0.852 \ \text{(using calculator)}$

20. $x' = -3 \cdot \sin(t)$, $y' = 4 \cdot \cos(t)$

$$L = \int_0^{2\pi} \sqrt{(-3 \cdot \sin(t))^2 + (4 \cdot \cos(t))^2} \ dt$$

$$= \int_0^{2\pi} \sqrt{9 \cdot \sin^2(t) + 16 \cdot \cos^2(t)} \ dt \ \approx 22.103 \ \text{(using calculator)}$$

21. $x' = -5 \cdot \sin(t)$, $y' = 2 \cdot \cos(t)$

$$L = \int_0^{2\pi} \sqrt{25 \cdot \sin^2(t) + 4 \cdot \cos^2(t)} \ dt \ \approx 23.018 \ \text{(using calculator)}$$

22. $x' = -t \cdot \sin(t) + \cos(t)$, $y' = t \cdot \cos(t) + \sin(t)$. Then

$(x')^2 + (y')^2 = t^2 + 1$ (check it) so $L = \int_0^{2\pi} \sqrt{1 + t^2}\ dt \approx 21.256$ (using calculator)

23. same x' and y' as in #22. $L = \int_{10}^{20} \sqrt{1 + t^2}\ dt \approx 150.346$ (using calculator)

24. $x' = R \cdot (1 - \cos(t))$ and $y' = R \cdot \sin(t)$: $(x')^2 + (y')^2 = 2R^2 \cdot (1 - \cos(t))$

$L = \int_0^{2\pi} \sqrt{2R^2 \cdot (1 - \cos(t))}\ dt = R\sqrt{2} \int_0^{2\pi} \sqrt{1 - \cos(t)}\ dt \approx 8R$ (Actually, $= 8R$)

25. 1 mile = 5,280 feet at 2π feet per revolution so 1 mile $= \dfrac{5280\ ft}{2\pi\ ft/rev} \approx 840.338$ revolutions.

From #24, total distance = (840.338 rev.)(8 feet/rev) ≈ 6722.705 feet (≈ 1.27 miles).

26. 2 (Why?)

27. $y' = 2x$: total length $= \int_0^4 \sqrt{1 + 4x^2}\ dx \approx 16.8186$ (using calculator)

(a) Find T so $\int_0^T \sqrt{1 + 4x^2}\ dx \approx \frac{1}{2}(16.8186) = 8.4093$.

         By experimenting on a calculator, $T \approx 2.77$.

(b) Find A so $\int_0^A \sqrt{1 + 4x^2}\ dx \approx \frac{1}{3}(16.8186) = 5.6062$ and

         B so $\int_0^B \sqrt{1 + 4x^2}\ dx \approx \frac{2}{3}(16.8186) = 11.2124$.

         By experimenting on a calculator, $A \approx 2.22$ and $B \approx 3.23$.

28. , 29. on your own.

30. (a) A: $SA_x = 2\pi(4)(4) = 32\pi$ B: $SA_x = 2\pi(2)(2) = 8\pi$

     (b) A: $SA_y = 2\pi(3)(4) = 24\pi$ B: $SA_y = 2\pi(6)(2) = 24\pi$

31. (a) A: $SA_x = 2\pi(3)(3) = 18\pi$ B: $SA_x = 2\pi(4)(\sqrt{3^2 + 4^2}\ ) = 40\pi$

     (b) A: $SA_y = 2\pi(2)(3) = 12\pi$ B: $SA_y = 2\pi(6.5)(\ 5\ ) = 65\pi$

33. (a) about $y = 1$: A: $SA = 2\pi(2)(3) = 12\pi$ B: $SA = 2\pi(3)(\ 5\ ) = 30\pi$

     (b) about $x = -2$: A: $SA = 2\pi(4)(3) = 24\pi$ B: $SA = 2\pi(8.5)(\ 5\ ) = 85\pi$

35. The largest surface area occurs when the midpoint is farthest from the line of rotation, the x–axis, when $\theta = 90^\circ$. Then $SA = 2\pi$(dist. to x–axis)(length) $= 2\pi(6)(2) = 24\pi$.

37. $y' = 6x^2$: $SA_y = \int_0^1 2\pi x \sqrt{1 + (y')^2}\ dx = \int_0^1 2\pi x \sqrt{1 + (6x^2)^2}\ dx$

$$= \int_0^1 2\pi x \sqrt{1 + 36x^4} \ dx \approx 10.207 \ \text{(using calculator)}$$

39. $y' = 4x$: $SA_x = \int_0^1 2\pi y \sqrt{1 + (y')^2} \ dx = \int_0^1 2\pi(2x^2)\sqrt{1 + 16x^2} \ dx \approx 13.306$ (using calculator)

41. $y' = 3x^2$: $SA_x = \int_0^2 2\pi y \sqrt{1 + (y')^2} \ dx = \int_0^2 2\pi(x^3)\sqrt{1 + 9x^4} \ dx \approx 203.046$ (using calculator)

43. $y' = 2x$: $SA_y = \int_0^2 2\pi x \sqrt{1 + (y')^2} \ dx = \int_0^2 2\pi x \sqrt{1 + 4x^2} \ dx \approx 36.177$ (using calculator)

or use a u–substitution with $u = 1 + 4x^2$: $SA_y = \int_{u=1}^{u=17} 2\pi u^{1/2} \frac{1}{8} \ du \approx 36.177$.

45–50. on your own

Section 5.3

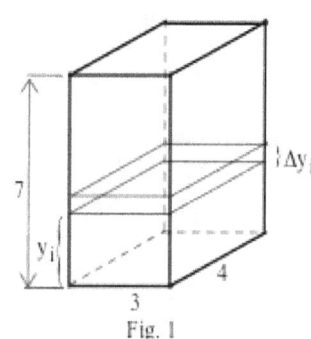
Fig. 1

1. Fig. 1. Weight = (volume)(density) = $(3 \cdot 4 \cdot \Delta y)(62.5)$,

distance raised = $7 - y_i$, so Work = $(750 \ \Delta y)(7 - y_i)$.

Total work = $\int_0^7 (7 - y)(750) \ dy = 750(7y - \frac{1}{2}y^2)\Big|_0^7 = 18{,}375$ ft–lbs.

3. Slice at height y: weight = $(5 \cdot 2 \cdot \Delta y)(60) = 600 \ \Delta y$ lb,

distance raised = $4 - y$ ft.

(a) Total work = $\int_0^4 (4 - y)(600) \ dy = 600(4y - \frac{1}{2}y^2)\Big|_0^4 = 4{,}800$ ft–lbs.

(b) The top 36 cubic feet of water corresponds to dimensions of 5 ft. long, 2 ft wide, and $\frac{36}{(5)(2)} = 3.6$ ft. high.

Weight of slice = $600 \ \Delta y$, distance raised = $4 - y$ (where y goes from 0.4 ft. to 4 ft.)

Total work = $\int_{0.4}^4 (4 - y)(600) \ dy = 600(4y - \frac{1}{2}y^2)\Big|_{0.4}^4 = 3{,}888$ ft–lbs.

(c) With 1 HP pump: it takes $\frac{4800}{33000}$ minutes ≈ 0.145 min ≈ 8.73 seconds to empty the tank.

With 1/2 HP pump: $\frac{8.73}{1/2} = 17.46$ seconds.

Both pumps do the same amount of work but in different times and thus have different power.

5. Slice at height y: weight = $(\pi \cdot r^2 \cdot \Delta y)(60) = (\pi \cdot 1^2 \cdot \Delta y)(60) = 60\pi \ \Delta y$ lb, distance raised = $6 - y$ ft.

(a) Total work = $\int_0^6 (6 - y)(60\pi) \ dy = 60\pi(6y - \frac{1}{2}y^2)\Big|_0^6 = 60\pi(18) \approx 3{,}393$ ft–lbs.

(b) weight of slice = $60\pi \ \Delta y$ lb, distance raised = $(2 + 6) - y = 8 - y$ ft.

$$\text{Total work} = \int_{6-1}^{6} (8-y)(60\pi)\ dy = 60\pi(8y - \tfrac{1}{2} y^2)\Big|_{5}^{6} = 150\pi \approx 471.2 \ \text{ft–lbs.}$$

(c) First find the **work** needed to empty the top 3 feet; using integration of part (a) with new limits.

$$\text{work} = \int_{3}^{6} (6-y)(60\pi)\ dy = 60\pi(6y - \tfrac{1}{2} y^2)\Big|_{3}^{6} = 60\pi(4.5) \approx 848 \ \text{ft–lbs.}$$

Knowing that 1 HP pump works at a rate of 33,000 ft–lbs/minute, then it takes

$$\frac{848}{33000} \approx 0.0257 \text{ minutes} \approx 1.54 \text{ seconds.}$$

7.   Fig. 2. Use similar triangles: slice at height y has

weight = (volume)(density) = $(7 \cdot \tfrac{3}{2} \cdot y \cdot \Delta y)(80) = 840y\ \Delta y$ and

distance raised = $4 - y$.

$$\text{Total work} = \int_{3}^{4} (4-y)(840y)\ dy = 840(2y^2 - \tfrac{1}{3} y^3)\Big|_{3}^{4} = 1,400 \ \text{ft–lbs.}$$

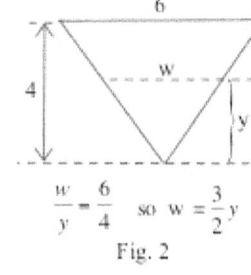

Fig. 2

8.   Fig. 3. Each slice (perpendicular to the y–axis) is a disk with

volume = $\pi(\text{radius})^2(\text{thickness}) = \pi (\tfrac{y}{2})^2 \Delta y$ so

weight = (volume)(density) = $(\pi (\tfrac{y}{2})^2 \Delta y)(40) = 10\pi y^2\ \Delta y$.

(a) Distance to top = $8- y$ so

$$\text{Total work} = \int_{0}^{8} (8-y)(10\pi y^2)\ dy = 10\pi \int_{0}^{8} 8y^2 - y^3\ dy$$

$$= 10\pi(\tfrac{8}{3} y^3 - \tfrac{1}{4} y^4)\Big|_{0}^{8} = \frac{10\pi}{12} \cdot 8^4 \approx 3,413.3\ \pi \approx 10,723 \text{ ft–lbs.}$$

(b) The limits of integration are now 6 to 8:

$$\text{work} = \int_{6}^{8} (8-y)(10\pi y^2)\ dy = 10\pi \int_{6}^{8} 8y^2 - y^3\ dy$$

$$= 10\pi(\tfrac{8}{3} y^3 - \tfrac{1}{4} y^4)\Big|_{6}^{8} \approx 893.3\ \pi \approx 2,806 \text{ ft–lbs.}$$

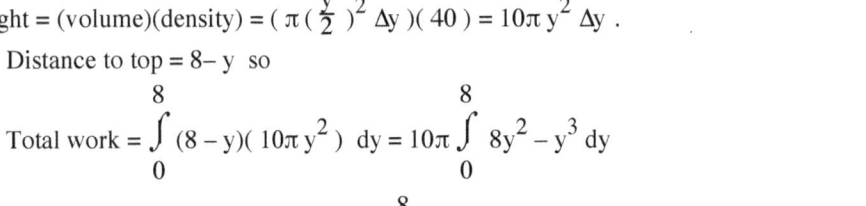

9.   Find H so $10\pi \int_{0}^{H} 8y^2 - y^3\ dy = \tfrac{1}{2}$ { total work done in 8(a) } $= \tfrac{1}{2}$ { 3,413.3

π } $\approx 1,706.6\pi$.

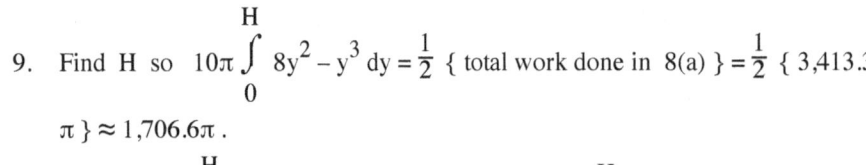

Fig. 3

$$10\pi \int_{0}^{H} 8y^2 - y^3\ dy = 10\pi\{ \tfrac{8}{3} y^3 - \tfrac{1}{4} y^4)\Big|_{0}^{H} = 10\pi H^3 \{ \tfrac{8}{3} - \tfrac{1}{4} H \} .$$

Some "exploring" with a calculator shows that if H ≈ 4.914 then $10\pi H^3 \{ \tfrac{8}{3} - \tfrac{1}{4} H \} \approx 1,706.6\pi$, the value we want. One person should remove the top 8 – 4.914 = 3.086 feet of grain and the other should remove the bottom 4.914 feet.

11. Fig. 4. Slice at height y:

$$\text{weight} = (\text{volume})(\text{density}) = (\pi r^2 \Delta y)(62.5)$$
$$= (\pi (\sqrt{2y})^2 \Delta y)(62.5) = 125\pi y \Delta y.$$

(a) Distance raised = 2 – y so

$$\text{work} = \int_0^2 (2-y)(125\pi y)\ dy = 125\pi \{\ y^2 - \tfrac{1}{3} y^3\ \}\Big|_0^2$$

$$= 125\pi (\ 4 - \tfrac{8}{3}\ ) \approx 523.6 \text{ ft–lbs.}$$

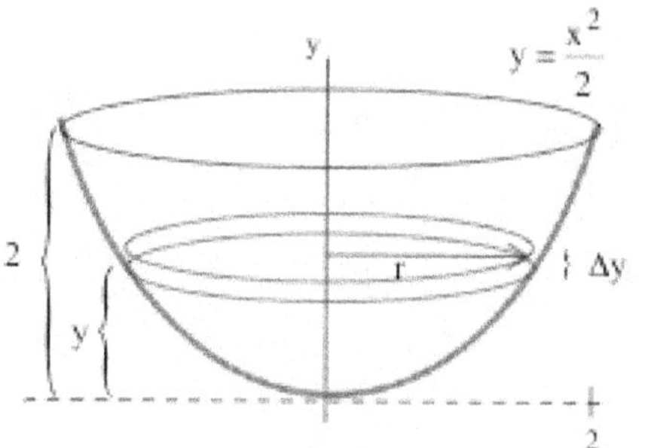

(b) The only change is the distance raised: distance = 5 – y.

$$\text{work} = \int_0^2 (5-y)(125\pi y)\ dy = 125\pi \{\ \tfrac{5}{2} y^2 - \tfrac{1}{3} y^3\ \}\Big|_0^2$$

$$= 125\pi (\ 10 - \tfrac{8}{3}\ ) \approx 2{,}879.8 \text{ ft–lbs.}$$

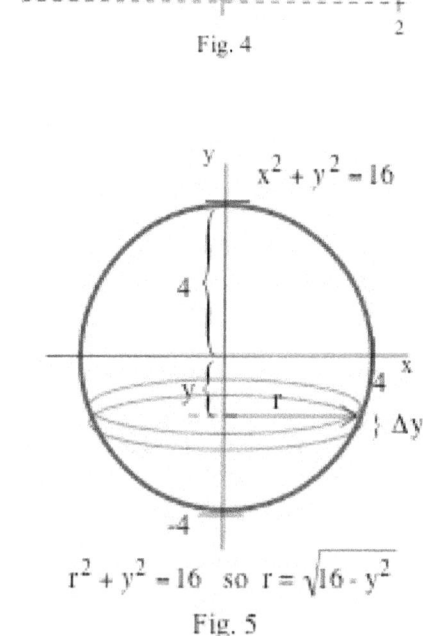

Fig. 4

13. Fig. 5. Slice at height y:

$$\text{weight} = (\text{volume})(\text{density}) = (\pi r^2 \Delta y)(62.5)$$
$$= (\pi (\sqrt{16-y^2})^2 \Delta y)(62.5) = 62.5\ \pi (\ 16 - y^2\ )\ \Delta y.$$

distance = 4 – y.

$$\text{work} = \int_{-4}^{-2} (4-y)(62.5\pi)(16-y^2)\ dy$$

$$= 62.5\pi \int_{-4}^{-2} 64 - 16y - 4y^2 + y^3\ dy = 62.5\pi \{\ 64y - 8y^2 - \tfrac{4}{3} y^3 + \tfrac{1}{4} y^4\}\Big|_{-4}^{-2}$$

$$\approx 62.5\pi (\ 89.33\ ) \approx 17{,}540\ \text{ft–lbs.}$$

$r^2 + y^2 = 16$ so $r = \sqrt{16 - y^2}$

Fig. 5

15. { work for (a) } > { work for (b) } > { work for (c) }. For container (c), most of the water is raised only a small distance, but for container (a), most of the water is raised a larger distance.

17. Work = area under the curve of force vs. position.

(a) work = $(10)(5) + \tfrac{1}{2}(2)(10) + (2)(10) + \tfrac{1}{2}(1)(10) = 85$ ft–lbs.

(b) work = $(2)(10) + (1)(10) + \tfrac{1}{2}(1)(10) = 35$ ft–lbs.

19. F = kx so 6x = kx and k = 6.

(a) work = $\int_0^3 6x\ dx = 3x^2\Big|_0^3 = 27$ in–oz.

(b) work = $\int_0^6 6x\ dx = 3x^2\Big|_0^6 = 108$ in–oz.

21. Work = { area under graph of force vs. length }. Use the graph in Fig. 26 to approximate the area.

(a) From $x = 23$ to $x = 33$, area $\approx (33 - 23 \text{ cm}) \{ \text{ avg. height of about } 15 \text{ g } \} = 150$ g–cm.

(b) From $x = 28$ to $x = 33$, area $\approx (33 - 28 \text{ cm}) \{ 15 \text{ g } \} + \frac{1}{2}(33 - 30) \{ 15 \text{ g}) = 97.5$ g–cm.

23. $F = kx$ so $3 = k \cdot 5$ and $k = 3/5$.

$$\text{work} = \int_{5}^{5+4} \frac{3}{5} x \ dx = \frac{3}{10} x^2 \Big|_{5}^{9} = 16.8 \text{ in–lbs.}$$

25. $\text{Work} = \int_{a}^{b} f(x) \ dx = \int_{R}^{R+h} \frac{R^2 P}{x^2} \ dx = \int_{300}^{300+h} \frac{(300)^2 (100)}{x^2} \ dx = 9 \cdot 10^6 \{ \frac{-1}{x} \} \Big|_{300}^{300+h}$

$$= 9 \cdot 10^6 \{ \frac{1}{300} - \frac{1}{300+h} \}$$

(a) $h = 100$: work $= 9 \cdot 10^6 \{ \frac{1}{300} - \frac{1}{400} \} \approx 7{,}500$ mile–pounds.

(b) $h = 200$: work $= 9 \cdot 10^6 \{ \frac{1}{300} - \frac{1}{500} \} \approx 12{,}000$ mile–pounds.

(c) $h = 300$: work $= 9 \cdot 10^6 \{ \frac{1}{300} - \frac{1}{600} \} \approx 15{,}000$ mile–pounds.

27. Assume you weigh P pounds.

$$\text{Work} = \int_{a}^{b} f(x) \ dx = \int_{R}^{R+h} \frac{R^2 P}{x^2} \ dx = \int_{4000}^{4000+h} \frac{(4000)^2 P}{x^2} \ dx$$

$$= 1.6 \cdot 10^7 P \{ \frac{-1}{x} \} \Big|_{4000}^{4000+h} = 1.6 \cdot 10^7 P \{ \frac{1}{4000} - \frac{1}{4000+h} \}$$

(a) $h = 200$: work $= 1.6 \cdot 10^7 P \{ \frac{1}{4000} - \frac{1}{4200} \}$ mile–pounds.

(b) $h = 400$: work $= 1.6 \cdot 10^7 P \{ \frac{1}{4000} - \frac{1}{4400} \}$ mile–pounds.

(c) $h = 10^6$: work $= 1.6 \cdot 10^7 P \{ \frac{1}{4000} - \frac{1}{4000+10^6} \}$ mile–pounds. What happens when h is really large?

29. $f(x) = \frac{-1}{kx^2}$ and we know that $f = 0.1$ when $x = 10$ so $0.1 = \frac{-1}{k10^2}$ and $k = -0.1$.

$$\text{work} = \int_{a}^{b} f(x) \ dx = \int_{a}^{b} \frac{-1}{(-0.1)x^2} \ dx = 10 \{ \frac{-1}{x} \} \Big|_{a}^{b} = 10 \{ \frac{1}{a} - \frac{1}{b} \}.$$

(a) from $x = 20$ to $x = 10$: work $= 10 \{ \frac{1}{10} - \frac{1}{20} \} = 0.5$ ft–lb.

(b) from $x = 10$ to $x = 1$: work $= 10 \{ \frac{1}{1} - \frac{1}{10} \} = 9$ ft–lb.

(c) from $x = 1$ to $x = 0.1$: work $= 10 \{ \frac{1}{0.1} - \frac{1}{1} \} = 90$ ft–lb.

31. $\text{Work} = \int_{t=a}^{t=b} f(t) \sqrt{ (dx/dt)^2 + (dy/dt)^2 } \ dt$

$$= \int_{0}^{2\pi} t \sqrt{ \sin^2(t) + \cos^2(t) } \ dt = \int_{0}^{2\pi} t \ dt = \frac{1}{2} t^2 \Big|_{0}^{2\pi} = 2 \pi^2 .$$

33. Work $= \int\limits_{t=a}^{t=b} f(t) \sqrt{(dx/dt)^2 + (dy/dt)^2} \ dt = \int\limits_{0}^{1} t\sqrt{4t^2 + 1} \ dt$.

(Use the substitution $u = 4t^2 + 1$ to find an antiderivative, and then evaluate.)

35. "Unroll" the region to get a triangle with base $= 2\pi(\text{radius}) = 2\pi(1) = 2\pi$ and height $= f(2\pi) = 2\pi$.

The "area" of the triangle is $\frac{1}{2}(\text{base})(\text{height}) = \frac{1}{2}(2\pi)(2\pi) = 2\pi^2$. (This is the same as problem 31.)

Section 5.4

1. (a) $M = 2 + 5 + 5 = 12$. $M_0 = (2)(4) + (5)(2) + (5)(6) = 48$. $\bar{x} = \dfrac{M_0}{M} = \dfrac{48}{12} = 4$.

(b) new $\bar{x} = 5 = \dfrac{M_0}{M} = \dfrac{(2)(4) + (5)(2) + (5)(6) + (8)(\,?\,)}{2 + 5 + 5 + 8} = \dfrac{48 + 8(\,?\,)}{20}$ so $? = \dfrac{100 - 48}{8} = 6.5$.

(c) $\bar{x} = 6 = \dfrac{M_0}{M} = \dfrac{48 + (\,?\,)(10)}{12 + ?}$ so $72 + 6(\,?\,) = 48 + 10(\,?\,)$ and $4(\,?\,) = 24$ so $? = 6$.

3. (a) $M = 2 + 5 + 5 = 12$. $\bar{x} = \dfrac{M_y}{M} = \dfrac{(2)(4) + (5)(2) + (5)(6)}{12} = \dfrac{48}{12} = 4$.

$\bar{y} = \dfrac{M_x}{M} = \dfrac{(2)(3) + (5)(4) + (5)(2)}{12} = \dfrac{36}{12} = 3$.

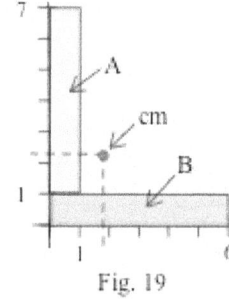

Fig. 19

(b) new $\bar{x} = 5 = \dfrac{48 + 10(\,?\,)}{12 + 10}$ so $? = \dfrac{110 - 48}{10} = 6.2$.

new $\bar{y} = 2 = \dfrac{36 + 10(?)}{12 + 10}$ so $? = \dfrac{44 - 36}{10} = 0.8$.

5. See Fig. 19. A: mass $= (1)(6)(\text{density}) = 6$. Center of mass $= (0.5, 4)$

B: mass $= (1)(6)(\text{density}) = 6$. Center of mass $= (3, 0.5)$

Total mass $= 6 + 6 = 12$. $M_y = (6)(0.5) + (6)(3) = 21$. $M_x = (6)(4) + (6)(0.5) = 27$.

$\bar{x} = \dfrac{M_y}{M} = \dfrac{21}{12} = 1.75$ and $\bar{y} = \dfrac{M_x}{M} = \dfrac{27}{12} = 2.25$.

Notice that the center of mass is not "in" the region.

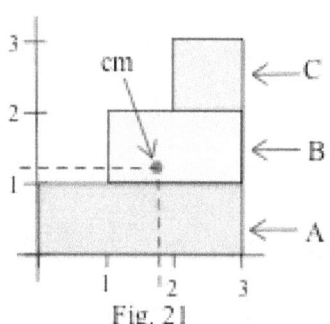

Fig. 21

7. See Fig. 21. A: mass $= (1)(3) = 3$. Center of mass $= (1.5, 0.5)$

B: mass $= (1)(2) = 2$. Center of mass $= (2, 1.5)$

C: mass $= (1)(1) = 1$. Center of mass $= (2.5, 2.5)$

Total mass $= 3 + 2 + 1 = 6$. $M_y = (3)(1.5) + (2)(2) + (1)(2.5) = 11$.

$M_x = (3)(0.5) + (2)(1.5) + (1)(2.5) = 7$.

$\bar{x} = \dfrac{M_y}{M} = \dfrac{11}{6} \approx 1.83$. $\bar{y} = \dfrac{M_x}{M} = \dfrac{7}{6} \approx 1.17$.

9. See Fig. 23. A: mass = (8)(4) = 32. Center of mass = (2 , 4)

   B: mass = $0.5\pi(2^2) \approx 6.28$.

       Center of mass $\approx (2, 8+0.4244(2)) = (2, 8.85)$

   C: mass = $0.5\pi(4^2) \approx 25.13$.

       Center of mass $\approx (4+0.4244(4), 4) = (5.70, 4)$

   Total mass ≈ 63.416. $M_y \approx (32)(2)+(6.28)(2)+(25.13)(5.70) = 219.801$.

   $M_x \approx (32)(4) + (6.28)(8.85) + (25.13)(4) = 284.098$.

   $\bar{x} = \dfrac{M_y}{M} \approx \dfrac{219.801}{63.416} \approx 3.47$. $\bar{y} = \dfrac{M_x}{M} = \dfrac{284.098}{63.416} \approx 4.48$.

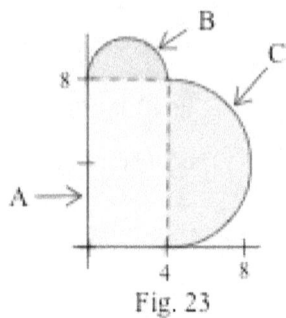

Fig. 23

11. Fig. 1. $\bar{x} = \dfrac{\displaystyle\int_0^3 x \cdot f(x)\, dx}{\displaystyle\int_0^3 f(x)\, dx} = \dfrac{\displaystyle\int_0^3 x \cdot x\, dx}{\displaystyle\int_0^3 x\, dx} = \dfrac{\left.\frac{1}{3}x^3\right|_0^3}{\left.\frac{1}{2}x^2\right|_0^3} = \dfrac{9}{4.5} = 2$

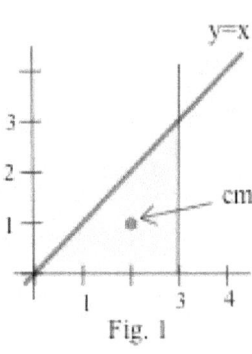

Fig. 1

$\bar{y} = \dfrac{\frac{1}{2}\displaystyle\int_0^3 (f(x))^2\, dx}{\displaystyle\int_0^3 f(x)\, dx} = \dfrac{\frac{1}{2}\displaystyle\int_0^3 x^2\, dx}{\displaystyle\int_0^3 x\, dx} = \dfrac{\left.\frac{1}{2}\frac{1}{3}x^3\right|_0^3}{\left.\frac{1}{2}x^2\right|_0^3} = \dfrac{4.5}{4.5} = 1$

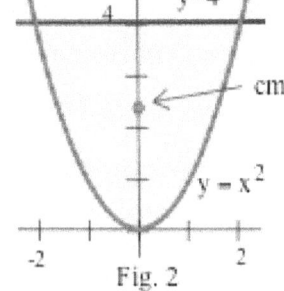

Fig. 2

13. Fig. 2. Because of symmetry about the y–axis, $\bar{x} = 0$. $\bar{y} = \dfrac{M_x}{M}$.

   $M_x = \frac{1}{2}\displaystyle\int_a^b \{ f^2(x) - g^2(x) \}\, dx = \frac{1}{2}\displaystyle\int_{-2}^2 \{ 4^2 - (x^2)^2 \}\, dx = \frac{1}{2} \{ 16x - \frac{1}{5}x^5$

   $\left.\}\right|_{-2}^2$

   $= \frac{1}{2} \{ (32-\frac{32}{5}) - (-32+\frac{32}{5}) \} \approx 25.6$

   $M = \displaystyle\int_a^b \{ 4 - x^2 \}\, dx = \left.\{ 4x - \frac{1}{3}x^3 \}\right|_{-2}^2 = \{ (8-\frac{8}{3}) - (-8+\frac{8}{3}) \} = \frac{32}{3} \approx 10.67$.

   $\bar{y} = \dfrac{M_x}{M} \approx \dfrac{25.6}{10.67} = 2.4$

15. Fig. 3. Because of symmetry about the y–axis, $\bar{x} = 0$. $\bar{y} = \dfrac{M_x}{M}$.

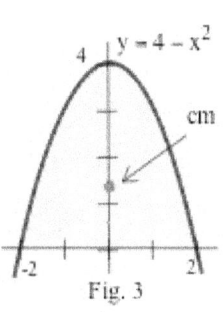

$$M_x = \frac{1}{2} \int_a^b \{ f^2(x) \} dx = \frac{1}{2} \int_{-2}^{2} (4 - x^2)^2 \, dx = \frac{1}{2} \{ 16x - \frac{8}{3} x^3 + \frac{1}{5} x^5 \} \Big|_{-2}^{2}$$

$$= \frac{1}{2} \{ (32 - \frac{64}{3} + \frac{32}{5}) - (-32 + \frac{64}{3} - \frac{32}{5}) \} \approx 17.07$$

$$M = \int_a^b \{ 4 - x^2 \} dx = \{ 4x - \frac{1}{3} x^3 \} \Big|_{-2}^{2} = \{ (8 - \frac{8}{3}) - (-8 + \frac{8}{3}) \} = \frac{32}{3} \approx 10.67 .$$

$$\bar{y} = \frac{M_x}{M} \approx \frac{17.07}{10.67} \approx 1.6$$

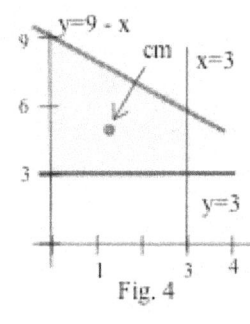

17. Fig. 4. $M = \int_0^3 f(x) - g(x) \, dx = \int_0^3 9 - x - 3 \, dx = 6x - \frac{1}{2} x^2 \Big|_0^3 = 13.5$

$$M_y = \int_0^3 x \cdot \{ f(x) - g(x) \} dx = \int_0^3 x \cdot (9 - x - 3) \, dx = 3x^2 - \frac{1}{3} x^3 \Big|_0^3 = 27 - 9 = 18.$$

$$M_x = \frac{1}{2} \int_a^b \{ f^2(x) - g^2(x) \} dx = \frac{1}{2} \int_0^3 \{ (9 - x)^2 - (3)^2 \} dx = \frac{1}{2} \int_0^3 \{ 81 - 18x + x^2 - 9 \} dx$$

$$= \frac{1}{2} \{ 72x - 9x^2 + \frac{1}{3} x^3 \} \Big|_0^3 = \frac{1}{2} \{ 216 - 81 + 9) = 72.$$

$$\bar{x} = \frac{M_y}{M} = \frac{18}{13.5} \approx 1.33 . \quad \bar{y} = \frac{M_x}{M} = \frac{72}{13.5} \approx 5.33 .$$

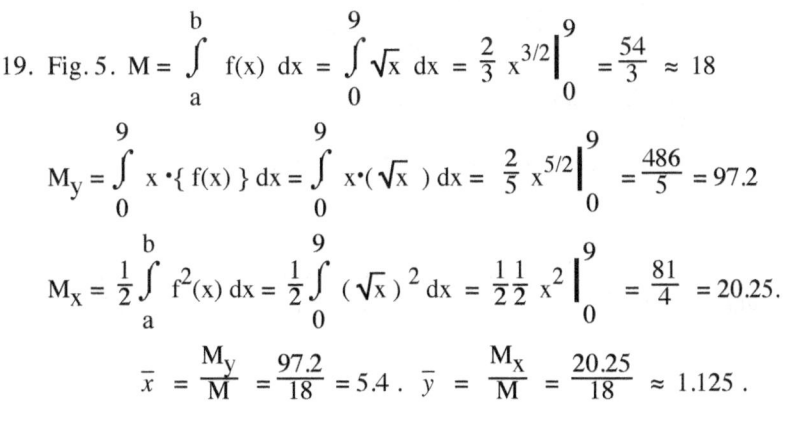

19. Fig. 5. $M = \int_a^b f(x) \, dx = \int_0^9 \sqrt{x} \, dx = \frac{2}{3} x^{3/2} \Big|_0^9 = \frac{54}{3} \approx 18$

$$M_y = \int_0^9 x \cdot \{ f(x) \} dx = \int_0^9 x \cdot (\sqrt{x}) \, dx = \frac{2}{5} x^{5/2} \Big|_0^9 = \frac{486}{5} = 97.2$$

$$M_x = \frac{1}{2} \int_a^b f^2(x) \, dx = \frac{1}{2} \int_0^9 (\sqrt{x})^2 \, dx = \frac{1}{2} \frac{1}{2} x^2 \Big|_0^9 = \frac{81}{4} = 20.25.$$

$$\bar{x} = \frac{M_y}{M} = \frac{97.2}{18} = 5.4 . \quad \bar{y} = \frac{M_x}{M} = \frac{20.25}{18} \approx 1.125 .$$

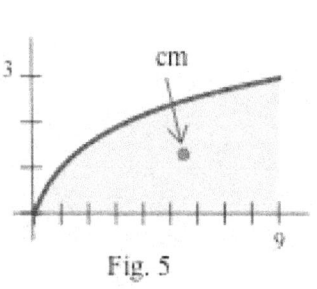

21. Fig. 6. $M = \int_a^b f(x) \, dx = \int_0^1 e - e^x \, dx = e \cdot x - e^x \Big|_0^1 = (e - e) - (0 - 1) = 1$

$$M_y = \int_0^1 x \cdot \{ f(x) \} \, dx = \int_0^1 x \cdot (e - e^x) \, dx = \left. e \cdot \frac{1}{2} x^2 - e^x (x-1) \right|_0^1$$

$$= (\frac{e}{2} - 0) - (0+1) \approx 0.359$$

$$M_x = \frac{1}{2} \int_a^b \{ f^2(x) - g^2(x) \} \, dx = \frac{1}{2} \int_0^1 e^2 - e^{2x} \, dx$$

$$= \frac{1}{2} \{ e^2 \cdot x - \frac{1}{2} e^{2x} \} \Big|_0^1 = \frac{1}{2} \{ e^2 - \frac{1}{2} e^2 \} - \frac{1}{2} \{ 0 - \frac{1}{2} \} \approx 2.097 .$$

$$\bar{x} = \frac{M_y}{M} = \frac{0.359}{1} = 0.359 . \qquad \bar{y} = \frac{M_x}{M} = \frac{2.097}{1} = 2.097 .$$

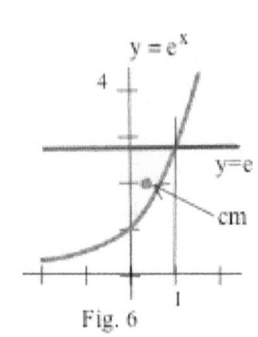

Fig. 6

23. (a) $h = \dfrac{M_0}{M} = \dfrac{(10)(6) + (\frac{x}{2})(\frac{x}{12})(60)}{10 + (\frac{x}{12})(60)} = \dfrac{60 + \frac{5}{2} x^2}{10 + 5x}$.

(b) Find minimum of h in part (a): calculate $\dfrac{dh}{dt}$, set $\dfrac{dh}{dt} = 0$ and solve for x to find critical points.

$$\frac{dh}{dt} = \frac{5x(10+5x) - 5(60 + \frac{5}{2} x^2)}{(10+5x)^2} .$$

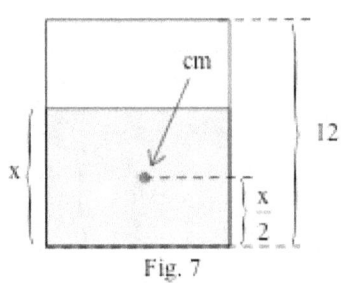

Fig. 7

If $\dfrac{dh}{dt} = 0$ then $5x(10+5x) - 5(60 + \frac{5}{2} x^2 = 0$ so $x = \dfrac{-4 \pm \sqrt{16+96}}{2} \approx 3.3$ inches.

25. Empty glass: weight $= 4$ oz., $\bar{y} = \frac{8}{2} = 4$ cm.

Full glass (liquid only): weight $= 20 - 4 = 16$ oz., $\bar{y} = 4$ cm, volume $= (4 \text{ cm})(4 \text{ cm})(8 \text{ cm}) = 128$ cm^3.

$$\text{density of liquid} = \frac{\text{weight}}{\text{volume}} = \frac{16 \text{ oz.}}{128 \text{ cm}^3} = \frac{1}{8} \frac{\text{oz.}}{\text{cm}^3}$$

Partially full glass (liquid only): **x = height of liquid**, $\bar{y} = \frac{x}{2}$ cm,

$$\text{weight} = (4 \text{ cm})(4 \text{ cm})(x \text{ cm})(\frac{1}{8} \frac{\text{oz.}}{\text{cm}^3}) = 2x \text{ oz}$$

h = height of cm of glass containing x cm of liquid $= \dfrac{(4 \text{ oz})(4 \text{ cm}) + (2x \text{ oz})(\frac{x}{2} \text{ cm})}{4 \text{ oz} + 2x \text{ oz}} = \dfrac{16 + x^2}{4 + 2x}$ cm .

27. and 29. On your own.

31. Center of gravity is 2 feet above the ground and is raised to a height of 10 feet: distance c.g. raised $= 8$ feet.

work $=$ (force)(distance) $= (300)(8) = 2,400$ ft–lb.

33. C.g. is raised $3 + 1 = 4$ feet. Work $= (250)(4) = 1,000$ ft–lb.

35. (a) Volume about x–axis = A·2π·y = 2^2·2π·4 = 32π ft^3.

   Surface area about x–axis = P·2π·y = 8·2π·4 = 64π ft^2.

   (b) Volume about y–axis = A·2π·x = 4·2π·3 = 24π ft^3.

   Surface area about y–axis = P·2π·x = 8·2π·3 = 48π ft^2.

   (c) Volume about line = A·2π·(dist. of c.g. from line) = 4·2π·2 = 16π ft^3.

   Surface area about line = P·2π·(dist of c.g. from line) = 8·2π·2 = 32π ft^2.

37. (a) Volume about x–axis = A·2π·y = $(\pi 2^2)$·2π·5 = 40π^2 ft^3.

   Surface area about x–axis = P·2π·y = 2π(2)·2π·5 = 40π^2 ft^2.

   (b) Volume about y–axis = A·2π·x = 4π·2π·3 = 24π^2 ft^3.

   Surface area about y–axis = P·2π·x = 4π·2π·3 = 24π^2 ft^2.

   (c) Volume about line = A·2π·(dist. of c.g. from line) = 4π·2π·3 = 24π^2 ft^3.

   Surface area about line = P·2π·(dist of c.g. from line) = 4π·2π·3 = 24π^2 ft^2.

39. Each rectangle has area = 8 ft^2, perimeter = 12 ft., and centroid 3 ft. from the line of rotation.

   Volume of each = 2π(radius)(area) = 2π(3 ft)(8 ft^2) = 48π ft^3 ≈ 150.8 ft^3 .

   Surface area of each = 2π(radius)(perimeter) = 2π(3 ft)(12 ft) = 72π ft^2 ≈ 226.2 ft^2 .

Section 5.5

Liquid Pressure

1. A: $\int_{2}^{6} d·x·(2)\, dx = 2d \int_{2}^{6} x\, dx = d\, x^2 \Big|_{2}^{6} = 32d.$

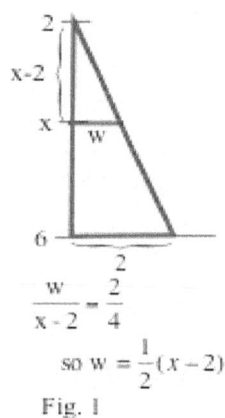

Fig. 1. B: $\int_{2}^{6} d·x·\frac{1}{2}(x-2)\, dx = \frac{d}{2} \int_{2}^{6} x^2 - 2x\, dx$

$= \frac{d}{2}(\frac{1}{3}x^3 - x^2) \Big|_{2}^{6} = \frac{56}{3} d .$

$\dfrac{w}{x-2} = \dfrac{2}{4}$

so $w = \frac{1}{2}(x-2)$

Fig. 1

2. Fig. 2. C: $\int_{2}^{6} d·x·\frac{1}{2}(6-x)\, dx = 13.33\, d$

   D: $\int_{4}^{6} d·x·(2)\, dx = 20\, d .$

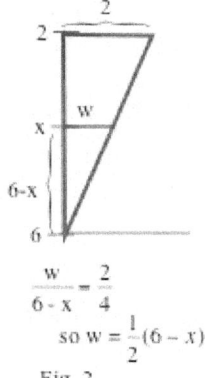

$\dfrac{w}{6-x} = \dfrac{2}{4}$

so $w = \frac{1}{2}(6-x)$

Fig. 2

3. Rectangular end: $\int_0^5 d{\cdot}x{\cdot}(4)\ dx = 4d \int_0^5 x\ dx = 2d\ x^2\Big|_0^5 = 50\ d$.

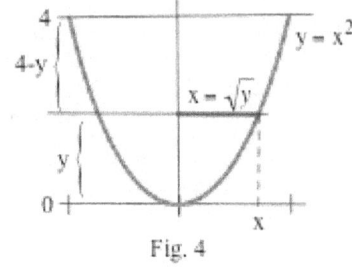

Fig. 3

Triangular end (Fig. 3): $\int_0^5 d{\cdot}x{\cdot}(\frac{4}{5})(5-x)\ dx = \frac{4}{5}\ d \int_0^4 5x - x^2\ dx$

$$= \frac{4}{5}\ d\ \{ \frac{5}{2}\ x^2 - \frac{1}{3}\ x^3)\Big|_0^5 = \frac{50}{3}\ d\ .$$

The total force is unchanged if the length is doubled.

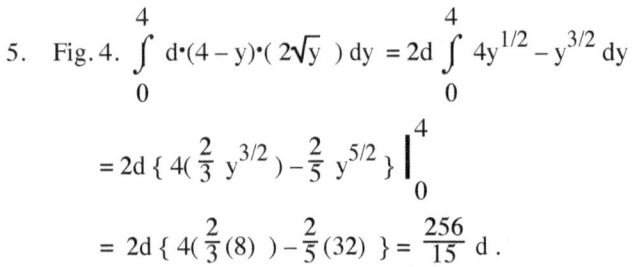

5. Fig. 4. $\int_0^4 d{\cdot}(4-y){\cdot}(2\sqrt{y})\ dy = 2d \int_0^4 4y^{1/2} - y^{3/2}\ dy$

$$= 2d\ \{ 4(\frac{2}{3}\ y^{3/2}) - \frac{2}{5}\ y^{5/2} \}\Big|_0^4$$

$$= 2d\ \{ 4(\frac{2}{3}(8)) - \frac{2}{5}(32) \} = \frac{256}{15}\ d\ .$$

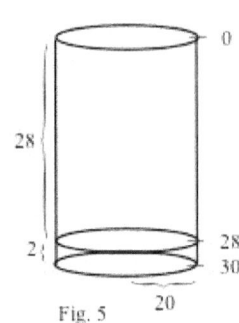

Fig. 4

6. All three have the same total force against their sides.

7. The one with the largest perimeter: not (a), probably (b)

9. Fig. 5. (a) $\int_{28}^{30} d{\cdot}x{\cdot}(40\pi)\ dx = 20\pi d\ x^2\Big|_{28}^{30} = 20\pi d\ (116) = 2{,}320\pi d$.

(b) $\int_{33}^{35} d{\cdot}x{\cdot}(40\pi)\ dx = 20\pi d\ x^2\Big|_{33}^{35} = 20\pi d\ (136) = 2{,}720\pi d$.

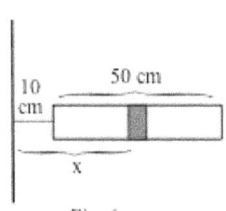

Fig. 5

Kinetic Energy

11. (a) $M = 20$ g. $v = 3$ rev/sec $= 3(\ 2\pi(15\text{cm})\ )/\text{sec} = 90\pi$ cm/sec.

$KE = \frac{1}{2}\ M\ v^2 = \frac{1}{2}(20\ \text{g})(\ 90\pi\ \text{cm/s}\ )^2 = 81{,}000\ \pi^2\ \text{ergs} \approx 799{,}438$ ergs.

(b) $M = 20$ g. $v = 3$ rev/sec $= 3(\ 2\pi(\mathbf{20}\text{cm})\ )/\text{sec} = 120\pi$ cm/sec.

$KE = \frac{1}{2}\ M\ v^2 = \frac{1}{2}(20\ \text{g})(\ 120\pi\ \text{cm/s}\ )^2 = 144{,}000\ \pi^2\ \text{ergs} \approx 1{,}421{,}223$ ergs.

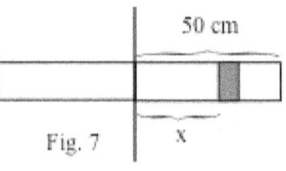

Fig. 6

13. Fig. 6. $KE_i = \frac{1}{2}(\ 3\ \Delta x\ )(\ 2{\cdot}2\pi x\ )^2 = 24\pi^2\ x^2\ \Delta x$

$KE = \int_{10}^{60} 24\pi^2\ x^2\ dx = 8\pi^2\ x^3\Big|_{10}^{60} = 1.72\ {\cdot}10^6\ \pi^2\ \text{ergs} \approx 1.698\ {\cdot}10^7$ ergs.

15. Fig. 7. $(20\ \text{g})/(100\ \text{cm}) = 0.2$ g/cm. $KE_i = \frac{1}{2}(\ 0.2\ \Delta x\ )(\ 2\pi x\ )^2 = 0.4\pi^2\ x^2\ \Delta x$

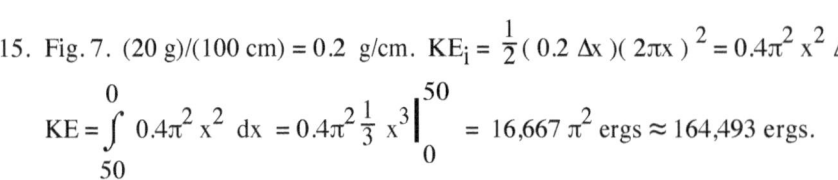

Fig. 7

$KE = \int_{50}^0 0.4\pi^2\ x^2\ dx = 0.4\pi^2\ \frac{1}{3}\ x^3\Big|_0^{50} = 16{,}667\ \pi^2\ \text{ergs} \approx 164{,}493$ ergs.

Total KE $= 2\{\ 16{,}667\ \pi^2\ \text{ergs}\ \} \approx 328{,}986$ ergs.

17. (a) $m_i = 2\pi\, r\, \Delta x = 2\pi\, x\, \Delta x$,

$v_i = 3$ rev/sec $= 3(\, 2\pi$ radians $)$/sec $= 6\pi x$ radians/sec.

$KE_i = \frac{1}{2}\, m_i\, (\, v_i\,)2 = \frac{1}{2}(\, 2\pi\, x\, \Delta x\,)(\, 6\pi x\,)^2 = 36\, \pi^3\, x^3\, \Delta x$

$KE = \int\limits_{1}^{3} 36\, \pi^3\, x^3\ dx = 36\, \pi^3\, (\frac{1}{4}\,)\, x^4\, \Big|_{1}^{3} = 9\, \pi^3\, (3^4 - 1^4) = 720\, \pi^3$

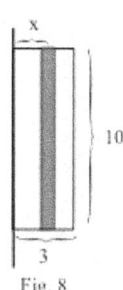

(b) $m_i = 4\pi\, x\, \Delta x$, $v_i = 3$ rev/sec $= 3(\, 2\pi$ radians $)$/sec $= 6\pi x$ radians/sec.

$KE_i = \frac{1}{2}\, m_i\, (\, v_i\,)2 = \frac{1}{2}(\, 4\pi\, x\, \Delta x\,)(\, 6\pi x\,)^2 = 72\, \pi^3\, x^3\, \Delta x$

$KE = \int\limits_{3}^{4} 72\, \pi^3\, x^3\ dx = 72\, \pi^3\, (\frac{1}{4}\,)\, x^4\, \Big|_{3}^{4} = 18\, \pi^3\, (4^4 - 3^4) = 3{,}150\, \pi^3$

Fig. 8

19. density $= 3$ g/cm^3 .

(a) Fig. 8. $m_i = ($area$)(\, 3$ g/cm$^3) = (10\, \Delta x)(\, 3$ g/cm$^3) = 30\, \Delta x.$

$v_i = 2$ rev/sec $= 2(\, 2\pi\, x$ radians $)$/sec $= 4\pi\, x$ radians/sec.

$KE_i = \frac{1}{2}\, m_i\, (\, v_i\,)2 = \frac{1}{2}(\, 30\, \Delta x\,)(\, 4\pi x\,)^2 = 240\, \pi^2\, x^2\, \Delta x$

$KE = \int\limits_{0}^{3} 240\, \pi^2\, x^2\ dx = 80\, \pi^2\, x^3\, \Big|_{0}^{3} = 18\, \pi^2\, (27) = 2{,}160\, \pi^2$

Total KE $= 2\{\, 2{,}160\, \pi^2\,) = 4{,}320\, \pi^2$

(b) Fig. 9. $m_i = ($area$)(\, 3$ g/cm$^3) = (6\, \Delta x)(\, 3$ g/cm$^3) = 18\, \Delta x.$ $v_i = 4\pi\, x$ radians/sec.

$KE_i = \frac{1}{2}\, m_i\, (\, v_i\,)2 = \frac{1}{2}(\, 18\, \Delta x\,)(\, 4\pi x\,)^2 = 144\, \pi^2\, x^2\, \Delta x$

$KE = \int\limits_{0}^{5} 144\, \pi^2\, x^2\ dx = 48\, \pi^2\, x^3\, \Big|_{0}^{5} = 48\, \pi^2\, (125) = 6{,}000\, \pi^2$

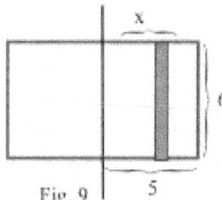

Total KE $= 2\{\, 6{,}000\, \pi^2\,) = 12{,}000\, \pi^2$

Fig. 9

Volumes using tubes: volume $= \int\limits_{a}^{b} 2\pi($radius$)($height$)($thickness$)$

21. Fig. 10. $\int\limits_{a}^{b} 2\pi($radius$)($height$)($thickness$)$

$= \int\limits_{0}^{1} 2\pi(\, x\,)(\sqrt{1 - x^2}\,)\ dx$ (put $u = 1 - x^2$, then $du = -2x\ dx$)

$= 2\pi \int\limits_{u=1}^{u=0} u^{1/2} \cdot \left(\frac{-1}{2} du\right) = -\pi\, (\frac{3}{2}\,)\, u^{3/2}\, \Big|_{u=1}^{u=0}$

$= \left(\frac{-2\pi}{3}\right)(0) - \left(\frac{-2\pi}{3}\right)(1) = \frac{2\pi}{3}$.

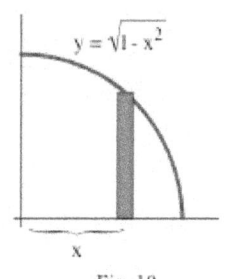

Fig. 10

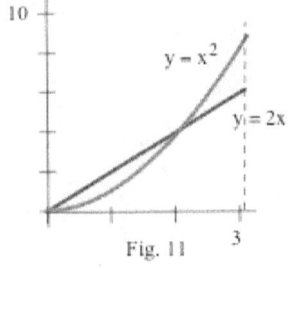

Fig. 11

23. Fig. 11. $\int_a^b 2\pi(\text{radius})(\text{height})(\text{thickness})$

$$= \int_0^2 2\pi(4-x)(2x-x^2)\,dx + \int_2^3 2\pi(4-x)(x^2-2x)\,dx$$

$$= 2\pi\int_0^2 -x^3+6x^2-8x\,dx + 2\pi\int_2^3 x^3-6x^2+8x\,dx = 2\pi\{4\} + 2\pi\{\frac{7}{4}\} \approx 36.13$$

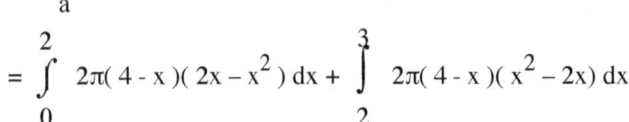

Fig. 12

25. Fig. 12. $\int_a^b 2\pi(\text{radius})(\text{height})(\text{thickness}) = \int_1^3 2\pi(5-x)(\frac{1}{x}-\frac{1}{3})\,dx$

$$= 2\pi\int_1^3 \frac{5}{x}-\frac{8}{3}+\frac{x}{3}\,dx = 2\pi\{5\ln(x)-\frac{8}{3}x+\frac{1}{6}x^2\}\Big|_1^3$$

$$= 2\pi\{-4+5\ln(3)\} = 9.38 \ .$$

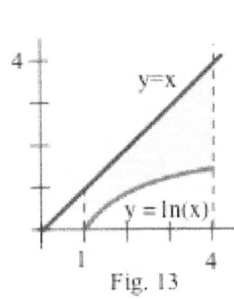

Fig. 13

27. Fig. 13. $\int_a^b 2\pi(\text{radius})(\text{height})(\text{thickness}) = \int_1^4 2\pi(x)(x-\ln(x))\,dx$

≈ 85.8 (Using calculator.)

29. Fig. 14. $\int_a^b 2\pi(\text{radius})(\text{height})(\text{thickness}) = \int_0^{3/\sqrt{2}} 2\pi(x)(\sqrt{9-x^2}-x)\,dx$

$=$ finish on your own.

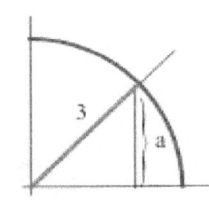

14

Voting

31. a votes for cand. A, b votes for cand. A, c votes for cand. C.

32. a votes for cand. B, b votes for cand. C, c votes for cand. C.

33. Fig. 39: B wins.   34. Fig. 40: A wins.

35. (a) Fig. 41a: A wins.

    (b) Fig. 41b: C wins.

    (c) Fig. 41c: B wins.

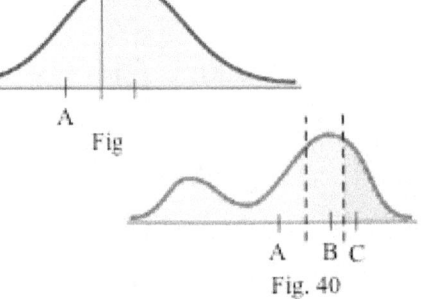

Fig

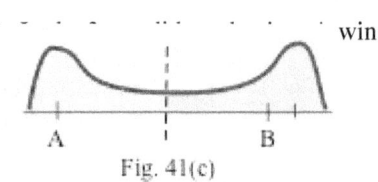

A   B C

Fig. 40

N[...] wins.

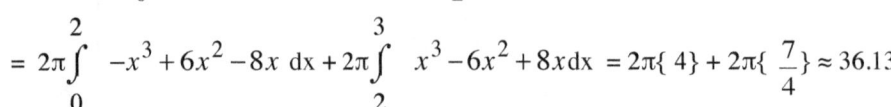

Fig. 41(a)   Fig. 41(b)   Fig. 41(c)

36. (a) B wins.   (b) B wins   (c) A wins   37. (a) A wins.   (b) A wins   (c) looks like C wins

38. (a) B wins.   (b) A wins   (c) looks like A wins   39. on your own

CHAPTER 8: IMPROPER INTEGRALS and INTEGRATION TECHNIQUES

Introduction

In previous sections we examined a variety of applications which require integrals, and we found antiderivatives of many important groups of functions: polynomials, some rational functions, the trigonometric functions, the logarithm functions and some exponential functions. With this information, it is easy to find antiderivatives of their sums and differences, but finding antiderivatives of their products, quotients, and compositions can still be quite difficult. This chapter introduces several techniques of integration which greatly expand the number and variety of functions you can integrate. The first section, however, does not discuss a technique for finding antiderivatives.

- Section 8.1 describes how to evaluate an **Improper Integral**, the integral of a function over an infinitely long interval or over a finite interval when the function is not bounded at one endpoint of the interval.

The rest of the chapter is devoted to finding antiderivatives, and the overall theme **transformation**, how to transform a new type of integrand into one we can integrate immediately or into one we can find in the tables at the end of the book. Our goal is to change the pattern of some new function or combination of functions into a pattern we recognize or can find in the tables.

- Section 8.2 **reviews** some of the most common patterns we have already encountered, and it emphasizes the powerful technique of **substitution**. When the substitution technique works, it is among the easiest and quickest to use.

The remaining four sections present additional techniques for finding antiderivatives. Each of the new techniques is very useful for finding antiderivatives of particular patterns and combinations of functions. There are more integration techniques besides the four presented here, but these four are the ones most commonly needed.

- Section 8.3 introduces a technique called **Integration By Parts** which is particularly useful for finding antiderivatives of products of functions. Integration by parts is the technique used to derive many of the integration formulas in the tables.

- Section 8.4 introduces an algebraic technique called **Partial Fraction Decomposition** for transforming difficult rational functions into sums of easier rational functions which can then be integrated using previous integration techniques.

- Section 8.5 introduces a technique called **Trigonometric Substitution** which is particularly useful for integrands containing sums and differences of squares of the forms $x^2 + a^2$, $x^2 - a^2$, and $a^2 - x^2$.

- Section 8.6 considers a variety of ways in which trigonometric identities and transformations can be used to find antiderivatives of some combinations of trigonometric functions.

Unfortunately, some functions simply do not have antiderivatives which are elementary combinations (sums, differences, products, quotients, roots, and compositions) of polynomials, rational functions, trigonometric, logarithmic and exponential functions, and none of the integration methods of this chapter will find their antiderivatives.

Historically, the integration techniques in this chapter and tables of antiderivatives were very important for people who needed to apply calculus and solve differential equations. Recently we have gained additional tools, computers and even calculators that can calculate the antiderivatives of many (but not all) functions. These electronic aides, like earlier tables of antiderivatives, can remove some computational difficulties on the way to an answer, but it is still up to you to understand and set up the problems and to interpret and use the answers. Now, perhaps more than ever, it is important that you master the concepts of calculus and understand how these concepts are related and are used. **The computer may help you get an answer once the problem has been understood and formulated in mathematical terms, but it is your understanding of the concepts that will enable you to formulate the problems so computers can help.**

Although the computational techniques in this chapter are less important than they were several years ago, they still contribute to understanding calculus and to recognizing patterns of functions and their derivatives and antiderivatives.

8.1   Improper Integrals

Our original development of the definite integral $\int_a^b f(x)\, dx$ used Riemann sums and assumed that

- the length of the interval of integration [a, b] was finite and

- that f(x) was defined and bounded at every point of the interval [a, b] (including the endpoints).

Sometimes, however, we need the value of an integral which does not satisfy one or both of these assumptions. In this section we extend the ideas of the definite integral to evaluate two types of **improper definite integrals**:

(1)   the length of the interval of integration is not finite

(2)   the integrand function is not bounded at a point of the interval of integration.

Example 1:   Represent each area as an improper definite integral.

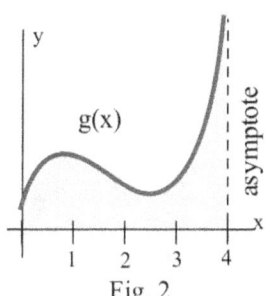

Fig. 2

(a) The area of the infinite region between

$f(x) = 1/x^2$ and the x–axis for $x \geq 1$ (Fig. 1)

(b) The area between g(x) and the x–axis for

$0 \leq x \leq 4$ (Fig. 2)

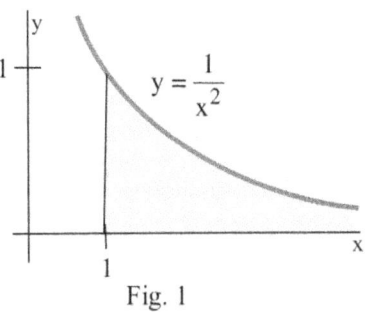

$y = \dfrac{1}{x^2}$

Fig. 1

Solution: (a) $\displaystyle\int_1^\infty \frac{1}{x^2}\, dx$ (b) $\displaystyle\int_0^4 g(x)\, dx$

Practice 1:   Represent each quantity as an improper definite integral.

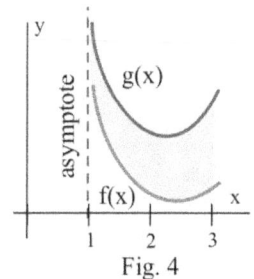

Fig. 4

(a) The volume swept out when the infinite region

between f(x) = 1/x and the x–axis is revolved about

the x–axis for $x \geq 4$ (Fig. 3)

(b) The area between the curves in Fig. 4 for

$1 \leq x \leq 3$.

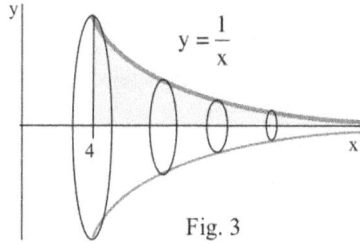

$y = \dfrac{1}{x}$

Fig. 3

General Strategy For Improper Integrals

Our general strategy for evaluating improper integrals is to shrink the interval of integration so we have a definite integral we can evaluate. Then as we let the interval grow to approach the interval of integration we want, the value of the integral on the growing intervals approaches the value of the improper integral. The value of the improper integral is the limiting value of the definite integrals as the interval grows to the interval we want, provided that the limit exists.

Infinitely Long Intervals of Integration

We evaluate an improper integral on an infinitely long interval by

- replacing the inifinitely long interval with a finite interval,

- evaluating the integral on the finite interval, and, finally,

- letting the finite interval grow longer and longer, approaching the interval we want.

Example 2:   Evaluate $\int_{1}^{\infty} \dfrac{1}{x^2}\ dx$.

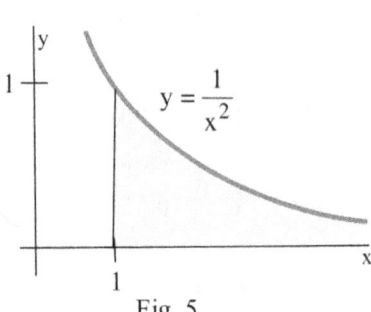

Fig. 5

Solution: The interval $[1, \infty)$ is infinitely long, but we can

evaluate the integral on the finite intervals $[1, 2], [1, 10], [1, 1000]$,

and in general $[1, C]$ (Fig. 5).

$$\int_{1}^{2} \frac{1}{x^2}\ dx \ = -\frac{1}{x}\ \Big|_{1}^{2}\ = \left(-\frac{1}{2}\right) - \left(-\frac{1}{1}\right) = 1 - \frac{1}{2} \ = \ \frac{1}{2}\ .$$

Similarly, $\int_{1}^{10} \dfrac{1}{x^2}\ dx\ = 1 - \dfrac{1}{10}\ = \ .9$, $\int_{1}^{1000} \dfrac{1}{x^2}\ dx\ = 1 - \dfrac{1}{1000}\ = \ .999$, and, in general,

$$\int_{1}^{C} \frac{1}{x^2}\ dx \ = -\frac{1}{x}\ \Big|_{1}^{C}\ = \left(-\frac{1}{C}\right) - \left(-\frac{1}{1}\right) = 1 - \frac{1}{C}\ .$$ As the value of C gets larger, the length of the

interval $[1, C]$ increases, and the value of $\int_{1}^{C} \dfrac{1}{x^2}\ dx$ approaches the value of $\int_{1}^{\infty} \dfrac{1}{x^2}\ dx$.

The value of $\int_{1}^{\infty} \dfrac{1}{x^2}\ dx$ is the limit of the values of $\int_{1}^{C} \dfrac{1}{x^2}\ dx$ as C approaches infinity:

$$\int_{1}^{\infty} \frac{1}{x^2}\ dx \ = \ \lim_{C \to \infty} \left\{ \int_{1}^{C} \frac{1}{x^2}\ dx \right\} \ = \ \lim_{C \to \infty} \left\{ 1 - \frac{1}{C} \right\} \ = \ 1\ .$$

We say that the improper integral $\int_{1}^{\infty} \dfrac{1}{x^2}\ dx$ is **convergent** and **converges to 1**.

If the following limits exist,

the value of $\displaystyle\int_a^\infty f(x)\ dx$ is defined to be the value of $\displaystyle\lim_{C\to\infty}\left\{\int_a^C f(x)\ dx\right\}$, and

the value of $\displaystyle\int_{-\infty}^b f(x)\ dx$ is defined to be the value of $\displaystyle\lim_{C\to-\infty}\left\{\int_C^b f(x)\ dx\right\}$.

In each case, first evaluate the proper integral and then take the limit.

If the limit is a finite number, we say the improper integral is **convergent.**

If the limit does not exist or if it is infinite, we say the improper integral is **divergent.**

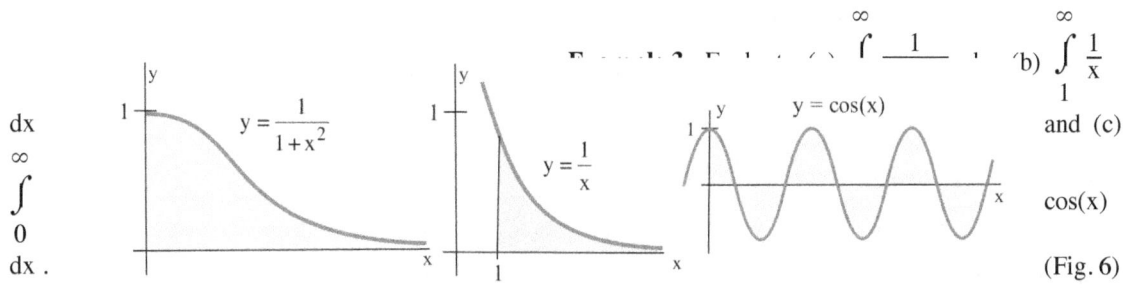

Fig. 6

Solution: (a) $\displaystyle\int_0^\infty \frac{1}{1+x^2}\ dx = \lim_{C\to\infty}\left\{\int_0^C \frac{1}{1+x^2}\ dx\right\} = \lim_{C\to\infty}\left\{ \arctan(x)\Big|_0^C \right\}$

$$= \lim_{C\to\infty}\left\{ \arctan(C) - \arctan(0) \right\} = \frac{\pi}{2} - 0 = \frac{\pi}{2}\ .$$

so we say that $\displaystyle\int_0^\infty \frac{1}{1+x^2}\ dx$ is **convergent.**

(b) $\displaystyle\int_1^\infty \frac{1}{x}\ dx = \lim_{C\to\infty}\left\{\int_1^C \frac{1}{x}\ dx\right\} = \lim_{C\to\infty}\left\{ \ln(x)\Big|_1^C \right\} = \lim_{C\to\infty}\left\{ \ln(C) - \ln(1) \right\} = \infty$

so we say the improper integral $\displaystyle\int_1^\infty \frac{1}{x}\ dx$ is **divergent.**

(c) $\int_0^\infty \cos(x)\ dx = \lim_{C\to\infty}\left\{\int_0^C \cos(x)\ dx\right\} = \lim_{C\to\infty}\left\{\sin(x)\Big|_0^C\right\}$

$= \lim_{C\to\infty}\left\{\sin(C) - \sin(0)\right\} = \lim_{C\to\infty}\sin(C)\ .$

As C increases the values of $\sin(C)$ oscilate between -1 and 1 and do not approach a single value

so the last limit does not exist and the improper integral $\int_1^\infty \cos(x)\ dx$ is **divergent**.

Practice 2: Evaluate (a) $\int_1^\infty \frac{1}{x^3}\ dx$ and (b) $\int_0^\infty \sin(x)\ dx$.

Functions Undefined At An Endpoint Of The Interval Of Integration

If the function we want to integrate is unbounded at one of the endpoints of an interval of finite length, we can shrink the interval so the function is bounded at both endpoints of the new, smaller interval, evaluate the integral over the smaller interval, and finally, let the smaller interval grow to approach the original interval.

Example 4:   Evaluate $\int_0^1 \frac{1}{\sqrt{x}}\ dx$. (Fig. 7)

Solution: The function $1/\sqrt{x}$ is not bounded at $x = 0$, the lower endpoint of

integration, but the function is bounded on the intervals $[.36, 1], [0.09, 1]$, and,

in general, on the interval $[C, 1]$ for any $C > 0$.

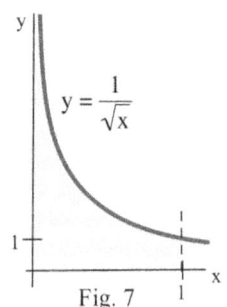

Fig. 7

$\int_{.36}^1 \frac{1}{\sqrt{x}}\ dx = 2\sqrt{x}\ \Big|_{.36}^1 = 2\sqrt{1} - 2\sqrt{.36} = 2 - 1.2 = 0.8$. Similarly,

$\int_{0.09}^1 \frac{1}{\sqrt{x}}\ dx = 2\sqrt{x}\ \Big|_{0.09}^1 = 2\sqrt{1} - 2\sqrt{0.09} = 1.4$ and $\int_C^1 \frac{1}{\sqrt{x}}\ dx = 2\sqrt{x}\ \Big|_C^1 = 2 - 2\sqrt{C}$.

As C approaches 0 from the right, the interval $[C, 1]$ approaches the interval $[0, 1]$ and the value of

$2 - 2\sqrt{C}$ approaches 2. We say that $\int_0^1 \frac{1}{\sqrt{x}}\ dx$ converges to 2 and write $\int_0^1 \frac{1}{\sqrt{x}}\ dx = 2$.

If f(x) is not bounded (not defined) at x = b,

$$\int_a^b f(x)\ dx \text{ is defined to be the value of } \lim_{C \to b^-} \left\{ \int_a^C f(x)\ dx \right\}$$

if this limit exists.

If f(x) is not bounded (not defined) at x = a,

$$\int_a^b f(x)\ dx \text{ is defined to be the value of } \lim_{C \to a^+} \left\{ \int_C^b f(x)\ dx \right\}.$$

if this limit exists.

In each case, first evaluate the proper integral and then take the limit.

If the limit is a finite number, we say the improper integral is **convergent**.

If the limit does not exist or if it is infinite, we say the improper integral is **divergent**.

Practice 3:   Show that (a) $\int_1^{10} \frac{1}{\sqrt{10-x}}\ dx = 6$ and (b) $\int_0^1 \frac{1}{x}\ dx$ is divergent.

If the function is unbounded at one or more points inside the interval of integration, we can split the original improper integral into several improper integrals on subintervals so the function is unbounded at one endpoint of each subinterval.

Testing For Convergence: The Comparison Test and P–Test

Sometimes the only thing that matters about an improper integral is whether or not it converges to a finite number. There are ways to determine its convergence even though we may not be able to or may not want to determine the exact value of the integral. In the remainder of this section we consider two methods for testing the convergence of an improper integral. Neither method gives us the value of the improper integral, but each enables us to determine whether some improper integrals are convergent. The Comparison Test For Integrals enables us to determine the convergence (or divergence) of some integrals by comparing them with some integrals we already know converge or diverge. The Comparison Test, however, requires that we know the convergence or divergence of the integrals we compare against, and the P–Test provides examples of known convergent and divergent integrals to use for this comparison.

We start with the P–Test in order to have some examples to use when we consider the Comparison Test.

P–Test for Integrals

For any a > 0, the improper integral $\displaystyle\int_a^\infty \frac{1}{x^p}\,dx$ $\begin{cases} \text{converges} & \text{if } p > 1 \\ \text{diverges} & \text{if } p \le 1 \end{cases}$

Proof: It is easiest to consider three cases: $p = 1, p > 1,$ and $p < 1$.

Case **p = 1**: Then $\displaystyle\int_a^\infty \frac{1}{x^p}\,dx = \int_a^\infty \frac{1}{x}\,dx = \lim_{C \to \infty}\left\{ \int_a^C \frac{1}{x}\,dx \right\}$

$$= \lim_{C \to \infty}\left\{ \ln(x) \Big|_a^C \right\} = \lim_{C \to \infty}\left\{ \ln(C) - \ln(a) \right\} = \infty \quad \text{so} \quad \int_a^\infty \frac{1}{x^p}\,dx \text{ diverges.}$$

For the other two cases, $p \neq 1$, so $\displaystyle\int_a^\infty \frac{1}{x^p}\,dx = \lim_{C \to \infty}\left\{ \int_a^C \frac{1}{x^p}\,dx \right\} = \lim_{C \to \infty}\left\{ \int_a^C x^{-p}\,dx \right\}$

$$= \lim_{C \to \infty}\left\{ \frac{1}{1-p} \cdot x^{1-p} \Big|_a^C \right\} = \lim_{C \to \infty}\left\{ C^{1-p} - a^{1-p} \right\}.$$

Case **p > 1**: Then $1 - p < 0$ so $\displaystyle\lim_{C \to \infty} C^{1-p} = 0$ and

$$\int_a^\infty \frac{1}{x^p}\,dx = \lim_{C \to \infty}\frac{1}{1-p}\left\{ C^{1-p} - a^{1-p} \right\} = -\frac{a^{1-p}}{1-p} = \frac{a^{1-p}}{p-1} \quad \text{, a finite number.}$$

Case **p < 1**: Then $1 - p > 0$ so $\displaystyle\lim_{C \to \infty} C^{1-p} = \infty$ and $\displaystyle\int_a^\infty \frac{1}{x^p}\,dx$ diverges.

Example 5: Determine the convergence or divergence of (a) $\displaystyle\int_5^\infty \frac{1}{x^2}\,dx$, (b) $\displaystyle\int_1^\infty \frac{1}{\sqrt{x}}\,dx$ and (c) $\displaystyle\int_1^8 \frac{1}{x^{1/3}}\,dx$.

Solution: (a) $\displaystyle\int_5^\infty \frac{1}{x^2}\,dx$ matches the form for the P–Test with $p = 2 > 1$, so the integral is convergent.

(b) $\displaystyle\int_1^\infty \frac{1}{\sqrt{x}}\,dx = \int_1^\infty \frac{1}{x^{1/2}}\,dx$ matches the form for the P–Test with $p = 1/2 < 1$ so the integral is divergent.

(c) $\displaystyle\int_1^8 \frac{1}{x^{1/3}}\,dx$ does not match the form for the P–Test because the interval $[1, 8]$ is finite.

$$\int_1^8 \frac{1}{x^{1/3}}\,dx = \int_1^8 x^{-1/3}\,dx = \frac{3}{2}x^{2/3}\Big|_1^8 = \frac{3}{2}\left\{ 8^{2/3} - 1^{2/3} \right\} = \frac{3}{2}\left\{ 4 - 1 \right\} = \frac{9}{2} \ .$$

The following Comparison Test enables us to determine the convergence or divergence of an improper integral of a new positive function by comparing the new function with functions whose improper integrals we already know converge or diverge.

Comparison Test for Integrals of Positive Functions

(a) If the new integral is smaller than one we know converges, then the new integral converges (Fig. 8):

$$\text{if } 0 \leq f(x) \leq g(x) \text{ and } \int_a^\infty g(x) \; dx \text{ converges, then } \int_a^\infty f(x) \; dx \text{ converges.}$$

(b) If the new integral is larger than one which diverges, then the new integral diverges (Fig. 9):

$$\text{if } f(x) \geq g(x) \geq 0 \text{ and } \int_a^\infty g(x) \; dx \text{ diverges, then } \int_a^\infty f(x) \; dx \text{ diverges.}$$

(c) If the new integral is larger than a convergent integral or smaller than a divergent integral,
then we can draw no immediate conclusion about the new integral — the new integral
may converge or diverge (Fig. 10).

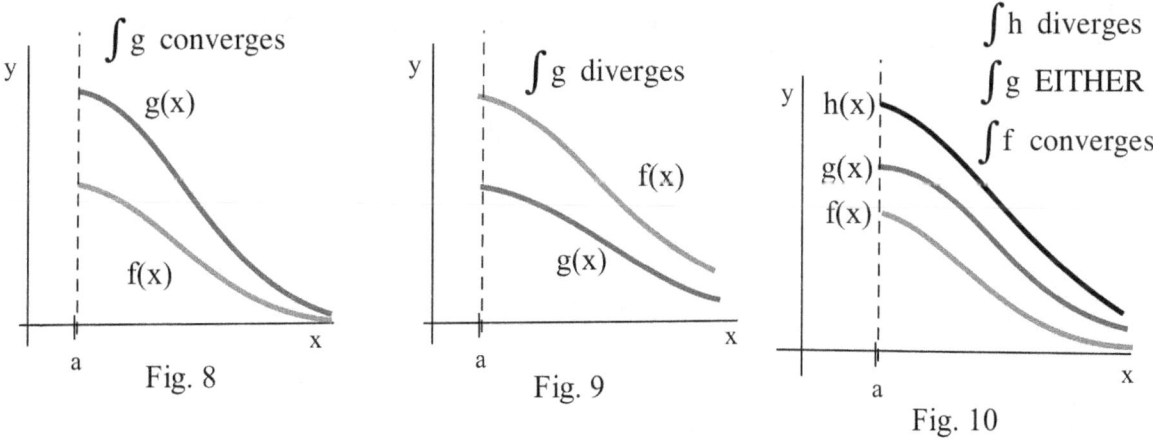

The proof is a straightforward application of the definition of the value of an improper integral and facts about limits.

Example 6:   Determine whether each of these integrals is convergent by comparing it with an appropriate
integral which you know converges or diverges.

$$\text{(a) } \int_1^\infty \frac{7}{x^3+5} \; dx \qquad \text{(b) } \int_1^\infty \frac{3+\sin(x)}{x^2} \; dx \qquad \text{(c) } \int_6^\infty \frac{9}{\sqrt{x-5}} \; dx$$

Solution: (a) The dominant power of $\dfrac{7}{x^3 + 5}$ is $\dfrac{1}{x^3}$ so we should compare with $\dfrac{1}{x^3}$.

$$\int_1^\infty \frac{7}{x^3 + 5}\ dx < \int_1^\infty \frac{7}{x^3}\ dx = 7\int_1^\infty \frac{1}{x^3}\ dx \ . \ \text{We know} \ \int_1^\infty \frac{1}{x^3}\ dx \ \text{is convergent}$$

by the P–Test $(p = 3 > 1)$, so we can conclude that $\displaystyle\int_1^\infty \frac{7}{x^3 + 5}\ dx$ is convergent.

(b) We know $0 < 3 + \sin(x) \le 4$ and the dominant term is $\dfrac{1}{x^2}$ so we should compare with $\dfrac{1}{x^2}$.

$$\int_1^\infty \frac{3 + \sin(x)}{x^2}\ dx \le \int_1^\infty \frac{4}{x^2}\ dx = 4\int_1^\infty \frac{1}{x^2}\ dx \ . \ \text{We know} \ \int_1^\infty \frac{1}{x^2}\ dx \ \text{is convergent}$$

by the P–Test $(p = 2 > 1)$, so we can conclude that $\displaystyle\int_1^\infty \frac{3 + \sin(x)}{x^2}\ dx$ is convergent.

(c) The dominant power of $\dfrac{9}{\sqrt{x - 5}}$ is $\dfrac{1}{\sqrt{x}}$ so we should compare with $\dfrac{1}{x^{1/2}}$.

$$\int_6^\infty \frac{9}{\sqrt{x - 5}}\ dx > \int_1^\infty \frac{9}{\sqrt{x}}\ dx = 9 \int_1^\infty \frac{1}{x^{1/2}}\ dx \ . \ \text{We know} \ \int_1^\infty \frac{1}{x^{1/2}}\ dx \ \text{is divergent}$$

by the P–Test $(p = 1/2 < 1)$, so we can conclude that $\displaystyle\int_6^\infty \frac{9}{\sqrt{x - 5}}\ dx$ is divergent.

PROBLEMS

In 1–21, use the definition of an improper integral to evaluate the given integral.

1. $\displaystyle\int_{10}^\infty \frac{1}{x^3}\ dx$

2. $\displaystyle\int_e^\infty \frac{5}{x \cdot \ln(x)^2}\ dx$

3. $\displaystyle\int_3^\infty \frac{2}{1 + x^2}\ dx$

4. $\displaystyle\int_1^\infty \frac{2}{e^x}\ dx$

5. $\displaystyle\int_e^\infty \frac{5}{x \cdot \ln(x)}\ dx$

6. $\displaystyle\int_0^\infty \frac{x}{1 + x^2}\ dx$

7. $\displaystyle\int_3^\infty \frac{1}{x - 2}\ dx$

8. $\displaystyle\int_3^\infty \frac{1}{(x - 2)^2}\ dx$

9. $\displaystyle\int_3^\infty \frac{1}{(x - 2)^3}\ dx$

10. $\int\limits_{3}^{\infty} \dfrac{1}{x+2} \, dx$

11. $\int\limits_{3}^{\infty} \dfrac{1}{(x+2)^2} \, dx$

12. $\int\limits_{3}^{\infty} \dfrac{1}{(x+2)^3} \, dx$

13. $\int\limits_{0}^{4} \dfrac{1}{\sqrt{x}} \, dx$

14. $\int\limits_{0}^{8} \dfrac{1}{\sqrt[3]{x}} \, dx$

15. $\int\limits_{0}^{16} \dfrac{1}{\sqrt[4]{x}} \, dx$

16. $\int\limits_{0}^{2} \dfrac{1}{\sqrt{2-x}} \, dx$

17. $\int\limits_{0}^{2} \dfrac{1}{\sqrt{4-x^2}} \, dx$

18. $\int\limits_{0}^{2} \dfrac{3x^2}{\sqrt{8-x^3}} \, dx$

19. $\int\limits_{-2}^{\infty} \sin(x) \, dx$

20. $\int\limits_{\pi}^{\infty} \sin(x) \, dx$

21. $\int\limits_{0}^{\pi/2} \tan(x) \, dx$

22. Example 3(b) showed that $\int\limits_{1}^{C} \dfrac{1}{x} \, dx$ grew arbitrarily large as C grew arbitrarily large, so no finite

amount of paint would cover the area bounded between the x–axis and the graph of f(x) = 1/x for x >
1 (Fig. 11a). Show that the volume obtained when the area in Fig. 11a is revolved about the
x–axis (Fig. 11b) is finite so the 3–dimensional trumpet–shaped region can be filled with a finite
amount of paint. Is there a contradiction here?

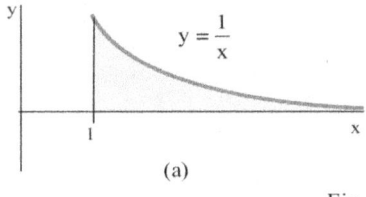 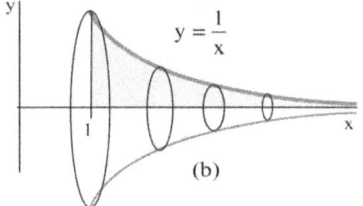

Fig. 11

23. In the **Lifting a Payload** discussion in Section 5.3, we determined that the amount of work needed to
lift a payload from the surface of a moon to an altitude of A above the moon's surface was

$$\text{work} = \int\limits_{R}^{R+A} \dfrac{R^2 P}{x^2} \, dx.$$

(a) Calculate the amount of work required to lift the payload to an altitude of R miles and 2R miles.

(b) Calculate the amount of work needed to lift the payload arbitrarily high. (Calculate the limit of
the work integral as "A → ∞.")

In problems 24–32, determine whether the improper integral is convergent or divergent. Do not evaluate
the integral.

24. $\int\limits_{3}^{\infty} \dfrac{7}{x^2+5} \, dx$

25. $\int\limits_{3}^{\infty} \dfrac{1}{x^3+x} \, dx$

26. $\int\limits_{7}^{\infty} \dfrac{1}{x-2} \, dx$

27. $\displaystyle\int_{3}^{\infty} \frac{7}{x + \ln(x)}\, dx$ 28. $\displaystyle\int_{3}^{\infty} \frac{1}{x^2 - 1}\, dx$ 29. $\displaystyle\int_{7}^{\infty} \frac{1 + \cos(x)}{x^2}\, dx$

30. The volume obtained when the area between the x–axis for $x \geq 1$ and the graph of $f(x) = \dfrac{\sin(x)}{x}$ (Fig. 12) is revolved about the x–axis.

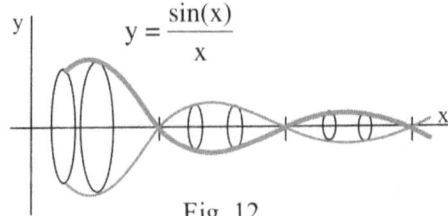

Fig. 12

31. (a) The volume obtained when the area between the positive x–axis ($x \geq 0$) and the graph of $f(x) = \dfrac{1}{x^2 + 1}$ (Fig. 13a) is revolved about the x–axis.

    (b) The volume obtained when the area between the positive x–axis ($x \geq 0$) and the graph of $f(x) = \dfrac{1}{x^2 + 1}$ (Fig. 13b) is revolved about the **y–axis**. (Use the method of "tubes" from section 5.5.)

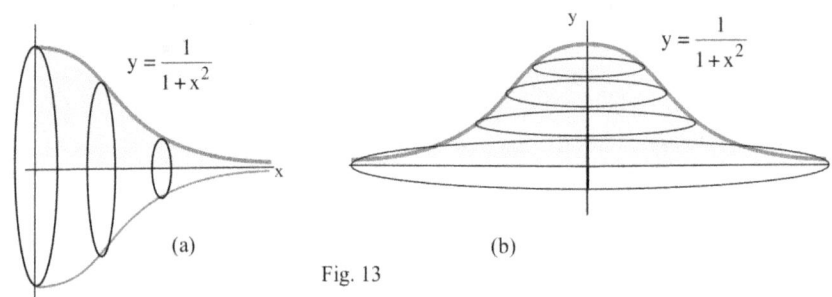

(a)   (b)

Fig. 13

32. (a) The volume obtained when the area between the positive x–axis ($x \geq 0$) and the graph of $f(x) = \dfrac{1}{e^x}$ is revolved about the x–axis.

    (b) The volume obtained when the area between the positive x–axis ($x \geq 0$) and the graph of $f(x) = \dfrac{1}{e^x}$ is revolved about the **y–axis**. (Use the method of "tubes" from section 5.5.

33. (a) Use Fig. 14a to help determine which is larger: $\displaystyle\int_{1}^{A} \frac{1}{x}\, dx$ or $\displaystyle\sum_{k=1}^{A-1} \frac{1}{k}$.

    (b) Use Fig. 14b to help determine which is larger: $\displaystyle\int_{1}^{A} \frac{1}{x}\, dx$ or $\displaystyle\sum_{k=2}^{A} \frac{1}{k}$.

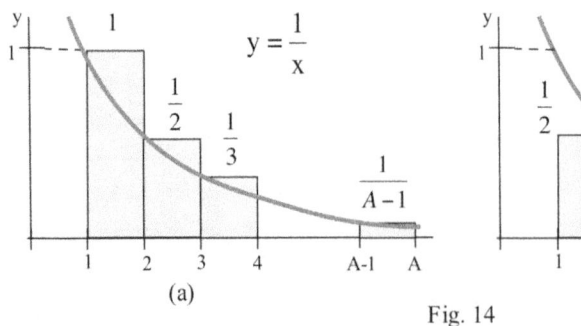 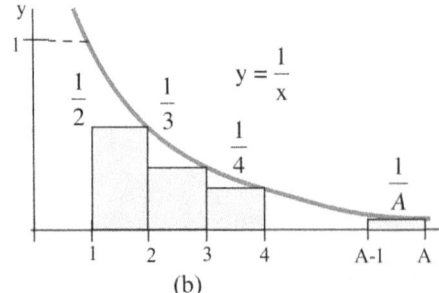

(a)   (b)

Fig. 14

34. (a) Use Fig. 15a to help determine which is larger: $\displaystyle\int_1^A \frac{1}{x^2}\,dx$ or $\displaystyle\sum_{k=1}^{A-1} \frac{1}{k^2}$.

(b) Use Fig. 15b to help determine which is larger: $\displaystyle\int_1^A \frac{1}{x^2}\,dx$ or $\displaystyle\sum_{k=2}^{A} \frac{1}{k^2}$.

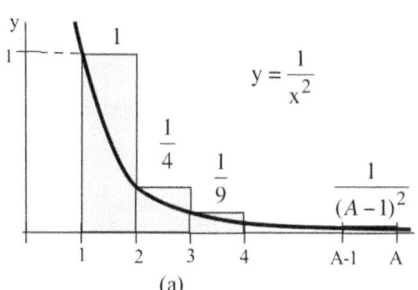

Fig. 15

Section 8.1   Practice Answers

Practice 1:   (a) $\displaystyle V = \int_4^\infty \pi\left(\frac{1}{x}\right)^2 dx$ (b) $\displaystyle A = \int_1^3 g(x) - f(x)\,dx$

Practice 2:   (a) $\displaystyle\int_1^\infty \frac{1}{x^3}\,dx = \lim_{A\to\infty}\int_1^A \frac{1}{x^3}\,dx = \lim_{A\to\infty}\left\{-\frac{1}{2x^2}\right\}\Big|_1^A = \lim_{A\to\infty}\frac{1}{2}-\frac{1}{2A^2} = \frac{1}{2}$.

(b) $\displaystyle\int_0^\infty \sin(x)\,dx = \lim_{A\to\infty}\int_1^A \sin(x)\,dx = \lim_{A\to\infty}\{-\cos(x)\}\Big|_0^A = \lim_{A\to\infty}\{1-\cos(A)\}$ DNE.

$\displaystyle\int_0^\infty \sin(x)\,dx$ is DIVERGENT (or DIVERGES).

Practice 3:   (a) $\displaystyle\int_1^{10} \frac{1}{\sqrt{10-x}}\,dx = \lim_{A\to10^-}\int_1^A \frac{1}{\sqrt{10-x}}\,dx = \lim_{A\to10^-}\left\{-2(10-x)^{1/2}\right\}\Big|_1^A$

$\displaystyle = \lim_{A\to10^-}\left\{-2(10-A)^{1/2} + 2(10-1)^{1/2}\right\} = \lim_{A\to10^-}\left\{6 + 2(10-A)^{1/2}\right\} = 6.$

(b) $\displaystyle\int_0^1 \frac{1}{x}\,dx = \lim_{B\to0^+}\int_B^1 \frac{1}{x}\,dx = \lim_{B\to0^+}\{\ln(x)\}\Big|_B^1 = \lim_{B\to0^+}\{\ln(1)-\ln(B)\} = -(-\infty) = \infty.$

$\displaystyle\int_0^1 \frac{1}{x}\,dx$ is DIVERGENT (or DIVERGES).

8.2   FINDING ANTIDERIVATIVES: A REVIEW

Success at integration is primarily a matter of recognizing standard patterns and being able to manipulate functions
into the form of these recognizable patterns. Integral tables such as the short one given below and the longer one
inside the covers of this book give antiderivatives for patterns of functions and their derivatives, but often it is
necessary to change the variable in order to see the pattern. For most people, skill with these patterns comes with
practice, and this section provides a variety of problems to review and hone your skills.

Table of Common Integral Formulas

Constant $\qquad\qquad \int k\, du = ku + C$

Powers $\qquad\qquad \int u^n\, du = \dfrac{u^{n+1}}{n+1} + C \quad (n \ne -1) \qquad\qquad \int u^{-1}\, du = \int \dfrac{1}{u}\, du = \ln|u| + C$

Trigonometric $\qquad \int \sin(u)\, du = -\cos(u) + C \qquad\qquad \int \cos(u)\, du = \sin(u) + C$

$\qquad\qquad\qquad \int \tan(u)\, du = -\ln|\cos(u)| + C \qquad\qquad \int \cot(u)\, du = \ln|\sin(u)| + C$

$\qquad\qquad\qquad \int \sec(u)\, du = \ln|\sec(u) + \tan(u)| + C \qquad \int \csc(u)\, du = -\ln|\csc(u) + \cot(u)| + C$

Exponential $\qquad\qquad \int e^u\, du = e^u + C \qquad\qquad\qquad \int a^u\, du = \dfrac{a^u}{\ln(a)} + C$

Inverse Trigonometric $\qquad \int \dfrac{1}{a^2 + u^2}\, du = \dfrac{1}{a}\arctan\!\left(\dfrac{u}{a}\right) + C$

$\qquad\qquad\qquad\qquad \int \dfrac{1}{\sqrt{a^2 - u^2}}\, du = \arcsin\!\left(\dfrac{u}{a}\right) + C$

$\qquad\qquad\qquad\qquad \int \dfrac{1}{|u|\sqrt{u^2 - a^2}}\, du = \dfrac{1}{a}\operatorname{arcsec}\!\left(\dfrac{u}{a}\right) + C$

The most generally useful and powerful integration technique, Changing the Variable, was introduced in
Chapter 4, and has been used extensively since then. The first problems at the end of this section provide
additional practice changing variables to calculate integrals. As we develop more complicated and more
specialized techniques for finding antiderivatives, your first thought should still be whether the integral can
be simplified by changing the variable.

A Particular Situation: An Irreducible Quadratic Denominator

Sometimes the appropriate change of variable is not obvious, and we may need to manipulate the function first.

Example 1:   Evaluate $\int \dfrac{18}{x^2 - 6x + 10}\ dx$.

Solution: When the denominator of a rational function is an irreducible quadratic $(ax^2 + bx + c$ with $b^2 - 4ac < 0)$, we can write it as the sum of two squares by using the algebraic technique of "completing the square:" $x^2 - 6x + 10 = (x - 3)^2 + 1$. Then

$$\int \dfrac{18}{x^2 - 6x + 10}\ dx = \int \dfrac{18}{(x - 3)^2 + 1}\ dx \quad \text{and the change of variable}\ u = x - 3\ \text{makes the}$$

arctangent pattern more evident. If $u = x - 3$, then $du = dx$ and

$$\int \dfrac{18}{x^2 - 6x + 10}\ dx = \int \dfrac{18}{u^2 + 1}\ du = 18 \arctan(\ u\ ) + C = 18 \arctan(\ x - 3\ ) + C.$$

Completing The Square: $x^2 + bx + c = (\ x + \dfrac{b}{2}\ )^2 + (\ c - \dfrac{b^2}{4}\ )$.

Practice 1:   Evaluate $\int \dfrac{5}{x^2 + 8x + 25}\ dx$ by completing the square and changing the variable.

Example 2:   Evaluate $\int \dfrac{6x}{x^2 - 6x + 10}\ dx$.

Solution: This would be an easy problem if the numerator was $6x - 18$. Then the numerator would be three times the derivative of the denominator and the pattern of the integral would be $\int \dfrac{3}{u}\ du$ with $u = x^2 - 6x + 10$. Fortunately, we can rewrite the numerator as $6x - 18 + 18$. Then

$$\int \dfrac{6x}{x^2 - 6x + 10}\ dx = \int \dfrac{6x - 18 + 18}{x^2 - 6x + 10}\ dx$$

$$= \int \dfrac{6x - 18}{x^2 - 6x + 10}\ dx + \int \dfrac{18}{x^2 - 6x + 10}\ dx$$

$$= \int \dfrac{3}{u}\ du + \int \dfrac{18}{w^2 + 1}\ dw \quad \text{where}\ u = x^2 - 6x + 10\ \text{and}\ w = x - 3$$

$$= 3 \cdot \ln|\ u\ | + 18 \cdot \arctan(\ w\ ) + C = 3 \cdot \ln(\ x^2 - 6x + 10\ ) + 18 \cdot \arctan(\ x - 3\ ) + C.$$

Practice 2:   Evaluate $\int \dfrac{4x+21}{x^2+8x+25}\,dx$. (Hint: if $u = x^2+8x+25$, then $2\,du = (4x+16)\,dx$.)

The "logarithm plus an arctangent" pattern of these answers is quite typical for integrals of linear functions divided by irreducible quadratics. If the quadratic denominator can be factored into a product of two linear factors, we use the technique discussed in Section 8.3, Partial Fraction Decomposition.

Problems

In problems 1 – 42, find the indefinite integrals and evaluate the definite integrals. In some of the problems a particular change of variable is suggested.

1. $\int 6x(x^2+7)^2\,dx$ $u = x^2+7$ 2. $\int 6x(x^2-1)^3\,dx$ $u = x^2-1$

3. $\int_2^4 \dfrac{6x}{\sqrt{x^2-3}}\,dx$ $u = x^2-3$ 4. $\int_0^\pi 12\cos(x)\cdot\{\sin(x)+2\}^2\,dx$ $u = \sin(x)+2$

5. $\int \dfrac{12x}{x^2+3}\,dx$ 6. $\int \dfrac{\cos(x)}{2+\sin(x)}\,dx$

7. $\int \sin(3x+2)\,dx$ 8. $\int \cos(x/5)\,dx$

9. $\int_0^1 e^x\sec^2(e^x+3)\,dx$ $u = e^x+3$ 10. $\int_0^{\pi/2} \cos(x)\sqrt{1+\sin(x)}\,dx$ $u = 1+\sin(x)$

11. $\int \dfrac{\ln(x)}{x}\,dx$ $u = \ln(x)$ 12. $\int \dfrac{\cos(\sqrt{x})}{\sqrt{x}}\,dx$ $u = \sqrt{x}$

13. $\int \cos(x)\cdot e^{\sin(x)}\,dx$ 14. $\int e^x\sin(e^x)\,dx$

15. $\int_1^3 \dfrac{5}{1+9x^2}\,dx$ $u = 3x$ 16. $\int_0^1 \dfrac{7}{1+(x+3)^2}\,dx$ $u = x+3$

17. $\int_1^2 \dfrac{1}{x^2}\cos(\tfrac{1}{x})\,dx$ $u = \dfrac{1}{x}$ 18. $\int_1^e \dfrac{\sec(2+\ln(x))}{x}\,dx$ $u = 2+\ln(x)$

19. $\int \dfrac{6\sin(x)\cos(x)}{5+\sin^2(x)}\,dx$ $u = 5+\sin^2(x)$ 20. $\int \dfrac{6\cos(x)}{5+\sin^2(x)}\,dx$ $u = \sin(x)$

21. $\int \dfrac{10}{2x+5}\,dx$ 22. $\int \dfrac{3}{8x+1}\,dx$

23. $\displaystyle\int_{1}^{3} \frac{20x}{5x^2 + 3}\ dx$

24. $\displaystyle\int_{1}^{5} \frac{4x}{x^2 + 9}\ dx$

25. $\displaystyle\int_{0}^{1} \frac{7}{(x+3)^2 + 4}\ dx \qquad u = x + 3$

26. $\displaystyle\int_{-2.1}^{-2.3} \frac{1}{\sqrt{1 - (x+2)^2}}\ dx \qquad u = x + 2$

27. $\displaystyle\int \frac{e^x}{1 + e^{2x}}\ dx$

28. $\displaystyle\int \frac{4x + 10}{x^2 + 5x + 9}\ dx \quad u = x^2 + 5x + 9$

29. $\displaystyle\int_{1}^{e} \frac{3}{x\cdot\{\,1 + \ln(x)\,\}}\ dx$

30. $\displaystyle\int_{0}^{1} \frac{e^x}{1 + e^x}\ dx$

31. $\displaystyle\int_{0}^{1} 2x\sqrt{1 - x^2}\ dx$

32. $\displaystyle\int_{0}^{3} \frac{2x}{\sqrt{5 + x^2}}\ dx$

33. $\displaystyle\int \cos(x)\,(1 + \sin(x)\,)^3\ dx$

34. $\displaystyle\int \cos(x)\,\sin^4(x)\ dx$

35. $\displaystyle\int_{1}^{e} \frac{\sqrt{\ln(x)}}{x}\ dx$

36. $\displaystyle\int_{1}^{2} e^x\sqrt{2 + e^x}\ dx$

37. $\displaystyle\int \frac{\sec^2(x)}{5 + \tan(x)}\ dx$

38. $\displaystyle\int \frac{6x}{(x^2 - 1)^3}\ dx$

39. $\displaystyle\int \tan(x - 5)\ dx$

40. $\displaystyle\int (x^3 + 3)^2\ dx$

41. $\displaystyle\int_{0}^{1} e^{5x}\ dx$

42. $\displaystyle\int \sec(2 + 3x)\ dx$

In problems 43 – 48, complete the square in the denominator, make the appropriate substitution, and integrate.

43. $\displaystyle\int \frac{7}{x^2 + 4x + 5}\ dx$

44. $\displaystyle\int \frac{3}{x^2 + 4x + 29}\ dx$

45. $\displaystyle\int \frac{2}{x^2 - 6x + 58}\ dx$

46. $\displaystyle\int \frac{11}{x^2 - 2x + 10}\ dx$

47. $\displaystyle\int \frac{3}{x^2 + 10x + 29}\ dx$

48. $\displaystyle\int \frac{5}{x^2 + 2x + 5}\ dx$

In problems 49 – 54, evaluate the first integral as a sum of two integrals.

49. $\displaystyle\int \frac{2x + 11}{x^2 + 4x + 5}\ dx = \int \frac{2x + 4}{x^2 + 4x + 5}\ dx + \int \frac{7}{x^2 + 4x + 5}\ dx$

50. $\displaystyle\int \frac{4x + 11}{x^2 + 4x + 5}\ dx = \int \frac{4x + 8}{x^2 + 4x + 5}\ dx + \int \frac{3}{x^2 + 4x + 5}\ dx$

51. $\displaystyle\int \frac{4x + 7}{x^2 - 6x + 10}\ dx = \int \frac{4x - 12}{x^2 - 6x + 10}\ dx + \int \frac{19}{x^2 - 6x + 10}\ dx$

52. $\int \dfrac{6x + 28}{x^2 + 10x + 34} \; dx$

53. $\int \dfrac{6x + 5}{x^2 - 4x + 13} \; dx$

54. $\int \dfrac{4x + 9}{x^2 + 6x + 13} \; dx$

Section 8.2   **PRACTICE Answers**

Practice 1: $\int \dfrac{5}{x^2 + 8x + 25} \; dx = \int \dfrac{5}{(x+4)^2 + 9} \; dx$ (Put $u = x + 4$, $du = dx$)

$= \int \dfrac{5}{(u)^2 + 9} \; du = \dfrac{5}{3} \arctan(\dfrac{u}{3}) + C = \dfrac{5}{3} \arctan(\dfrac{x+4}{3}) + C$.

Practice 2: $\int \dfrac{4x + 21}{x^2 + 8x + 25} \; dx = \int \dfrac{4x + 16}{x^2 + 8x + 25} \; dx + \int \dfrac{5}{x^2 + 8x + 25} \; dx = (a) + (b)$

(a) Put $u = x^2 + 8x + 25$. Then $du = (2x + 8) \; dx$ so $2 \; du = (4x + 16) \; dx$.

$\int \dfrac{4x + 16}{x^2 + 8x + 25} \; dx = \int \dfrac{2}{u} \; du = 2 \cdot \ln| \, u \, | + C = 2 \cdot \ln(\, x^2 + 8x + 25 \,) + C$

(b) $\int \dfrac{5}{x^2 + 8x + 25} \; dx = \dfrac{5}{3} \arctan(\dfrac{x+4}{3}) + C$. (see Practice 1)

Therefore, $\int \dfrac{4x + 21}{x^2 + 8x + 25} \; dx = \mathbf{2 \cdot ln(\, x^2 + 8x + 25 \,) + \dfrac{5}{3} \, arctan(\dfrac{x+4}{3}) + C}$.

8.3   INTEGRATION BY PARTS

Integration by parts is an integration method which enables us to find antiderivatives of some new functions such as ln(x) and arctan(x) as well as antiderivatives of products of functions such as $x^2 \cdot \ln(x)$ and $e^x \cdot \sin(x)$. It is the method used to derive many of the general integral formulas in the Table of Integrals. The Integration By Parts Formula for integrals comes from the Product Rule for derivatives.

For functions u = u(x) and v = v(x), the Product Rule for derivatives is

$$\frac{d(\,uv\,)}{dx} = u\,\frac{dv}{dx} + v\,\frac{du}{dx} \quad \text{or, in the form using differentials, } d(\,uv\,) = u\,dv + v\,du \;.$$

Algebraically solving for u dv , we have u dv = d(uv) – v du which can then be integrated to give

$$\int u\,dv = \int d(\,uv\,) - \int v\,du = uv - \int v\,du\;.$$

This last formula is called the Integration By Parts Formula, and it enables us to find antiderivatives for many functions which we have not been able to integrate using the substitution method. In practice, the Integration By Parts Formula allows us to exchange the problem of finding one integral, $\int u\,dv$, for the problem of finding a different integral, $\int v\,du$. This trade of one integral for another may not look very useful, but we can often arrange the exchange so we trade a difficult integral for a much easier one.

INTEGRATION BY PARTS FORMULA

If   u, v, u', and v' are continuous functions,

then $\int u\,dv = u\cdot v - \int v\,du$.

For definite integrals, the Integration By Parts Formula is

$$\int_a^b u\,dv = u\cdot v \Big|_a^b - \int_a^b v\,du = \left\{ u(b)\cdot v(b) - u(a)\cdot v(a) \right\} - \int_a^b v\,du\;.$$

Example 1:   Use Integration By Parts to evaluate $\int x\cdot\cos(x)\,dx$

and $\int_0^{\pi} x\cdot\cos(x)\,dx$. (Fig. 1)

Solution: Our first step is to write this integral in the form of the

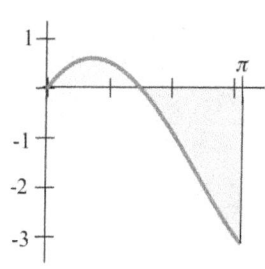

Fig. 1: $y = x \cdot \cos(x)$

Integration By Parts Formula, $\int \mathbf{u} \, dv$. If we put $\mathbf{u} = x$, then we **must** have $dv = \cos(x) \, dx$ so that

u dv completely represents the integrand x•cos(x) . In order to use integration by parts, we also need

to calculate du and v:

Since $\mathbf{u} = x$ and $dv = \cos(x) \, dx$

then $du = dx$ and $v = \sin(x)$.

Putting these pieces into the Integration By Parts Formula, we have

$$\int x\cdot\cos(x) \, dx \;=\; x\cdot\sin(x) \;-\; \int \sin(x) \, dx \;=\; x\cdot\sin(x) + \cos(x) + C \;.$$

(To check this result, differentiate x•sin(x) + cos(x) to verify that its derivative is x•cos(x) .)

$$\int_0^{\pi} x\cdot\cos(x) \, dx \;=\; x\cdot\sin(x) + \cos(x) \Big|_0^{\pi} \;=\; \{\pi\cdot\sin(\pi) + \cos(\pi)\} - \{0\cdot\sin(0) + \cos(0)\} = -1 - 1 = -2 \;.$$

The Integration By Parts formula allowed us to exchange the problem of evaluating $\int x\cdot\cos(x) \, dx$

for the much easier problem of evaluating $\int \sin(x) \, dx$.

Practice 1:   Use the Integration By Parts Formula on $\int x\cdot\cos(x) \, dx$ with the choice $u = \cos(x)$

and $dv = x \, dx$. Why does this lead to a poor exchange?

Example 2:   Use integration by parts to evaluate $\int x\cdot e^{3x} \, dx$ and $\int_0^1 x\cdot e^{3x} \, dx$. (Let $u = x$.)

Solution: Let $u = x$. Then $dv = e^{3x} \, dx$ so $du = dx$ and $v = \frac{1}{3} e^{3x}$.

Using the Integration By Parts Formula, we get

$$\int x\cdot e^{3x} \, dx \;=\; x\cdot\frac{1}{3} \cdot e^{3x} \;-\; \int \frac{1}{3} \cdot e^{3x} \, dx \;=\; \frac{x}{3} \cdot e^{3x} - \frac{1}{9} \cdot e^{3x} \;+ C \;.$$

$$\int_0^1 x\cdot e^{3x} \, dx \;=\; \frac{x}{3} \cdot e^{3x} - \frac{1}{9} \cdot e^{3x} \Big|_0^1 \;=\; \left\{ \frac{1}{3} \cdot e^3 - \frac{1}{9} \cdot e^3 \right\} - \left\{ 0 - \frac{1}{9} \right\} = \frac{2}{9} \cdot e^3 + \frac{1}{9} \;.$$

In this Example, it is valid to choose $u = e^{3x}$ and $dv = x \, dx$, but that choice results in an

integral that is more difficult than the original one. If we put $u = e^{3x}$ and $dv = x \, dx$, then

$du = 3e^{3x} \, dx$ and $v = \frac{x^2}{2}$, and the Integration By Parts Formula gives

$$\int x \cdot e^{3x}\, dx = e^{3x} \cdot \frac{x^2}{2} - \int \frac{x^2}{2}\, 3e^{3x}\, dx\ .$$

We end up exchanging the integral $\int x \cdot e^{3x}\, dx$ for the more difficult integral $\int \frac{x^2}{2}\, 3e^{3x}\, dx$.

Practice 2:   Evaluate $\int x \cdot \sin(x)\, dx$ and $\int x \cdot e^{5x}\, dx$. (In each integral, let $u = x$.)

Once we have chosen u and dv to represent the integrand as $u\, dv$, we need to calculate du and v. The "du" calculation is usually easy, but finding v from dv can be difficult for some choices of dv. In practice, you need to select u so dv is a simple enough part of the integrand so you can find v, the antiderivative of dv.

Example 3:   Evaluate $\int 2x \cdot \ln(x)\, dx$.

Solution: The choice $u = 2x$ seems fine until we go a little further with the process. If $u = 2x$, then $dv = \ln(x)\, dx$ and we need to find du and v. Finding $du = 2\, dx$ is simple, but then we have the difficult problem of finding an antiderivative v for our choice $dv = \ln(x)\, dx$.

In this Example, the choice $u = \ln(x)$ results in easier calculations.

Let $u = \ln(x)$. Then $dv = 2x\, dx$

so $du = \frac{1}{x}\, dx$ and $v = x^2$.

Then the Integration By Parts Formula gives

$$\int 2x \cdot \ln(x)\, dx \quad = \ln(x)\, x^2 - \int x^2 \cdot \frac{1}{x}\, dx$$

$$= x^2 \cdot \ln(x) - \int x\, dx \quad = \quad x^2 \cdot \ln(x) - \frac{x^2}{2} + C\ .$$

If you can not find a v for your original choice of dv, try a different choice for u and dv.

Integration by parts also enables us to evaluate the integrals of the inverse trigonometric functions and of the logarithm.

Example 4:   Evaluate $\int \arctan(x)\, dx$.

Solution: Let $u = \arctan(x)$. Then $dv = dx$

so $du = \frac{1}{1 + x^2}\, dx$ and $v = x$.

Then $\int \arctan(x)\,dx = x\arctan(x) - \int x{\cdot}\dfrac{1}{1+x^2}\,dx$. We can evaluate the new integral

$\int x{\cdot}\dfrac{1}{1+x^2}\,dx$ by changing the variable using $w = 1 + x^2$. Then $dw = 2x\,dx$, so

$\int x{\cdot}\dfrac{1}{1+x^2}\,dx = \int \dfrac{1}{2}\dfrac{1}{w}\,dw = \dfrac{1}{2}\,\ln|\,w\,| = \dfrac{1}{2}\,\ln|\,1+x^2\,|$. Putting this all together,

$$\int \arctan(x)\,dx = x\arctan(x) - \dfrac{1}{2}\,\ln|\,1+x^2\,| + C .$$

Practice 3:   Evaluate $\int \ln(x)\,dx$ and $\displaystyle\int_{1}^{e} \ln(x)\,dx$.

Notes: 1.   Once u is chosen, then dv is completely determined: dv = rest of the integrand.

2.   Since we need to find an antiderivative of dv to get v, pick u and dv so an antiderivative v can be found for the chosen dv.

3.   The Integration By Parts Formula allows us to trade one integral for another one.

   (a) If the new integral is more difficult than the original integral, then we have made a poor choice of u and dv . Try a different choice for u and dv or try a different technique.

   (b) To evaluate the new integral $\int v\,du$ we may need to use substitution, integration by

parts again, or some other technique such as the ones discussed later in this chapter.

More General Uses of Integration By Parts

The Integration By Parts Formula is also used to derive many of the entries in the Table of Integrals. For some integrands such as $x^n{\cdot}\ln(x)$, the result is simply a function, an antiderivative of the integrand. For some integrands such as $\sin^n(x)$, the result is a **reduction formula**, a formula which still contains an integral, but the new integrand is the sine function raised to a smaller power, $\sin^{n-2}(x)$. By repeatedly applying the reduction formula, we can evaluate the integral of sine raised to any positive integer power.

General Patterns

Example 5:   Evaluate $\int x^n{\cdot}\ln(x)\,dx$ for $n \ne -1$.

Solution: Let $u = \ln(x)$. Then $dv = x^n\,dx$

so $du = \dfrac{1}{x}\,dx$ and $v = \dfrac{1}{n+1}\,x^{n+1}$.

Then $\int x^n \ln(x)\ dx = \frac{1}{n+1} \cdot x^{n+1} \cdot \ln(x) - \int \frac{1}{n+1} \cdot x^{n+1} \cdot \frac{1}{x}\ dx$.

But $\int \frac{1}{n+1} \cdot x^{n+1} \cdot \frac{1}{x}\ dx = \frac{1}{n+1} \int x^n\ dx = \frac{1}{n+1} \cdot \frac{1}{n+1} \cdot x^{n+1} = \frac{x^{n+1}}{(n+1)^2}$, so

$\int x^n \cdot \ln(x)\ dx = \frac{x^{n+1}}{n+1} \cdot \ln(x) - \frac{x^{n+1}}{(n+1)^2} + C = \frac{x^{n+1}}{n+1} \left\{ \ln(x) - \frac{1}{n+1} \right\} + C$ for $n \ne -1$.

Practice 4:   Use the **result** of Example 5 to evaluate $\int x^2 \cdot \ln(x)\ dx$ and $\int \ln(x)\ dx$.

Reduction Formulas

Sometimes the general pattern still contains an integral, but a simpler one with a smaller exponent. In that case we can reuse the reduction pattern until the resulting integral is simple enough to integrate completely.

Example 6:   Evaluate $\int x^n e^x\ dx$ and use the result to evaluate $\int x^2 e^x\ dx$.

Solution: Put $u = x^n$. Then $dv = e^x\ dx$,

so $du = n\ x^{n-1}\ dx$ and $v = e^x$.

The Integration By Parts Formula gives $\int x^n e^x\ dx = x^n e^x - n \int x^{n-1} e^x\ dx$, a reduction

formula since we have reduced the power of x by 1 and have succeeded in trading the integral $\int x^n e^x\ dx$ for the "reduced" integral $\int x^{n-1} e^x\ dx$.

$\int x^2 e^x\ dx$ has the form of the general pattern $\int x^n e^x\ dx$ with $n = 2$, so

$$\int x^2 e^x\ dx = x^2 e^x - 2 \int x^1 e^x\ dx .$$

Using the pattern on $\int x^1 e^x\ dx$ with $n = 1$, we have

$$\int x^2 e^x\ dx = x^2 \cdot e^x - 2 \int x^1 e^x\ dx$$

$$= x^2 \cdot e^x - 2\left\{ x \cdot e^x - \int e^x\ dx \right\}$$

$$= x^2 \cdot e^x - 2x \cdot e^x + 2e^x + C \text{ or } e^x \left\{ x^2 - 2x + 2 \right\} + C .$$

Practice 5:   Derive the reduction formula $\int x^n \cdot \sin(x)\ dx = -x^n \cdot \cos(x) + n \int x^{n-1} \cdot \cos(x)\ dx$.

The Reappearing Integral

Sometimes the integral we are trying to evaluate shows up on both sides of the equation during our calculations in such a way that we can solve for the integral algebraically.

Example 7:   Evaluate $\int e^X \cos(x)\ dx$.

Solution: Let $u = e^X$. Then $dv = \cos(x)\ dx$

so $du = e^X\ dx$ and $v = \sin(x)$.

Then $\int e^X \cos(x)\ dx = e^X \sin(x) - \int e^X \sin(x)\ dx$. The new integral does not look any easier than

the original one, but lets try to evaluate the new integral using integration by parts again.

To evaluate $\int e^X \sin(x)\ dx$. , let $u = e^X$ and $dv = \sin(x)\ dx$. Then $du = e^X\ dx$ and $v = -\cos(x)$ so

$$\int e^X \sin(x)\ dx = -e^X \cos(x) + \int e^X \cos(x)\ dx.$$

Putting this result back into the original problem, we get

$$\int e^X \cos(x)\ dx \qquad = e^X \sin(x) - \int e^X \sin(x)\ dx = e^X \sin(x) - \left\{ -e^X \cos(x) + \int e^X \cos(x)\ dx \right\}$$

$$= e^X \sin(x) + e^X \cos(x) - \int e^X \cos(x)\ dx .$$

The integral of $e^X \cos(x)$ appears on each side of this last equation, and we can algebraically solve for it to get

$$2 \int e^X \cos(x)\ dx = e^X \sin(x) + e^X \cos(x) \text{ , and finally,}$$

$$\int e^X \cos(x)\ dx = \tfrac{1}{2} \left\{ e^X \sin(x) + e^X \cos(x) \right\} + C .$$

Practice 6:   Derive the formula $\int e^X \sin(x)\ dx = \tfrac{1}{2} \left\{ e^X \sin(x) - e^X \cos(x) \right\} + C$.

PROBLEMS

In problems 1–6, a function u or dv is given. Find the piece u or dv which is not given, calculate du and v, and apply the Integration by Parts Formula.

1. $\int 12x \cdot \ln(x)\, dx$ $u = \ln(x)$ 2. $\int x \cdot e^{-x}\, dx$ $u = x$

3. $\int x^4 \ln(x)\, dx$ $dv = x^4\, dx$ 4. $\int x \cdot \sec^2(3x)\, dx$ $dv = \sec^2(3x)\, dx$

5. $\int x \cdot \arctan(x)\, dx$ $dv = x\, dx$ 6. $\int x \cdot (5x + 1)^{19}\, dx$ $u = x$

In problems 7– 24 , evaluate the integrals.

7. $\displaystyle\int_0^1 \frac{x}{e^{3x}}\, dx$ 8. $\displaystyle\int_0^1 10x \cdot e^{3x}\, dx$ 9. $\int x \cdot \sec(x) \cdot \tan(x)\, dx$

10. $\displaystyle\int_0^\pi 5x \cdot \sin(2x)\, dx$ 11. $\displaystyle\int_{\pi/3}^{\pi/2} 7x \cdot \cos(3x)\, dx$ 12. $\int 6x \cdot \sin(x^2 + 1)\, dx$

13. $\int 12x \cdot \cos(3x^2)\, dx$ 14. $\int x^2 \cos(x)\, dx$ 15. $\displaystyle\int_1^3 \ln(2x + 5)\, dx$

16. $\int x^3 \ln(5x)\, dx$ 17. $\displaystyle\int_1^e (\ln(x))^2\, dx$ 18. $\displaystyle\int_1^e \sqrt{x} \cdot \ln(x)\, dx$

19. $\int \arcsin(x)\, dx$ 20. $\int x^2 e^{5x}\, dx$ 21. $\int x \cdot \arctan(3x)\, dx$

22. $\int x \ln(x + 1)\, dx$ 23. $\displaystyle\int_1^2 \frac{\ln(x)}{x}\, dx$ 24. $\displaystyle\int_1^2 \frac{\ln(x)}{x^2}\, dx$

These reduction formulas can all be derived using integration by parts. In problems 25–30 , use them to help evaluate the integrals. (These are entries 19, 20 and 23 in the Table of Integrals with a = 1.)

$$\int \sin^n(x)\, dx = \frac{1}{n}\left\{ -\sin^{n-1}(x) \cdot \cos(x) + (n-1) \int \sin^{n-2}(x)\, dx \right\} + C$$

$$\int \cos^n(x)\, dx = \frac{1}{n}\left\{ \cos^{n-1}(x) \cdot \sin(x) + (n-1) \int \cos^{n-2}(x)\, dx \right\} + C$$

$$\int \sec^n(x)\, dx = \frac{1}{n-1}\left\{ \sec^{n-2}(x) \cdot \tan(x) + (n-2) \int \sec^{n-2}(x)\, dx \right\} + C$$

25. (a) $\int \sin^3(x)\,dx$ (b) $\int \sin^4(x)\,dx$ (c) $\int \sin^5(x)\,dx$

26. (a) $\int \cos^3(x)\,dx$ (b) $\int \cos^4(x)\,dx$ (c) $\int \cos^5(x)\,dx$

27. (a) $\int \sec^3(x)\,dx$ (b) $\int \sec^4(x)\,dx$ (c) $\int \sec^5(x)\,dx$

28. $\int \sin^3(5x - 2)\,dx$ 29. $\int \cos^3(2x + 3)\,dx$ 30. $\int \sec^3(7x - 1)\,dx$

31. $\int x \cdot (2x + 5)^{19}\,dx$ can be evaluated using integration by parts or a change of variable. (a) Evaluate the integral using integration by parts with $u = x$ and $dv = (2x + 5)^{19}\,dx$. (b) Evaluate the integral using change of variable with $u = 2x + 5$. (c) Which method is easier?

32. $\int \dfrac{x}{\sqrt{1 + x}}\,dx$ can be evaluated using integration by parts or using a change of variable. (a) Evaluate the integral using integration by parts with $u = x$ and $dv = \dfrac{1}{\sqrt{1 + x}}\,dx$. (b) Evaluate the integral using change of variable with $u = 1 + x$.

33. (a) Before evaluating the integrals, which do you think is larger, $\int_0^1 x \cdot \sin(x)\,dx$ or $\int_0^1 \sin(x)\,dx$? Why?

(b) Evaluate $\int_0^1 x \cdot \sin(x)\,dx$ and $\int_0^1 \sin(x)\,dx$. Was your prediction in part (a) correct?

34. (a) Before evaluating the integrals, which do you think is larger, $\int_0^\pi x \cdot \sin(x)\,dx$ or $\int_0^\pi \sin(x)\,dx$? Why?

(b) Evaluate $\int_0^\pi x \cdot \sin(x)\,dx$ and $\int_0^\pi \sin(x)\,dx$. Was your prediction in part (a) correct?

35. In Fig. 2, the volume swept out when region A is revolved about the x–axis is $\int_{x=1}^{e} \pi\,(\ln(x))^2\,dx$

(using the disk method), and the volume swept out when region B is revolved about the x–axis is

$\int_{y=0}^{1} 2\pi y \cdot e^y\,dy$ (using the tube method).

(a) Before evaluating the integrals, which volume do you think is larger? Why?

(b) Evaluate the integrals. Was your prediction in part (a) correct?

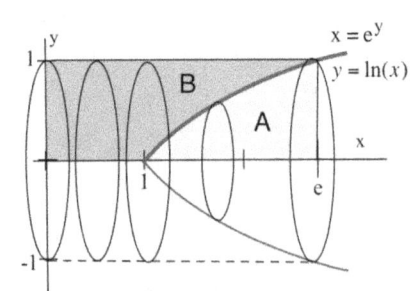

Fig. 2: Comparing volumes

36. Use the tube method to calculate the volume when the region between the x–axis and the graph of
   $y = \sin(x)$ for $0 \le x \le \pi$ is rotated about the y–axis.

37. We derived the Integration by Parts Formula analytically, but the formula also has a geometric interpretation.
   In Fig. 3, let D be the large rectangle formed by the regions A, B, and C so we have the area
   equation

   $\qquad$ (area of C) = (area of D) – (area of A) – (area of B).

   (a) Represent the area of the large rectangle D as
       a function of u_2 and v_2.

   (b) Represent the area of the small rectangle (region A)
       as a function of u_1 and v_1.

   (c) Represent the area of region C as an integral with
       respect to the variable u.

   (d) Represent the area of region B as an integral with
       respect to the variable v.

   (e) Rewrite the area equation using the representations
       in parts (a) – (d). This result should look very familiar.

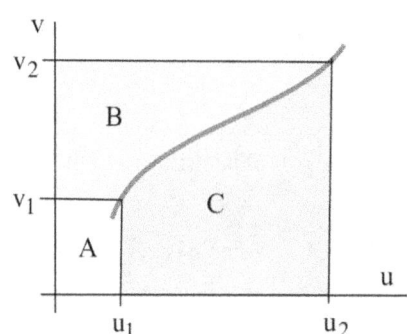

D is the region obtained by putting
together regions A, B and C

Fig. 3: Graphic version of Integration by Parts

38. $\displaystyle\int x\cdot(\ln(x))^2\, dx$

39. $\displaystyle\int x^2\cdot\arctan(x)\, dx$

40. $\displaystyle\int_0^1 e^{-x}\sin(x)\, dx$

41. $\displaystyle\int_0^1 \frac{\cos(x)}{e^x}\, dx$

42. $\displaystyle\int \sin(\ln(x))\, dx$

43. $\displaystyle\int \cos(\ln(x))\, dx$

44. $\displaystyle\int e^{3x}\sin(x)\, dx$

45. $\displaystyle\int e^x\cos(3x)\, dx$

46. Use integration by parts to evaluate $\displaystyle\int \sec^3(x)\, dx$.

47. Derive a reduction formula for $\displaystyle\int x^n e^{ax}\, dx$.

48. Derive a reduction formula for $\displaystyle\int x^n \sin(ax)\, dx$.

49. Derive a reduction formula for $\displaystyle\int x\cdot(\ln(x))^n\, dx$.

50. Suppose f and f ' are continuous and bounded on the interval $[0, 2\pi]$ ($|f(x)| < M$ and $|f '(x)| < M$

for all $0 \le x \le 2\pi$). The nth Fourier Sine Coefficient of f is defined as the value of

$$S_n = \int_0^{2\pi} f(x) \cdot \sin(nx) \, dx \ .$$

(a) Use the Integration by Parts Formula with $u = f(x)$ and $dv = \sin(nx) \, dx$ to represent the formula

for S_n in a different way.

(b) Use the new representation of S_n in part (a) to determine what happens to the values of

S_n when n is very large ($n \to \infty$). (Hint: $| f '(x) \cdot \cos(nx) | \le | f '(x)| \cdot |\cos(nx)| < M \cdot 1 = M$.)

(c) What happens to the values of the nth Fourier Cosine Coefficient $C_n = \int_0^{2\pi} f(x) \cdot \cos(nx) \, dx$

when n is very large.

Section 8.3   PRACTICE Answers

Practice 1: $\int_a^b x \cdot \cos(x) \, dx$. Put $u = \cos(x)$ and $dv = x \, dx$. Then $du = -\sin(x) \, dx$ and $v = \frac{1}{2} x^2$.

$$uv - \int v \, du \ = \ (\cos(x))(\frac{1}{2} x^2) - \int (\frac{1}{2}x^2)(- \sin(x)) \ dx$$

$$= \ \frac{1}{2} x^2 \cos(x) + \frac{1}{2} \int x^2 \sin(x) \, dx \ .$$

The last integral is worse than the original integral.

Practice 2:   (a) $\int_a^b x \cdot \sin(x) \, dx$. Put $u = x$ and $dv = \sin(x) \, dx$. Then $du = dx$ and $v = -\cos(x) \, dx$.

$$uv - \int v \, du \ = \ (x)(-\cos(x)) - \int - \cos(x) \ dx \ = \ - x \cdot \cos(x) + \sin(x) + C \ .$$

(b) $\int_a^b x \cdot e^{5x} \, dx$. Put $u = x$ and $dv = e^{5x} \, dx$. Then $du = dx$ and $v = \frac{1}{5} e^{5x}$.

$$uv - \int v \, du \ = \ (x)(\frac{1}{5} e^{5x}) - \int \frac{1}{5} e^{5x} \ dx \ = \ \frac{1}{5} x \cdot e^{5x} - \frac{1}{25} e^{5x} + C \ .$$

Practice 3: $\int \ln(x) \, dx$. Put $u = \ln(x)$ and $dv = dx$. Then $du = \frac{1}{x} \, dx$ and $v = x$.

$$uv - \int v \, du \ = \ \ln(x) \cdot x - \int x \cdot \frac{1}{x} \ dx \ = \ x \cdot \ln(x) - \int 1 \, dx = x \cdot \ln(x) - x + C \ .$$

$$\int_1^e \ln(x) \, dx = x \cdot \ln(x) - x \Big|_1^e \ = \ \{ e \cdot \ln(e) - e \} - \{ 1 \cdot \ln(1) - 1 \} = \{ e - e \} - \{ 0 - 1 \} = 1 \ .$$

Practice 4:   Example 5: $\int x^n \ln(x)\, dx = \dfrac{x^{n+1}}{n+1} \left\{ \ln(x) - \dfrac{1}{n+1} \right\} + C$.

(a) n = 2: $\int x^2 \ln(x)\, dx = \dfrac{x^3}{3} \left\{ \ln(x) - \dfrac{1}{3} \right\} + C$.

(b) n = 0: $\int \ln(x)\, dx = \dfrac{x^1}{1} \left\{ \ln(x) - \dfrac{1}{1} \right\} + C = x{\cdot}\ln(x) - x + C$.

Practice 5: $\int x^n \sin(x)\, dx$. Put $u = x^n$ and $dv = \sin(x)\, dx$. Then $du = nx^{n-1}\, dx$ and $v = -\cos(x)$.

$$uv - \int v\, du \;=\; (\,x^n\,)(-\cos(x)\,) \;-\; \int -\cos(x)\, nx^{n-1}\, dx$$

$$=\; -x^n \cos(x) \;+\; n \int x^{n-1} \cos(x)\, dx .$$

Practice 6:   (Similar to Example 7)

$\int e^x \sin(x)\, dx$. Put $u = e^x$ and $dv = \sin(x)\, dx$. Then $du = e^x\, dx$ and $v = -\cos(x)\, dx$.

$$uv - \int v\, du \;=\; (\,e^x\,)(-\cos(x)\,) \;-\; \int -\cos(x)\, e^x\, dx$$

$$=\; -e^x \cos(x) + \int e^x \cos(x)\, dx .$$

(For the last integral, put $u = e^x$, $dv = \cos(x)\, dx$, $du = e^x\, dx$, $v = \sin(x)\, dx$)

$$=\; -e^x \cos(x) + \int e^x \cos(x)\, dx \;=\; -e^x \cos(x) + \left\{ e^x \sin(x) - \int e^x \sin(x)\, dx \right\}$$

So $2 \int e^x \sin(x)\, dx = \; -e^x \cos(x) + e^x \sin(x)$

and $\int e^x \sin(x)\, dx = \dfrac{1}{2} \left\{ -e^x \cos(x) + e^x \sin(x) \right\} + C$.

8.4   PARTIAL FRACTION DECOMPOSITION

Rational functions (polynomials divided by polynomials) and their integrals are important in mathematics and applications, but if you look through a table of integral formulas, you will find very few formulas for their integrals. Partly that is because the general formulas are rather complicated and have many special cases, and partly it is because they can all be reduced to just a few cases using the algebraic technique discussed in this section, Partial Fraction Decomposition.

In algebra you learned to add rational functions to get a single rational function. Partial Fraction Decomposition is a technique for reversing that procedure to "decompose" a single rational function into a **sum** of simpler rational functions. Then the integral of the single rational function can be evaluated as the sum of the integrals of the simpler functions.

Example 1:   Use the algebraic decomposition $\dfrac{17x-35}{2x^2-5x} = \dfrac{7}{x} + \dfrac{3}{2x-5}$ to evaluate $\int \dfrac{17x-35}{2x^2-x}$ dx .

Solution: The decomposition allows us to exchange the original integral for two much easier ones:

$$\int \frac{17x-35}{2x^2-5x} \ dx \ = \ \int \frac{7}{x} + \frac{3}{2x-5} \ dx \ \ = \int \frac{7}{x} \ dx + \int \frac{3}{2x-5} \ dx$$

$$= \ 7\ln| \ x \ | + \frac{3}{2} \ \ln| \ 2x-5 \ | \ + C \ .$$

Practice 1:   Use the algebraic decomposition $\dfrac{7x-11}{3x^2-8x-3} = \dfrac{4}{3x+1} + \dfrac{1}{x-3}$ to evaluate $\int \dfrac{7x-11}{3x^2-8x-3}$ dx .

The Example illustrates how to use a "decomposed" fraction with integrals, but it does not show how to achieve the decomposition. The algebraic basis for the Partial Fraction Decomposition technique is that every polynomial can be factored into a product of linear factors $ax+b$ and irreducible quadratic factors $ax^2 + bx + c$ (with $b^2 - 4ac < 0$). These factors may not be easy to find , and they will typically be more complicated than the examples in this section, but every polynomial has such factors. Before we apply the Partial Fraction Decomposition technique, the fraction must have the following form:

(i)   (the degree of the numerator) $<$ (degree of the denominator)

(ii) The denominator has been factored into a product of linear factors and irreducible quadratic factors.

If assumption (i) is not true, we can use polynomial division until we get a remainder which has a smaller degree than the denominator. If assumption (ii) is not true, we simply cannot use the Partial Fraction Decomposition technique.

Example 2:   Put each fraction into a form for Partial Fraction Decomposition:

$$\text{(a)}\ \frac{2x^2 + 4x - 6}{x^2 - 2x} \qquad \text{(b)}\ \frac{3x^3 - 3x^2 - 9x + 8}{x^2 - x - 6} \qquad \text{(c)}\ \frac{7x^2 + 12x - 12}{x^3 - 4x}$$

Solution: (a) $\dfrac{2x^2 + 4x - 6}{x^2 - 2x} = 2 + \dfrac{8x - 6}{x^2 - 2x} = 2 + \dfrac{8x - 6}{x(x - 2)}$

(b) $\dfrac{3x^3 - 3x^2 - 9x + 8}{x^2 - x - 6} = 3x + \dfrac{9x + 8}{x^2 - x - 6} = 3x + \dfrac{9x + 8}{(x + 2)(x - 3)}$

(c) $\dfrac{7x^2 + 12x - 12}{x^3 - 4x} = \dfrac{7x^2 + 12x - 12}{x(x + 2)(x - 2)}$

Distinct Linear Factors

If the denominator can be factored into a product of distinct linear factors, then the original fraction can be written as the **sum** of fractions of the form $\dfrac{\text{number}}{\text{linear factor}}$. Our job is to find the values of the numbers in the numerators, and that typically requires solving a system of equations.

Example 3: Find values for A and B so $\dfrac{17x - 35}{x(2x - 5)} = \dfrac{A}{x} + \dfrac{B}{2x - 5}$.

Solution: We can combine the two terms on the right by putting them over the common denominator $x(2x - 5)$. Multiplying the $\dfrac{A}{x}$ term by $\dfrac{2x - 5}{2x - 5}$ and multiplying the $\dfrac{B}{2x-5}$ term by $\dfrac{x}{x}$, we have

$$\frac{A}{x} \cdot \frac{2x - 5}{2x - 5} + \frac{B}{2x - 5} \cdot \frac{x}{x} = \frac{A2x - 5A + Bx}{x(2x - 5)} = \frac{(2A + B)x - 5A}{x(2x - 5)} \ .$$

Since $\dfrac{(2A + B)x - 5A}{x(2x - 5)} = \dfrac{17x - 35}{x(2x - 5)}$, the coefficients of like terms in the numerators must be equal:

coefficients of x: $2A + B = 17$

constant terms: $-5A = -35$

Solving this system of two equations with two unknowns, we get A = 7 and B = 3 so

$$\frac{17x - 35}{x(2x - 5)} = \frac{7}{x} + \frac{3}{2x - 5} \ .$$

As a check, add $\frac{7}{x}$ and $\frac{3}{2x-5}$ and verify that the sum is $\frac{17x-35}{x(2x-5)}$.

Practice 2: Find values of A and B so $\frac{6x-7}{(x+3)(x-2)} = \frac{A}{x+3} + \frac{B}{x-2}$.

In general, there is one unknown coefficient for each distinct linear factor of the denominator. However, if the number of distinct linear factors is large, we would need to solve a large system of equations for the unknowns.

Example 4:   Find values for A, B, and C so $\frac{2x^2+7x+9}{x(x+1)(x+3)} = \frac{A}{x} + \frac{B}{x+1} + \frac{C}{x+3}$.

Solution: $\frac{A}{x} + \frac{B}{x+1} + \frac{C}{x+3} = \frac{A}{x}\frac{(x+1)(x+3)}{(x+1)(x+3)} + \frac{B}{(x+1)}\frac{x(x+3)}{x(x+3)} + \frac{C}{(x+3)}\frac{x(x+1)}{x(x+1)}$

$$= \frac{A(x+1)(x+3) + Bx(x+3) + Cx(x+1)}{x(x+1)(x+3)}$$

$$= \frac{(A+B+C)x^2 + (4A+3B+C)x + (3A)}{x(x+1)(x+3)} = \frac{2x^2+7x+9}{x(x+1)(x+3)} .$$

The coefficients of the like terms in the numerators must be equal:

coefficients of x^2:   A + B + C = 2

coefficients of x: 4A + 3B + C = 7

constant terms: 3A   = 9   so A = 3, B = –2, and C = 1 .

Finally, $\frac{2x^2+7x+9}{x(x+1)(x+3)} = \frac{3}{x} + \frac{-2}{x+1} + \frac{1}{x+3}$.

Practice 3:   Use the result of Example 4 to evaluate $\int \frac{2x^2+7x+9}{x(x+1)(x+3)} \, dx$.

Practice 4:   How large would the system be for a Partial Fraction Decomposition of $\frac{something}{5^{th} \ degree \ polynomial}$?

The next two subsections describe how to decompose fractions whose denominators contain irreducible quadratic factors and repeated factors. We will not discuss why the suggestions work except to note that they provide enough, but not too many, unknown coefficients for the decomposition.

Distinct Irreducible Quadratic Factors

If the factored denominator includes a distinct irreducible quadratic factor, then the Partial Fraction Decomposition **sum** contains a fraction of the form of a linear polynomial with unknown coefficients divided by the irreducible quadratic factor:

$$\frac{\text{linear polynomial}}{\text{irreducible quadratic factor}} \quad \text{or} \quad \frac{Ax + B}{\text{irreducible quadratic factor}} \cdot$$

Once again we will solve a system of equations to find the values of the unknown coefficients A and B.

Example 5:   Find values for A, B, and C so $\dfrac{x^2 + 3x - 15}{(x^2 + 2x + 5)x} = \dfrac{Ax + B}{x^2 + 2x + 5} + \dfrac{C}{x}$.

Solution: $\dfrac{Ax + B}{x^2 + 2x + 5} + \dfrac{C}{x}$ $= \dfrac{Ax + B}{x^2 + 2x + 5}\left(\dfrac{x}{x}\right) + \dfrac{C}{x}\left(\dfrac{x^2 + 2x + 5}{x^2 + 2x + 5}\right)$

$$= \frac{Ax^2 + Bx + Cx^2 + 2Cx + 5C}{x(x^2 + 2x + 5)}$$

$$= \frac{(A + C)x^2 + (B + 2C)x + 5C}{x(x^2 + 2x + 5)} = \frac{x^2 + 3x - 15}{(x^2 + 2x + 5)x} \cdot$$

Then A + C = 1, B + 2C = 3, and 5C = –15 so C = – 3, B = 9, and A = 4.

In general, there are 2 unknown coefficients for each distinct irreducible quadratic factor of the denominator. We would start the decomposition of

$$\frac{6x^3 + 36x^2 + 50x + 53}{(x^2 + 4)(x^2 + 4x + 5)} \quad \text{by writing it as the sum} \quad \frac{Ax + B}{x^2 + 4} + \frac{Cx + D}{x^2 + 4x + 5} \cdot$$

We would finish this decomposition by solving the system of 4 equations with 4 unknowns, A + C = 6, 4A + B + D = 36, 5A + 4B + 4C = 50, and 5B + 4D = 53 to get A = 6, B = 5, C = 0, and D = 7.

Repeated Factors

If the factored denominator contains a linear factor raised to a power (greater than one), then we need to start the decomposition with several terms. There should be one term with one unknown coefficient for each power of the linear factor. For example,

$$\frac{\text{something}}{(x + 1)(x - 2)^3} = \frac{A}{x + 1} + \frac{B}{x - 2} + \frac{C}{(x - 2)^2} + \frac{D}{(x - 2)^3} \cdot$$

Similarly, if the factored denominator contains an irreducible quadratic factor raised to a power greater than one), then we need to start the decomposition with several terms. There should be one term with two unknown coefficients for each power of the irreducible quadratic. For example,

$$\frac{something}{x^2(x^2+9)^3} = \frac{A}{x} + \frac{B}{x^2} + \frac{Cx+D}{x^2+9} + \frac{Ex+F}{(x^2+9)^2} + \frac{Gx+H}{(x^2+9)^3} \ .$$

This leads to a system of 8 equations with 8 unknowns.

Example 6:   Decompose $\dfrac{-4x^2+5x+3}{x(x-1)^2}$ and evaluate $\displaystyle\int \dfrac{-4x^2+5x+3}{x(x-1)^2} \, dx$.

Solution: $\dfrac{-4x^2+5x+3}{x(x-1)^2} = \dfrac{A}{x} + \dfrac{B}{x-1} + \dfrac{C}{(x-1)^2}$

$$= \frac{A}{x}\left(\frac{(x-1)^2}{(x-1)^2}\right) + \frac{B}{x-1}\left(\frac{x(x-1)}{x(x-1)}\right) + \frac{C}{(x-1)^2}\left(\frac{x}{x}\right)$$

$$= \frac{A(x-1)^2 + Bx(x-1) + Cx}{x(x-1)^2}$$

$$= \frac{(A+B)x^2 + (-2A-B+C)x + A}{x(x-1)^2} = \frac{-4x^2+5x+3}{x(x-1)^2} \ .$$

Then $A + B = -4$, $-2A - B + C = 5$, and $A = 3$ so $A = 3$, $B = -7$, and $C = 4$. Finally,

$$\int \frac{-4x^2+5x+3}{x(x-1)^2} \, dx = \int \frac{3}{x} + \frac{-7}{x-1} + \frac{4}{(x-1)^2} \, dx = 3\ln|x| - 7\ln|x-1| + \frac{-4}{x-1} + C.$$

Practice 5:   Decompose $\dfrac{2x^2+27x+85}{(x+5)^2}$ and evaluate $\displaystyle\int \dfrac{2x^2+27x+85}{(x+5)^2} \, dx$.

The primary use of the partial fraction technique in this course is to put rational functions in a form that is easier to integrate, but this algebraic technique can also be used to simplify the differentiation of some rational functions. The next example illustrates the use of partial fractions to make a differentiation problem easier.

Example 7:   For $f(x) = \dfrac{2x+13}{x^2+x-2}$, calculate $f'(x)$, $f''(x)$, and $f'''(x)$.

Solution: You already know how to calculate these derivatives using the quotient rule, but that process is

rather tedious for the second and third derivatives here. Instead, we can use the partial fraction technique to

rewrite f as $f(x) = \dfrac{5}{x-1} - \dfrac{3}{x+2} = 5(x-1)^{-1} - 3(x+2)^{-1}$. Then the derivatives are very

straightforward:

$$f'(x) = -5(x-1)^{-2} + 3(x+2)^{-2} ,$$

$$f''(x) = 10(x-1)^{-3} - 6(x+2)^{-3} , \text{ and}$$

$$f'''(x) = -30(x-1)^{-4} + 18(x+2)^{-4} .$$

Practice 6:   Use the partial fraction decomposition of $g(x) = \dfrac{9x+1}{x^2-2x-3}$ to calculate $g'(x)$,

$g''(x)$, and $g^{(4)}(x)$.

PROBLEMS

In problems 1 – 12, decompose the fractions.

1. $\dfrac{7x+2}{x(x+1)}$ 2. $\dfrac{7x+9}{(x+3)(x-1)}$ 3. $\dfrac{11x+25}{x^2+9x+8}$ 4. $\dfrac{3x+7}{x^2-1}$

5. $\dfrac{2x^2+15x+25}{x^2+5x}$ 6. $\dfrac{3x^3+3x^2}{x^2+x-2}$ 7. $\dfrac{6x^2+9x-15}{x(x+5)(x-1)}$ 8. $\dfrac{6x^2-x-1}{x^3-x}$

9. $\dfrac{8x^2-x+3}{x^3+x}$ 10. $\dfrac{9x^2+13x+15}{x^3+2x^2-3x}$ 11. $\dfrac{11x^2+23x+6}{x^2(x+2)}$ 12. $\dfrac{6x^2+14x-9}{x(x+3)^2}$

In problems 13 – 30, evaluate the integrals.

13. $\int \dfrac{3x+13}{(x+2)(x-5)} \, dx$ 14. $\int \dfrac{2x+11}{(x-7)(x-2)} \, dx$ 15. $\int_{2}^{5} \dfrac{2}{x^2-1} \, dx$

16. $\int_{1}^{3} \dfrac{5x^2+5x+3}{x^3+x} \, dx$ 17. Integrate the functions in problems 1 – 4.

18.   Integrate the functions in problems 5 – 8.

19. $\int \dfrac{2x^2+5x+3}{x^2-1} \, dx$ 20. $\int \dfrac{2x^2+19x+22}{x^2+x-12} \, dx$ 21. $\int \dfrac{3x^2+19x+24}{x^2+6x+5} \, dx$

22. $\int \dfrac{7x^2+8x-2}{x^2+2x} \, dx$ 23. $\int \dfrac{3x^2-1}{x^3-x} \, dx$ 24. $\int \dfrac{x^4+5x^3+x-15}{x^2+5x} \, dx$

25. $\int \dfrac{x^3 + 3x^2 - 4x + 30}{x^2 + 3x - 10}\ dx$

26. $\int \dfrac{2x + 5}{(x + 1)^2}\ dx$

27. $\int \dfrac{12x^2 + 19x - 6}{x^3 + 3x^2}\ dx$

28. $\int \dfrac{7x^3 + x^2 + 7x + 10}{x^4 + 2x^3}\ dx$

29. $\int \dfrac{7x^2 + 3x + 7}{x^3 + x}\ dx$

30. $\int \dfrac{7x^2 - 4x + 4}{x^3 + 1}\ dx$

31. Integrals are very sensitive to small changes in the integrand. Evaluate

  (a) $\int \dfrac{1}{x^2 + 2x + 2}\ dx$

  (b) $\int \dfrac{1}{x^2 + 2x + 1}\ dx$

  (c) $\int \dfrac{1}{x^2 + 2x + 0}\ dx$.

32. Evaluate   (a) $\int \dfrac{1}{x^2 - 6x + 8}\ dx$

  (b) $\int \dfrac{1}{x^2 - 6x + 9}\ dx$

  (c) $\int \dfrac{1}{x^2 - 6x + 10}\ dx$.

33. Use the partial fraction decomposition of the functions in problems 1 and 2 to calculate their first and second derivatives.

34. Use the partial fraction decomposition of the functions in problems 3 and 4 to calculate their first and second derivatives.

35. Use the partial fraction decomposition of the functions in problems 5 and 6 to calculate their first and second derivatives.

The following two applications involve a type of differential equation which can be solved by separating the variables and using a partial fraction decomposition to help calculate the antiderivatives. The same type of differential equation is also used to model the spread of rumors and diseases as well as some populations and chemical reactions.

Logistic Growth: The growth rate of many different populations depends not only on the number of individuals (leading to exponential growth) but also on a "carrying capacity" of the environment. If

  x is the population at time t and the growth rate of x is proportional to the **product** of the population and the carrying capacity M minus the population, then the growth rate is described by the differential equation

$$\frac{dx}{dt} = k \cdot x \cdot (M - x)$$

where k and M are constants for a given species in a given environment.

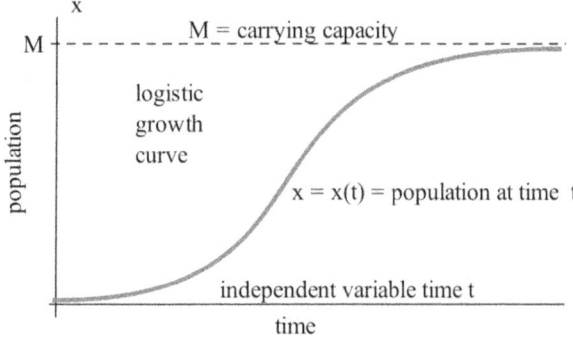

Fig. 1: Logistic growth curve

36. Let $k = 1$ and $M = 100$, and assume the initial population is $x(0) = 5$.

    (a) Solve the differential equation $\frac{dx}{dt} = x(100 - x)$ for x.

    (b) Graph the population $x(t)$ for $0 \le t \le 20$.

    (c) When will the population be 20? 50? 90? 100?

    (d) What is the population after a "long" time? (Find the limit, as t becomes arbitrarily large, of x.)

    (e) Explain the shape of the graph in (a) in terms of a population of bacteria.

    (f) When is the growth rate largest? (Maximize dx/dt.)

    (g) What is the population when the growth rate is largest?

37. Let $k = 1$ and $M = 100$, and assume the initial population is $x(0) = 150$.

    (a) Solve the differential equation $\frac{dx}{dt} = x(100 - x)$ for x and graph $x(t)$ for $0 \le t \le 20$.

    (b) When will the population be 120? 110? 100?

    (c) What is the population after a "long" time? (Find the limit, as t becomes arbitrarily large, of x.)

    (d) Explain the shape of the graph in (a) in terms of a population of bacteria.

38. Let k and M be positive constants, and assume the initial population is $x(0) = x_0$.

    (a) Solve the differential equation $\frac{dx}{dt} = k \cdot x \cdot (M - x)$ for x.

    (b) What is the population after a "long" time? (Find the limit, as t becomes arbitrarily large, of x.)

    (c) When is the growth rate largest? (Maximize dx/dt.)

    (d) What is the population when the growth rate is largest?

Chemical Reaction: In some chemical reactions, a new material X is formed from materials A and B, and the rate at which X forms is proportional to the **product** of the amount of A and the amount of B remaining in the solution. Let x represent the amount of material X present at time t, and assume that the reaction begins with a grams of A, b grams of B, and no material X ($x(0) = 0$). Then the rate of formation of material X can be described by the differential equation

$$\frac{dx}{dt} = k(a - x)(b - x).$$

39. Solve the differential equation $\frac{dx}{dt} = k(a - x)(b - x)$ for x if $k = 1$ and the reaction begins with

    (i) 7 grams of A and 5 grams of B, and (ii) 6 grams of A and 6 grams of B.

40. Solve the differential equation $\frac{dx}{dt} = k(a - x)(b - x)$ for x if $k = 1$ and the reaction begins with

    (i) a grams of A and b grams of B with $a \ne b$, and (ii) c grams of A and c grams of B ($c \ne 0$).

Section 8.4   **PRACTICE Answers**

Practice 1: $\int \dfrac{7x-11}{3x^2-8x-3}\ dx\ =\ \int \dfrac{4}{3x+1}\ dx\ +\ \int \dfrac{1}{x-3}\ dx$

$$=\tfrac{4}{3}\cdot \ln|3x+1|+\ln|x-3|+C\ .$$

Practice 2: $\dfrac{6x-7}{(x+3)(x-2)}=\dfrac{A}{x+3}+\dfrac{B}{x-2}=\dfrac{A(x-2)+B(x+3)}{(x+3)(x-2)}=\dfrac{(A+B)x+(-2A+3B)}{(x+3)(x-2)}\ .$

This gives us the system: $A+B=6$ and $-2A+3B=-7$ so (solving) **A = 5** and **B = 1**.

$$\dfrac{6x-7}{(x+3)(x-2)}=\dfrac{5}{x+3}+\dfrac{1}{x-2}\ .$$

Practice 3: From Example 4,

$$\int \dfrac{2x^2+7x+7}{x(x+1)(x+3)}\ dx\ =\ \int \dfrac{3}{x}+\dfrac{-2}{x+1}+\dfrac{1}{x+3}\ dx$$

$$=3\cdot\ln|x|-2\cdot\ln|x+1|+\ln|x+3|+C\ .$$

Practice 4: If the 5th degree polynomial can be factored into a product of 5 distinct linear terms, then we would have

$$\dfrac{\text{something}}{5^{th}\ \text{degree polynomial}}=\dfrac{A}{1^{st}\ \text{term}}+\dfrac{B}{2^{nd}\ \text{term}}+\ldots+\dfrac{E}{5^{th}\ \text{term}}\ .$$

Practice 5: $\dfrac{2x^2+27x+85}{(x+5)^2}=\dfrac{2x^2+27x+85}{x^2+10x+25}=2+\dfrac{7x+35}{(x+5)^2}=2+\dfrac{7}{x+5}\ .$

Then $\int \dfrac{2x^2+27x+85}{(x+5)^2}\ dx\ =\ \int 2+\dfrac{7}{x+5}\ dx\ =\ \mathbf{2x+7\cdot\ln|x+5|+C}\ .$

Practice 6: $g(x)=\dfrac{9x+1}{(x-3)(x+1)}=\dfrac{A}{x-3}+\dfrac{B}{x+1}=\dfrac{A(x+1)+B(x-3)}{(x-3)(x+1)}=\dfrac{(A+B)x+(A-3B)}{(x-3)(x+1)}\ .$

This gives us the system $A+B=9$ and $A-3B=1$ so (solving) $A=7$ and $B=2$.

$g(x)=\dfrac{7}{x-3}+\dfrac{2}{x+1}=7(x-3)^{-1}+2(x+1)^{-1}\ .$ Then

$g'(x)=-7(x-3)^{-2}-2(x+1)^{-2}\ ,$

$g''(x)=14(x-3)^{-3}+4(x+1)^{-3}\ ,$

$g'''(x)=-42(x-3)^{-4}-12(x+1)^{-4}\ ,$ and

$g''''(x)=168(x-3)^{-5}+48(x+1)^{-5}\ .$

8.5 Trigonometric Substitution — Another Change of Variable

Changing the variable is a very powerful technique for finding antiderivatives, and by now you have probably found a lot of integrals by setting u = something. This section also involves a change of variable, but for more specialized patterns, and the change is more complicated. Another difference from previous work is that instead of setting u equal to a function of x we will be replacing x with a function of θ.

The next three examples illustrate the typical steps involved making trigonometric substitutions. After these examples, we examine each step in more detail and consider how to make the appropriate decisions.

Example 1:   In the expression $\sqrt{9 - x^2}$ replace x with $3\sin(\theta)$ and simplify the result.

Solution: Replacing x with $3\sin(\theta)$, $\sqrt{9 - x^2}$ becomes

$$\sqrt{9 - (3\sin(\theta))^2} = \sqrt{9 - 9\sin^2(\theta)} = \sqrt{9 \cdot (1 - \sin^2(\theta))} = 3\cos(\theta) .$$

Example 2:   Evaluate $\int \sqrt{9 - x^2}\ dx$ using the change of variable $x = 3\sin(\theta)$ and then use the antiderivative to evaluate $\int_0^3 \sqrt{9 - x^2}\ dx$.

Solution: If $x = 3\sin(\theta)$, then $dx = 3\cos(\theta)\ d\theta$ and $\sqrt{9 - x^2} = 3\cos(\theta)$. With this change of variable, the integral becomes

$$\int \sqrt{9 - x^2}\ dx \ = \int 3\cos(\theta)\ 3\cos(\theta)\ d\theta = 9\int \cos^2(\theta)\ d\theta = 9\left\{ \frac{\theta}{2} - \frac{\sin(2\theta)}{4} \right\} + C$$

$$= 9\left\{ \frac{\theta}{2} - \frac{2\sin(\theta)\cos(\theta)}{4} \right\} + C = \frac{9}{2}\left\{ \theta - \sin(\theta)\cos(\theta) \right\} + C$$

This antiderivative, a function of the variable θ, can be converted back to a function of the variable x. Since $x = 3\sin(\theta)$ we can solve for θ to get $\theta = \arcsin(x/3)$. Replacing θ with arcsin(x/3) in the antiderivative, we get

$$\frac{9}{2}\left\{ \theta - \sin(\theta)\cos(\theta) \right\} + C = \frac{9}{2}\left\{ \textbf{arcsin(x/3)} - \sin(\textbf{arcsin(x/3)})\cos(\textbf{arcsin(x/3)}) \right\} + C$$

$$= \frac{9}{2}\left\{ \arcsin\left(\frac{x}{3}\right) + \frac{x}{3}\frac{\sqrt{9 - x^2}}{3} \right\} + C = \frac{9}{2}\arcsin\left(\frac{x}{3}\right) + \frac{1}{2}x\sqrt{9 - x^2} + C.$$

Using this antiderivative, we can evaluate the definite integral:

$$\int_0^3 \sqrt{9-x^2}\ dx = \frac{9}{2}\ \text{arcsin}(\frac{x}{3}) + \frac{x}{2}\sqrt{9-x^2}\ \Big|_0^3$$

$$= \{\frac{9}{2}\ \text{arcsin}(\frac{3}{3}) + \frac{3}{2}\sqrt{9-3^2}\ \} - \{\frac{9}{2}\ \text{arcsin}(\frac{0}{3}) + \frac{0}{2}\sqrt{9-0^2}\ \} = \frac{9\pi}{4}\ .$$

Example 3:   The definite integral $\int_0^3 \sqrt{9-x^2}\ dx$ represents the area of what region?

Solution: The area of one fourth of the circle of radius 3 which lies in the first

quadrant (Fig. 1). The area of this quarter circle is

$$\frac{\text{area of whole circle}}{4} = \frac{1}{4}\ \pi\, r^2 = \frac{1}{4}\ \pi\, 3^2 = \frac{9}{4}\ \pi$$

which agrees with the value found in the previous example.

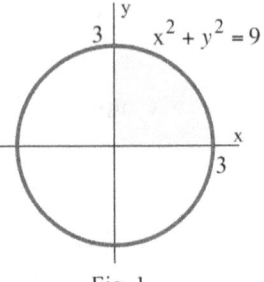

Fig. 1

Each Trigonometric Substitution involves four major steps:

1.   Choose which substitution to make, x = a trigonomteric function of θ .

2.   Rewrite the original integral in terms of θ and $d\theta$.

3.   Find an antiderivative of the new integral.

4.   Write the antiderivative in step 3 in terms of the original variable x.

The rest of this section discusses each of these steps. The first step requires you to make a decision. Then
the other three steps follow from that decision. For most students, the key to success with the
Trigonometric Substitution technique is to THINK TRIANGLES.

Step 1: Choosing the substitution

The first step requires that you make a decision, and the pattern of the familiar Pythagorean Theorem can
help you make the correct choice.

Pythagorean Theorem: $(\text{side})^2 + (\text{side})^2 = (\text{hypotenuse})^2$ or $(\text{side})^2 = (\text{hypotenuse})^2 - (\text{side})^2$.

The pattern $\mathbf{3^2 + x^2}$ matches the Pythagorean pattern if 3 and

x are sides of a right triangle. For a right triangle with sides 3

and x (Fig. 2), we know $\tan(\theta) = \text{opposite/adjacent} = x/3$ so

$\mathbf{x = 3\ tan(\theta)}$.

$\sqrt{3^2 + x^2}$

x

θ

$\theta = \text{arctan}\left(\frac{x}{3}\right)$

3

Fig. 2

The pattern $3^2 - x^2$ matches the Pythagorean pattern if 3 is the hypotenuse and x is a side of a right triangle (Fig. 3). Then $\sin(\theta)$ = opposite/hypotenuse = x/3 so **x = 3 sin(θ)**.

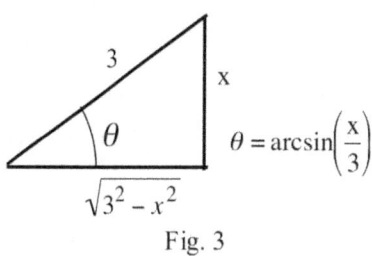

Fig. 3

The pattern $x^2 - 3^2$ matches the Pythagorean pattern if x is the hypotenuse and 3 is a side of a right triangle (Fig. 4). Then $\sec(\theta)$ = hypotenuse/adjacent = x/3 so **x = 3 sec(θ)**.

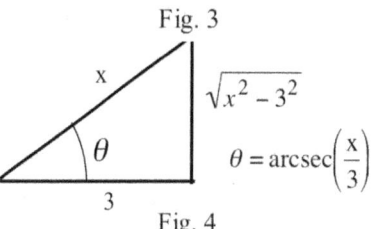

Fig. 4

Once the choice has been made for the substitution, then several things follow automatically:

  dx   can be calculated by differentiating x with respect to θ ,

  θ   can be found by solving the substitution equation for θ ,

    (if x = 3 tan(θ) then tan(θ) = x/3 so θ = arctan(x/3)) , and

the patterns $3^2 + x^2$, $3^2 - x^2$, and $x^2 - 3^2$ can be simplified using algebra and the trigonometric identities $1 + \tan^2(\theta) = \sec^2(\theta)$, $1 - \sin^2(\theta) = \cos^2(\theta)$, and $\sec^2(\theta) - 1 = \tan^2(\theta)$.

These results are collected in the table below.

$3^2 + x^2$ (Fig. 2)	$3^2 - x^2$ (Fig. 3)	$x^2 - 3^2$ (Fig. 4)
Put **x = 3 tan(θ)** .	Put **x = 3 sin(θ)** .	Put **x = 3 sec(θ)** .
Then $dx = 3 \sec^2(\theta)\, d\theta$	Then $dx = 3 \cos(\theta)\, d\theta$	Then $dx = 3 \sec(\theta)\tan(\theta)\, d\theta$
$\theta = \arctan(\frac{x}{3})$	$\theta = \arcsin(\frac{x}{3})$	$\theta = \operatorname{arcsec}(\frac{x}{3})$
$3^2 + x^2$ $= 3^2 + 3^2 \tan^2(\theta)$ $= 3^2(1 + \tan^2(\theta))$ $= 3^2 \sec^2(\theta)$	$3^2 - x^2$ $= 3^2 - 3^2 \sin^2(\theta)$ $= 3^2(1 - \sin^2(\theta))$ $= 3^2 \cos^2(\theta)$	$x^2 - 3^2$ $= 3^2 \sec^2(\theta) - 3^2$ $= 3^2(\sec^2(\theta) - 1)$ $= 3^2 \tan^2(\theta)$

Example 4:   For the patterns $16 - x^2$ and $5 + x^2$, (a) decide on the appropriate substitution for x,

(b) calculate dx and θ, and (c) use the substitution to simplify the pattern.

Solution: $16 - x^2$: This matches the Pythagorean pattern if 4 is a hypotenuse and x is the side of a right

triangle. Then $\sin(\theta) = $ opposite/hypotenuse $ = x/4$ so $\mathbf{x = 4 \sin(\theta)}$. For $x = 4 \sin(\theta)$, dx

$= 4 \cos(\theta) \, d\theta$ and $\theta = \arcsin(x/4)$. Finally,

$$16 - x^2 = 16 - (4 \sin(\theta))^2 = 16 - 16\sin^2(\theta) = 16(1 - \sin^2(\theta)) = 16 \cos^2(\theta) .$$

$5 + x^2$: This matches the Pythagorean pattern if x and $\sqrt{5}$ are the sides of a right triangle. Then

$\tan(\theta) = $ opposite/adjacent $ = x/\sqrt{5}$ so $\mathbf{x = \sqrt{5} \tan(\theta)}$. For $x = \sqrt{5} \tan(\theta)$, $dx = \sqrt{5}$

$\sec^2(\theta) \, d\theta$ and $\theta = \arctan(x/\sqrt{5})$. Finally,

$$5 + x^2 = 5 + (\sqrt{5} \tan(\theta))^2 = 5 + 5 \tan^2(\theta) = 5(1 + \tan^2(\theta)) = 5 \sec^2(\theta) .$$

Practice 1:   For the patterns $25 + x^2$ and $x^2 - 13$, (a) decide on the appropriate substitution for x,

(b) calculate dx and θ, and (c) use the substitution to simplify the pattern.

Step 2: Rewriting the integral in terms of θ and $d\theta$

Once we decide on the appropriate substitution, calculate dx , and simplify the the pattern, then the second

step is very straightforward.

Example 5:   Use the substitution $x = 5 \tan(\theta)$ to rewrite the integral $\int \dfrac{1}{\sqrt{25 + x^2}}$ dx in terms of θ and $d\theta$.

Solution: Since $x = 5 \tan(\theta)$, then $dx = 5 \sec^2(\theta) \, d\theta$ and

$$25 + x^2 = 25 + (5 \tan(\theta))^2 = 25 + 25 \tan^2(\theta) = 25\{ 1 + \tan^2(\theta)) = 25 \sec^2(\theta) . \text{ Finally,}$$

$$\int \frac{1}{\sqrt{25 + x^2}} \; dx = \int \frac{1}{\sqrt{25 \sec^2(\theta)}} \; 5 \sec^2(\theta) \, d\theta = \int \frac{5 \sec^2(\theta)}{5 \sec(\theta)} \; d\theta = \int \sec(\theta) \; d\theta .$$

Practice 2:   Use the substitution $x = 5 \sin(\theta)$ to rewrite the integral $\int \dfrac{1}{\sqrt{25 - x^2}}$ dx in terms of θ and $d\theta$.

Steps 3 & 4:   Finding an antiderivative of the new integral & writing the answer in terms of x

After changing the variable, the new integral typically involves trigonometric functions and we can use any

of our previous methods (a change of variable, integration by parts, a trigonometric identity, or the integral

tables) to find an antiderivative.

Once we have an antiderivative, usually a trigonometric function of θ, we can replace θ with the appropriate inverse trigonometric function of x and simplify. Since the antiderivatives commonly contain trigonometric functions, we frequently need to simplify a trigonometric function of an inverse trigonometric function, and it is **very** helpful to refer back to the right triangle we used at the beginning of the substitution process.

Example 6:   By replacing x with $5\tan(\theta)$, $\int \dfrac{1}{\sqrt{25 + x^2}}\,dx$

becomes $\int \sec(\theta)\,d\theta$. Evaluate $\int \sec(\theta)\,d\theta$ and write the

resulting antiderivative in terms of the variable x.

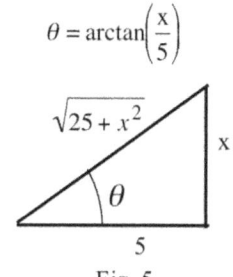

Fig. 5

Solution: $x = 5\tan(\theta)$ so $\theta = \arctan(x/5)$ (Fig. 5). Then

$$\int \sec(\theta)\,d\theta = \ln|\sec(\theta) + \tan(\theta)| + C \quad = \ln|\sec(\textbf{arctan(x/5)}) + \tan(\textbf{arctan(x/5)})| + C$$

By referring to the right triangle in Fig. 5, we see that

$$\sec(\arctan(x/5)) = \frac{\sqrt{25 + x^2}}{5} \quad \text{and} \quad \tan(\arctan(x/5)) = \frac{x}{5} \quad \text{so}$$

$$\ln|\sec(\arctan(x/5)) + \tan(\arctan(x/5))| + C = \ln\left|\frac{\sqrt{25 + x^2}}{5} + \frac{x}{5}\right| + C.$$

Putting these pieces together, we have $\displaystyle\int \frac{1}{x\sqrt{25 + x^2}}\,dx = \ln\left|\frac{\sqrt{25 + x^2}}{5} + \frac{x}{5}\right| + C.$

Practice 3:   Show that by replacing x with $3\sin(\theta)$, $\displaystyle\int \frac{1}{x^2\sqrt{9 - x^2}}\,dx$ becomes $\dfrac{1}{9}\displaystyle\int \csc^2(\theta)\,d\theta$.

Evaluate $\dfrac{1}{9}\displaystyle\int \csc^2(\theta)\,d\theta$ and write the resulting antiderivative in terms of the variable x.

Sometimes it is useful to "complete the square" in an irreducible quadratic to make the pattern more obvious.

Example 7:   Rewrite $x^2 + 2x + 26$ by completing the square and evaluate $\displaystyle\int \frac{1}{\sqrt{x^2 + 2x + 26}}\,dx$.

Solution: $x^2 + 2x + 26 = (x + 1)^2 + 25$ so $\displaystyle\int \frac{1}{\sqrt{x^2 + 2x + 26}}\,dx = \int \frac{1}{\sqrt{(x + 1)^2 + 25}}\,dx$.

Put $u = x + 1$. Then $du = dx$, and

$$\int \frac{1}{\sqrt{x^2 + 2x + 26}} \, dx = \int \frac{1}{\sqrt{(x+1)^2 + 25}} \, dx$$

$$= \int \frac{1}{\sqrt{u^2 + 25}} \, du$$

$$= \ln \left| \frac{\sqrt{25 + u^2}}{5} + \frac{u}{5} \right| + C \quad \text{(using the result of Example 6)}$$

$$= \ln \left| \frac{\sqrt{25 + (x+1)^2}}{5} + \frac{x+1}{5} \right| + C = \ln \left| \frac{\sqrt{x^2 + 2x + 26}}{5} + \frac{x+1}{5} \right| + C$$

THINK TRIANGLES. The first and last steps of the method (choosing the substitution and writing the answer interms of x) are easier if you understand the triangles (Figures 2, 3, and 4) and have drawn the appropriate triangle for the problem. Of course, you also need to practice the method.

PROBLEMS

In problems 1–6, (a) make the given substitution and simplify the result, and (b) calculate dx.

1. $x = 3 \cdot \sin(\theta)$ in $\dfrac{1}{\sqrt{9 - x^2}}$

2. $x = 3 \cdot \tan(\theta)$ in $\dfrac{1}{\sqrt{x^2 + 9}}$

3. $x = 3 \cdot \sec(\theta)$ in $\dfrac{1}{\sqrt{x^2 - 9}}$

4. $x = 6 \cdot \sin(\theta)$ in $\dfrac{1}{36 - x^2}$

5. $x = \sqrt{2} \cdot \tan(\theta)$ in $\dfrac{1}{\sqrt{2 + x^2}}$

6. $x = \sec(\theta)$ in $\dfrac{1}{x^2 - 1}$

In problems 7–12,   (a) solve for θ as a function of x,

(b) replace θ in $f(\theta)$ with you result in part (a), and   (c) simplify.

7. $x = 3 \cdot \sin(\theta)$, $f(\theta) = \cos(\theta) \cdot \tan(\theta)$

8. $x = 3 \cdot \tan(\theta)$, $f(\theta) = \sin(\theta) \cdot \tan(\theta)$

9. $x = 3 \cdot \sec(\theta)$, $f(\theta) = \sqrt{1 + \sin^2(\theta)}$

10. $x = 5 \cdot \sin(\theta)$, $f(\theta) = \dfrac{\cos(\theta)}{1 + \sec(\theta)}$

11. $x = 5 \cdot \tan(\theta)$, $f(\theta) = \dfrac{\cos^2(\theta)}{1 + \cot(\theta)}$

12. $x = 5 \cdot \sec(\theta)$, $f(\theta) = \cos(\theta) + 7 \cdot \tan^2(\theta)$

In problems 13–36, evaluate the integrals. (More than one method works for some of the integrals.)

13. $\int \dfrac{1}{x^2\sqrt{9-x^2}} \ dx$

14. $\int \dfrac{x^2}{\sqrt{9-x^2}} \ dx$

15. $\int \dfrac{1}{\sqrt{x^2+49}} \ dx$

16. $\int \dfrac{1}{\sqrt{x^2+1}} \ dx$

17. $\int \sqrt{36-x^2} \ dx$

18. $\int \sqrt{1-36x^2} \ dx$

19. $\int \dfrac{1}{\sqrt{36+x^2}} \ dx$

20. $\int \dfrac{1}{x\sqrt{25-x^2}} \ dx$

21. $\int \dfrac{1}{\sqrt{49+x^2}} \ dx$

22. $\int \dfrac{\sqrt{25-x^2}}{x^2} \ dx$

23. $\int \dfrac{x}{\sqrt{25-x^2}} \ dx$

24. $\int \dfrac{1}{x^2+49} \ dx$

25. $\int \dfrac{x}{x^2+49} \ dx$

26. $\int \dfrac{1}{49x^2+25} \ dx$

27. $\int \dfrac{1}{(x^2-9)^{3/2}} \ dx$

28. $\int \dfrac{1}{(4x^2-1)^{3/2}} \ dx$

29. $\int \dfrac{5}{2x\sqrt{x^2-25}} \ dx$

30. $\int \dfrac{1}{x\sqrt{3-x^2}} \ dx$

31. $\int \dfrac{1}{25-x^2} \ dx$

32. $\int \dfrac{1}{a^2+x^2} \ dx$

33. $\int \dfrac{1}{\sqrt{a^2+x^2}} \ dx$

34. $\int \dfrac{1}{x\cdot\sqrt{a^2+x^2}} \ dx$

35. $\int \dfrac{1}{x^2\sqrt{a^2+x^2}} \ dx$

36. $\int \dfrac{1}{(a^2+x^2)^{3/2}} \ dx$

In problems 37–42, first complete the square, make the appropriate substitutions, and evaluate the integral.

37. $\int \dfrac{1}{\sqrt{(x+1)^2+9}} \ dx$

38. $\int \dfrac{1}{\sqrt{(x+3)^2+1}} \ dx$

39. $\int \dfrac{1}{x^2+10x+29} \ dx$

40. $\int \dfrac{1}{x^2-4x+13} \ dx$

41. $\int \dfrac{1}{\sqrt{x^2+4x+3}} \ dx$

42. $\int \dfrac{1}{\sqrt{x^2-6x-16}} \ dx$

Section 8.5   Practice Answers

Practice 1: $25 + x^2$:   (a) Put $x = 5 \cdot \tan(\theta)$

(b) Then $dx = 5 \cdot \sec^2(\theta) \, d\theta$ and $\theta = \arctan(x/5)$

(c) $25 + x^2 = 25 + 25 \cdot \tan^2(\theta) = 25(1 + \tan^2(\theta)) = 25 \cdot \sec^2(\theta)$

$x^2 - 13$:   (a) Put $x = \sqrt{13} \cdot \sec(\theta)$

(b) Then $dx = \sqrt{13} \cdot \sec(\theta) \cdot \tan(\theta) \, d\theta$ and $\theta = \text{arcsec}(x/\sqrt{13})$

(c) $x^2 - 13 = 13 \cdot \sec^2(\theta) - 13 = 13(\sec^2(\theta) - 1) = 13 \cdot \tan^2(\theta)$

Practice 2: $x = 5 \cdot \sin(\theta)$ so $dx = 5 \cdot \cos(\theta) \, d\theta$ and $25 - x^2 = 25(1 - \sin^2(\theta)) = 25 \cdot \cos^2(\theta)$.

Then $\displaystyle \int \frac{1}{\sqrt{25 - x^2}} \, dx = \int \frac{1}{\sqrt{25 - \sin^2(\theta)}} \, 5 \cdot \cos(\theta) \, d\theta$

$\displaystyle = \int \frac{1}{\sqrt{25 \cdot \cos^2(\theta)}} \, 5 \cdot \cos(\theta) \, d\theta = \int 1 \, d\theta = \theta + C = \arcsin(x/5) + C$

Practice 3: $x = 3 \cdot \sin(\theta)$ so $dx = 3 \cdot \cos(\theta) \, d\theta$ and $9 - x^2 = 9(1 - \sin^2(\theta)) = 9 \cdot \cos^2(\theta)$.

Then $\displaystyle \int \frac{1}{x^2 \sqrt{9 - x^2}} \, dx = \int \frac{1}{9 \cdot \sin^2(\theta) \sqrt{9 - \sin^2(\theta)}} \, 3 \cdot \cos(\theta) \, d\theta = \int \frac{1}{9 \cdot \sin^2(\theta) \sqrt{9 \cdot \cos^2(\theta)}} \, 3 \cdot \cos(\theta) \, d\theta$

$\displaystyle = \frac{1}{9} \int \csc^2(\theta) \, d\theta = -\frac{1}{9} \cot() + C = -\frac{1}{9} \cot(\arcsin(x/3)) + C = -\frac{1}{9} \frac{\sqrt{9 - x^2}}{x} + C.$

8.6 Integrals of Trigonometric Functions

There are an overwhelming number of combinations of trigonometric functions which appear in integrals, but fortunately they fall into a few patterns and most of their integrals can be found using reduction formulas and tables of integrals. This section examines some of the patterns of these combinations and illustrates how some of their integrals can be derived.

Products of Sine and Cosine: $\int \sin(ax)\cdot\sin(bx)\ dx$, $\int \cos(ax)\cdot\cos(bx)\ dx$, $\int \sin(ax)\cdot\cos(bx)\ dx$

All of these integrals are handled by referring to the trigonometric identities for sine and cosine of sums and differences:

$$\sin(A + B) = \sin(A)\cos(B) + \cos(A)\sin(B)$$

$$\sin(A - B) = \sin(A)\cos(B) - \cos(A)\sin(B)$$

$$\cos(A + B) = \cos(A)\cos(B) - \sin(A)\sin(B)$$

$$\cos(A - B) = \cos(A)\cos(B) + \sin(A)\sin(B)$$

By adding or subtracting the appropriate pairs of identities, we can write the various products such as $\sin(ax)\cos(bx)$ as a sum or difference of single sines or cosines. For example, by adding the first two identities we get $2\sin(A)\cos(B) = \sin(A + B) + \sin(A - B)$ so $\sin(A)\cos(B) = \frac{1}{2}\{ \sin(A+B) + \sin(A-B) \}$. Using this last identity, the integral of $\sin(ax)\cos(bx)$ for $a \neq b$ is relatively easy:

$$\int \sin(ax)\cos(bx)\ dx = \int \frac{1}{2}\{ \sin((a+b)x) + \sin((a-b)x) \}\ dx = \frac{1}{2}\left\{ \frac{-\cos((a-b)x)}{a - b} + \frac{-\cos((a+b)x)}{a + b} \right\} + C.$$

The other integrals of products of sine and cosine follow in a similar manner.

If $a \neq b$, then

$$\int \sin(ax)\cdot\sin(bx)\ dx = \frac{1}{2}\left\{ \frac{\sin((a-b)x)}{a - b} - \frac{\sin((a+b)x)}{a + b} \right\} + C$$

$$\int \cos(ax)\cdot\cos(bx)\ dx = \frac{1}{2}\left\{ \frac{\sin((a-b)x)}{a - b} + \frac{\sin((a+b)x)}{a + b} \right\} + C$$

$$\int \sin(ax)\cdot\cos(bx)\ dx = \frac{-1}{2}\left\{ \frac{\cos((a-b)x)}{a - b} + \frac{\cos((a+b)x)}{a + b} \right\} + C$$

If $a = b$, we have patterns we have already used.

$$\int \sin^2(ax)\, dx \qquad = \frac{x}{2} - \frac{\sin(2ax)}{4a} + C = \frac{x}{2} - \frac{\sin(ax)\cdot\cos(ax)}{2a} + C$$

$$\int \cos^2(ax)\, dx \qquad = \frac{x}{2} + \frac{\sin(2ax)}{4a} + C = \frac{x}{2} + \frac{\sin(ax)\cdot\cos(ax)}{2a} + C$$

$$\int \sin(ax)\cdot\cos(ax)\, dx \quad = \frac{\sin^2(ax)}{2a} + C = \frac{1 - \cos(2ax)}{4a} + C$$

The first and second of these integral formulas follow from the identities $\sin^2(ax) = \dfrac{1 - \cos(2ax)}{2}$ and $\cos^2(ax) = \dfrac{1 + \cos(2ax)}{2}$, and the third can be derived by changing the variable to $u = \sin(ax)$.

Powers of Sine and Cosine Alone: $\int \sin^n(x)\, dx$, $\int \cos^n(x)\, dx$

All of these antiderivatives can be found using integration by parts or the reduction formulas (formulas 19 and 20 in the integral tables) which were derived using integration by parts. For small values of m and n it is just as easy to find the antiderivatives directly.

Even Powers of Sine or Cosine Alone

For **even** powers of sine or cosine, we can successfully reduce the size of the exponent by repeatedly applying the identities $\sin^2(x) = \dfrac{1 - \cos(2x)}{2}$ and $\cos^2(x) = \dfrac{1 + \cos(2x)}{2}$.

Example 1: Evaluate $\int \sin^4(x)\, dx$.

Solution: $\sin^4(x) = \{ \sin^2(x) \}^2 = \{ \frac{1}{2} [1 - \cos(2x)] \}^2 = \frac{1}{4} \{ 1 - 2\cos(2x) + \cos^2(2x) \}$ so

$$\int \sin^4(x)\, dx = \int \frac{1}{4} \{ 1 - 2\cos(2x) + \cos^2(2x) \}\, dx$$

$$= \frac{1}{4} \left\{ x + \sin(2x) + \frac{x}{2} + \frac{\sin(2x)\cos(2x)}{2} \right\} + C .$$

Practice 1: Evaluate $\int \cos^4(x)\, dx$.

Odd Powers of Sine or Cosine Alone

For odd powers of sine or cosine we can split off one factor of sine or cosine, reduce the remaining even exponent using the identities $\sin^2(x) = 1 - \cos^2(x)$ or $\cos^2(x) = 1 - \sin^2(x)$, and finally integrate by changing the variable.

Example 2:   Evaluate $\int \sin^5(x)\, dx$.

Solution: $\sin^5(x) = \sin^4(x)\sin(x) = \{\, \sin^2(x)\, \}^2 \sin(x) \qquad = \{\, 1 - \cos^2(x)\, \}^2 \sin(x)$

$$= \{\, 1 - 2\cos^2(x) + \cos^4(x)\, \}\sin(x)\,.$$

Then $\int \sin^5(x)\, dx = \int \sin(x)\, dx - 2\int \cos^2(x)\sin(x)\, dx + \int \cos^4(x)\sin(x)\, dx$.

The first integral is easy, and the last two can be evaluated by changing the variable to $u = \cos(x)$:

$$\int \sin^5(x)\, dx = -\cos(x) - 2\left\{\, -\frac{\cos^3(x)}{3}\, \right\} + \left\{\, -\frac{\cos^5(x)}{5}\, \right\} + C\,.$$

Practice 2:   Evaluate $\int \cos^5(x)\, dx$.

Patterns for $\int \sin^m(x) \cos^n(x)\ dx$

If the exponent of sine is odd, we can split off one factor $\sin(x)$ and use the identity $\sin^2(x) = 1 - \cos^2(x)$ to rewrite the remaining even power of sine in terms of cosine. Then the change of variable $u = \cos(x)$ makes all of the integrals straightforward.

Example 3:   Evaluate $\int \sin^3(x) \cos^6(x)\ dx$.

Solution: $\sin^3(x)\cos^6(x) = \sin(x)\,\mathbf{\sin^2(x)}\cos^6(x) = \sin(x)\,\{\, \mathbf{1 - \cos^2(x)}\, \}\cos^6(x)$

$$= \sin(x)\cos^6(x) - \sin(x)\cos^8(x)\,.$$

Then $\int \sin^3(x)\cos^6(x)\ dx = \int \sin(x)\cos^6(x) - \sin(x)\cos^8(x)\ dx$ (put $u = \cos(x)$)

$$= -\frac{\cos^7(x)}{7} + \frac{\cos^9(x)}{9} + C\,.$$

Practice 3:   Evaluate $\int \sin^3(x) \cos^4(x)\ dx$.

If the **exponent of cosine is odd**, we can split off one factor $\cos(x)$ and use the identity $\cos^2(x) = 1 - \sin^2(x)$ to rewrite the remaining even power of cosine in terms of sine. Then the change of variable $u = \sin(x)$ makes all of the integrals straightforward.

If **both exponents are even**, we can use the identities $\sin^2(x) = \frac{1}{2}(1 - \cos(2x))$ and $\cos^2(x) = \frac{1}{2}(1 + \cos(2x))$ to rewrite the integral in terms of powers of $\cos(2x)$ and then proceed with integrating even powers of cosine.

Powers of Secant and Tangent Alone: $\int \sec^n(x)\,dx$, $\int \tan^n(x)\,dx$

All of the integrals of powers of secant and tangent can be evaluated by knowing

$$\int \sec(x)\,dx \ = \ \ln|\sec(x) + \tan(x)| + C \ \text{ and}$$

$$\int \tan(x)\,dx = -\ln|\cos(x)| + C \ = \ \ln|\sec(x)| + C$$

and then using the reduction formulas

$$\int \sec^n(x)\,dx \ = \ \frac{\sec^{n-2}(x)\cdot\tan(x)}{n-1} \ + \ \frac{n-2}{n-1} \int \sec^{n-2}(x)\,dx \ \text{ and}$$

$$\int \tan^n(x)\,dx \ = \ \frac{\tan^{n-1}(x)}{n-1} \ - \ \int \tan^{n-2}(x)\,dx \ .$$

Example 4:   Evaluate $\int \sec^3(x)\,dx$.

Solution: Using the reduction formula with $n = 3$,

$$\int \sec^3(x)\,dx \ = \ \frac{\sec(x)\cdot\tan(x)}{2} \ + \ \frac{1}{2} \int \sec(x)\,dx \ = \ \frac{\sec(x)\cdot\tan(x)}{2} \ + \ \frac{1}{2} \ln|\sec(x) + \tan(x)| + C.$$

Practice 4:   Evaluate $\int \tan^3(x)\,dx$ and $\int \sec^5(x)\,dx$.

Patterns for $\int \sec^m(x)\cdot\tan^n(x)\,dx$

The patterns for evaluating $\int \sec^m(x)\cdot\tan^n(x)\,dx$ are similar to those for $\int \sin^m(x)\cdot\cos^n(x)\,dx$ because we treat the even and odd powers differently and we use the identities $\tan^2(x) = \sec^2(x) - 1$ and $\sec^2(x) = \tan^2(x) + 1$.

If the **exponent of secant is even**, factor off $\sec^2(x)$, replace the other even powers (if any) of secant using $\sec^2(x) = \tan^2(x) + 1$, and make the change of variable $u = \tan(x)$ (then $du = \sec^2(x)\,dx$).

If the **exponent of tangent is odd**, factor off $\sec(x)\tan(x)$, replace the remaining even powers (if any) of tangent using $\tan^2(x) = \sec^2(x) - 1$, and make the change of variable $u = \sec(x)$ (then $du = \sec(x)\tan(x)\,dx$).

If the **exponent of secant is odd and the exponent of tangent is even**, replace the even powers of tangent using $\tan^2(x) = \sec^2(x) - 1$. Then the integral contains only powers of secant, and we can use the patterns for integrating powers of secant alone.

Example 5:   Evaluate $\int \sec(x)\cdot\tan^2(x)\,dx$.

Solution: Since the exponent of secant is odd and and the exponent of tangent is even, we can use the

last method mentions: replace the even powers of tangent using $\tan^2(x) = \sec^2(x) - 1$. Then

$$\int \sec(x)\cdot\tan^2(x)\,dx \;=\; \int \sec(x)\cdot\{\,\sec^2(x) - 1\,\}\,dx$$

$$=\; \int \sec^3(x) - \sec(x)\,dx \;=\; \int \sec^3(x)\,dx \;-\; \int \sec(x)\,dx$$

$$=\; \Big\{\;\frac{\sec(x)\cdot\tan(x)}{2} \;+\; \frac{1}{2}\,\ln|\sec(x) + \tan(x)|\,\Big\} \;-\; \ln|\sec(x) + \tan(x)| \;+\;C$$

$$=\; \frac{\sec(x)\cdot\tan(x)}{2} \;-\; \frac{1}{2}\,\ln|\sec(x) + \tan(x)| \;+\;C.$$

Practice 5:   Evaluate $\int \sec^4(x)\cdot\tan^2(x)\,dx$.

Wrap Up

Even if you use tables of integrals (or computers) for most of your future work, it is important to realize that most of the integral formulas can be derived from some basic facts using the techniques we have discussed in this and earlier sections.

PROBLEMS

Evaluate the integrals. (More than one method works for some of the integrals.)

1. $\int \sin^2(3x)\,dx$ 2. $\int \cos^2(5x)\,dx$ 3. $\int e^x\cdot\sin(e^x)\cdot\cos(e^x)\,dx$

4. $\int \frac{1}{x}\cdot\sin^2(\ln(x))\,dx$ 5. $\int_0^\pi \sin^4(3x)\,dx$ 6. $\int_0^\pi \cos^4(5x)\,dx$

7. $\displaystyle\int_0^\pi \sin^3(7x) \; dx$ 8. $\displaystyle\int_0^\pi \cos^3(5x) \; dx$ 9. $\displaystyle\int \sin(7x)\cdot\cos(7x) \; dx$

10. $\displaystyle\int \sin(7x)\cdot\cos^2(7x) \; dx$ 11. $\displaystyle\int \sin(7x)\cdot\cos^3(7x) \; dx$ 12. $\displaystyle\int \sin^2(3x)\cdot\cos(3x) \; dx$

13. $\displaystyle\int \sin^2(3x)\cdot\cos^2(3x) \; dx$ 14. $\displaystyle\int \sin^2(3x)\cdot\cos^3(3x) \; dx$ 15. $\displaystyle\int \sec^2(5x)\cdot\tan(5x) \; dx$

16. $\displaystyle\int \sec^2(3x)\cdot\tan^2(3x) \; dx$ 17. $\displaystyle\int \sec^3(3x)\cdot\tan(3x) \; dx$ 18. $\displaystyle\int \sec^3(5x)\cdot\tan^2(5x) \; dx$

The definite integrals of various combinations of sine and cosine on the interval $[0, 2\pi]$ exhibit a number of interesting patterrns. For now these patterns are simply curiousities and a source of additional problems for practice, but the patterns are very important as the foundation for an applied topic, Fourier Series, that you may encounter in more advanced courses.

The next three problems ask you to show that the definite integral on $[0, 2\pi]$ of $\sin(mx)$ multiplied by almost any other combination of $\sin(nx)$ or $\cos(nx)$ is 0. The only nonzero value comes when $\sin(mx)$ is multiplied by itself.

19. Show that if m and n are integers with $m \neq n$, then $\displaystyle\int_0^{2\pi} \sin(mx)\cdot\sin(nx) \; dx = 0$.

20. Show that if m and n are integers, then $\displaystyle\int_0^{2\pi} \sin(mx)\cdot\cos(nx) \; dx = 0$. (Consider $m = n$ and $m \neq n$.)

21. Show that if $m \neq 0$ is an integer, then $\displaystyle\int_0^{2\pi} \sin(mx)\cdot\sin(mx) \; dx = \pi$.

22. Suppose $P(x) = 5\cdot\sin(x) + 7\cdot\cos(x) - 4\cdot\sin(2x) + 8\cdot\cos(2x) - 2\cdot\sin(3x)$. (This is called a trigonometric polynomial.) Use the **results** of problems 19–21 to quickly evaluate

    (a) $a_1 = \dfrac{1}{\pi}\displaystyle\int_0^{2\pi} \sin(1x)\cdot P(x) \; dx$ (b) $a_2 = \dfrac{1}{\pi}\displaystyle\int_0^{2\pi} \sin(2x)\cdot P(x) \; dx$

    (c) $a_3 = \dfrac{1}{\pi}\displaystyle\int_0^{2\pi} \sin(3x)\cdot P(x) \; dx$ (d) $a_4 = \dfrac{1}{\pi}\displaystyle\int_0^{2\pi} \sin(4x)\cdot P(x) \; dx$

    (e)   Describe how the values of a_i are related to the coeffiecients of $P(x)$.

    (f)   Make up your own trigonometric polynomial $P(x)$ and see if your description in part (e) holds for the a_i values calculated from the new $P(x)$.

(g)   Just by knowing the a_i values we can "rebuild" part of $P(x)$. Find a similar method for getting the coefficients of the cosine terms of $P(x)$: b_i = ??

23. Show that if n is a positive, **odd** integer, then $\int_{0}^{2\pi} \sin^n(x)\ dx = 0$.

24. It is straightforward (using formula 19 in the integral table) to show that $\int_{0}^{2\pi} \sin^2(x)\ dx = \pi$, $\int_{0}^{2\pi}$

$\sin^4(x)\ dx = \frac{3}{4}\ \pi$, and $\int_{0}^{2\pi} \sin^6(x)\ dx = \frac{5}{6}\frac{3}{4}\ \pi$.   (a)   Evaluate $\int_{0}^{2\pi} \sin^8(x)\ dx$.

(b)   Predict the value of $\int_{0}^{2\pi} \sin^{10}(x)\ dx$ and then evalaute the integral.

Section 8.6   Practice Answers

Practice 1: $\int \cos^4(x)\ dx$ $\{$ Use $\cos^2(x) = \frac{1}{2}(1 + \cos(2x))$ $\}$

$$= \int \cos^2(x) \cdot \cos^2(x)\ dx\ \ = \int \frac{1}{2}(1 + \cos(2x))\frac{1}{2}(1 + \cos(2x))\ dx$$

$$= \frac{1}{4} \int 1 + 2\cos(2x) + \cos^2(2x)\ dx\ \ = \frac{1}{4} \int 1 + 2\cos(2x) + \frac{1}{2}\{1 + \cos(4x)\}\ dx$$

$$= \frac{1}{4} \int \frac{3}{2} + 2\cos(2x) + \frac{1}{2}\cos(4x)\ dx\ \ = \frac{3}{8}\ x + \frac{1}{4}\ \sin(2x) + \frac{1}{32}\ \sin(4x) + C\ .$$

Practice 2: $\int \cos^5(x)\ dx\ =\ \int \cos^2(x) \cdot \cos^2(x) \cdot \cos(x)\ dx\ =\ \int (1 - \sin^2(x))(1 - \sin^2(x))\cos(x)\ dx$

$$= \int \{1 - 2\sin^2(x) + \sin^4(x)\}\cos(x)\ dx$$

$$= \int \cos(x)\ dx\ -\ 2 \int \sin^2(x) \cdot \cos(x)\ dx\ +\ \int \sin^4(x) \cdot \cos(x)\ dx \quad (\text{Use } u = \sin(x),\ du = \cos(x)\ dx\ )$$

$$= \sin(x) - \frac{2}{3}\ \sin^3(x) + \frac{1}{5}\ \sin^5(x) + C\ .$$

Practice 3: $\int \sin^3(x) \cdot \cos^4(x)\ dx\ =\ \int \sin(x) \cdot \sin^2(x) \cdot \cos^4(x)\ dx\ =\ \int \sin(x) \cdot (1 - \cos^2(x)) \cdot \cos^4(x)\ dx$

$$= \int \sin(x) \cdot \cos^4(x)\ dx\ -\ \int \sin(x) \cdot \cos^6(x)\ dx \quad (\text{Use } u = \cos(x),\ du = -\sin(x)\ dx\ )$$

$$= -\frac{1}{5}\ \cos^5(x) + \frac{1}{7}\ \cos^7(x) + C$$

Practice 4: $\int \tan^3(x)\,dx = \frac{1}{2}\tan^2(x) - \int \tan(x)\,dx = \frac{1}{2}\tan^2(x) - \ln|\sec(x)| + C$.

$$\int \sec^5(x)\,dx = \frac{1}{2}\sec^3(x)\cdot\tan(x) + \frac{3}{4}\int \sec^3(x)\,dx$$

$$= \frac{1}{2}\sec^3(x)\cdot\tan(x) + \frac{3}{4}\left\{\frac{1}{2}\sec(x)\cdot\tan(x) + \frac{1}{2}\int \sec(x)\,dx\right\}$$

$$= \frac{1}{2}\sec^3(x)\cdot\tan(x) + \frac{3}{8}\sec(x)\cdot\tan(x) + \frac{3}{8}\ln|\sec(x) + \tan(x)| + C.$$

Practice 5: $\int \sec^4(x)\cdot\tan^2(x)\,dx = \int \sec^2(x)\cdot\sec^2(x)\cdot\tan^2(x)\,dx$

$$= \int \sec^2(x)\cdot(\tan^2(x) + 1)\cdot\tan^2(x)\,dx$$

$$= \int \sec^2(x)\cdot\tan^4(x)\,dx + \int \sec^2(x)\cdot\tan^2(x)\,dx \quad (\text{Use } u = \tan(x),\ du = \sec^2(x)\,dx\ )$$

$$= \frac{1}{5}\tan^5(x) + \frac{1}{3}\tan^3(x) + C .$$

PROBLEM ANSWERS Chapter Eight

Section 8.1

1. $\int = \lim\limits_{A\to\infty}\{-\dfrac{1}{2x^2}\Big|_{10}^{A}\} = \lim\limits_{A\to\infty}\{(-\dfrac{1}{2A^2})-(-\dfrac{1}{200})\} = \lim\limits_{A\to\infty}\{\dfrac{1}{200}-\dfrac{1}{2A^2}\} = \dfrac{1}{200}$.

3. $\int = \lim\limits_{A\to\infty}\{2\!\cdot\!\arctan(x)\Big|_{3}^{A}\} = \lim\limits_{A\to\infty}\{2\!\cdot\!\arctan(A) - 2\!\cdot\!\arctan(3)\} = 2(\dfrac{\pi}{2}) - 2\!\cdot\!\arctan(3) \approx 0.644$.

5. Use $u = \ln(x)$.

$\int = \lim\limits_{A\to\infty}\{5\!\cdot\!\ln(\ln(x))\Big|_{e}^{A}\} = \lim\limits_{A\to\infty}\{5\!\cdot\!\ln(\ln(A)) - 5\!\cdot\!\ln(\ln(e))\} = \lim\limits_{A\to\infty}\{5\!\cdot\!\ln(\ln(A)) - 0\} = \infty$.

$\int$ DIVERGES.

7. $\int = \lim\limits_{A\to\infty}\{\ln(x-2)\Big|_{3}^{A}\} = \lim\limits_{A\to\infty}\{\ln(A-2) - \ln(3-2)\} = \lim\limits_{A\to\infty}\{\ln(A-2) - 0\} = \infty$. $\int$ DIVERGES.

9. $\int = \lim\limits_{A\to\infty}\{\dfrac{-1}{2(x-2)^2}\Big|_{3}^{A}\} = \lim\limits_{A\to\infty}\{\dfrac{-1}{2(A-2)^2} - \dfrac{-1}{2(3-2)^2}\} = \lim\limits_{A\to\infty}\{\dfrac{-1}{2(A-2)^2} + \dfrac{1}{2}\} = \dfrac{1}{2}$.

11. $\int = \lim\limits_{A\to\infty}\{\dfrac{-1}{x+2}\Big|_{3}^{A}\} = \lim\limits_{A\to\infty}\{\dfrac{-1}{A+2} - \dfrac{-1}{3+2}\} = \lim\limits_{A\to\infty}\{\dfrac{-1}{A+2} + \dfrac{1}{5}\} = \dfrac{1}{5}$.

13. $\int = \lim\limits_{A\to\infty}\{2\sqrt{x}\Big|_{A}^{4}\} = \lim\limits_{A\to\infty}\{2\sqrt{4} - 2\sqrt{A}\} = 4$.

15. $\int = \lim\limits_{A\to0^{+}}\{\dfrac{4}{3}x^{3/4}\Big|_{A}^{16}\} = \lim\limits_{A\to0^{+}}\{\dfrac{4}{3}(16)^{3/4} - \dfrac{4}{3}A^{3/4}\} = \lim\limits_{A\to0^{+}}\{\dfrac{32}{3} - \dfrac{4}{3}A^{3/4}\} = \dfrac{32}{3}$.

17. $\int = \lim\limits_{A\to2^{-}}\{\arcsin(\dfrac{x}{2})\Big|_{0}^{A}\} = \lim\limits_{A\to2^{-}}\{\arcsin(\dfrac{A}{2}) - \arcsin(\dfrac{0}{2})\} = \dfrac{\pi}{2} - 0 = \dfrac{\pi}{2}$.

19. $\int = \lim\limits_{A\to\infty}\{-\cos(x)\Big|_{-2}^{A}\} = \lim\limits_{A\to\infty}\{-\cos(A) - -\cos(-2)\} = \lim\limits_{A\to\infty} -0.416 - \cos(A)$ DNE. $\int$ DIVERGES.

21. $\int = \lim\limits_{A\to\pi/2}\{-\ln|\cos(x)|\,\Big|_{0}^{A}\} = \lim\limits_{A\to\pi/2}\{-\ln|\cos(A)| - -\ln|\cos(0)|\} = \lim\limits_{A\to\pi/2} 0 - \ln|\cos(A)|$ DNE.

$\int$ DIVERGES.

23. (a) $A = R$: $\displaystyle\int_R^{2R} \frac{R^2 P}{x^2}\, dx = R^2 P\left(\frac{-1}{x}\right)\Big|_R^{2R} = R^2 P\left(\frac{-1}{2R} - \frac{-1}{R}\right) = \frac{1}{2}\, RP$.

    $A = 2R$: $\displaystyle\int_R^{3R} \frac{R^2 P}{x^2}\, dx = R^2 P\left(\frac{-1}{x}\right)\Big|_R^{3R} = R^2 P\left(\frac{-1}{3R} - \frac{-1}{R}\right) = \frac{2}{3}\, RP$.

   (b) $\displaystyle\lim_{A \to \infty} \int_R^{R+A} \frac{R^2 P}{x^2}\, dx = \lim_{A \to \infty} R^2 P\left(\frac{-1}{x}\right)\Big|_R^{R+A} = \lim_{A \to \infty} R^2 P\left(\frac{-1}{R+A} - \frac{-1}{R}\right) = R^2 P\left(\frac{1}{R}\right) = RP.$

25. $\displaystyle\int_3^{\infty} \frac{1}{x(x^2+1)}\, dx = \int_3^{\infty} \frac{1}{x^3+x}\, dx < \int_3^{\infty} \frac{1}{x^3}\, dx$ which converges by the p–test.

   Therefore, $\displaystyle\int_3^{\infty} \frac{1}{x(x^2+1)}\, dx$ converges.

27. Fof $x > 0$, $\ln(x) < x$ so $x + \ln(x) < 2x$ and $\dfrac{7}{x + \ln(x)} > \dfrac{7}{2x}$.

   Then $\displaystyle\int_3^{\infty} \frac{7}{2x}\, dx = \frac{7}{2}\int_3^{\infty} \frac{1}{x}\, dx$ which diverges by the p–test. Therefore, $\displaystyle\int_3^{\infty} \frac{7}{x + \ln(x)}\, dx$ diverges.

29. $-1 \le \cos(x) \le 1$ so $0 \le 1 + \cos(x) \le 2$ and $0 \le \dfrac{1 + \cos(x)}{x^2} \le \dfrac{2}{x^2}$. Then

   $\displaystyle\int_7^{\infty} \frac{1 + \cos(x)}{x^2}\, dx \le \int_7^{\infty} \frac{2}{x^2}\, dx$ which converges by the p–test. Therefore, $\displaystyle\int_7^{\infty} \frac{1 + \cos(x)}{x^2}\, dx$ converges.

31. $V = \displaystyle\int_0^{\infty} \pi\left(\frac{1}{x^2+1}\right)^2 dx < \pi\int_0^{\infty} \frac{1}{x^4}\, dx$ which converges by the p–test.

   Therefore, $V = \displaystyle\int_0^{\infty} \pi\left(\frac{1}{x^2+1}\right)^2 dx$ converges.

33. (a) $\displaystyle\int_1^{A} \frac{1}{x}\, dx < \sum_{k=1}^{A-1} \frac{1}{k}$ (b) $\displaystyle\int_1^{A} \frac{1}{x}\, dx > \sum_{k=2}^{A} \frac{1}{k}$

Section 8.2

1. $u = x^2 + 7$, $du = 2x\, dx$, $3\, du = 6x\, dx$: $\displaystyle\int = \int u^2\, 3\, du = u^3 + C = (x^2+7)^3 + C$.

3. $u = x^2 - 3$, $du = 2x\, dx$, $3\, du = 6x\, dx$: $\displaystyle\int = \int \frac{1}{\sqrt{u}}\, 3\, du = 3 \cdot 2\sqrt{u} = 6\sqrt{x^2-3}\,\Big|_2^4 = 6\sqrt{13} - 6\sqrt{1} \approx 15.6$

5. $u = x^2 + 3$: $\int = 6 \ln(x^2 + 3) + C$.   7. $u = 3x + 2$: $\int = -\frac{1}{3} \cos(3x + 2) + C$.

9. $u = e^x + 3, du = e^x\, dx$: $\int = \int \sec^2(u)\, du = \tan(u) = \tan(e^x + 3)\Big|_0^1 = \tan(e + 3) - \tan(1+3) \approx -1.79$

11. $u = \ln(x), du = \frac{1}{x}\, dx$: $\int = \int u\, du = \frac{1}{2} u^2 + C = \frac{1}{2}(\ln(x))^2 + C$.

13. $u = \sin(x),\ du = \cos(x)\, dx$: $\int = \int e^u\, du = e^u + C = e^{\sin(x)} + C$.

15. $u = 3x, du = 3\, dx$: $\int = \int \frac{5}{1 + u^2}\, du = \frac{5}{3} \arctan(u) = \frac{5}{3} \arctan(3x)\Big|_1^3 = \frac{5}{3}\arctan(9) - \frac{5}{3}\arctan(3) \approx$

0.35

17. $u = \frac{1}{x}, du = \frac{-1}{x^2}\, dx$: $\int = \int -\cos(u)\, du = -\sin(u) = -\sin(\frac{1}{x})\Big|_1^2 = (-\sin(\frac{1}{2})) - (-\sin(1)) \approx 0.36$

19. $u = 5 + \sin^2(x),\ du = 2 \cdot \sin(x) \cdot \cos(x)\, dx$: $\int = \int \frac{1}{u}\ 3\, du = 3 \ln|u| + C = 3 \ln|5 + \sin^2(x)| + C$.

21. $\int = 5 \ln|2x + 5| + C$.

23. $\int = 2 \ln|5x^2 + 3|\Big|_1^3 = 2 \ln|48| - 2 \ln|8| = 2 \ln|\frac{48}{8}| = \ln(36) \approx 3.58$

25. $\int = \frac{7}{2} \arctan(\frac{x+3}{2})\Big|_0^1 = \frac{7}{2} \arctan(2) - \frac{7}{2} \arctan(1.5) \approx 0.44$

27. $u = e^x, du = e^x\, dx$: $\int = \int \frac{1}{1 + u^2}\, du = \arctan(u) + C = \arctan(e^x) + C$.

29. $u = 1 + \ln(x),\ du = \frac{1}{x}\, dx$: $\int = \int \frac{3}{u}\, du = 3 \ln|u| = 3 \ln|1 + \ln(x)|\Big|_1^e = 3 \ln|2| \approx 2.08$

31. $u = 1 - x^2,\ du = -2x\, dx$: $\int = \int -\sqrt{u}\, du = \frac{-2}{3} u^{3/2} = \frac{-2}{3}(1 - x^2)^{3/2}\Big|_0^1 = \frac{-2}{3}(0)^{3/2} - \frac{-2}{3}(1)^{3/2} = \frac{2}{3}$

33. $u = 1 + \sin(x),\ du = \cos(x)\, dx$: $\int = \int u^3\, du = \frac{1}{4} u^4 + C = \frac{1}{4}(1 + \sin(x))^4 + C$.

35. $u = \ln(x), du = \frac{1}{x}\, dx$: $\int = \int \sqrt{u}\, du = \frac{2}{3} u^{3/2} = \frac{2}{3}(\ln(x))^{3/2}\Big|_1^e = \frac{2}{3}(\ln(e))^{3/2} - \frac{2}{3}(\ln(1))^{3/2} = \frac{2}{3}$.

37. $u = 5 + \tan(x)$, $du = \sec^2(x)\,dx$: $\int = \int \frac{1}{u}\,du = \ln|u| + C = \ln|5 + \tan(x)| + C$.

39. $u = x - 5$, $du = dx$: $\int = \int \tan(u)\,du = \ln|\sec(u)| + C = \ln|\sec(x - 5)| + C$.

41. $u = 5x$, $du = 5\,dx$: $\int = \int \frac{1}{5} e^u\,du = \frac{1}{5} e^u = \frac{1}{5} e^{5x}\Big|_0^1 = \frac{1}{5} e^5 - \frac{1}{5} e^0 \approx 29.48$

43. $\int = \int \frac{7}{(x+2)^2 + 1}\,dx = 7\cdot\arctan(x+2) + C$.   45. $\int = \int \frac{2}{(x-3)^2 + 49}\,dx = \frac{2}{7}\cdot\arctan(\frac{x-3}{7}) + C$.

47. $\int = \int \frac{3}{(x+5)^2 + 4}\,dx = \frac{3}{2}\cdot\arctan(\frac{x+5}{2}) + C$.

49. $\int = \ln|x^2 + 4x + 5| + 7\cdot\arctan(x+2) + C$.   51. $\int = 2\cdot\ln|x^2 - 6x + 10| + 19\cdot\arctan(x-3) + C$.

53. $\int = \int \frac{6x - 12}{x^2 - 4x + 13}\,dx + \int \frac{17}{(x-2)^2 + 9}\,dx = 3\cdot\ln|x^2 - 4x + 13| + \frac{17}{3}\cdot\arctan(\frac{x-2}{3}) + C$.

Section 8.3

1. $\int 12x\cdot\ln(x)\,dx$ $u = \ln(x)$. Then $dv = 12x\,dx$, $du = \frac{1}{x}\,dx$, and $v = 6x^2$.

    $= uv - \int v\,du = \ln(x)\cdot 6x^2 - \int 6x^2 \cdot \frac{1}{x}\,dx = 6x^2 \ln(x) - \int 6x\,dx = \mathbf{6x^2\ln(x) - 3x^2 + C}$.

3. $\int x^4 \ln(x)\,dx$ $dv = x^4\,dx$. Then $u = \ln(x)$, $du = \frac{1}{x}\,dx$, and $v = \frac{1}{5} x^5$.

    $= uv - \int v\,du = \ln(x)\cdot\frac{1}{5} x^5 - \int \frac{1}{5} x^5 \cdot \frac{1}{x}\,dx = \frac{1}{5} x^5 \cdot\ln(x) - \int \frac{1}{5} x^4\,dx = \mathbf{\frac{1}{5} x^5 \cdot ln(x) - \frac{1}{25} x^5 + C}$.

5. $\int x\cdot\arctan(x)\,dx$ $dv = x\,dx$. Then $u = \arctan(x)$, $du = \frac{1}{1 + x^2}\,dx$, and $v = \frac{1}{2} x^2$.

    $= uv - \int v\,du = \arctan(x)\cdot\frac{1}{2} x^2 - \int \frac{1}{2} x^2 \cdot \frac{1}{1 + x^2}\,dx = \frac{1}{2} x^2\cdot\arctan(x) - \frac{1}{2} \int 1 - \frac{1}{1 + x^2}\,dx$

    $= \frac{1}{2} x^2\cdot\arctan(x) - \frac{1}{2} \{x - \arctan(x)\} + C = \mathbf{\frac{1}{2} x^2 \cdot arctan(x) - \frac{1}{2} \cdot x + \frac{1}{2} \cdot arctan(x) \} + C}$.

7. $\int_0^1 \frac{x}{e^{3x}}\,dx = \int_0^1 x\cdot e^{-3x}\,dx$. Put $u = x$. Then $dv = e^{-3x}\,dx$, $du = dx$, and $v = \frac{-1}{3} e^{-3x}$.

    $= uv - \int v\,du = x\cdot\frac{-1}{3} e^{-3x} - \int \frac{-1}{3} e^{-3x}\,dx = x\cdot\frac{-1}{3} e^{-3x} - \frac{1}{9} e^{-3x}\Big|_0^1$

    $= \{1\cdot\frac{-1}{3} e^{-3(1)} - \frac{1}{9} e^{-3(1)}\} - \{0\cdot\frac{-1}{3} e^{-3(0)} - \frac{1}{9} e^{-3(0)}\} = \mathbf{\frac{1}{9} - \frac{4}{9} e^{-3}}$.

9. $\int x \cdot \sec(x) \cdot \tan(x)\, dx$.   Put $u = x$. Then $dv = \sec(x) \cdot \tan(x)\, dx$, $du = dx$, and $v = \sec(x)$.

$= uv - \int v\, du = x \cdot \sec(x) - \int \sec(x)\, dx = \mathbf{x \cdot \sec(x) - \ln|\sec(x) + \tan(x)| + C}$.

11. $\displaystyle\int_{\pi/3}^{\pi/2} 7x \cdot \cos(3x)\, dx$ Put $u = 7x$. Then $dv = \cos(3x)\, dx$, $du = 7\, dx$, and $v = \frac{1}{3}\sin(3x)$.

$= uv - \int v\, du = 7x \cdot \frac{1}{3}\sin(3x) - \int \frac{1}{3}\sin(3x) \cdot 7\, dx = \frac{7}{3} x \cdot \sin(3x) + \frac{7}{9}\cos(3x)\ \Big|_{\pi/3}^{\pi/2}$

$= \left\{ \frac{7}{3} \cdot \frac{\pi}{2} \cdot \sin(3 \cdot \frac{\pi}{2}) + \frac{7}{9}\cos(3 \cdot \frac{\pi}{2}) \right\} - \left\{ \frac{7}{3} \cdot \frac{\pi}{3} \cdot \sin(3 \cdot \frac{\pi}{3}) + \frac{7}{9}\cos(3 \cdot \frac{\pi}{3}) \right\} \approx -2.887$.

13. $\int 12x \cdot \cos(3x^2)\, dx$. **Use u–substitution!** Put $u = 3x^2$. Then $du = 6x\, dx$ and $2\, du = 12x\, dx$.

$\int = \int \cos(u) \cdot 2\, du = 2 \cdot \sin(u) + C = \mathbf{2 \cdot \sin(3x^2) + C}$.

15. $\displaystyle\int_{1}^{3} \ln(2x + 5)\, dx$.   Put $u = \ln(2x + 5)$. Then $dv = dx$, $du = \frac{2}{2x + 5}\, dx$, and $v = x$.

$= uv - \int v\, du = \ln(2x + 5) \cdot x - \int x \cdot \frac{2}{2x + 5}\, dx = x \cdot \ln(2x + 5) - \int 1 - \frac{5}{2x + 5}\, dx$

$= x \cdot \ln(2x + 5) - \left\{ x - \frac{5}{2} \cdot \ln|2x + 5| \right\}\Big|_{1}^{3} = \left\{ 3 \cdot \ln(11) - 3 + \frac{5}{2} \cdot \ln|11| \right\} - \left\{ 1 \cdot \ln(7) - 1 + \frac{5}{2} \cdot \ln|7| \right\}$

$= \frac{11}{2} \cdot \ln(11) - \frac{7}{2} \cdot \ln(7) - 2 \approx \mathbf{4.38}$.

17. $\displaystyle\int_{1}^{e} (\ln(x))^2\, dx$.   Put $u = (\ln(x))^2$. Then $dv = dx$, $du = 2 \cdot \ln(x) \cdot \frac{1}{x}\, dx$, and $v = x$.

$= uv - \int v\, du = (\ln(x))^2 x - \int x \cdot 2 \cdot \ln(x) \cdot \frac{1}{x}\, dx$

$= x \cdot (\ln(x))^2 - \int 2 \cdot \ln(x)\, dx = x \cdot (\ln(x))^2 - 2\left\{ x \cdot \ln(x) - x \right\}\Big|_{1}^{e}$

$= \left\{ e(\ln(e))^2 - 2e \cdot \ln(e) + 2e \right\} - \left\{ 1(\ln(1))^2 - 2 \cdot \ln(1) + 2 \right\} = e - 2 \approx \mathbf{0.718}$.

19. $\int \arcsin(x)\, dx$.   Put $u = \arcsin(x)$. Then $dv = dx$, $du = \frac{1}{\sqrt{1 - x^2}}\, dx$, and $v = x$.

$= uv - \int v\, du = \arcsin(x) \cdot x - \int x \cdot \frac{1}{\sqrt{1 - x^2}}\, dx = x \cdot \arcsin(x) - \int \frac{x}{\sqrt{1 - x^2}}\, dx$ (use u–sub with $u = 1 - x^2$)

$= x \cdot \arcsin(x) - \int \frac{1}{\sqrt{u}}\left(\frac{-1}{2}\right) du = x \cdot \arcsin(x) + \sqrt{u} + C = \mathbf{x \cdot \arcsin(x) + \sqrt{1 - x^2} + C}$.

21. $\int x \cdot \arctan(3x)\, dx$.   Put $u = \arctan(3x)$. Then $dv = x\, dx$, $du = \dfrac{3}{1+9x^2}\, dx$, and $v = \dfrac{1}{2} x^2$.

$= uv - \int v\, du = \arctan(x) \cdot \dfrac{1}{2} x^2 - \int \dfrac{1}{2} x^2 \cdot \dfrac{3}{1+9x^2}\, dx = \dfrac{1}{2} x^2 \cdot \arctan(3x) - \dfrac{1}{2} \int \dfrac{1}{3} - \dfrac{1/3}{1+9x^2}\, dx$

$= \dfrac{1}{2} x^2 \cdot \arctan(3x) - \dfrac{1}{2} \{ \dfrac{x}{3} - \dfrac{1}{9} \cdot \arctan(3x) \} + C = \dfrac{1}{2} \mathbf{x^2 \cdot arctan(3x)} - \dfrac{1}{6} \mathbf{x} + \dfrac{1}{18} \mathbf{arctan(3x)} \} + \mathbf{C}.$

23. $\displaystyle\int_1^2 \dfrac{\ln(x)}{x}\, dx.$ **Use u–substitution!** Put $u = \ln(x)$. Then $du = \dfrac{1}{x}\, dx$.

$\displaystyle\int = \int u\, du = \dfrac{1}{2} u^2 = \dfrac{1}{2}(\ln(x))^2 \Big|_1^2 = \dfrac{1}{2}(\ln(2))^2 - \dfrac{1}{2}(\ln(1))^2 = \dfrac{1}{2}(\ln(2))^2 \approx \mathbf{0.240}$.

25. (a) $\int \sin^3(x)\, dx = \dfrac{1}{3} \{ -S^2 \cdot C + 2 \int S\, dx \} = \dfrac{1}{3} \{ -S^2 \cdot C - 2C \} + K$

$= \dfrac{1}{3} \{ -\sin^2(x) \cdot \cos(x) - 2 \cdot \cos(x) \} + K$.

(b) $\int \sin^4(x)\, dx = \dfrac{1}{4} \{ -S^3 \cdot C + 3 \int S^2\, dx \} = \dfrac{1}{4} \{ -S^4 \cdot C + 3[\dfrac{1}{2}(-SC + x)] \} + K$

$= \dfrac{1}{4} \{ -S^3 \cdot C - \dfrac{3}{2} SC + \dfrac{3}{2} x \} + K$

$= \dfrac{1}{4} \{ -\sin^3(x) \cdot \cos(x) - \dfrac{3}{2} \sin(x) \cdot \cos(x) + \dfrac{3}{2} x \} + K$ (c) on your own.

27. (a) $\int \sec^3(x)\, dx = \dfrac{1}{2} \{ \sec(x)\tan(x) + \int \sec(x)\, dx \} = \dfrac{1}{2} \{ \mathbf{sec(x) \cdot tan(x) + ln| sec(x) + tan(x) |} \} + \mathbf{K}$

(b) and (c)   on your own.

29. $\int \cos^3(2x + 3)\, dx$.   First do a substitution: $u = 2x + 3$. Then $du = 2\, dx$ and $dx = \dfrac{1}{2} du$.

$= \int \dfrac{1}{2} \cos^3(u)\, du = \dfrac{1}{2} \{ \dfrac{1}{3}(C^2 \cdot S + 2 \int C\, du) \} = \dfrac{1}{6} \{ C^2 \cdot S + 2S \} + K$

$= \dfrac{1}{6} \{ \cos^2(u) \cdot \sin(u) + 2\sin(u) \} + K = \dfrac{1}{6} \{ \mathbf{cos^2(2x + 3) \cdot sin(2x + 3) + 2sin(2x + 3)} \} + \mathbf{K}$

31. $\int x \cdot (2x + 5)^{19}\, dx$.

(a) By parts: put $u = x$. Then $dv = (2x + 5)^{19}\, dx$, $du = dx$, and $v = \dfrac{1}{40}(2x + 5)^{20}$.

$\int = uv - \int v\, du = x \cdot \dfrac{1}{40}(2x + 5)^{20} - \int \dfrac{1}{40}(2x + 5)^{20}\, dx$

$= x \cdot \dfrac{1}{40}(2x + 5)^{20} - \dfrac{1}{40}\dfrac{1}{42}(2x + 5)^{21} + C.$

(b) Substitution: put $u = 2x + 5$. Then $du = 2\, dx$ and $dx = \dfrac{1}{2} du$. Also, $x = \dfrac{1}{2}(u - 5)$.

$\int = \int \dfrac{1}{2}(u - 5) \cdot u^{19} \dfrac{1}{2}\, du = \dfrac{1}{4} \int u^{20} - 5u^{19}\, du = \dfrac{1}{4} \{ \dfrac{1}{21} u^{21} - \dfrac{5}{20} u^{20} \} + C$

$= \dfrac{1}{84}(2x + 5)^{21} - \dfrac{5}{80}(2x + 5)^{20} + C.$

The answers (antiderivatives) in parts (a) and (b) look different, but you can check that the derivative of each answer is $x \cdot (2x + 5)^{19}$.

33. (a) Make an informed prediction.

(b) $\int_0^1 \sin(x)\,dx = -\cos(x)\Big|_0^1 = (-\cos(1)) - (-\cos(0)) = \cos(0) - \cos(1) \approx 1 - 0.54 = \mathbf{0.46}$.

$\int_0^1 x\cdot\sin(x)\,dx = -x\cdot\cos(x) + \sin(x)\Big|_0^1 = \{-1\cdot\cos(1) + \sin(1)\} - \{-0\cdot\cos(0) + \sin(0)\} = \sin(1) - \cos(1) \approx \mathbf{0.30}$.

35. (a) Make an informed prediction.

(b) See problem 17: $V_{x-axis} = \int_{x=1}^e \pi(\ln(x))^2\,dx = \pi(e - 2) \approx \mathbf{2.257}$.

$V_{y-axis} = \int_{y=0}^1 2\pi\cdot y\cdot e^y\,dy = 2\pi \int_{y=0}^1 y\cdot e^y\,dy$ (use integration by parts with $u = y$, $dv = e^y\,dy$: see Example 2)

$= 2\pi(y\cdot e^y - e^y)\Big|_0^1 = 2\pi(1\cdot e^1 - e^1) - 2\pi(0\cdot e^0 - e^0) = 2\pi(0) - 2\pi(-1) = 2\pi \approx \mathbf{6.283}$.

37. On your own.

39. $\int x^2\cdot\arctan(x)\,dx$.   Put $u = \arctan(x)$. Then $dv = x^2\,dx$, $du = \dfrac{1}{x^2+1}\,dx$, and $v = \dfrac{1}{3}x^3$.

$= uv - \int v\,du = \arctan(x)\cdot\dfrac{1}{3}x^3 - \int \dfrac{1}{3}x^3\,\dfrac{1}{x^2+1}\,dx$

$= \dfrac{1}{3}x^3\cdot\arctan(x) - \dfrac{1}{3}\int x - \dfrac{x}{x^2+1}\,dx$ (dividing x^3 by x^2+1)

$= \dfrac{1}{3}x^3\cdot\arctan(x) - \dfrac{1}{3}\{\dfrac{1}{2}x^2 - \dfrac{1}{2}\ln(x^2+1)\} + C$

$-\dfrac{1}{3}x^3\cdot\mathbf{arctan(x)} - \dfrac{1}{6}x^2 + \dfrac{1}{6}\ln(x^2+1)\} + C$.

40 – 50. On your own.

Section 8.4

1. $= \dfrac{A}{x} + \dfrac{B}{x+1} = \dfrac{2}{x} + \dfrac{5}{x+1}$ 3. $= \dfrac{A}{x+1} + \dfrac{B}{x+8} = \dfrac{2}{x+1} + \dfrac{9}{x+8}$

5. Divide first: $\dfrac{2x^2 + 15x + 25}{x^2 + 5x} = 2 + \dfrac{5x + 25}{x^2 + 5x} = 2 + \dfrac{5(x+5)}{x(x+5)} = 2 + \dfrac{5}{x}$.

7. $\dfrac{6x^2 + 9x - 15}{x(x+5)(x-1)} = \dfrac{A}{x} + \dfrac{B}{x+5} + \dfrac{C}{x-1} = \dfrac{A(x^2 + 4x - 5) + B(x^2 - x) + C(x^2 + 5x)}{x(x+5)(x-1)}$.

Solving x^2: $A + B + C = 6$
x: $4A - B + 5C = 9$
k: $-5A = -15$ we get $A = 3$, $B = 3$, and $C = 0$ so

$$\frac{6x^2 + 9x - 15}{x(x + 5)(x - 1)} = \frac{3}{x} + \frac{3}{x+5} + \frac{0}{x-1} = \frac{3}{x} + \frac{3}{x+5} \ .$$

9. $\dfrac{8x^2 - x + 3}{x(x^2 + 1)} = \dfrac{A}{x} + \dfrac{Bx + C}{x^2 + 1} = \dfrac{A(x^2 + 1) + x(Bx + C)}{x(x^2 + 1)}$.

Solving x^2:   A +   B   = 8
            x: C   = -1
            k: A   = 3   we get A = 3, B = 5, and C = -1 so

$$\frac{8x^2 - x + 3}{x(x^2 + 1)} = \frac{3}{x} + \frac{5x - 1}{x^2 + 1} \ .$$

11. $\dfrac{11x^2 + 23x + 6}{x^2(x + 2)} = \dfrac{A}{x} + \dfrac{B}{x^2} + \dfrac{C}{x + 2} = \dfrac{A(x)(x + 2) + B(x + 2) + C(x^2)}{x^2(x + 2)}$.

Solving x^2:   A +   C   = 11
            x: 2A +   B   = 23
            k: 2B   = 6   we get A = 10, B = 3, and C = 1 so

$$\frac{11x^2 + 23x + 6}{x^2(x + 2)} = \frac{10}{x} + \frac{3}{x^2} + \frac{1}{x + 2} \ .$$

13. $\displaystyle\int \frac{3x + 13}{(x + 2)(x - 5)} \ dx = \int \frac{-1}{x + 2} + \frac{4}{x - 5} \ dx = -\ln|\,x + 2\,| + 4\cdot\ln|\,x - 5\,| + C$.

15. $\displaystyle\int_2^5 \frac{2}{x^2 - 1} \ dx = \int_2^5 \frac{-1}{x + 1} + \frac{1}{x - 1} \ dx = -\ln|\,x+1\,| + \ln|\,x - 1\,|$

$$= \ln\left|\frac{x - 1}{x + 1}\right|\ \Big|_2^5 = \ln\left(\frac{4}{6}\right) - \ln\left(\frac{1}{3}\right) \approx \mathbf{0.693} \ .$$

17. (1) $\displaystyle\int \frac{2}{x} + \frac{5}{x + 1} \ dx = \mathbf{2\cdot ln|\,x\,| + 5\cdot ln|\,x + 1\,| + C}$

   (2) $\displaystyle\int \frac{3}{x + 3} + \frac{4}{x - 1} \ dx = \mathbf{3\cdot ln|\,x + 3\,| + 4\cdot ln|\,x - 1\,| + C}$

19. $\displaystyle\int \frac{2x^2 + 5x + 3}{x^2 - 1} \ dx = \int 2 + \frac{5}{x - 1} \ dx = \mathbf{2x + 5\cdot ln|\,x - 1\,| + C}$.

21. $\displaystyle\int \frac{3x^2 + 19x + 24}{x^2 + 6x + 5} \ dx = \int 3 + \frac{2}{x + 1} + \frac{-1}{x + 5} \ dx = \mathbf{3x + 2\cdot ln|\,x + 1\,| - ln|\,x + 5\,| + C}$.

23. $\displaystyle\int \frac{3x^2 - 1}{x^3 - x} \ dx$. Use u–substitution with $u = x^3 - x$. Then $du = 3x^2 - 1$ so

$\displaystyle\int = \int \frac{1}{u} \ du = \ln|\,u\,| + C = \ln|\,x^3 - x|\ + C$. A partial fraction decomposition also works but takes longer.

25. $\int \dfrac{x^3 + 3x^2 - 4x + 30}{x^2 + 3x - 10}\ dx = \int\ x + \dfrac{6}{x-2}\ dx = \dfrac{1}{2}\ x^2 + 6\ln|\ x - 2\ | + C$.

27. $\int \dfrac{12x^2 + 19x - 6}{x^3 + 3x^2}\ dx = \int \dfrac{7}{x} + \dfrac{-2}{x^2} + \dfrac{5}{x+3}\ dx = 7\cdot\ln|\ x\ | + \dfrac{2}{x} + 5\cdot\ln|\ x+3\ | + C$.

29. $\int \dfrac{7x^2 + 3x + 7}{x^3 + x}\ dx = \int \dfrac{7}{x} + \dfrac{3}{x^2 + 1}\ dx = 7\cdot\ln|\ x\ | + 3\cdot\arctan(\ x\ ) + C$.

31. (a) $\int \dfrac{1}{x^2 + 2x + 2}\ dx = \int \dfrac{1}{(x+1)^2 + 1}\ dx = \arctan(\ x+1\ ) + C$.

   (b) $\int \dfrac{1}{x^2 + 2x + 1}\ dx = \int \dfrac{1}{(x+1)^2}\ dx = -(\ x+1\ )^{-1} + C = \dfrac{-1}{x+1} + C$.

   (c) $\int \dfrac{1}{x^2 + 2x + 0}\ dx = \int \dfrac{1}{x(x+2)}\ dx = \int \dfrac{1/2}{x} + \dfrac{-1/2}{x+2}\ dx = \dfrac{1}{2}\ \ln|\ x\ | - \dfrac{1}{2}\ \ln|\ x+2\ | + C$.

33. Prob. 1: $f(x) = \dfrac{2}{x} + \dfrac{5}{x+1} = 2\cdot(x)^{-1} + 5\cdot(x+1)^{-1}$. Then

   $f'(x) = -2\cdot(x)^{-2} - 5\cdot(x+1)^{-2}$ and $f''(x) = 4\cdot(x)^{-3} + 10\cdot(x+1)^{-3}$.

   Prob. 3: $g(x) = \dfrac{3}{x+3} + \dfrac{4}{x-1} = 3\cdot(x+3)^{-1} + 4\cdot(x-1)^{-1}$. Then

   $g'(x) = -3\cdot(x+3)^{-2} - 4\cdot(x-1)^{-2}$ and $g''(x) = 6\cdot(x+3)^{-3} + 8\cdot(x-1)^{-3}$.

35. Prob. 5: $f(x) = \dfrac{2x^2 + 15x + 15}{x^2 + 5x} = 2 + \dfrac{5}{x}$ so $f'(x) = \dfrac{-5}{x^2}$ and $f''(x) = \dfrac{10}{x^3}$.

   Prob. 6: On your own.

37. (a) Solve $\dfrac{dx}{dt} = x(100 - x)$.

   Separate the variables: $\dfrac{1}{x(100 - x)}\ dx = dt$.

   Use partial fractions: $\dfrac{1}{x(100 - x)} = \dfrac{A}{x} + \dfrac{B}{100 - x} = \dfrac{0.01}{x} + \dfrac{0.01}{100 - x}$ (solving $-A+B = 0$ and $100A = 1$)

   so $\left\{ \dfrac{0.01}{x} + \dfrac{0.01}{100 - x} \right\} dx = dt$ and, integrating, $\int \dfrac{0.01}{x} + \dfrac{0.01}{100 - x}\ dx = \int 1\ dt$. Then

   $0.01\cdot\ln|\ x\ | - 0.01\cdot\ln|\ 100-x\ | = t + C$ so $\ln|\ \dfrac{x}{100 - x}\ | = 100t + K$. ($K = 100C$ is a constant)

   Using the initial condition $x(0) = 150$: $\ln|\ \dfrac{150}{100-150}\ | = \ln(\ 3\ ) = 100\cdot(0) + K$ so $K = \ln(\ 3\ )$.

   Finally, $\ln|\ \dfrac{x}{100 - x}\ | = 100\cdot t + \ln(3)$ so $|\ \dfrac{x}{100 - x}\ | = e^{100t}\cdot e^{\ln(3)} = 3\cdot e^{100t}$ and

   $x = \dfrac{-300\cdot e^{100t}}{1 - 3\cdot e^{100t}}$. Graph this on your own.

(b)   In this part it is easier to use the form $| \frac{x}{100-x} | = 3 \cdot e^{100t}$.

Put $x = 120$ and solve $| \frac{120}{100-120} | = 3 \cdot e^{100t}$: $6 = 3 \cdot e^{100t}$ so $t = \frac{1}{100} \ln(2) \approx \mathbf{0.0069}$.

Put $x = 110$ and solve $| \frac{110}{100-110} | = 3 \cdot e^{100t}$: $11 = 3 \cdot e^{100t}$ so $t = \frac{1}{100} \ln(\frac{11}{3}) \approx \mathbf{0.013}$.

Put $x = 100$. Then $| \frac{x}{100-x} |$ is undefined (division by 0) so $x(t)$ is never equal to 100 .

(c)   limit (as t becomes arbitrarily large) of $x = \frac{-300 \cdot e^{100t}}{1 - 3 \cdot e^{100t}}$ is $\frac{-300}{-3} = 100$.

(d)   The population $x(t)$ is decling to the "carrying capacity" $M = 100$ of the enviroment.

39. (a)   Solve $\frac{dx}{dt} = (7-x)(5-x)$ by separating the variables and using partial fractions to rewrite the fraction.

$\frac{7-x}{5-x} = \frac{7}{5} e^{2kt}$ and $x(t) = \frac{7 \cdot e^{2kt} - 7}{\frac{7}{5} \cdot e^{2kt} - 1}$. (As t gets big, x approaches 5.)

(b)   Solve $\frac{dx}{dt} = (6-x)(6-x) = (6-x)2$ by separating the variables: $\frac{1}{(6-x)^2}$ dx = dt and integrating.

$\frac{1}{6-x} = t + C$ $(C = \frac{1}{6})$ so $x = \frac{6t+1}{t + \frac{1}{6}} = \frac{36t+6}{6t+1}$. (As t gets big, x approaches 6.)

Section 8.5

1. $x = 3 \cdot \sin(\theta)$ (a) $9 - x^2 = 9 - 9\sin^2(\theta) = 9(1 - \sin^2(\theta)) = 9\cos^2(\theta)$ so $\frac{1}{\sqrt{9-x^2}} = \frac{1}{3\cos(\theta)}$.

(b) $dx = 3\cos(\theta) \, d\theta$

3. $x = 3 \cdot \sec(\theta)$ (a) $x^2 - 9 = 9\sec^2(\theta) - 9 = 9(\sec^2(\theta) - 1) = 9\tan^2(\theta)$ so $\frac{1}{\sqrt{x^2-9}} = \frac{1}{3\tan(\theta)}$.

(b) $dx = 3\sec(\theta)\tan(\theta) \, d\theta$

5. $x = \sqrt{2} \, \tan(\theta)$ (a) $2 + x^2 = 2 + 2\tan^2(\theta) = 2(1 + \tan^2(\theta)) = 2\sec^2(\theta)$ so $\frac{1}{\sqrt{2+x^2}} = \frac{1}{\sqrt{2} \sec(\theta)}$.

(b) $dx = \sqrt{2} \, \sec^2(\theta) \, d\theta$

7. $x = 3 \cdot \sin(\theta)$ (a) $\theta = \arcsin(x/3)$

(b) & (c) $f(\theta) = \cos(\theta) \cdot \tan(\theta) = \cos(\arcsin(x/3)) \cdot \tan(\arcsin(x/3)) = \frac{\sqrt{9-x^2}}{3} \cdot \frac{x}{\sqrt{9-x^2}} = \frac{x}{3}$.

9. $x = 3 \cdot \sec(\theta)$ (a) $\theta = \text{arcsec}(x/3)$

(b) & (c) $f(\theta) = \sqrt{1 + \sin^2(\theta)} = \sqrt{1 + \sin^2(\text{arcsec}(x/3))} = \sqrt{1 + \left(\frac{\sqrt{x^2-9}}{x}\right)^2}$

11. $x = 5 \cdot \tan(\theta)$ (a) $\theta = \arctan(x/5)$

(b) & (c) $f(\theta) = \dfrac{\cos^2(\theta)}{1 + \cot(\theta)} = \dfrac{\left(\dfrac{5}{\sqrt{25 + x^2}} \right)^2}{1 + \left(\dfrac{5}{x} \right)} = \dfrac{\dfrac{25}{25 + x^2}}{1 + \dfrac{5}{x}}$

13. Same as Practice 3. Sorry.

15. $x = 7 \cdot \tan(\theta)$. $dx = 7 \sec^2(\theta)\, d\theta$. $x^2 + 49 = 49 \tan^2(\theta) + 49 = 49(\tan^2(\theta) + 1) = 49 \sec^2(\theta)$.

$\displaystyle \int \frac{1}{\sqrt{x^2 + 49}}\, dx = \int \frac{1}{\sqrt{49 \sec^2(\theta)}}\, 7 \sec^2(\theta)\, d\theta$

$\displaystyle = \int \sec(\theta)\, d\theta = \ln| \sec(\theta) + \tan(\theta) | + C = \ln| \sec(\arctan(x/7)) + \tan(\arctan(x/7)) | + C$

$= \ln\left| \dfrac{\sqrt{x^2 + 49}}{7} + \dfrac{x}{7} \right| + C$.

17. $x = 6 \cdot \sin(\theta)$. $dx = 6 \cos(\theta)\, d\theta$. $36 - x^2 = 36 \cos^2(\theta)$.

$\displaystyle \int \sqrt{36 - x^2}\, dx = \int \sqrt{36 - \cos^2(\theta)}\ 6 \cos(\theta)\, d\theta = 36 \int \cos^2(\theta)\, d\theta =$ (use Table #14)

$= 36 \left\{ \dfrac{1}{2} \theta + \dfrac{1}{2} \sin(\theta) \cdot \cos(\theta) \right\} + C = 36 \left\{ \dfrac{1}{2} \arcsin(x/6) + \dfrac{1}{2} \left(\dfrac{x}{6} \right) \cdot \left(\dfrac{\sqrt{36 - x^2}}{6} \right) \right\} + C$.

19. $x = 6 \cdot \tan(\theta)$. $dx = 6 \cdot \sec^2(\theta)\, d\theta$. $36 + x^2 = 36 \sec^2(\theta)$.

$\displaystyle \int \sqrt{36 + x^2}\, dx = \int \frac{1}{\sqrt{36 + x^2}}\, 6 \cdot \sec^2(\theta)\, d\theta = \int \sec(\theta)\, d\theta =$ (use Table #11)

$= \ln| \sec(\theta) + \tan(\theta) | + C = \ln\left| \dfrac{\sqrt{36 + x^2}}{6} + \dfrac{x}{6} \right| + C$ or $\ln| \sqrt{36 + x^2} + x | + K$.

21. Similar to 19: $x = 7 \cdot \tan(\theta)$. $\displaystyle \int \frac{1}{\sqrt{49 + x^2}}\, dx = \ln| \sqrt{49 + x^2} + x | + C$.

23. $x = 5 \cdot \sin(\theta)$. $\displaystyle \int \ = -5 \cos(\arcsin(x/5)) + C = -5\, \dfrac{\sqrt{25 - x^2}}{5} + C = -\sqrt{25 - x^2} + C$.

25. $x = 7 \cdot \tan(\theta)$. $\displaystyle \int \ = -\ln| \cos(\arctan(x/7)) | + C = -\ln\left| \dfrac{7}{\sqrt{49 + x^2}} \right| + C$ (now some algebra)

$= -\ln| 7 | + \ln| \sqrt{49 + x^2} | + C = \dfrac{1}{2} \ln| 49 + x^2 | + K$.

(A u–substitution with $u = 49 + x^2$ is **much** easier.)

27. $x = 3 \cdot \sec(\theta)$. $\int = \frac{1}{9} \int \frac{\cos(\theta)}{\sin^2(\theta)} d\theta = (\text{put } u = \sin(\theta)) \frac{-1}{9} \frac{1}{\sin(\theta)} + C = -\csc(\theta) + C = \frac{-1}{9} \frac{x}{\sqrt{x^2 + 9}} + C$.

29. $x = 5 \cdot \sec(\theta)$. $\int = \frac{1}{2} \theta + C = \frac{1}{2} \text{arcsec}(\frac{x}{5}) + C$.

31. $x = 5 \cdot \sin(\theta)$. $\int = \frac{1}{5} \ln| \sec(\theta) + \tan(\theta) | + C = \frac{1}{5} \ln\left| \frac{5}{\sqrt{25 - x^2}} + \frac{x}{\sqrt{25 - x^2}} \right| + C$

$$= (\text{after lots of algebra}) \ \frac{1}{10} \ln\left| \frac{5 + x}{5 - x} \right| + C.$$

33. Similar to 19. $\int = \ln\left| \frac{\sqrt{a^2 + x^2}}{a} + \frac{x}{a} \right| + C = \ln\left| \sqrt{a^2 + x^2} + x \right| + K$.

35. $x = a \cdot \tan(\theta)$. $\int = \frac{-1}{a^2} \frac{\sqrt{a^2 + x^2}}{x} + C$.

37. $x + 1 = u$. Then $u = 3 \cdot \tan(\theta)$. $\int = \ln| \sec(\theta) + \tan(\theta) | + C = \ln\left| \frac{\sqrt{u^2 + 9}}{3} + \frac{u}{3} \right| + C$

$$= \ln| \sqrt{u^2 + 9} + u | - \mathbf{ln(3)} + C = \ln| \sqrt{u^2 + 9} + u | + \mathbf{K} = \ln| \sqrt{(x+1)^2 + 9} + (x+1) | + K.$$

39. $\frac{1}{2} \arctan(\frac{x+5}{2}) + C$ 41. $\ln| (x+2) + \sqrt{x^2 + 4x + 3} | + C$.

Section 8.6

1. $\int \sin^2(3x) dx = \frac{x}{2} - \frac{\sin(6x)}{12} + C = \frac{x}{2} - \frac{\sin(3x) \cdot \cos(3x)}{6} + C$.

3. Put $u = \sin(e^x)$. $\int = \frac{1}{2} \sin^2(e^x) + C$. (If you put $w = \cos(e^x)$ then $\int = -\frac{1}{2} \cos^2(e^x) + C$.)

5. $\frac{3}{8} \pi$ 7. $\frac{4}{21}$

9. $\frac{1}{14} \sin^2(7x) + C$ or $-\frac{1}{14} \cos^2(7x) + C$

11. Put $u = \cos(7x)$. $\int = -\frac{1}{28} \cos^4(7x) + C$.

13. $\int \sin^2(3x)\cos^2(3x)\,dx = \int \frac{1}{2}(1-\cos(6x))\frac{1}{2}(1+\cos(6x))\;dx$

$$= \frac{1}{4}\int 1-\cos^2(6x)\;dx = \frac{x}{4} - \frac{1}{4}\int \cos^2(6x)\;dx$$

$$= \frac{x}{4} - \frac{1}{4}\left\{\frac{x}{2} + \frac{\sin(12x)}{24}\right\} + C.$$

15. $\int = \frac{1}{10}\tan^2(5x) + C.$
17. $\int = \frac{1}{9}\sec^3(3x) + C.$

19. $m \ne n.$ $\displaystyle\int_0^{2\pi} \sin(mx)\sin(nx)\,dx = \frac{1}{2}\left\{\frac{\sin((m-n)x)}{m-n} - \frac{\sin((m+n)x)}{m+n}\right\}\Big|_0^{2\pi}$

$$= \frac{1}{2}\left\{\frac{\sin((m-n)2\pi)}{m-n} - \frac{\sin((m+n)2\pi)}{m+n}\right\} - \frac{1}{2}\left\{\frac{\sin((m-n)0)}{m-n} - \frac{\sin((m+n)0)}{m+n}\right\}$$

$$= \frac{1}{2}\{0-0\} - \frac{1}{2}\{0-0\} = 0.$$

21. $\displaystyle\int_0^{2\pi} \sin(mx)\sin(mx)\,dx = \int_0^{2\pi} \sin^2(mx)\,dx = \frac{x}{2} - \frac{\sin(mx)\cos(mx)}{2m}\Big|_0^{2\pi}$

$$= \left\{\frac{2\pi}{2} - \frac{\sin(m2\pi)\cos(m2\pi)}{2m}\right\} - \left\{\frac{0}{2} - \frac{\sin(0)\cos(0)}{2m}\right\} = \pi.$$

23. On your own.

9.1 POLAR COORDINATES

The rectangular coordinate system is immensely useful, but it is not the only way to assign an address to a point in the plane and sometimes it is not the most useful. In many experimental situations, our location is fixed and we or our instruments, such as radar, take readings in different directions (Fig. 1); this information can be graphed using rectangular coordinates (e.g., with the angle on the horizontal axis and the measurement on the vertical axis). Sometimes, however, it is more useful to plot the information in a way similar to the way in which it was collected, as magnitudes along radial lines (Fig. 2). This system is called the Polar Coordinate System.

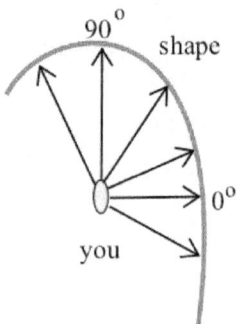

angle	distance
$0°$	120 ft
$30°$	140
$60°$	230
$90°$	270
$120°$	200

Numerical Data

Taking Measurements

Fig. 1

Rectangular Coordinate Graph of Data

In this section we introduce polar coordinates and examine some of their uses. We start with graphing points and functions in polar coordinates, consider how to change back and forth between the rectangular and polar coordinate systems, and see how to find the slopes of lines tangent to polar graphs. Our primary reasons for considering polar coordinates, however, are that they appear in applications, and that they provide a "natural" and easy way to represent some kinds of information.

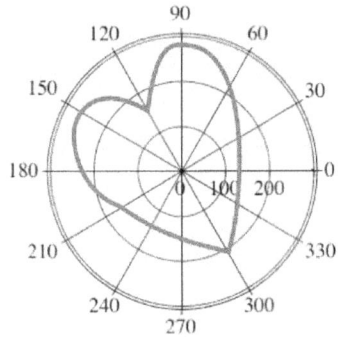

Polar coordinate graph of data

Fig. 2

Example 1: SOS! You've just received a distress signal from a ship located at A on your radar screen (Fig. 3). Describe its location to your captain so your vessel can speed to the rescue.

Solution: You could convert the relative location of the other ship to rectangular coordinates and then tell your captain to go due east for 7.5 miles and north for 13 miles,

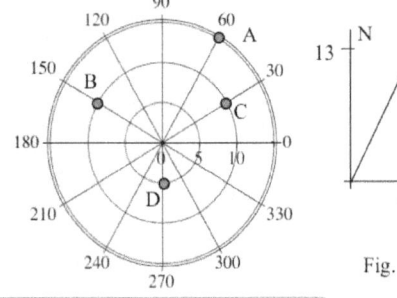

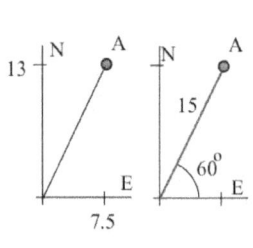

Fig. 3

but that certainly is not the quickest way to reach the other ship. It is better to tell the captain to sail for 15

miles in the direction of 60°. If the distressed ship was at B on the radar screen, your vessel should sail

for 10 miles in the direction 150°. (Real radar screens have 0° at the top of the screen, but the convention

in mathematics is to put 0° in the direction of the positive x–axis and to measure positive angles

counterclockwise from there. And a real sailor speaks of "bearing" and "range" instead of direction and

magnitude.)

Practice 1:   Describe the locations of the ships at C and D in

Fig. 3 by giving a distance and a direction to those

ships from your current position at the center of the

radar screen.

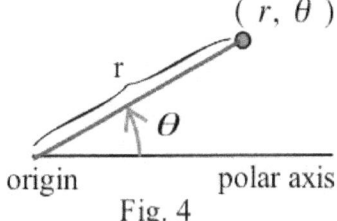

Fig. 4

Points in Polar Coordinates

To construct a polar coordinate system we need a starting point (called the **origin** or **pole**) for the magnitude

measurements and a starting direction (called the **polar axis**) for the angle measurements (Fig. 4).

A polar coordinate pair for a point P in the plane is an ordered pair (r, θ) where r is the directed distance

along a radial line from O to P, and θ is the angle formed by the polar axis and the segment OP (Fig. 4).

The angle θ is positive when the angle of the radial line OP is measured counterclockwise from the polar

axis, and θ is negative when measured clockwise.

Degree or Radian Measure for θ? Either degree or radian measure can be used for the angle in the polar

coordinate system, but when we differentiate and integrate trigonometric functions of θ we will want all of the

angles to be given in radian measure. From now on, we will primarily use radian measure. You should

assume that all angles are given in radian measure unless the units " ° " ("degrees") are shown.

Example 2:   Plot the points with the given polar coordinates: $A(2, 30°), B(3, \pi/2), C(-2, \pi/6)$,

and $D(-3, 270°)$.

Solution: To find the location of A, we look along the ray that makes an angle of 30° with the polar axis, and

then take two steps in that direction (assuming 1 step = 1 unit). The locations of

A and B are shown in Fig. 5.

To find the location of C, we look along the ray which makes an angle of $\pi/6$ with

the polar axis, and then we take two steps backwards since $r = -2$ is negative.

Fig. 6 shows the locations of C and D.

Notice that the points B and D have different addresses, $(3, \pi/2)$ and $(-3, 270°)$,

but the same location.

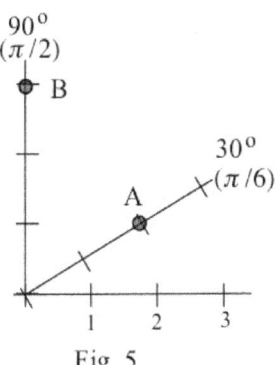

Fig. 5

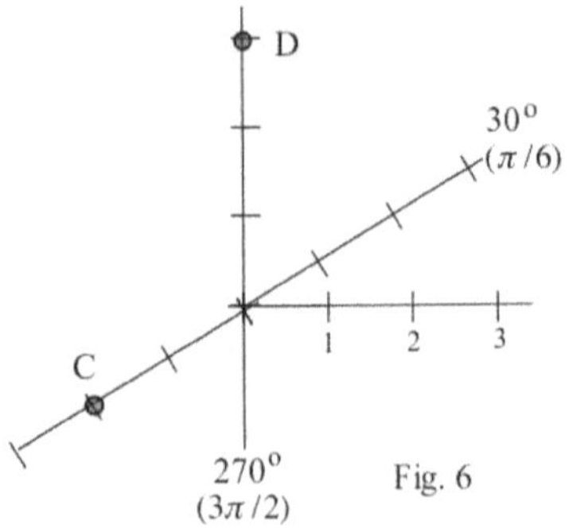

Practice 2:   Plot the points with the given polar coordinates: A(2, π/2),

B(2, –120°), C(–2, π/3), D(–2, –135°), and E(2, 135°). Which two

points coincide?

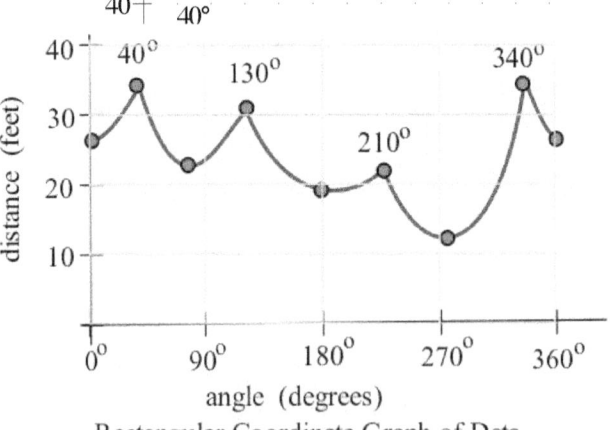

Fig. 6

Each polar coordinate pair (r,θ) gives the location of one point, but each

location has lots of different addresses in the polar coordinate system: the polar

coordinates of a point are not unique. This nonuniqueness of addresses comes

about in two ways. First, the angles θ, θ ± 360°, θ ± 2·360°, . . . all

describe the same radial line (Fig. 7), so the polar coordinates

(r, θ), (r, θ ± 360°), (r, θ ± 2·360°) , . . . all locate the same point.

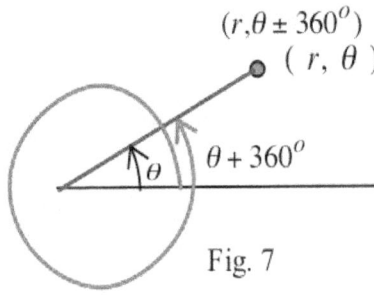

$(r,\theta \pm 360^{o})$

(r, θ)

$\theta + 360^{o}$

Fig. 7

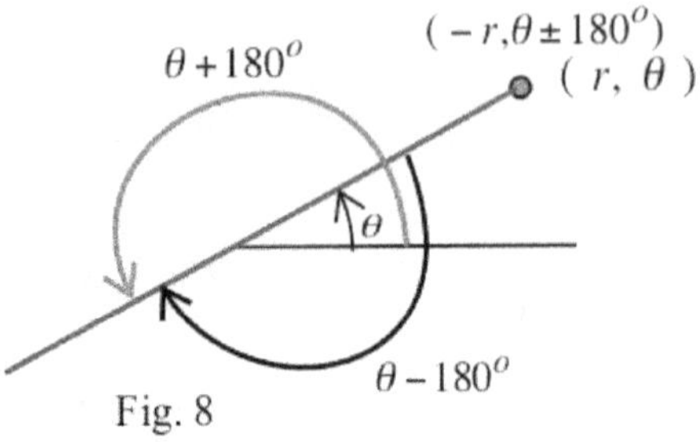

Secondly, the angle θ ± 180° describes

the radial line pointing in exactly the

opposite direction from the radial line

described by the angle θ (Fig. 8), so the

polar coordinates (r, θ) and (–r, θ± 180°) locate the same point. A polar

coordinate pair gives the location of exactly one point, but the location of one

point is described by many (an infinite number) different polar coordinate pairs.

Note: In the rectangular coordinate system we use (x, y) and y = f(x): first variable independent and

second variable dependent. In the polar coordinate system we use (r, θ) and r = f(θ): first variable

dependent and second variable independent, a reversal from the rectangular coordinate usage.

Practice 3:   Table 1 contains measurements to the edge of a plateau taken by a remote sensor which crashed

on the plateau. Fig. 9 shows the data plotted in

rectangular coordinates. Plot the data in polar

angle	distance		angle	distance
0°	28 feet		150°	22 feet
20°	30		180°	18
40°	36		210°	21
60°	27		230°	13
80°	24		270°	10
100°	24		330°	18
130°	30		340°	30

Table 1

coordinates and determine the shape of the

top of the plateau.

Rectangular Coordinate Graph of Data

Fig. 9

Graphing Functions in the Polar Coordinate System

In the rectangular coordinate system, we have worked with functions given by tables of data, by graphs, and by formulas. Functions can be represented in the same ways in polar coordinates.

* If a function is given by a table of data, we can graph the function in polar coordinates by plotting individual points in a polar coordinate system and connecting the plotted points to see the shape of the graph. By hand, this is a tedious process; by calculator or computer, it is quick and easy.

* If the function is given by a rectangular coordinate graph of magnitude as a function of angle, we can read coordinates of points on the rectangular graph and replot them in polar coodinates. In essence, as we go from the rectangular coordinate graph to the polar coordinate graph we "wrap" the rectangular graph around the "pole" at the origin of the polar coordinate system. (Fig. 10)

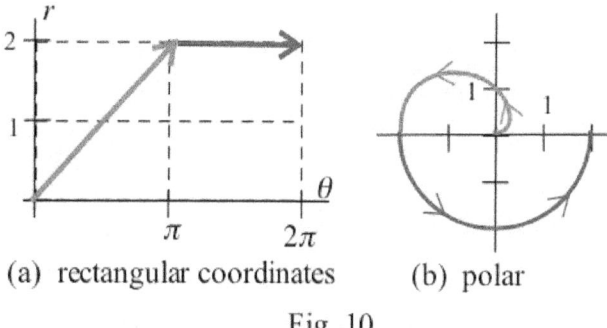

(a) rectangular coordinates   (b) polar

Fig. 10

* If the function is given by a formula, we (or our calculator) can graph the function to help us obtain information about its behavior. Typically, a graph is created by evaluating the function at a lot of points and then plotting the points in the polar coordinate system. Some of the following examples illustrate that functions given by simple formulas may have rather exotic graphs in the polar coordinate system.

If a function is already given by a polar coordinate graph, we can use the graph to answer questions about the behavior of the function. It is usually easy to locate the maximum value(s) of r on a polar coordinate graph, and, by moving counterclockwise around the graph, we can observe where r is increasing, constant, or decreasing.

Example 3:   Graph $r = 2$ and $r = \pi - \theta$ in the polar coordinate system for $0 \le \theta \le 2\pi$.

Solution: $r = 2$: In every direction θ, we simply move 2 units along the radial line and plot a point. The resulting polar graph (Fig. 11b) is a circle centered at the origin with a radius of 2. In the rectangular coordinate system, the graph of a constant $y = k$ is a horizontal line. In the polar coordinate system, the graph of a constant $r = k$ is a circle with radius $|k|$.

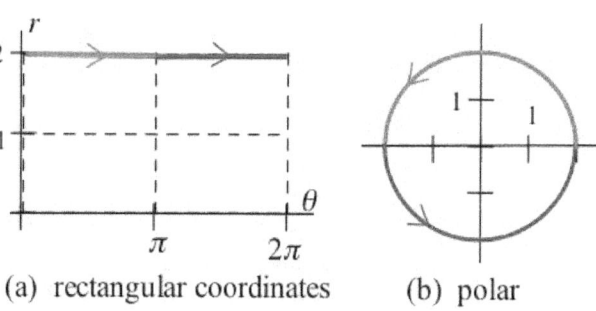

(a) rectangular coordinates   (b) polar

Fig. 11

r = π – θ: The rectangular coordinate graph of r = π – θ is

    shown in Fig. 12a. If we read the values of r and θ

    from the rectangular coordinate graph and plot them

    in polar coordinates, the result is the shape in Fig.

    12b. The different line thicknesses are used in the

    figures to help you see which values from the

    rectangular graph became which parts of the loop in

    the polar graph.

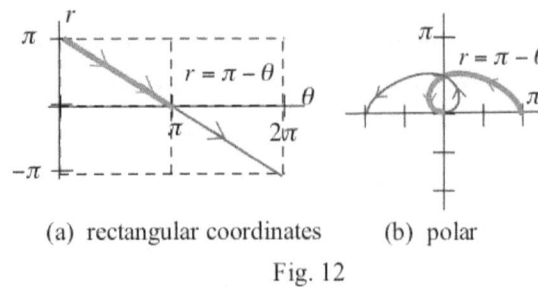

(a) rectangular coordinates   (b) polar

Fig. 12

Practice 4:   Graph r = –2 and r = cos(θ) in the polar coordinate system.

Example 4:   Graph r = θ and r = 1 + sin(θ) in the polar coordinate system.

Solution: r = θ: The rectangular coordinate graph of

    r = θ is a straight line (Fig. 13a). If we read the

    values of r and θ from the rectangular

    coordinate graph and plot them in polar

    coordinates, the result is the spiral, called an

    Archimedean spiral, in Fig. 13b.

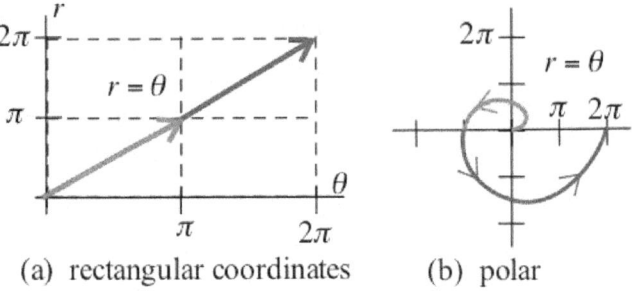

(a) rectangular coordinates   (b) polar

Fig. 13

    r = 1 + sin(θ): The rectangular coordinate graph

    of r = 1 + sin(θ) is shown in Fig. 14a, and it is

    the graph of the sine curve shifted up 1 unit. In

    polar coordinates, the result of adding 1 to sine

    is much less obvious and is shown in Fig. 14b.

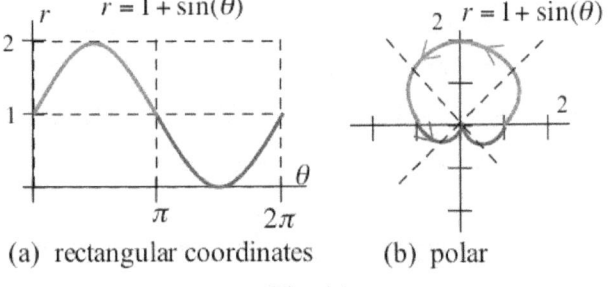

(a) rectangular coordinates   (b) polar

Fig. 14

angle (radians)	distance (meters)
0	3.0
π/6	1.9
π/4	1.7
π/3	1.6
π/2	2.0

Table 2

Practice 5:   Plot the points in Table 2 in the polar coordinate system and connect them with a smooth curve. Describe the shape of the graph in words.

Fig. 15 shows the effects of adding various constants to the rectangular and polar graphs of $r = \sin(\theta)$. In rectangular coordinates the result is a graph shifted up or down by k units. In polar coordinates, the result **may** be a graph with an entirely different shape (Fig. 16).

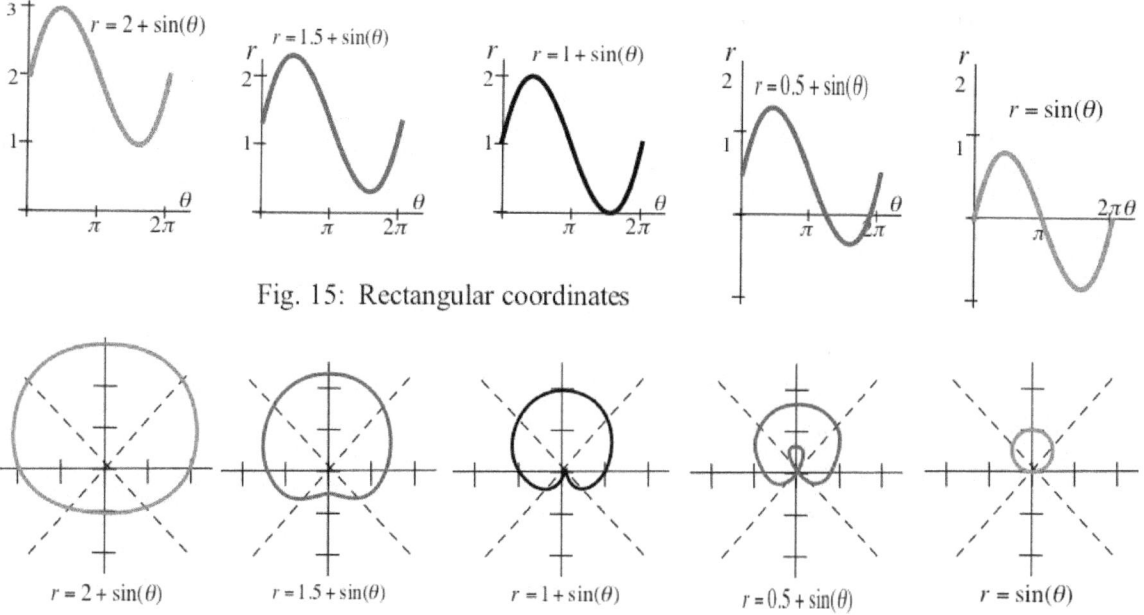

Fig. 15: Rectangular coordinates

Fig. 16: Polar coordinates

Fig. 17 shows the effects of adding a constant to the independent variable in rectangular coordinates, and the result is a horizontal shift of the original graph. In polar coordinates, Fig. 18, the result is a rotation of the original graph. Generally it is difficult to find formulas for rotated figures in rectangular coordinates, but rotations are easy in polar coordinates.

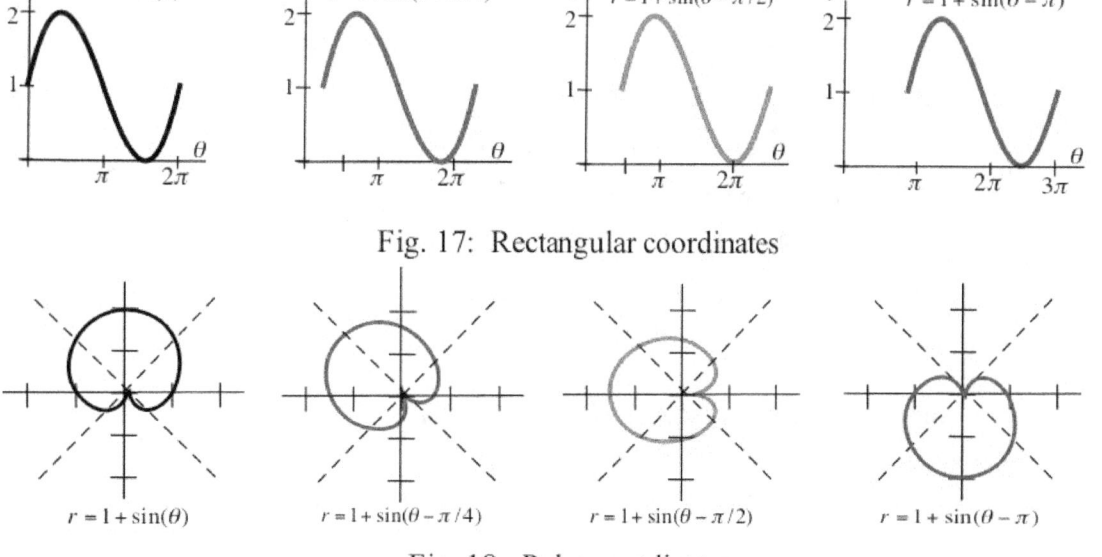

Fig. 17: Rectangular coordinates

Fig. 18: Polar coordinates

The formulas and names of several functions with exotic shapes in polar coordinates are given in the problems. Many of them are difficult to graph "by hand," but by using a graphing calculator or computer you can enjoy the shapes and easily examine the effects of changing some of the constants in their formulas.

Converting Between Coordinate Systems

Sometimes both rectangular and polar coordinates are needed in the same application, and it is necessary to change back and forth between the systems. In such a case we typically place the two origins together and align the polar axis with the positive x–axis. Then the conversions are straightforward exercises using trigonometry and right triangles (Fig. 19).

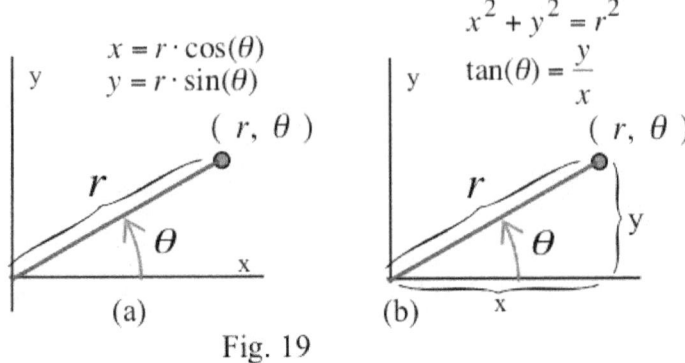

$$x = r \cdot \cos(\theta)$$
$$y = r \cdot \sin(\theta)$$
(r, θ)

(a)

$$x^2 + y^2 = r^2$$
$$\tan(\theta) = \frac{y}{x}$$
(r, θ)

(b)

Fig. 19

Polar to Rectangular (Fig. 19a)	Rectangular to Polar (Fig. 19b)
$x = r \cdot \cos(\theta)$	$r^2 = x^2 + y^2$
$y = r \cdot \sin(\theta)$	$\tan(\theta) = \frac{y}{x}$ (if $x \neq 0$)

Example 5:   Convert (a) the polar coordinate point $P(7, 0.4)$ to rectangular coordinates, and (b) the rectangular coordinate point $R(12, 5)$ to polar coordinates.

Solution: (a) $r = 7$ and $\theta = 0.4$ (Fig. 20) so $x = r \cdot \cos(\theta) = 7 \cdot \cos(0.4) \approx 7(0.921) = 6.447$ and
$y = 7 \cdot \sin(0.4) \approx 7(0.389) = 2.723$.

  (b) $x = 12$ and $y = 5$ so $r^2 = x^2 + y^2 = 144 + 25 = 169$ and
  $\tan(\theta) = y/x = 5/12$ so we can take $r = 13$ and
  $\theta = \arctan(5/12) \approx 0.395$. The polar coordinate addresses
  $(13, 0.395 \pm n \cdot 2\pi)$ and $(-13, 0.395 \pm (2n+1) \cdot \pi)$ give the location
  of the same point.

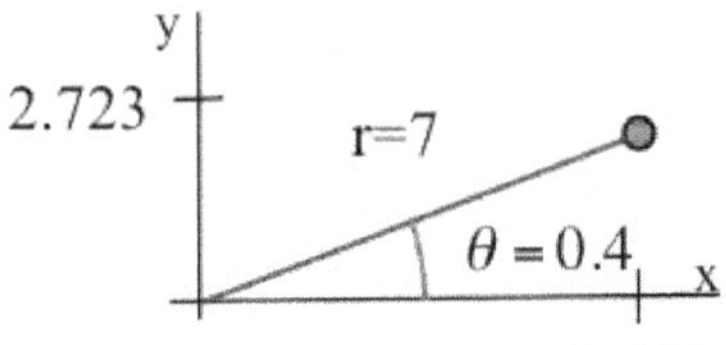

2.723

r=7

$\theta = 0.4$

Fig. 20   6.447

The conversion formulas can also be used to convert function equations from one system to the other.

Example 6:   Convert the rectangular coordinate linear equation $y = 3x + 5$ (Fig. 21) to a polar coordinate equation.

Solution: This simply requires that we replace x with r·cos(θ) and y with

r·sin(θ). Then

$y = 3x + 5$ becomes $r \cdot \sin(\theta) = 3r \cdot \cos(\theta) + 5$

so $r \cdot (\sin(\theta) - 3\cos(\theta)) = 5$ and $r = 5/(\sin(\theta) - 3\cos(\theta))$. This final

representation is valid only for θ such that $\sin(\theta) - 3\cos(\theta) \neq 0$.

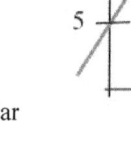

Fig. 21

Practice 6:   Convert the polar coordinate equation $r^2 = 4r \cdot \sin(\theta)$ to a rectangular

coordinate equation.

Example 7:   **Robotic Arm**: A robotic arm has a hand at the end of a 12 inch

long forearm which is connected to an 18 inch long upper arm

(Fig. 22). Determine the position of the hand, relative to the

shoulder, when $\theta = 45°$ ($\pi/4$) and $\phi = 30°$ ($\pi/6$).

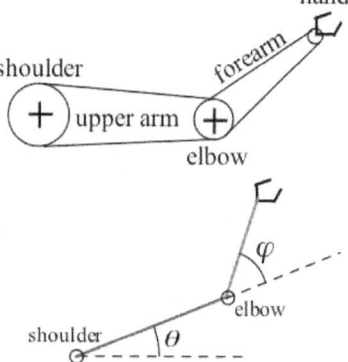

Fig. 22: Robot arm

Solution: The hand is $12 \cdot \cos(\pi/4 + \pi/6) \approx 3.1$ inches to the right of the

elbow (Fig. 23) and $12\sin(\pi/4 + \pi/6) \approx 11.6$

inches above the elbow. Similarly, the elbow is

$18 \cdot \cos(\pi/4) \approx 12.7$ inches to the right of the

shoulder and $18 \cdot \sin(\pi/4) \approx 12.7$ inches above the shoulder. Finally, the hand is

approximately $3.1 + 12.7 = 15.8$ inches to the right of the shoulder and approximately

$11.6 + 12.7 = 24.3$ inches above the shoulder. In polar coordinates, the hand is

approximately 29 inches from the shoulder, at an angle of about 57° (about 0.994

radians) above the horizontal.

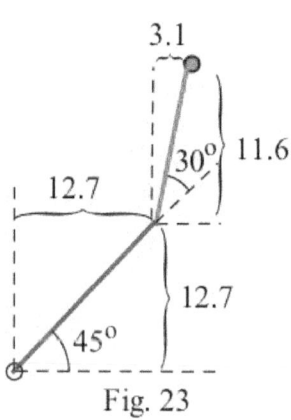

Fig. 23

Practice 7:   Determine the position of the hand, relative to the shoulder,

when $\theta = 30°$ and $\phi = 45°$.

Graphing Functions in Polar Coordinates on a Calculator or Computer

Some calculators and computers are programmed to graph polar functions simply by keying in the formula for

r, either as a function of θ or of t, but others are only designed to display rectangular coordinate graphs.

However, we can graph polar functions on most of them as well by using the rectangular to polar conversion

formulas, selecting the parametric mode (and the radian mode) on the calculator, and then graphing the

resulting parametric equations in the rectangular coordinate system:

To graph $r = r(\theta)$ for θ between 0 and 3π,

define $x(t) = r(t) \cdot \cos(t)$ and $y(t) = r(t) \cdot \sin(t)$

and graph the parametric equations $x(t), y(t)$ for t taking values from 0 to 9.43.

Which Coordinate System Should You Use?

There are no rigid rules. Use whichever coordinate system is easier or more "natural" for the problem or data you have. Sometimes it is not clear which system to use until you have graphed the data both ways, and some problems are easier if you switch back and forth between the systems.

Generally, the polar coordinate system is easier if

* the data consists of measurements in various directions (radar)
* your problem involves locations in relatively featureless locations (deserts, oceans, sky)
* rotations are involved.

Typically, the rectangular coordinate system is easier if

* the data consists of measurements given as functions of time or location (temperature, height)
* your problem involves locations in situations with an established grid (a city, a chess board)
* translations are involved.

PROBLEMS

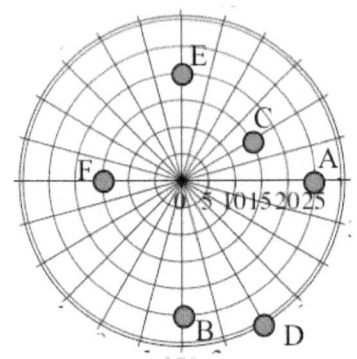

1. Give the locations in polar coordinates (using radian measure) of the points labeled A, B, and C in Fig. 24.

2. Give the locations in polar coordinates (using radian measure) of the points labeled D, E, and F in Fig. 24.

Fig. 24

3. Give the locations in polar coordinates (using radian measure) of the points labeled A, B, and C in Fig. 25.

4. Give the locations in polar coordinates (using radian measure) of the points labeled D, E, and F in Fig. 25.

In problems 5–8, plot the points A – D in polar coordinates, connect the dots by line segments in order (A to B to C to D to A), and name the approximate shape of the resulting figure.

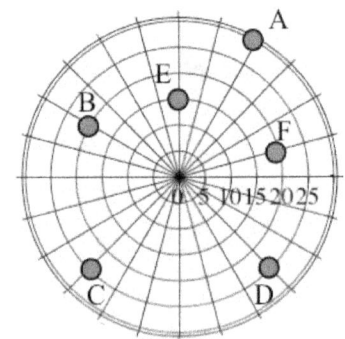

5. A(3, 0°), B(2, 120°), C(2, 200°), and D(2.8, 315°).

6. A(3, 30°), B(2, 130°), C(3, 150°), and D(2, 280°).

Fig. 25

7. A(2, 0.175), B(3, 2.269), C(2, 2.618), and D(3, 4.887).

8. A(3, 0.524), B(2, 2.269), C(3, 2.618), and D(2, 4.887).

In problems 9–14, the rectangular coordinate graph of a function $r = r(\theta)$ is shown. Sketch the polar coordinate graph of $r = r(\theta)$.

9. The graph in Fig. 26.   10. The graph in Fig. 27.   11. The graph in Fig. 28.

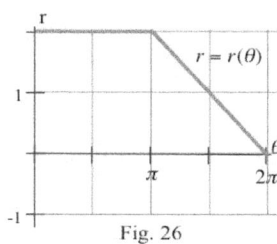
Fig. 26

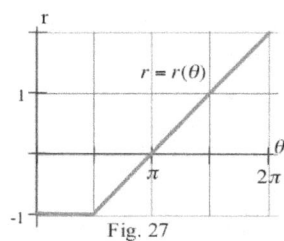
Fig. 27

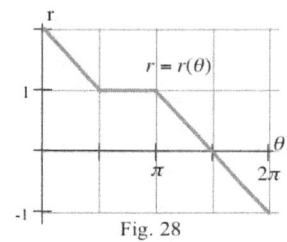
Fig. 28

12. The graph in Fig. 29.   13. The graph in Fig. 30.   14. The graph in Fig. 31.

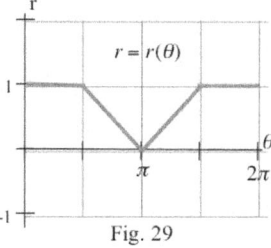
Fig. 29

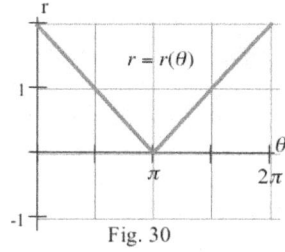
Fig. 30

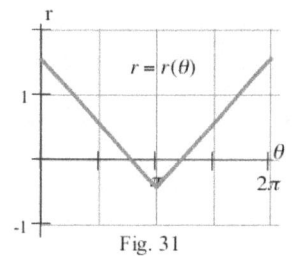
Fig. 31

15. The rectangular coordinate graph of $r = f(\theta)$ is shown in Fig. 32.

    (a) Sketch the rectangular coordinate graphs of $r = 1 + f(\theta)$, $r = 2 + f(\theta)$, and $r = -1 + f(\theta)$.

    (b) Sketch the polar coordinate graphs of $r = f(\theta)$, $r = 1 + f(\theta)$, $r = 2 + f(\theta)$, and $r = -1 + f(\theta)$.

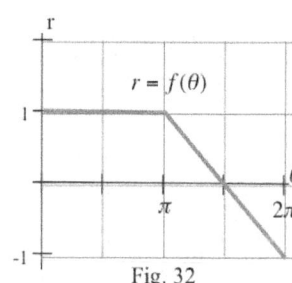
Fig. 32

16. The rectangular coordinate graph of $r = g(\theta)$ is shown in Fig. 33.

    (a) Sketch the rectangular coordinate graphs of $r = 1 + g(\theta)$, $r = 2 + g(\theta)$, and $r = -1 + g(\theta)$.

    (b) Sketch the polar coordinate graphs of $r = g(\theta)$, $r = 1 + g(\theta)$, $r = 2 + g(\theta)$, and $r = -1 + g(\theta)$.

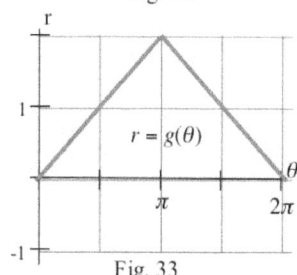
Fig. 33

17. The rectangular coordinate graph of $r = f(\theta)$ is shown in Fig. 34.

    (a) Sketch the rectangular coordinate graphs of $r = 1 + f(\theta)$, $r = 2 + f(\theta)$, and $r = -1 + f(\theta)$.

    (b) Sketch the polar coordinate graphs of $r = f(\theta)$, $r = 1 + f(\theta)$, $r = 2 + f(\theta)$, and $r = -1 + f(\theta)$.

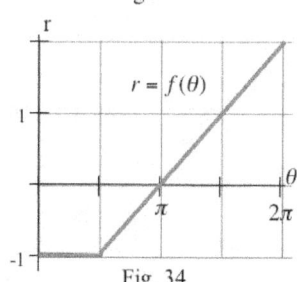
Fig. 34

18. The rectangular coordinate graph of $r = g(\theta)$ is shown in Fig. 35.

    (a) Sketch the rectangular coordinate graphs of $r = 1 + g(\theta)$,

        $r = 2 + g(\theta)$, and $r = -1 + g(\theta)$.

    (b) Sketch the polar coordinate graphs of $r = g(\theta)$, $r = 1 + g(\theta)$,

        $r = 2 + g(\theta)$, and $r = -1 + g(\theta)$.

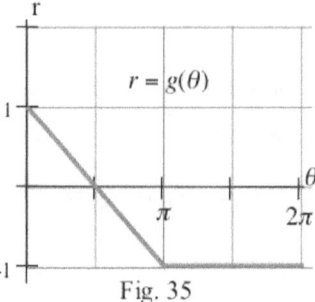

Fig. 35

19. Suppose the rectangular coordinate graph of $r = f(\theta)$ has the horizontal asymptote $r = 3$ as θ grows arbitrarily large. What does that tell us about the polar coordinate graph of $r = f(\theta)$ for large values of θ?

20. Suppose the rectangular coordinate graph of $r = f(\theta)$ has the vertical asymptote $\theta = \pi/6$: $\lim\limits_{\theta \to \pi/6} f(\theta) = +\infty$.

    What does that tell us about the polar coordinate graph of $r = f(\theta)$ for values of θ near $\pi/6$?

A computer or graphing calculator is recommended for the problems marked with a * .

In problems 21–40, graph the functions in polar coordinates for $0 \le \theta \le 2\pi$.

21. $r = -3$	22. $r = 5$	23. $\theta = \pi/6$	24. $\theta = 5\pi/3$
25. $r = 4 \cdot \sin(\theta)$	26. $r = -2 \cdot \cos(\theta)$	27. $r = 2 + \sin(\theta)$	28. $r = -2 + \sin(\theta)$
29. $r = 2 + 3 \cdot \sin(\theta)$	30. $r = \sin(2\theta)$	*31. $r = \tan(\theta)$	*32. $r = 1 + \tan(\theta)$

*33. $r = \dfrac{3}{\cos(\theta)}$ *34. $r = \dfrac{2}{\sin(\theta)}$ *35. $r = \dfrac{1}{\sin(\theta) + \cos(\theta)}$ 36. $r = \dfrac{\theta}{2}$

37. $r = 2 \cdot \theta$	38. $r = \theta^2$	39. $r = \dfrac{1}{\theta}$	40. $r = \sin(2\theta) \cdot \cos(3\theta)$

*41. $r = \sin(m\theta) \cdot \cos(n\theta)$ produces lovely graphs for various small integer values of m and n. Go exploring with a graphic calculator to find values of m and n which result in shapes you like.

*42. Graph $r = \dfrac{1}{1 + 0.5 \cdot \cos(\theta + a)}$, $0 \le \theta \le 2\pi$, for $a = 0, \pi/6, \pi/4$, and $\pi/2$. How are the graphs related?

*43. Graph $r = \dfrac{1}{1 + 0.5 \cdot \cos(\theta - a)}$, $0 \le \theta \le 2\pi$, for $a = 0, \pi/6, \pi/4$, and $\pi/2$. How are the graphs related?

*44. Graph $r = \sin(n\theta)$, $0 \le \theta \le 2\pi$, for $n = 1, 2, 3$, and 4 and count the number of "petals" on each graph. Predict the number of "petals" for the graphs of $r = \sin(n\theta)$ for $n = 5, 6$, and 7, and then test your prediction by creating those graphs.

*45. Repeat the steps in problem 44 but using $r = \cos(n\theta)$.

In problems 46–49, convert the rectangular coordinate locations to polar coordinates.

46. $(0, 3), (5, 0)$, and $(1, 2)$ 47. $(-2, 3), (2, -3)$, and $(0, -4)$.

48. $(0, -2), (4, 4)$, and $(3, -3)$ 49. $(3, 4), (-1, -3)$, and $(-7, 12)$.

In problems 50–53, convert the polar coordinate locations to rectangular coordinates.

50. $(3, 0), (5, 90°)$, and $(1, \pi)$ 51. $(-2,3), (2,-3)$, and $(0,-4)$.

52. $(0,3), (5,0)$, and $(1,2)$ 53. $(2,3), (-2,-3)$, and $(0,4)$.

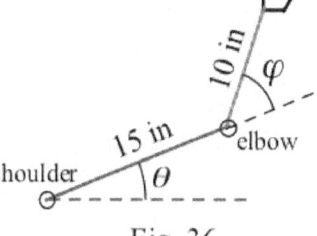

Problems 54–60 refer to the robotic arm in Fig. 36.

54.   Determine the position of the hand, realtive to the shoulder, when $\theta = 60°$ and $\phi = -45°$.

Fig. 36

55.   Determine the position of the hand, relative to the shoulder, when $\theta = -30°$ and $\phi = 30°$.

56.   Determine the position of the hand, relative to the shoulder, when $\theta = 0.6$ and $\phi = 1.2$.

57.   Determine the position of the hand, relative to the shoulder, when $\theta = -0.9$ and $\phi = 0.4$.

58.   Suppose the robot's shoulder can pivot so that $-\pi/2 \le \theta \le \pi/2$, but the elbow is broken and ϕ is always $0°$. Sketch the points the hand can reach.

59.   Suppose the robot's shoulder can pivot so that $-\pi/2 \le \theta \le \pi/2$, and the elbow can pivot so that $-\pi/2 \le \phi \le \pi/2$. Sketch the points the hand can reach.

60.   Suppose the robot's shoulder can pivot so that $-\pi/2 \le \theta \le \pi/2$, and the elbow can pivot completely so $-\pi \le \phi \le \pi$. Sketch the points the hand can reach.

*61.   Graph $r = \dfrac{1}{1 + a \cdot \cos(\theta)}$ for $0 \le \theta \le 2\pi$ and $a = 0.5, 0.8, 1, 1.5$, and 2. What shapes do the various values of a produce?

*62.   Repeat problem 61 with $r = \dfrac{1}{1 + a \cdot \sin(\theta)}$.

Some Exotic Curves (and Names)

Many of the following curves were discovered and named even before polar coordinates were invented. In most cases the path of a point moving on or around some object is described. You may enjoy using your calculator to graph some of these curves or you can invent your own exotic shapes. (An inexpensive source for these shapes and names is A Catalog Of Special Plane Curves by J. Dennis Lawrence, Dover Publications, 1972, and the page references below are to that book)

Some Classics:

Cissoid ("like ivy") of Diocles (about 200 B.C.): $r = a \sin(\theta) \cdot \tan(\theta)$ p. 98

Right Strophoid ("twisting") of Barrow (1670): $r = a(\sec(\theta) - 2\cos(\theta))$ p. 101

Trisectrix of Maclaurin (1742): $r = a \sec(\theta) - 4a \cos(\theta)$ p. 105

Lemniscate ("ribbon") of Bernoulli (1694): $r^2 = a^2 \cos(2\theta)$ p. 122

Conchoid ("shell") of Nicomedes (225 B.C.): $r = a + b \cdot \sec(\theta)$ p. 137

Hippopede ("horse fetter") of Proclus (about 75 B.C.): $r^2 = 4b(a - b \sin^2(\theta))$ p. 144 b = 3, a = 1, 2, 3, 4

Devil's Curve of Cramer (1750): $r^2(\sin^2(\theta) - \cos^2(\theta)) = a^2 \sin^2(\theta) - b^2 \cos^2(\theta)$ p. 151 a= 2, b=3

Nephroid ("kidney") of Freeth: $r = a \cdot (1 + 2 \sin(\frac{\theta}{2}))$ p. 175   a = 3

Some of our own: (Based on their names, what shapes do you expect for the following curves?)

Piscatoid of Pat (1992): $r = \dfrac{1}{\cos(\theta)} - 3\cos(\theta)$ for $-1.1 \le \theta \le 1.1$ Window x: (–2, 1) and y: (–1, 1)

Kermitoid of Kelcey (1992) :

   $r = 2.5 \cdot \sin(2\theta) \cdot (\theta - 4.71) \cdot INT(\theta/\pi) + \{ 5 \cdot \sin^3(\theta) - 3 \cdot \sin^9(\theta)\} \cdot \{ 1 - INT(\theta/\pi) \}$ for $0 \le \theta \le 2\pi$
   Window x: (–3, 3) and y: (–1, 4)

Bovine Oculoid: $r = 1 + INT(\theta/(2\pi))$ for $0 \le \theta \le 6\pi$ (≈ 18.85) Window x: (–5, 5) and y: (–4, 4)

A Few Reference Facts

The polar form of the linear equation $Ax + By + C = 0$ is $r \cdot (A \cdot \cos(\theta) + B \cdot \sin(\theta)) + C = 0$

The equation of the line through the polar coordinate points (r_1, θ_1) and (r_2, θ_2) is

   $r \cdot \{r_1 \cdot \sin(\theta - \theta_1) + r_2 \cdot \sin(\theta_2 - \theta) \} = r_1 \cdot r_2 \cdot \sin(\theta_2 - \theta_1)$

The graph of $r = a \cdot \sin(\theta) + b \cdot \cos(\theta)$ is a circle through the origin with center $(b/2, a/2)$ and radius $\frac{1}{2}\sqrt{a^2 + b^2}$. (Hint: multiply each side by r, and then convert to rectangular coordinates.)

The equations $r = \dfrac{1}{1 \pm a \cdot \cos(\theta)}$ and $r = \dfrac{1}{1 \pm a \cdot \sin(\theta)}$ are conic sections with one focus at the origin.

If $a < 1$, the denominator is **never** 0 for $0 \le \theta < 2\pi$ and the graph is an **ellipse**.

If $a = 1$, the denominator is 0 for **one** value of $\theta, 0 \le \theta < 2\pi$, and the graph is a **parabola**.

If $a > 1$, the denominator is 0 for **two** values of $\theta, 0 \le \theta < 2\pi$, and the graph is a **hyperbola**.

Section 9.1   **PRACTICE Answers**

Practice 1:   Point C is at a distance of 10 miles in
the direction 30°. D is 5 miles at 270°.

Practice 2:   The points are plotted in Fig. 37.

Practice 3:   See Fig. 38.

The top of the plateau is roughly rectangular.

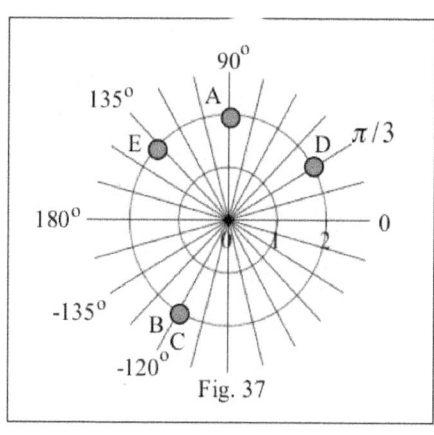

Fig. 37

Practice 4:   The graphs are shown in Figs. 39 and 40.

Note that the graph of $r = \cos(\theta)$ traces out a
circle **twice**; once as θ goes from 0 to π, and
a second time as θ goes from π to 2π.

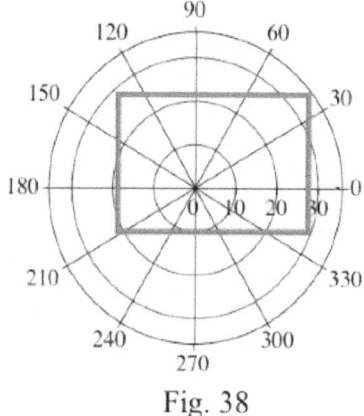

Fig. 38

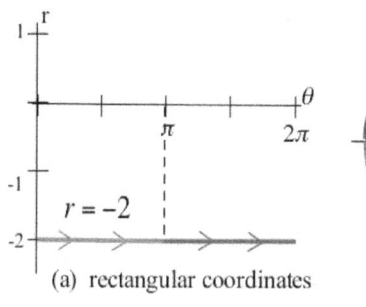

$r = -2$

(a) rectangular coordinates

Fig. 39

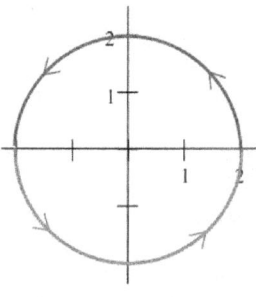

(b) polar coordinates

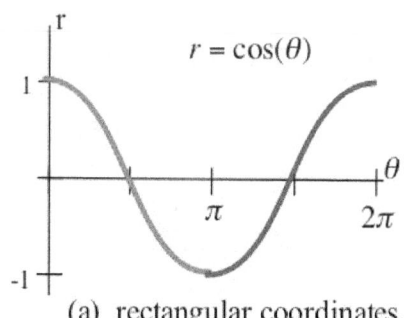

(a) rectangular coordinates

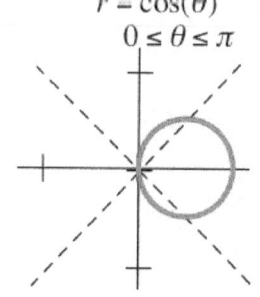

(b) polar coordinates

Fig. 40

Practice 5:   The points are plotted in Fig. 41.

The points (almost) lie on a straight line.

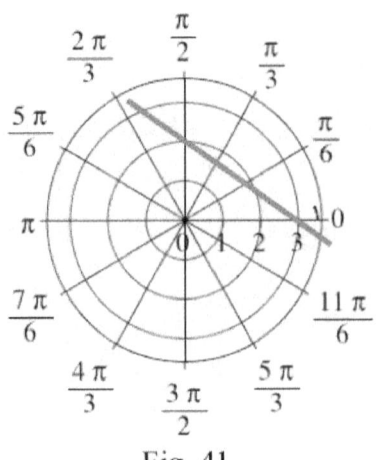

Fig. 41

Practice 6: $r^2 = x^2 + y^2$ and $r \cdot \sin(\theta) = y$ so $r^2 = 4r \cdot \sin(\theta)$ becomes $x^2 + y^2 = 4y$.

Putting this last equation into the standard form for a circle (by completing the square) we have $x^2 + (y-2)^2 = 4$, the equation of a circle with center at $(0, 2)$ and radius 2.

Practice 7:   See Fig. 42.

For point A, the "elbow," relative to O, the "shoulder:"

$$x = 18 \cdot \cos(30^o) \approx 15.6 \text{ inches} \text{ and } y = 18 \cdot \sin(30^o) = 9 \text{ inches.}$$

For point B, the "hand," relative to A:

$$x = 12 \cdot \cos(75^o) \approx 3.1 \text{ inches} \text{ and } y = 12 \cdot \sin(75^o) \approx 11.6 \text{ inches.}$$

Then the retangular coordinate location of the B relative to O is

$$x \approx 15.6 + 3.1 = 18.7 \text{ inches} \text{ and } y \approx 9 + 11.6 = 20.6 \text{ inches.}$$

The polar coordinate location of B relative to O is

$$r = \sqrt{x^2 + y^2} \approx 27.8 \text{ inches} \text{ and } \theta \approx 47.7^o \text{ (or 0.83 radians)}$$

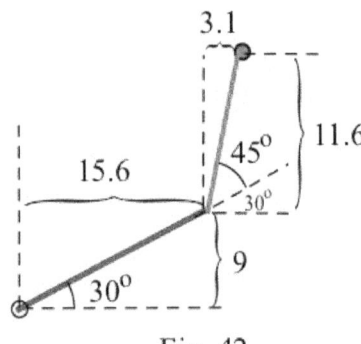

Fig. 42

9.2   CALCULUS IN THE POLAR COORDINATE SYSTEM

The previous section introduced the polar coordinate system and discussed how to plot points, how to create graphs of functions (from data, a rectangular graph, or a formula), and how to convert back and forth between the polar and rectangular coordinate systems. This section examines calculus in polar coordinates: rates of changes, slopes of tangent lines, areas, and lengths of curves. The results we obtain may look different, but they all follow from the approaches used in the rectangular coordinate system.

Polar Coordinates and Derivatives

In the rectangular coordinate system, the derivative dy/dx measured both the rate of change of y with respect to x and the slope of the tangent line. In the polar coordinate system two different derivatives commonly appear, and it is important to distinguish between them.

$\dfrac{dr}{d\theta}$ measures the **rate of change** of r with respect to θ.

The sign of $\dfrac{d\,r}{d\theta}$ tells us whether r is increasing or decreasing as θ increases.

$\dfrac{dy}{dx}$ measures the **slope** $\dfrac{\Delta y}{\Delta x}$ **of the tangent line** to the polar graph of r.

We can use our usual rules for derivatives to calculate the derivative of a polar coordinate equation r with respect to θ, and dr/dθ tells us how r is changing with respect to (increasing) θ. For example, if dr/dθ > 0 then the directed distance r is increasing as θ increases (Fig. 1). However, dr/dθ is **NOT** the slope of the line tangent to the polar graph of r. For the simple spiral r = θ (Fig. 2), $\dfrac{d\,r}{d\theta}$ = 1 > 0 for all values of θ; but the slope of the tangent line, $\dfrac{dy}{dx}$, may be positive (at A and C) or negative (at B and D).

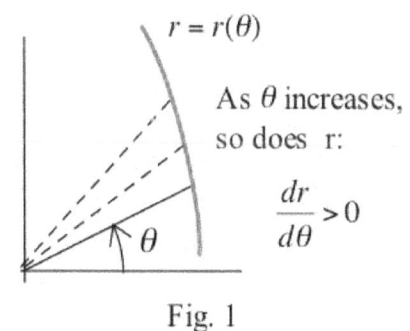

r = r(θ)

As θ increases, so does r:

$\dfrac{dr}{d\theta} > 0$

Fig. 1

Similarly, $\dfrac{dx}{d\theta}$ is the rate of change of the x-coordinate of the graph with respect to (increasing) θ , and $\dfrac{dy}{d\theta}$ is the rate of change of the y-coordinate of the graph with respect to (increasing) θ . The values of the derivatives dy/dθ and dx/dθ depend on the location on the graph. They will also be used to calculate the slope $\dfrac{dy}{dx}$ of the tangent line, and also to express the formula for arc length in polar coordinates.

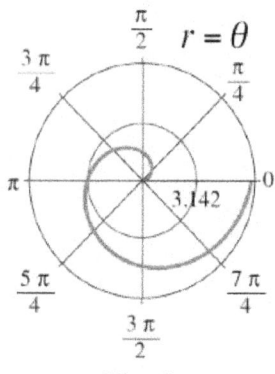

Fig. 2

Example 1:   State whether the values of dr/dθ, dx/dθ , dy/dθ , and dy/dx are + (positive),

– (negative), 0 (zero), or U (undefined) at the points A and B on the graph in Fig. 3.

Solution: The values of the derivatives at A and B are given in Table 1.

Practice 1:   Fill in the rest of Table 1 for points labeled C and D.

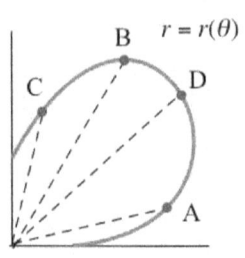

Fig. 3

When r is given by a formula we can calculate dy/dx , the slope of

the tangent line, by using the polar–rectangular conversion formulas

and the Chain Rule. By the Chain Rule $\dfrac{dy}{d\theta} = \dfrac{dy}{dx}\dfrac{dx}{d\theta}$, so we can

solve for $\dfrac{dy}{dx}$ by dividing each side of the equation by $\dfrac{dx}{d\theta}$.

Point	$\dfrac{dx}{d\theta}$	$\dfrac{dy}{d\theta}$	$\dfrac{dr}{d\theta}$	$\dfrac{dy}{dx}$
A	+	+	+	+
B	−	0	−	0
C				
D				

Table 1

Then the slope $\dfrac{dy}{dx}$ of the line tangent to the polar coordinate graph of r(θ) is

(1) $$\frac{dy}{dx} = \frac{\dfrac{dy}{d\theta}}{\dfrac{dx}{d\theta}} = \frac{\dfrac{d\,(r\cdot\sin(\theta)\,)}{d\theta}}{\dfrac{d\,(r\cdot\cos(\theta)\,)}{d\theta}} \ .$$

Since r is a function of θ, r = r(θ), we may use the product rule and the Chain Rule

for derivatives to calculate each derivative and to obtain

(2) $$\frac{dy}{dx} = \frac{r\cdot\cos(\theta) + r\,'\cdot\sin(\theta)}{-r\cdot\sin(\theta) + r\,'\cdot\cos(\theta)} \quad (\text{with } r\,' = dr/d\theta) \ .$$

The result in (2) is difficult to remember, but the starting point (1) and derivation are straightforward.

Example 2:   Find the slopes of the lines tangent to the spiral r = θ (shown in Fig. 2) at the points

P(π/2, π/2) and Q(π, π) .

Solution: y = r·sin(θ) = θ·sin(θ) and x = r·cos(θ) = θ·cos(θ) so

$$\frac{dy}{dx} = \frac{\dfrac{d\,(r\cdot\sin(\theta))}{d\theta}}{\dfrac{d\,(r\cdot\cos(\theta))}{d\theta}} = \frac{\dfrac{d\,(\theta\cdot\sin(\theta))}{d\theta}}{\dfrac{d\,(\theta\cdot\cos(\theta))}{d\theta}} = \frac{\theta\cdot\cos(\theta) + 1\cdot\sin(\theta)}{-\theta\cdot\sin(\theta) + 1\cdot\cos(\theta)} \ .$$

At the point P, θ = π/2 and r = π/2 so $\dfrac{dy}{dx} = \dfrac{\frac{\pi}{2}\cdot 0 + 1\cdot(1)}{-\frac{\pi}{2}\cdot(1) + 1\cdot(0)} = -\dfrac{2}{\pi} \approx -0.637$.

At the point Q, θ = π and r = π so $\dfrac{dy}{dx} = \dfrac{\pi\cdot(-1) + 1\cdot(0)}{-\pi\cdot(0) + 1\cdot(-1)} = \dfrac{-\pi}{-1} = \pi \approx 3.142$.

The function   r = θ is steadily increasing, but the slope of the line tangent to the polar graph can

negative or positive or zero or even undefined (where?).

Practice 2:   Find the slopes of the lines tangent to the cardioid

r = 1 – sin(θ) (Fig. 4)   when   θ = 0, π/4, and π/2.

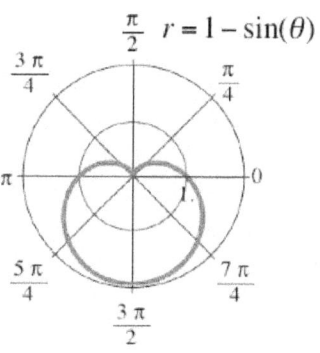

Fig. 4

Areas in Polar Coordinates

The patterns for calculating areas in rectangular and polar coordinates

look different, but they are derived in the same way: partition the area

into pieces, calculate areas of the pieces, add the small areas together to

get a Riemann sum, and take the limit of the Riemann sum to get a definite

integral. The major difference is the shape of the pieces: we use thin

rectangular pieces in the rectangular system and thin sectors (pieces of pie)

in the polar system. The formula we need for the area of a sector can be

found by using proportions (Fig. 5):

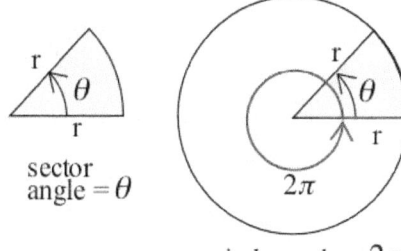

Fig. 5

$$\frac{\text{area of sector}}{\text{area of whole circle}} = \frac{\text{sector angle}}{\text{angle of whole circle}} = \frac{\theta}{2\pi}$$

so (area of sector) $= \frac{\theta}{2\pi}$ (area of whole circle) $= \frac{\theta}{2\pi}(\pi r^2) = \frac{1}{2} r^2 \theta$.

Figures 6 and 7 refer to the area discussion after the figures.

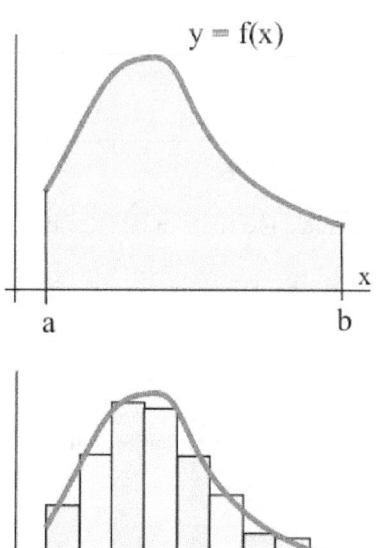

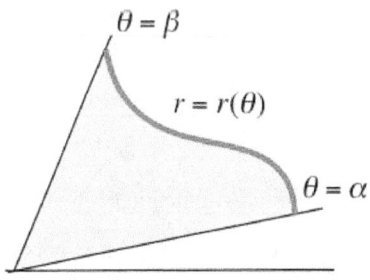

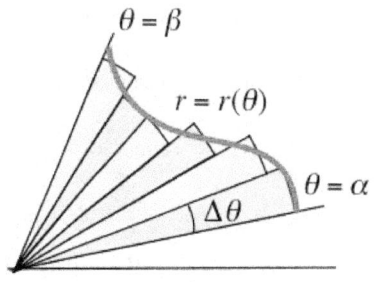

Fig. 6 Area with rectangular
coordinates

Fig. 7: Area with polar
coordinates

Area in Rectangular Coordinates (Fig. 6)	**Area in Polar Coordinates** (Fig. 7)
Partition the domain x of the rectangular coordinate function into small pieces of width Δx .	Partition the domain θ of the polar coordinate function into small pieces of angular width $\Delta\theta$.
Build rectangles on each piece of the domain.	Build "nice" shapes (pieces of pie shaped sectors) along each piece of the domain.
Calculate the area of each piece (rectangle): $\text{area}_i = (\text{base}_i)\cdot(\text{height}_i) = f(x_i)\cdot\Delta x_i$.	Calculate the area of each piece (sector): $\text{area}_i = \frac{1}{2}(\text{radius}_i)^2(\text{angle}_i) = \frac{1}{2}r_i^2\,\Delta\theta_i$.
Approximate the total area by adding the small areas together, a Riemann sum: $\text{total area} \approx \sum \text{area}_i = \sum f(x_i)\cdot\Delta x_i$.	Approximate the total area by adding the small areas together, a Riemann sum: $\text{total area} \approx \sum \text{area}_i = \sum \frac{1}{2}r_i^2\,\Delta\theta_i$.
The limit of the Riemann sum is a definite integral: $\text{Area} = \displaystyle\int_{x=a}^{b} f(x)\,dx$.	The limit of the Riemann sum is a definite integral: $\text{Area} = \displaystyle\int_{\theta=\alpha}^{\beta} \frac{1}{2}r^2(\theta)\,d\theta$.

If r is a continuous function of θ , then the limit of the Riemann sums is a finite number, and we have a formula for the area of a region in polar coordinates.

Area In Polar Coordinates

The area of the region bounded by a continuous function $r(\theta)$ and radial lines at

angles $\theta = \alpha$ and $\theta = \beta$ is

$$\text{area} = \int_{\theta=\alpha}^{\beta} \frac{1}{2}r^2(\theta)\,d\theta .$$

Example 3:   Find the area inside the cardioid $r = 1 + \cos(\theta)$. (Fig. 8)

Solution: This is a straightforward application of the area formula.

$$\text{Area} = \int_{\theta=0}^{2\pi} \frac{1}{2}(1 + \cos(\theta))^2 \, d\theta = \frac{1}{2} \int_{\theta=0}^{2\pi} \{1 + 2\cdot\cos(\theta) + \cos^2(\theta)\} \, d\theta$$

$$= \frac{1}{2} \left\{ \theta + 2\cdot\sin(\theta) + \frac{1}{2}\left[\theta + \frac{1}{2} \sin(2\theta) \right] \right\} \Big|_0^{2\pi}$$

$$= \frac{1}{2} \{ [2\pi + 0 + \frac{1}{2}(2\pi + 0)] - [0 + 0 + 0] \} = \frac{3}{2}\pi .$$

We could also have used the symmetry of the region and determined this area by integrating from 0 to π (Fig. 9) and multiplying the result by 2.

Practice 3:   Find the area inside one "petal" of the rose $r = \sin(3\theta)$. (Fig. 10)

We can also calculate the area between curves in polar coordinates.

The area of the region (Fig. 11) between the continuous curves

$r_1(\theta) \leq r_2(\theta)$ for $\alpha \leq \theta \leq \beta$ is

$$\int_{\theta=\alpha}^{\beta} \frac{1}{2} r_2^2(\theta) \, d\theta = \int_{\theta=\alpha}^{\beta} \frac{1}{2} r_1^2(\theta) \, d\theta = \int_{\theta=\alpha}^{\beta} \frac{1}{2}\{ r_2^2(\theta) - r_1^2(\theta)\} \, d\theta .$$

It is a good idea to sketch the graphs of the curves to help determine the endpoints of integration.

Example 4:   Find the area of the shaded region in Fig. 12.

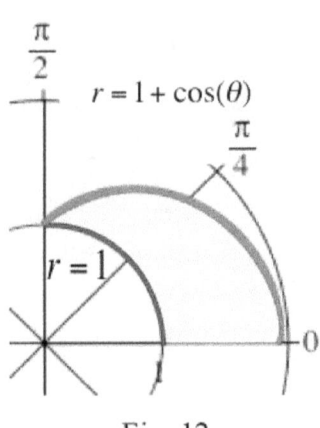

Solution: A_1 = area between the circle and the origin = $\int\limits_{\theta=0}^{\pi/2} \frac{1}{2}\, 1^2 \, d\theta$

$$= \frac{1}{2}\, \theta \Big|_0^{\pi/2} \quad = \frac{\pi}{4} \approx 0.785 \ .$$

A_2 = area between the cardioid and the origin = $\int\limits_{\theta=0}^{\pi/2} \frac{1}{2}(1 + \cos(\theta))^2 \, d\theta$

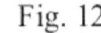

Fig. 12

$$= \frac{3}{4}\, \theta + \sin(\theta) + \frac{1}{8}\, \sin(2\theta) \Big|_0^{\pi/2} \quad = \{\, \frac{3\pi}{8} + 1 + 0 \,\} - \{\, 0 + 0 + 0 \,\} \approx$$
2.178 .

The area we want is $A_2 - A_1 = 1 + \frac{3\pi}{8} - \frac{\pi}{4} = 1 + \frac{\pi}{8} \approx 1.393$.

Practice 4:   Find the area of the region outside the cardioid $1 + \cos(\theta)$ and

inside the circle $r = 2$. (Fig. 13)

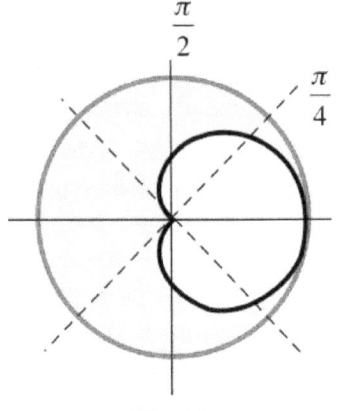

Fig. 13

Arc Length in Polar Coordinates

The patterns for calculating the lengths of curves in rectangular and polar coordinates look different, but
they are derived from the Pythagorean Theorem and the same sum we used in Section 5.2 (Fig. 14):

$$\text{length} \approx \sum \sqrt{ (\Delta x)^2 + (\Delta y)^2 } = \sum \sqrt{ (\frac{\Delta x}{\Delta\theta})^2 + (\frac{\Delta y}{\Delta\theta})^2 } \ \ \Delta\theta \ .$$

If x and y are differentiable functions of θ, then as $\Delta\theta$ approaches 0, $\Delta x/\Delta\theta$
approaches $dx/d\theta$, $\Delta y/\Delta\theta$ approaches $dy/d\theta$, and the Riemann sum approaches
the definite integral

$$\text{length} = \int\limits_{\theta=\alpha}^{\beta} \sqrt{ (\frac{dx}{d\theta})^2 + (\frac{dy}{d\theta})^2 } \ \ d\theta \ .$$

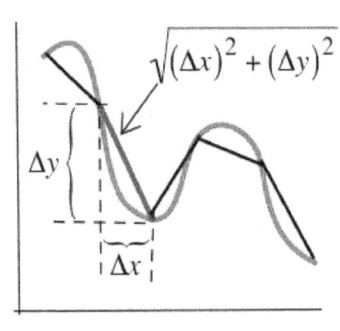

Fig. 14

Replacing x with $r \cdot \cos(\theta)$ and y with $r \cdot \sin(\theta)$, we have $dx/d\theta = -r \cdot \sin(\theta) + r\,' \cdot \cos(\theta)$ and

$dy/d\theta = r \cdot \cos(\theta) + r' \cdot \sin(\theta)$. Then $(dx/d\theta)^2 + (dy/d\theta)^2$ inside the square root simplifies to $r^2 + (r')^2$ and we have a more useful form of the integral for arc length in polar coordinates.

Arc Length

If r is a differentiable function of θ for $\alpha \leq \theta \leq \beta$, then the length of the graph of r is

$$\text{Length} = \int_{\theta=\alpha}^{\beta} \sqrt{(r)^2 + \left(\frac{dr}{d\theta}\right)^2} \ d\theta \ .$$

Problems

Derivatives

In problems 1–4, fill in the table for each graph with + (positive), – (negative), 0 (zero), or U (undefined) for each derivative at each labeled point.

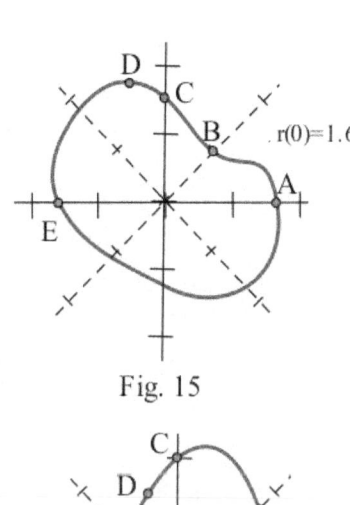

Fig. 15

1. Use Fig. 15. Point

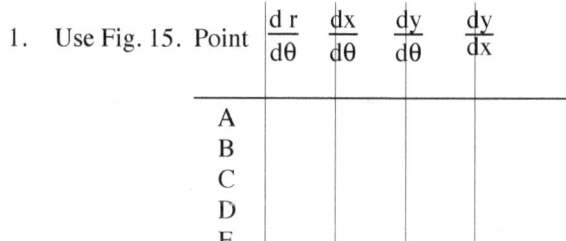

Point	$\frac{dr}{d\theta}$	$\frac{dx}{d\theta}$	$\frac{dy}{d\theta}$	$\frac{dy}{dx}$
A				
B				
C				
D				
E				

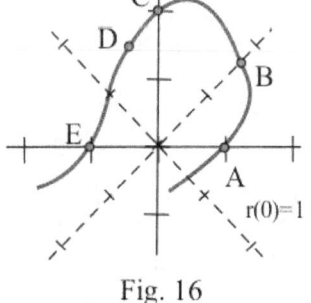

Fig. 16

2. Use Fig. 16. Point

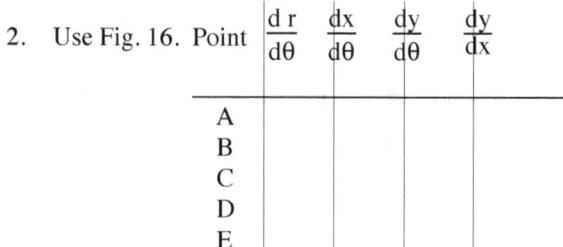

Point	$\frac{dr}{d\theta}$	$\frac{dx}{d\theta}$	$\frac{dy}{d\theta}$	$\frac{dy}{dx}$
A				
B				
C				
D				
E				

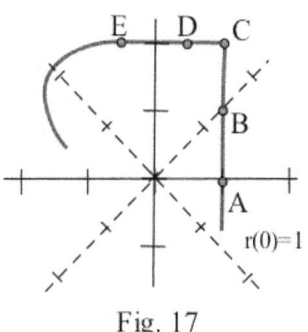

Fig. 17

3. Use Fig. 17. Point

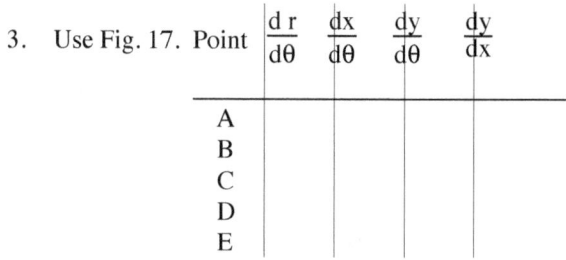

Point	$\frac{dr}{d\theta}$	$\frac{dx}{d\theta}$	$\frac{dy}{d\theta}$	$\frac{dy}{dx}$
A				
B				
C				
D				
E				

4.   Use Fig. 18. Point

Point	$\dfrac{dr}{d\theta}$	$\dfrac{dx}{d\theta}$	$\dfrac{dy}{d\theta}$	$\dfrac{dy}{dx}$
A				
B				
C				
D				
E				

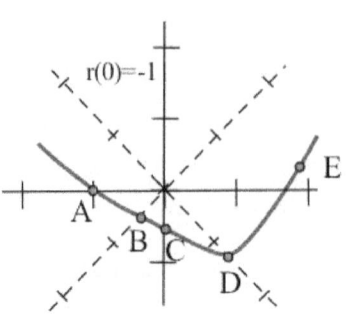

Fig. 18

In problems 5–8, sketch the graph of the polar coordinate function

$r = r(\theta)$ for $0 \le \theta \le 2\pi$, label the points with the given polar coordinates on the graph, and calculate the

values of $\dfrac{dr}{d\theta}$ and $\dfrac{dy}{dx}$ at the points with the given polar coordinates.

5. $r = 5$ at $A(5, \pi/4), B(5, \pi/2),$ and $C(5, \pi)$.

6. $r = 2 + \cos(\theta)$ at $A(2 + \dfrac{\sqrt{2}}{2}, \pi/4), B(2, \pi/2),$ and $C(1, \pi)$.

7. $r = 1 + \cos^2(\theta)$ at $A(2, 0), B(3/2, \pi/4),$ and $C(1, \pi/2)$.

8. $r = \dfrac{6}{2 + \cos(\theta)}$ at $A(2, 0), B(3, \pi/2),$ and $C(\dfrac{24 - 6\sqrt{2}}{7}, \pi/4) \approx (2.216, \pi/4)$.

9.   Graph $r = 1 + 2 \cdot \cos(\theta)$ for $0 \le \theta \le 2\pi$, and show that the graph goes through the origin when
     $\theta = 2\pi/3$ and $\theta = 4\pi/3$. Calculate dy/dx when $\theta = 2\pi/3$ and $\theta = 4\pi/3$. How can a curve have two
     different tangent lines (and slopes) when it goes through the origin?

10. Graph the cardiod $r = 1 + \sin(\theta)$ for $0 \le \theta \le 2\pi$.
     (a) At what points on the cardioid does $dx/d\theta = 0$? (b) At what points on the cardiod does $dy/d\theta = 0$?
     (c) At what points on the cardioid does $dr/d\theta = 0$? (d) At what points on the cardiod does $dy/dx = 0$?

11. Show that if a polar coordinate graph goes through the origin when the angle is θ_0 (and if $dr/d\theta$
     exists and does not equal 0 there), then the slope of the tangent line at the origin is $\tan(\theta_0)$.
     (Suggestion: Evaluate formula (2) for dy/dx at the point $(0, \theta_0)$.)

Areas

In problems 12–20, represent each area as a definite integral. Then evaluate the integral exactly or using Simpson's rule (with n = 100).

12. The area of the shaded region in Fig. 19.

13. The area of the shaded region in Fig. 20.

14. The area of the shaded region in Fig. 21.

15. The area in the first quadrant outside the circle $r = 1$ and inside the cardiod $r = 1 + \cos(\theta)$.

16. The region in the second quadrant bounded by $r = \theta$ and $r = \theta^2$.

17. The area inside one "petal" of the graph of (a) $r = \sin(3\theta)$ and (b) $r = \sin(5\theta)$.

18. The area (a) inside the "peanut" $r = 1.5 + \cos(2\theta)$ and (b) inside $r = a + \cos(2\theta)$ (a > 1).

19. The area inside the circle $r = 4 \cdot \sin(\theta)$.

20. The area of the shaded region in Fig. 22.

21. Goat and Square Silo: (This problem does not require calculus.)

    One end of a 40 foot long rope is attached to the middle of a wall of a 20 foot square silo, and the other end is tied to a goat (Fig. 23).

    (a) Sketch the region that the goat can reach.

    (b) Find the area of the region that the goat can reach.

    (c) Can the goat reach more area if the rope is tied to the corner of the silo?

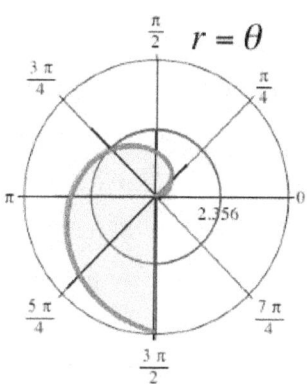

Fig. 19

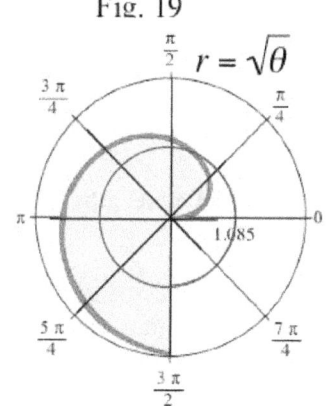

Fig. 20

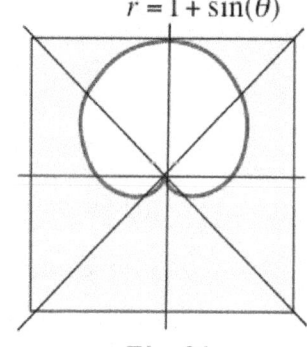

Fig. 21

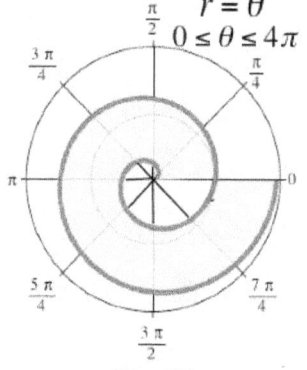

Fig. 22

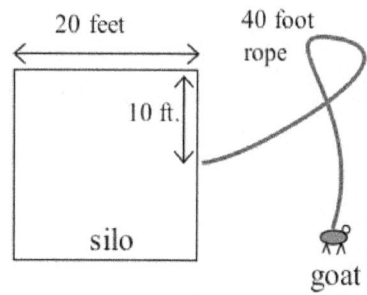

Fig. 23

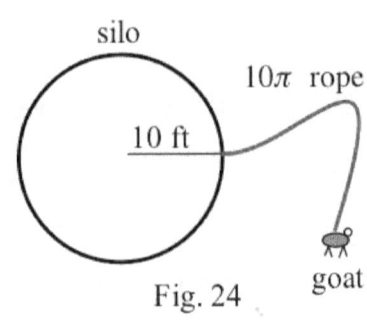

silo

10π rope

10 ft

Fig. 24

goat

22. **Goat and Round Silo:** One end of a 10π foot long rope is attached to the wall of a round silo that has a radius of 10 feet, and the other end is tied to a goat (Fig. 24).

   (a) Sketch the region the goat can reach.

   (b) Justify that the area of the region in Fig. 25 as the goat goes around the silo from having θ feet of rope taut against the silo to having $\theta + \Delta\theta$ feet taut against the silo is approximately

$$\frac{1}{2}(10\pi - 10{\cdot}\theta)^{2}\,\Delta\theta.$$

   (c) Use the result from part (b) to help calculate the area of the region that the goat can reach.

Arc Lengths

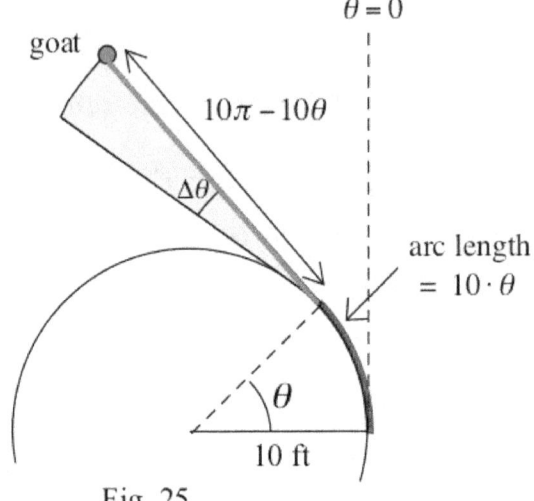

goat

$\theta = 0$

$10\pi - 10\theta$

$\Delta\theta$

arc length $= 10 \cdot \theta$

θ

10 ft

Fig. 25

In problems 23–29, represent the length of each curve as a definite integral. Then evaluate the integral exactly or using your calculator.

23. The length of the spiral $r = \theta$ from $\theta = 0$ to $\theta = 2\pi$.

24. The length of the spiral $r = \theta$ from $\theta = 2\pi$ to $\theta = 4\pi$.

25. The length of the cardiod $r = 1 + \cos(\theta)$.

26. The length of $r = 4{\cdot}\sin(\theta)$ from $\theta = 0$ to $\theta = \pi$.

27. The length of the circle $r = 5$ from $\theta = 0$ to $\theta = 2\pi$.

28. The length of the "peanut" $r = 1.2 + \cos(2\theta)$.

29. The length (a) of one "petal" of the graph of $r = \sin(3\theta)$ and (b) of one "petal" of $r = \sin(5\theta)$.

30. Assume that r is a differentiable function of θ. Verify that $\{ \frac{dx}{d\theta} \}^{2} + \{ \frac{dy}{d\theta} \}^{2} = \{ r \}^{2} + \{ \frac{dr}{d\theta} \}^{2}$ by replacing x with $r{\cdot}\cos(\theta)$ and y with $r{\cdot}\sin(\theta)$ in the left side of the equation, differentiating, and then simplifying the result to obtain the right side of the equation.

Section 9.2   **PRACTICE Answers**

Practice 1:   The values are shown in Fig. 26.

Point	$\dfrac{dx}{d\theta}$	$\dfrac{dy}{d\theta}$	$\dfrac{dr}{d\theta}$	$\dfrac{dy}{dx}$
A	+	+	+	+
B	−	0	−	0
C	−	−	−	+
D	−	+	+	−

Fig. 26

Practice 2: $r = 1 - \sin(\theta)$ and $r' = -\cos(\theta)$.

$$\frac{dy}{dx} = \frac{\dfrac{dy}{d\theta}}{\dfrac{dx}{d\theta}} = \frac{r \cdot \cos(\theta) + r' \cdot \sin(\theta)}{-r \cdot \sin(\theta) + r' \cdot \cos(\theta)}$$

$$= \frac{(1 - \sin(\theta)) \cdot \cos(\theta) + (-\cos(\theta)) \cdot \sin(\theta)}{-(1 - \sin(\theta)) \cdot \sin(\theta) + (-\cos(\theta)) \cdot \cos(\theta)} = \frac{\cos(\theta) - 2 \cdot \sin(\theta) \cdot \cos(\theta)}{-\sin(\theta) + \sin^2(\theta) - \cos^2(\theta)} \quad .$$

When $\theta = 0$, $\dfrac{dy}{dx} = \dfrac{1 - 0}{-0 + 0 - 1} = -1$.

When $\theta = \dfrac{\pi}{4}$, $\dfrac{dy}{dx} = \dfrac{(1/\sqrt{2}) - 2(1/\sqrt{2})(1/\sqrt{2})}{-(1/\sqrt{2}) + (1/\sqrt{2})^2 - (1/\sqrt{2})^2} = \dfrac{1/\sqrt{2} - 1}{-1/\sqrt{2} + \frac{1}{2} - \frac{1}{2}} = \sqrt{2} - 1 \approx 0.414$.

When $\theta = \dfrac{\pi}{2}$, $\dfrac{dy}{dx} = \dfrac{0 - 0}{-1 + 1 - 0}$ which is undefined. Why does this result make sense in terms

of the graph of the cardioid $r = 1 - \sin(\theta)$?

Practice 3:   One "petal" of the rose $r = \sin(3\theta)$ is swept out as θ goes from 0 to $\pi/3$ (see Fig. 10) so

the endpoints of the area integral are 0 and $\pi/3$.

$$\text{area} = \int_{\theta=\alpha}^{\beta} \frac{1}{2} r^2(\theta) \, d\theta = \int_{\theta=0}^{\pi/3} \frac{1}{2} \{ \sin(3\theta) \}^2 \, d\theta \qquad \text{(then using integral table entry \#13)}$$

$$= \frac{1}{2} \left\{ \frac{1}{2} \theta - \frac{1}{4(3)} \sin(2 \cdot 3\theta) \right\} \Bigg|_0^{\pi/3} = \frac{1}{2} \left\{ \left[\frac{\pi}{6} - \frac{1}{12}(0) \right] - [0 - 0] \right\} = \frac{\pi}{12} \approx 0.262 .$$

Practice 4:   The area we want in Fig. 13 is

$\{\text{area of circle}\} - \{\text{area of cardioid from Example 3}\} = \pi(2)^2 - \frac{3}{2}\pi = \frac{5}{2}\pi \approx 7.85$.

9.3   PARAMETRIC EQUATIONS

Some motions and paths are inconvenient, difficult or impossible for us to describe
by a single function or formula of the form y = f(x).

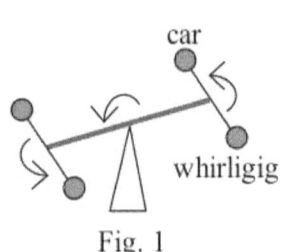

car

whirligig

Fig. 1

- A rider on the "whirligig" (Fig. 1) at the carnival goes in circles at the end
  of a rotating bar.
- A robot delivering supplies in a factory (Fig. 2) needs to avoid obstacles.
- A fly buzzing around the room (Fig. 3) and a molecule in a solution follow
  erratic paths.
- A stone caught in the tread of a rolling wheel has a smooth path with some
  sharp corners (Fig. 4).

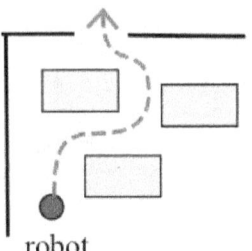

robot

Fig. 2

Parametric equations provide a way to describe all of these motions and paths. And
parametric equations generalize easily to describe paths and motions in 3 dimensions.

fly

Fig. 3

Parametric equations were used briefly in earlier sections (2.5: Applications of the Chain
Rule and 5.2: Arc Length). In those sections the equations were always given. In this
section we look at functions given parametrically by data, graphs, and
formulas and examine how to build formulas to describe some motions
parametrically. The last curve in this section is the cycloid, one of the
most famous curves in mathematics. The next section considers
calculus with parametric equations: slopes of tangent lines, arc lengths,
and areas.

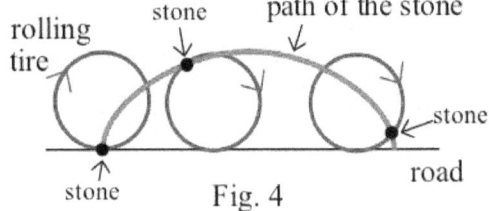

rolling
tire

stone

path of the stone

stone

road

stone   Fig. 4

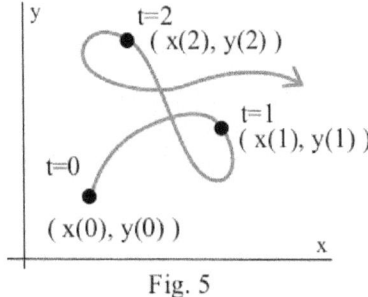

Fig. 5

Parametric equations describe the location of a point (x,y) on a graph or path as
a function of a single independent variable t, a "parameter" often representing
time. In 2 dimensions, the coordinates x and y are functions of the variable t:
$x = x(t)$ and $y = y(t)$ (Fig. 5). In 3 dimensions, the z coordinate is also a
function of t: $z = z(t)$. With parametric equations we can also analyze the
forces acting on an object separately in each coordinate direction and then
combine the results to see the overall behavior of the object. Parametric
equations often provide an easier way to understand and build equations for complicated motions.

Graphing Parametric Equations

The data for creating a parametric equation graph can be given as a table of values, as graphs of (t, x(t))
and (t, y(t)), or as formulas for x and y as functions of t.

Example 1:   Table 1 is a record of the location of a roller coaster car relative to its starting location. Use the data to sketch a graph of the car's path for the first 7 seconds.

t	x(t)	y(t)	t	x(t)	y(t)
0	0	70	7	90	55
1	30	20	8	105	85
2	70	50	9	125	100
3	60	75	10	130	80
4	30	70	11	150	65
5	32	35	12	180	75
6	60	15	13	200	30

Table 1

Solution: Figure 6 is a plot of the (x, y) locations of the car for t = 0 to 7 seconds. The points are connected by a smooth curve to show a possible path of the car.

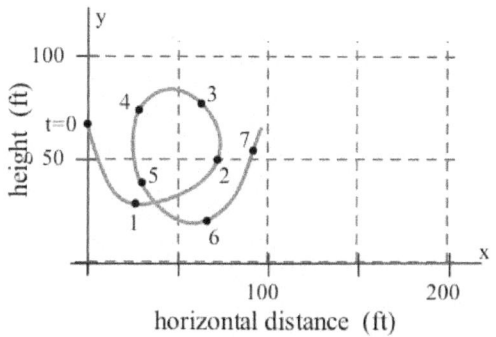

Fig. 6: Roller coaster for 0≤t≤7 seconds

Practice 1:   Use the data in Table 1 to sketch the path of the roller coaster for the next 6 seconds.

Note: Clearly the graph in Fig. 6 is not the graph of a function y = f(x). But every y = f(x) function has an easy parametric representation by setting x(t) = t and y(t) = f(t).

Sometimes a parametric graph can show patterns that are not clearly visible in individual graphs.

Example 2:   Figures 7a and 7b are graphs of the populations of rabbits and foxes on an island. Use these graphs to sketch a parametric graph of rabbits (x–axis) versus foxes (y–axis) for 0 ≤ t ≤ 10 years.

Solution: The separate rabbit and fox population graphs give us information about each population separately, but the parametric graph helps us see the effects of the interaction between the rabbits and the foxes more clearly.

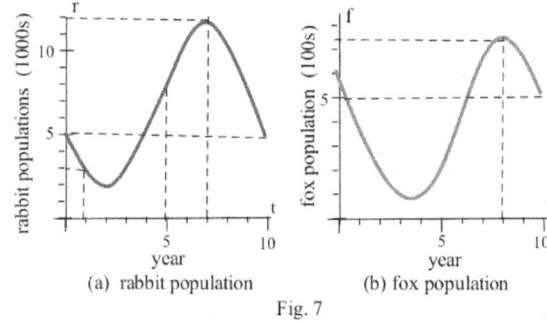

Fig. 7

For each time t we can read the rabbit and fox populations from the separate graphs (e.g., when t = 1, there are approximately 3000 rabbits and 400 foxes so x ≈ 3000 and y ≈ 400) and then combine this information to plot a single point on the parametric graph. If we repeat this process for a large number of values of t, we get a graph (Fig. 8) of the "motion" of the rabbit and fox populations over a period of time, and we can ask questions about why the populations might show this behavior.

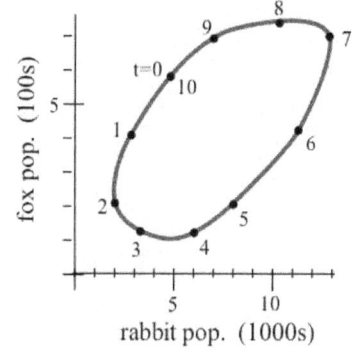

Fig. 8

The type of graph in Fig. 8 is very common for "predator–prey" interactions. Some two–species populations tend to approach a "steady state" or "fixed point" (Fig. 9). However, many two–species population graphs tend to cycle over a period of time as in Fig. 9.

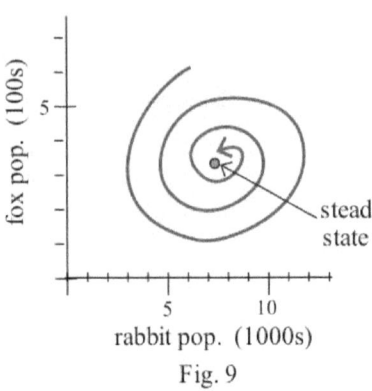

Fig. 9

Practice 2:   What would it mean if the rabbit–fox parametric equation graph hit the horizontal axis as in Fig. 10?

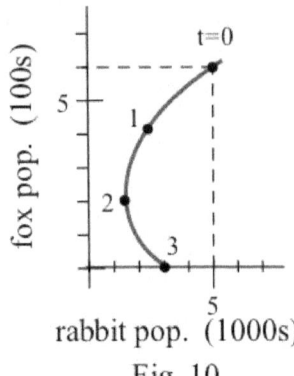

rabbit pop. (1000s)

Fig. 10

Example 3:   Graph the pair of parametric equations x(t) = 2t – 2 and y(t) = 3t + 1.

Solution: Table 2 shows the values of x and y for several values of t. These points are plotted in Fig. 11, and the graph appears to be a straight line.

t	x(t)	y(t)
0	–2	1
1	0	4
2	2	7
–1	–4	–2

Table 2

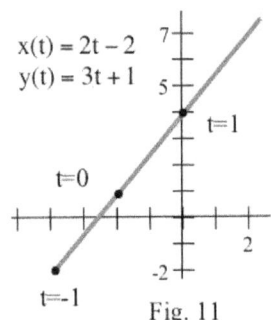

x(t) = 2t – 2
y(t) = 3t + 1

Fig. 11

Usually it is not possible to write y as a simple function of x, but in this case we can do so. By solving x = 2t – 2 for $t = \frac{1}{2} x + 1$ and then replacing the t in the equation y = 3t + 1,

we get $y = 3t + 1 = 3\{ \frac{1}{2} x + 1 \} + 1 = \frac{3}{2} x + 4$, a linear function of x.

Practice 3:   Graph the pair of parametric equations x(t) = 3 – t and $y(t) = t^2 + 1$. Write y as a function of x alone and identify the shape of the graph.

Example 4:   Graph the pair of parametric equations x(t) = 3·cos(t) and y(t) = 2·sin(t) for $0 \le t \le 2\pi$,

and show that these equations satisfy the relation $\frac{x^2}{9} + \frac{y^2}{4} = 1$ for all values of t.

Solution: The graph, an ellipse, is shown in Fig. 12.

$$\frac{x^2}{9} + \frac{y^2}{4} = \frac{3^2 \cdot \cos^2(t)}{9} + \frac{2^2 \cdot \sin^2(t)}{4} = \cos^2(t) + \sin^2(t) = 1.$$

Practice 4:   Graph the pair of parametric equations x(t) = sin(t) and y(t) = 5·cos(t) for $0 \le t \le 2\pi$, and show that these equations

satisfy the relation $\frac{x^2}{1} + \frac{y^2}{25} = 1$ for all values of t.

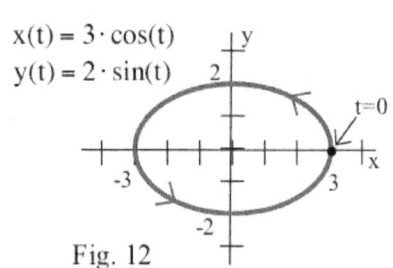

x(t) = 3 · cos(t)
y(t) = 2 · sin(t)

Fig. 12

Example 5:   Describe the motion of a point whose position is

$x(t) = -R \cdot \sin(t)$ and $y(t) = -R \cdot \cos(t)$.

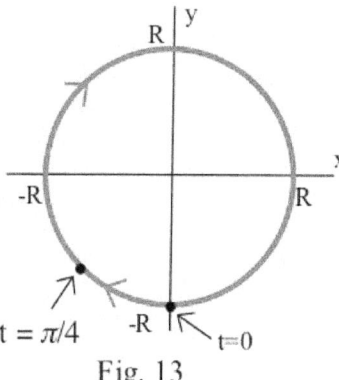

Fig. 13

Solution: The point starts at $x(0) = -R \cdot \sin(0) = 0$ and $y(0) = -R \cdot \cos(0) = -R$.

By plotting $x(t)$ and $y(t)$ for several other values of t (Fig. 13), we can see

that the point is rotating clockwise around the origin. Since

$x^2(t) + y^2(t) = R^2 \sin^2(t) + R^2 \cos^2(t) = R^2$, we know the point is

always on the circle of radius R which is centered at the origin.

Practice 5:   The path of each parametric equation given below is a circle with radius 1 and center at the

origin. If an object is located at the point (x, y) at time t seconds:

(a) Where is the object at t = 0? (b) Is the object traveling clockwise or counterclockwise around the

circle? (c) How long does it take the object to make 1 revolution?

A: $x = \cos(2t), y = \sin(2t)$ B: $x = -\cos(3t), y = \sin(3t)$ C: $x = \sin(4t), y = -\cos(4t)$

Putting Motions Together

If we know how an object moves horizontally and how it moves vertically, then we can put these motions

together to see how it moves in the plane.

If an object is thrown straight upward with an initial velocity of A feet per second, then its height after t

seconds is $y(t) = A \cdot t - \frac{1}{2} g \cdot t^2$ feet where $g = 32$ feet/second^2 is the downward acceleration of gravity

(Fig. 14a). If an object is thrown horizontally with an initial velocity of B feet per second, then its

horizontal distance from the starting place after t seconds is $x(t) = B \cdot t$ feet (Fig. 14b).

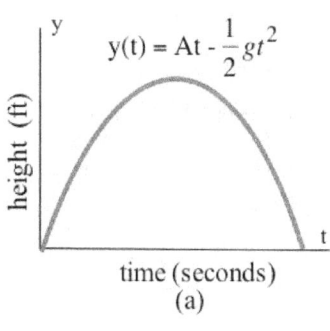

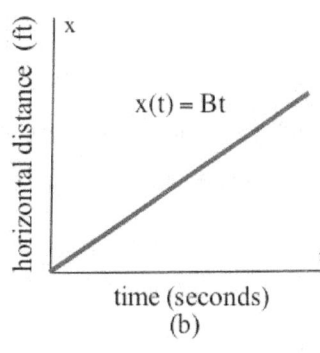

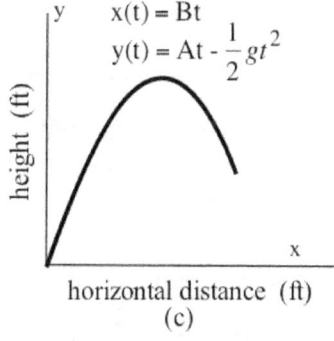

Fig, 14

Example 6:   Write an equation for the position at time t (Fig. 14c) of an object thrown at an angle of

30° with the ground (horizontal) with an initial velocity 100 feet per second.

Solution: If the object travels 100 feet along a line at an angle of

30° to the horizontal ground (Fig. 15), then it travels

100·sin(30°) = 50 feet upward and 100·cos(30°) ≈ 86.6

feet sideways, so A = 50 and B = 86.6 . The position

of the object at time t is

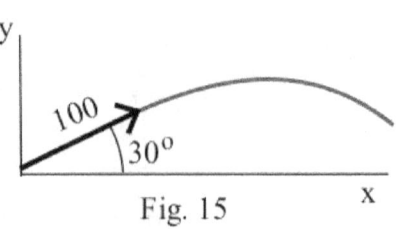

Fig. 15

$$y(t) = 50{\cdot}t - \frac{1}{2}\,g\,t^2 \ \text{ and } \ x(t) = 86.6{\cdot}t \ .$$

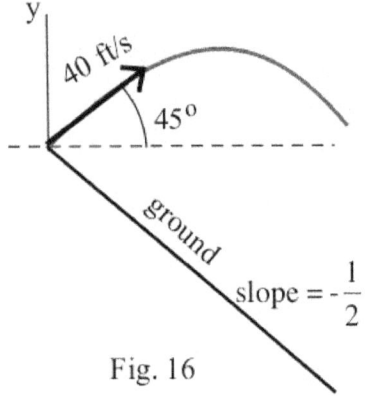

Fig. 16

Practice 6:   A ball is thrown upward at an angle of 45° (Fig. 16) with an

initial velocity of 40 ft/sec.

(a) Write the parametric equations for the position of the ball as

a function of time.

(b) Use the parametric equations to find when and then where

the ball will hit the sloped ground. (Suggestion: set

$y(t) = -0.5x(t)$ from part (a) and solve for t. Then use that

value of t to evaluate x(t) and y(t).)

Sometimes the location or motion of an object is measured by an instrument which is in motion itself (e.g.,
tracking a pod of migrating whales from a moving ship), and we want to determine the path of the object
independent of the location of the instrument. In that case, the "absolute" location of the object with
respect to the origin is the sum of the relative location of the object (pod of whales) with respect to the
instrument (ship) and the location of the instrument (ship) with respect to the origin. The same approach
works for describing the motion of linked objects such as connected gears.

Example 7:   **Carnival Ride**   The car (Fig. 17) makes one counterclockwise revolution (r = 8 feet)

about the pivot point A every 2 seconds and the long arm (R = 20 feet) makes one counterclockwise

revolution about its pivot point (the origin) every 5 seconds. Assume that the ride begins with the

two arms along the positive x–axis and sketch the path you think the car will follow. Find a pair of

parametric equations to describe the position of the car at time t.

Solution: The position of the car relative to its pivot point A is

$$x_c(t) = 8{\cdot}\cos(\frac{2\pi}{2}\,t) \ \text{ and } \ y_c(t) = 8{\cdot}\sin(\frac{2\pi}{2}\,t).$$

The position of the pivot point A relative to the origin is

$$x_p(t) = 20{\cdot}\cos(\frac{2\pi}{5}\,t) \ \text{ and } \ y_p(t) = 20{\cdot}\sin(\frac{2\pi}{5}\,t), \text{ so the}$$

location of the car, relative to the origin, is

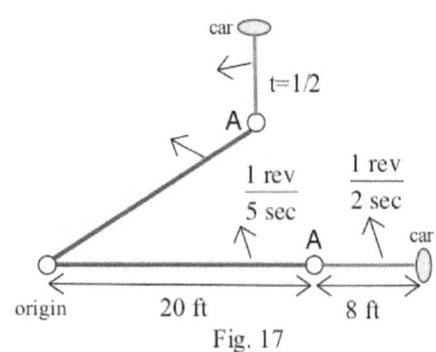

Fig. 17

$$x(t) = x_p(t) + x_c(t) = 20 \cdot \cos(\tfrac{2\pi}{5} t) + 8 \cdot \cos(\tfrac{2\pi}{2} t) \text{ and}$$

$$y(t) = y_p(t) + y_c(t) = 20 \cdot \sin(\tfrac{2\pi}{5} t) + 8 \cdot \sin(\tfrac{2\pi}{2} t) .$$

Use a graphing calculator to graph the path of the car for 5 seconds.

Example 8:   **Cycloid**   A light is attached to the edge of a wheel of radius R which is rolling along a level road (Fig. 18). Find parametric equations to describe the location of the light.

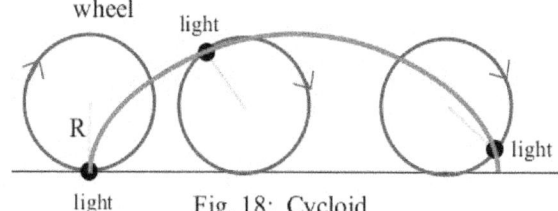

Fig. 18: Cycloid

Solution: We can describe the location of the axle of the wheel, the location of the light relative to the axle, and then put the results together to get the location of the light.

The axle of the wheel is always R inches off the ground, so the y coordinate of the axle is $y_a(t) = R$

(Fig. 19). When the wheel has rotated t radians about its axle, the wheel has rolled a distance of $R \cdot t$ along the road, and the x coordinate of the axle is $x_a(t) = R \cdot t$.

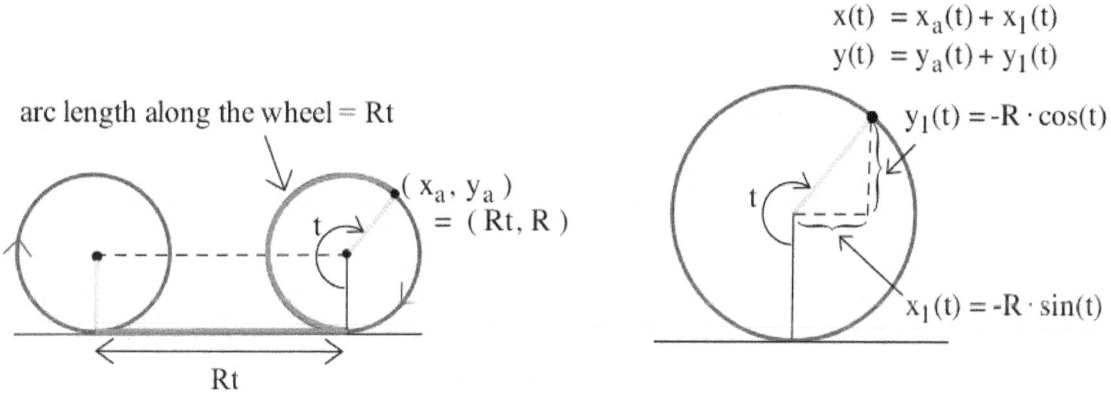

arc length along the wheel = Rt

(x_a, y_a)
$= (Rt, R)$

Rt

$$x(t) = x_a(t) + x_l(t)$$
$$y(t) = y_a(t) + y_l(t)$$

$$y_l(t) = -R \cdot \cos(t)$$

$$x_l(t) = -R \cdot \sin(t)$$

Fig. 19

The position of the light relative to the axle is $x_l(t) = -R \cdot \sin(t)$ and $y_l(t) = -R \cdot \cos(t)$ (see Example 3) so the position of the light is

$$x(t) = x_a(t) + x_l(t) = R \cdot t - R \cdot \sin(t) = R \cdot \{ t - \sin(t) \} \text{ and}$$

$$y(t) = y_a t) + y_l(t) = R - R \cdot \cos(t) = R \cdot \{ 1 - \cos(t) \} .$$

This curve is called a **cycloid**, and it is one of the most famous and interesting curves in mathematics. Many great mathematicians and physicists (Mersenne, Galileo, Newton, Bernoulli, Huygens, and others) examined the cycloid, determined its properties, and used it in physical applications.

Practice 7:   A light is attached r units from the axle of an R inch radius wheel

(r < R) that is rolling along a level road (Fig. 20). Use the approach of the

solution to Example 8 to find parametric equations to describe the location

of the light. The resulting curve is called an curate cycloid.

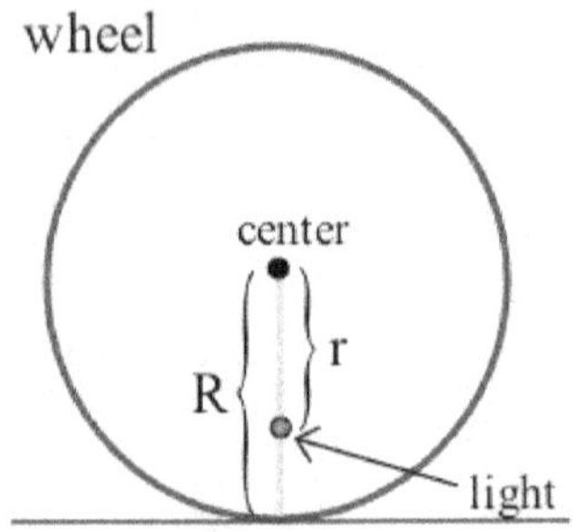

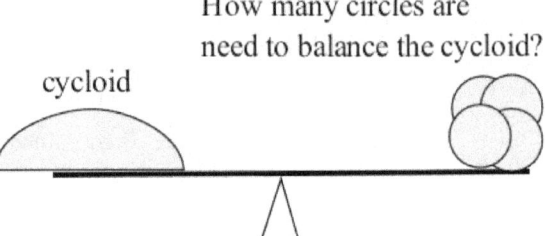

Fig. 20

The cycloid, the path of a point on a rolling circle, was studied in the early 1600's by

Mersenne (1588–1648) who thought the path might be part of an ellipse (it isn't).

In 1634 Roberval determined the parametric form of the cycloid and found the area under the cycloid as did

Descartes and Fermat. This was done before Newton (1642–1727) was even born; they used various

specialized geometric approaches to solve the area problem. About the same time Galileo determined the

area experimentally by cutting a cycloid region from a sheet of lead and balancing it against a number of

circular regions (with the same radius as the circle which

generated the cycloid) cut from the same material. How

many circles do you think balanced the cycloid region's

area (Fig. 21)?

How many circles are need to balance the cycloid?

cycloid

However, the most amazing properties of the cycloid involve

motion along a cycloid–shaped path, and their discovery had

Fig. 21

to wait for Newton and the calculus. These calculus–based properties are discussed at the end of the next

section.

PROBLEMS

For problems 1–4, use the data in each table to create three graphs:

(a) (t, x(t)), (b) (t, y(t)), and (c) the parametric graph (x(t), y(t)).

(Connect the points with straight

line segments to create the graph.)

1. Use Table 3.   2. Use Table 4.

3. Use Table 5.   4. Use Table 6.

t	x(t)	y(t)
0	2	1
1	2	0
2	–1	0
3	1	–1

Table 3

t	x(t)	y(t)
0	0	1
1	1	1
2	1	–1
3	2	0

Table 4

t	x(t)	y(t)
0	1	2
1	–1	–1
2	1	2
3	0	2

Table 5

t	x(t)	y(t)
0	0	1
1	–1	0
2	0	–2
3	3	1

Table 6

For problems 5–8, use the data in the given graphs of $(t, x(t))$ and $(t, y(t))$ to sketch the parametric graph $(x(t), y(t))$.

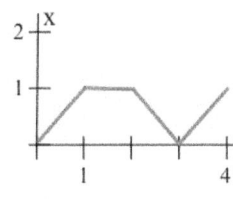

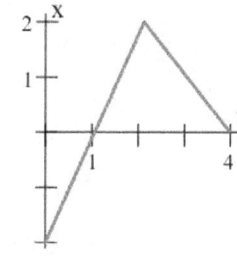

5.   Use x and y from Fig. 22.

6.   Use x and y from Fig. 23.

7.   Use x and y from Fig. 24.

8.   Use x and y from Fig. 25.

9.   Graph $x(t) = 3t - 2$, $y(t) = 1 - 2t$. What shape is this graph?

10. Graph $x(t) = 2 - 3t$, $y(t) = 3 + 2t$.
     What shape is this graph?

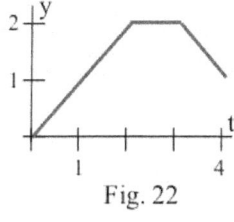

Fig. 22

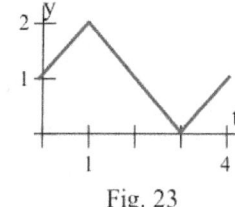

Fig. 23

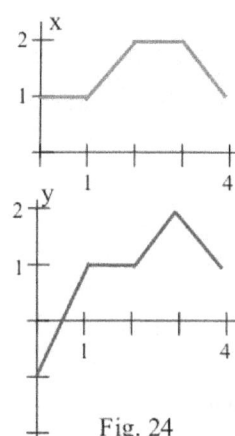

Fig. 24

11. Calculate the slope of the line through the points
     $P = (x(0), y(0))$ and $Q = (x(1), y(1))$ for $x(t) = at + b$
     and $y(t) = ct + d$.

12. Graph $x(t) = 3 + 2 \cdot \cos(t)$, $y(t) = -1 + 3 \cdot \sin(t)$ for
     $0 \le t \le 2\pi$. Describe the shape of the graph.

13. $x(t) = -2 + 3 \cdot \cos(t)$, $y(t) = 1 - 4 \cdot \sin(t)$ for $0 \le t \le 2\pi$.
     Describe the shape of the graph.

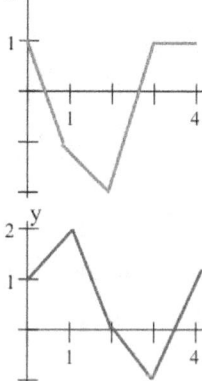

Fig. 25

14. Graph (a) $x(t) = t^2$, $y(t) = t$, (b) $x(t) = \sin^2(t)$, $y(t) = \sin(t)$, and
     (c) $x(t) = t, y(t) = \sqrt{t}$. Describe the similarities and the differences among
     these graphs.

15. Graph (a) $x(t) = t$, $y(t) = t$, (b) $x(t) = \sin(t)$, $y(t) = \sin(t)$, and (c) $x(t) = t^2, y(t) = t^2$.
     Describe the similarities and the differences among these graphs.

16. Graph $x(t) = (4 - \frac{1}{t})\cos(t)$, $y(t) = (4 - \frac{1}{t})\sin(t)$ for $t \ge 1$. Describe the behavior of the graph.

17. Graph $x(t) = \frac{1}{t} \cdot \cos(t)$, $y(t) = \frac{1}{t} \cdot \sin(t)$ for $t \ge \pi/4$. Describe the behavior of the graph.

18. Graph $x(t) = t + \sin(t)$, $y(t) = t^2 + \cos(t)$ for $0 \le t \le 2\pi$. Describe the behavior of the graph.

Problems 19–22 refer to the rabbit–fox population graph shown
in Fig. 26 which shows several different population cycles
depending on the various numbers of rabbits and foxes. Wildlife
biologists sometimes try to control animal populations by
"harvesting" some of the animals, but it needs to be done with
care. The thick dot on the graph is the fixed point for this two–
species population.

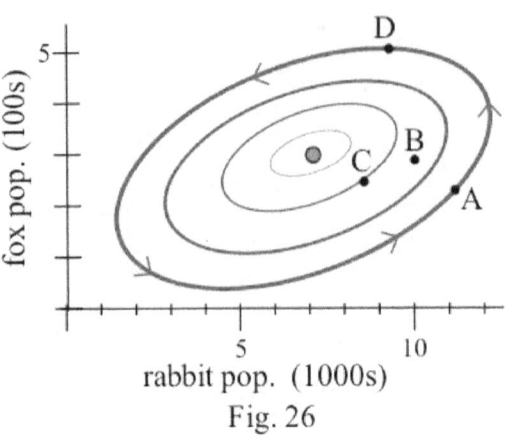
Fig. 26

19. Suppose there are currently 11,000 rabbits and 200 foxes
    (point A on the graph), and 1,000 rabbits are "harvested" (removed from the population). Does the
    harvest shift the populations onto a cycle closer to or farther from the fixed point?

20. Suppose there are currently 10,000 rabbits and 300 foxes (point B on the graph), and 100 foxes are
    "harvested." Does the harvest shift the populations onto a cycle closer to or farther from the fixed point?

21. Suppose there are currently 8,000 rabbits and 250 foxes (point C on the graph), and 1,000 rabbits die
    during a hard winter. Does the wildlife biologist need to take action to maintain the population
    balance? Justify your response.

22. Suppose there are currently 9,000 rabbits and 500 foxes (point D on the graph), and 2,000 rabbits die
    during a hard winter. Does the wildlife biologists need to take action to maintain the population
    balance? Justify your response.

23. Suppose x and y are functions of the form $x(t) = a \cdot t + b$ and $y(t) = c \cdot t + d$ with $a \neq 0$ and $c \neq 0$. Write
    y as a function of x alone and show that the parametric graph (x, y) is a straight line. What is the slope
    of the resulting line?

24. The parametric equations given in (a) – (e) all satisfy $x^2 + y^2 = 1$, and, for $0 \leq t \leq 2\pi$, the path of each
    object is a circle with radius 1 and center at the origin. Explain how the motions of the objects **differ**.
    (a) $x(t) = \cos(t), y(t) = \sin(t),$ (b) $x(t) = \cos(-t), y(t) = \sin(-t),$ (c) $x(t) = \cos(2t), y(t) = \sin(2t),$
    (d) $x(t) = \sin(t), y(t) = \cos(t),$ and   (e) $x(t) = \cos(t + \pi/2), y(t) = \sin(t + \pi/2)$

25. From a tall building you observe a person is walking along a straight path while twirling a light (parallel
    to the ground) at the end of a string. (a) If the person is walking slowly, sketch the path of the light. (b)
    How would the graph change if the person was running? (c) Sketch the path for a person walking
    (running) along a parabolic path.

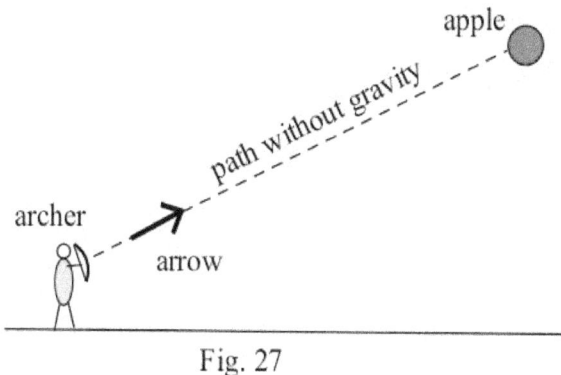

Fig. 27

26. William Tell and the Falling Apple: William Tell is aiming directly at an apple, and releases the arrow at exactly the same instant that the apple stem breaks. In a world without gravity (or air resistance), the apple remains in place after the stem breaks, and the arrow flies in a straight line to hit the apple (Fig. 27). Sketch the path of the apple and the arrow in a world with gravity (but still no air). Does the arrow still hit the apple? Why or why not?

27. Find the radius R of a circle which generates a cycloid starting at the point $(0,0)$ and

(a) passing through the point $(10\pi, 0)$ on its first complete revolution $(0 \le t \le 2\pi)$..

(b) passing through the point $(5, 2)$ on its first complete revolution. (A calculator is helpful here.)

(c) passing through the point $(2, 3)$ on its first complete revolution. (A calculator is helpful here.)

(d) passing through the point $(4\pi, 8)$ on its first complete revolution.

The Ferris Wheel and the Apple (problems 28 – 30).

28. Your friends are on the Ferris wheel illustrated in Fig. 28, and at time t seconds, their location is given parametrically as

$$(-20\sin(\frac{2\pi}{15} t), 30 - 20\cos(\frac{2\pi}{15} t)).$$

(a) Is the Ferris wheel turning clockwise of counterclockwise?

(b) How many seconds does it take the Ferris wheel to make a revolution?

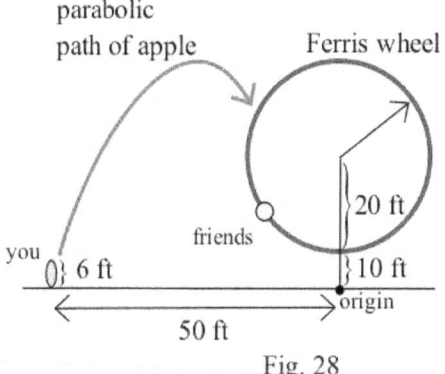

Fig. 28

29. You are 50 feet to the left of the Ferris wheel in problem 28, and you toss an apple from a height of 6 feet above the ground at an angle of 45°. Write parametric equations for the location of the apple (relative to the origin in Fig. 29) at time t if

(a) its initial velocity is 30 feet per second, and (b) its initial velocity is V feet per second.

30. Help — the Ferris wheel won't stop! To keep your friends on the Ferris wheel in problem 28 from getting too hungry, you toss an apple to them (at time t = 0). Write an equation for the distance between the apple and your friends at time t. Somehow, find a value for the initial velocity V of the apple so that it comes close enough for your friend to catch it, within 2 feet. (Note: A calculator or computer is probably required for this problem.)

Section 9.3   **PRACTICE Answers**

Practice 1:   A possible path for the car is shown

in Fig. 29.

Practice 2:   If the (rabbit, fox) parametric graph touches

the horizontal axis, then there are 0 foxes: the foxes are

extinct.

Practice 3: $x = 3 - t$ and $y = t^2 + 1$.

Then $t = 3 - x$ and $y = (3 - x)^2 + 1 = x^2 - 6x + 10$.

The graph in Fig. 30 is parabola, opening upward, with

vertex at $(3, 1)$.

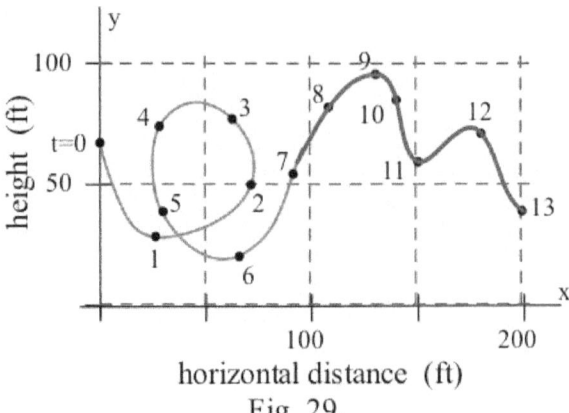

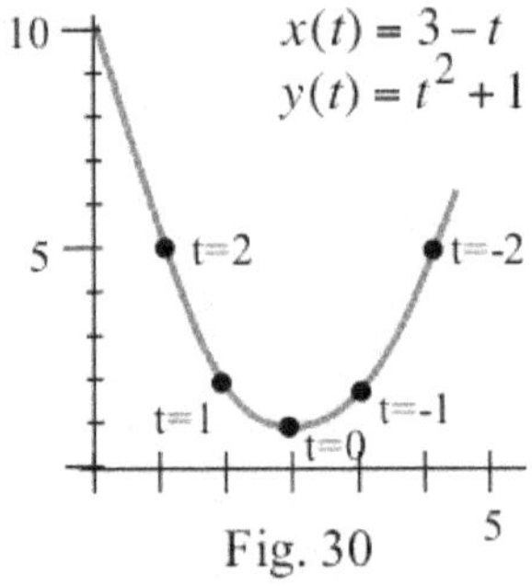

horizontal distance (ft)

Fig. 29

Practice 4:   The parametric graph of $x(t) = \sin(t)$ and $y(t) = 5\cos(t)$

is shown in Fig. 31. For all t,

$$\frac{x^2}{1} + \frac{y^2}{25} = \frac{\sin^2(t)}{1} + \frac{25\cos^2(t)}{25} = \sin^2(t) + \cos^2(t) = 1.$$

Practice 5:   A: Starts at $(1, 0)$, travels counterclockwise, and

takes $2\pi/2 = \pi$ seconds to make one revolution.

B: Starts at $(-1, 0)$, travels clockwise, and

takes $2\pi/3$ seconds to make one revolution.

C: Starts at $(0, -1)$, travels counterclockwise, and

takes $2\pi/4 = \pi/2$ seconds to make one revolution.

Fig. 30

Practice 6:   (a) $x(t) = 40 \cdot \cos(45^\circ) \cdot t$, $y(t) = 40 \cdot \sin(45^\circ) \cdot t - 16t^2$.

(b)   Let $A = 40 \cdot \sin(45^\circ) = 40 \cdot \cos(45^\circ) \approx 28.284$.

Then the ball is at $x(t) = At$ and $y(t) = At - 16t^2$.

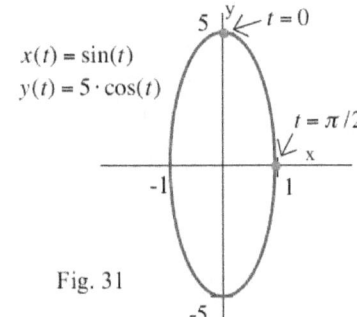

Fig. 31

Along the ground line, $y = -\frac{1}{2}x$ so the ball intersects the ground when $y(t) = -\frac{1}{2}x(t)$: $At - 16t^2 = -\frac{1}{2}At$.

When $t \neq 0$, we can solve $At - 16t^2 = -\frac{1}{2}At$ for $t = \frac{3}{32}A$.

Putting $t = \frac{3}{32}A$ into the equations for the location of the ball, we have

$$x(\frac{3}{32}A) = A \cdot (\frac{3}{32}A) = \frac{3}{32}A^2 \text{ and } y(\frac{3}{32}A) = A \cdot (\frac{3}{32}A) - 16(\frac{3}{32}A)^2 = -\frac{3}{64}A^2.$$

The ball hits the ground after $t = \frac{3}{32}A = \frac{3}{32} \cdot 40 \cdot \sin(45^\circ) \approx \textbf{2.652 seconds}$.

The ball hits the ground at the location $\mathbf{x} = \frac{3}{32}A^2 = \textbf{75 feet}$ and $\mathbf{y} = -\frac{3}{64}A^2 = \mathbf{-37.5\ feet}$.

Practice 7:   Axle: $x_a = R \cdot t$ and $y_a = R$. Light relative to the axle: $x_l = -r \cdot \sin(t)$ and $y_l = -r \cdot \cos(t)$.

Then $x(t) = x_a + x_l = R \cdot t - r \cdot \sin(t)$ and $y(t) = y_a + y_l = R - r \cdot \cos(t)$.

9.4   CALCULUS AND PARAMETRIC EQUATIONS

The previous section discussed parametric equations, their graphs, and some of their uses for visualizing and analyzing information. This section examines some of the ideas and techniques of calculus as they apply to parametric equations: slope of a tangent line, speed, arc length, and area. Slope, speed, and arc length were considered earlier (in optional parts of sections 2.5 and 5.2), and the presentation here is brief. The material on area is new and is a variation on the Riemann sum development of the integral. This section ends with a presentation of some of the properties of the cycloid.

Slope (also see section 2.5)

If $x(t)$ and $y(t)$ are differentiable functions of t, then the derivatives dx/dt and dy/dt measure the rates of change of x and y with respect to t: dx/dt and dy/dt tell how fast each variable is changing. The derivative dy/dx measures the slope of the line tangent to the parametric graph $(x(t), y(t))$. To calculate dy/dx we need to use the Chain Rule:

$$\frac{dy}{dt} = \frac{dy}{dx} \cdot \frac{dx}{dt} .$$

Dividing each side of the Chain Rule by $\frac{dx}{dt}$, we have $\frac{dy}{dx} = \frac{dy/dt}{dx/dt}$.

Slope with Parametric Equations

If $x(t)$ and $y(t)$ are differentiable functions of t and $\frac{dx}{dt} \neq 0$,

then   the **slope** of the line tangent to the parametric graph is $\frac{dy}{dx} = \frac{dy/dt}{dx/dt}$.

Example 1:   The location of an object is given by the parametric equations $x(t) = t^3 + 1$ feet and $y(t) = t^2 + t$ feet at time t seconds.

(a) Evaluate $x(t)$ and $y(t)$ at $t = -2, -1, 0, 1,$ and 2, and then graph the path of the object for $-2 \leq t \leq 2$.

(b) Evaluate dy/dx for $t = -2, -1, 0, 1,$ and 2. Do your calculated values for dy/dx agree with the shape of your graph in part (a)?

t	x	y	dy/dx
-2	-7	2	-3/12 = -1/4
-1	0	0	-1/3
0	1	0	undefined
1	2	2	3/3 = 1
2	9	6	5/12

Table 1

Solution: (a)   When $t = -2$,

$x = (-2)^3 + 1 = -7$ and

$y = (-2)^2 + (-2) = 2$. The other values   for x

and y are given in Table 1.

The graph of (x, y) is shown in Fig. 1.

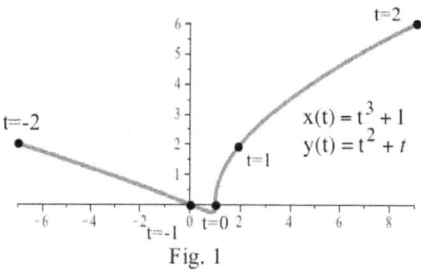

Fig. 1

(b) $dy/dt = 2t + 1$ and $dx/dt = 3t^2$ so $\dfrac{dy}{dx} = \dfrac{2t + 1}{3t^2}$. When $t = -2$, $\dfrac{dy}{dx} = \dfrac{-3}{12}$. The other values for dy/dx are given in Table 1.

Practice 1: Find the equation of the line tangent to the graph of the parametric equations in Example 1 when $t = 3$.

An object can "visit" the same location more than once, and a parametric graph can go through the same point more than once.

Example 2: Fig. 2 shows the x and y coordinates of an object at time t.

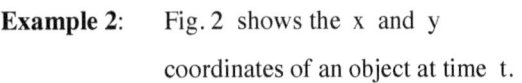

Fig. 2

(a) Sketch the parametric graph $(x(t), y(t))$, the position of the object at time t.

(b) Give the coordinates of the object when $t = 1$ and $t = 3$.

(c) Find the slopes of the tangent lines to the parametric graph when $t = 1$ and $t = 3$.

Solution: (a) By reading the x and y values on the graphs in Fig. 2, we can plot points on the parametric graph. The parametric graph is shown in Fig. 3.

(b) When $t = 1$, $x = 2$ and $y = 2$ so the parametric graph goes through the point (2,2).
When $t = 3$, the parametric graph goes through the same point (2,2).

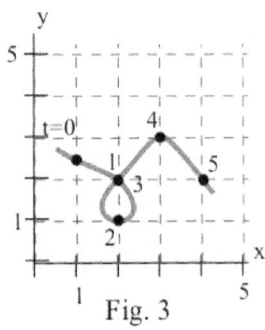

Fig. 3

(c) When $t = 1$, $dy/dt \approx -1$ and $dx/dt \approx +1$ so $\dfrac{dy}{dx} = \dfrac{dy/dt}{dx/dt} \approx \dfrac{-1}{+1} = -1$.

When $t = 3$, $dy/dt \approx +1$ and $dx/dt \approx +1$ so $\dfrac{dy}{dx} \approx \dfrac{+1}{+1} = +1$.

These values agree with the appearance of the parametric graph in Fig. 3.

The object goes through the point (2,2) twice (when t=1 and t=3), but it is traveling in a different direction each time.

Practice 2: (a) Estimate the slopes of the lines tangent to the parametric graph when $t = 2$ and $t = 5$.

(b) When does $y'(t) = 0$ in Fig. 2?

(c) When does the parametric graph in Fig. 3 have a maximum? A minimum?

(d) How are the maximum and minimum points on a parametric graph related to the derivatives of $x(t)$ and $y(t)$?

Speed

If we know how fast an object is moving in the x direction (dx/dt) and how fast in the y direction (dy/dt), it is straightforward to determine the speed of the object, how fast it is moving in the xy –plane.

If, during a short interval of time Δt, the object's position changes Δx in the x direction and Δy in the y direction (Fig. 4), then the object has moved $\sqrt{(\Delta x)^2 + (\Delta y)^2}$ in Δt time. Then

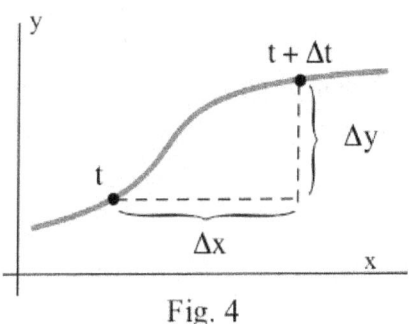

Fig. 4

$$\text{average speed} = \frac{\text{distance moved}}{\text{time change}} = \frac{\sqrt{(\Delta x)^2 + (\Delta y)^2}}{\Delta t}$$

$$= \sqrt{\left(\frac{\Delta x}{\Delta t}\right)^2 + \left(\frac{\Delta y}{\Delta t}\right)^2} \; .$$

If x(t) and y(t) are differentiable functions of t, and if we take the limit of the average speed as Δt approaches 0, then

$$\text{speed} = \lim_{\Delta t \to 0} \{\text{average speed}\} = \lim_{\Delta t \to 0} \sqrt{\left(\frac{\Delta x}{\Delta t}\right)^2 + \left(\frac{\Delta y}{\Delta t}\right)^2} = \sqrt{\left(\frac{dx}{dt}\right)^2 + \left(\frac{dy}{dt}\right)^2} \; .$$

Speed with Parametric Equations

If   an object is located at $(x(t), y(t))$ at time t, and $x(t)$ and $y(t)$ are differentiable functions of t ,

then   the **speed** of the object is $\sqrt{\left(\frac{dx}{dt}\right)^2 + \left(\frac{dy}{dt}\right)^2}$.

Example 3:   At time t seconds an object is located at (cos(t) feet, sin(t) feet) in the plane. Sketch the path of the object and show that it is travelling at a constant speed.

Solution: The object is moving in a circular path (Fig. 5). $dx/dt = -\sin(t)$ feet/second and $dy/dt = \cos(t)$ feet/second so at all times the speed of the object is

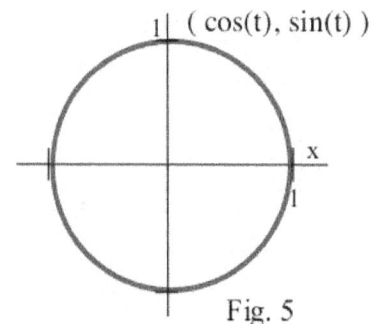

Fig. 5

$$\sqrt{\left(\frac{dx}{dt}\right)^2 + \left(\frac{dy}{dt}\right)^2} \qquad = \sqrt{(-\sin(t))^2 + (\cos(t))^2}$$

$$= \sqrt{\sin^2(t) + \cos^2(t)} = \sqrt{1} = 1 \text{ foot per second.}$$

Practice 3:   Is the object in Example 2 traveling faster when $t = 1$ or when $t = 3$? When $t = 1$ or

when $t = 2$?

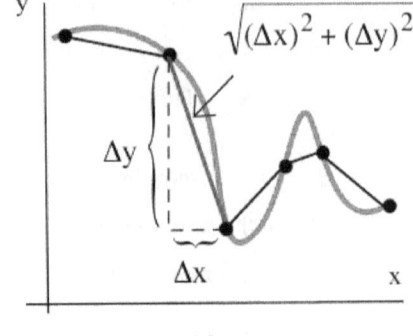

Arc Length (also see section 5.2)

In section 5.2 we approximated the total length L of a curve by
partitioning the curve into small pieces (Fig. 6), approximating the length
of each piece using the distance formula, and then adding the lengths of
the pieces together to get

$$L \approx \sum \sqrt{(\Delta x)^2 + (\Delta y)^2}$$

Fig. 6

$$= \sum \sqrt{(\tfrac{\Delta x}{\Delta x})^2 + (\tfrac{\Delta y}{\Delta x})^2} \;\; \Delta x \;\;, \text{a Riemann sum.}$$

As Δx approaches 0, the Riemann sum approaches the definite integral

$$L = \int_{x=a}^{b} \sqrt{1 + (\tfrac{dy}{dx})^2} \;\; dx \;.$$

A similar approach also works for parametric equations, but in this case we factor out a Δt from the
original summation:

$$L \approx \sum \sqrt{(\Delta x)^2 + (\Delta y)^2} \;\; = \sum \sqrt{(\tfrac{\Delta x}{\Delta t})^2 + (\tfrac{\Delta y}{\Delta t})^2} \;\; \Delta t \;\; \text{(a Riemann sum)}$$

$$\longrightarrow \int_{t=a}^{b} \sqrt{(\tfrac{dx}{dt})^2 + (\tfrac{dy}{dt})^2} \;\; dt \;\; \text{as} \;\; \Delta t \to 0.$$

Arc Length with Parametric Equations

If $x(t)$ and $y(t)$ are differentiable functions of t

then   the length of the parametric graph from $(x(a), y(a))$ to $(x(b), y(b))$ is

$$L = \int_{t=a}^{t=b} \sqrt{(\tfrac{dx}{dt})^2 + (\tfrac{dy}{dt})^2} \;\; dt$$

Example 4:   Find the length of the cycloid

$x = R(\,t - \sin(t)\,)$ $y = R(\,1 - \cos(t)\,)$

for $0 \le t \le 2\pi$. (Fig. 7)

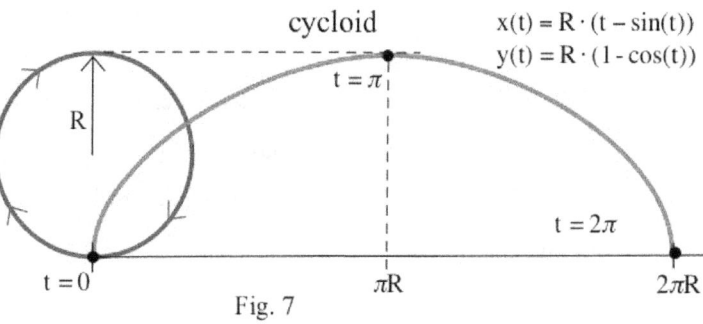

cycloid $x(t) = R \cdot (t - \sin(t))$
$y(t) = R \cdot (1 - \cos(t))$

Fig. 7

Solution: Since $dx/dt = R(\,1 - \cos(t)\,)$ and

$dy/dt = R \cdot \sin(t)$,

$$L = \int_{t=a}^{b} \sqrt{(\tfrac{dx}{dt})^2 + (\tfrac{dy}{dt})^2}\ dt\ =$$

$$\int_{t=0}^{2\pi} \sqrt{(\,R(\,1-\cos(t)\,)\,)^2 + (\,R \cdot \sin(t)\,)^2}\ dt$$

$$= R \int_{t=0}^{2\pi} \sqrt{1 - 2\cos(t) + \cos^2(t) + \sin^2(t)}\ dt\ =\ R \int_{t=0}^{2\pi} \sqrt{2 - 2 \cdot \cos(t)}\ dt\ .$$

By replacing θ with $t/2$ in the formula $\sin^2(\theta) = \dfrac{1 - \cos(2\theta)}{2}$ we have $\sin^2(t/2) = \dfrac{1 - \cos(t)}{2}$

so $2 - 2 \cdot \cos(t) = 4 \cdot \sin^2(t/2)$, and the integral becomes

$$L = R \int_{t=0}^{2\pi} 2 \cdot \sin(\,t/2\,)\ dt\ =\ 2R\{\,-2 \cdot \cos(\,t/2\,)\,\}\ \Big|_0^{2\pi}\ =\ 2R\,\{\,-2 \cdot \cos(\pi) + 2 \cdot \cos(0)\,\}\ =\ \mathbf{8R}\ .$$

The length of a cycloid arch is 8 times the radius of the rolling circle that generated the cycloid.

Practice 4:   Represent the length of the ellipse $x = 3 \cdot \cos(t)$

$y = 2 \cdot \sin(t)$ for for $0 \le t \le 2\pi$ (Fig. 8). as a definite

integral. Use a calculator to approximate the value of

the integral.

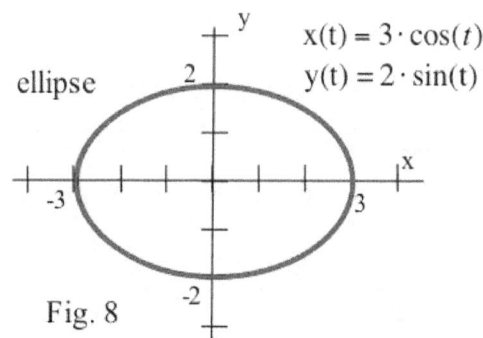

ellipse $x(t) = 3 \cdot \cos(t)$
$y(t) = 2 \cdot \sin(t)$

Fig. 8

Area

When we first discussed area and developed the definite integral,
we approximated the area of a positive function y (Fig. 9) by
partitioning the domain $a \leq x \leq b$ into pieces of length Δx, finding
the areas of the thin rectangles, and approximating the
total area by adding the little areas together:

$$A \approx \sum y \, \Delta x \quad \text{(a Riemann sum)}.$$

As Δx approached 0, the Riemann sum approached the definite

integral $\displaystyle\int_{x=a}^{x=b} y \; dx$.

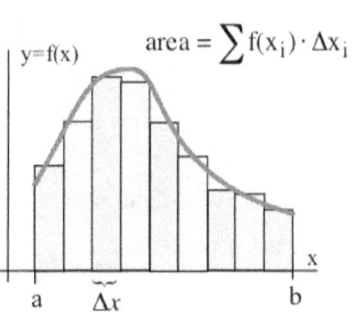

$$\text{area} = \sum f(x_i) \cdot \Delta x_i$$

$y=f(x)$

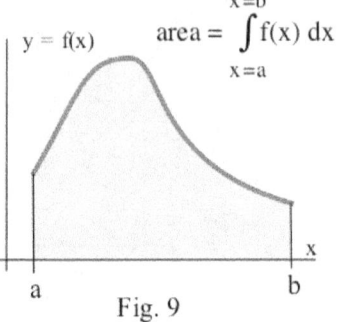

$$\text{area} = \int_{x=a}^{x=b} f(x) \, dx$$

$y = f(x)$

Fig. 9

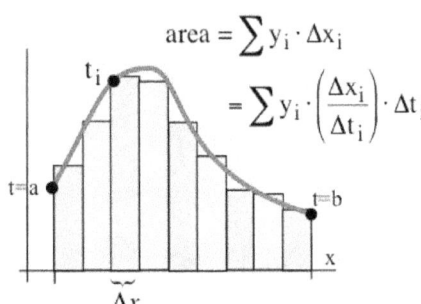

$$\text{area} = \sum y_i \cdot \Delta x_i$$
$$= \sum y_i \cdot \left(\frac{\Delta x_i}{\Delta t_i}\right) \cdot \Delta t_i$$

For parametric equations, the independent variable is t and the domain is
an interval [a, b].

If x is an increasing function of t, then a partition of the t–interval [a, b]
into pieces of length Δt induces a partition along the x–axis (Fig. 10),
and we can use the induced partition
of the x–axis to approximate the total area by

$$A \approx \sum y \, \Delta x = \sum y \, \frac{\Delta x}{\Delta t} \, \Delta t \quad \text{which approaches the definite}$$

integral $\displaystyle A = \int_{t=a}^{t=b} y \cdot \left(\frac{dx}{dt}\right) dt$ as Δt approaches 0.

$y = f(x)$ $\displaystyle \text{area} = \int_{t=a}^{t=b} y \cdot \left(\frac{dx}{dt}\right) dt$

Fig. 10

Area with Parametric Equations

If   y and dx/dt do not change sign for $a \leq t \leq b$,

then   the **area** between the graph (x , y) and the x–axis is $A = \left| \displaystyle\int_{t=a}^{t=b} y \cdot \left(\frac{dx}{dt}\right) dt \right|$.

The requirement that y not change sign for a ≤ t ≤ b is to prevent the parametric graph from being above the x–axis sometimes and below the x–axis sometimes. The requirement that dx/dt not change sign for a ≤ t ≤ b is to prevent the graph from "turning around" (Fig. 11). If either of those situations occurs, some of the area is evaluated as positive and some of the area is evaluated as negative.

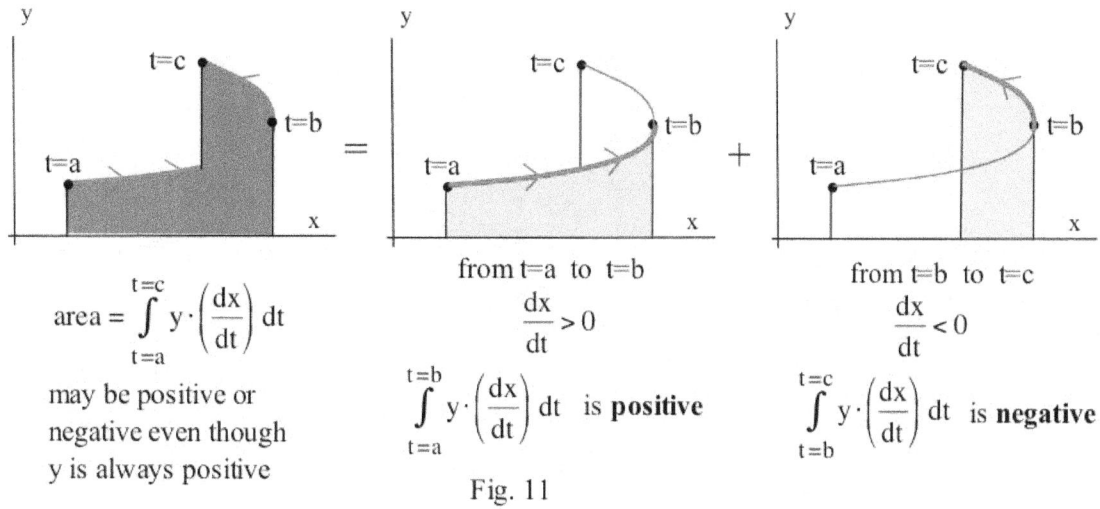

$$area = \int_{t=a}^{t=c} y \cdot \left(\frac{dx}{dt}\right) dt$$

may be positive or
negative even though
y is always positive

from t=a to t=b

$$\frac{dx}{dt} > 0$$

$$\int_{t=a}^{t=b} y \cdot \left(\frac{dx}{dt}\right) dt \text{ is \textbf{positive}}$$

from t=b to t=c

$$\frac{dx}{dt} < 0$$

$$\int_{t=b}^{t=c} y \cdot \left(\frac{dx}{dt}\right) dt \text{ is \textbf{negative}}$$

Fig. 11

Example 5:   Find the area of the ellipse $x = a \cdot \cos(t), y = b \cdot \sin(t)$

(a,b > 0) in the first quadrant (Fig. 12).

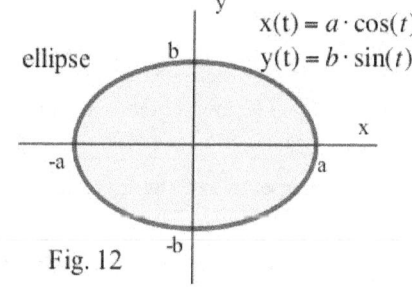

ellipse

$$x(t) = a \cdot \cos(t)$$
$$y(t) = b \cdot \sin(t)$$

Fig. 12

Solution: The derivative $dx/dt = -a \cdot \sin(t)$, and in the first quadrant

$0 \le t \le \pi/2$. Then the

$$\text{area of the ellipse in first quadrant} = \left| \int_{t=a}^{b} y \cdot \left(\frac{dx}{dt}\right) dt \right|$$

$$= \left| \int_{t=0}^{\pi/2} \{ b \cdot \sin(t) \} \cdot \left(-a \cdot \sin(t) \right) dt \right|$$

$$= \left| -ab \int_{t=0}^{\pi/2} \sin^2(t) \, dt \right| = ab \int_{t=0}^{\pi/2} \sin^2(t) \, dt \quad (\text{ replace } \sin^2(t) \text{ with } \frac{1 - \cos(2t)}{2})$$

$$= \frac{1}{2} ab \int_{t=0}^{\pi/2} 1 - \cos(2t) \, dt = \frac{1}{2} ab \{ t - \frac{1}{2} \cdot \sin(2t) \} \Big|_{0}^{\pi/2} = \frac{1}{4} ab\pi .$$

The area of the whole ellipse is $4\{ \frac{1}{4} ab\pi \} = \boldsymbol{\pi ab}$.

If a = b, the ellipse is a circle with radius r = a = b, and its area is πr^2 as expected.

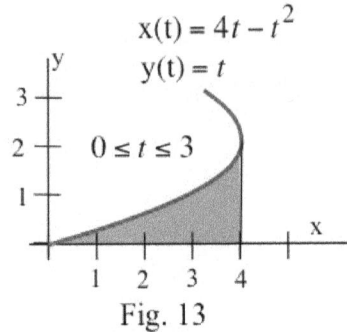

$x(t) = 4t - t^2$

$y(t) = t$

$0 \le t \le 3$

Fig. 13

Practice 5:   Let $x(t) = 4t - t^2$ and $y(t) = t$ (Fig. 13).

(a) Represent the shaded area in Fig. 13 as an integral and evaluate the integral.

(b) Evaluate $\displaystyle\int_{t=0}^{t=3} t \cdot (4 - 2t)\, dt$. Does this value represent an area?

Area under a Cycloid: (Fig. 7) For all $t \ge 0$, $x = R(t - \sin(t)) \ge 0$, $y = R(1 - \cos(t)) \ge 0$, and

$dx/dt = R(1 - \cos(t)) \ge 0$ so we can use the area formula. Then

$$\text{area} = \left| \int_{t=a}^{b} y \cdot \left(\frac{dx}{dt} \right) dt \right| = \left| \int_{t=0}^{2\pi} \{ R(1 - \cos(t)) \} \cdot (R(1 - \cos(t)))\, dt \right|$$

$$= R^2 \int_{t=0}^{2\pi} 1 - 2 \cdot \cos(t) + \cos^2(t)\, dt \quad (\text{replace } \cos^2(t) \text{ with } \frac{1 + \cos(2t)}{2} \text{ and integrate })$$

$$= R^2 \left\{ t - 2 \cdot \sin(t) + \frac{1}{2} t + \frac{1}{4} \sin(2t) \right\} \Big|_{0}^{2\pi} = R^2 \{ 2\pi + \pi \} = 3\pi R^2 .$$

The area under one arch of a cycloid is 3 times the area of the circle that generates the cycloid.

Properties of the Cycloid

Suppose you and a friend decide to have a contest to see who can build a slide that gets a person from point A to point B (Fig. 14) in the shortest time. What shape should you make your slide — a straight line, part of a circle, or something else? Assuming that the slide is frictionless and that the only acceleration is due to gravity, John Bernoulli showed that the **shortest time** ("brachistochrone" for "brachi" = short and "chrone" = time) path is a cycloid that starts at A that also goes through the point B. Fig. 15 shows the cycloid paths for A and B as well as the cycloid paths for two other "finish" points, C and D.

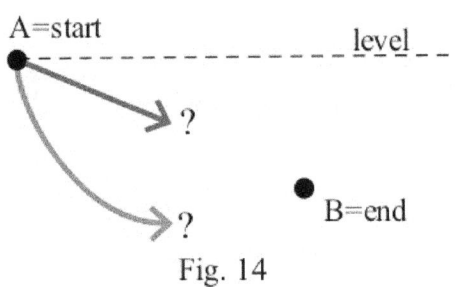

Fig. 14

Even before Bernoulli solved the brachistochrone problem, the astronomer (physicist, mathematician) Huygens was trying to design an accurate pendulum clock. On a standard pendulum clock (Fig. 16), the path of the bob is part of a circle, and the period of the swing depends on the displacement angle of the bob. As friction slows the bob, the displacement angle gets smaller and the clock slows

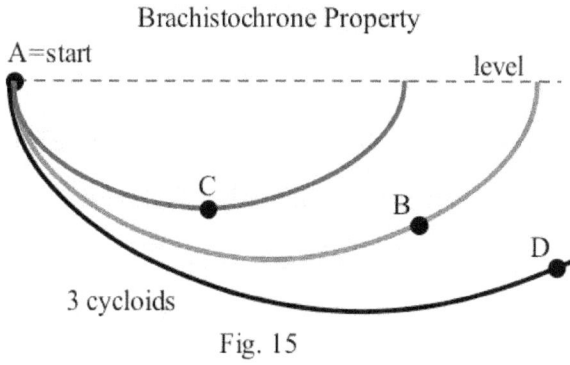

Brachistochrone Property

3 cycloids

Fig. 15

down. Huygens designed a clock (Fig. 17) whose bob swung in a curve so that the period of the swing did not depend on the displacement angle. The curve Huygens found to solve the **same time** ("tautochrone" for "tauto" = same and "chrone" = time) problem was the cycloid. Beads strung on a wire in the shape of a cycloid (Fig. 18) reach the bottom in the same amount of time, no matter where along the wire (except the bottom point) they are released.

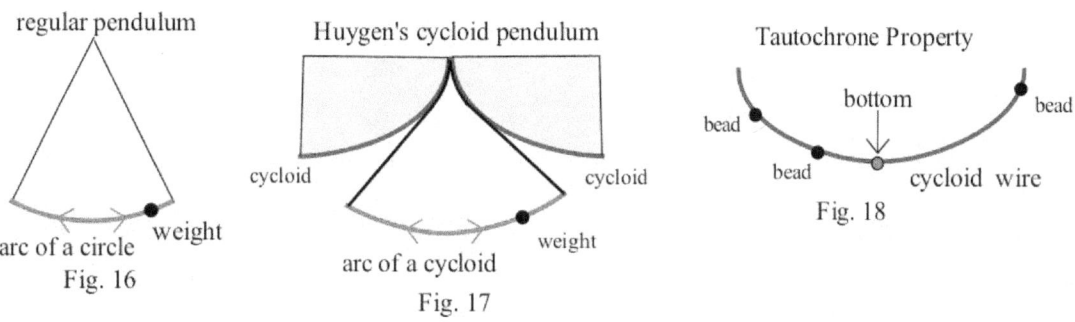

Fig. 16

Fig. 17

Fig. 18

The brachistochane and tautochrone problems are examples from a field of mathematics called the Calculus of Variations. Typical optimization problems in calculus involve finding a point or number that maximizes or minimizes some quantity. Typical optimization problems in the Calculus of Variations involve finding the curve or function that maximizes or minimizes some quantity. For example, what curve or shape with a given length encloses the greatest area? (Answer: a circle) Modern applications of Calculus of Variations include finding routes for airliners and ships to minimize travel time or fuel consumption depending on prevailing winds or currents.

PROBLEMS

Slope

For problems 1–8, (a) sketch the parametric graph (x,y) ,

(b) find the slope of the line tangent to the parametric graph at the given values of t, and

(c) find the points (x,y) at which dy/dx is either 0 or undefined.

1. $x(t) = t - t^2$, $y(t) = 2t + 1$ at t = 0, 1, and 2 .

2. $x(t) = t^3 + t$, $y(t) = t^2$ at t = 0, 1, and 2.

3. $x(t) = 1 + \cos(t)$, $y(t) = 2 + \sin(t)$ at t = 0, π/4, and π/2 .

4. $x(t) = 1 + 3 \cdot \cos(t)$, $y(t) = 2 + 2 \cdot \sin(t)$ at t = 0, π/4, π/2, and π.

5. $x(t) = \sin(t)$, $y(t) = \cos(t)$ at t = 0, π/4, π/2, and 17.3 .

6. $x(t) = 3 + \sin(t)$, $y(t) = 2 + \sin(t)$ at t = 0, π/4, π/2, and 17.3 .

7. $x(t) = \ln(t)$, $y(t) = 1 - t^2$ at t = 1, 2, and e .

8. $x(t) = \arctan(t)$, $y(t) = e^t$ at t = 0, 1, and 2 .

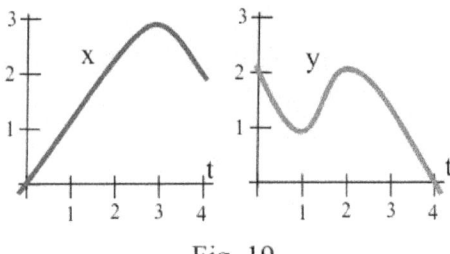

Fig. 19

In problems 9–12, the graphs of x(t) and y(t) are given.

Use this graphical information to estimate

(a) the slope of the line tangent to the parametric graph

at t = 0, 1, 2, and 3 , and

(b) the points (x,y) at which dy/dx is either 0 or undefined.

9. x(t) and y(t) in Fig. 19.

10. x(t) and y(t) in Fig. 20.

11. x(t) and y(t) in Fig. 21.

12. x(t) and y(t) in Fig. 22.

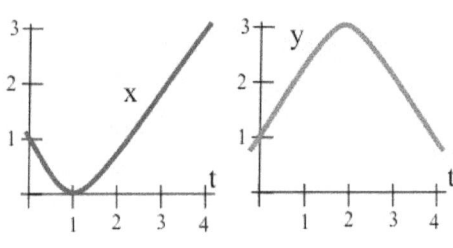

Fig. 20

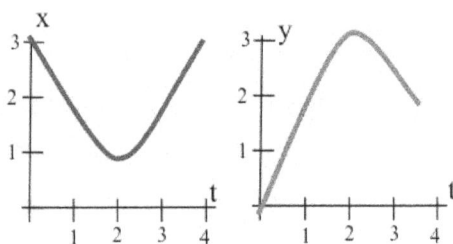

Fig. 21

Speed

For problems 13– 20, the locations x(t) and y(t) (in feet)

of an object are given at time t seconds. Find the speed of

the object at the given times.

13. $x(t) = t - t^2$, $y(t) = 2t + 1$ at t = 0, 1, and 2 .

14. $x(t) = t^3 + t$, $y(t) = t^2$ at t = 0, 1, and 2.

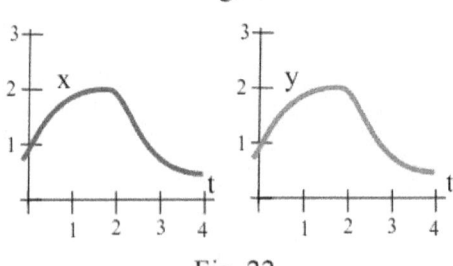

Fig. 22

15. $x(t) = 1 + \cos(t)$, $y(t) = 2 + \sin(t)$ at $t = 0$, $\pi/4$, $\pi/2$, and π .

16. $x(t) = 1 + 3 \cdot \cos(t)$, $y(t) = 2 + 2 \cdot \sin(t)$ at $t = 0$, $\pi/4$, $\pi/2$, and π.

17. x and y in Fig. 19 at $t = 0, 1, 2, 3$, and 4.

18. x and y in Fig. 20 at $t = 0, 1, 2$, and 3.

19. x and y in Fig. 21 at $t = 0, 1, 2$, and 3.

20. x and y in Fig. 22 at $t = 0, 1, 2$, and 3.

21. At time t seconds an object is located at the point $x(t) = R \cdot (t - \sin(t))$, $y(t) = R \cdot (1 - \cos(t))$ (in feet).

   (a) Find the speed of the object at time t.   (b) At what time is the object traveling fastest?

   (c) Where is the object on the cycloid when it is traveling fastest?

22. At time t seconds an object is located at the point $x(t) = 5 \cdot \cos(t)$, $y(t) = 2 \cdot \sin(t)$ (in feet).

   (a) Find the speed of the object at time t.   (b) At what time is the object traveling fastest?

   (c) Where is the object on the ellipse when it is traveling fastest?

Arc Length

For problems 23–28, (a) represent the arc length of each parametric function as a definite integral, and

(b) evaluate the integral (if necessary, use your calculator's **fnInt()** feature to evaluate the integral).

23. $x(t) = t - t^2$, $y(t) = 2t + 1$ for $t = 0$ to 2 .

24. $x(t) = t^3 + t$, $y(t) = t^2$ for $t = 0$ to 2 .

25. $x(t) = 1 + \cos(t)$, $y(t) = 2 + \sin(t)$ for $t = 0$ to π .

26. $x(t) = 1 + 3 \cdot \cos(t)$, $y(t) = 2 + 2 \cdot \sin(t)$ for $t = 0$ to π .

27. x and y in Fig. 23 for $t = 1$ to 3 .

28. x and y in Fig. 22 for $t = 0$ to 2 .

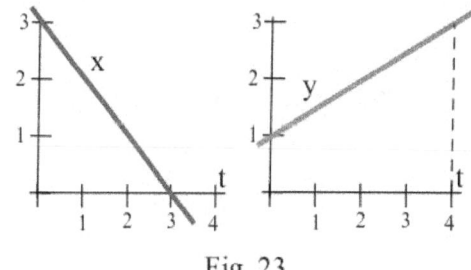
Fig. 23

Area

For problems 29–34, (a) represent the area of each region as a definite integral, and

(b) evaluate the integral (if necessary, use your calculator's **fnInt()** feature to evaluate the integral).

29. $x(t) = t^2$, $y(t) = 4t^2 - t^4$ for $0 \le t \le 2$.

30. $x(t) = 1 + \sin(t)$, $y(t) = 2 + \sin(t)$ for $0 \le t \le \pi$.

31. $x(t) = t^2$, $y(t) = 1 + \cos(t)$ for $0 \le t \le 2$.

32. $x(t) = \cos(t)$, $y(t) = 2 - \sin(t)$ for $0 \le t \le \pi/2$.

33. "Cycloid" with a square wheel: Find the area under one "arch" of the path of a point on the corner of a "rolling" square that has sides of length R. (This problem does not require calculus.)

34. The region bounded between the x–axis and the curate cycloid $x(t) = R \cdot t - r \cdot \sin(t)$, $y(t) = R - r \cdot \cos(t)$ for $0 \le t \le 2\pi$.

Section 9.4   **PRACTICE Answers**

Practice 1: $\frac{dy}{dt} = 2t + 1$, so when $t = 3$, $\frac{dy}{dt} = 7$. $\frac{dx}{dt} = 3t^2$, so when $t = 3$, $\frac{dx}{dt} = 27$.

Finally, $\frac{dy}{dx} = \frac{dy/dt}{dx/dt}$ so when $t = 3$, $\frac{dy}{dx} = \frac{7}{27}$.

When $t = 3$, $x = 28$ and $y = 12$ so the equation of the tangent line is $y - 12 = \frac{7}{27}(x - 28)$.

Practice 2:   (a) When $t = 2$, $dy/dx \approx 0$. When $t = 5$, $dy/dx \approx -1$.

(b) In Fig. 2, $\frac{dy}{dt} = 0$ when $t \approx 2$ and $t \approx 4$.

(c) In Fig. 3, a minimum occurs when $t \approx 2$ and a maximum when $t \approx 4$.

(d) If the parametric graph has a maximum or minimum at $t = t^*$, then dy/dt is either 0 or undefined when $t = t^*$.

Practice 3:   When $t = 1$, speed $= \sqrt{(dx/dt)^2 + (dy/dt)^2} \approx \sqrt{(1)^2 + (-1)^2} = \sqrt{2} \approx 1.4$ ft/sec.

When $t = 2$, speed $= \sqrt{(dx/dt)^2 + (dy/dt)^2} \approx \sqrt{(-1)^2 + (0)^2} = \sqrt{1} = 1$ ft/sec.

When $t = 3$, speed $= \sqrt{(dx/dt)^2 + (dy/dt)^2} \approx \sqrt{(1)^2 + (1)^2} = \sqrt{2} \approx 1.4$ ft/sec.

Practice 4:   Length $= \int_{t=0}^{2\pi} \sqrt{(-3 \sin(t))^2 + (2 \cos(t))^2} \; dt$

≈ 15.87 (using my calculator's **fnInt()** feature)

Practice 5:   (a) $A = \int_{t=0}^{2} t \cdot (4 - 2t) \; dt = 2 \cdot t^2 - \frac{2}{3} t^3 \Big|_{0}^{2} = \{ 8 - \frac{16}{3} \} - \{ 0 - 0 \} = \frac{8}{3}$.

(b) $\int_{t=0}^{3} t \cdot (4 - 2t) \; dt = 2 \cdot t^2 - \frac{2}{3} t^3 \Big|_{0}^{3} = \{ 18 - 18 \} - \{ 0 - 0 \} = 0$.

This integral represents {shaded area in Fig. 13 } – { area from $t = 2$ to $t = 3$ }.

9.4 $\frac{1}{2}$ BEZIER CURVES — Getting the shape you want

Historically, parametric equations were often used to model the motion of objects, and that is the approach we have seen so far. But more recently, as computers became more common in design work and manufacturing, a need arose to efficiently find formulas for shapes such as airplane wings and automobile bodies and even letters of the alphabet that designers or artists had created.

One simple but inefficient method for describing and storing the shape of a curve is to measure the location and save the coordinates of hundreds or thousands of points along the curve. This result is called a "bitmap" of the shape. However, bitmaps typically require a large amount of computer memory, and when the bitmap is reconverted from stored coordinates back into a graphic image, originally smooth curves often appear jagged (Fig. 1). Also, when these bitmapped shapes are stretched or rotated, the new location of every one of the points must be calculated, a relatively slow process.

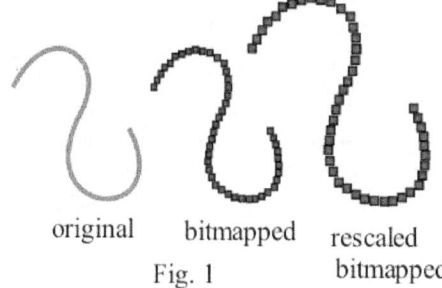

original   bitmapped   rescaled
                                    bitmapped
Fig. 1

A second method, still simple but more efficient than bitmaps, is to store fewer points along the curve, but to automatically connect consecutive points with line segments (Fig. 2). Less computer memory is required since fewer coordinates are stored, and stretches and rotations are calculated more quickly since the new locations of fewer points are needed. This method is commonly used in computer graphics to store and redraw surfaces (Fig. 3). Sometimes instead of saving the coordinates of each point, a "vector" is used to describe how to get to the next point from the previous point, and the result is a "vector map" of the curve. The major drawback of this method is that the stored and redrawn curve consists of straight segments and corners even though the original curve may have been smooth.

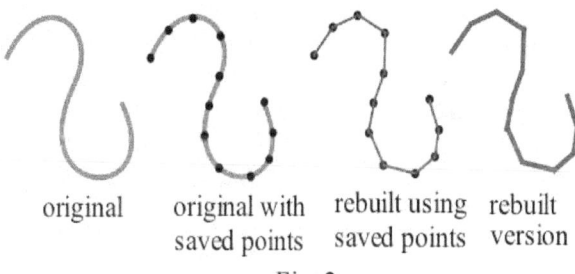

original   original with   rebuilt using   rebuilt
             saved points   saved points   version
Fig. 2

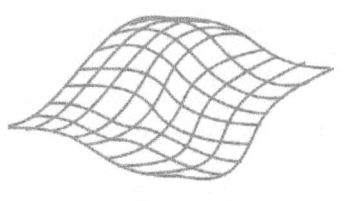

original surface

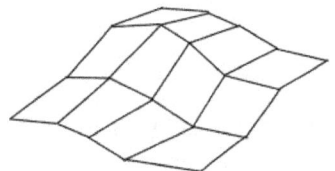

surface rebuilt using
points and line segments

Fig. 3

The primary building block for curves and surfaces represented as line segments is the line segment given by parametric equations.

Example 1:   Show that the parametric equations

$$x(t) = (1 - t){\cdot}x_0 + t{\cdot}x_1 \quad \text{and}$$

$$y(t) = (1 - t){\cdot}y_0 + t{\cdot}y_1$$

for $0 \leq t \leq 1$ go through the points $P_0 = (x_0, y_0)$ and $P_1 = (x_1, y_1)$ (Fig. 4) and that the slope is

$$\frac{d\,y}{d\,x} = \frac{y_1 - y_0}{x_1 - x_0}$$

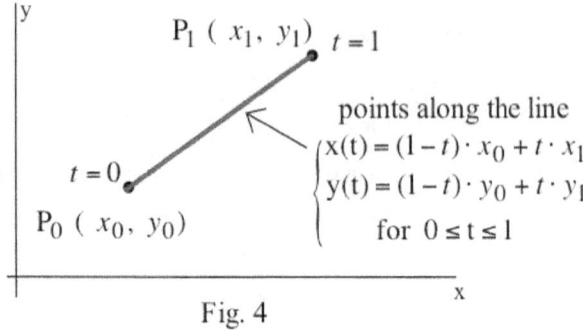

Fig. 4

for all values of t between 0 and 1.

We typically abbreviate the pair of parametric equations as $P(t) = (1 - t){\cdot}P_0 + t{\cdot}P_1$.

Solution: $x(\mathbf{0}) = (1 - \mathbf{0}){\cdot}x_0 + \mathbf{0}{\cdot}x_1 = x_0$ and $y(\mathbf{0}) = (1 - \mathbf{0}){\cdot}y_0 + \mathbf{0}{\cdot}y_1 = y_0$ so $P(\mathbf{0}) = P_0$.

$x(\mathbf{1}) = (1 - \mathbf{1}){\cdot}x_0 + \mathbf{1}{\cdot}x_1 = x_1$ and $y(\mathbf{1}) = (1 - \mathbf{1}){\cdot}y_0 + \mathbf{1}{\cdot}y_1 = y_1$ so $P(\mathbf{1}) = P_1$.

$$\frac{d\,y(t)}{d\,t} = -y_0 + y_1 \quad \text{and} \quad \frac{d\,x(t)}{d\,t} = -x_0 + x_1$$

so $\dfrac{d\,y}{d\,x} = \dfrac{dy/dt}{dx/dt} = \dfrac{y_1 - y_0}{x_1 - x_0}$, the slope of the line segment from P_0 to P_1 .

Practice 1:   Use the pattern of Example 1 to write parametric equations for the line segments that

(a) connect $(1,2)$ to $(5,4)$, (b) connect $(5,4)$ to $(1,2)$, and (c) connect $(6,-2)$ to $(3,1)$.

Bezier Curves

One solution to the problem of efficiently saving and redrawing a smooth curve was independently developed in the 1960s by two French automobile engineers, Pierre Bezier (pronounced "bez–ee–ay") who worked for Renault automobile company and P. de Casteljau who worked for Citroen. Originally, the solutions were considered industrial secrets, but Bezier's work was eventually published first. The curves that result using Bezier's method are called Bezier curves. The method of Bezier curves allow us to efficiently store information about smooth (and not–so–smooth) shapes and to quickly stretch, rotate and distort these shapes. Bezier curves are now commonly used in computer–aided design work and in most computer drawing programs. They are also used to specify the shapes of letters of the alphabet in different fonts. By using this method, a computer and a laser printer can have many different fonts in many different sizes available without using a large amount of memory. (Bezier curves were used to produce most of the graphs in this book.)

Here we define Bezier curves and examine some of their properties. At the end of this section some optional material describes the mathematical construction of Bezier curves.

Definition: Bezier Curve

The Bezier curve $B(t)$ defined for the four points $P_0, P_1, P_2,$
and P_3 (Fig. 5) is

$$B(t) = (1-t)^3 \cdot P_0 + 3(1-t)^2 \cdot t \cdot P_1 + 3(1-t) \cdot t^2 \cdot P_2 + t^3 \cdot P_3$$

$$\text{for } 0 \le t \le 1:$$

$$x(t) = (1-t)^3 \cdot x_0 + 3(1-t)^2 \cdot t \cdot x_1 + 3(1-t) \cdot t^2 \cdot x_2 + t^3 \cdot x_3 \quad \text{and}$$

$$y(t) = (1-t)^3 \cdot y_0 + 3(1-t)^2 \cdot t \cdot y_1 + 3(1-t) \cdot t^2 \cdot y_2 + t^3 \cdot y_3 \ .$$

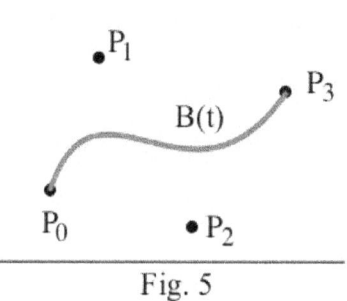

Fig. 5

The four points $P_0, P_1, P_2,$ and P_3 are called **control points** for the Bezier curve. Fig. 6 shows Bezier curves for several sets of control points. The dotted lines connecting the control points in Fig. 6 are shown to help illustrate the relationship between the graph of $B(t)$ and the control points.

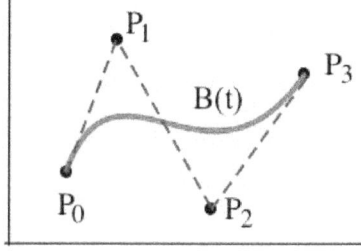

Example 2: Plot the points $P_0 = (0,3), P_1 = (1,5), P_2 = (3,-1),$ and $P_3 = (4,0),$ and determine the equation of the Bezier curve for these control points. Then graph the Bezier curve.

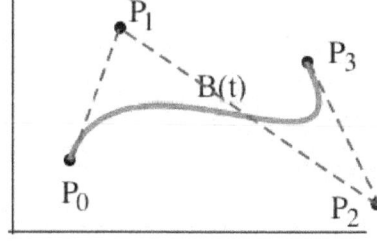

Solution:

$$x(t) = (1-t)^3 \cdot 0 + 3(1-t)^2 \cdot t \cdot 1 + 3(1-t) \cdot t^2 \cdot 3 + t^3 \cdot 4 = -2t^3 + 3t^2 + 3t \ .$$

$$y(t) = (1-t)^3 \cdot 3 + 3(1-t)^2 \cdot t \cdot 5 + 3(1-t) \cdot t^2 \cdot (-1) + t^3 \cdot 0$$

$$= 15t^3 - 24t^2 + 6t + 3 \ .$$

The control points and the graph of $B(t) = (\, x(t), y(t)\,)$ are shown in Fig. 7.

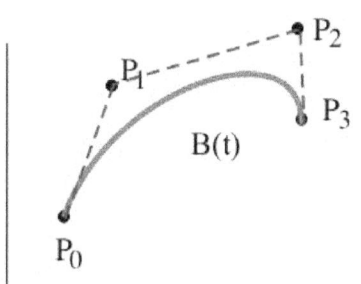

Fig. 6: Several Bezier curves
as P_2 is moved

Practice 2: Plot the points $P_0 = (0, 4),$
$P_1 = (1, 2), P_2 = (4, 2),$ and
$P_3 = (4, 4),$ and determine the equation of
the Bezier curve for these control points.
Then graph the Bezier curve.

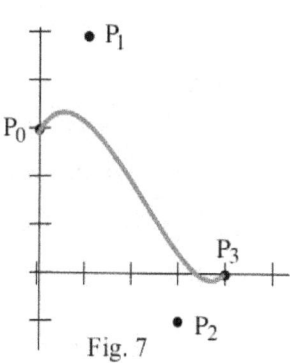

Fig. 7

Properties of Bezier Curves

Bezier curves have a number of properties that make them particularly useful for design work, and some of them are stated below. These properties are verified at the end of this section.

(1) $B(0) = P_0$ and $B(1) = P_3$ so the Bezier curve goes through the points P_0 and P_3.

This property guarantees that $B(t)$ goes through specified points. If we want two Bezier curves to fit together, it is important that the value at the end of one curve matches the starting value of the next curve. This property guarantees that we can control the values of the Bezier curves at their endpoints by choosing appropriate values for the control points P_0 and P_3.

(2) $B(t)$ is a cubic polynomial.

This is an important property because it guarantees that $B(t)$ is continuous and differentiable at each point so its graph is connected and smooth at each point. It also guarantees that the graph of $B(t)$ does not "wiggle" too much between control points.

(3) $B'(0) =$ slope of the line segment from P_0 to P_1; $B'(1) =$ slope of the line segment from P_2 to P_3.

This is an important property because it means we can match the ending slope of one curve with the starting slope of the next curve to result in a smooth connection. We can see in Fig. 6 that the dotted line from P_0 to P_1 is tangent to the graph of $B(t)$ at the point P_0.

(4) For $0 \le t \le 1$, the graph of $B(t)$ is in the region whose corners are the control points.

Visually, property (4) means that if we put a rubber band around the four control points $P_0, P_1, P_2,$ and P_3 (Fig. 8), then the graph of $B(t)$ will be inside the rubber banded region. This is an important property of Bezier curves because it guarantees that the graph of $B(t)$ does not get too far from the four control points.

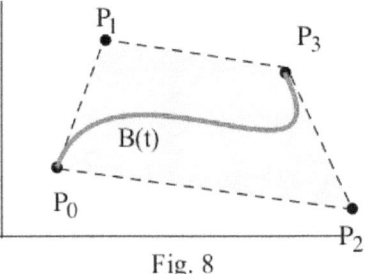

Fig. 8

Example 3:   Find a formula for a Bezier curve that goes through the points $(1, 1)$ and $(0, 0)$ and is shaped like an "S."

Solution: Since we want the curve to begin at the point $(1, 1)$ and end at $(0, 0)$ we can put $P_0 = (1, 1)$ and $P_3 = (0, 0)$. A little experimentation with values for P_1 and P_2 indicates that $P_1 = (-1, 2)$ and $P_2 = (2, -1)$ gives a mediocre "S" shape (Fig. 9). Then the formula for $B(t)$ is

$$B(t) = (1 - t)^3 \cdot P_0 + 3(1 - t)^2 \cdot t \cdot P_1 + 3(1 - t) \cdot t^2 \cdot P_2 + t^3 \cdot P_3$$

for $0 \le t \le 1$, and

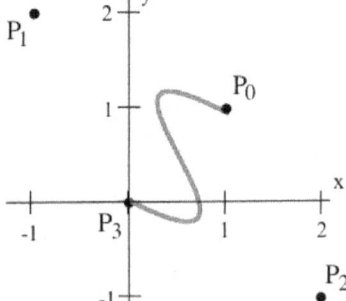

Fig. 9

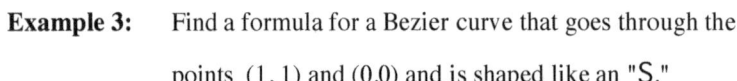
$$x(t) = (1 - t)^3 \cdot x_0 + 3(1 - t)^2 \cdot t \cdot x_1 + 3(1 - t) \cdot t^2 \cdot x_2 + t^3 \cdot x_3 = (1-t)^3 \cdot 1 + 3(1-t)^2 \cdot t \cdot (-1) + 3(1-t) \cdot t^2 \cdot 2 + t^3 \cdot 0$$

and

$$y(t) = (1-t)^3 \cdot y_0 + 3(1-t)^2 \cdot t \cdot y_1 + 3(1-t) \cdot t^2 \cdot y_2 + t^3 \cdot y_3 = (1-t)^3 \cdot 1 + 3(1-t)^2 \cdot t \cdot 2 + 3(1-t) \cdot t^2 \cdot (-1) + t^3 \cdot 0$$

Certainly other values of P_1 and P_2 can give similar shapes.

Practice 3:   Find a formula for a Bezier curve that goes through the points $(0, 0)$ and $(0,1)$ and is shaped like a "C."

Example 4:   Find a formula for a Bezier curve that goes through the point $(0, 5)$ with a slope of 2 and through the point $(6, 1)$ with a slope of 3.

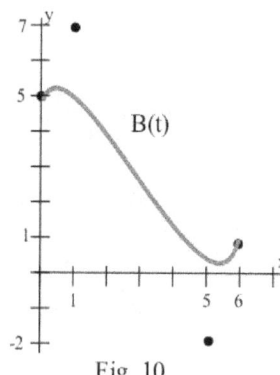

Fig. 10

Solution: Since we want to curve to begin at $(0,5)$ and end at $(6,1)$, we put $P_0 = (0,5)$ and $P_3 = (6,1)$. To get the slopes we need to pick P_1 so the slope of the line segment from P_0 to P_1 is 2: going "over 1 and up 2" to get $P_1 = (1,7)$ works fine as do several other points. Similarly, to get the right slope at P_3 we can go "back 1 and down 3" to get $P_2 = (5,-2)$. The Bezier curve for $P_0 = (0,5), P_1 = (1,7), P_2 = (5,-2)$, and $P_3 = (6,1)$ is

$$x(t) = (1-t)^3 \cdot 0 + 3(1-t)^2 \cdot t \cdot 1 + 3(1-t) \cdot t^2 \cdot 5 + t^3 \cdot 6 = -6t^3 + 9t^2 + 3t \text{ and}$$
$$y(t) = (1-t)^3 \cdot 5 + 3(1-t)^2 \cdot t \cdot 7 + 3(1-t) \cdot t^2 \cdot (-2) + t^3 \cdot 1 = 23t^3 - 33t^2 + 6t + 5.$$

The graph of this B(t) is shown in Fig. 10.

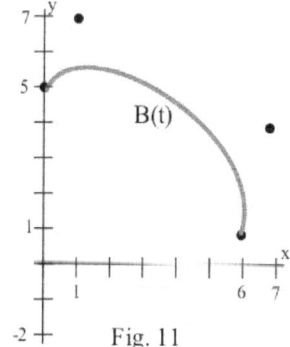

Fig. 11

Keeping the previous values of P_0, P_1 and P_3, we could pick P_2 by going "over 1 and up 3" to $P_2 = (7,4)$, and the graph of the B(t) for this choice of P_2 is shown in Fig. 11. The graph for B(t) when $P_1 = (2,9)$ and $P_2 = (5.5, -0.5)$ is shown in Fig. 12. These, and other choices of P_1 and P_2 satisfy the conditions specified in the problem: the choice of which one you use depends on the other properties of the shape that you want the curve to have.

Practice 4:   Find a formula for a Bezier curve that goes through the point $(1, 3)$ with a slope of -2 and through the point $(5, 3)$ with a slope of -1.

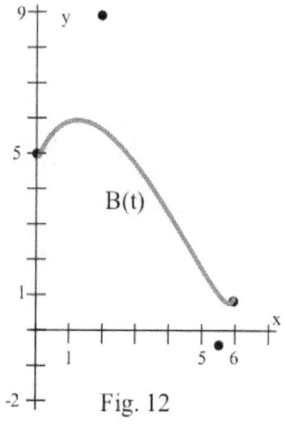

Fig. 12

Example 5:   Find formulas for a pair of Bezier curves so that the first starts at the point $A = (0,3)$ with slope 2, the second ends at the point $C = (7,2)$ with slope 1, and the curves connect at the point $B = (4,6)$ with slope 0.

Solution: For the first Bezier curve B(t) take $P_0 = A = (0,3), P_1 = (1,5)$ (to get the slope 2), $P_3 = B = (4,6)$, and $P_2 = (3,6)$ (to get the slope 0 at the connecting point). Then, for $0 \le t \le 1$, $x(t) = (1-t)^3 \cdot 0 + 3(1-t)^2 \cdot t \cdot 1 + 3(1-t) \cdot t^2 \cdot 3 + t^3 \cdot 4$ and

$y(t) = (1-t)^3 \cdot 3 + 3(1-t)^2 \cdot t \cdot 5 + 3(1-t) \cdot t^2 \cdot 6 + t^3 \cdot 6$.

For the second Bezier curve $C(t)$ take $P_0 = B = (4,6)$, $P_1 = (5,6)$ (to get the slope 0 at

the connecting point), $P_3 = C = (7,2)$, and $P_2 = (6,1)$ (to get the slope 1). Then,

for $0 \le t \le 1$, $x(t) = (1-t)^3 \cdot 4 + 3(1-t)^2 \cdot t \cdot 5 + 3(1-t) \cdot t^2 \cdot 6 + t^3 \cdot 7$ and

$y(t) = (1-t)^3 \cdot 6 + 3(1-t)^2 \cdot t \cdot 6 + 3(1-t) \cdot t^2 \cdot 1 + t^3 \cdot 2$.

Fig. 13 shows the graphs of B(t) and
C(t) and illustrates how they connect,
continuously and smoothly, at the
common point $(4,6)$.

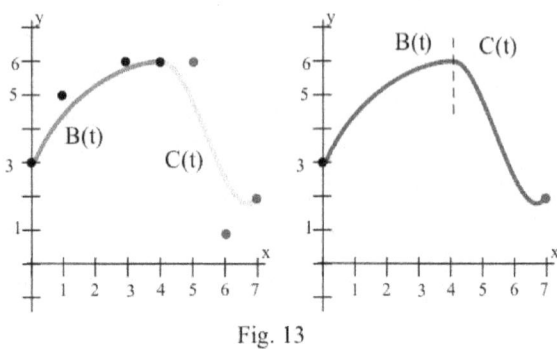

Fig. 13

Using Bezier Curves

In practice, Bezier curves are usually used in
computer design or manufacturing programs, and
the user of Bezier curves does not have to know the mathematics behind them. But the program creator does!

Typically a designer sketches a crude shape for an object and then moves certain points to locations specified by
the plans. Sometimes the designer adds additional points along the curve to "fix" the location of the curve.
These "fixed" points along the curve become the endpoints P_0 and P_3 for each of the sections of the curve
that will be described by a Bezier formula. Then, for each section of the curve, the designer visually
experiments with different locations of the interior control points P_1 and P_2 to get the shape "just right."

Meanwhile, the computer program adjusts the formulas for the Bezier curves based on the current locations of
the control points for each section, and, when the design is complete, saves the locations of the control points.

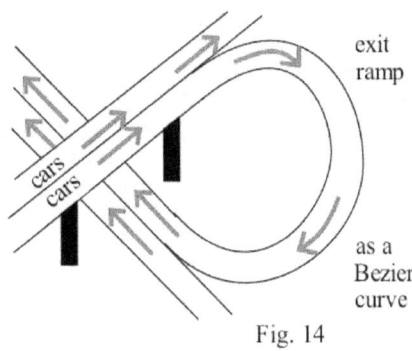

exit
ramp

as a
Bezier
curve

Fig. 14

You may never need to calculate the formulas for Bezier curves
(outside of a mathematics class), but if you do any computer–aided
design work you will certainly be using these curves. And the ideas
and formulas for Bezier curves in two dimensions extend very easily
and naturally to describe paths in three dimensions such as the route of
a highway exit ramp (Fig. 14) or the path of a hydraulic hose for the
landing gear of an airplane.

Mathematical Construction of Bezier Curves & Verifications of Their Properties

In order to use and program Bezier curves we don't need to know where the formulas came from, but their construction is a beautiful piecing–together of simple geometric ideas.

The first idea is the parametric representation of a line segment from point
P_A to P_B as $L(t) = (1 - t) \cdot P_A + t \cdot P_B$ for $0 \le t \le 1$.

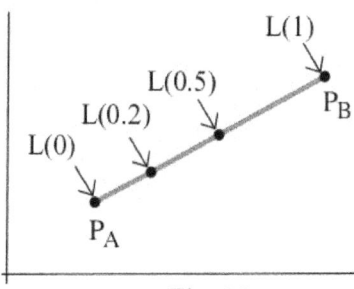
Fig. 15

This parametric pattern for a line is used in the construction of a Bezier curve.

When $t = 0$, the point is $L(0) = P_A$. When $t = 1$, $L(1) = P_B$. When $t = 0.5$, $L(0.5)$ is the midpoint of the line from P_A to P_B (Fig. 15). When $t = 0.2$, $L(0.2)$ is 20% of the way along the line L from P_A to P_B.

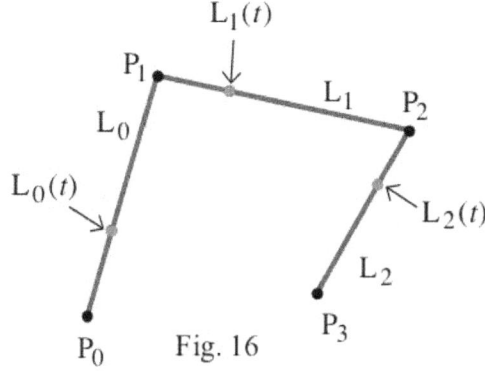
Fig. 16

To construct the Bezier curve for the four control points P_0, P_1, P_2, and P_3 we start by fixing a value of t between 0 and 1. Then we find the point $L_0(t)$ along the parametric line from P_0 to P_1, the point $L_1(t)$ along the parametric line from P_1 to P_2, and the point $L_2(t)$ along the parametric line from P_2 to P_3 (Fig. 16):

$$L_0(t) = (1-t) \cdot P_0 + t \cdot P_1,$$
$$L_1(t) = (1-t) \cdot P_1 + t \cdot P_2,$$
$$L_2(t) = (1-t) \cdot P_2 + t \cdot P_3 .$$

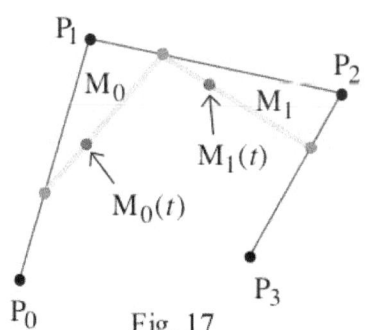
Fig. 17

For the fixed value of t, we then find the point $M_0(t)$ along the parametric line from the point $L_0(t)$ to the point $L_1(t)$, and the point $M_1(t)$ along the parametric line from the point $L_1(t)$ to the point $L_2(t)$ (Fig. 17):

$$M_0(t) = (1 - t) \cdot L_0(t) + t \cdot L_1(t)$$
$$M_1(t) = (1 - t) \cdot L_1(t) + t \cdot L_2(t) .$$

For the same fixed value of t, we finally find the point $B(t)$ along the parametric line from the point $M_0(t)$ to the point $M_1(t)$ (Fig. 18):

$$B(t) = (1 - t) \cdot M_0(t) + t \cdot M_1(t) .$$

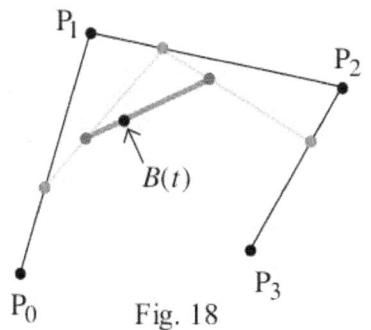
Fig. 18

As the variable t takes on different values between 0 and 1, the points $L_0(t)$, $L_1(t)$, and $L_2(t)$ move along the lines connecting P_0, P_1, P_2, and P_3 . Similarly, the points $M_0(t)$ and $M_1(t)$ move along the lines connecting $L_0(t)$, $L_1(t)$, and $L_2(t)$, and the point $B(t)$ moves along the line connecting $M_0(t)$ and $M_1(t)$. It is all quite dynamic.

Fig. 19 shows these points and lines for several values of t between 0 and 1.

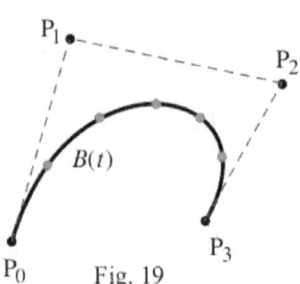

We can use the previous geometric construction to obtain the formula for B(t) given in the definition of Bezier curves by working backwards from $B(t) = (1 - t) \cdot M_0(t) + t \cdot M_1(t)$:

Fig. 19

$$
\begin{aligned}
B(t) &= (1 - t) \cdot M_0 + t \cdot M_1 \\
&= (1 - t) \cdot \{ (1 - t) \cdot L_0 + t \cdot L_1 \} + t \cdot \{ (1 - t) \cdot L_1 + t \cdot L_2 \} \qquad \text{replacing } M_0 \text{ and } M_1 \text{ in terms} \\
&\qquad\qquad\qquad\qquad\qquad\qquad\qquad\qquad\qquad\qquad\qquad \text{of } L_0, L_1 \text{ and } L_2 \\
&= (1 - t)^2 \cdot L_0 + 2(1 - t) \cdot t \cdot L_1 + t^2 \cdot L_2 \qquad\qquad \text{simplifying} \\
&= (1 - t)^2 \cdot \{ (1 - t) \cdot P_0 + t \cdot P_1 \} + 2(1 - t) \cdot t \cdot \{ (1 - t) \cdot P_1 + t \cdot P_2 \} + t^2 \cdot \{ (1 - t) \cdot P_2 + t \cdot P_3 \} \\
&\qquad \text{replacing } L_0, L_1 \text{ and } L_2 \text{ in terms of } P_0, P_1, P_2 \text{ and } P_3 \\
&= (1 - t)^3 \cdot P_0 + 3(1 - t)^2 \cdot t \cdot P_1 + 3(1 - t) \cdot t^2 \cdot P_2 + t^3 \cdot P_3 \qquad \text{simplifying}
\end{aligned}
$$

Verifications of Properties (1) – (4)

Property (1) is easy to verify by evaluating B(0) and B(1):

$$B(0) = (1 - 0)^3 \cdot P_0 + 3(1 - 0)^2 \cdot 0 \cdot P_1 + 3(1 - 0) \cdot t^2 \cdot P_2 + 0^3 \cdot P_3 = P_0 . \text{ Similarly,}$$

$$B(1) = (1 - 1)^3 \cdot P_0 + 3(1 - 1)^2 \cdot 1 \cdot P_1 + 3(1 - 1) \cdot 1^2 \cdot P_2 + 1^3 \cdot P_3 = P_3 .$$

Property (2) is clear from the defining formula for B(t), or we can expand the powers of $1 - t$ and t and collect the similar terms to rewrite B(t) as

$$B(t) = (-P_0 + 3P_1 - 3P_2 + P_3) \cdot t^3 + (3P_0 - 6P_1 + 3P_2) \cdot t^2 + (-3P_0 + 3P_1) \cdot t + (P_0).$$

Property (3) can be verified using the rewritten form from Property 2,

$$x(t) = (-x_0 + 3x_1 - 3x_2 + x_3) \cdot t^3 + (3x_0 - 6x_1 + 3x_2) \cdot t^2 + (-3x_0 + 3x_1) \cdot t + (x_0) \text{ and}$$

$$y(t) = (-y_0 + 3y_1 - 3y_2 + y_3) \cdot t^3 + (3y_0 - 6y_1 + 3y_2) \cdot t^2 + (-3y_0 + 3y_1) \cdot t + (y_0).$$

Then $\dfrac{d\,x(t)}{d\,t} = 3(-x_0 + 3x_1 - 3x_2 + x_3) \cdot t^2 + 2(3x_0 - 6x_1 + 3x_2) \cdot t + (-3x_0 + 3x_1)$ and

$\dfrac{d\,y(t)}{d\,t} = 3(-y_0 + 3y_1 - 3y_2 + y_3) \cdot t^2 + 2(3y_0 - 6y_1 + 3y_2) \cdot t + (-3y_0 + 3y_1)$.

When $t = 0$, $\dfrac{d\,x(t)}{d\,t} = -3x_0 + 3x_1 = -3(x_1 - x_0)$ and $\dfrac{d\,x(t)}{d\,t} = -3(y_1 - y_0)$ so

$$\frac{d\,B(t)}{d\,t} = \frac{d\,y(t)/dt}{d\,x(t)/dt} = \frac{-3(y_1 - y_0)}{-3(x_1 - x_0)} = \frac{y_1 - y_0}{x_1 - x_0} = \text{slope of the line from } P_0 \text{ to } P_1.$$

The verification that B '(1) equals the slope of the line segment from P_2 to P_3 is similar:

evaluate $x'(1), y'(1)$ and $B'(1) = \dfrac{y'(1)}{x'(1)}$.

We will not verify Property (4) here, but it follows from the fact that each point on the Bezier curve is a "weighted average" of the four control points. For $0 \le t \le 1$, each of the coefficients $(1-t)^3$, $3(1-t)^2 \cdot t$, $3(1-t) \cdot t^2$ and t^3 is between (or equal to) 0 and 1, and they always add up to 1 (just expand the powers and add them to check this statement).

Problems

In problems 1 – 6, pairs of points, P_A and P_B, are given. In each problem (a) sketch the line segment L from P_A and P_B, and (b) plot the locations of the points $L(0.2)$, $L(0.5)$, and $L(0.9)$ on the line segment in part (a). Finally, (c) determine the equation of the line segment $L(t)$ from P_A and P_B and graph it.

1. P_A and P_B are given in Fig. 20.   2. P_A and P_B are given in Fig. 21.

3. P_A and P_B are given in Fig. 22.   4. P_A and P_B are given in Fig. 23.

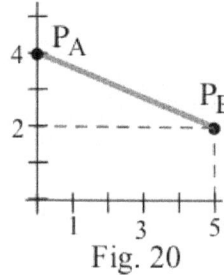
Fig. 20

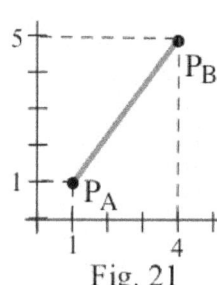
Fig. 21

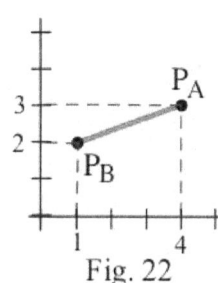
Fig. 22

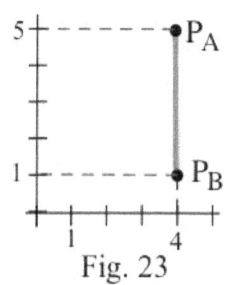
Fig. 23

5. $P_A = (1, 4)$ and $P_B = (5, 1)$.   6. $P_A = (8, 5)$ and $P_B = (4, 3)$.

7. Show that the parametric equations for the line segment given in Example 1,
$$x(t) = (1 - t) \cdot x_0 + t \cdot x_1 \text{ and } y(t) = (1 - t) \cdot y_0 + t \cdot y_1 \text{ for } 0 \le t \le 1,$$

is equivalent to the parametric equations
$$x(t) = x_0 + t \cdot \Delta x \text{ and } y(t) = y_0 + t \cdot \Delta y \text{ for } 0 \le t \le 1 \text{ where } \Delta x = x_1 - x_0 \text{ and } \Delta y = y_1 - y_0.$$

In problems 8 – 13, find the parametric equations for a Bezier curve with the given control points or the given properties.

8. $P_0 = (1, 0)$, $P_1 = (2, 3)$, $P_2 = (5, 2)$, $P_3 = (6, 3)$ 9. $P_0 = (0, 5)$, $P_1 = (2, 3)$, $P_2 = (1, 4)$, $P_3 = (4, 2)$

10. $P_0 = (5, 1)$, $P_1 = (3, 3)$, $P_2 = (3, 5)$, $P_3 = (2, 1)$ 11. $P_0 = (6, 5)$, $P_1 = (6, 3)$, $P_2 = (2, 5)$, $P_3 = (2, 0)$

12. $P_0 = (0, 1)$, $P_3 = (4, 1)$ and $B'(0) = 1$, $B'(1) = -3$ 13. $P_0 = (5, 1)$, $P_3 = (1, 3)$ and $B'(0) = 2$, $B'(1) = 3$

In problems 14 – 17, sets of control points P_0, P_1, P_2, and P_3 are shown. Sketch a reasonable Bezier curve for the given control points.

14. P_0, P_1, P_2, and P_3 are given in Fig. 24.   15. P_0, P_1, P_2, and P_3 are given in Fig. 25.

16. P_0, P_1, P_2, and P_3 are given in Fig. 26.   17. P_0, P_1, P_2, and P_3 are given in Fig. 27.

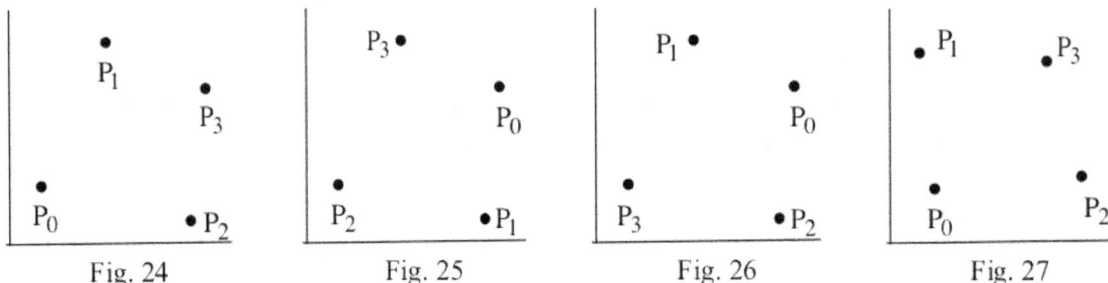

Fig. 24   Fig. 25   Fig. 26   Fig. 27

In problems 18 – 21, sets of control points P_0, P_1, P_2, and P_3 are shown as well as a curve C(t). For each problem explain why we can be certain that C(t) is NOT the Bezier curve for the given control points (state which property or properties of Bezier curves C(t) does not have).

18. P_0, P_1, P_2, and P_3 are given in Fig. 28.   19. P_0, P_1, P_2, and P_3 are given in Fig. 29.

20. P_0, P_1, P_2, and P_3 are given in Fig. 30.   21. P_0, P_1, P_2, and P_3 are given in Fig. 31.

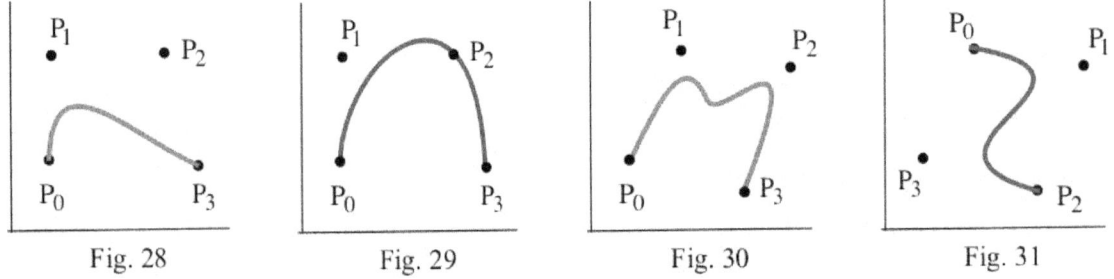

Fig. 28   Fig. 29   Fig. 30   Fig. 31

In problems 22 – 23, find a pair of Bezier curves that satisfy the given conditions.

22. B(0) = (0, 5), B '(0) = 2, B(1) = C(0) = (3, 1), B '(1) = C '(0) = 2, C(1) = (6, 2), and C '(1) = 4.

23. B(0) = (0, 5), B '(0) = 2, B(1) = C(0) = (3, 1), B '(1) = C '(0) = 2, C(1) = (6, 2), and C '(1) = 4

Some Applications of Bezier Curves

The following Applications illustrate just a few of the wide variety of design applications of Bezier curves. This combination of differentiation and algebra is very powerful.

Applications

For each of the following applications, write the equation of a Bezier curve $B(t)$ that satisfies the requirements of the application, and then use your calculator/computer to graph $B(t)$.
(Typically in these applications, the starting and ending points, P_0 and P_3, are specified, but several choices of the control points P_1 and P_2 meet the requirements of the application. Select P_1 and P_2 so the resulting graph of $B(t)$ satisfies the requirements of the application and is also "visually pleasing.")

1. You have been hired to design an escalator for a shopping mall, and the design requirements are that the entrance and exit of the escalator must be horizontal (Fig. 32), the total rise is 20 feet, and the total run is 30 feet.

    (a) Find the equation of a Bezier curve $B(t)$ that meets these design requirements and graph it.
       (Suggestion: Place the origin at the lower left end of the escalator.)

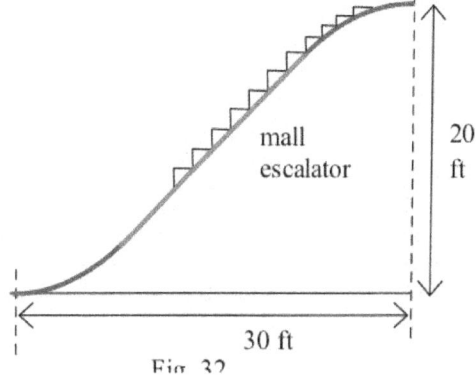

Fig. 32

    (b) In practice, the middle section of an escalator is straight, and each end consists of a curved section (a Bezier curve) that smoothly converts our horizontal motion to motion along the straight section and then to horizontal motion again for our exit. Find the equation of a Bezier curve that models the curve at the entrance to the up escalator (Fig. 33), and the equation of the straight line section.

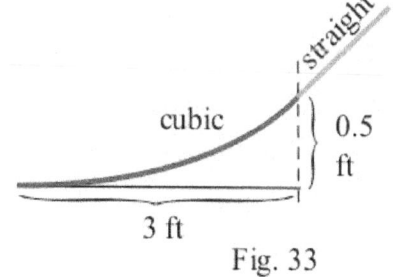

Fig. 33

2. To design the exit ramp on a highway (Fig. 34), you need to find the equation of a Bezier curve so that at the beginning of the exit the elevation is 20 feet with a slope of -0.05 (about $3°$), and 600 feet later, measured horizontally, the elevation is 0 feet and the ramp is horizontal.

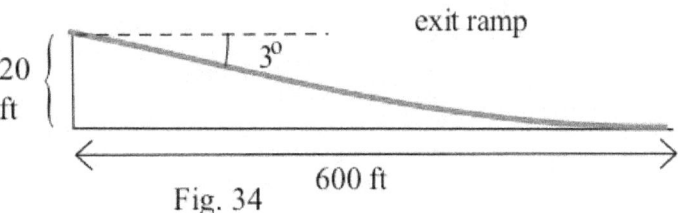

Fig. 34

3.   Find a Bezier curve that describes the final 60 feet
of the ski jump shown in Fig. 35.

4.   Find a Bezier curve that describes the left half of
the arch shown in Fig. 36.

5.   Find two Bezier curves B(t) and C(t) that describe the
pieces of the curve for the top half of the hull of the
Concordia Yawl that is 40 feet long, has a beam of 10
feet, and a transom width of 2 feet (Fig. 37).

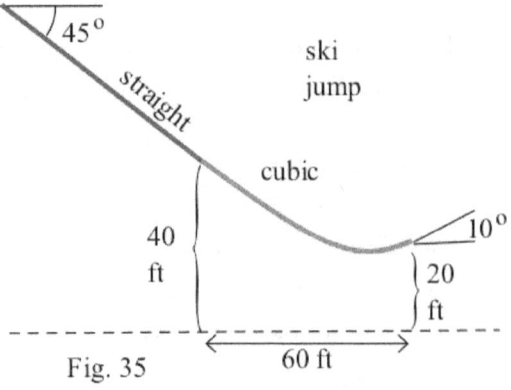

Fig. 35

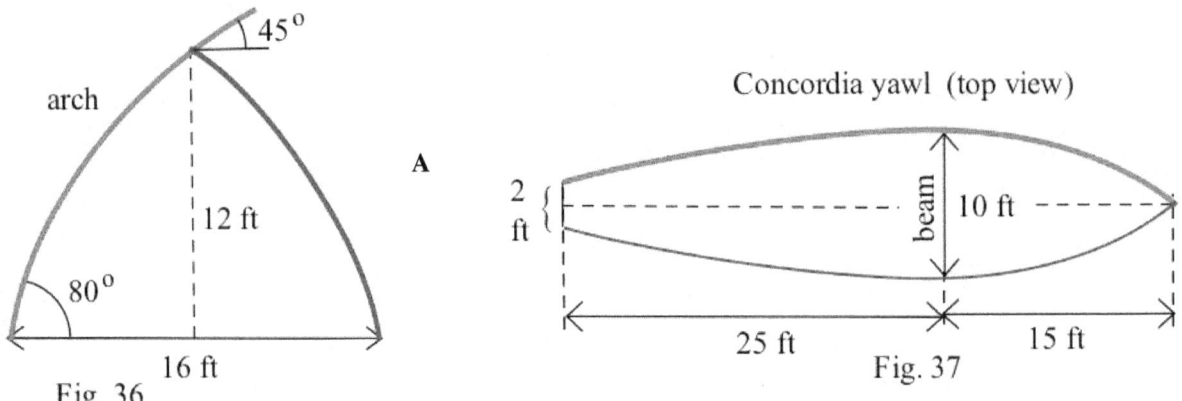

Fig. 36

Fig. 37

Final Note: The examples and applications given here have consisted of only one or two Bezier
curves, and they are intended only as an introduction to the ideas and techniques of fitting Bezier
curves to particular situations. But these ideas extend very nicely to curves that require pieces of
several different Bezier curves for a good fit (Fig. 38) and even to curves and surfaces in three
dimensions. Unfortunately, the systems of equations tend to grow very large for these extended
applications.

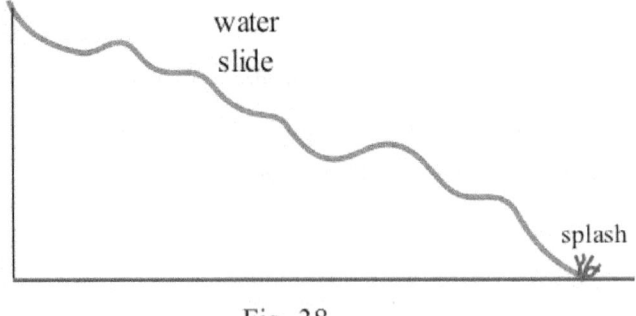

Fig. 38

Practice Answers

Practice 1: (a) $x(t) = (1 - t) \cdot x_0 + t \cdot x_1 = (1 - t) \cdot 1 + t \cdot 5 = 1 + 4t$,

$y(t) = (1 - t) \cdot y_0 + t \cdot y_1 = (1 - t) \cdot 2 + t \cdot 4 = 2 + 2t$.

(b) $x(t) = 5 - 4t,\ y(t) = 4 - 2t$

(c) $x(t) = 6 - 3t,\ y(t) = -2 + 3t$

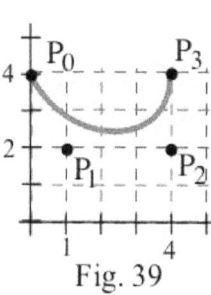

Fig. 39

Practice 2: $P_0 = (0, 4), P_1 = (1, 2),\ P_2 = (4, 2),$ and $P_3 = (4, 4)$. Then

$x(t) = (1 - t)^3 \cdot 0 + 3(1 - t)^2 \cdot t \cdot 1 + 3(1 - t) \cdot t^2 \cdot 4 + t^3 \cdot 4 = -5t^3 + 6t^2 + 3t$.

$y(t) = (1 - t)^3 \cdot 4 + 3(1 - t)^2 \cdot t \cdot 2 + 3(1 - t) \cdot t^2 \cdot 2 + t^3 \cdot 4 = 6t^2 - 6t + 4$.

The control points and the graph of $B(t) = (x(t), y(t))$ are shown in Fig. 39.

Practice 3: Take $P_0 = (0, 0)$ and $P_3 = (0, 1)$ to get the correct endpoints.

Take $P_1 = (-1, -1)$ and $P_2 = (-1, 2)$ to get a "C" shape. Then

$x(t) = (1 - t)^3 \cdot 0 + 3(1 - t)^2 \cdot t \cdot (-1) + 3(1 - t) \cdot t^2 \cdot (-1) + t^3 \cdot 0$

$y(t) = (1 - t)^3 \cdot 0 + 3(1 - t)^2 \cdot t \cdot (-1) + 3(1 - t) \cdot t^2 \cdot 2 + t^3 \cdot 1$.

The control points and the graph of $B(t) = (x(t), y(t))$ are shown in Fig. 40.

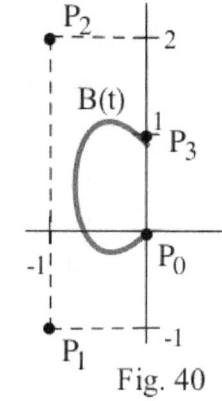

Fig. 40

Practice 4: Take $P_0 = (1, 3)$ and $P_3 = (5, 3)$ to get the correct endpoints.

Take $P_1 = (2, 1)$ and $P_2 = (4, 4)$ to get the correct slopes. Then

$x(t) = (1 - t)^3 \cdot 1 + 3(1 - t)^2 \cdot t \cdot 2 + 3(1 - t) \cdot t^2 \cdot 4 + t^3 \cdot 5$

$y(t) = (1 - t)^3 \cdot 3 + 3(1 - t)^2 \cdot t \cdot 1 + 3(1 - t) \cdot t^2 \cdot 4 + t^3 \cdot 3$.

The control points and the graph of

$B(t) = (x(t), y(t))$ are shown in Fig. 41.

Other choices for P_1 and P_2 can also yield correct slopes, and then we

have different formulas for $x(t)$ and $y(t)$

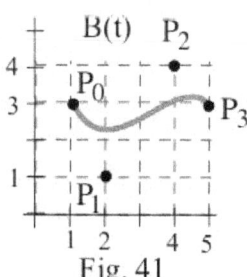

Fig. 41

9.5   CONIC SECTIONS

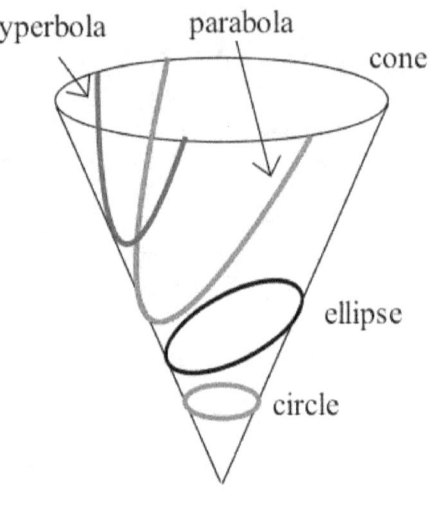

Fig. 1

The conic sections are the curves obtained when a cone is cut by a plane (Fig. 1). They have attracted the interest of mathematicians since the time of Plato, and they are still used by scientists and engineers. The early Greeks were interested in these shapes because of their beauty and their representations by sets of points that met certain distance definitions (e.g., the circle is the set of points at a fixed distance from a given point). Mathematicians and scientists since the 1600s have been interested in the conic sections because the planets, moons, and other celestial objects follow paths that are (approximately) conic sections, and the reflective properties of the conic sections are useful for designing telescopes and other instruments. Finally, the conic sections give the **complete** answer to the question, "what is the shape of the graph of the general quadratic equation $Ax^2 + Bxy + Cy^2 + Dx + Ey + F = 0$?"

This section discusses the "cut cone" and distance definitions of the conic sections and shows their standard equations in rectangular coordinate form. The section ends with a discussion of the discriminant, an easy way to determine the shape of the graph of any standard quadratic equation $Ax^2 + Bxy + Cy^2 + Dx + Ey + F = 0$. Section 9.6 examines the polar coordinate definitions of the conic sections, some of the reflective properties of the conic sections, and some of their applications.

Cutting A Cone

When a (right circular double) cone is cut by a plane, only a few shapes are possible, and these are called the conic sections (Fig. 1). If the plane makes an angle of θ with the horizontal, and $\theta < \alpha$, then the set of points is an ellipse (Fig. 2). When $\theta = 0 < \alpha$, we have a circle, a special case of an ellipse (Fig. 3). If $\theta = \alpha$, a parabola is formed (Fig. 4), and if $\theta > \alpha$, a hyperbola is formed (Fig. 5). When the plane goes through the vertex of the cone, degenerate conics are formed: the degenerate ellipse ($\theta < \alpha$) is a point, the degenerate parabola ($\theta = \alpha$) is a line, and a degenerate hyperbola ($\theta > \alpha$) is a pair of intersecting lines.

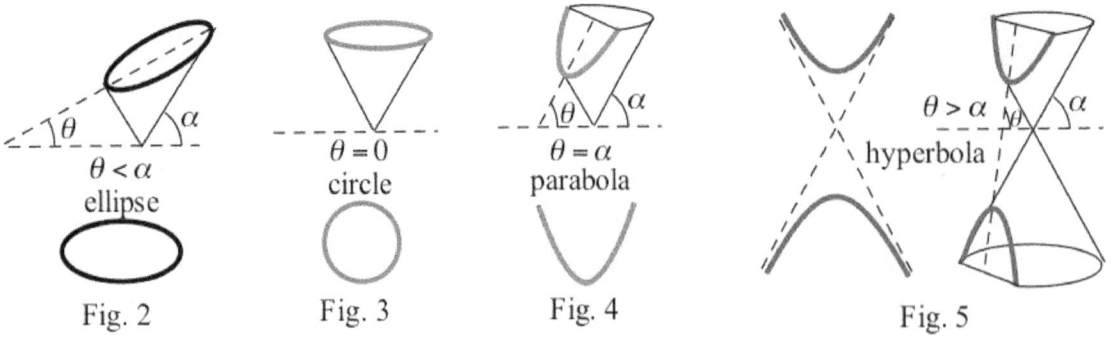

Fig. 2   Fig. 3   Fig. 4   Fig. 5

The conic sections are lovely to look at, but we will not use the conic sections as pieces of a cone because the "cut cone" definition of these shapes does not easily lead to formulas for them. To determine formulas for the conic sections it is easier to use alternate definitions of these shapes in terms of distances of points from fixed points and lines. Then we can use the formula for distance between two points and some algebra to derive formulas for the conic sections.

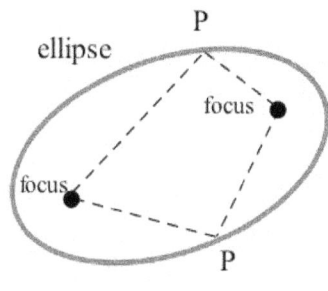

Fig. 6

The Ellipse

Ellipse: An ellipse is the set of all points P for which the **sum** of the
             distances from P to two fixed points (called foci) **is a constant**:
             dist(P, one focus) + dist(P, other focus) = constant. (Fig. 6)

Example 1:   Find the set of points whose distances from the foci $F_1 = (4,0)$
                 and $F_2 = (-4, 0)$ add up to 10.

Solution: If the point $P = (x, y)$ is on the ellipse, then the distances $PF_1 = \sqrt{(x-4)^2 + y^2}$ and
            $PF_2 = \sqrt{(x+4)^2 + y^2}$ must total 10 so we have the equation

$$PF_1 + PF_2 = \sqrt{(x-4)^2 + y^2} + \sqrt{(x+4)^2 + y^2} = 10 \quad \text{(Fig. 7)}$$

Moving the second radical to the right side of the equation, squaring both
sides, and simplifying, we get

$$4x + 25 = 5\sqrt{(x+4)^2 + y^2} \quad .$$

Squaring each side again and simplifying, we have $225 = 9x^2 + 25y^2$ so,
after dividing each side by 225,

$$\frac{x^2}{25} + \frac{y^2}{9} = 1 \; .$$

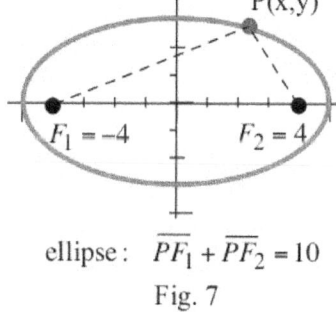

ellipse : $\overline{PF_1} + \overline{PF_2} = 10$

Fig. 7

Practice 1:   Find the set of points whose distances from the foci $F_1 = (3,0)$ and $F_2 = (-3, 0)$ add up to 10.

Using the same algebraic steps as in Example 1, it can be shown (see the Appendix at the end of the problems) that the set of points $P = (x,y)$ whose distances from the foci $F_1 = (c,0)$ and $F_2 = (-c, 0)$ add up to 2a (a > c) is described by the formula

$$\frac{x^2}{a^2} + \frac{y^2}{b^2} = 1 \; \text{ where } \; b^2 = a^2 - c^2 \; .$$

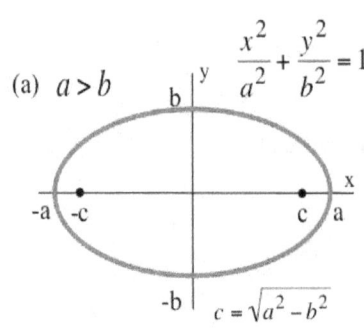

(a) $a > b$

$\dfrac{x^2}{a^2} + \dfrac{y^2}{b^2} = 1$

$c = \sqrt{a^2 - b^2}$

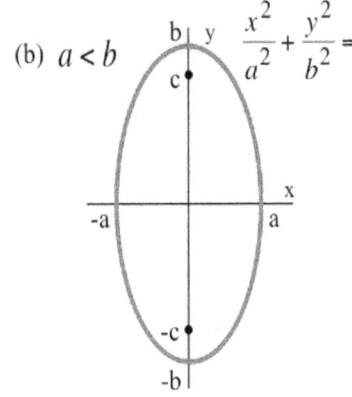

(b) $a < b$

$\dfrac{x^2}{a^2} + \dfrac{y^2}{b^2} = 1$

Fig. 8

Ellipse

The standard formula for an ellipse is $\dfrac{x^2}{a^2} + \dfrac{y^2}{b^2} = 1$.

a = b: The ellipse is a circle.

a > b: (Fig. 8a)   The vertices are at $(\pm a, 0)$ on the x–axis,

the foci are at $(\pm c, 0)$ with $c = \sqrt{a^2 - b^2}$, and

for any point P on the ellipse,

dist(P, one focus) + dist(P, other focus) = 2a.

The length of the semimajor axis is a.

a < b: (Fig. 8b)   The vertices are at $(0, \pm b)$ on the y–axis,

the foci are at $(0, \pm c)$ with $c = \sqrt{b^2 - a^2}$, and

for any point P on the ellipse,

dist(P, one focus) + dist(P, other focus) = 2b.

The length of the semimajor axis is b.

Practice 2:   Use the information in the box to determine the vertices, foci, and length of the semimajor

axis of the ellipse $\dfrac{x^2}{169} + \dfrac{y^2}{25} = 1$.

The Parabola

Parabola: A parabola is the set of all points P for which the distance from P

to a fixed point (focus) **is equal to** the distance from P to a

fixed line (directrix): dist(P, focus) = dist(P, directrix). (Fig. 9)

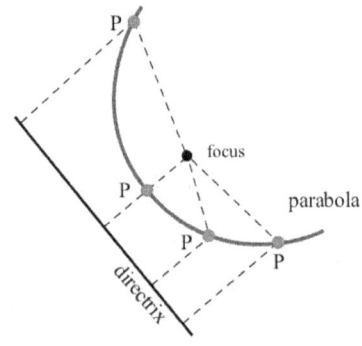

dist(P, focus) = dist(P, directix)

Fig. 9

Example 2:   Find the set of points P = (x,y) whose distance from the

focus F = (4,0) equals the distance from the directrix x = −1.

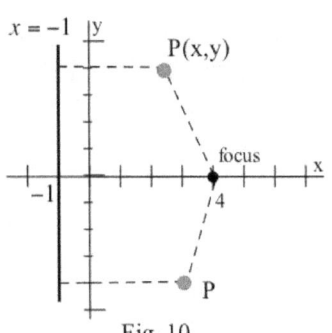

Fig. 10

Solution: The distance $PF = \sqrt{(x-4)^2 + y^2}$, and the distance from P to

to

the directrix (Fig. 10) is x+1. If these two distances are equal then

we have the equation $\sqrt{(x-4)^2 + y^2} = x + 1$.

Squaring each side,

$$(x-4)^2 + y^2 = (x+1)^2$$

so $x^2 - 8x + 16 + y^2 = x^2 + 2x + 1$.

This simplifies to $x = \frac{1}{10} y^2 + \frac{3}{2}$, the equation

of a parabola opening to the right (Fig. 11).

Practice 3:   Find the set of points P = (x,y)

whose distance from the focus F = (0,2)

equals the distance from the directrix y = –2..

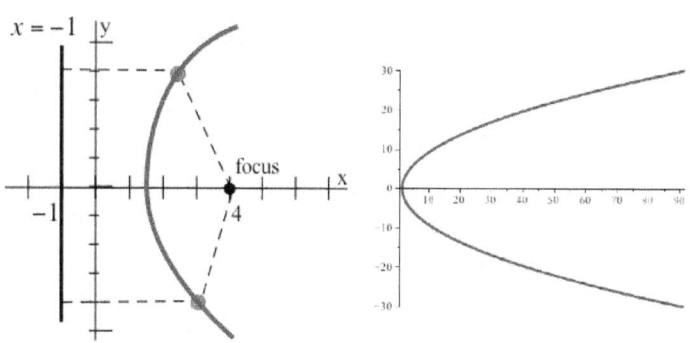

parabola: $x = \frac{3}{2} + \frac{1}{10} y^2$ (two scales)

Fig. 11

(a) $y = ax^2$ $(a > 0)$

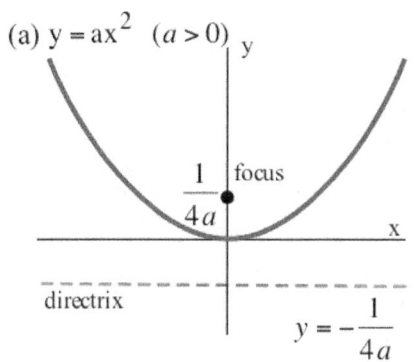

(b) $x = ay^2$ $(a > 0)$

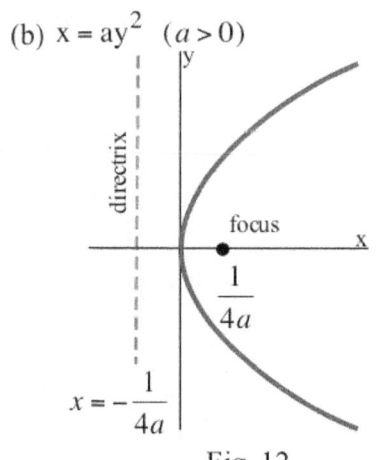

Fig. 12

Parabola

The standard parabola $y = ax^2$ opens around the y–axis

(Fig. 12a) with vertex = (0,0),   focus = $(0 , \frac{1}{4a})$, and

directrix $y = -\frac{1}{4a}$.

The standard parabola $x = ay^2$ opens around the x–axis

(Fig. 12b) with vertex = (0,0),   focus = $(\frac{1}{4a} , 0)$, and

directrix $x = -\frac{1}{4a}$.

Proof for the case $y = ax^2$:

The set of points p = (x,y) that are equally distant from the

focus

$F = (0 , \frac{1}{4a})$ and the directrix $y = -\frac{1}{4a}$ satisfy the distance

equation

PF = PD so

$\sqrt{ x^2 + (y - \frac{1}{4a})^2 }$ = $(y + \frac{1}{4a})$. Squaring each side, we have

$x^2 + (y - \frac{1}{4a})^2$ = $(y + \frac{1}{4a})^2$ and $x^2 + y^2 - \frac{2}{4a} y + \frac{1}{16a^2}$ = $y^2 + \frac{2}{4a} y + \frac{1}{16a^2}$.

Then $x^2 = \frac{2}{4a} y + \frac{2}{4a} y = \frac{1}{a} y$ and, finally, $y = ax^2$.

Practice 4: Prove that the set of points P= (x,y) that are equally distant from the focus

$$F = (\frac{1}{4a} , 0) , \text{ and directrix } x = -\frac{1}{4a} \text{ satisfy the equation } x = ay^2 .$$

Hyperbola

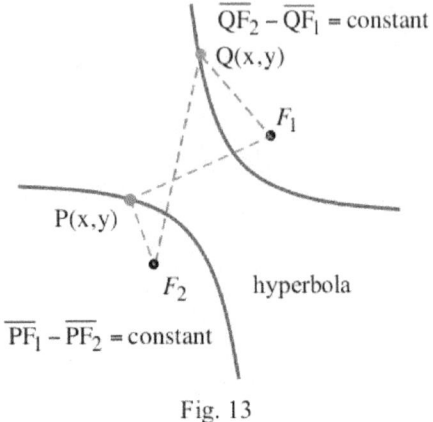

$$\overline{QF_2} - \overline{QF_1} = \text{constant}$$

Hyperbola: A hyperbola is the set of all points P for which the

difference of the distances from P to two fixed points

(foci) **is a constant**:

dist(P, one focus) – dist(P, other focus) = constant. (Fig. 13)

Example 3: Find the set of points for which the **difference** of

the distances from the points to the foci $F_1 = (5,0)$

and $F_2 = (-5, 0)$ is always 8.

$$\overline{PF_1} - \overline{PF_2} = \text{constant}$$

Fig. 13

Solution: If the point P = (x, y) is on the hyperbola, then the

difference of the distances $PF_1 = \sqrt{(x-5)^2 + y^2}$ and $PF_2 = \sqrt{(x+5)^2 + y^2}$ is 8 so we have

the equation $PF_1 - PF_2 = \sqrt{(x-5)^2 + y^2} - \sqrt{(x+5)^2 + y^2} = 8$ (Fig. 14).

Moving the second radical to the right side of the equation, squaring both sides, and simplifying, we get

$$5x + 16 = -4\sqrt{(x+5)^2 + y^2} .$$

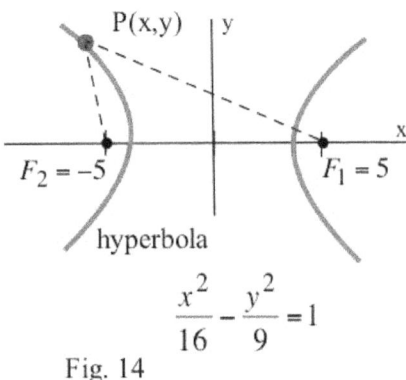

$$PF_1 - PF_2 = 8 \quad \text{or} \quad PF_2 - PF_1 = 8$$

Squaring each side again and simplifying, we have $9x^2 - 16y^2 = 144 .$

After dividing each side by 144, $\frac{x^2}{16} - \frac{y^2}{9} = 1.$

If we start with the difference $PF_2 - PF_1 = 8$, we have the equation

$$\sqrt{(x + 5)^2 + y^2} - \sqrt{(x - 5)^2 + y^2} = 8 .$$

Solving this equation, we again get $9x^2 - 16y^2 = 144$ and

Fig. 14

$$\frac{x^2}{16} - \frac{y^2}{9} = 1.$$

$$\frac{x^2}{16} - \frac{y^2}{9} = 1 .$$

Using the same algebraic steps as in the Example 3, it can be shown (see the Appendix at the end of the

problems) that the set of points P = (x,y) whose distances from the foci $F_1 = (c,0)$ and $F_2 = (-c, 0)$

differ by 2a (a < c) is described by the formula

$$\frac{x^2}{a^2} - \frac{y^2}{b^2} = 1 \text{ where } b^2 = c^2 - a^2 .$$

(a) hyperbola $\dfrac{x^2}{a^2} - \dfrac{y^2}{b^2} = 1$

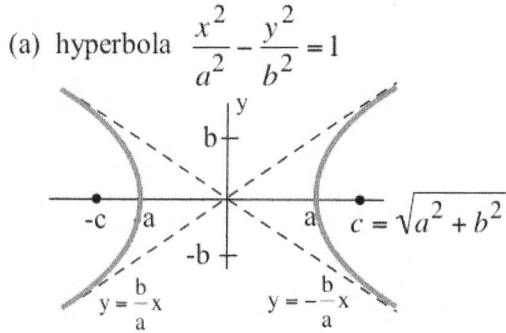

(b) hyperbola $\dfrac{y^2}{b^2} - \dfrac{x^2}{a^2} = 1$

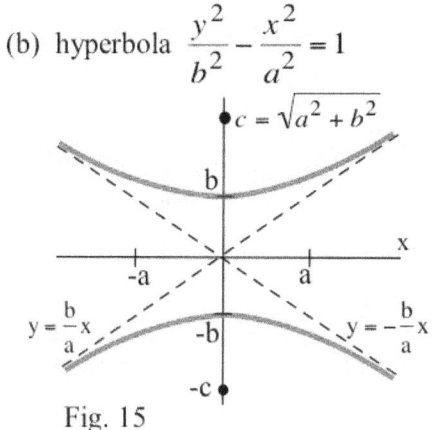

Fig. 15

Hyperbola

The standard hyperbola $\dfrac{x^2}{a^2} - \dfrac{y^2}{b^2} = 1$

opens around the x–axis (Fig. 15a)

with vertices at $(\pm a, 0)$, foci at $(\pm\sqrt{a^2 + b^2}, 0)$,

and linear asymptotes $y = \pm\dfrac{b}{a} x$.

The standard hyperbola $\dfrac{y^2}{b^2} - \dfrac{x^2}{a^2} = 1$

opens around the y–axis (Fig. 15b)

with vertices at $(0, \pm b)$, foci at $(0, \pm\sqrt{a^2 + b^2})$,

and linear asymptotes $y = \pm\dfrac{b}{a} x$.

Practice 5:　Graph the hyperbolas $\dfrac{x^2}{25} - \dfrac{y^2}{16} = 1$ and $\dfrac{y^2}{25} - \dfrac{x^2}{16} = 1$ and find the linear

asymptotes for each hyperbola.

Visually distinguishing the conic sections

If you only observe a small part of the graph of a conic section, it may be impossible to determine which conic section it is, and you may need to look at more of its graph. Near a vertex or in small pieces, all of the conic sections can be quite similar in appearance, but on a larger graph the ellipse is easy to distinguish from the other two. On a large graph, the hyperbola and parabola can be distinguished by noting that the hyperbola has two linear asymptotes and the parabola has no linear asymptotes.

The General Quadratic Equation and the Discriminant

Every equation that is quadratic in the variables x or y or both can be written in the form

$$Ax^2 + Bxy + Cy^2 + Dx + Ey + F = 0 \text{ where } A \text{ through } F \text{ are constants.}$$

The form $Ax^2 + Bxy + Cy^2 + Dx + Ey + F = 0$ is called the **general quadratic equation**.

In particular, each of the conic sections can be written in the form of a general quadratic equation by clearing all fractions and collecting all of the terms on one side of the equation. What is perhaps surprising is that the graph of a general quadratic equation is always a conic section or a degenerate form of a conic section. Usually the graph of a general quadratic equation is not centered at the origin and is not symmetric about either axis, but the shape is always an ellipse, parabola, hyperbola, or degenerate form of one of these.

Even more surprising, a quick and easy calculation using just the coefficients A, B, and C of the general quadratic equation tells us the shape of its graph: ellipse, parabola, or hyperbola. The value obtained by this simple calculation is called the discriminant of the general quadratic equation.

Discriminant

The **discriminant** of the the general quadratic form $Ax^2 + Bxy + Cy^2 + Dx + Ey + F = 0$

is the value $\mathbf{B^2 - 4AC}$.

Example 4:   Write each of the following in its general quadratic form and calculate its discriminant.

(a) $\dfrac{x^2}{25} + \dfrac{y^2}{9} = 1$ (b) $3y + 7 = 2x^2 + 5x + 1$ (c) $5x^2 + 3 = 7y^2 - 2xy + 4y + 8$

Solution: (a) $9x^2 + 25y^2 - 225 = 0$ so $A = 9, C = 25, F = -225,$ and $B = D = E = 0$. $B^2 - 4AC = -900$.

(b) $2x^2 + 5x - 3y - 6 = 0$ so $A = 2, D = 5, E = -3, F = -6,$ and $B = C = 0$. $B^2 - 4AC = 0$.

(c) $5x^2 + 2xy - 7y^2 - 4y - 5 = 0$ so $A = 5, B = 2, C = -7, D = 0, E = -4$ and $F = -5$.
$B^2 - 4AC = 4 - 4(5)(-7) = 144$.

Practice 6:   Write each of the following in its general quadratic form and calculate its discriminant.

(a) $1 = \dfrac{x^2}{36} - \dfrac{y^2}{9}$ (b) $x = 3y^2 - 5$ (c) $\dfrac{x^2}{16} + \dfrac{(y-2)^2}{25} = 1$

One very important property of the discriminant is that it is invariant under translations and rotations, its value does not change even if the graph is rigidly translated around the plane and rotated. When a graph is shifted or rotated or both, its general quadratic equation changes, but the discriminant of the new quadratic equation is the same value as the discriminant of the original quadratic equation. And we can determine the shape of the graph simply from the sign of the discriminant.

Quadratic Shape Theorem

The graph of the general quadratic equation $Ax^2 + Bxy + Cy^2 + Dx + Ey + F = 0$ is

     an ellipse if $B^2 - 4AC < 0$ (degenerate forms: one point or no points)

     a parabola if $B^2 - 4AC = 0$ (degenerate forms: two lines, one line, or no points)

     a hyperbola if $B^2 - 4AC > 0$ (degenerate form: pair of intersecting lines).

The proofs of this result and of the invariance of the discriminant under translations and rotations are "elementary" and just require a knowledge of algebra and trigonometry, but they are rather long and are very computational. A proof of the invariance of the discriminant under translations and rotations and of the Quadratic Shape Theorem is given in the Appendix after the problem set.

Example 5:   Use the discriminant to determine the shapes of the graphs of the following equations.

     (a) $x^2 + 3xy + 3y^2 = -7y - 4$ (b) $4x^2 + 4xy + y^2 = 3x - 1$ (c) $y^2 - 4x^2 = 0$.

Solution: (a) $B^2 - 4AC = 3^2 - 4(1)(3) = -3 < 0$. The graph is an ellipse.

           (b) $B^2 - 4AC = 4^2 - 4(4)(1) = 0$. The graph is a parabola.

           (c) $B^2 - 4AC = 0^2 - 4(-4)(1) = 16 > 0$. The graph is a hyperbola — actually a degenerate

               hyperbola. The graph of $0 = y^2 - 4x^2 = (y + 2x)(y - 2x)$ consists of the two lines

               $y = -2x$ and $y = 2x$.

Practice 7:   Use the discriminant to determine the shapes of the graphs of the following equations.

     (a) $x^2 + 2xy = 2y^2 + 4x + 3$ (b) $y^2 + 2x^2 = xy - 3y + 7$ (c) $2x^2 - 4xy = 3 + 5y - 2y^2$.

Sketching Standard Ellipses and Hyperbolas

The graphs of general ellipses and hyperbolas require plotting lots of points (a computer or calculator can help), but it is easy to sketch good graphs of the standard ellipses and hyperbolas. The steps for doing so are given below.

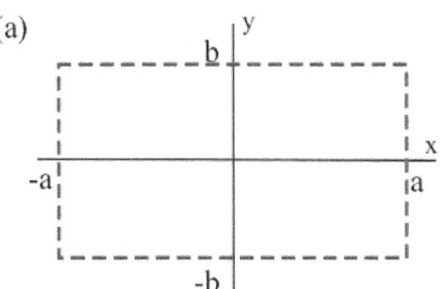

Graphing the Standard Ellipse $\dfrac{x^2}{a^2} + \dfrac{y^2}{b^2} = 1$

1. Sketch short vertical line segments at the points $(\pm a, 0)$ on the x–axis and short horizontal line segments at the points $(0, \pm b)$ on the y–axis (Fig. 16a). Draw a rectangle whose sides are formed by extending the line segments.

2. Use the tangent line segments in step 1 as guide to sketching the ellipse (Fig. 16b). The graph of the ellipse is always inside the rectangle except at the 4 points that touch it.

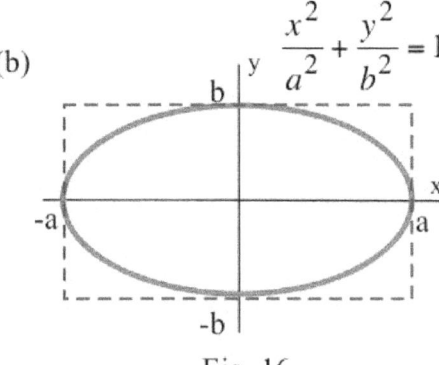

Fig. 16

Graphing the Standard Hyperbolas $\dfrac{x^2}{a^2} - \dfrac{y^2}{b^2} = 1$ and $\dfrac{y^2}{b^2} - \dfrac{x^2}{a^2} = 1$

1. Sketch the rectangle that intersects the x–axis at the points $(\pm a, 0)$ and the y–axis at the points $(0, \pm b)$. (Fig. 17a)

2. Draw the lines which go through the origin and the corners of the rectangle from step 1. (Fig. 17b) These lines are the asymptotes of the hyperbola.

3. For $\dfrac{x^2}{a^2} - \dfrac{y^2}{b^2} = 1$, plot the points $(\pm a, 0)$ on the hyperbola, and use the asymptotes from step 2 as a guide to sketching the rest of the hyperbola. (Fig. 17c)

3'. For $\dfrac{y^2}{b^2} - \dfrac{x^2}{a^2} = 1$, plot the points $(0, \pm b)$ on the hyperbola, and use the asymptotes from step 2 as a guide to sketching the rest of the hyperbola. (Fig. 17d)

The graph of the hyperbola is always outside the rectangle except at the 2 points which touch it.

(a)

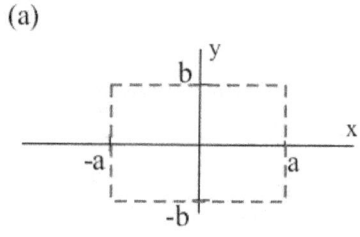

(b)

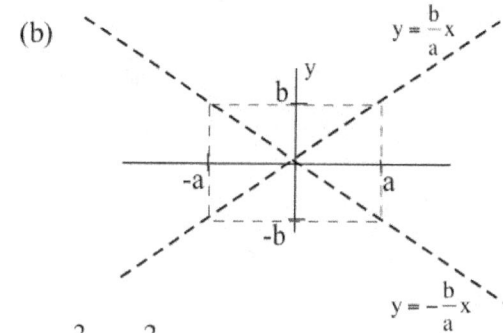

(c) $\dfrac{x^2}{a^2} - \dfrac{y^2}{b^2} = 1$

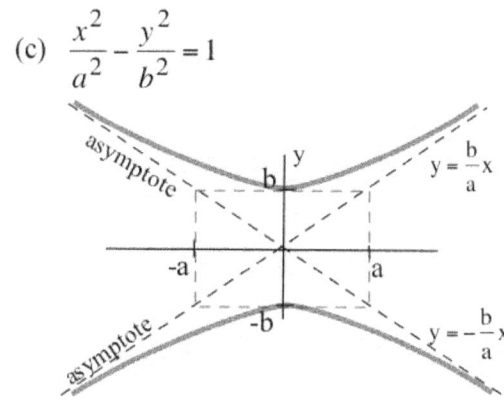

(d) $\dfrac{y^2}{b^2} - \dfrac{x^2}{a^2} = 1$

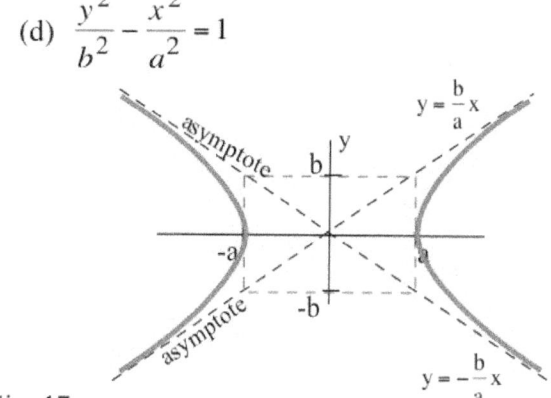

Fig. 17

Symmetry of the Conic Sections

Symmetry properties of the conic sections can simplify the task of graphing them. A parabola has one line of symmetry, so once we have graphed half of a parabola we can get the other half by folding along the line of symmetry. An ellipse and a hyperbola each have two lines of symmetry, so once we have graphed one fourth of an ellipse or hyperbola we can get the rest of the graph by folding along each line of symmetry.

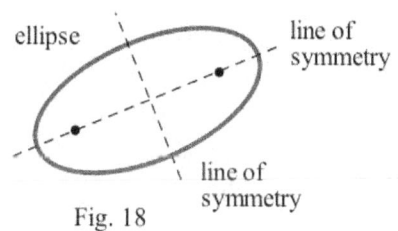

Fig. 18

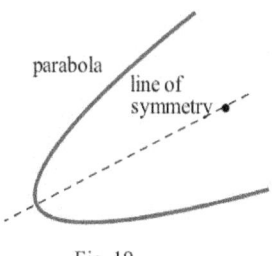

Fig. 19

- The parabola is symmetric about the line through the focus and the vertex (Fig. 18).
- The ellipse is symmetric about the line through the two foci. It is also symmetric about the perpendicular bisector of the line segment through the two foci (Fig. 19).
- The hyperbola is symmetric about the line through the two foci and about the perpendicular bisector of the line segment through the two foci (Fig. 20).

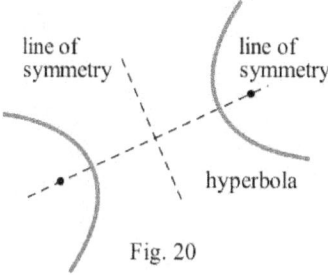

Fig. 20

The Conic Sections as "Shadows of Spheres"

There are a lot of different shapes at the beach on a sunny day, even conic sections. Suppose we have a sphere resting on a flat surface and a point radiating light.

- If the point of light is higher than the top of the sphere, then the shadow of the sphere is an ellipse (Fig. 21).
- If the point of light is exactly the same height as the top of the sphere, then the shadow of the sphere is a parabola (Fig. 22).
- If the point of light is lower than the top of the sphere, then the shadow of the sphere is one branch of a hyperbola (Fig. 23).

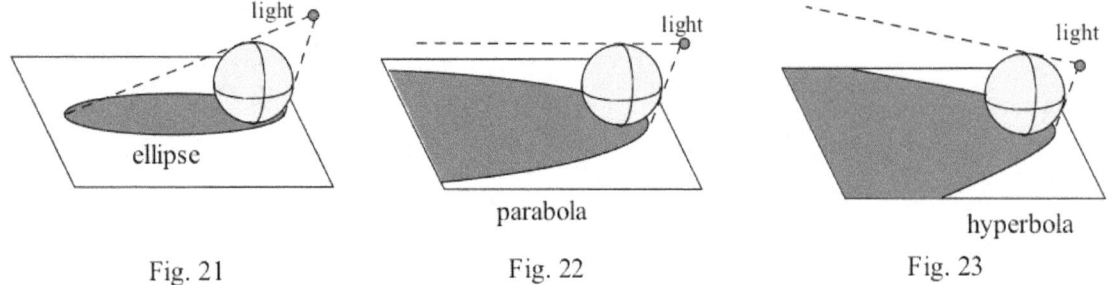

Fig. 21   Fig. 22   Fig. 23

PROBLEMS

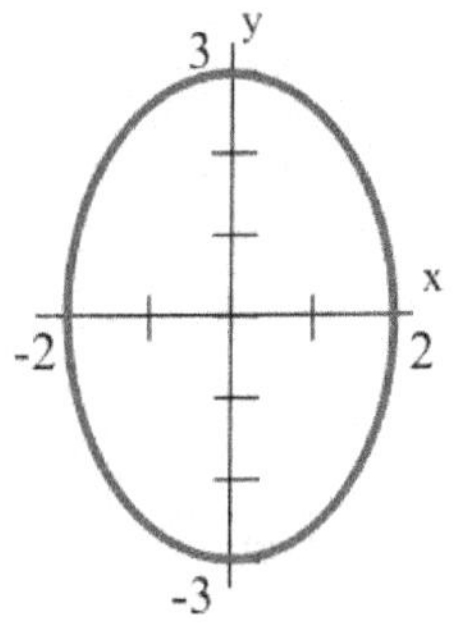

1. What is the shape of the graph of the set of points whose distances from (6,0) and (–6,0) always add up to 20? Find an equation for the graph.

2. What is the shape of the graph of the set of points whose distances from (2,0) and (–2,0) always add up to 20? Find an equation for the graph.

Fig. 24

3. What is the shape of the graph of the set of points whose distance from the point (0,5) is equal to the distance from the point to the line y = –5? Find an equation for the graph.

4. What is the shape of the graph of the set of points whose distance from the point (2,0) is equal to the distance from the point to the line x = –4? Find an equation for the graph.

5. Give the standard equation for the ellipse in Fig. 24.

6. Give the standard equation for the ellipse in Fig. 25.

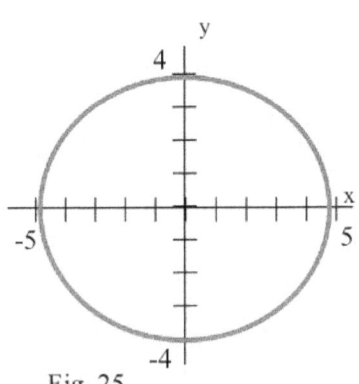

Fig. 25

7. What lines are linear asymptotes for the hyperbola $4x^2 - 9y^2 = 36$, and where are the foci?

8. What lines are linear asymptotes for the hyperbola $25x^2 - 4y^2 = 100$, and where are the foci?

9. What lines are linear asymptotes for the hyperbola $5y^2 - 3x^2 = 15$, and where are the foci?

10. What lines are linear asymptotes for the hyperbola $5y^2 - 3x^2 = 120$, and where are the foci?

In problems 11–16, rewrite each equation in the form of the general quadratic equation $Ax^2 + Bxy + Cy^2 + Dx + Ey + F = 0$ and then calculate the value of the discriminant. What is the shape of each graph?

11. (a) $\dfrac{x^2}{4} + \dfrac{y^2}{25} = 1$ (b) $\dfrac{x^2}{a^2} + \dfrac{y^2}{b^2} = 1$ 12. (a) $\dfrac{x^2}{4} - \dfrac{y^2}{25} = 1$ (b) $\dfrac{x^2}{a^2} - \dfrac{y^2}{b^2} = 1$

13. $x + 2y = 1 + \dfrac{3}{x - y}$ 14. $y = \dfrac{5 + 2y - x^2}{4x + 5y}$

15. $x = \dfrac{7x - 3 - 2y^2}{2x + 4y}$ 16. $x = \dfrac{2y^2 + 7x - 3}{2x + 5y}$

Problems 17–20 illustrate that a small change in the value of just **one** coefficient in the quadratic equation $Ax^2 + Bxy + Cy^2 + Dx + Ey + F = 0$ can have a dramatic effect on the shape of the graph. Determine the shape of the graph for each formula.

17. (a) $2x^2 + \mathbf{3}xy + 2y^2 + (\textit{terms for } x, y, \textit{ and a constant}) = 0$.
    (b) $2x^2 + \mathbf{4}xy + 2y^2 + (\textit{terms for } x, y, \textit{ and a constant}) = 0$.
    (c) $2x^2 + \mathbf{5}xy + 2y^2 + (\textit{terms for } x, y, \textit{ and a constant}) = 0$.
    (d) What are the shapes if the coefficients of the xy term are $3.99, 4,$ and 4.01?

18. (a) $\mathbf{1}x^2 + 4xy + 2y^2 + (\textit{terms for } x, y, \textit{ and a constant}) = 0$.
    (b) $\mathbf{2}x^2 + 4xy + 2y^2 + (\textit{terms for } x, y, \textit{ and a constant}) = 0$.
    (c) $\mathbf{3}x^2 + 4xy + 2y^2 + (\textit{terms for } x, y, \textit{ and a constant}) = 0$.
    (d) What are the shapes if the coefficients of the x^2 term are $1.99, 2,$ and 2.01?

19. (a) $x^2 + 4xy + \mathbf{3}y^2 + (\textit{terms for } x, y, \textit{ and a constant}) = 0$.
    (b) $x^2 + 4xy + \mathbf{4}y^2 + (\textit{terms for } x, y, \textit{ and a constant}) = 0$.
    (c) $x^2 + 4xy + \mathbf{5}y^2 + (\textit{terms for } x, y, \textit{ and a constant}) = 0$.
    (d) What are the shapes if the coefficients of the y^2 term are $3.99, 4,$ and 4.01?

20. Just changing a single sign can also dramatically change the shape of the graph.

   (a) $x^2 + 2xy + y^2 + (terms\ for\ x,\ y,\ and\ a\ constant) = 0.$

   (b) $x^2 + 2xy - y^2 + (terms\ for\ x,\ y,\ and\ a\ constant) = 0.$

21. Find the volume obtained when the region enclosed by the ellipse $\dfrac{x^2}{2^2} + \dfrac{y^2}{5^2} = 1$ is rotated

   (a) about the x–axis, and (b) about the y–axis.

22. Find the volume obtained when the region enclosed by the ellipse $\dfrac{x^2}{a^2} + \dfrac{y^2}{b^2} = 1$ is rotated

   (a) about the x–axis, and (b) about the y–axis.

23. Find the volume obtained when the region enclosed by the hyperbola $\dfrac{x^2}{2^2} - \dfrac{y^2}{5^2} = 1$ and the vertical

   line $x = 10$ is rotated (a) about the x–axis, and (b) about the y–axis.

24. Find the volume obtained when the region enclosed by the

   hyperbola $\dfrac{x^2}{a^2} - \dfrac{y^2}{b^2} = 1$ and the vertical line $x = L$

   (Fig. 26) is rotated (a) about the x–axis, and (b) about

   the y–axis. (Assume $a < L$.)

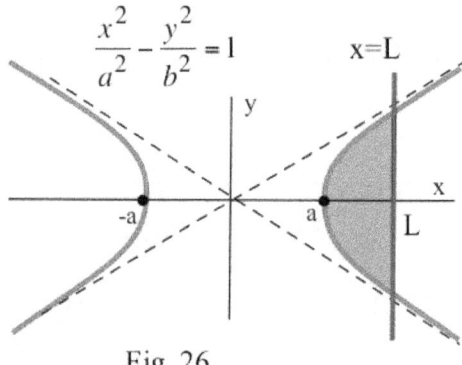

Fig. 26

25. Find the ratio of the area of the shaded parabolic region in
   Fig. 27 to the area of the rectangular region.

26. Find the ratio of the volumes obtained when the parabolic
   and rectangular regions in Fig. 27 are rotated about the y–
   axis.

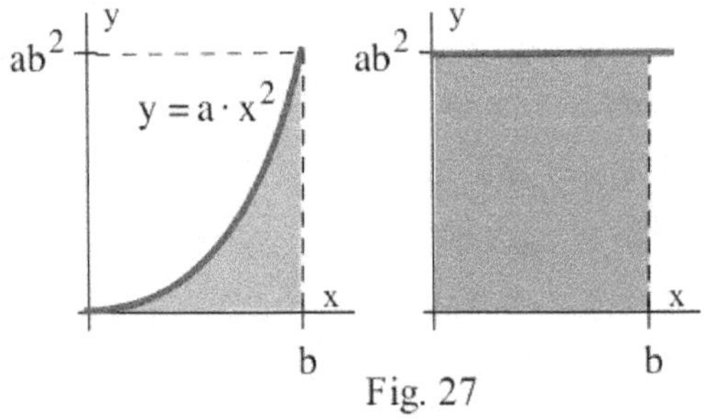

Fig. 27

String Constructions of Ellipses, Parabolas, and Hyperbolas   (Optional)

All of the conic sections can be drawn with the help of some pins and string, and the directions and figures show how it can be done. For each conic section, you are asked to determine and describe why each construction produces the desired shape.

Ellipse: Pin the two ends of the string to a board so the string is not taut. Put the point of a pencil in the bend in the string (Fig. 28), and, keeping the string taut, draw a curve.

27. How is the distance between the vertices of the ellipse related to the length of the string?

28. Explain why this method produces an ellipse, a set of points whose distances from the two fixed points (foci) always sum to a constant. What is the constant?

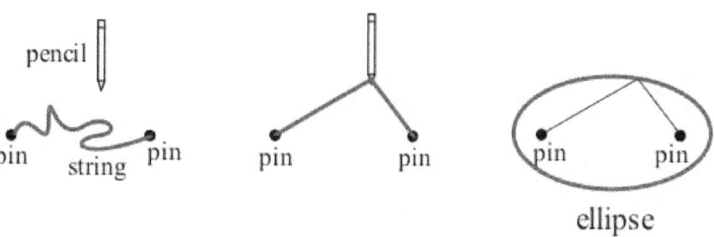

Fig. 28

29. What happens to the shape of the ellipse as the two foci are moved closer together (and the piece of string stays the same length)? Draw several ellipses using the same piece of string and different fixed points, and describe the results.

Parabola: Pin one end of the string to a board and the other end to the corner of a T–square bar that is the same length as the string. Put the point of a pencil in the bend in the string (Fig. 29) and keep the string taut. As the T–square is slid sideways, the pencil draws a curve.

30. Explain why this method produces a parabola, a set of points whose distance from a fixed point (one end of the string) is equal to the distance from a fixed line (the edge of the table).

31. What happens if the length of the string is slightly shorter than the length of the T–square bar? Draw several curves with several slightly shorter pieces of string and describe the results. What shapes are the curves?

32. Find a way to use pins, string and a pencil to sketch the graph of a hyperbola.

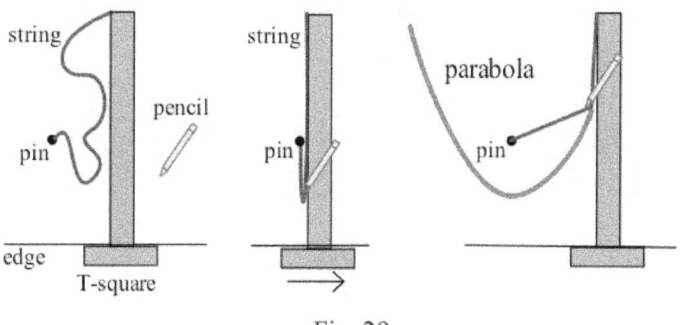

Fig. 29

Section 9.5   **PRACTICE Answers**

Practice 1: $F_1 = (3,0)$, $F_2 = (-3,0)$, and $P = (x,y)$. We want $\text{dist}(F_1, P) + \text{dist}(F_2, P) = 10$ so

$$\text{dist}((x,y),(3,0)) + \text{dist}((x,y),(-3,0)) = 10 \text{ and}$$

$$\sqrt{(x-3)^2 + y^2} + \sqrt{(x+3)^2 + y^2} = 10.$$

Moving the second radical to the right side and squaring, we get

$$(x-3)^2 + y^2 = 100 - 20\sqrt{(x+3)^2 + y^2} + (x+3)^2 + y^2 \text{ and}$$

$$x^2 - 6x + 9 + y^2 = 100 - 20\sqrt{(x+3)^2 + y^2} + x^2 + 6x + 9 + y^2 \text{ so}$$

$$-12x - 100 = -20\sqrt{(x+3)^2 + y^2}.$$

Dividing each side by -2 and then squaring, we have

$$36x^2 + 600x + 2500 = 100(x^2 + 6x + 9 + y^2) \text{ so}$$

$$1600 = 64x^2 + 100y^2 \text{ and}$$

$$1 = \frac{64x^2}{1600} + \frac{100y^2}{1600} = \frac{x^2}{25} + \frac{y^2}{16}.$$

Practice 2: $a = 13$ and $b = 5$ so the vertices of the ellipse are $(13, 0)$ and $(-13, 0)$. The value of c is $\sqrt{169 - 25} = 12$ so the foci are $(12, 0)$ and $(-12,0)$. The length of the semimajor axis is 13.

Practice 3: $\text{dist}(P, \text{focus}) = \text{dist}(P, \text{directrix})$ so $\text{dist}((x,y),(0,2)) = \text{dist}((x,y), \text{line } y=-2)$:

$$\sqrt{(x-0)^2 + (y-2)^2} = y + 2.$$

Squaring, we get $x^2 + y^2 - 4y + 4 = y^2 + 4y + 4$ so $x^2 = 8y$ or $y = \frac{1}{8}x^2$.

Practice 4:   This is similar to Practice 3: $\text{dist}(P, \text{focus}) = \text{dist}(P, \text{directrix})$ so

$$\text{dist}((x,y),(\tfrac{1}{4a}, 0)) = \text{dist}((x,y), \text{line } x = -\tfrac{1}{4a}). \text{ Then}$$

$$\sqrt{(x - \tfrac{1}{4a})^2 + (y - 0)^2} = x + \frac{1}{4a}. \text{ Squaring each side we get}$$

$$x^2 - 2x\frac{1}{4a} + \frac{1}{16a^2} + y^2 = x^2 + 2x\frac{1}{4a} + \frac{1}{16a^2} \text{ so } y^2 = \frac{1}{a}x \text{ and } x = ay^2.$$

Practice 5:   The graphs are shown in Fig. 30. Both hyperbolas have the same linear asymptotes:

$y = \frac{4}{5} \, x$ and $y = -\frac{4}{5} \, x$.

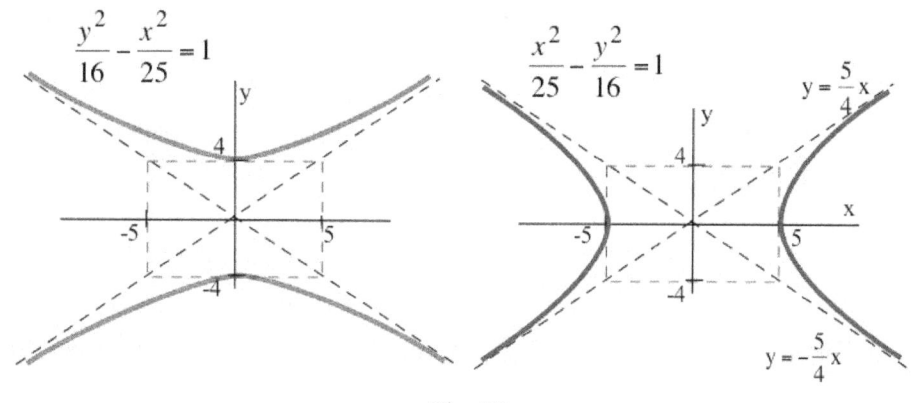

Fig. 30

Practice 6:   (a) $324 = 9x^2 - 36y^2$ so $9x^2 - 36y^2 - 324 = 0$.

A = 9, B = 0, and C = –36 so D = 0 – 4(9)(–36) = **576**.

(b) $0x^2 + 0xy + 3y^2 - x - 5 = 0$.

A = 0, B = 0, and C = 3 so D = 0 – 4(0)(3) = **0**.

(c) $25x^2 + 16(y-2)^2 = 400$ so $25x^2 + 16y^2 - 64y + 48 - 400 = 0$.

A = 25, B = 0, and C = 16 so D = 0 – 4(25)(16) = **–1600**.

Practice 7:   (a) $x^2 + 2xy - 2y^2 - 4x - 3 = 0$.

A = 1, B = 2, C = –2 so D = 4 – 4(1)(–2) = 12 > 0: hyperbola.

(b) $2x^2 - 1xy + 1y^2 + 3y - 7 = 0$.

A = 2, B = –1, C = 1 so D = 1 – 4(2)(1) = –7 < 0: ellipse.

(c) $2x^2 - 4xy + 2y^2 - 5y - 3 = 0$.

A = 2, B = –4, and C = 2 so D = 16 – 4(2)(2) = 0: parabola.

Appendix for 9.5: Conic Sections

Deriving the Standard Forms from Distance Definitions of the Conic Sections

Ellipse

> Ellipse   An ellipse is the set of all points P so the **sum** of the distances of P from two fixed points
> (called foci) **is a constant**.

If F_1 and F_2 are the foci (Fig. 40), then for every point P on the ellipse, the distance from P to F_1 PLUS
the distance from P to F_2 is a constant: $PF_1 + PF_2 = $ constant. If the center of the ellipse is at the origin
and the foci lie on the x–axis at $F_1 = (c, 0)$ and $F_2 = (-c, 0)$, we can translate the words into a formula:

$$PF_1 + PF_2 = \text{constant} \quad \text{becomes} \quad \sqrt{(x-c)^2 + y^2} \;+\; \sqrt{(x+c)^2 + y^2} \;=\; 2a .$$

(Calling the constant 2a simply makes some of the later algebra easier.)

By moving the second radical to the right side of the equation, squaring each side, and simplifying, we get

$$\sqrt{(x-c)^2 + y^2} \;=\; 2a - \sqrt{(x+c)^2 + y^2}$$

$$(x-c)^2 + y^2 \;=\; 4a^2 - 4a\sqrt{(x+c)^2 + y^2} \;+\; (x+c)^2 + y^2$$

$$\text{so } x^2 - 2xc + c^2 + y^2 \;=\; 4a^2 - 4a\sqrt{(x+c)^2 + y^2} \;+\; x^2 + 2xc + c^2 + y^2$$

$$\text{and } xc + a^2 \;=\; a\sqrt{(x+c)^2 + y^2} \;.$$

Squaring each side again and simplifying, we get

$$(xc + a^2)^2 \;=\; a^2 \{ (x+c)^2 + y^2 \} \quad \text{so } x^2c^2 + 2xca^2 + a^4 \;=\; a^2x^2 + 2xca^2 + a^2c^2 + a^2y^2$$

$$\text{and } a^2(a^2 - c^2) = x^2(a^2 - c^2) + y^2a^2 \;.$$

Finally, dividing each side by $a^2(a^2 - c^2)$, we get $\dfrac{x^2}{a^2} + \dfrac{y^2}{a^2 - c^2} = 1$.

By setting $b^2 = a^2 - c^2$, we have $\dfrac{x^2}{a^2} + \dfrac{y^2}{b^2} = 1$, the standard form of the ellipse.

Hyperbola

Hyperbola: A hyperbola is the set of all points P so the **difference** of the distances of P from the two

fixed points (foci) **is a constant**.

If F_1 and F_2 are the foci (Fig. 9), then for every point P on the hyperbola, the distance from P to F_1
MINUS the distance from P to F_2 is a constant: $PF_1 - PF_2 = $ constant (Fig. 42). If the center of the
hyperbola is at the origin and the foci lie on the x–axis at $F_1 = (c, 0)$ and $F_2 = (-c, 0)$, we can translate the
words into a formula:

$$PF_1 - PF_2 = \text{constant becomes } \sqrt{(x-c)^2 + y^2} - \sqrt{(x+c)^2 + y^2} = 2a.$$

(Calling the constant 2a simply makes some of the later algebra easier.)

The algebra which follows is very similar to that used for the ellipse.

Moving the second radical to the right side of the equation, squaring each side, and simplifying, we get

$$\sqrt{(x-c)^2 + y^2} = 2a + \sqrt{(x+c)^2 + y^2}$$

$$(x-c)^2 + y^2 = 4a^2 + 4a\sqrt{(x+c)^2 + y^2} + (x+c)^2 + y^2$$

$$\text{so } x^2 - 2xc + c^2 + y^2 = 4a^2 + 4a\sqrt{(x+c)^2 + y^2} + x^2 + 2xc + c^2 + y^2$$

$$\text{and } xc + a^2 = -a\sqrt{(x+c)^2 + y^2} .$$

Squaring each side again and simplifying, we get

$$(xc + a^2)^2 = a^2\{(x+c)^2 + y^2\} \text{ so } x^2c^2 + 2xca^2 + a^2 = a^2x + 2xca^2 + a^2c^2 + a^2y^2$$

$$\text{and } x^2(c^2 - a^2) - y^2a^2 = a^2(c^2 - a^2) .$$

Finally, dividing each side by $a^2(c^2 - a^2)$, we get $\dfrac{x^2}{a^2} + \dfrac{y^2}{c^2 - a^2} = 1$.

By setting $b^2 = c^2 - a^2$, we have $\dfrac{x^2}{a^2} - \dfrac{y^2}{b^2} = 1$, the standard form of the hyperbola.

Invariance Properties of the Discriminant

The **discriminant** of $Ax^2 + Bxy + Cy^2 + Dx + Ey + F = 0$ is $d = \mathbf{B^2 - 4AC}$.

The Discriminant $\mathbf{B^2 - 4AC}$ is invariant under translations (shifts):

If a point (x, y) is shifted h units up and k units to the right, then the coordinates of the new point are $(x',$
$y') = (x+h, y+k)$. To show that the discriminant of $Ax^2 + Bxy + Cy^2 + Dx + Ey + F = 0$ is invariant under
translations, we need to show that the discriminant of $Ax^2 + Bxy + Cy^2 + Dx + Ey + F = 0$ and the
discriminant of $A(x')^2 + B(x')(y') + C(y')^2 + Dx' + Ey' + F = 0$ are equal for $x' = x + h$ and $y' = y + k$.

The discriminant of $Ax^2 + Bxy + Cy^2 + Dx + Ey + F = 0$ is equal to $B^2 - 4AC$.
Replacing x' with x+h and y' with y+k,

$$A(x')^2 + B(x')(y') + C(y')^2 + Dx' + Ey' + F$$
$$= A(x+h)^2 + B(x+h)(y+k) + C(y+k)^2 + D(x+h) + E(y+k) + F$$
$$= A(x^2 + 2xh + h^2) + B(xy + xk + yh + hk) + C(y^2 + 2yk + k^2) + D(x+h) + E(y+k) + F$$
$$= Ax^2 + Bxy + Cy^2 + (2Ah + Bk + D)x + (Bh + 2Ck + E)y + (Ah^2 + Bhk + Ck^2 + Dh + Ek + F).$$

The discriminant of this final formula is $B^2 - 4AC$, the same as the discriminant of
$Ax^2 + Bxy + Cy^2 + Dx + Ey + F$. In fact, a translation does not change the values of the coefficients of
x^2, xy, and y^2 (the values of A, B, and C) so the discriminant is unchanged.

The Discriminant $\mathbf{B^2 - 4AC}$ is invariant under rotation by an angle θ:

If a point (x, y) is rotated about the origin by an angle of θ, then the coordinates of the new point are
$(x', y') = (x \cdot \cos(\theta) - y \cdot \sin(\theta), x \cdot \sin(\theta) + y \cdot \cos(\theta))$. To show that the discriminant of
$Ax^2 + Bxy + Cy^2 + Dx + Ey + F = 0$ is invariant under rotations, we need to show that the discriminant of
$Ax^2 + Bxy + Cy^2 + Dx + Ey + F = 0$ and the discriminant of $A(x')^2 + B(x')(y') + C(y')^2 + Dx' + Ey' + F = 0$
are equal when $x' = x \cdot \cos(\theta) - y \cdot \sin(\theta)$ and $y' = x \cdot \sin(\theta) + y \cdot \cos(\theta)$.

The discriminant of $Ax^2 + Bxy + Cy^2 + Dx + Ey + F = 0$ is equal to $B^2 - 4AC$.
Replacing x' with $x \cdot \cos(\theta) - y \cdot \sin(\theta) = x \cdot c - y \cdot s$ and y' with $x \cdot \sin(\theta) + y \cdot \cos(\theta) = x \cdot s + y \cdot c$

$$A(x')^2 + B(x')(y') + C(y')^2 + Dx' + Ey' + F = 0$$
$$= A(xc - ys)^2 + B(xc - ys)(xs + yc) + C(xs + yc)^2 + \ldots (\text{terms without } x^2, xy, \text{ and } y^2)$$
$$= A(x^2c^2 - 2xysc + y^2s^2) + B(x^2sc - xys^2 + xyc^2 - y^2sc) + C(x^2s^2 + 2xysc + y^2c^2) + \ldots$$
$$= (Ac^2 + Bsc + Cs^2)x^2 + (-2Asc - Bs^2 + Bc^2 + 2Csc)xy + (As^2 + Bsc + Cc^2)y^2 + \ldots$$

Then $A' = Ac^2 + Bsc + Cs^2$, $B' = -2Asc - Bs^2 + Bc^2 + 2Csc$, and $C' = As^2 + Bsc + Cc^2$, so the new
discriminant is

$$(B')^2 - 4(A')(C')$$

$$= (-2Asc - Bs^2 + Bc^2 + 2Csc)^2 - 4(Ac^2 + Bsc + Cs^2)(As^2 + Bsc + Cc^2)$$

$$= \{ s^4(B^2) + s^3c(4AB - 4BC) + s^2c^2(4A^2 - 8AC - 2B^2 + 4C^2) + sc^3(-4AB + 4BC) + c^4(B^2) \}$$

$$\quad - 4\{ s^4(AC) + s^3c(AB - BC) + s^2c^2(A^2 - B^2 + C^2) + sc^3(-AB + BC) + c^4(AC) \}$$

$$= s^4(B^2 - 4AC) + s^2c^2(2B^2 - 8AC) + c^4(B^2 - 4AC)$$

$$= (B^2 - 4AC)(s^4 + 2s^2c^2 + c^4) = (B^2 - 4AC)(s^2 + c^2)(s^2 + c^2) = (B^2 - 4AC) \text{, the original discriminant.}$$

The invariance of the discriminant under translation and rotation shows that any conic section can be translated so its "center" is at the origin and rotated so its axis is the x–axis without changing the value of the discriminant: the value of the discriminant depends strictly on the shape of the curve, not on its location or orientation. When the axis of the conic section is the x–axis, the standard quadratic equation

$$Ax^2 + Bxy + Cy^2 + Dx + Ey + F = 0$$

does not have an xy term (B=0) so we only need to investigate the reduced form

$$Ax^2 + Cy^2 + Dx + Ey + F = 0.$$

1)　A = C = 0 (discriminant d=0). A straight line. (special case: no graph)

2)　A = C ≠ 0　(d<0).　　A circle. (special cases: a point or no graph)

3)　A = 0, C ≠ 0　or　A ≠ 0, C = 0　(d=0):　A parabola. (special cases: 2 lines, 1 line, or no graph)

4)　A and C both positive or both negative　(d<0):　An Ellipse. (special cases: a point or no graph)

5)　A and C have opposite signs (d>0):　　A Hyperbola.　　(special case: a pair of intersecting lines)

9.6　PROPERTIES OF THE CONIC SECTIONS

This section presents some of the interesting and important properties of the conic sections that can be proven using calculus. It begins with their reflection properties and considers a few ways these properties are used today. Then we derive the polar coordinate form of the conic sections and use that form to examine one of the reasons conic sections are still extensively used: the paths of planets, satellites, comets, baseballs, and even subatomic particles are often conic sections. The section ends with a specialized examination of elliptical orbits. To understand and describe the motions of the universe, at telescopic and microscopic levels, we need conic sections!

Reflections on the Conic Sections

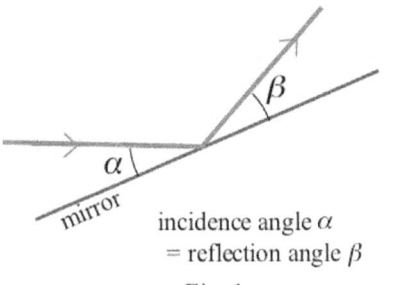

incidence angle α
= reflection angle β

Fig. 1

This discussion of reflection assumes that the angle of incidence of a light ray or billiard ball is equal to the angle of reflection of the ray or ball. The assumption is valid for light rays and mirrors (Fig. 1) but is not completely valid for balls: the spin of the ball before it hits the wall may make the reflection angle smaller than, greater than, or equal to the incidence angle.

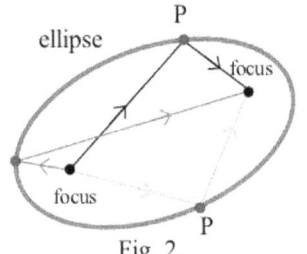

Fig. 2

Reflection Property of an Ellipse

An elliptical mirror reflects light from one focus to the other focus (Fig. 2) and all of the light rays take the same amount of time to be reflected to the other focus.

Outline of a proof: We can assume that the ellipse is oriented so its equation is

$$\frac{x^2}{a^2} + \frac{y^2}{b^2} = 1 \quad \text{(Fig. 3) and } a > b > 0.$$

Then the foci are at the points $F_1 = (-c, 0)$ and $F_2 = (c, 0)$ with $c = \sqrt{a^2 - b^2}$.

To show that the light rays from one focus are always reflected to the other focus, we need to show that angle α, the angle between the ray from F_1 and the tangent line to the ellipse, is equal to angle β, the angle between the tangent to the ellipse and the ray to F_2 .

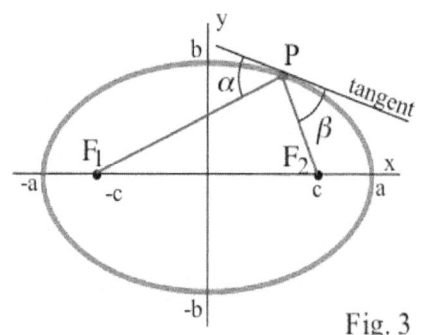

m_1 = slope of F_1P
m_2 = slope of F_2P
m_3 = slope of tangent line at P

Fig. 3

The most direct way to show that $\alpha = \beta$ is to start by calculating the slopes m_1 = slope of the line from F_1 to P, m_2 = the slope of the line from P to F_2, and m_3 = the slope of the tangent line:

$$m_1 = \frac{y}{x+c} \quad , \quad m_2 = \frac{y}{x-c} \quad , \text{and, by implicit differentiation, } m_3 = \frac{-x}{y} \frac{b^2}{a^2} \quad .$$

Then, from section 0.2, we know that $\tan(\alpha) = \frac{m_3 - m_1}{1 + m_1 m_3}$ and $\tan(\beta) = \frac{m_2 - m_3}{1 + m_2 m_3}$ so we just need to evaluate $\tan(\alpha)$ and $\tan(\beta)$ and show that they are equal. Since the process is algebraically tedious and not very enlightening, it has been relegated to an Appendix after the problem set.

Since straight paths from one focus to the ellipse and back to the other focus all have the same length (the definition of an ellipse), all of the light rays from the one focus take the same amount of time to reach the other focus. If a small stone is dropped into an elliptical pool at one focus (Fig. 4), then the waves radiate in all directions, reflect off the sides of the pool to the other focus and create a splash there because they all arrive at the same time. Similarly, if a room is in the shape of an (half) ellipsoid of revolution (Fig. 5), then the sound waves from a whisper at one focus will bounce off the walls and all arrive at the same time at the other focus where an eavesdropper can hear the conversation.

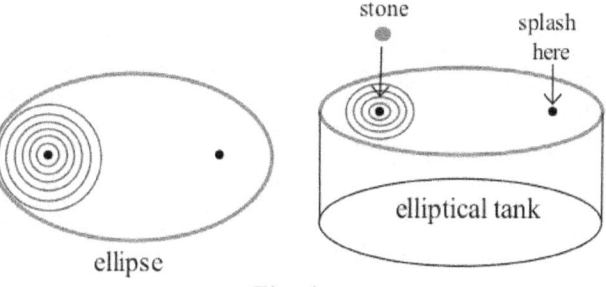

Fig. 4

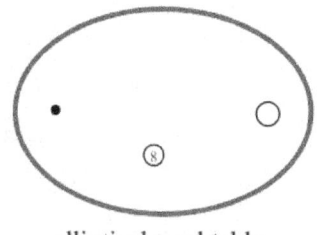

elliptical pool table

Fig. 6

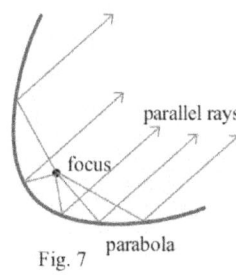

elliptical dome

whisper here   hear it here

Fig. 5

Practice 1:   What simple directions ensure that a ball shot from anywhere on an elliptical pool table (Fig. 6) will bounce off one wall and go into the single hole located at a focus of the ellipse?

Reflection Property of a Parabola

A parabolic mirror reflects light from the focus in a line parallel to the axis of the parabola (Fig. 7). In reverse, incoming light rays parallel to the axis are reflected to the focus.

Fig. 7   parabola

Outline of a proof: If the parabolic mirror is given by $x = ay^2$ (Fig. 8), then its focus is at $F = (\frac{1}{4a}, 0)$ and the parabola is symmetric with respect to the x–axis. To prove the reflection property of the parabola, we need to show that $\alpha = \beta$ or,

since α and β are both acute angles, that $\tan(\alpha) = \tan(\beta)$.

From Fig. 8 we calculate that

$$m_1 = 0, \quad m_2 = \frac{y}{x - \frac{1}{4a}} = \frac{4ay}{4ax - 1} = \frac{4ay}{4a^2y^2 - 1} \quad , \text{and, by implicit}$$

differentiation, $m_3 = \frac{1}{2ay}$.

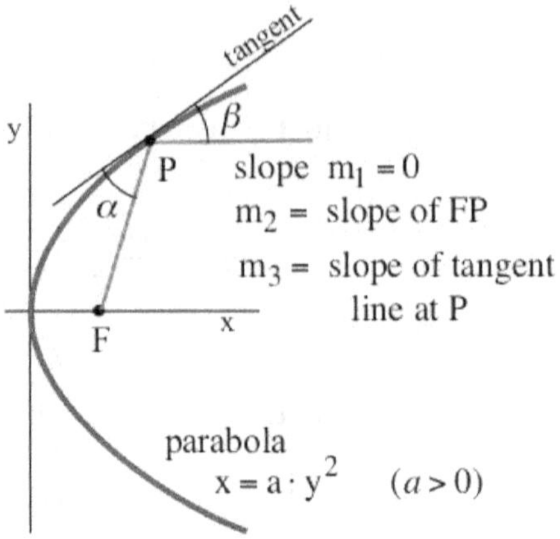

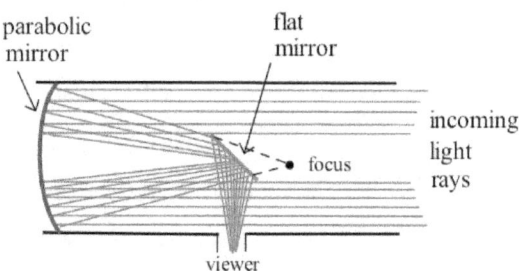

P   slope $m_1 = 0$
$m_2 = $ slope of FP
$m_3 = $ slope of tangent
          line at P

parabola
    $x = a \cdot y^2$ $(a > 0)$

Fig. 8

Since $m_1 = 0$, we know that $\tan(\alpha) = \frac{m_3 - m_1}{1 + m_1 m_3} = m_3 = \frac{1}{2ay}$.

An elementary but tedious algebraic argument shows that $\tan(\beta) = \frac{m_2 - m_3}{1 + m_2 m_3}$ simplifies to the same value

as $\tan(\alpha)$ so $\tan(\alpha) = \tan(\beta)$ and $\alpha = \beta$.

Because of this reflection property, the parabola is used in a

variety of instruments and devices. Mirrors in reflecting

telescopes are parabolic (Fig. 9) so that the dim incoming

(parallel) light rays from distant stars are all reflected to an

eyepiece at the mirror's focus for viewing. Similarly, radio

telescopes use a parabolic surface to collect weak signals. A

well known scientific supply company sells an 18 inch

diameter parabolic reflector "ideal for a broad range of

applications including solar furnaces, solar energy collectors,

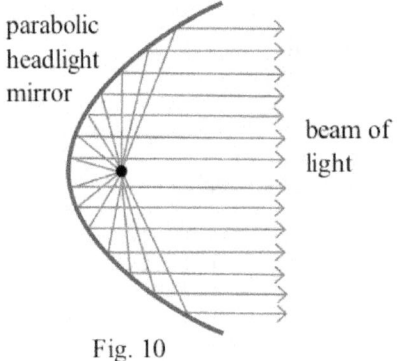

Fig. 9: Reflecting telescope

and parabolic and directional microphones." For outgoing light, flashlights and automobile headlights use

(almost) parabolic mirrors so a light source set at the focus of the mirror creates a tight beam of light (Fig. 10).

parabolic
headlight
mirror

beam of
light

Fig. 10

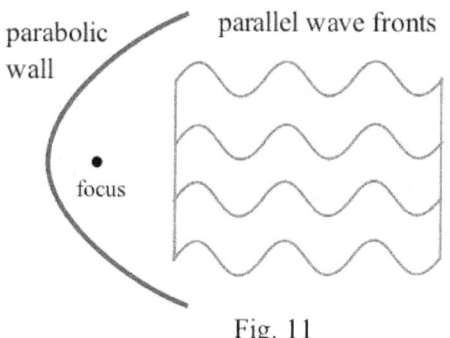

parabolic wall

parallel wave fronts

• focus

Fig. 11

Not only are incoming parallel rays reflected to the same point, but they reach that point at the same time. More precisely, if two objects start at the same distance from the y–axis and travel parallel to the x–axis, they both travel the same distance to reach the focus. If they are traveling at the same speed, they reach the focus together. An incoming linear wave front (Fig. 11) is reflected by a parabolic wall to create a splash at the focus. A small stone dropped into a wave tank at the focus of a parabola creates a linear outgoing wave. This "same distance" property of the reflection is something you can prove.

Practice 2:   An object starts at the point (p,q), travels to the left until it encounters the parabola $x = ay^2$ (Fig. 12) and then goes straight to the focus at $(\frac{1}{4a}, 0)$. Show that the total distance traveled, L_1 plus L_2 , equals $p + \frac{1}{4a}$ so the total distance is the same for all values of q. (Assume that $p > aq^2$ so the starting point is to the left of the parabola.)

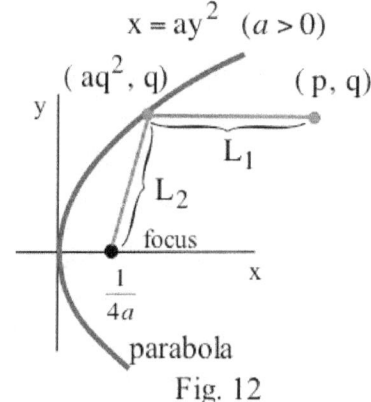

$x = ay^2$ $(a > 0)$

(aq^2, q) (p, q)

y L_1

L_2

focus

$\frac{1}{4a}$ x

parabola

Fig. 12

The hyperbola also has a reflection property, but it is less useful than those for ellipses and parabolas.

Reflection Property of an Hyperbola

An hyperbolic mirror reflects light aimed at one focus to the other focus (Fig. 13).

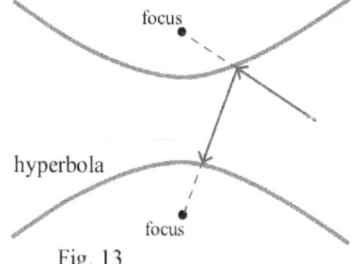

focus •

hyperbola

focus •

Fig. 13

Polar Coordinate Forms for the Conic Sections

In the rectangular coordinate system, the graph of the general quadratic equation $Ax^2 + Bxy + Cy^2 + Dx + Ey + F = 0$ is always a conic section, and the value of the discriminant $B^2 - 4AC$ tells us which type. In the polar coordinate system, an even simpler function describes all of the conic section shapes, and a single parameter in that function tells us the shape of the graph.

For $e \geq 0$, the polar coordinate graphs of $r = \dfrac{k}{1 \pm e \cdot \cos(\theta)}$ and $r = \dfrac{k}{1 \pm e \cdot \sin(\theta)}$

are conic sections with one focus at the origin.

If $e < 1$, the graph is an ellipse. (If $e = 0$, the graph is a circle.)

If $e = 1$, the graph is a parabola.

If $e > 1$, the graph is an hyperbola.

The number e is called the **eccentricity** of the conic section.

Fig. 14 shows graphs of $r = \dfrac{1}{1 + e \cdot \cos(\theta)}$ for several values of e.

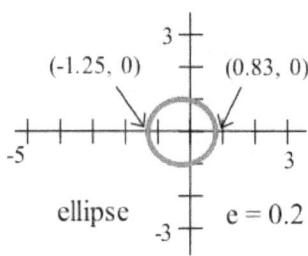

ellipse $e = 0.2$

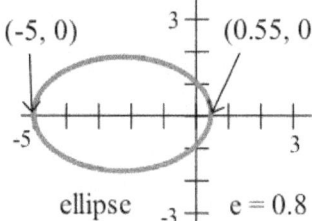

ellipse $e = 0.8$

$$r = \frac{1}{1 + e \cdot \cos(\theta)}$$

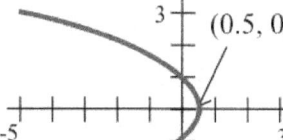

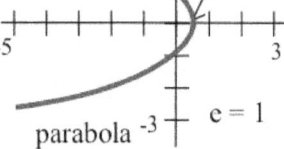

parabola $e = 1$

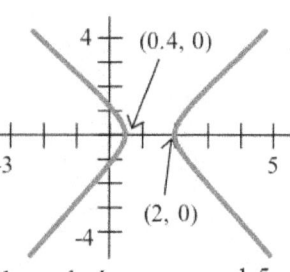

hyperbola $e = 1.5$

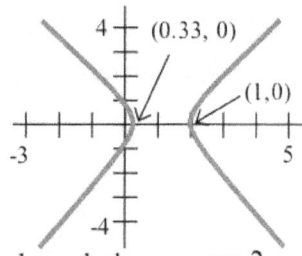

hyperbola $e = 2$

Fig. 14

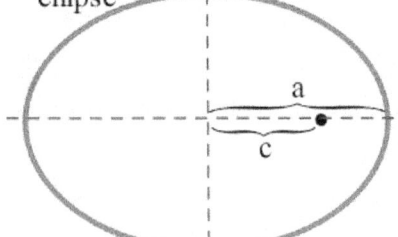

eccentricity $= \dfrac{c}{a} < 1$

Fig. 15

For an ellipse, the eccentricity $= \dfrac{\text{dist(center, focus)}}{\text{dist(center, vertex)}}$ (Fig. 15) .

If the eccentricity of an ellipse is close to zero, then the ellipse is "almost" a circle. If the eccentricity of an ellipse is close to 1, the ellipse is rather "narrow."

For a hyperbola, the eccentricity = $\dfrac{\text{dist(center, focus)}}{\text{dist(center, vertex)}}$ (Fig. 16) .

If the eccentricity of a hyperbola is close to 1, then the hyperbola is "narrow." If the eccentricity of a hyperbola is very large, the hyperbola "opens wide."

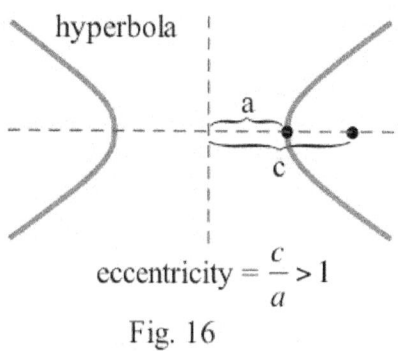

eccentricity $= \dfrac{c}{a} > 1$

Fig. 16

Proof: The proof uses a strategy common in mathematics: move the problem into a system we know more about. In this case we move the problem from the polar coordinate system to the rectangular coordinate system, put the resulting equation into the form of a general quadratic equation, and then use the discriminant to determine the shape of the graph.

If $r = \dfrac{k}{1 + e{\cdot}\cos(\theta)}$, then $r + e{\cdot}r{\cdot}\cos(\theta) = k$. Replacing r with $\sqrt{x^2 + y^2}$ and $r{\cdot}\cos(\theta)$ with x, we get $\sqrt{x^2 + y^2} + e{\cdot}x = k$ and $\sqrt{x^2 + y^2} = k - e{\cdot}x$.

Squaring each side and collecting all of the terms on the right gives the equivalent general quadratic equation

$$(1 - e^2)x^2 + y^2 + 2kex - k = 0 \text{ so } A = 1 - e^2, B = 0, \text{ and } C = 1.$$

The discriminant of this general quadratic equation is $B^2 - 4AC = 0 - 4(1 - e^2)(1) = 4(e^2 - 1)$ so

if $e < 1$, then $B^2 - 4AC < 0$ and the graph is an ellipse,

if $e = 1$, then $B^2 - 4AC = 0$ and the graph is an parabola, and

if $e > 1$, then $B^2 - 4AC > 0$ and the graph is an hyperbola.

The graph of $r = \dfrac{k}{1 + e{\cdot}\cos(\theta)}$ is a conic section, and the value of the eccentricity tells which shape the graph has. We will not prove that one focus of the conic section is at the origin, but it's true.

Practice 3:   Graph $r = \dfrac{k}{1 + (0.8){\cdot}\cos(\theta)}$ for $k = 0.5, 1, 2,$ and 3. What effect does the value of k

have on the graph?

Subtracting a constant α from θ rotates a polar coordinate graph counterclockwise about the origin by an angle of α, but does not change the shape of the graph, so the graphs of

$$r = \frac{k}{1 + e \cdot \cos(\theta - \alpha)}$$
are all conic sections whose shapes depend on the size of the parameter e.

In particular, the polar graphs of $r = \dfrac{1}{1 - e \cdot \cos(\theta)}$, $r = \dfrac{1}{1 + e \cdot \sin(\theta)}$, and $r = \dfrac{1}{1 - e \cdot \sin(\theta)}$

are conic sections, rotations about the origin of the graphs of $r = \dfrac{1}{1 + e \cdot \cos(\theta)}$, since

$-\cos(\theta) = \cos(\theta - \pi)$, $\sin(\theta) = \cos(\theta - \pi/2)$, and $-\sin(\theta) = \cos(\theta - -\pi/2)$. Fig. 17 shows several of these graphs for $e = 0.8$.

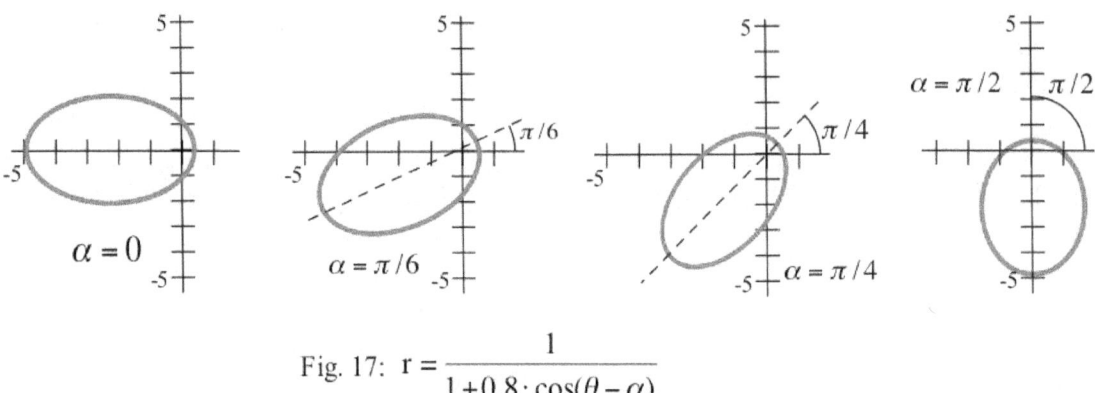

Fig. 17: $r = \dfrac{1}{1 + 0.8 \cdot \cos(\theta - \alpha)}$

The Path of Every Object in the Universe is a Conic Section (not really)

Rather than engage in endless philosophical discussion about how the planets ought to move, Tycho Brahe (1546–1601) had a better idea about how to find out how they actually do move — collect data! Even before the invention of the telescope, he built an observatory, and with the aid of devices like protractors he carefully cataloged the positions of the planets for 20 years. Just before his death, he passed this accumulated data to Johannes Kepler to edit and publish. From these remarkable data, Kepler deduced his three laws of planetary motion, the first of which says **each planet moves in an elliptical orbit with the sun at one focus** (Fig. 18).
From Kepler's laws, Newton was able to deduce that a force, gravity, held the planets in orbits and that the force varied inversely as the square of the distance between the planet and the sun. From this "inverse square" fact it can be shown that the position of one object (e.g., a planet) with respect to another (e.g., the sun) is given by the polar coordinate formula

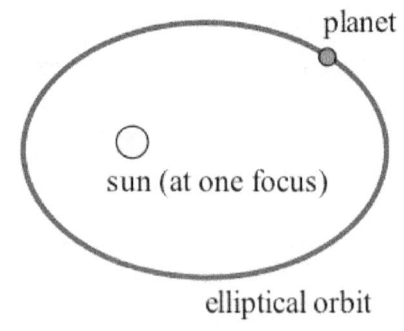

planet

sun (at one focus)

elliptical orbit

Fig. 18

$$r = \frac{h}{1 + (h-1)\cdot\cos(\theta)} \qquad \text{where} \qquad h = \frac{r_0 \cdot v_0{}^2}{GM}$$

r_0 = initial distance (m), v_0 = initial velocity (m/s)

G = universal gravition constant ($6.7 \times 10^{-11}\ \frac{N\ m^2}{kg^2}$)

M = mass of the sun or "other object" (kg) .

You should recognize the pattern of this formula as the pattern for the conic sections with eccentricity

$$e = |h - 1| = \left| \frac{r_0 \cdot v_0{}^2}{GM} - 1 \right|.$$

In a "two body" universe, all motion paths are conic sections, and the shape of the conic section is determined by the value of $r_0 \cdot v_0{}^2$. If the object is far away or moving very rapidly or both, then $r_0 \cdot v_0{}^2 > 2GM$, the eccentricity is greater than one, and the path is a hyperbola. If the objects are relatively close and/or moving slowly, then $r_0 \cdot v_0{}^2 < 2GM$ (the situation with each planet and the sun), the eccentricity is less than one and the path is an ellipse. If $r_0 \cdot v_0{}^2 = 2GM$, then the eccentricity is 1 and the path is a parabola. If $r_0 \cdot v_0{}^2 = GM$, then the eccentricity is 0 and the path is a circle. It is rare to encounter values of r_0 and v_0 so $r_0 \cdot v_0{}^2$ exactly equals GM or 2GM.

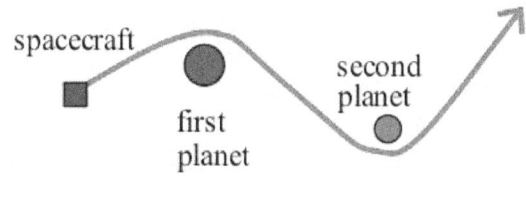

Fig. 19

The universe obviously contains more than two bodies and the paths of most objects are not conic sections, but there are still important situations in which the force on an object is due almost entirely to the gravitational attraction between it and one other body. For example, scientists and engineers use the position formula to determine the orbital position and velocity needed to put a satellite into an orbit with the desired eccentricity, and the position formula was used to help calculate how close Voyager 2 should come to Jupiter and Saturn so the gravity of those planets could be used to change the path of Voyager 2 to a hyperbola and send it on to other planets (Fig. 19). The conic sections even appear at less grand scales.

In a vacuum, the path of a thrown baseball (or bat) is a parabola, unless it is thrown hard enough to achieve an elliptical orbit. And at the subatomic scale, Rutherford (1871 – 1937) discovered that alpha particles shot toward the nucleus of an atom are repelled away from the nucleus along hyperbolic paths (Fig. 20). Conic sections are everywhere.

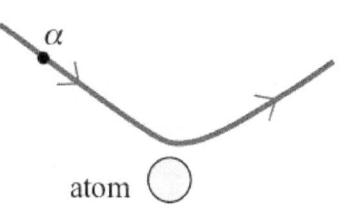

Fig. 20

Elliptical Orbits

When a planet orbits a sun, the orbit is an ellipse, and we can use information about ellipses to calculate information about these orbits.

The position of a planet in elliptical orbit around a sun is given by the polar equation

$$r = \frac{k}{1 + e \cdot \cos(\theta)}$$

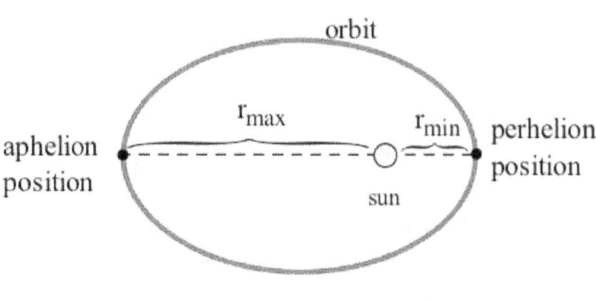

Fig. 21

for some value of the eccentricity $e < 1$ (Fig. 21). The planet is closest (at perhelion) to the sun when

$\theta = 0$ and this minimum distance is $r_{min} = \frac{k}{1+e}$. The greatest distance (at aphelion) occurs when

$\theta = \pi$ and that distance is $r_{max} = \frac{k}{1-e}$. If the width of the ellipse (technically, the length of the major axis) is 2a, then

$$2a = r_{min} + r_{max} = \frac{k}{1+e} + \frac{k}{1-e} = \frac{2k}{1-e^2}$$

so $k = a(1 - e^2)$ and the position of the planet is given by

$$r = \frac{a(1-e^2)}{1 + e \cdot \cos(\theta)} \quad \text{with } r_{min} = \frac{a(1-e^2)}{1+e} = a(1-e) \text{ and } r_{max} = \frac{a(1-e^2)}{1-e} = a(1+e).$$

Example 4:   We want to put a satellite in an elliptical orbit around the earth (radius ≈ 6360 km) so the maximum height of the satellite is 20,000 km and the minimum height is 10,000 km (Fig. 22). Find the eccentricity of the orbit and give a polar formula for its position.

Solution: r_{max} = maximum height plus the radius of the earth = 26,360 km and r_{min} = minimum

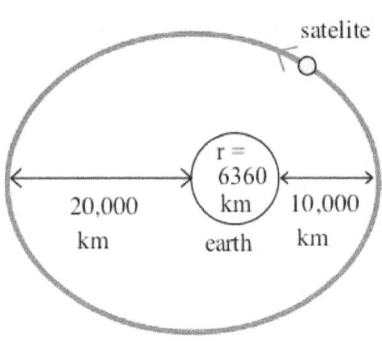

Fig. 22

height plus the radius of the earth = 16,360 km so $a(1 + e) = 26,360$ and $a(1 - e) = 16,360$. Dividing these last two quantities, we have

$$\frac{r_{max}}{r_{min}} = \frac{26360}{16360} = \frac{a(1+e)}{a(1-e)} = \frac{1+e}{1-e} \text{ and } e = \frac{10000}{42720} \approx \textbf{0.234}$$

Using $r_{min} = a(1 - e) = a(1 - 0.234) = 16360$ we have $a \approx 21358$.

Finally, $r = \dfrac{a(1-e^2)}{1 + e \cdot \cos(\theta)} = \dfrac{\textbf{20189}}{\textbf{1 + (0.234)} \cdot \textbf{cos}(\theta)}$.

Practice 4:   Pluto's orbit has an eccentricity of 0.2481 and its semimajor axis is 5,909 million

kilometers. Find the minimum and maximum distance of Pluto from the sun during one orbit. The

orbit of Neptune has an eccentricity of 0.0082 with a semimajor axis of 4,500 million kilometers. Is

Neptune ever farther from the sun than Pluto?

PROBLEMS

Reflection properties

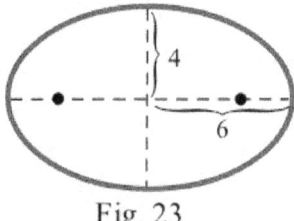

Fig. 23

1.   For the ellipse in Fig. 23, how far does a ball travel as it moves from

one focus to any point on the ellipse and on to the other focus?

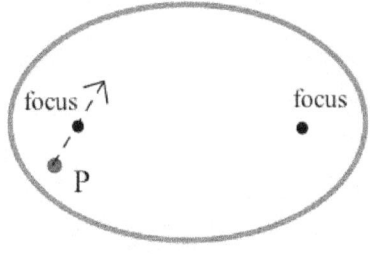

Fig. 24

2.   In Fig. 24 a ball rolls from point P over the the focus at A and keeps

rolling. Sketch the path of the ball for the first 5 bounces it makes off of the

ellipse. What does the path of the ball look like after "a long time?"

3.   In Fig. 25 a ball rolls from point P toward the the focus at A

and bounces off of the hyperbola. Sketch the path of the ball for

the first 5 bounces it makes off of the hyperbola. What does the

path of the ball look like after "a long time?"

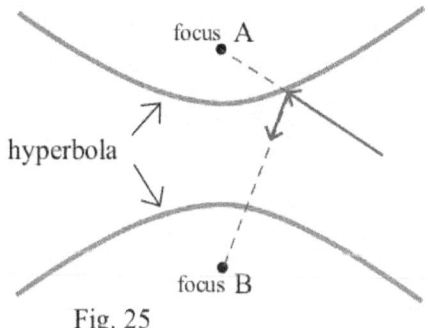

Fig. 25

4.   An explosion is set off at one focus inside a very strong

ellipsoidal shell. What might happen to a piece of graphite located at the other focus?

5.   A straight wave front is approaching a parabolic jetty. Why wouldn't you want your boat to be at the

focus of the parabola?

6.   The members of a marching band are grouped near point A (Fig. 26),

the focus of a parabola. At a signal from the director, the band

members each march (at the same speed) in different directions

toward the parabola, immediately turn and then march due west.

What shape will the formation have after all of the marchers have

made their turns?

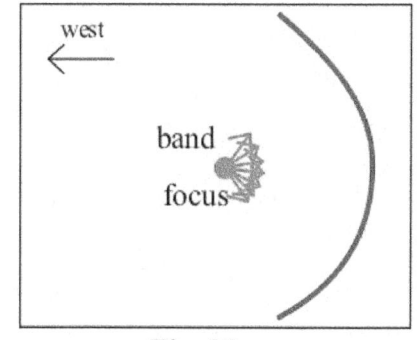

Fig. 26

7. A reflecting telescope is built with a parabolic mirror and a
    hyperbolic mirror (Fig. 27) so F_1 is the focus of the parabola
    and F_1 and F_2 are the foci of the hyperbola. Trace the paths
    of the incoming parallel light rays a, b, and c as they reflect off
    both mirrors. Why is the eyepiece located at F_2 ?

8. Make a slight design change in the telescope in Fig. 27 so
    the eyepiece can be located at the point off to the side of the
    parabolic mirror.

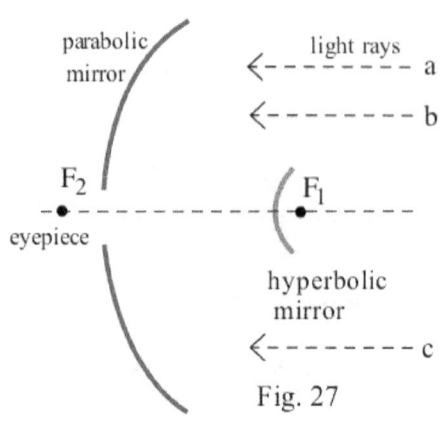

Fig. 27

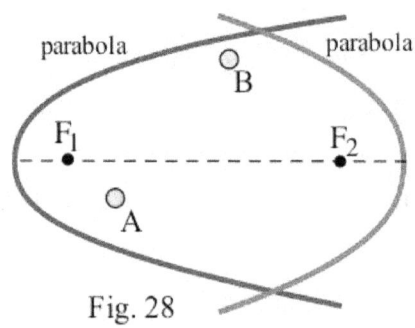

Fig. 28

9. The billiards table in Fig. 28 consists of parts of two
    parabolas: the right parabola has focus F_1 and the left
    parabola has focus F_2 . (a) Determine a strategy for
    shooting the balls located at A and B to make a two–cushion
    (i.e., two bounce) shot into the hole located at F_2 . (b) Are
    there any places on the table where your strategy in part (a)
    does not work? (Unfortunately for my ability to win at

billiards, the angle of incidence does not necessarily equal the angle of
reflection. But assume they are equal for these problems.)

10. The billiards table in Fig. 29 consists of parts of two ellipses: the short
    ellipse has foci F_1 and F_2 and the tall ellipse has foci F_2 and F_3 .

    (a) Determine a strategy for shooting the balls located at A and B to
    make a two–cushion (i.e., two bounce) shot into the hole located at F_3 .

    (b) Are there any places on the table where your strategy in part (a)
    does not work?

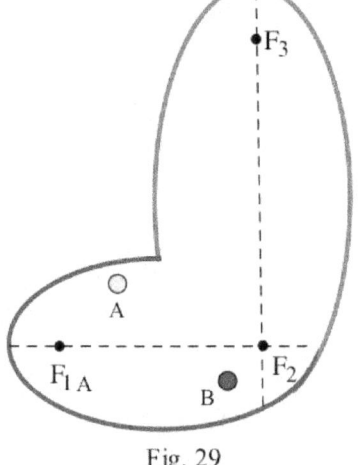

Fig. 29

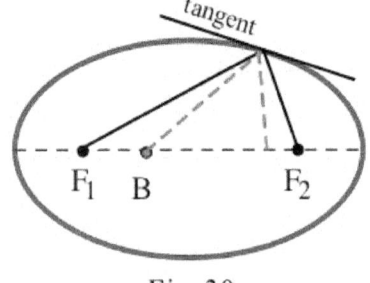

Fig. 30

11. Use Fig. 30 to help explain geometrically
    why a ball located on the major axis of an
    ellipse between the two foci is always
    reflected back to a point on the major axis
    between the two foci.

12. Is a rectangular reflection path (Fig. 31) possible for the ellipse

    (a) $\dfrac{x^2}{25} + \dfrac{y^2}{16} = 1$? (b) $\dfrac{x^2}{a^2} + \dfrac{y^2}{b^2} = 1$?

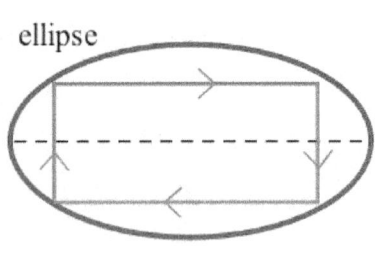

Fig. 31

Polar forms

In problems 13–18, determine the eccentricity of each conic section, identify the shape, and determine where it crosses the x and y axes.

13. $r = \dfrac{11}{3 + 5 \cdot \cos(\theta)}$

14. $r = \dfrac{8}{7 + 3 \cdot \sin(\theta + \pi/6)}$

15. $r = \dfrac{1}{2 + 2 \cdot \sin(\theta - \pi/3)}$

16. $r = \dfrac{-4}{3 - 3 \cdot \cos(\theta)}$

17. $r = \dfrac{17}{7 - 5 \cdot \cos(\theta + 3\pi)}$

18. $r = \dfrac{3}{4 - 2 \cdot \sin(\theta + \pi/11)}$

In problems 19–24, sketch each ellipse and determine the coordinates of the foci. (Reminder: In the standard polar coordinate form used here, one focus is always at the origin.)

19. $r = \dfrac{6}{2 + \cos(\theta)}$

20. $r = \dfrac{6}{2 + \sin(\theta)}$

21. $r = \dfrac{12}{3 - \sin(\theta)}$

22. $r = \dfrac{12}{3 - \cos(\theta)}$

23. $r = \dfrac{3}{2 + \sin(\theta - \pi/4)}$

24. $r = \dfrac{3}{2 + \cos(\theta + \pi/4)}$

In problems 25 and 26, represent the length and area of each ellipse as definite integrals and use Simpson's rule with n = 100 to approximate the values of the integrals.

25. $r = \dfrac{1}{1 + 0.5 \cdot \cos(\theta)}$

26. $r = \dfrac{1}{1 + 0.9 \cdot \cos(\theta)}$

Conic section paths:

Problems 27–30 refer to the two objects in Fig. 32. Determine the shape of the path of object B. (Object A has mass 10^{19} kg, r = 10^5 m)

27. v = 17.6 m/s   28. v = 115 m/s

29. v = 120 m/s

30. Determine a velocity for B so that the path is a parabola.

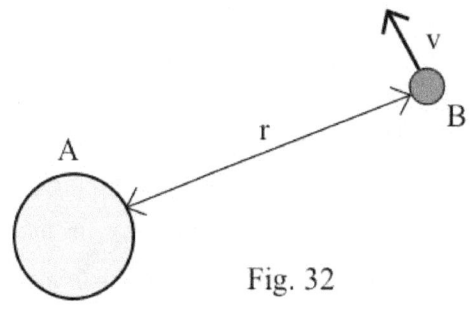

Fig. 32

31. An object at a distance r from the center of a planet of mass M (Fig. 33) has velocity v. Determine conditions on v as a function of r, M, and G so the resulting path is (a) circular,

(b) elliptical, (c) parabolic, and

(d) hyperbolic.

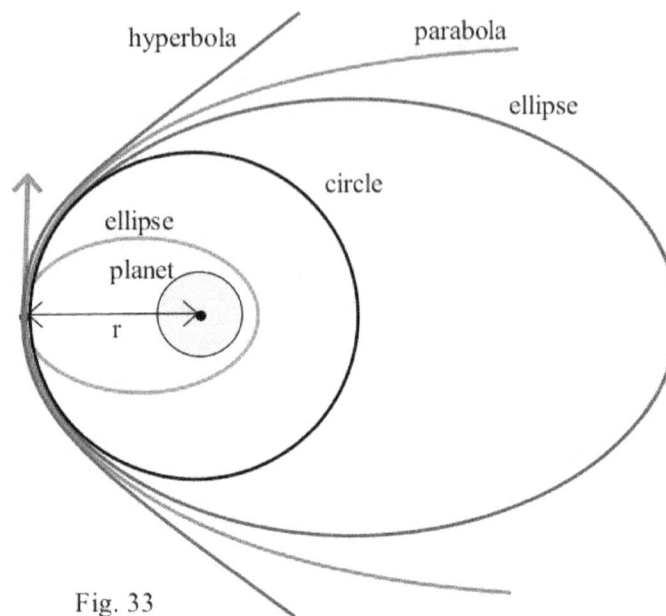

Fig. 33

Problems 32–34 refer to Earth and moon orbits: use

$$G = 6.7\text{x}10^{-11} \text{ N·m}^2/\text{kg}^2,$$

$$M_{\text{Earth}} = 5.98\text{x}10^{24} \text{ kg},$$

$$r_{\text{Earth}} = 6.360\text{x}10^{6} \text{ m},$$

$$M_{\text{moon}} = 7\text{x}10^{22} \text{ kg} \text{ , and}$$

$$r_{\text{moon}} = 1.738\text{x}10^{6} .$$

32. At the Earth's surface, determine a velocity so that the resulting path is circular. This is the minimum velocity needed to achieve "orbit," a very low orbit.

33. At the Earth's surface, determine a velocity so that the resulting path is parabolic. This is the minimum velocity needed to escape from orbit, and is called the "escape velocity".

34. At the moon's surface, determine the minimum orbital velocity and the (minimum) escape velocity.

Elliptical orbits

35. We want to put a satellite into orbit around Earth so the maximum altitude of the satellite is 1000 km and the minimum altitude is 800 km. Find the eccentricity of this orbit and give a polar coordinate formula for its position.

36. The Earth follows an elliptical orbit around the sun, and this ellipse has a semimajor axis of $149.6\text{x}10^{6}$ km and an eccentricity of 0.017. (a) Determine the maximum and minimum distances of the Earth from the sun. (b) How far apart are the two foci of this ellipse?

37. Determine the altitude needed for an Earth satellite to make one orbit on a circular path every 24 hours ("geosyncronous"). (Since the orbit is circular and the satellite makes one orbit every 24 hours, you can determine the velocity v (in m/s) as a function of the distance r from the center of the Earth. Since the orbit is circular, e = 0 and $r·v^2 = GM$.)

Section 9.6   **PRACTICE Answers**

Practice 1:   "Shoot the ball toward the left focus." Since the table is an ellipse, the ball will roll over the

focus, hit the wall, and be reflected into the hole at the other focus.

Practice 2: L_1 = distance from (p, q) to $(a \cdot q^2, q)$ = $p - a \cdot q^2$.

L_2 = distance from $(a \cdot q^2, q)$ to $(\frac{1}{4a}, 0)$

$$= \sqrt{ (a \cdot q^2 - \frac{1}{4a})^2 + (q - 0)^2 } \quad = \sqrt{ a^2 q^4 - 2aq^2(\frac{1}{4a}) + \frac{1}{16a^2} + q^2 }$$

$$= \sqrt{ a^2 q^4 - \frac{1}{2}q^2 + \frac{1}{16a^2} + q^2 }$$

$$= \sqrt{ a^2 q^4 + \frac{1}{2}q^2 + \frac{1}{16a^2} } \quad = \sqrt{ (aq^2 + \frac{1}{4a})^2 } \quad = aq^2 + \frac{1}{4a} \; .$$

Therefore, $L_1 + L_2 = (p - a \cdot q^2) + (aq^2 + \frac{1}{4a}) = p + \frac{1}{4a} \; .$

Practice 3:   The graphs of $r = \dfrac{k}{1 + (0.8) \cdot \cos(\theta)}$

for $k = 0.5, 1,$ and 2 are shown in Fig. 34.

Each graph is an ellipse with eccentricity 0.8 and one

focus at the origin. The value of k determines the

size of the ellipse. The larger the magnitude of k, the

larger the ellipse.

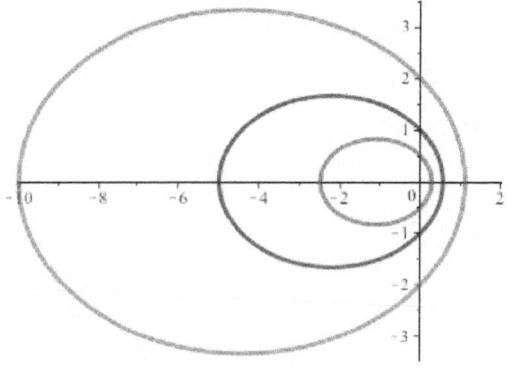

Fig. 34

Practice 4:   For Pluto: e = 0.2481 and semimajor axis length = 5,909 km (a = 5,909 km).

r_{min} = a(1 − e) = 5909(0.7519) ≈ 4443 km.

r_{max} = a(1 + e) = 5909(1.2481) ≈ 7375 km.

For Neptune: e = 0.0082 and a = 4,500 km.

r_{min} = a(1 − e) = 4500(0.9918) ≈ 4463 km, a distance closer than Pluto at its closest!

r_{max} = a(1 + e) = 4500(1.0082) ≈ 4537 km.

In fact, Neptune is the farthest planet from the sun between January 1979 and March 1999. Now

that Pluto has been reclassified as a "dwarf plane," Neptune is always the farthest planet from the

sun. Poor Pluto.

Appendix: Reflection Property of the Ellipse

Let $P = (x,y)$ be on the ellipse $\dfrac{x^2}{a^2} + \dfrac{y^2}{b^2} = 1$ with foci at $(-c, 0)$ and $(c, 0)$ for $c = \sqrt{a^2 - b^2}$.

Then the slopes (Fig. 40) are $m_1 =$ slope from P to $(-c, 0) = \dfrac{y}{x+c}$, $m_2 =$ slope from P to $(c, 0) = \dfrac{y}{x-c}$,

and $m_3 =$ slope of tangent line to ellipse at $(x,y) = \dfrac{-x}{y}\dfrac{b^2}{a^2}$ (by implicit differentiation) .

We know that $\tan(\alpha) = \dfrac{m_3 - m_1}{1 + m_1 m_3}$ and $\tan(\beta) = \dfrac{m_2 - m_3}{1 + m_2 m_3}$, and we want to show that $\alpha = \beta$ or,

equivalently, that $\tan(\alpha) = \tan(\beta)$.

$$\tan(\alpha) = \frac{m_3 - m_1}{1 + m_1 m_3} \quad = \quad \frac{\dfrac{-x}{y}\dfrac{b^2}{a^2} - \dfrac{y}{x+c}}{1 + \dfrac{y}{x+c}\dfrac{-x}{y}\dfrac{b^2}{a^2}} \qquad \text{multiply top \& bottom by } ya^2(x+c)$$

$$= \quad \frac{-xb^2(x+c) - y^2 a^2}{ya^2(x+c) - xyb^2} \qquad = \quad \frac{-x^2 b^2 - xb^2 c - y^2 a^2}{xya^2 + ya^2 c - xyb^2}$$

$$= \quad \frac{-xb^2 c - (x^2 b^2 + y^2 a^2)}{xy(a^2 - b^2) + ya^2 c} \qquad x^2 b^2 + y^2 a^2 = a^2 b^2 \ \text{ and } \ a^2 - b^2 = c^2$$

$$= \quad \frac{-xb^2 c - a^2 b^2}{xyc^2 + ya^2 c} \quad = \quad \frac{-b^2}{yc}\frac{xc + a^2}{xc + a^2} \quad = \quad \frac{-b^2}{yc} \ .$$

Similarly, $\tan(\beta) = \dfrac{m_2 - m_3}{1 + m_2 m_3} \quad = \quad \dfrac{\dfrac{y}{x-c} - \dfrac{-x}{y}\dfrac{b^2}{a^2}}{1 + \dfrac{y}{x-c}\dfrac{-x}{y}\dfrac{b^2}{a^2}}$ multiply top \& bottom by $ya^2(x-c)$

$$= \quad \frac{y^2 a^2 + xb^2(x-c)}{ya^2(x-c) - xyb^2} \qquad = \quad \frac{x^2 b^2 - xb^2 c + y^2 a^2}{xya^2 - ya^2 c - xyb^2}$$

$$= \quad \frac{-xb^2 c + (x^2 b^2 + y^2 a^2)}{xy(a^2 - b^2) - ya^2 c} \qquad\qquad (\ x^2 b^2 + y^2 a^2 = a^2 b^2 \ \text{ and } \ a^2 - b^2 = c^2 \ )$$

$$= \quad \frac{-xb^2 c + a^2 b^2}{xyc^2 - ya^2 c} \quad = \quad \frac{-b^2}{yc}\frac{xc - a^2}{xc - a^2} \quad = \quad \frac{-b^2}{yc} \ = \ \tan(\alpha). \ \ (\text{Yes!})$$

Reflection Property of the Parabola $x = ay^2$ **with focus** $(\frac{1}{4a}, 0)$

The slopes of the line segments in Fig. 41 are m_1 = slope of "incoming" ray = 0,

m_2 = slope from P to focus = $\dfrac{y}{x - \frac{1}{4a}}$ = $\dfrac{4ay}{4ax - 1}$ = $\dfrac{4ay}{4a^2y^2 - 1}$, and, by implicit differentiation,

m_3 = slope of the tangent line at (x,y) = $\dfrac{1}{2ay}$.

We know that $\quad \tan(\alpha) = \dfrac{m_3 - m_1}{1 + m_1 m_3}\quad$ and $\tan(\beta) = \dfrac{m_2 - m_3}{1 + m_2 m_3}$, and we want to show that $\alpha = \beta$

or, equivalently, that $\tan(\alpha) = \tan(\beta)$.

$\tan(\alpha) = \dfrac{m_3 - m_1}{1 + m_1 m_3}\qquad = m_3 = \dfrac{1}{2ay}$

$\tan(\beta) = \dfrac{m_2 - m_3}{1 + m_2 m_3}\qquad = \dfrac{\dfrac{4ay}{4a^2y^2 - 1} - \dfrac{1}{2ay}}{1 + \dfrac{4ay}{4a^2y^2 - 1}\dfrac{1}{2ay}}\qquad$ multiply top & bottom by $(4a^2y^2 - 1)(2ay)$

$\qquad = \dfrac{8a^2y^2 - 4a^2y^2 + 1}{(4a^2y^2 - 1)(2ay) + 4ay} = \dfrac{4a^2y^2 + 1}{8a^3y^3 - 2ay + 4ay}$

$\qquad = \dfrac{4a^2y^2 + 1}{8a^3y^3 + 2ay}\qquad = \dfrac{4a^2y^2 + 1}{2ay(4a^2y^2 + 1)}\qquad = \dfrac{1}{2ay} = \tan(\alpha)$. (Yes!)

There are other, more geometric ways to prove this result.

Chapter Nine

Section 9.1   Odd Answers

1.   A: $(25, 0)$ B: $(25, 3\pi/2)$ C: $(15, \pi/6)$ 3.   A: $(30, \pi/3)$ B: $(20, 5\pi/6)$ C: $(25, 5\pi/4)$

5.   Graph is given. Shape is almost
     rectangular.

7.   Graph is given.

9.   Graph is given.

11. Graph is given.

13. Graph is given.

15. Graphs are given.

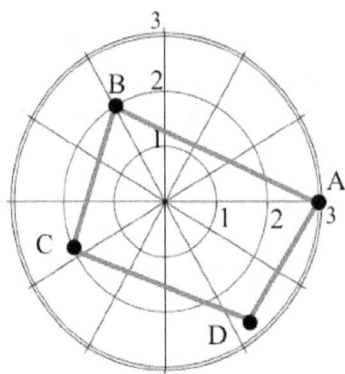

Prob. 5

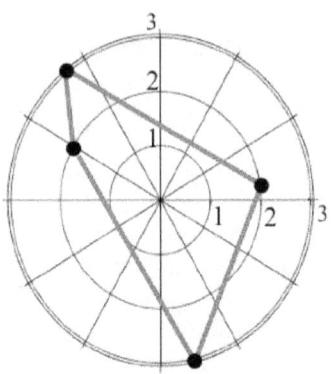

Prob. 7

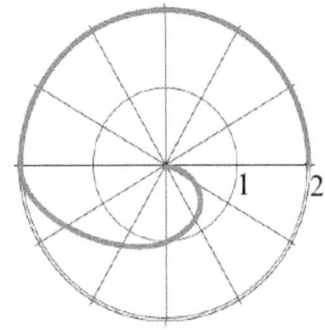

Prob. 9

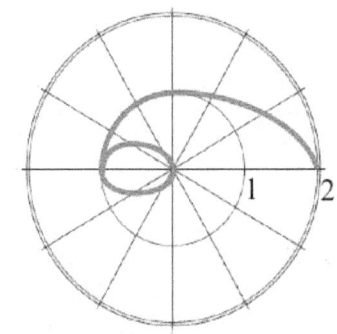

Prob. 11

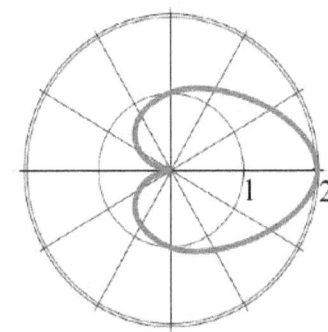

Prob. 13

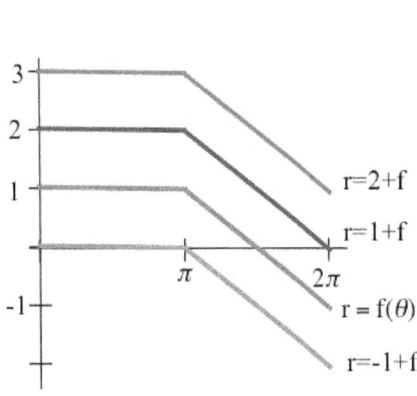

Prob. 15

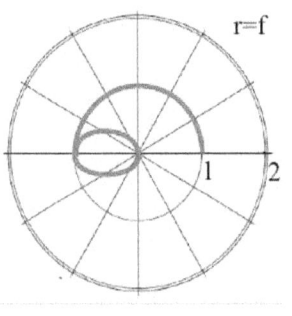

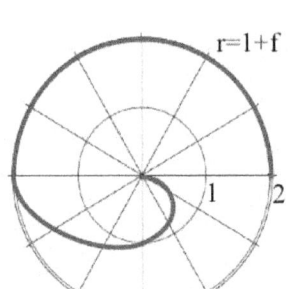

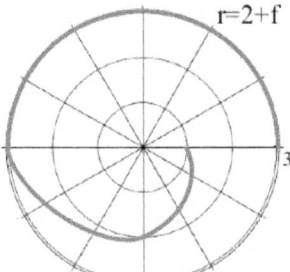

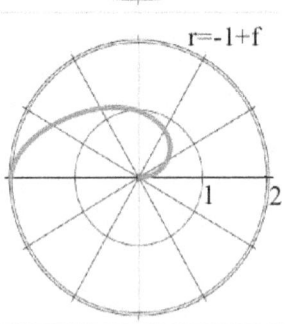

17. Graphs are given.

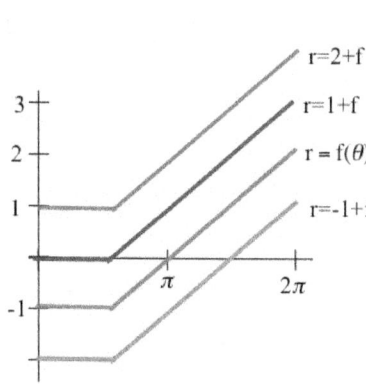

Prob. 17

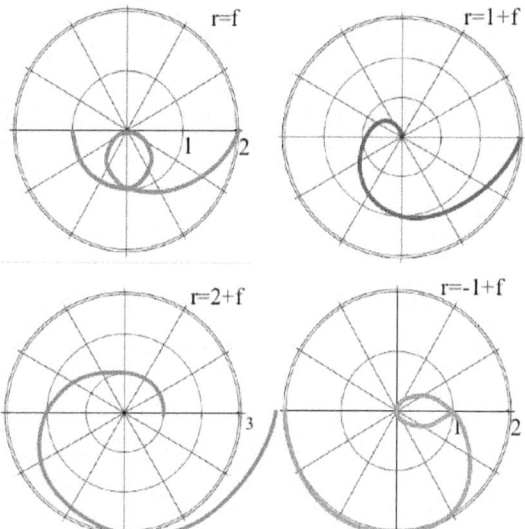

19. The polar graph will approach (spiral in or out to) a
    circle of radius 3.

21. Circle centered at origin with radius 3.

23. Line through the origin making an
    angle of $\pi/6$ with the x–axis.

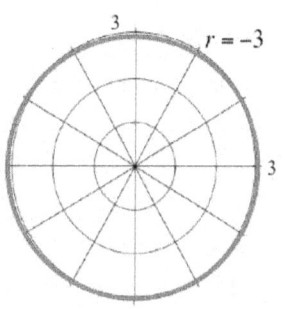

Prob. 21

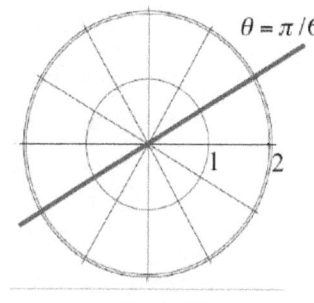

Prob. 23

25. A circle sitting atop the x–axis,
    touching the origin.

27. Graph is given.

29. Graph is given.

31. Graph is given.

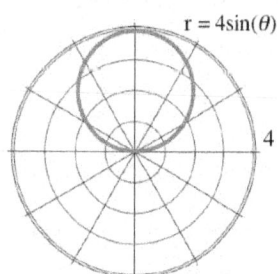

Prob. 25

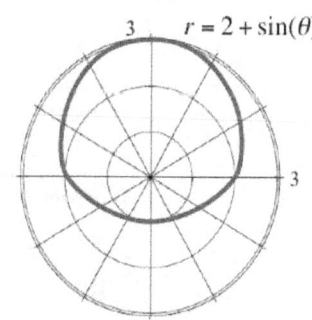

Prob. 27

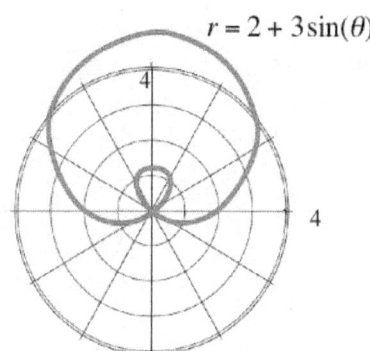

Prob. 29

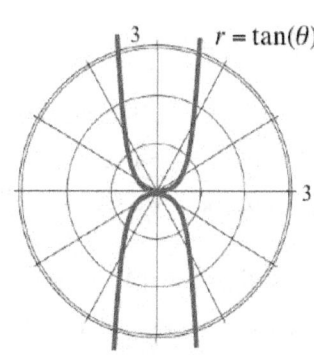

Prob. 31

33. Graph is given. A vertical line through the
    rectangular coordinate point (3,0).

35. Graph is given. A line: in rectangular coordinates
    y = 1 − x.

37. Graph is given: a fast growing spiral.

39. Graph is given.

41. {m = 1, n = 2}, {m = 2, n = 4},
    {m = 3, n = 3}, {m = 4, n = 4}
    all have pleasing shapes. Find
    some others for yourself.

43. The graphs are given. The graph
    for a = 0 is rotated π/6, π/4,
    and π/2 counterclockwise about
    the origin as a = π/6, π/4, and
    π/2 respectively.

45.   n   1   2   3   4   5   6   7   8
      # petals   1   4   3   8   5   12 7   16

47. Rectangular (−2, 3) is polar ($\sqrt{13}$, 2.159).
    Rect (2, −3) is polar ($\sqrt{13}$, 5.300) or (−$\sqrt{13}$, 2.159).
    Rect (0, −4) is polar (−4, π/2) or (4, 3π/2).

49. Rectangular (3, 4) is polar (5, 0.927).
    Rect (−1, −3) is polar ($\sqrt{10}$, 4.391).
    Rect (−7, 12) is polar ($\sqrt{193}$, −1.043) or ($\sqrt{193}$, −1.043 + π) ≈ ($\sqrt{193}$, 2.099).

51. Polar (−2, 3) is rect. (1.98, −0.282).   Polar (2, −3) is rect. (−1.98, −0.282).   Polar (0, −4) is rect. (0, 0).

53. Polar (2, 3) is rect. (−1.98, 0.282).   Polar (−2, −3) is rect. (1.98, 0.282).   Polar (0, 4) is rect. (0, 0).

55. x = 15·cos(−30°) + 10·cos(0°) ≈ 22.99 inches to the right of the shoulder.
    y = 15·sin(−30°) + 10·sin(0°) ≈ −7.5 inches or 7.5 inches **below** the shoulder.
    Polar location of hand: (24.18, −18.07°) or (24.18, −0.32).

57. x = 15·cos(−0.9) + 10·cos(−0.5) ≈ 18.10 inches to the right of the shoulder.
    y = 15·sin(−0.9) + 10·sin(−0.5) ≈ 16.54 inches **below** the shoulder. Polar is (24.52, -0.74).

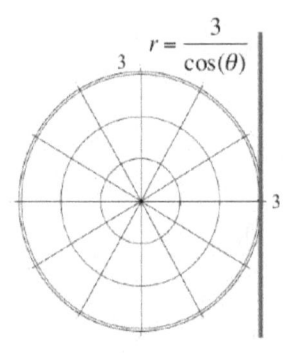

$$r = \frac{3}{\cos(\theta)}$$

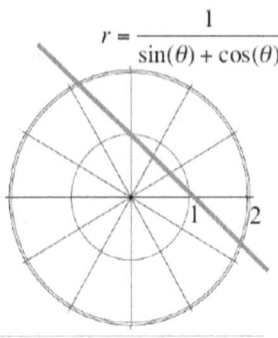

$$r = \frac{1}{\sin(\theta) + \cos(\theta)}$$

Prob. 33   Prob. 35

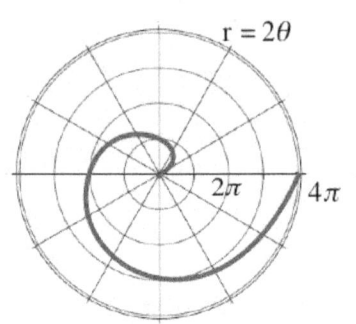

r = 2θ

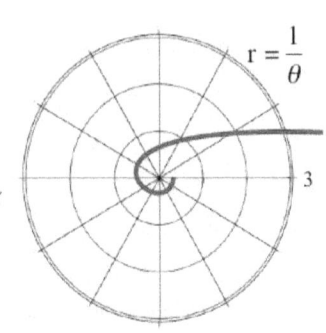

$$r = \frac{1}{\theta}$$

Prob. 37   Prob. 39

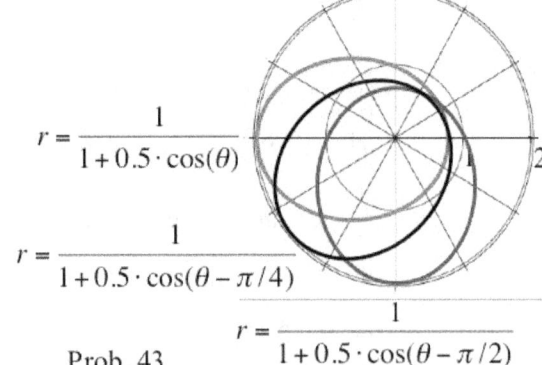

$$r = \frac{1}{1 + 0.5 \cdot \cos(\theta)}$$

$$r = \frac{1}{1 + 0.5 \cdot \cos(\theta - \pi/4)}$$

Prob. 43

$$r = \frac{1}{1 + 0.5 \cdot \cos(\theta - \pi/2)}$$

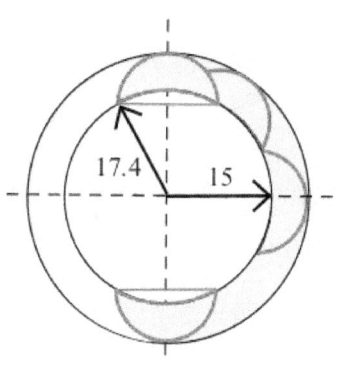

Prob. 59

59. Figure is given: $-\pi/2 \leq \theta \leq \pi/2$ and $-\pi/2 \leq \phi \leq \pi/2$. The robot's hand can reach the shaded region.

60. Figure is given: $-\pi/2 \leq \theta \leq \pi/2$ and $-\pi \leq \phi \leq \pi$. The robot's hand can reach the shaded region.

61. On your own.

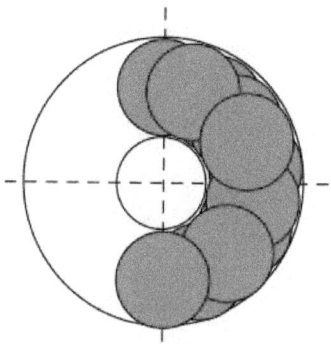

Prob. 60

Section 9.2   Odd Answers

1.	Point	$dr/d\theta$	$dx/d\theta$	$dy/d\theta$	dy/dx
	A	−	−	+	−
	B	0	−	+	−
	C	+	−	+	−
	D	+	−	0	0
	E	−	+	−	−

2.	Point	$dr/d\theta$	$dx/d\theta$	$dy/d\theta$	dy/dx
	A	+	+	+	+
	B	+	−	+	−
	C	−	−	−	+
	D	−	−	−	+
	E	+	−	−	+

3.	Point	$dr/d\theta$	$dx/d\theta$	$dy/d\theta$	dy/dx
	A	0	0	+	Und.
	B	+	0	+	Und.
	C	U	U	U	U
	D	−	−	0	0
	E	+	−	0	0

4.	Point	$dr/d\theta$	$dx/d\theta$	$dy/d\theta$	dy/dx
	A	−	+	−	−
	B	−	+	−	−
	C	+	+	−	−
	D	+	+	0	0
	E	+	+	+	+

5. The polar graph is a circle centered at the origin with radius 5.

   At A: $dr/d\theta = 0$, $dy/dx = -1$. At B: $dr/d\theta = 0$, $dy/dx = 0$.

   At C: $dr/d\theta = 0$, $dy/dx = $ Und.

7. The graph is given.

   At A: $dr/d\theta = 0$, $dy/dx = $ Und. At B: $dr/d\theta = -1$, $dy/dx = -1/5$.

   At C: $dr/d\theta = 0$, $dy/dx = 0$.

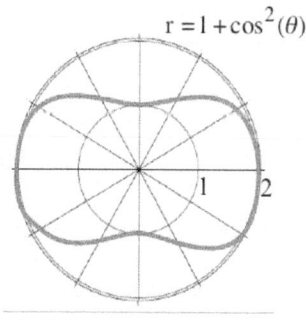

$r = 1 + \cos^2(\theta)$

Prob. 7

9. When $\theta = 2\pi/3$, $dy/dx = -\sqrt{3}$. When $\theta = 4\pi/3$, $dy/dx = +\sqrt{3}$.

11. We have $\dfrac{dy}{dx} = \dfrac{r \cdot \cos(\theta) + \dfrac{dr}{d\theta} \cdot \sin(\theta)}{-r \cdot \sin(\theta) + \dfrac{dr}{d\theta} \cdot \cos(\theta)}$,

   so if $r(\theta) = 0$ (and $\dfrac{dr}{d\theta}$ exists and $\dfrac{dr}{d\theta} \neq 0$)

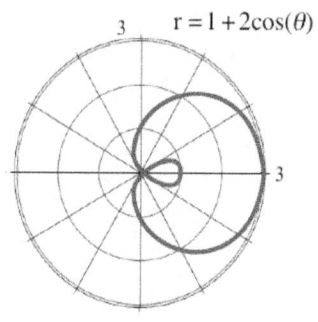

$r = 1 + 2\cos(\theta)$

Prob. 9

$$\text{then } \frac{dy}{dx} = \frac{\frac{dr}{d\theta} \cdot \sin(\theta)}{\frac{dr}{d\theta} \cdot \cos(\theta)} = \tan(\theta) .$$

13. $\frac{9}{16} \pi^2 \approx 5.552$

15. {area of cardioid in first quad.} – {area of circle in first quad.} = $\{ 1 + \frac{3\pi}{8} \} - \{ \frac{\pi}{4} \} = 1 + \frac{\pi}{8} \approx 1.393$. This problem is worked out in Example 4.

17. 3–petal: $A = \frac{1}{2} \int_0^{\pi/3} \sin^2(3\theta) \, d\theta = \frac{\pi}{12} \approx 0.262$.   5–petal: $A = \frac{1}{2} \int_0^{\pi/5} \sin^2(5\theta) \, d\theta = \frac{\pi}{20} \approx 0.157$.

19. $A = \pi(2)^2 = 4\pi \approx 12.566$

21. (b) add several semicirular regions to get $\frac{1}{2} \pi(40)^2 + \frac{1}{2} \pi(30)^2 + \frac{1}{2} \pi(10)^2 = 1300\pi \, \text{ft}^2 \approx 4{,}084.1 \, \text{ft}^2$

   (c) $\frac{3}{4} \pi(40)^2 + \frac{1}{2} \pi(20)^2 = 1400\pi \, \text{ft}^2 \approx 4{,}398.2 \, \text{ft}^2 > 4{,}084.1 \, \text{ft}^2$

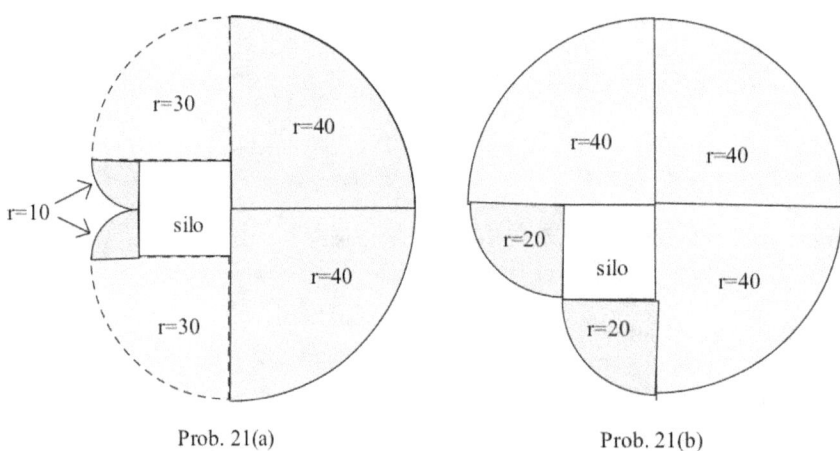

Prob. 21(a)   Prob. 21(b)

23. $r = \theta$, $dr/d\theta = 1$, $L = \int_0^{2\pi} \sqrt{\theta^2 + 1} \, d\theta \approx 21.256$ (using Simpson's rule with n = 20).

25. $r = 1 + \cos(\theta)$, $dr/d\theta = -\sin(\theta)$,

$$L = \int_0^{2\pi} \sqrt{\{1 + \cos(\theta)\}^2 + \{-\sin(\theta)\}^2} \, d\theta = \int_0^{2\pi} \sqrt{2 + 2 \cdot \cos(\theta)} \, d\theta = \int_0^{2\pi} \sqrt{4 \cdot \cos^2(\theta/2)} \, d\theta$$

$$= 2 \int_0^{2\pi} |\cos(\theta/2)| \, d\theta = 4 \int_0^{\pi} \cos(\theta/2) \, d\theta = 8 \sin(\theta/2) \Big|_0^{\pi} = 8 .$$

(Simpson's rule with n= 20 gives the same result.)

27. 10π

29. $r = \sin(3\theta)$, $dr/d\theta = 3 \cdot \cos(3\theta)$, $L = \int\limits_{0}^{\pi/3} \sqrt{\sin^2(3\theta) + \{3 \cdot \cos(3\theta)\}^2} \; d\theta \approx 2.227$ (Simpson, n = 20).

Section 9.3   Odd Answers

1. Graph is given.   3. Graph is given.

5. Graph is given.   7. Graph is given.

9. Graph is given: a straight line.

11. $(x(0), y(0)) = (b, d)$ and

$(x(1), y(1)) = (a+b, c+d)$ so

slope $= \dfrac{(c+d) - d}{(a+b) - b} = \dfrac{c}{a}$.

13. Graph is given.

15. Graphs are given.

(a) is the entire line $y = x$.

(b) is the "half–line" $y = x$ for $x \geq 0$.

(c) is the line segment $y = x$ for

$-1 \leq x \leq 1$: the location oscillates

along the line between $(-1,-1)$ and $(1,1)$.

All of these graphs satisfy the same relationship

between x and y, $y = x$, but the graphs cover

different parts of the graph of $y = x$ (different

domains).

17. The graph is given. The graph begins

at ($\dfrac{2\sqrt{2}}{\pi}$, $\dfrac{2\sqrt{2}}{\pi}$) when $t = \pi/4$,

and then it spirals counterclockwise

around and in toward the origin.

As t increases from $\pi/4$, the radial

distance of a point on the graph

decreases, approaching 0.

19. closer to

21. No. The new values give a point very close to the fixed point for these

populations so the system will be in good balance.

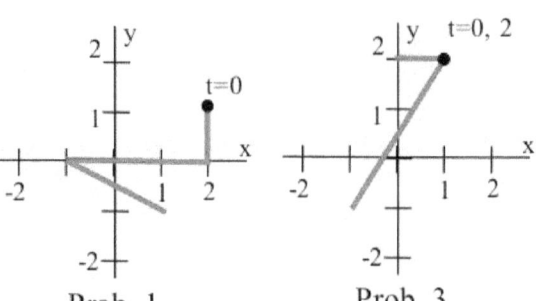

Prob. 1   Prob. 3

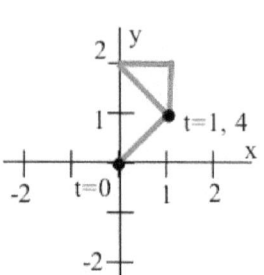

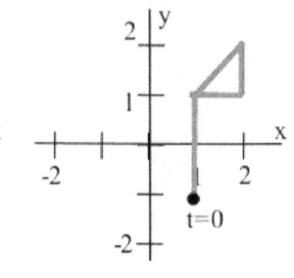

Prob. 5   Prob. 7

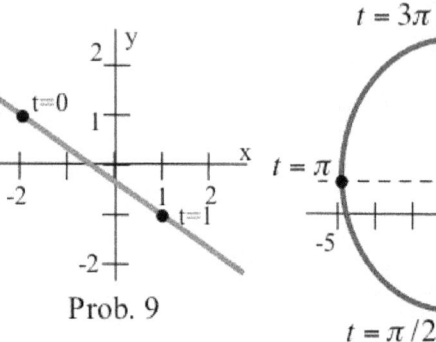

Prob. 9   Prob. 13

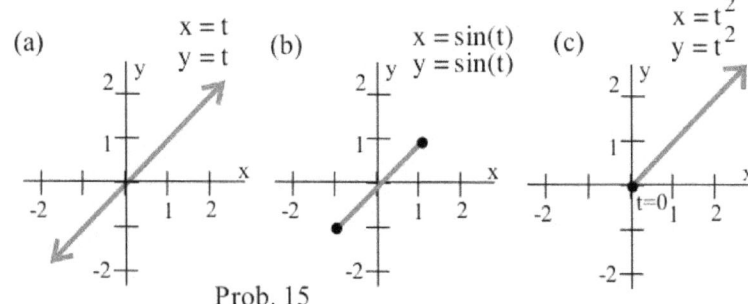

Prob. 15

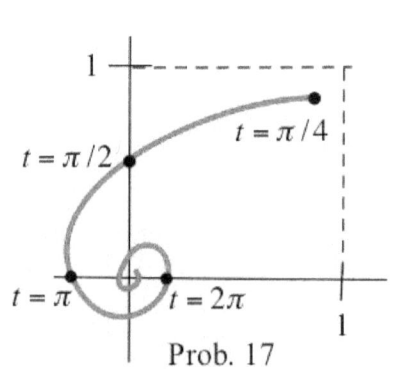

Prob. 17

23. Solve $x = at + b$ for t and substitute this value of t into the equation $y = ct + d$. Then $t = \dfrac{x - b}{a}$ and

$y = c\left(\dfrac{x - b}{a}\right) + d = \dfrac{c}{a}x + \left(d - \dfrac{bc}{a}\right)$, the equation of a straight line with slope $\dfrac{c}{a}$. The slope $\dfrac{c}{a}$ of y as a

function of x is the slope of y as a function of t divided by the slope of x as a function of t.

25. Sketches of some possible paths are given.

    (a) walking slowly   (b) running (like a stretched spring)   (c) walking along a parabolic path

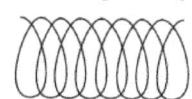   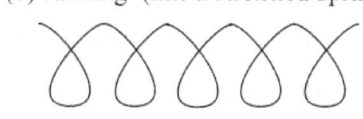   

parabola

Prob. 25

27. (a) $x = R(t - \sin(t))$, $y = R(1 - \cos(t))$. If $t = 2\pi$, then $(x, y) = (2\pi R, 0) = (10\pi, 0)$ so $R = 5$.

    (b) Set $x = R(t - \sin(t)) = 5$ and $y = R(1 - \cos(t)) = 2$, divide y by x to eliminate R, solve (graphically

    or using Newton's method or some other way) to get $t \approx 3.820$. Substitute this value into the equation

    for x or y and solve for $R \approx 1.124$.

    (c) Set $x = R(t - \sin(t)) = 2$ and $y = R(1 - \cos(t)) = 3$, divide y by x to eliminate R, solve (graphically

    or using Newton's method or some other way) to get $t \approx 1.786$. Substitute this value into the equation

    for x or y and solve for $R \approx 2.472$.

    (d) Set $x = R(t - \sin(t)) = 4\pi$ and $y = R(1 - \cos(t)) = 8$ and solve to get $t = \pi$ and $R = 4$.

29. (a) $x = -50 + 30(\dfrac{1}{\sqrt{2}})\cdot t = -50 + (15\sqrt{2})\cdot t$ feet and

    $y = 6 + 30(\dfrac{1}{\sqrt{2}})\cdot t - 16t^2 = 6 + (15\sqrt{2})\cdot t - 16t^2$ feet.

    (b) $x = -50 + V(\dfrac{\sqrt{2}}{2})\cdot t$ feet and $y = 6 + V(\dfrac{\sqrt{2}}{2})\cdot t - 16t^2$ feet.

Section 9.4   Odd Answers

1. (a) The graph is given.

    (b) $dx/dt = 1 - 2t$, $dy/dt = 2$. When $t = 0$, $dy/dx = 2$. When $t = 1$, $dy/dx = -2$.

    When $t = 2$, $dy/dx = -2/3$.

    (c) dy/dx is never 0. dy/dx is undefined when $t = 1/2$: at $(x, y) = (1/4, 2)$.

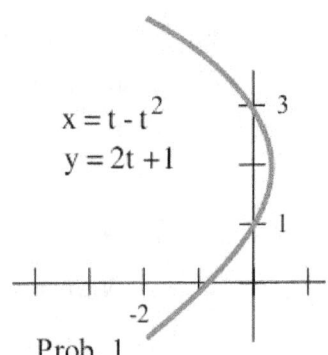

$x = t - t^2$

$y = 2t + 1$

Prob. 1

3.   (a)   The graph is given.

     (b) $dx/dt = -\sin(t)$, $dy/dt = \cos(t)$. When $t = 0$, dy/dx is undefined.

        When $t = \pi/4$, $dy/dx = -1$. When $t = \pi/2$, $dy/dx = 0$.

     (c) $dy/dx = 0$ whenever $t = (k + \frac{1}{2})\pi$ for k an integer. dy/dx is

        undefined whenever $t = k\pi$ for k an integer.

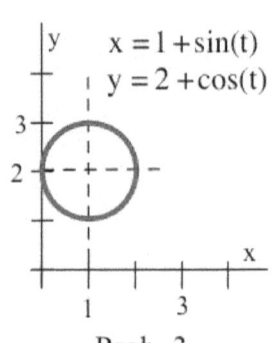

Prob. 3

5.   (a)   The graph is given.

     (b) $dx/dt = \cos(t)$, $dy/dt = -\sin(t)$. When $t = 0$, $dy/dx = 0$. When $t = \pi/4$,

        $dy/dx = -1$. When $t = \pi/2$, dy/dx is undefined.

        When $t = 17.3$, $dy/dx \approx 47.073$.

     (c) $dy/dx = 0$ whenever $t = k\pi$ for k an integer. dy/dx is undefined

        whenever $t = (k + \frac{1}{2})\pi$ for k an integer.

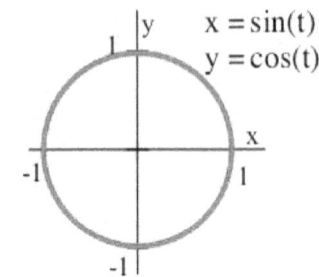

Prob. 5

7.   (a)   The graph is given.

     (b) $dx/dt = 1/t$, $dy/dt = -2t$. When $t = 1$, $dy/dx = -2$.

        When $t = 2$, $dy/dx = -8$. When $t = e$, $dy/dx = -2e^2$.

     (c)   The function is only defined for $t > 0$, and for all $t > 0$ the

        slope of the tangent line dy/dx is defined and is not equal to 0.

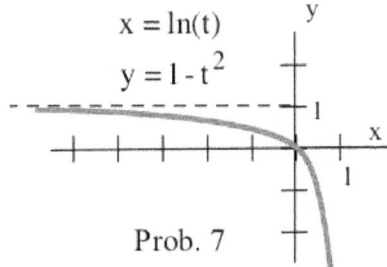

Prob. 7

9.   (a) $m_0 \approx \frac{-1}{1} = -1$. $m_1 \approx \frac{0}{1} = 0$. $m_2 \approx \frac{0}{1} = 0$. m_3 is undefined.

     (b) $dy/dx = 0$ when $t = 1$ and $t = 2$.

11.   (a) $m_0 \approx \frac{2}{-1} = -2$. $m_1 \approx \frac{1}{-1} = -1$. m_2 is undefined. $m_3 = \frac{-1}{1} = -1$.

     (b) dy/dx is undefined when $t = 2$.

13. $dx/dt = 1 - 2t$, $dy/dt = 2$ so $v = \sqrt{(1 - 2t)^2 + (2)^2} = \sqrt{4t^2 - 4t + 5}$.

     $v_0 = \sqrt{5} \approx 2.24$ ft/s , $v_1 = \sqrt{5} \approx 2.24$ ft/s , $v_2 = \sqrt{13} \approx 3.61$ ft/s .

15. $dx/dt = -\sin(t)$, $dy/dt = \cos(t)$ so

     $v = \sqrt{(-\sin(t))^2 + (\cos(t))^2} = \sqrt{\sin^2(t) + \cos^2(t)} = 1$ ft/s for all values of t.

     $v_0 = v_{\pi/4} = v_{\pi/2} = v_\pi = 1$ ft/s.

17. $v_0 \approx \sqrt{(1)^2 + (-1)^2} = \sqrt{2} \approx 1.41$ ft/s. $v_1 \approx \sqrt{(1)^2 + (0)^2} = 1$ ft/s.

     $v_2 \approx \sqrt{(1)^2 + (0)^2} = 1$ ft/s. $v_3 \approx \sqrt{(0)^2 + (-1)^2} = 1$ ft/s.

     $v_4 \approx \sqrt{(-1)^2 + (-1)^2} = \sqrt{2} \approx 1.41$ ft/s.

19. $v_0 \approx \sqrt{(-1)^2 + (2)^2} = \sqrt{5} \approx 2.24$ ft/s. $v_1 \approx \sqrt{(-1)^2 + (1)^2} = \sqrt{2} \approx 1.41$ ft/s.

$v_2 \approx \sqrt{(0)^2 + (0)^2} = 0$ ft/s. $v_3 \approx \sqrt{(1)^2 + (-1)^2} = \sqrt{2} \approx 1.41$ ft/s.

21. (a) $dx/dt = R(1 - \cos(t))$ and $dy/dt = R\sin(t)$ so

$v = \sqrt{R^2(1 - \cos(t))^2 + R^2(\sin(t))^2} = |R| \sqrt{2} \sqrt{1 - \cos(t)}$ ft/s.

(b) v is maximum when $\cos(t) = -1$, when $t = (2k+1)\pi$ seconds for k an integer.

(c) $v_{max} = 2R$ ft/s.

23. $L = \int_0^2 \sqrt{(1 - 2t)^2 + 2^2} \; dt \approx 4.939$ (using a calculator).

25. π (half the circumference of the circle).

27. $x = 3 - t, y = 1 + \frac{1}{2} t$. $L = \int_1^3 \sqrt{(-1)^2 + (1/2)^2} \; dt = 2 \cdot \frac{\sqrt{5}}{2} = \sqrt{5} \approx 2.24$.

Alternately, the graph of (x, y) is a straight line, and we can calculate the distance from $(x(1), y(1)) = (2, 1.5)$ to the point $(x(3), y(3)) = (0, 2.5)$: distance $= \sqrt{2^2 + 1^2} = \sqrt{5}$.

29. $y = 4t^2 - t^4, dx/dt = 2t$. $A = \int_0^2 (4t^2 - t^4) \cdot 2t \; dt = 2t^4 - \frac{1}{3} t^6 \Big|_0^2 = 32 - \frac{64}{3} = \frac{32}{3}$.

31. $y = 1 + \cos(t), dx/dt = 2t$. $A = \int_0^2 (1 + \cos(t)) \cdot 2t \; dt \approx 4.805$ (using a calculator).

33. See Fig. 24. $A = \frac{\pi R^2}{4} + \frac{R^2}{2} + \frac{\pi R^2}{2} + \frac{R^2}{2} + \frac{\pi R^2}{4} = \pi R^2 + R^2 = R^2 (\pi + 1)$.

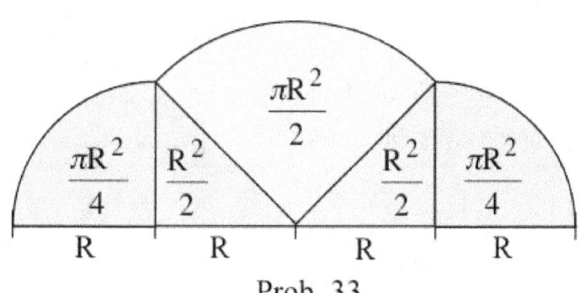

Prob. 33

Section 9.4.5 Odd Answers

1. $P_A = (0,4)$, $P_B = (5,2)$. $x(t) = 0 + 5t = (1-t)\cdot 0 + t\cdot 5$ and $y(t) = 4 + (-2)t = (1-t)\cdot 4 + t\cdot 2$.

3. $P_A = (4,3)$, $P_B = (1,2)$. $x(t) = 4 - 3t = (1-t)\cdot 4 + t\cdot 1$ and $y(t) = 3 - 1t = (1-t)\cdot 3 + t\cdot 2$.

5. $P_A = (1,4)$, $P_B = (5,1)$. $x(t) = 1 + 4t = (1-t)\cdot 1 + t\cdot 5$ and $y(t) = 4 + (-3)t = (1-t)\cdot 4 + t\cdot 1$

7.   If we start with the equation $x(t) = x_0 + t\cdot \Delta x$ and replace Δx with $x_1 - x_0$ then

$x(t) = x_0 + t\cdot(x_1 - x_0) = x_0 + t\cdot x_1 - t\cdot x_0 = (1-t)\cdot x_0 + t\cdot x_1$ which is the pattern we wanted.

The algebra for y(t) is similar.

For problems 9-13 the Bezier pattern is

$$B(t) = (1-t)^3\cdot P_0 + 3(1-t)^2 t\cdot P_1 + 3(1-t)t^2\cdot P_2 + t^3\cdot P_3$$

9. $x(t) = (1-t)^3\cdot 0 + 3(1-t)^2 t\cdot 2 + 3(1-t)t^2\cdot 1 + t^3\cdot 4$

$y(t) = (1-t)^3\cdot 5 + 3(1-t)^2 t\cdot 3 + 3(1-t)t^2\cdot 4 + t^3\cdot 2$

11. $x(t) = (1-t)^3\cdot 6 + 3(1-t)^2 t\cdot 6 + 3(1-t)t^2\cdot 2 + t^3\cdot 2$

$y(t) = (1-t)^3\cdot 5 + 3(1-t)^2 t\cdot 3 + 3(1-t)t^2\cdot 5 + t^3\cdot 0$

13. $P_0 = (5,1)$ and $B'(0) = 2$ tells us that P_1 could be (5+1, 1+2) or (5+2, 1+4) or (5+h, 1 + 2h)

We pick $P_1 = (6,3)$

$P_3 = (1,3)$ and $B'(1) = 3$ tells us P_2 could be (1+1,3+3) or (1+h, 3+3h).

We pick $P_2 = (2,6)$.

$$x(t) = (1-t)^3\cdot 5 + 3(1-t)^2 t\cdot 6 + 3(1-t)t^2\cdot 2 + t^3\cdot 1$$
$$y(t) = (1-t)^3\cdot 1 + 3(1-t)^2 t\cdot 3 + 3(1-t)t^2\cdot 6 + t^3\cdot 3$$

14. See Fig. 20.   15. See Fig. 21.

16. See Fig. 22.   17. See Fig. 23.

18. Violates Property (3): B'(1) does not equal the slope of the segment from P_2 to P_3

19. Violates Property (3) at B(1). Also violates Property (4) since the B(t) graph goes outside the "rubber band" around the 4 control points.

20. Violates Property (2). Since the graph of B(t) has 3 "turns" then B(t) is not a cubic polynomial (which can only have 2 turns).

21. Violates Property (1): B(1) = P_2 instead of P_3.

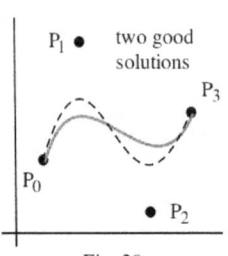

Fig. 20

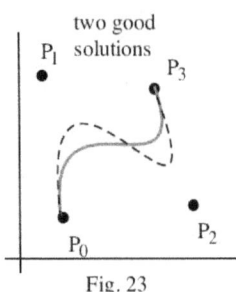

Fig. 21

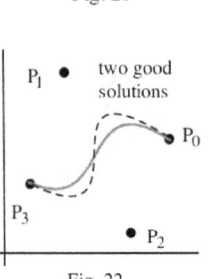

Fig. 22

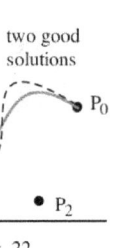

Fig. 23

Section 9.5   Odd Answers

1. An ellipse. Figure is given Since $b^2 + 6^2 = 10^2$, $b = 8$. $c = \sqrt{a^2 - b^2}$ so

    $6 = \sqrt{a^2 - 64}$ and $a = 10$. The ellipse is given by $\dfrac{x^2}{10^2} + \dfrac{y^2}{8^2} = 1$.

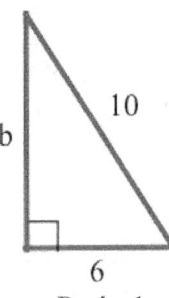

Prob. 1

3. A parabola with focus at $(0,5)$ and vertex at $(0,0)$.

    The parabola has an equation of the form $y = ax^2$

    with $\dfrac{1}{4a} = 5$ so $a = \dfrac{1}{20}$. The parabola is given

    by $y = \dfrac{1}{20} x^2$.

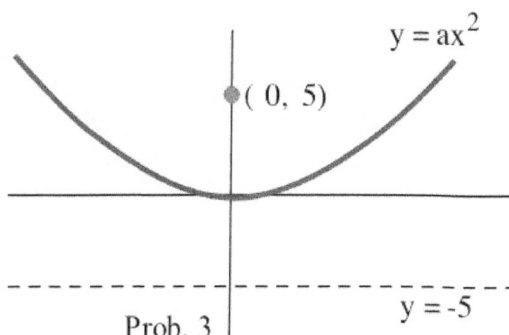

Prob. 3

5. $\dfrac{x^2}{4} + \dfrac{y^2}{9} = 1$.

7. Linear asymptotes: $y = \dfrac{2}{3} x$ and $y = -\dfrac{2}{3} x$.

    (Set $4x^2 - 9y^2 = 0$ and solve for y.)

    $c^2 = a^2 + b^2 = 4 + 9$ so $c = \sqrt{13}$: foci are

    at $(\sqrt{13}, 0)$ and $(-\sqrt{13}, 0)$.

9. Linear asymptotes: $y = \sqrt{\dfrac{3}{5}} x$ and $y = -\sqrt{\dfrac{3}{5}} x$. Foci: $(0, \sqrt{8})$ and $(0, -\sqrt{8})$.

11. (a) $25x^2 + 4y^2 + (-100) = 0$: discriminant $= (0)^2 - 4(25)(4) = -400 < 0$. The graph is an ellipse.

    (b) $b^2 x^2 + a^2 y^2 + (a^2 b^2) = 0$: discriminant $= (0)^2 - 4(a^2)(b^2) = -4a^2 b^2 < 0$. The graph is an ellipse.

13. $x^2 + xy - 2y^2 - x + y - 3 = 0$: discriminant $= (1)^2 - 4(1)(-2) = 9 > 0$. The graph is a hyperbola.

15. $2x^2 + 4xy + 2y^2 - 7x + 3 = 0$: discriminant $= (4)^2 - 4(2)(2) = 0$. The graph is a parabola.

17. (a) $B^2 - 4AC = 9 - 4(2)(2) < 0$: ellipse.

    (b) $B^2 - 4AC = 16 - 4(2)(2) = 0$: parabola.

    (c) $B^2 - 4AC = 25 - 4(2)(2) > 0$: hyperbola.

    (d) same answers as for parts (a), (b), and (c).

19. (a) $B^2 - 4AC = 16 - 4(1)(3) > 0$: hyperbola.

    (b) $B^2 - 4AC = 16 - 4(1)(4) = 0$: parabola.

    (c) $B^2 - 4AC = 16 - 4(1)(5) < 0$: ellipse.

    (d) same answers as for parts (a), (b), and (c).

21. Figure is shown.

(a) About the x–axis: $V = 2\int_0^2 \pi y^2\, dx = 2\pi \int_0^2 25(1 - \frac{x^2}{4})\, dx$

$= 50\pi \{ x - \frac{x^3}{12} \} \Big|_0^2 = \frac{200\pi}{3}$.

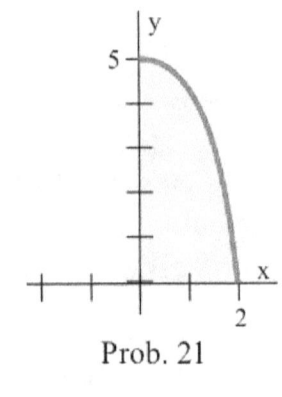

Prob. 21

(b) About the y–axis: $V = 2\int_0^5 \pi x^2\, dy$

$= 2\pi \int_0^5 4(1 - \frac{y^2}{25})\, dx = 8\pi \{ y - \frac{y^3}{75} \} \Big|_0^5 = \frac{80\pi}{3}$.

23. Figure is shown.

(a) About the x–axis: $V = \int_2^{10} \pi y^2\, dx$

$= \pi \int_2^{10} 25(\frac{x^2}{4} - 1)\, dx = 25\pi \{ \frac{x^3}{12} - x \} \Big|_2^{10}$

$= 25\pi \{ (\frac{1000}{12} - 10) - (\frac{8}{12} - 2) \} = \frac{5600\pi}{3} \approx 5864.3$.

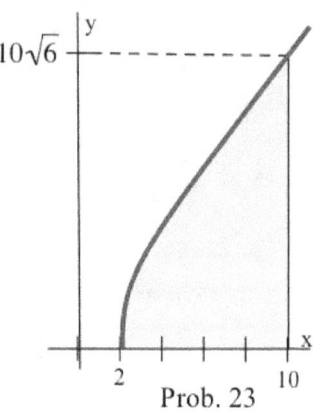

Prob. 23

(b) About the y–axis: $V = 2\pi \int_0^{10\sqrt6} \{ 100 - 4(\frac{y^2}{25} + 1) \}\, dy = 2\pi \{ 96y - \frac{4y^3}{75} \} \Big|_0^{10\sqrt6}$

$= 20\sqrt6\, \pi \{ 96 - \frac{2400}{75} \} = \frac{3840\sqrt6\, \pi}{3} \approx 9850$.

25. $A_{parabolic} = \int_a^b ax^2\, dx = \frac{1}{3} ax^3 \Big|_a^b = \frac{1}{3} ab^3$. $A_{rectangular} = ab^2 \cdot b = ab^3$. $\frac{A_{parabolic}}{A_{rectangular}} = \frac{1}{3}$.

27. The length of the string is the distance between the vertices.

29. When the pins (foci) are far apart, the ellipse tends to be long and narrow, cigar shaped. As the pins are moved closer together, the ellipse becomes more rounded and circular. In the limit, with the pins together, the ellipse is a cirle with diameter equal to the length of the string.

31. The curves are parabolas As the string is shortened, the vertex is moved up and the parabola narrows, approaching a vertical ray in the limit as the length of the string nears the vertical distance from the pin to the corner of the T–square where the string is attached.

Section 9.6　　　Odd Answers

1. 12 units, independent of where it bounces off of the ellipse.

3. After a "long time," the ball oscillates (almost)
along a line between the vertices of the hyperbola.

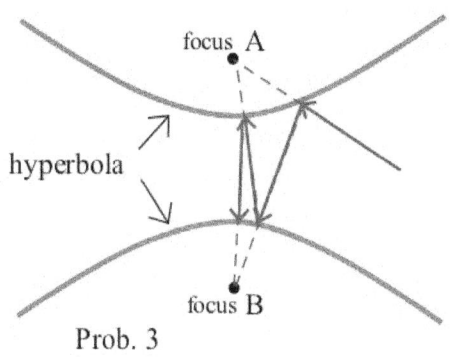

focus A

hyperbola

focus B

Prob. 3

5. All of the energy along the wave front is reflected to the focus
of the parabolic jetty at the same time　The boat would be
struck by a wave of considerable force.

7. The traced rays are shown. The eyepiece is located
at F_2 because all of the incoming light is focused there.

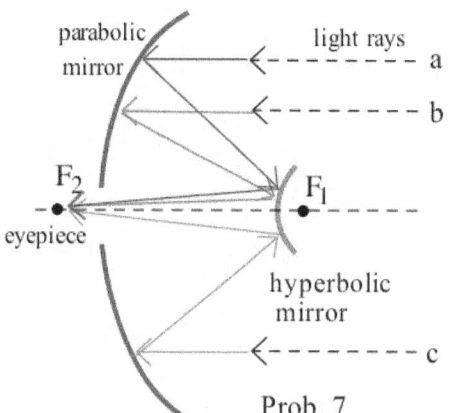

parabolic mirror

light rays

a

b

F_2

eyepiece

F_1

hyperbolic mirror

c

Prob. 7

9. (a) Roll the ball toward the focus F1 . The paths of A and
B are shown.

(b) The strategy in part (a) does not work for a ball in the
shaded region. Why not?

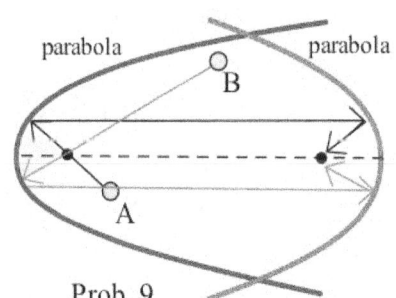

parabola　　　　　　　parabola

B

A

Prob. 9

11. At any point E on the ellipse, the angle of
incidence equals the angle of reflection so the angles α and
β are equal and the angle a and b are equal. Since
{angle a} > {angle α} we have that {angle b} > {angle β}
and the ball is reflected to a point C between the two foci.

13. $r = \dfrac{11/3}{1 + (5/3)\cos(\theta)}$　so　$e = 5/3 > 1$　and the graph is a

hyperbola. The hyperbola crosses the x–axis when $\theta = 0$
and π: at the points $(11/8, 0)$ and $(11/2, 0)$. It crosses
the y–axis when $\theta = \pi/2$ and $3\pi/2$: at the points $(0, 11/3)$
and $(0, -11/3)$.

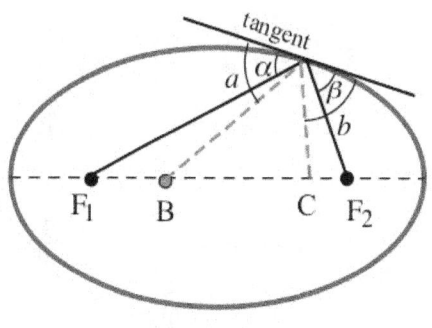

tangent

α

a

β

b

F_1　B　　　C　F_2

Prob. 11

15. $r = \dfrac{1/2}{1 + 1 \cdot \sin(\theta - \pi/3)}$　so　$e = 1$　and the graph is a parabola.

The parabola crosses the x–axis when $\theta = 0$ and π :
approximately
at the points $(3.73 , 0)$ and $(-0.27 , 0)$. It crosses the y–axis
when $\theta = \pi/2$ and $3\pi/2$: at the points $(0, 1/3)$ and $(0, -1)$.

17. $r = \dfrac{17/7}{1-(5/7)\cdot\cos(\theta+3\pi)}$ so e = 5/7 ≤ 1 and the graph is an ellipse.

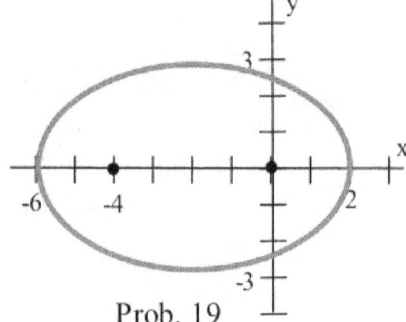

Prob. 19

The ellipse crosses the x–axis when θ = 0 and π: approximately at the

points (1.42 , 0) and (−8.5 , 0). It crosses the y–axis when θ = π/2

and 3π/2: at the points (0, 2.43) and (0, −2.43).

19. One focus is at (0,0) and, by symmetry, the other

focus is at (−4, 0).

21. One focus is at (0,0) and, by symmetry, the other

focus is at (0, 3).

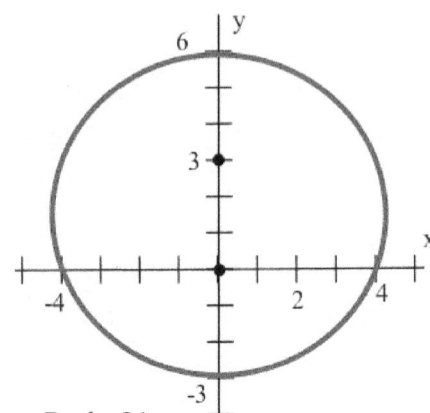

Prob. 21

23. See Fig. 41. This one is more difficult because the ellipse is tilted,

but we can still use the symmetry of the ellipse and the fact that it

is tilted at an angle of π/4 to the x–axis. One focus is at (0,0) and

the other focus is at (1.41 , −1.41).

25. $r = \dfrac{1}{1 + 0.5\cdot\cos(\theta)}$.

$\dfrac{dr}{d\theta} = -(1 + 0.5\cdot\cos(\theta))^{-2}\cdot\{-0.5\cdot\sin(\theta)\} = \dfrac{\sin(\theta)}{2 + \cos(\theta)}$.

Length $= \displaystyle\int_0^{2\pi} \sqrt{(r)^2 + (dr/d\theta)^2}\ d\theta \approx 7.659$

Area $= \displaystyle\int_0^{2\pi} \frac{1}{2}\, r^2(\theta)\, d \approx 4.8368$

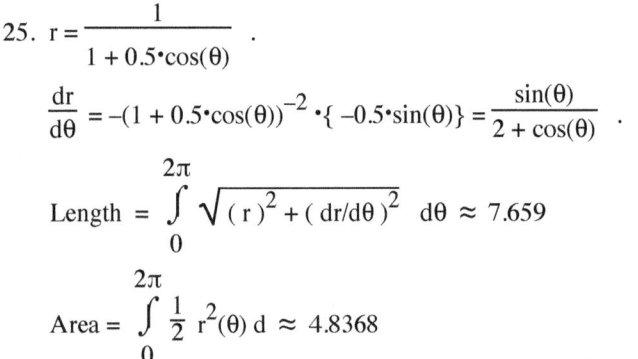

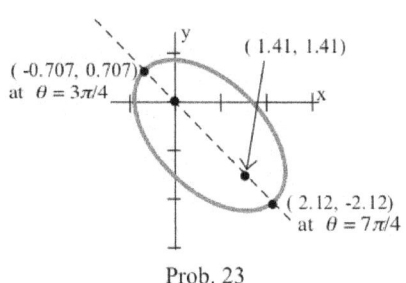

Prob. 23

Both integrals were approximated using Simpson's rule with n = 20)

27. $h = \dfrac{r_o v_o^2}{GM} = \dfrac{10^5\,(17.6)^2}{(6.7)(10^{-11})(10^{19})} = \dfrac{3.097 \times 10^7}{6.7 \times 10^8} \approx 4.62 \times 10^{-2}$ so e = | h − 1 | ≈ 0.95 < 1

and the path is an ellipse (but a long narrow ellipse such as a comet might have).

29. $h = \dfrac{r_o v_o^2}{GM} = \dfrac{10^5\,(120)^2}{(6.7)(10^{-11})(10^{19})} = \dfrac{1.44 \times 10^9}{6.7 \times 10^8} \approx 2.149$ so e = | h − 1 | ≈ 1.149 > 1 and

the path is a hyperbola.

31. $e = \dfrac{r \cdot v^2}{GM} - 1$.

    (a) The path is circular if $e = 0$, so $\dfrac{r \cdot v^2}{GM} - 1 = 0$ and $v = \sqrt{\dfrac{GM}{r}}$.

    (b) The path is elliptical if $e < 1$, so $\dfrac{r \cdot v^2}{GM} - 1 < 1$ and

$$v < \sqrt{\dfrac{2GM}{r}} = \sqrt{2}\sqrt{\dfrac{GM}{r}} = \sqrt{2} \cdot \{ \text{circular velocity} \}.$$

    (c) The path is parabolic if $e = 1$, so $v = \sqrt{\dfrac{2GM}{r}} = \sqrt{2}\sqrt{\dfrac{GM}{r}} = \sqrt{2} \cdot \{ \text{circular velocity} \}.$

    (d) The path is hyperbolic if $e > 1$, so $v > \sqrt{\dfrac{2GM}{r}} = \sqrt{2}\sqrt{\dfrac{GM}{r}} = \sqrt{2} \cdot \{ \text{circular velocity} \}.$

33. We want $v = \sqrt{\dfrac{2GM}{r}} = \sqrt{\dfrac{2(6.7 \times 10^{-11})(5.98 \times 10^{24})}{6.36 \times 10^6}} \approx \sqrt{1.26 \times 10^8} \approx 11{,}225$ m/s or

approximately 25,110 miles per hour.

35. $r_{max} = 1000 + r_{earth} \approx 7.36 \times 10^6$. $r_{min} = 800 + r_{earth} \approx 7.16 \times 10^6$.

    $2a = r_{min} + r_{max} \approx 14.520 \times 10^6$ so $a \approx 7.26 \times 10^6$.

    Also, $r_{max} = a(1 + e)$ so $7.36 \times 10^6 = 7.26 \times 10^6 (1 + e)$ and $e \approx 0.01377$.

    Finally, $k = a(1 - e^2) \approx 7.26 \times 10^6 (1 - (0.01377)^2) \approx 7.2586 \times 10^6$ so

$$r = \dfrac{k}{1 + e \cdot \cos(\theta)} \approx \dfrac{7.2586 \times 10^6}{1 + (0.01377) \cdot \cos(\theta)} .$$

37. On your own.

10.0   INTRODUCTION TO SEQUENCES AND SERIES

Chapter 10 is an introduction to two special topics in calculus, sequences and series. The main idea underlying this chapter is that **polynomials are easy**, and that even the hard functions such as sin(x) and log(x) can be represented as "big polynomials."

Polynomials are easy. It is easy to do arithmetic (evaluate, add, subtract, multiply, and even divide) with polynomials. It is easy to do calculus (differentiate and integrate) with polynomials. And, strangely enough, every polynomial is <u>completely</u> determined by its value and the values of all of its derivatives at x = 0: if we know the values of   P(0), P'(0), P''(0), etc., we can determine a formula for P(x) that is valid for all x.

Unfortunately, many of the important functions we need for applications (sin, cos, exp, log) are not polynomials: sin and cos have too many wiggles; exp grows too fast; and log has an asymptote. However, even these "hard" functions are "almost" polynomials and share many properties with polynomials:

For some values of x (to be specified in later sections), many important functions can be represented as "big polynomials" called power series:

$$\sin(x) = x - \frac{x^3}{2 \cdot 3} + \frac{x^5}{2 \cdot 3 \cdot 4 \cdot 5} - \frac{x^7}{2 \cdot 3 \cdot 4 \cdot 5 \cdot 6 \cdot 7} + \ldots + (-1)^n \frac{x^{2n+1}}{(2n+1)!} + \ldots \qquad \text{for } n = 0, 1, 2, \ldots$$

$$\cos(x) = 1 - \frac{x^2}{2} + \frac{x^4}{2 \cdot 3 \cdot 4} - \frac{x^6}{2 \cdot 3 \cdot 4 \cdot 5 \cdot 6} + \ldots + (-1)^n \frac{x^{2n}}{(2n)!} + \ldots$$

the " . . . " at the end means

$$\exp(x) = 1 + x + \frac{x^2}{2} + \frac{x^3}{2 \cdot 3} + \frac{x^4}{2 \cdot 3 \cdot 4} + \ldots + \frac{x^n}{n!} + \ldots$$

the pattern of the terms continues

$$\frac{1}{1-x} = 1 + x + x^2 + x^3 + x^4 + \ldots + x^n + \ldots$$

"forever"

In this chapter we examine

*   what it means to sum an **infinite** number of terms,
*   how to do algebra and calculus with series ("big polynomials"),
*   how to represent functions as series ("big polynomials"), and
*   how to use such series ("big polynomials") to calculate derivatives, integrals, and even solve differential equations.

First, however, we need to lay a foundation, and that foundation is the study of lists of numbers, their properties and behavior.

Section 10.1 focuses on this foundation material. It introduces lists of numbers, called sequences, and examines some specific sequences we will need later. It also introduces the idea of the convergence of a sequence and examines some ways we can determine whether or not a sequence converges.

Sections 10.2 to 10.7 focus on what it means to add up an infinite number of numbers, an infinite series, and on how we can determine whether the resulting sum is a finite number.

Sections 10.8 to 10.11 generalize the idea of an infinite series of numbers to infinite series that contain a variable. These series that contain powers of a variable are called power series. These sections discuss how we can represent and approximate functions such as $\sin(x)$ and e^x with power series, how accurate these approximations are for commonly needed functions, and how we can use them with derivatives and integrals.

PROBLEMS

These problems illustrate, at an elementary level, some of the problems and concepts we will examine more deeply in this chapter. They are intended to start you thinking in certain ways that are useful and necessary for Chapter 10.

Patterns in lists of numbers

For problems 1 – 6, the first four numbers a_1, a_2, a_3, and a_4 in a list are given. (a) Write the next two numbers in the list, (b) write a formula for the 5th number a_5 in the list and (c) write a formula for the nth number a_n in the list.

1. $2, 4, 8, 16,$ ___ , ___

2. $3, 9, 27, 81,$ ___ , ___

3. $-1, +1, -1, +1,$ ___ , ___

4. $1, 1/2, 1/3, 1/4,$ ___ , ___

5. $1, 2, 6, 24,$ ___ , ___

6. $1, 4, 9, 16,$ ___ , ___

For problems 7 – 11, evaluate each of the four given numbers and write the next two numbers in the list.

7. $1, 1 + 1/2, 1 + 1/2 + 1/3, 1 + 1/2 + 1/3 + 1/4,$ _____ , _____

8. $1, 1 + 1/2, 1 + 1/2 + 1/4, 1 + 1/2 + 1/4 + 1/6,$ _____ , _____

9. $1, 1 - 1/2, 1 - 1/2 + 1/4, 1 - 1/2 + 1/4 - 1/8,$ _____ , _____

10. $1, 1 + 2, 1 + 2 + 4, 1 + 2 + 4 + 8,$ ___ , ___

11. $1, 1 - 1, 1 - 1 + 1, 1 - 1 + 1 - 1,$ ___ , ___

Lists and graphs

For problems 12 – 15, (a) fill in the next two entries in the table and (b) graph the function for $x = 1, 2, ... 6$. These particular functions are defined only for integer values of x.

12.

x	f(x)
1	2
2	4
3	8
4	16
5	
6	

13.

x	g(x)
1	−1
2	+1
3	−1
4	+1
5	
6	

14.

x	s(x)
1	1 + 1/2
2	1 + 1/2 + 1/3
3	1 + 1/2 + 1/3 + 1/4
4	1 + 1/2 + 1/3 + 1/4 + 1/5
5	
6	

15.

x	t(x)
1	1 − 1/2
2	1 − 1/2 + 1/4
3	1 − 1/2 + 1/4 − 1/8
4	1 − 1/2 + 1/4 − 1/8 + 1/16
5	
6	

Polynomials and sine, cosine, and the exponential function

16. (a) Fill in the table for $P(x) = x - \dfrac{x^3}{2 \cdot 3}$ and $\sin(x)$.

    (b) Graph $y = P(x)$ and $y = \sin(x)$ for $-2 \le x \le 2$.

    (c) Repeat parts (a) and (b) for $P(x) = x - \dfrac{x^3}{2 \cdot 3} + \dfrac{x^5}{2 \cdot 3 \cdot 4 \cdot 5}$.

x	P(x)	sin(x)	| P(x) − sin(x) |
0			
0.1			
0.2			
0.3			
1.0			
2.0			

17. (a) Fill in the table for $P(x) = 1 - \dfrac{x^2}{2}$ and $\cos(x)$.

    (b) Graph $y = P(x)$ and $y = \cos(x)$ for $-2 \le x \le 2$.

    (c) Repeat parts (a) and (b) for $P(x) = 1 - \dfrac{x^2}{2} + \dfrac{x^4}{2 \cdot 3 \cdot 4}$.

x	P(x)	cos(x)	| P(x) − cos(x) |
0			
0.1			
0.2			
0.3			
1.0			
2.0			

18. (a) Fill in the table for $P(x) = 1 + x + \dfrac{x^2}{2.}$ and e^x.

    (b) Graph $y = P(x)$ and $y = e^x$. for $-2 \le x \le 2$.

    (c) Repeat parts (a) and (b) for $P(x) = 1 + x + \dfrac{x^2}{2} + \dfrac{x^3}{3}$

x	P(x)	e^x	| P(x) − e^x |
0			
0.1			
0.2			
0.3			
1.0			
2.0			

Polynomials and their values at x = 0

These problems illustrate how we can determine a formula for a polynomial when we know the values of the polynomial and its derivatives at $x = 0$.

In problems 19 – 24, $P(x) = Ax + B$ is a linear polynomial, and the values of $P(0)$ and $P'(0)$ are given. Find the values of A and B and write a formula for $P(x)$.

19. $P(0) = 5$, $P'(0) = 3$ 20. $P(0) = -2$, $P'(0) = 7$ 21. $P(0) = 4$, $P'(0) = -1$

22. $P(0) = 8$, $P'(0) = 5$ 23. $P(0) = 4$, $P'(0) = 0$ 24. $P(0) = -3$, $P'(0) = -2$

25. How are the values of A and B related to the values of $P(0)$ and $P'(0)$?

In problems 26 – 31, $P(x) = Ax^2 + Bx + C$ is a quadratic polynomial, and the values of $P(0)$, $P'(0)$, and $P''(0)$ are given. Find the values of A, B, and C and write a formula for $P(x)$.

26. $P(0) = 5$, $P'(0) = 3$, $P''(0) = 4$ 27. $P(0) = -2$, $P'(0) = 7$, $P''(0) = 6$

28. $P(0) = 4$, $P'(0) = -1$, $P''(0) = -2$ 29. $P(0) = 8$, $P'(0) = 5$, $P''(0) = 10$

30. $P(0) = 4$, $P'(0) = 0$, $P''(0) = -4$ 31. $P(0) = -3$, $P'(0) = -2$, $P''(0) = 4$

32. How are the values of A, B, and C related to the values of $P(0)$, $P'(0)$, and $P''(0)$?

In problems 33 – 38, $P(x) = Ax^3 + Bx^2 + Cx + D$ is a cubic polynomial, and the values of $P(0)$, $P'(0)$, $P''(0)$, and $P'''(0)$ are given. Find the values of A, B, C, and D and write a formula for $P(x)$.

33. $P(0) = 5$, $P'(0) = 3$, $P''(0) = 4$, $P'''(0) = 6$ 34. $P(0) = -2$, $P'(0) = 7$, $P''(0) = 6$, $P'''(0) = 18$

35. $P(0) = 4$, $P'(0) = -1$, $P''(0) = -2$, $P'''(0) = -12$ 36. $P(0) = 8$, $P'(0) = 5$, $P''(0) = 10$, $P'''(0) = 12$

37. $P(0) = 4$, $P'(0) = 0$, $P''(0) = -4$, $P'''(0) = 36$ 38. $P(0) = -3$, $P'(0) = -2$, $P''(0) = 4$, $P'''(0) = 36$

39. How are the values of A, B, C, and D related to the values of $P(0)$, $P'(0)$, $P''(0)$, and $P'''(0)$?

10.1　SEQUENCES

Sequences play important roles in several areas of theoretical and applied mathematics. As you study additional mathematics you will encounter them again. In this course, however, their role is primarily as a foundation for our study of series ("big polynomials"). In order to understand how and where it is valid to represent a function such as sine as a series, we need to examine what it means to add together an infinite number of values. And in order to understand this infinite addition we need to analyze lists of numbers (called sequences) and determine whether or not the numbers in the list are converging to a single value. This section examines sequences, how to represent sequences graphically, what it means for a sequence to converge, and several techniques to determine if a sequence converges.

Example 1:　A person places $100 in an account that pays 8% interest at the end of each year. How much will be in the account at the end of 1 year, 2 years, 3 years, and n years?

Solution:　After one year, the total is the principal plus the interest: $100 + (.08)100 = (1.08) \cdot 100 = \$108$.

At the end of the second year, the amount is 108% of the amount at the start of the second year:

$$(1.08) \{ (1.08)100 \} = (1.08)^2 \cdot 100 = \$116.64 .$$

At the end of the third year, the amount is 108% of the amount at the start of the third year:

$$(1.08) \{ (1.08)^2 100 \} = (1.08)^3 \cdot 100 = \$125.97 .$$

These results are shown in Fig. 1. In general, at the end of the nth year, the amount in the account is $(1.08)^n \cdot 100$ dollars.

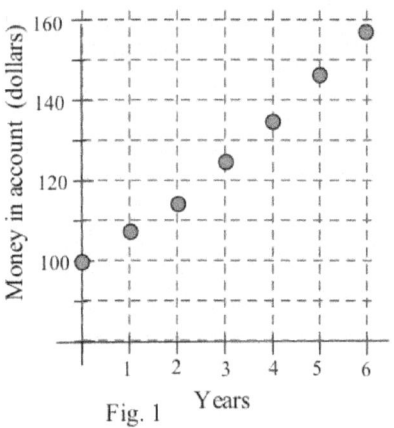

Fig. 1　Years

Practice 1:　A layer of protective film transmits two thirds of the light that reaches that layer. How much of the incoming light is transmitted through 1 layer, 2 layers, 3 layers, and n layers? (Fig. 2)

The Example and Practice each asked for a list of numbers in a definite order: a first number, then a second number, and so on. Such a list of numbers in a definite order is called a **sequence**. An infinite sequence is one that just keeps going and has no last number. Often the pattern of a sequence is clear from the first few numbers, but in order to precisely specify a sequence, a rule for finding the value of the nth term , a_n ("a sub n") , in the sequence is usually given.

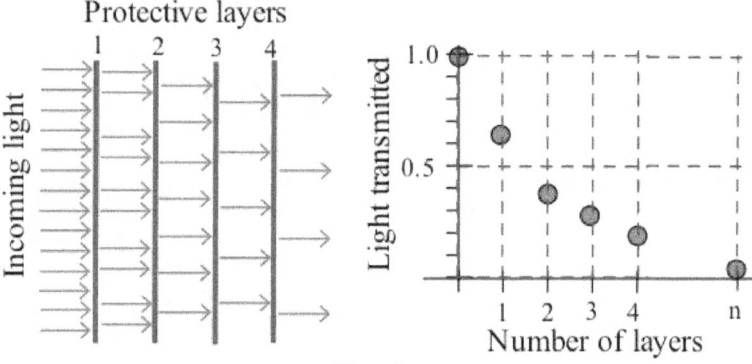

Fig. 2

Example 2:   List the next two numbers in each sequence and give a rule for calculating the nth number, a_n:

(a) $1, 4, 9, 16, \ldots$ (b) $-1, 1, -1, 1, \ldots$ (c) $\frac{1}{2}, \frac{1}{4}, \frac{1}{8}, \frac{1}{16}, \ldots$.

Solution: (a) $a_5 = 25, a_6 = 36$, and $a_n = n^2$. (b) $a_5 = -1, a_6 = 1$, and $a_n = (-1)^n$.

(c) $a_5 = \frac{1}{32}, a_6 = \frac{1}{64}$, and $a_n = (\frac{1}{2})^n = \frac{1}{2^n}$.

Practice 2:   List the next two numbers in each sequence and give a rule for calculating the nth number, a_n:

(a) $1, \frac{1}{2}, \frac{1}{3}, \frac{1}{4}, \ldots$ (b) $\frac{-1}{2}, \frac{1}{4}, \frac{-1}{8}, \frac{1}{16}, \ldots$ (c) $2, 2, 2, 2, \ldots$

Definition and Notation

Since a sequence gives a single value for each integer n, a sequence is a function, but a function whose domain is restricted to the integers.

Definition

A **sequence** is a function whose domain is all integers greater than or equal to a starting integer.

Most of our sequences will have a starting integer of 1, but sometimes it is convenient to start with 0 or another integer value.

Notation: The symbol a_n represents a single number called the nth term.

The symbol $\{a_n\}$ represents the entire sequence of numbers, the set of all terms.

The symbol $\{rule\}$ represents the sequence generated by the rule.
The symbol $\{a_n\}_{n=3}$ represents the sequence that starts with $n = 3$.

Because sequences are functions, we can add, subtract, multiply, and divide them, and we can combine them with other functions to form new sequences. We can also graph sequences, and their graphs can sometimes help us describe and understand their behavior.

Example 3:   For the sequences given by $a_n = 3 - \frac{1}{n}$ and $b_n = \frac{1}{2^n}$, graph the points (n, a_n) and (n, b_n) for

n = 1 to 5. Calculate the first 5 terms of $c_n = a_n + b_n$ and graph the points (n, c_n).

Solution: $c_1 = (3 - \frac{1}{1}) + (\frac{1}{2^1}) = 2.5, c_2 = 2.75, c_3 \approx 2.792$,

$c_4 = 2.8125, c_5 = 2.83125$. The graphs of $(n, a_n), (n, b_n)$, and (n, c_n) are shown in Fig. 3.

Practice 3:   For a_n and b_n in the previous example, calculate the first 5

terms of $c_n = a_n - b_n$ and $d_n = (-1)^n b_n$ and graph the

points (n, c_n) and (n, d_n).

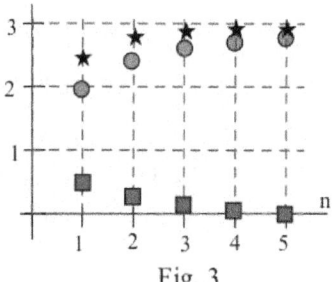

Fig. 3

Recursive Sequences

A recursive sequence is a sequence defined by a rule that gives each new term in the sequence as a combination of some of the previous terms. We already encountered a recursive sequence when we studied Newton's Method for approximating roots of a function (Section 2.7). Newton's method for finding the roots of a function generates a recursive sequence $\{ x_1, x_2, x_3, x_4, ... \}$, as do successive iterations of a function and other operations.

Example 4:   Let $f(x) = x^2 - 4$. Take $x_1 = 3$ and apply Newton's method (Section 2.7) to calculate x_2 and x_3. Give a rule for x_n.

Solution: $f(x) = x^2 - 4$ so $f'(x) = 2x$, and, by Newton's method,

$$x_2 = x_1 - \frac{f(x_1)}{f'(x_1)} = 3 - \frac{f(3)}{f'(3)} = 3 - \frac{5}{6} = \frac{13}{6} \approx 2.1667.$$

$$x_3 = x_2 - \frac{f(x_2)}{f'(x_2)} = \frac{13}{6} - \frac{f(13/6)}{f'(13/6)} = \frac{13}{6} - \frac{25}{156} = \frac{313}{156} \approx 2.0064.$$

In general, $x_n = x_{n-1} - \frac{f(x_{n-1})}{f'(x_{n-1})} = x_{n-1} - \frac{(x_{n-1})^2 - 4}{2x_{n-1}}$.

The terms $x_1, x_2, \ldots$ approach the value 2, one solution of $x^2 - 4 = 0$. The sequence $\{ x_n \}$ is a recursive sequence since each term x_n is defined as a function of the previous term x_{n-1}.

Practice 4:   Let $f(x) = 2x - 1$, and define $a_n = f(f(f(\ldots f(a_0) \ldots)))$ where the function is applied n times. Put $a_0 = 3$ and calculate $a_1, a_2,$ and a_3. Note that a_n can be defined recursively as $a_n = f(a_{n-1})$.

Example 5:   Let $a_n = 1/2^n$, and define a second sequence $\{ s_n \}$ by the rule that s_n is the **sum** of the first n terms of a_n. Calculate the values of s_n for n = 1 to 5.

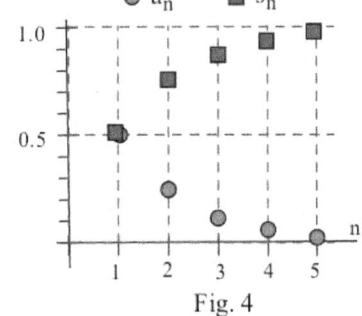

Solution: $s_1 = a_1 = 1/2$, $s_2 = a_1 + a_2 = 1/2 + 1/4 = 3/4$,

$s_3 = a_1 + a_2 + a_3 = 1/2 + 1/4 + 1/8 = 7/8$, $s_4 = 15/16$,

and $s_5 = 31/32$. (Fig. 4)

Fig. 4

You should notice two patterns in these sums.

First, it appears that $s_n = (2^n - 1)/2^n$.

Second, you can simplify the addition process: each term s_n is the sum of the previous term s_{n-1} and the a_n term: $s_n = s_{n-1} + a_n$. We will meet this second pattern again in the next section.

Practice 5:   Let $b_0 = 0$ and, for n > 0, define $b_n = b_{n-1} + 1/3^n$. Calculate b_n for n = 1 to 4.

Limits of Sequences: Convergence

Since sequences are discrete functions defined only on integers, some calculus ideas for continuous functions are not applicable to sequences. One type of limit, however, is used: the limit as n approaches infinity. Do the values a_n eventually approach (or equal) some number?

We say that the limit of a sequence $\{ a_n \}$ is L if the terms a_n are arbitrarily close to L for sufficiently large values of n — the terms at the beginning of the sequence can be any values, but for large values of n, the a_n terms are all close to L . The following definition puts this idea more precisely.

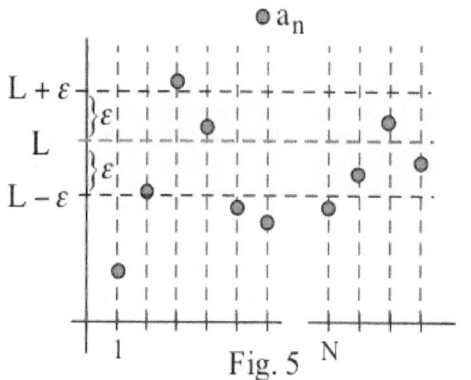

Definition

$$\lim_{n\to\infty} \mathbf{a_n} = \mathbf{L} \quad \text{if for any } \varepsilon > 0 \ \text{("epsilon} > 0")$$

there is an index N (typically depending on ε)

so that a_n is within ε of L whenever

n is larger than N :

$$n > N \text{ implies } | a_n - L | < \varepsilon \quad \text{(Fig. 5)}$$

Fig. 5

If a sequence has a finite limit L , we say that the sequence "converges to L." If a sequence does not have a finite limit, we say the sequence "diverges." Typically a sequence diverges because its terms grow infinitely large (positively or negatively) or because the terms oscillate and do not approach a single number.

Example 7:   For $a_n = 3 + 1/n^2$ show that $\lim_{n\to\infty} a_n = 3$.

Fig. 6 $N = \dfrac{1}{\sqrt{\varepsilon}}$

Solution: We need to show that for any positive ε , there is a number N so that the distance from a_n to L , $| a_n - L |$, is less than ε whenever n is larger than N. For this particular sequence and limit we need to show that for any positive ε, there is a number N so that (Fig. 6)

$$| (3 + 1/n^2) - 3 | < \varepsilon \text{ whenever } n > N.$$

To determine what N might be, we solve the inequality

$$| (3 + 1/n^2) - 3 | < \varepsilon \text{ for n in terms of } \varepsilon:$$

$$| 1/n^2 | < \varepsilon \text{ so } 1/\varepsilon < n^2 \text{ and } n > 1/\sqrt{\varepsilon} .$$

So for **any** positive ε, we can take $N = 1/\sqrt{\varepsilon}$ (or the next larger integer) . Then for n > N we know that

$$n > 1/\sqrt{\varepsilon} \text{ so } 1/n^2 < \varepsilon \text{ and } | (3 + 1/n^2) - 3 | < \varepsilon .$$

Practice 6:   For $a_n = (n+1)/n$ show that $\lim\limits_{n\to\infty} a_n = 1$.   (Fig. 7)

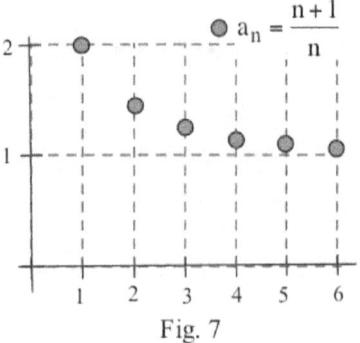

The limit of a sequence, as n approaches infinity, depends only on the

behavior of the terms of the sequence for large values of n (the "tail end")

and not on the values of the first few (or few thousand) terms. As a

consequence, we can insert or delete any **finite** number of terms without

changing the convergence behavior of the sequence.

Fig. 7

Sequences are functions, so limits of sequences share many properties with limits of other functions, and

we state only a few of them.

Uniqueness Theorem

    If a sequence converges to a limit, then the limit is unique.

    A sequence can not converge to two different values.

A proof of the Uniqueness Theorem is given after the problem set.

Sometimes it is useful to replace a sequence $\{\, a_n \,\}$, a

function whose domain is integers, with a function f

whose domain is the real numbers so $a_n = f(n)$. If

$f(x)$ has a limit as "$x \to \infty$," as x gets arbitrarily

large, then $f(n)$ has the same limit as "$n \to \infty$"

(Fig. 8). This replacement of "x" with "n" allows us

to use earlier results about functions, particularly

L'Hopital's Rule, to calculate limits of sequences.

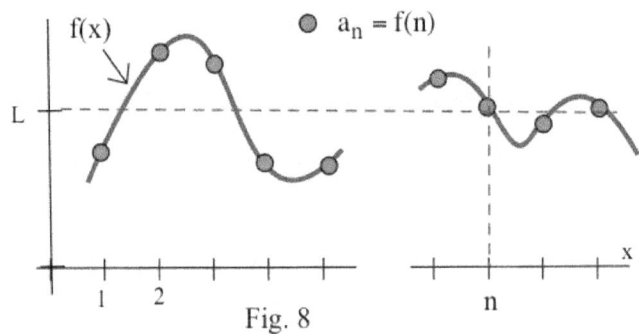

Fig. 8

Theorem: If $a_n = f(n)$ and $\lim\limits_{x\to\infty} f(x) = L$

      then $\{\, a_n \,\}$ converges to L: $\lim\limits_{n\to\infty} a_n = L$

Example 8:   Calculate $\lim\limits_{n\to\infty} \left(1 + \dfrac{2}{n} \right)^n$.

Solution: The terms of the sequence are $a_n = (1 + \frac{2}{n})^n$, so we can define $f(x) = (1 + \frac{2}{x})^x$ by

    replacing the integer values n with real number values x. Then $a_n = f(n)$, and we can use

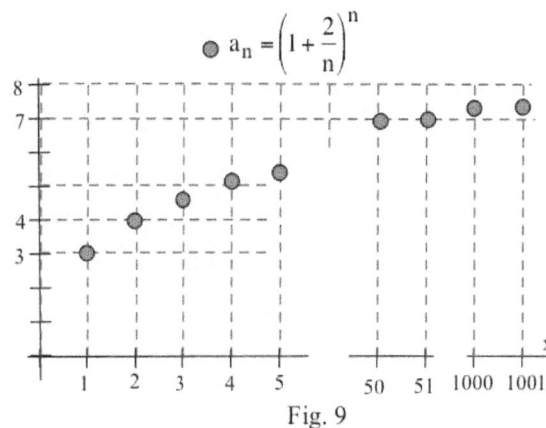
Fig. 9

L'Hopital's rule to get $\lim\limits_{x \to \infty} \left(1 + \dfrac{2}{x}\right)^x = e^2$ (Section 3.7,

Example 6).   Finally, we can conclude that

$$\lim_{n \to \infty} \left(1 + \frac{2}{n}\right)^n = e^2 \approx 7.389 \quad \text{(Fig. 9)}.$$

Practice 7:   Calculate $\lim\limits_{n \to \infty} \dfrac{\ln(n)}{n}$.

A **subsequence** is an infinite set of terms from a sequence that

occur in the same order as they appear in the original sequence. The sequence of even integers $\{\, 2, 4, 6, ..$

$.\,\}$ is a subsequence of the sequence of all positive integers $\{\, 1, \mathbf{2}, 3, \mathbf{4}, \ldots \,\}$. The sequence of reciprocals

of primes $\{\, 1/2, 1/3, 1/5, 1/7, \ldots \,\}$ is a subsequence of the sequence of the reciprocals of all positive

integers $\{\, 1, \mathbf{1/2}, \mathbf{1/3}, 1/4, \mathbf{1/5}, \ldots \,\}$. Subsequences inherit some properties from their original sequences.

Subsequence Theorem

    Every subsequence of a convergent sequence converges to the same limit as the original sequence:

        if $\lim\limits_{n \to \infty} a_n = L$ and $\{\, b_n \,\}$ is a subsequence of $\{\, a_n \,\}$,

        then $\lim\limits_{n \to \infty} b_n = L$.   (Fig. 10)

    If the sequence $\{\, a_n \,\}$ does not converge, then the subsequence $\{\, b_n \,\}$ may or may not converge.

Corollary: If two subsequences of the same sequence converge to two different limits, then the

             original sequence diverges.

Example 9:   Show that the sequence $\left\{\, \dfrac{(-1)^n n}{n+1} \,\right\}$ diverges.

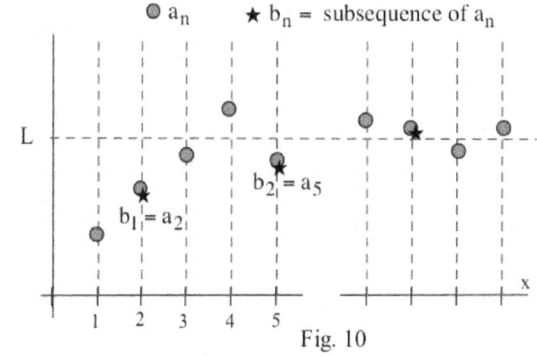
Fig. 10

Solution: If n is even, then the even terms $a_n = \dfrac{(-1)^{\text{even}} n}{n+1} =$

$\dfrac{n}{n+1}$ converge to $+1$ so the subsequence of even terms

converges to 1.

If n is odd, then the odd terms $a_n = \dfrac{(-1)^{\text{odd}} n}{n+1} = \dfrac{-n}{n+1}$

converge to -1 so the subsequence of odd terms converges to -1. Finally, since the two subsequences

converge to different values, we can conclude that the original sequence $\left\{\, \dfrac{(-1)^n n}{n+1} \,\right\}$ diverges.

Practice 8:   Show that the sequence $\{\, \sin(\, n\pi/2 \,) \,\}$ diverges.

Bounded and Monotonic Sequences

A sequence $\{a_n\}$ is **bounded above** if there is a value A so that $a_n \leq A$ for all values of n: A is called an **upper bound** of the sequence (Fig. 11). Similarly, $\{a_n\}$ is **bounded below** if there is a value B so that $B \leq a_n$ for all n: B is called a **lower bound** of the sequence. A sequence is **bounded** if it has an upper bound and a lower bound. All of the terms of a bounded sequence are between (or equal to) the upper and lower bounds (Fig. 12)

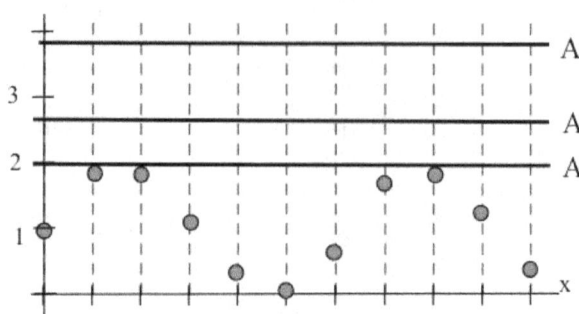

Fig. 11: Several upper bounds of the sequence $a_n = 1 + \sin(n)$

A **monotonically increasing** sequence is a sequence in which each term is greater than or equal to the previous term, $a_1 \leq a_2 \leq a_3 \leq \ldots$ (Fig. 13); a **monotonically decreasing** sequence is one in which each term is less than or equal to the previous term, $a_1 \geq a_2 \geq a_3 \geq \ldots$ (Fig. 14). A monotonic sequence does not oscillate: if one term is larger than a previous term and another term is smaller than a previous term, then the sequence is not monotonic increasing or decreasing.

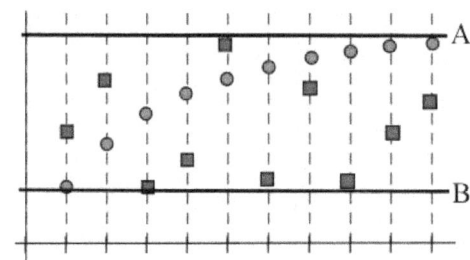

Fig. 12: Two bounded sequences

There are three basic ways to show that a sequence is monotonically increasing:

    (i)   by showing that $a_{n+1} \geq a_n$ for all n,

    (ii)   by showing that all the a_n are positive and

$$\frac{a_{n+1}}{a_n} \geq 1 \text{ for all } n, \text{ or}$$

    (iii)   by showing that $a_n = f(n)$ for integer values n

        and $f'(x) \geq 0$ for all $x > 0$.

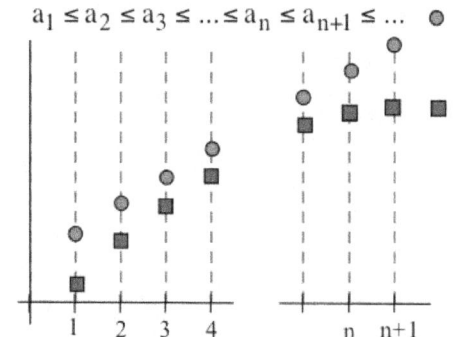

$$a_1 \leq a_2 \leq a_3 \leq \ldots \leq a_n \leq a_{n+1} \leq \ldots$$

Two monotonic increasing sequences

Fig. 13

Practice 9: List three ways you can show that a sequence is monotonically decreasing.

Example 10: Show that the sequence $a_n = \dfrac{2^n}{n!}$ is monotonically decreasing by showing that $\dfrac{a_{n+1}}{a_n} \leq 1$ for all n.

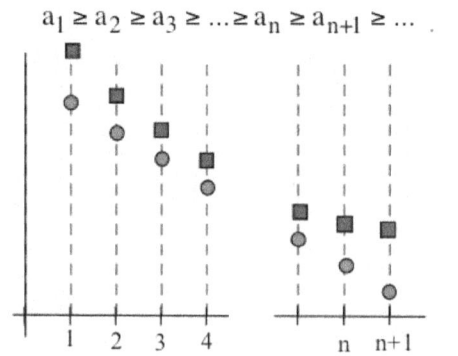

$$a_1 \geq a_2 \geq a_3 \geq \ldots \geq a_n \geq a_{n+1} \geq \ldots$$

Two monotonic decreasing sequences

Fig. 14

Solution: $a_n = \dfrac{2^n}{n!} = \dfrac{2^n}{1 \cdot 2 \cdot 3 \cdot \ldots \cdot n}$ and $a_{n+1} = \dfrac{2^{n+1}}{(n+1)!} = \dfrac{2^n \cdot 2}{1 \cdot 2 \cdot 3 \cdot \ldots \cdot n \cdot (n+1)}$.

Then $\dfrac{a_{n+1}}{a_n} = \dfrac{2^n \cdot 2}{1 \cdot 2 \cdot 3 \cdot \ldots \cdot n \cdot (n+1)} \cdot \dfrac{1 \cdot 2 \cdot 3 \cdot \ldots \cdot n}{2^n}$ by inverting and multiplying

$= \dfrac{2^n \cdot 2}{2^n} \cdot \dfrac{1 \cdot 2 \cdot 3 \cdot \ldots \cdot n}{1 \cdot 2 \cdot 3 \cdot \ldots \cdot n \cdot (n+1)}$ by reorganizing the top and bottom

$= \dfrac{2}{n+1} \leq 1$ for all positive integers n.

Practice 10:   Show that $\left\{ \left(\frac{2}{3} \right)^n \right\}$ is monotonically decreasing.

Since the behavior of a monotonic sequence is so regular, it is usually easy to determine if a monotonic sequence has a finite limit: all we need to do is show that it is bounded.

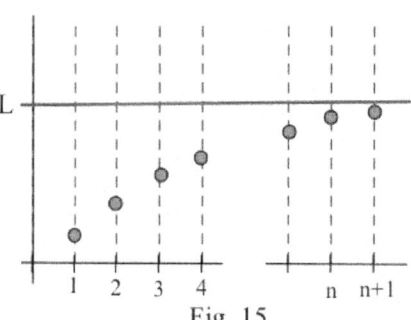

Fig. 15

Monotone Convergence Theorem

If   a monotonic sequence is bounded,

then   the sequence converges. (Fig. 15)

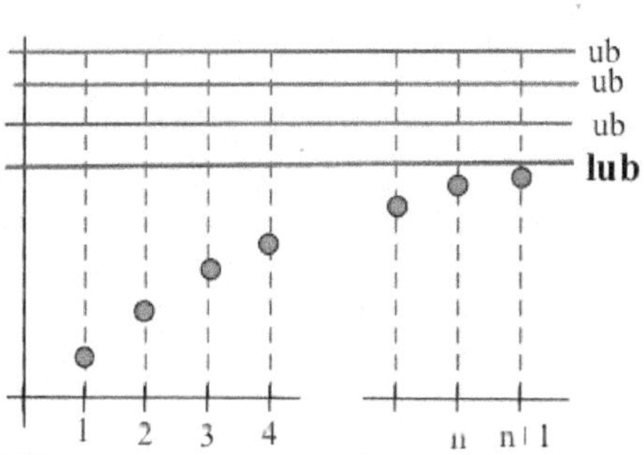

Fig. 16

Idea for a proof for a monotonically increasing sequence that is bounded above:

If the sequence $\{ a_n \}$ is bounded above, then $\{ a_n \}$ has an infinite number of upper bounds (Fig. 16), and each of these upper bounds is larger than every a_n. If L is the smallest of the upper bounds (the least upper bound) of $\{ a_n \}$, then there is a value a_N as close as we want to L (otherwise there would be an upper bound smaller than L). Finally, if a_N is close to L, then the later values, a_n with n ≥ N , are even closer to L because $\{ a_n \}$ is monotonically increasing, so L is the limit of $\{ a_n \}$.

Cauchy and other mathematicians accepted this theorem on intuitive and geometric grounds similar to the "idea for a proof" given above, but later mathematicians felt more rigor was needed. However, even the mathematician Dedekind who supplied much of that rigor recognized the usefulness of geometric intuition.

"Even now such resort to geometric intuition in a first presentation of differential calculus, I regard as exceedingly useful, from a didactic standpoint, and indeed indispensable if one does not wish to lose too much time." (Dedekind, Essays on the Theory of Numbers, 1901 (Dover, 1963), pp. 1–2)

PROBLEMS   In problems $1 - 6$, find a rule which describes the given numbers in the sequence.

1. $1, 1/4, 1/9, 1/16, 1/25, \ldots$ 2. $1, 1/8, 1/27, 1/64, 1/125, \ldots$ 3. $0, 1/2, 2/3, 3/4, 4/5, \ldots$

4. $-1, 1/3, -1/9, 1/27, -1/81, \ldots$ 5. $1/2, 2/4, 3/8, 4/16, 5/32, \ldots$ 6. $7, 7, 7, 7, 7, \ldots$

(Bonus: O, T, T, F, F, S, S, E, ? , ?)

In problems $7 - 18$, calculate the first 6 terms (starting with $n = 1$) of each sequence and graph these terms.

7. $\left\{ 1 - \dfrac{2}{n} \right\}$ 8. $\left\{ 3 + \dfrac{1}{n^2} \right\}$ 9. $\left\{ \dfrac{n}{2n-1} \right\}$ 10. $\left\{ \dfrac{\ln(n)}{n} \right\}$

11. $\left\{ 3 + \dfrac{(-1)^n}{n} \right\}$ 12. $\left\{ 4 + (-1)^n \right\}$ 13. $\left\{ (-1)^n \dfrac{n-1}{n} \right\}$ 14. $\left\{ \cos(n\pi/2) \right\}$

15. $\left\{ \dfrac{1}{n!} \right\}$ 16. $\left\{ \dfrac{n+1}{n!} \right\}$ 17. $\left\{ \dfrac{2^n}{n!} \right\}$ 18. $\left\{ \left(1 + \dfrac{1}{n}\right)^n \right\}$

In problems $19 - 24$, calculate the first 10 terms (starting with $n = 1$) of each sequence.

19. $a_1 = 2$ and $a_{n+1} = -a_n$ 20. $b_1 = 3$ and $b_{n+1} = 1/b_n$

21. $\left\{ \sin\left(\dfrac{2\pi n}{3} \right) \right\}$ 22. $a_1 = 2, a_2 = 3$, and, for $n \geq 3$, $a_n = a_{n-1} - a_{n-2}$

23. c_n = the sum of the first n positive integers

24. d_n = the sum of the first n prime numbers (2 is the first prime)

In problems $25 - 28$, state whether each sequence appears to be converging or diverging. If you think the sequence is converging, mark its limit as a value on the vertical axis. (***Important Note:*** *The behavior of a sequence can change drastically after awhile, and the first terms have **no** influence on whether or not the sequence converges. However, sometimes the first few terms are the only values we have, and we need to reach a **tentative** conclusion based on those values.*)

25. Sequences A and B in Fig. 17.   26. Sequences C and D in Fig. 18.

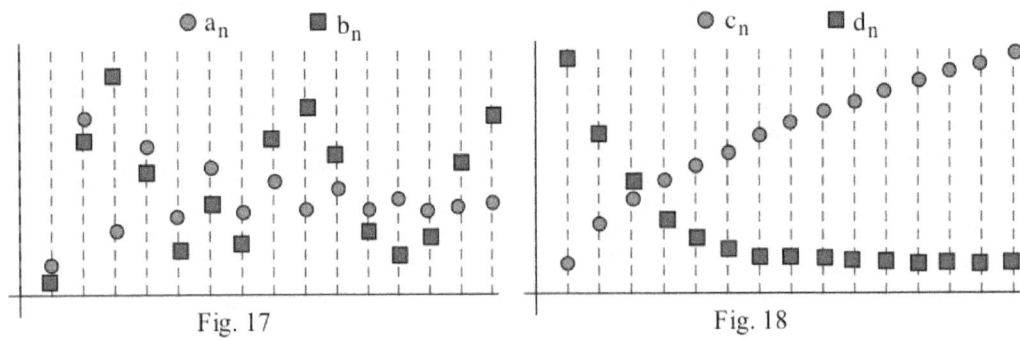

Fig. 17   Fig. 18

27. Sequences E and F in Fig. 19.   28. Sequences G and H in Fig. 20.

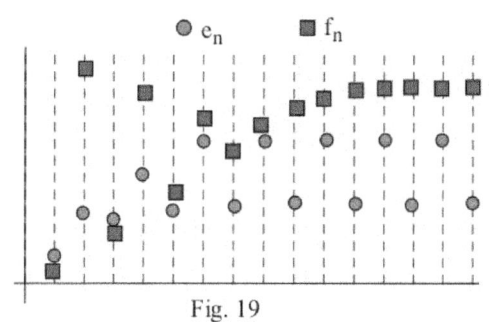

Fig. 19

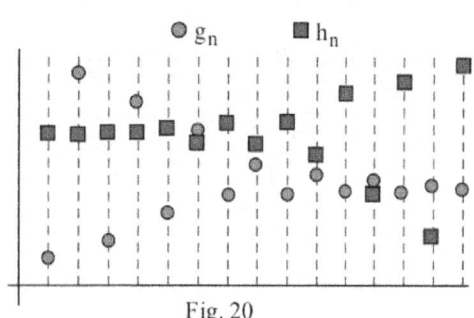

Fig. 20

In problems 29 – 43, state whether each sequence converges or diverges. If the sequence converges, find its limit.

29. $\left\{ 1 - \frac{2}{n} \right\}$
30. $\left\{ n^2 \right\}$
31. $\left\{ \frac{n^2}{n+1} \right\}$
32. $\left\{ 3 + \frac{1}{n^2} \right\}$

33. $\left\{ \frac{n}{2n-1} \right\}$
34. $\left\{ \frac{\ln(n)}{n} \right\}$
35. $\left\{ \ln(3 + \frac{7}{n}) \right\}$
36. $\left\{ 3 + \frac{(-1)^{n+1}}{n} \right\}$

37. $\left\{ 4 + (-1)^n \right\}$
38. $\left\{ (-1)^n \frac{n-1}{n} \right\}$
39. $\left\{ \frac{1}{n!} \right\}$
40. $\left\{ (1 + \frac{3}{n})^n \right\}$

41. $\left\{ (1 - \frac{1}{n})^n \right\}$
42. $\left\{ \frac{\sqrt{n} - 3}{\sqrt{n} + 3} \right\}$
43. $\left\{ \frac{(n+2)(n-5)}{n^2} \right\}$

In problems 44 – 47, prove that the sequence converges to the given limit by showing that for any $\varepsilon > 0$, you can find an N which satisfies the conditions of the definition of convergence.

44. $\lim_{n \to \infty} 2 - \frac{3}{n} = 2$

45. $\lim_{n \to \infty} \frac{3}{n^2} = 0$

46. $\lim_{n \to \infty} \frac{7}{n+1} = 0$

47. $\lim_{n \to \infty} \frac{3n-1}{n} = 3$

In problems 48 – 53, use subsequences to help determine whether the sequence converges or diverges. If the sequence converges, find its limit.

48. $\left\{ (-1)^n 3 \right\}$
49. $\left\{ \frac{1}{n^{\text{th}} \text{ prime}} \right\}$
50. $\left\{ (-1)^n \frac{n+1}{n} \right\}$

51. $\left\{ (-2)^n (\frac{1}{3})^n \right\}$
52. $\left\{ (1 + \frac{1}{3n})^{3n} \right\}$
53. $\left\{ (1 + \frac{5}{n^2})^{(n^2)} \right\}$

In problems 54 – 58, calculate $a_{n+1} - a_n$ and use that value to determine whether $\{ a_n \}$ is monotonic increasing, monotonic decreasing, or neither.

54. $\left\{ \frac{3}{n} \right\}$
55. $\left\{ 7 - \frac{2}{n} \right\}$
56. $\left\{ \frac{n-1}{2n} \right\}$
57. $\left\{ 2^n \right\}$
58. $\left\{ 1 - \frac{1}{2^n} \right\}$

In problems 59 – 63, calculate a_{n+1} / a_n and use that value to determine whether each sequence is monotonic increasing, monotonic decreasing, or neither.

59. $\left\{ \dfrac{n+1}{n!} \right\}$ 60. $\left\{ \dfrac{n}{n+1} \right\}$ 61. $\left\{ \left(\dfrac{5}{4}\right)^n \right\}$ 62. $\left\{ \dfrac{n^2}{n!} \right\}$ 63. $\left\{ \dfrac{n}{e^n} \right\}$

In problems 64 – 68, use derivatives to determine whether each sequence is monotonic increasing, monotonic decreasing, or neither.

64. $\left\{ \dfrac{n+1}{n} \right\}$ 65. $\left\{ 5 - \dfrac{3}{n} \right\}$ 66. $\left\{ n \cdot e^{-n} \right\}$ 67. $\left\{ \cos(1/n) \right\}$ 68. $\left\{ \left(1 + \dfrac{1}{n}\right)^3 \right\}$

In problems 69 – 73, show that each sequence is monotonic.

69. $\left\{ \dfrac{n+3}{n!} \right\}$ 70. $\left\{ \dfrac{n}{n+1} \right\}$ 71. $\left\{ 1 - \dfrac{1}{2^n} \right\}$ 72. $\left\{ \sin(1/n) \right\}$ 73. $\left\{ \dfrac{n+1}{e^n} \right\}$

74. The Fibonacci sequence (after Leonardo Fibonacci (1170–1250) who used it to model a population of rabbits) is obtained by setting the first two terms equal to 1 and then defining each new term as the sum of the two previous terms: $a_n = a_{n-1} + a_{n-2}$ for $n \geq 3$. (a) Write the first 7 terms of this sequence. (b) Calculate the successive ratios of the terms, a_n/a_{n-1}. (These ratios approach the "golden mean," approximately 1.618)

75. Heron's method for approximating roots: To approximate the square root of a positive number N, put $a_1 = N$ and let $a_{n+1} = \dfrac{1}{2}\left(a_n + \dfrac{N}{a_n}\right)$. Then $\{ a_n \}$ converges to $\sqrt{N}$. Calculate a_1 through a_4 for $N = 4, 9,$ and 5. (Heron's method is equivalent to Newton's method applied to the function $f(x) = x^2 - N$.)

76. Hailstone Sequence   For the initial or "seed" value h_0, define the hailstone sequence by the rule

$$h_n = \begin{cases} 3 \cdot h_{n-1} + 1 & \text{if } h_{n-1} \text{ is odd} \\ \dfrac{1}{2} \cdot h_{n-1} & \text{if } h_{n-1} \text{ is even} . \end{cases}$$

Define the **length** of the sequence to be the first value of n so that $h_n = 1$. If the seed value is $h_0 = 3$, then $h_1 = 3(3) + 1 = 10, h_2 = (10)/2 = 5, h_3 = 3(5) + 1 = 16, h_4 = 16/2 = 8, h_5 = 4, h_6 = 2,$ and $h_7 = 1$ so the **length** of the hailstone sequence is 7 for the seed value $h_0 = 3$.

(a) Find the length of the hailstone sequence for each seed value from 2 to 10.

(b) Find the length of the hailstone sequence for a seed value $h_0 = 2^n$.

** (c) Open question (no one has been able to answer the this question): Is the length of the hailstone sequence finite for every seed value?

(This is called the hailstone sequence because for some seed values, the terms of the sequence rise and drop just like the path of a hailstone as it forms. This sequence is attributed to Lothar Collatz and (c) is also called Ulam's conjecture, Syracuse's problem, Kakutani's problem and Hasse's algorithm. "The 3n+1 sequence has probably consumed more CPU time than any other number theoretic conjecture," says Gaston Gonnett of Zurich.)

77. Negative Eugenics: Suppose that individuals with the gene combination "aa" do not reproduce and those with the combinations "aA" and "AA" do reproduce. When the initial proportion of individuals with "aa" is $a_0 = p$ (typically a small number), then the proportion of individuals with "aa" in the kth generation is $a_k = \dfrac{p}{kp + 1}$. Use this formula for a_k to answer the following questions.

   (a) If 2% of a population initially have the combination "aa" and these individuals do not reproduce, then how many generations will it take for the proportion of individuals with "aa" to drop to 1% ?

   (b) In general, find the number of generations until the proportion of individuals with "aa" is half of the initial proportion.

   ("Negative eugenics" is a strategy in which individuals with an undesirable trait are prevented from reproducing. It is not an effective strategy for traits carried by recessive genes (the above example) which are uncommon (p small) in a species which reproduces slowly (people). Mathematics shows that the social strategy of sterilizing people with some undesirable trait, as proposed in the early 20th century, won't effectively reduce the trait in the population.)

78. The fractional part of a number is the number minus its integer part: $x - \text{INT}(x)$. The sequence of fractional parts of multiples of a number x is the sequence with terms $a_n = n{\cdot}x - \text{INT}(n{\cdot}x)$. The behavior of the sequence of fractional parts of the multiples of a number is one way in which rational numbers differ from irrational numbers.

   (a) Let $a_n = nx - \text{INT}(nx)$ be the fractional part of the nth multiple of x . Calculate a_1 through a_6 for $x = 1/3$. These are the fractional parts of the first 6 multiples of 1/3.

   (b) Calculate the fractional parts of the first 9 multiples of 3/4, and 2/5 .

   (c) Calculate the fractional parts of the first 5 multiples of π.

* (d) Let $a_n = n{\cdot}\pi - \text{INT}(n{\cdot}\pi)$ be the fractional part of the nth multiple of π. Is it possible for two different multiples of π to have the same fractional part? (Suggestion: Assume the answer is yes and derive a contradiction. Assume that $a_n = a_m$ for some $m \neq n$, and derive the contradiction that $\pi = \dfrac{\text{INT}(n\pi) - \text{INT}(m\pi)}{n-m}$. Why is this a contradiction?)

An Alternate Way to Visualize Sequences and Convergence

A sequence is a function, and we have graphed sequences in the xy plane in the same way we graphed other functions: since $a_n = f(n)$, we plotted the point (n, a_n). If the sequence $\{a_n\}$ converges to L, then the points (n, a_n) are eventually (for big values of n) close to or on the horizontal line $y = L$.

We can also graph a sequence $\{a_n\}$ in one dimension, on the x–axis. For each value of n, we plot the point $x = a_n$. Then the graph of $\{a_n\}$ consists of a collection of points on the x–axis. Fig. 21 shows the one dimensional graphs of $a_n = \dfrac{1}{n}$, $b_n = 2 + \dfrac{(-1)^n}{n}$, and $c_n = (-1)^n$.

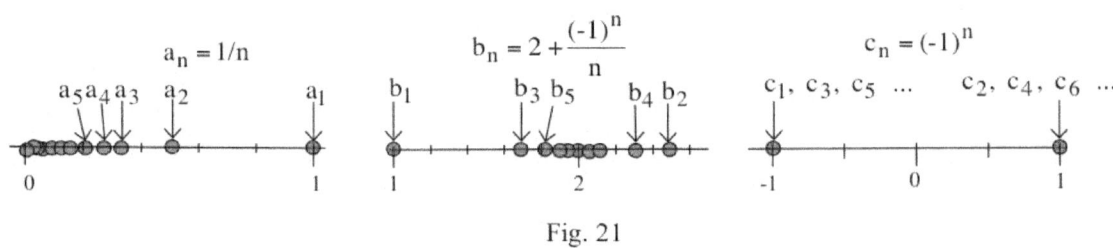

Fig. 21

If $\{a_n\}$ converges to L, then the points $x = a_n$ are eventually (for big values of n) close to or on the point $x = L$. If we build a narrow box, with width $2\varepsilon > 0$, and center the box at the point $x = L$, then all of the points a_n will fall into the box once n is larger than some value N.

79. Suppose that the sequence $\{a_n\}$ converges to 3 and that you place a single grain of sand at each point $x = a_n$ on the x–axis. Describe the likely result (a) after a few grains have been placed and (b) after a lot (thousands or millions) of grains have been placed.

80. Suppose the sequence $\{a_n\}$ converges to 3, $\{b_n\}$ converges to 1, and that you place a single grain of sand at each point $(x, y) = (a_n, b_n)$ on the xy–plane. Describe the likely result (a) after a few grains have been placed and (b) after a lot (thousands or millions) of grains have been placed.

81. Suppose that $a_n = \sin(n)$ for positive integers n. If you place a single grain of sand at each point $x = a_n$ on the x–axis. Describe the likely result (a) after a few grains have been placed and (b) after a lot (thousands or millions) of grains have been placed. (c) Do two grains ever end up on the same point?

82. Suppose that $a_n = \cos(n)$, and $b_n = \sin(n)$ for positive integers n. If you place a single grain of sand at each point $(x, y) = (a_n, b_n)$ on the xy–plane. Describe the likely result (a) after a few grains have been placed and (b) after a lot (thousands or millions) of grains have been placed.

Practice Answers

Practice 1: One layer transmits $2/3$ of the original light. Two layers transmit $(\frac{2}{3})(\frac{2}{3}) = (\frac{2}{3})^2$ of the original light. Three layers transmit $(\frac{2}{3})^3$, and, in general, n layers transmit $(\frac{2}{3})^n$ of the original light.

Practice 2:   (a) $1, 1/2, 1/3, 1/4, 1/5, 1/6, \ldots, 1/n, \ldots$

             (b) $-1/2, 1/4, -1/8, 1/16, -1/32, 1/64, \ldots, (-1/2)^n$ or $(-1)^n (1/2)^n, \ldots$

             (c) $2, 2, 2, 2, 2, \ldots, 2, \ldots$

Practice 3: $c_n = a_n - b_n$: $c1 = (3 - \frac{1}{1}) - (\frac{1}{2}) = \frac{3}{2} = 1.5$, $c_2 = (3 - \frac{1}{2}) - (\frac{1}{2^2}) = \frac{9}{4} = 2.25$,

$c_3 = (3 - \frac{1}{3}) - (\frac{1}{2^3}) = \frac{61}{24} \approx 2.542$, $c_4 = (3 - \frac{1}{4}) - (\frac{1}{2^4}) = \frac{43}{16} \approx 2.687$,

$c_5 = (3 - \frac{1}{5}) - (\frac{1}{2^5}) = \frac{443}{160} \approx 2.769$

$d_n = (-1)^n b_n$: $d_1 = (-1)^1 (\frac{1}{2}) = -\frac{1}{2}$, $d_2 = (-1)^2 (\frac{1}{2^2}) = \frac{1}{4}$, $d_3 = (-1)^3 (\frac{1}{2^3}) = -\frac{1}{8}$,

$d_4 = \frac{1}{16}$, $d_2 - \frac{1}{32}$

Practice 4: $a_0 = 3$: $a_1 = f(a_0) = 2(3) - 1 = 5$, $a_2 = f(a_1) = 2(5) - 1 = 9$, $a_3 = f(a_2) = 2(9) - 1 = 17$

Practice 5: $b_1 = b_0 + \frac{1}{3} = 0 + \frac{1}{3} = \frac{1}{3}$, $b_2 = b_1 + \frac{1}{9} = \frac{4}{9}$, $b_3 = b_2 + \frac{1}{27} = \frac{13}{27}$, $b_4 = b_3 + \frac{1}{81} = \frac{40}{81}$

Practice 6: For $a_n = (n+1)/n$ show that $\lim\limits_{n \to \infty} a_n = 1$:

We need to show that for any positive ε, there is a number N so that the distance from $a_n = \frac{n+1}{n}$ to L, $|a_n - L|$, is less than ε whenever n is larger than N.

For this particular $|a_n - L| = |\frac{n+1}{n} - 1| = |\frac{n}{n} + \frac{1}{n} - 1| = |\frac{1}{n}|$.

To determine what N might be, we solve the inequality

$|\frac{1}{n}| < \varepsilon$ for n in terms of ε: $|1/n| < \varepsilon$ so $1/\varepsilon < n$ and $n > 1/\varepsilon$.

So for **any** positive ε, we can take $N = 1/\varepsilon$ (or any larger number).

Then $n > N = 1/\varepsilon$ implies that $\varepsilon > \frac{1}{n} = |\frac{1}{n}| = |\frac{n+1}{n} - 1| = |a_n - L|$.

Practice 7: $\lim\limits_{n\to\infty} \dfrac{\ln(n)}{n}$: Since $\lim\limits_{n\to\infty} \ln(x) = \infty$ and $\lim\limits_{n\to\infty} x = \infty$, we can use L'Hopital's rule

(Section 3.7). $\mathbf{D}(\ln(x)) = \dfrac{1}{x}$ and $\mathbf{D}(x) = 1$, so $\lim\limits_{n\to\infty} \dfrac{\ln(n)}{n} = \lim\limits_{x\to\infty} \dfrac{1/x}{1} = 0$.

Then $\lim\limits_{x\to\infty} \dfrac{\ln(x)}{x} = 0$ and $\lim\limits_{n\to\infty} \dfrac{\ln(n)}{n} = 0$.

Practice 8:   We can show that $a_n = \{ \sin(\frac{\pi n}{2}) \}$ diverges by finding two subsequences b_n and c_n of

a_n so that the subsequences b_n and c_n converge to different limiting values.

Let $\{ b_n \}$ consist of the terms $\{ a_1 , a_5 , a_9 , \ldots , a_{4n-3} , \ldots \} = \{ \sin(\pi/2), \sin(5\pi/2),$

$\sin(9\pi/2), \ldots , \sin(\frac{(4n-3)\pi}{2}), \ldots \} = \{ 1, 1, 1, \ldots , 1 , \ldots \}$. Then $\lim\limits_{n\to\infty} b_n = 1$.

Let $\{ c_n \}$ consist of the terms $\{ a_2 , a_4 , a_6 , \ldots , a_{2n} , \ldots \} = \{ \sin(2\pi/2), \sin(4\pi/2),$

$\sin(6\pi/2), \ldots , \sin(\frac{2n\pi}{2}), \ldots \} = \{ 0, 0, 0, \ldots , 0 , \ldots \}$. Then $\lim\limits_{n\to\infty} c_n = 0$.

Since the subsequences $\{ b_n \}$ and $\{ c_n \}$ have different limits, we can conclude that the

original sequence $\{ a_n \}$ diverges. (Note: Many other pairs of subsequences also work in

place of the $\{ b_n \}$ and $\{ c_n \}$ that we used.)

Practice 9:   (i)   by showing that $a_{n+1} \le a_n$ for all n,

(ii)   by showing that all the a_n are positive and $\dfrac{a_{n+1}}{a_n} \le 1$ for all n, or

(iii)   by showing that $a_n = f(n)$ for integer values n and $f'(x) \le 0$ for all $x \ge 1$.

Practice 10:   Show that $\left\{ (\frac{2}{3})^n \right\}$ is monotonically decreasing.

Using method (i) of Practice 9: $(\frac{2}{3}) < 1$ so multiplying each side by $(\frac{2}{3})^n > 0$ we

have $(\frac{2}{3})(\frac{2}{3})^n < 1(\frac{2}{3})^n$ and $(\frac{2}{3})^{n+1} < (\frac{2}{3})^n$ so $a_{n+1} < a_n$.

We could have used method (ii) of Practice 9 instead: $a_{n+1} = (\frac{2}{3})^{n+1}$ and $a_n = (\frac{2}{3})^n$ so

$\dfrac{a_{n+1}}{a_n} = \dfrac{(2/3)^{n+1}}{(2/3)^n} = \dfrac{2}{3} < 1$ so $\left\{ (\frac{2}{3})^n \right\}$ is monotonically decreasing.

Appendix: Proof of the Uniqueness Theorem

The proof starts by assuming that the limit is not unique. Then we show that this assumption of nonuniqueness leads to a contradiction so the assumption is false and the limit is unique.

Suppose that a sequence $\{ a_n \}$ converges to two different limits L_1 and L_2. Then the distance between L_1 and L_2 is $d = |L_1 - L_2| > 0$. Since $\{ a_n \}$ converges to L_1, then for any $\varepsilon > 0$ there is an N_1 so that $n > N_1$ implies $|a_n - L_1| < \varepsilon$. Take $\varepsilon = d/3$. Then there is an N_1 so $n > N_1$ implies $|a_n - L_1| < \varepsilon = d/3$. Similarly, there is an N_2 so that $n > N_2$ implies $|a_n - L_2| < \varepsilon = d/3$. Finally, if n is larger than both N_1 and N_2, then both conditions are satisfied and

$$
\begin{aligned}
d = |L_1 - L_2| \ &= \ |(a_n - L_2) - (a_n - L_1)| && \text{by adding } 0 = a_n - a_n \text{ inside the absolute value} \\
&\leq |a_n - L_2| + |a_n - L_1| && \text{by the Triangle Inequality} \\
&< \ d/3 \ + \ d/3 && \text{since } |a_n - L_1| < \varepsilon = d/3 \text{ and } |a_n - L_2| < \varepsilon = d/3 \\
&= 2d/3.
\end{aligned}
$$

We have found that $0 < d < \frac{2}{3} d$, a contradiction, and we can conclude the assumption, $L_1 \neq L_2$, is false. A sequence **can not** converge to two different values.

(Note: *We chose $\varepsilon = d/3$ because it "works" for our purpose by leading to a contradiction. Several other choices, any ε less than $d/2$, also lead to the contradiction. Since the definition says "for **any** $\varepsilon > 0$," we picked one we wanted.*)

10.2   INFINITE SERIES

Our goal in this section is to add together the numbers in a sequence. Since it would take a "very long time" to add together the infinite number of numbers, we first consider finite sums, look for patterns in these finite sums, and take limits as more and more numbers are included in the finite sums.

What does it mean to add together an infinite number of terms? We will define that concept carefully in this section. Secondly, is the sum of all the terms a finite number? In the next few sections we will examine a variety of techniques for determining whether an infinite sum is finite. Finally, if we know the sum is finite, can we determine the value of the sum? The difficulty of finding the exact value of the sum varies from very easy to very, very difficult.

Example 1:   A golf ball is thrown 9 feet straight up into the air, and on each bounce it rebounds to two thirds of its previous height (Fig. 1). Find a sequence whose terms give the distances the ball travels during each successive bounce. Represent the **total** distance traveled by the ball as a sum.

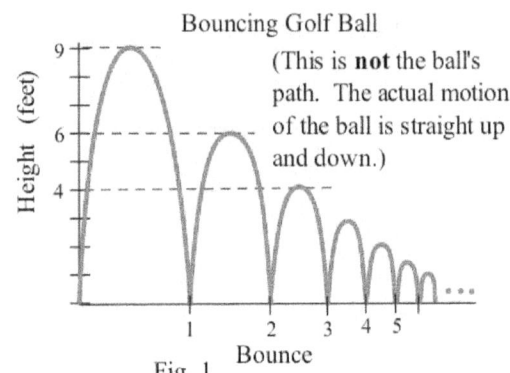

Bouncing Golf Ball
(This is **not** the ball's path. The actual motion of the ball is straight up and down.)

Fig. 1

Solution: The heights of the successive bounces are 9 feet,

$(\frac{2}{3})\cdot 9$ feet, $(\frac{2}{3})\cdot[(\frac{2}{3})\cdot 9]$ feet, $(\frac{2}{3})^3\cdot 9$ feet, and so forth. On each bounce, the ball rises and falls so the distance traveled is twice the height of that bounce:

$$18 \text{ feet}, (\frac{2}{3})\cdot 18 \text{ feet}, (\frac{2}{3})\cdot(\frac{2}{3})\cdot 18 \text{ feet}, (\frac{2}{3})^3\cdot 18 \text{ feet}, (\frac{2}{3})^4\cdot 18 \text{ feet}, \ldots .$$

The total distance traveled is the sum of the bounce–distances:

$$\text{total distance} \quad = \quad 18 + (\frac{2}{3})\cdot 18 + (\frac{2}{3})\cdot(\frac{2}{3})\cdot 18 + (\frac{2}{3})^3\cdot 18 + (\frac{2}{3})^4\cdot 18 + \ldots$$

$$= \quad 18\{ 1 + \frac{2}{3} + (\frac{2}{3})^2 + (\frac{2}{3})^3 + (\frac{2}{3})^4 + \ldots \}$$

At the completion of the first bounce the ball has traveled 18 feet. After the second bounce, it has traveled 30 feet, a total of 38 feet after the third bounce, $43\frac{1}{3}$ feet after the fourth, and so on. With a calculator and some patience, we see that after the 20th bounce the ball has traveled a total of approximately 53.996 feet, after the 30th bounce approximately 53.99994 feet, and after the 40th bounce approximately 53.9999989 feet.

Practice 1:   A tennis ball is thrown 10 feet straight up

into the air, and on each bounce it rebounds to 40%

of its previous height. Represent the total distance

traveled by the ball as a sum, and find the total distance

traveled by the ball after the completion of its third

bounce. (Fig. 2)

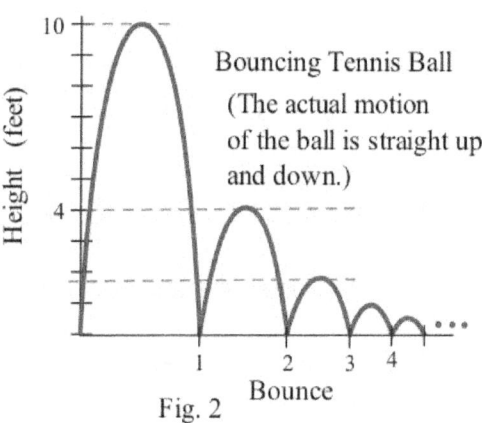

Fig. 2

Infinite Series

The infinite sums in the Example and Practice are called infinite series, and they are the objects we will

start to examine in this section.

Definitions

An **infinite series** is an expression of the form

$$a_1 + a_2 + a_3 + a_4 + \ldots \quad \text{or} \quad \sum_{k=1}^{\infty} a_k \; .$$

The numbers $a_1, a_2, a_3, a_4, \ldots$ are called the **terms** of the series. (Fig. 3)

Example 2:   Represent the following series using the sigma notation. (a) $1 + 1/3 + 1/9 + 1/27 + \ldots$,

(b) $-1 + 1/2 - 1/3 + 1/4 - 1/5 + \ldots$, (c) $18(2/3 + 4/9 + 8/27 + 16/81 + \ldots)$

(d) $0.777 \ldots = 7/10 + 7/100 + 7/1000 + \ldots$, and (e) $0.222\ldots$

Solution: (a) $1 + 1/3 + 1/9 + 1/27 + \ldots = \displaystyle\sum_{k=0}^{\infty} (\tfrac{1}{3})^k$ or $\displaystyle\sum_{k=1}^{\infty} (\tfrac{1}{3})^{k-1}$

(b) $-1 + 1/2 - 1/3 + 1/4 - 1/5 + \ldots = \displaystyle\sum_{k=1}^{\infty} (-1)^k \tfrac{1}{k}$ (c) $18 \displaystyle\sum_{k=1}^{\infty} (\tfrac{2}{3})^k$

(d) $0.777 \ldots = 7/10 + 7/100 + 7/1000 + \ldots = \displaystyle\sum_{k=1}^{\infty} \tfrac{7}{10^k}$ (e) $\displaystyle\sum_{k=1}^{\infty} \tfrac{2}{10^k}$

Practice 2:   Represent the following series using the sigma notation. (a) $1 + 2 + 3 + 4 + \ldots$,

(b) $-1 + 1 - 1 + 1 - \ldots$ (c) $2 + 1 + 1/2 + 1/4 + \ldots$

(d) $1/2 + 1/4 + 1/6 + 1/8 + 1/10 + \ldots$ (e) $0.111\ldots$

In order to determine if the infinite series adds up to a finite value, we examine the sums as more and more

terms are added.

Definition

The **partial sums** s_n of the infinite series $\displaystyle\sum_{k=1}^{\infty} a_k$

    are the numbers

        $s_1 = a_1,$

        $s_2 = a_1 + a_2,$

        $s_3 = a_1 + a_2 + a_3,$

        . . .

In general, $\displaystyle s_n = a_1 + a_2 + a_3 + \ldots + a_n = \sum_{k=1}^{n} a_k$

    or, recursively, as $s_n = s_{n-1} + a_n$.

The partial sums form the **sequence of partial sums** $\{ s_n \}$.

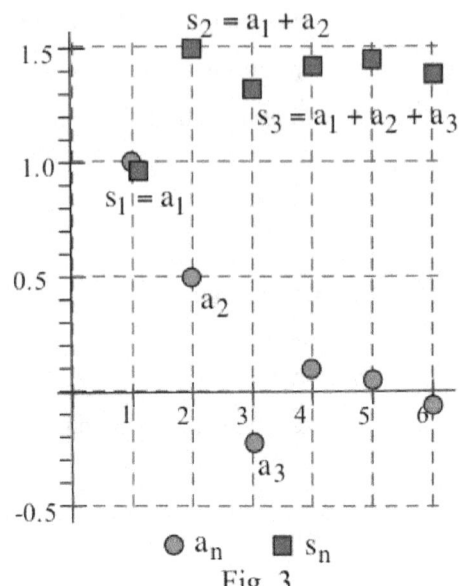

Fig. 3

Example 3:   Calculate the first 4 partial sums for the following series.

$$\text{(a) } 1 + 1/2 + 1/4 + 1/8 + 1/16 + \ldots, \text{ (b) } \sum_{k=1}^{\infty} (-1)^k, \text{ and (c) } \sum_{n=1}^{\infty} \frac{1}{n} .$$

Solution: (a) $s_1 = 1, s_2 = 1 + 1/2 = 3/2, s_3 = 1 + 1/2 + 1/4 = 7/4, s_4 = 1 + 1/2 + 1/4 + 1/8 = 15/8$

        It is usually easier to use the recursive version of s_n :

        $s_3 = s_2 + a_3 = 3/2 + 1/4 = 7/4$ and $s_4 = s_3 + a_4 = 7/4 + 1/8 = 15/8$.

        (b) $s_1 = (-1)^1 = -1,\ s_2 = s_1 + a_2 = -1 + (-1)^2 = 0, s_3 = s_2 + a_3 = 0 + (-1)^3 = -1, s_4 = 0.$

        (c) $s_1 = 1, s_2 = 3/2, s_3 = 11/6, s_4 = 25/12$.

Practice 3:   Calculate the first 4 partial sums for the following series.

$$\text{(a) } 1 - 1/2 + 1/4 - 1/8 + 1/16 - \ldots, \text{ (b) } \sum_{k=1}^{\infty} \left(\frac{1}{3}\right)^k, \text{ and (c) } \sum_{n=2}^{\infty} \frac{(-1)^n}{n} .$$

If we know the values of the partial sums s_n, we can recover the values of the terms a_n used to build the s_n.

Example 4:   Suppose $s_1 = 2.1, s_2 = 2.6, s_3 = 2.84$, and $s_4 = 2.87$ are the first partial sums of $\displaystyle\sum_{k=1}^{\infty} a_k$. Find

        the values of the first four terms of $\{ a_n \}$.

Solution: $s_1 = a_1$ so $a_1 = 2.1$. $s_2 = a_1 + a_2$ so $2.6 = 2.1 + a_2$ and $a_2 = 0.5$.

        Similarly, $s_3 = a_1 + a_2 + a_3$ so $2.84 = 2.1 + 0.5 + a_3$ and $a_3 = 0.24$. Finally, $a_4 = 0.03$.

An alternate solution method starts with $a_1 = s_1$ and then uses the fact that $s_n = s_{n-1} + a_n$ so $a_n = s_n - s_{n-1}$. Then

$$a_2 = s_2 - s_1 = 2.6 - 2.1 = 0.5 \ .$$

$$a_3 = s_3 - s_2 = 2.84 - 2.6 = 0.24, \text{ and}$$

$$a_4 = s_4 - s_3 = 2.87 - 2.84 = 0.03 \ .$$

Practice 4:   Suppose $s_1 = 3.2$, $s_2 = 3.6$, $s_3 = 3.5$, $s_4 = 4$, $s_{99} = 7.3$, $s_{100} = 7.6$, and $s_{101} = 7.8$ are partial

sums of $\displaystyle\sum_{k=1}^{\infty} a_k$. Find the values of a_1, a_2, a_3, a_4 , and a_{100} .

Example 5:   Graph the first five **terms** of the series $\displaystyle\sum_{k=1}^{\infty} (\frac{1}{2})^k$. Then graph the first five **partial sums**.

Solution: $a_1 = \left(\dfrac{1}{2}\right)^1 = \dfrac{1}{2}, \ a_2 = \left(\dfrac{1}{2}\right)^2 = \dfrac{1}{4}, \ a_3 = \left(\dfrac{1}{2}\right)^3 = \dfrac{1}{8}, \ a_4 = \dfrac{1}{16}, \ a_4 = \dfrac{1}{32}$

$s_1 = \dfrac{1}{2}, \ s_2 = \dfrac{1}{2} + \dfrac{1}{4} = \dfrac{3}{4}, \ s_3 = \dfrac{1}{2} + \dfrac{1}{4} + \dfrac{1}{8} = \dfrac{7}{8}, \ s_4 = \dfrac{15}{16}, \text{ and } s_5 = \dfrac{31}{32}.$

These values are graphed in Fig. 4.

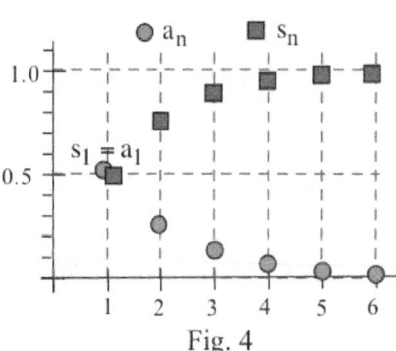

Fig. 4

Practice 5:   Graph the first five terms of the series

$\displaystyle\sum_{k=1}^{\infty} (-\frac{1}{2})^k$. Then graph the first five partial sums.

Convergence of a Series

The convergence of a series is defined in terms of the behavior of the sequence of partial sums. If the partial sums, the sequence obtained by adding more and more of the terms of the series, approaches a finite number, we say the **series** converges to that finite number. If the sequence of partial sums diverges (does not approach a single finite number), we say that the **series** diverges.

Definitions

Let $\{ s_n \}$ be the sequence of partial sums

of the series $\displaystyle\sum_{k=1}^{\infty} a_k$: $s_n = \displaystyle\sum_{k=1}^{n} a_k$

If $\{ s_n \}$ is a convergent sequence,   (Fig. 5)

we say the series $\displaystyle\sum_{k=1}^{\infty} \mathbf{a_k}$ **converges**.

If the sequence of partial sums $\{ s_n \}$ converges to A ,

we say the series $\displaystyle\sum_{k=1}^{\infty} \mathbf{a_k}$ **converges to A**

or **the sum of the series is A** ,

and we write $\displaystyle\sum_{k=1}^{\infty} \mathbf{a_k} = \mathbf{A}$.

If the sequence of partial sums $\{ s_n \}$ diverges,

we say the series $\displaystyle\sum_{k=1}^{\infty} \mathbf{a_k}$ **diverges**.

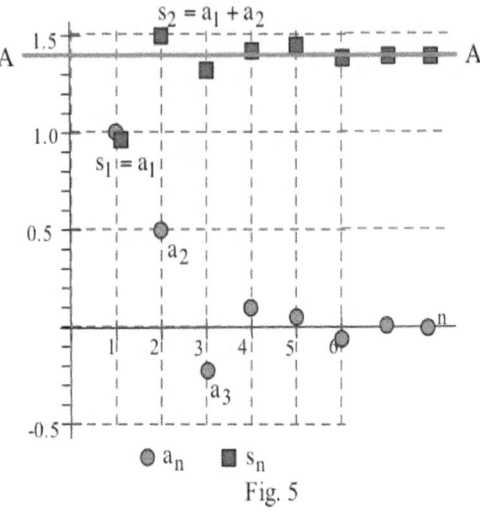

Fig. 5

Example 6:   In the next section we present a method for determining that the nth partial sum

of $\displaystyle\sum_{k=1}^{\infty} (\tfrac{1}{2})^k$ is $s_n = \dfrac{2^n - 1}{2^n} = 1 - \dfrac{1}{2^n}$.   Use this result to evaluate the limit of $\{ s_n \}$.

Does the series $\displaystyle\sum_{k=1}^{\infty} (\tfrac{1}{2})^k$ converge? If so, to what value?

Solution: $\displaystyle\lim_{n\to\infty} s_n = \lim_{n\to\infty} 1 - \dfrac{1}{2^n} = 1$ (Fig. 6), so $\displaystyle\sum_{k=1}^{\infty} (\tfrac{1}{2})^k$ converges to 1: $\displaystyle\sum_{k=1}^{\infty} (\tfrac{1}{2})^k = 1$.

Practice 6:   The nth partial sum of $\displaystyle\sum_{k=1}^{\infty} (-\tfrac{1}{2})^k$ is $s_n = -\dfrac{1}{3} + \dfrac{1}{3} \cdot (-\tfrac{1}{2})^n$.   Use this result to

evaluate the limit of $\{ s_n \}$. Does the series $\displaystyle\sum_{k=1}^{\infty} (-\tfrac{1}{2})^k$ converge? If so, to what value?

The next theorem says that if a series converges, then the terms of the series must approach 0. When a series is convergent then the partial sums s_n approach a finite limit (Fig. 3) so all of the s_n must be close to that limit when n is large. Then s_n and s_{n-1} must be close to each other (why?) and $a_n = s_n - s_{n-1}$ must be close to 0.

Theorem: If the series $\displaystyle\sum_{k=1}^{\infty} a_k$ converges, then $\displaystyle\lim_{k \to \infty} a_k = 0$.

We can NOT use this theorem to conclude that a series converges. If the terms of the series do approach 0, then the series may or may not converge — more information is needed to draw a conclusion. However, an alternate form of the theorem, called the nth Term Test for Divergence, is **very useful for quickly concluding that some series diverge.**

---

n[th] Term Test for Divergence of a Series

If   the terms a_n of a series do not approach 0 (as "n→∞")

then   the series diverges:

$$\text{if } \lim_{n \to \infty} a_n \neq 0, \text{ then } \sum_{n=1}^{\infty} a_n \text{ diverges.}$$

Example 7:   Which of these series diverge **by the nth Term Test?**

(a) $\displaystyle\sum_{n=1}^{\infty} (-1)^n$ (b) $\displaystyle\sum_{n=1}^{\infty} (\frac{3}{4})^n$ (c) $\displaystyle\sum_{n=1}^{\infty} (1 + \frac{1}{n})^n$ (d) $\displaystyle\sum_{n=1}^{\infty} \frac{1}{n}$

Solution: (a) $a_n = (-1)^n$ oscillates between -1 and $+1$ and does not approach 0. $\displaystyle\sum_{n=1}^{\infty} (-1)^n$ **diverges.**

(b) $a_n = (\frac{3}{4})^n$ approaches 0 so $\displaystyle\sum_{n=1}^{\infty} (\frac{3}{4})^n$ **may or may not** converge.

(c) $a_n = (1 + \frac{1}{n})^n$ approaches $e \neq 0$, so $\displaystyle\sum_{n=1}^{\infty} (1 + \frac{1}{n})^n$ **diverges.**

(d) $a_n = \frac{1}{n}$ approaches 0 so $\displaystyle\sum_{n=1}^{\infty} \frac{1}{n}$ **may or may not** converge.

We can be certain that (a) and (c) diverge. We don't have enough information yet to decide about (b) and (d). (In the next section we show that (b) converges and (d) diverges.)

Practice 7:   Which of these series diverge by the nth Term Test?

(a) $\displaystyle\sum_{n=1}^{\infty} (-0.9)^n$ (b) $\displaystyle\sum_{n=1}^{\infty} (1.1)^n$ (c) $\displaystyle\sum_{n=1}^{\infty} \sin(n\pi)$ (d) $\displaystyle\sum_{n=1}^{\infty} \frac{1}{\sqrt{n}}$

New Series From Old

If we know about the convergence of a series, then we also know about the convergence of several related series.

- Inserting or deleting a "few" terms, any **finite** number of terms, does not change the convergence or divergence of a series. The insertions or deletions typically change the sum (the limit of the partial sums), but they do not change whether or not the series converges. (Inserting or deleting an infinite number of terms can change the convergence or divergence.)

- Multiplying each term in a series by a nonzero constant does not change the convergence or divergence of a series:

    Suppose $c \neq 0$. $\displaystyle\sum_{n=1}^{\infty} a_n$ converges if and only if $\displaystyle\sum_{n=1}^{\infty} c \cdot a_n$ converges.

- Term–by–term addition and subtraction of the terms of two convergent series result in convergent series. (Term by term multiplication and division of series do not have such nice results.)

Theorem:

If $\displaystyle\sum_{n=1}^{\infty} a_n$ and $\displaystyle\sum_{n=1}^{\infty} b_n$ converge with $\displaystyle\sum_{n=1}^{\infty} a_n = A$ and $\displaystyle\sum_{n=1}^{\infty} b_n = B$,

then $\displaystyle\sum_{n=1}^{\infty} C \cdot a_n = C \cdot A$,

$\displaystyle\sum_{n=1}^{\infty} (a_n + b_n) = A + B$, and

$\displaystyle\sum_{n=1}^{\infty} (a_n - b_n) = A - B$.

The proofs of these statements follow directly from the definition of convergence of a series and from results about convergence of sequences (of partial sums).

PROBLEMS

In problems 1 – 6, rewrite each sum using sigma notation starting with $k = 1$.

1. $1 + \frac{1}{2} + \frac{1}{3} + \frac{1}{4} + \frac{1}{5} + ...$

2. $1 + \frac{1}{4} + \frac{1}{9} + \frac{1}{16} + \frac{1}{25} + ...$

3. $\frac{2}{3} + \frac{2}{6} + \frac{2}{9} + \frac{2}{12} + \frac{2}{15} + \frac{2}{18} + ...$

4. $\sin(1) + \sin(8) + \sin(27) + \sin(64) + \sin(125) + ...$

5. $(-\frac{1}{2}) + (\frac{1}{4}) + (-\frac{1}{8}) + (\frac{1}{16}) + (-\frac{1}{32}) + ...$

6. $(-\frac{1}{3}) + (\frac{1}{9}) + (-\frac{1}{27}) + (\frac{1}{81}) + (-\frac{1}{243}) + ...$

In problems 7 – 14, calculate and graph the first four partial sums s_1 to s_4 of the given series $\sum\limits_{n=1}^{\infty} a_n$.

7. $\sum\limits_{n=1}^{\infty} n^2$

8. $\sum\limits_{n=1}^{\infty} (-1)^n$

9. $\sum\limits_{n=1}^{\infty} \frac{1}{n+2}$

10. $\sum\limits_{n=1}^{\infty} \left\{ \frac{1}{n} - \frac{1}{n+1} \right\}$

11. $\sum\limits_{n=1}^{\infty} \frac{1}{2^n}$

12. $\sum\limits_{n=1}^{\infty} (-\frac{1}{2})^n$

In problems 13 – 18, the first five partial sums s_1 to s_5 are given. Find the first four terms a_1 to a_4 of the series.

13. $s_1 = 3$, $s_2 = 2$, $s_3 = 4$, $s_4 = 5$, $s_5 = 3$

14. $s_1 = 3$, $s_2 = 5$, $s_3 = 4$, $s_4 = 6$, $s_5 = 5$

15. $s_1 = 4$, $s_2 = 4.5$, $s_3 = 4.3$, $s_4 = 4.8$, $s_5 = 5$

16. $s_1 = 4$, $s_2 = 3.7$, $s_3 = 3.9$, $s_4 = 4.1$, $s_5 = 4$

17. $s_1 = 1, s_2 = 1.1, s_3 = 1.11, s_4 = 1.111, s_5 = 1.1111$

18. $s_1 = 1, s_2 = 0.9, s_3 = 0.93, s_4 = 0.91, s_5 = 0.92$

In problems 19 – 28 , represent each repeating decimal as a series using the sigma notation.

19. 0.888 ...

20. 0.333 ...

21. 0.555 ...

22. 0.111 ...

23. 0.aaa ...

24. 0.232323 ...

25. 0.171717 ...

26. 0.838383 ...

27. 0.070707 ...

28. 0.ababab ...

29. Find a pattern for a fraction representation of the repeating decimal 0.abcabcabc ... .

30. A golf ball is thrown 20 feet straight up into the air, and on each bounce it rebounds to 60% of its previous height. Represent the total distance traveled by the ball as a sum.

31. A "super ball" is thrown 15 feet straight up into the air, and on each bounce it rebounds to 80% of its previous height. Represent the total distance traveled by the ball as a sum.

32. Each special washing of a pair of overalls removes 80% of the radioactive particles attached to the overalls. Represent, as a sequence of numbers, the percent of the original radioactive particles that remain after each washing.

33. Each week, 20% of the argon gas in a container leaks out of the container. Represent, as a sequence of numbers, the percent of the original argon gas that remains in the container at the end of the 1st, 2nd, 3rd, and n^{th} weeks.

34. Eight people are going on an expedition by horseback through desolate country. The people and scientific equipment (fishing gear) require 12 horses, and additional horses are needed to carry food for the horses. Each horse can carry enough food to feed 2 horses for the trip. Represent the number of horses needed to carry food as a sum. (Start of a solution: The original 12 horses will require 6 new horses to carry their food. The 6 new horses require 3 additional horses to carry their food. The 3 additional horses require another 1.5 horses to carry food for them, etc.)

Which of the series in problems 35 – 43, definitely diverge by the nth Term Test? What can we conclude about the other series in these problems?

35. $\displaystyle\sum_{n=1}^{\infty} \left(\frac{1}{4}\right)^n$

36. $\displaystyle\sum_{n=1}^{\infty} \frac{7}{n}$

37. $\displaystyle\sum_{n=1}^{\infty} \left(\frac{4}{3}\right)^n$

38. $\displaystyle\sum_{n=1}^{\infty} \left(-\frac{7}{4}\right)^2$

39. $\displaystyle\sum_{n=1}^{\infty} \frac{\sin(n)}{n}$

40. $\displaystyle\sum_{n=1}^{\infty} \frac{\ln(n)}{n}$

41. $\displaystyle\sum_{n=1}^{\infty} \cos\left(\frac{1}{n}\right)$

42. $\displaystyle\sum_{n=1}^{\infty} \frac{n^2-20}{n^2+4}$

43. $\displaystyle\sum_{n=1}^{\infty} \frac{n^2-20}{n^5+4}$

Practice Answers

Practice 1:   The heights of the bounces are $10, (0.4)\cdot 10, (0.4)\cdot(0.4)\cdot 10, (0.4)^3\cdot 10, \ldots$ so the distances traveled (up and down) by the ball are $20, (0.4)\cdot 20, (0.4)\cdot(0.4)\cdot 20, (0.4)^3\cdot 20, \ldots$

The total distance traveled is

$$20 + (0.4)\cdot 20 + (0.4)^2\cdot 20 + (0.4)^3\cdot 20 + \ldots = 20\{ 1 + 0.4 + (0.4)^2 + (0.4)^3 + \ldots \} = 20 \sum_{k=0}^{\infty} (0.4)^k$$

After 3 bounces the ball has traveled $20 + (0.4)(20) + (0.4)^2(20) = 20 + 8 + 3.2 = 31.2$ feet.

Practice 2:   (a) $\displaystyle\sum_{k=1}^{\infty} k$ (b) $\displaystyle\sum_{k=1}^{\infty} (-1)^k$

(c) $2(1 + \frac{1}{2} + \frac{1}{4} + ...) = 2 \sum_{k=0}^{\infty} (\frac{1}{2})^k$ or $\sum_{k=0}^{\infty} (\frac{1}{2})^{k-1}$ or $\sum_{k=1}^{\infty} (\frac{1}{2})^{k-2}$

(d) $\sum_{k=1}^{\infty} \frac{1}{2k}$ (e) $\frac{1}{10} + \frac{1}{100} + \frac{1}{1000} + ... = \frac{1}{10} + \frac{1}{10^2} + \frac{1}{10^3} + ... = \sum_{k=1}^{\infty} (\frac{1}{10})^k$ or $\sum_{k=1}^{\infty} \frac{1}{10^k}$

Practice 3:   (a) Partial sums: $1, 1/2, 3/4, 5/8$ (b) $\frac{1}{3} + \frac{1}{9} + \frac{1}{27} + ...$; partial sums: $\frac{1}{3}, \frac{4}{9}, \frac{13}{27}, \frac{40}{81}$

(c) $\frac{1}{2} - \frac{1}{3} + \frac{1}{4} - ...$;   partial sums: $\frac{1}{2}, \frac{1}{6}, \frac{10}{24} = \frac{5}{12}, \frac{13}{60}$

Practice 4: $a_1 = s_1 = 3.2$, $a_2 = s_2 - s_1 = (3.6) - (3.2) = 0.4$, $a_3 = s_3 - s_2 = (3.5) - (3.6) = -0.1$,

$a_4 = s_4 - s_3 = (4) - (3.5) = 0.5$, $a_{100} = s_{100} - s_{99} = (7.6) - (7.3) = 0.3$

Practice 5: $a_1 = -1/2, a_2 = 1/4, a_3 = -1/8, a_4 = 1/16, a_5 = -1/32$

$s_1 = a_1 = -\frac{1}{2}$, $s_2 = a_1 + a_2 = -\frac{1}{2} + \frac{1}{4} = -\frac{1}{4}$,

$s_3 = s_2 + a_3 = -\frac{1}{4} - \frac{1}{8} = -\frac{3}{8} \approx -0.375$,

$s_4 = s_3 + a_4 = -\frac{3}{8} + \frac{1}{16} = -\frac{5}{16} \approx -0.3125$,

$s_5 = s_4 + a_5 = -\frac{5}{16} - \frac{1}{32} = -\frac{11}{32} \approx -0.34375$,

The graphs of a_n and s_n are shown in Fig. 6.

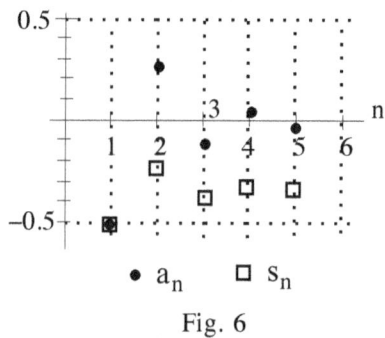

Fig. 6

Practice 6: $\lim_{n \to \infty} s_n = \lim_{n \to \infty} -\frac{1}{3} + \frac{1}{3} \cdot (-\frac{1}{2})^n = -\frac{1}{3} + \frac{1}{3} \cdot 0$

$= -\frac{1}{3}$.

The limit, as "$n \to \infty$" , of s_n is a finite number, so $\sum_{k=1}^{\infty} (-\frac{1}{2})^k$ **converges to $-\frac{1}{3}$** .

Practice 7:   (a) $\sum_{n=1}^{\infty} (-0.9)^n$: $a_n = (-0.9)^n$ approaches 0 so $\sum_{n=1}^{\infty} (-0.9)^n$ **may converge**.

(b) $\sum_{n=1}^{\infty} (1.1)^n$: $a_n = (1.1)^n$ "approaches infinity" so $\sum_{n=1}^{\infty} (1.1)^n$ **diverges**.

(c) $\sum_{n=1}^{\infty} \sin(n\pi) = 0 + 0 + 0 + ...$: $\sin(n\pi) = 0$ "approaches 0" so $\sum_{n=1}^{\infty} \sin(n\pi)$ **may converge**.

(d) $\sum_{n=1}^{\infty} \frac{1}{\sqrt{n}}$: $a_n = \frac{1}{\sqrt{n}}$ approaches 0 so $\sum_{n=1}^{\infty} \frac{1}{\sqrt{n}}$ **may converge**.

(Later in this chapter we will show that series (a) and (c) converge and series (d) diverges.)

10.3 GEOMETRIC AND HARMONIC SERIES

This section uses ideas from Section 10.2 about series and their convergence to investigate some special

types of series. Geometric series are very important and appear in a variety of applications. Much of the early

work in the 17th century with series focused on geometric series and generalized them. Many of the ideas used

later in this chapter originated with geometric series. It is easy to determine whether a geometric series converges

or diverges, and when one does converge, we can easily find its sum. The harmonic series is important as an

example of a divergent series whose terms approach zero. A final type of series, called "telescoping," is discussed

briefly. Telescoping series are relatively uncommon, but their partial sums exhibit a particularly nice pattern.

Geometric Series: $\displaystyle\sum_{k=0}^{\infty} C \cdot r^{k} = C + C \cdot r + C \cdot r^2 + C \cdot r^3 + \ldots$

Example 1: Bouncing Ball: A "super ball" is thrown 10 feet

straight up into the air. On each bounce, it rebounds to

four fifths of its previous height (Fig. 1) so the sequence

of heights is 10 feet, 8 feet, 32/5 feet, 128/25 feet, etc.

(a) How far does the ball travel (up and down) during

its nth bounce? (b) Use a sum to represent the total distance

traveled by the ball.

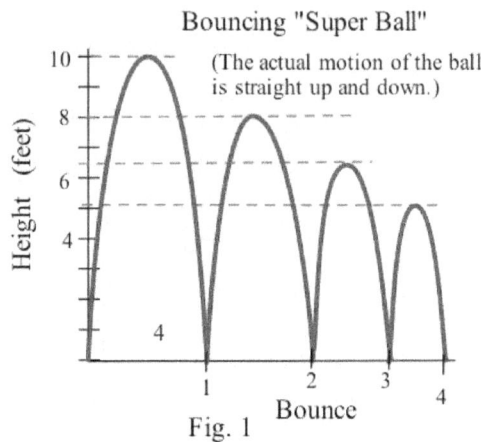

Bouncing "Super Ball"

(The actual motion of the ball is straight up and down.)

Fig. 1

Solution: Since the ball travels up and down on each bounce,

the distance traveled during each bounce is twice the height

of the ball on that bounce so $d_1 = 2(10 \text{ feet}) = 20$ feet,

$d_2 = 16$ feet, $d_3 = 64/5$ feet, and, in general, $d_n = \frac{4}{5} \cdot d_{n-1}$. Looking at these values in another way,

$$d_1 = 20, \quad d_2 = \frac{4}{5} \cdot (20), \quad d_3 = \frac{4}{5} d_2 = \frac{4}{5} \cdot \frac{4}{5} \cdot 20 = (\frac{4}{5})^2 (20), \quad d_4 = \frac{4}{5} \cdot d_3 = \frac{4}{5} \cdot ((\frac{4}{5})^2 \cdot (20)) = (\frac{4}{5})^3 \cdot (20),$$

and, in general, $d_n = (\frac{4}{5})^{n-1} \cdot (20)$.

In theory, the ball bounces up and down forever, and the total distance traveled by the ball is the sum of the

distances traveled during each bounce (an up and down flight):

(first bounce) + (second bounce) + (third bounce) + (forth bounce) + . . .

$$= 20 + \frac{4}{5}(20) + (\frac{4}{5})^2 (20) + (\frac{4}{5})^3 (20) + \ldots$$

$$= 20 \cdot (1 + \frac{4}{5} + (\frac{4}{5})^2 + (\frac{4}{5})^3 + \ldots) = 20 \cdot \sum_{k=0}^{\infty} (\frac{4}{5})^k \; .$$

Practice 1:   Cake: Three calculus students want to share a small square cake equally, but they go about it in a

rather strange way.   First they cut the cake into 4 equal square pieces, each person takes one square, and one

square is left (Fig. 2).   Then they cut the leftover piece into 4 equal square pieces, each person takes one square

and one square is left.   And they keep repeating this process. (a) What fraction of the total cake does each

person "eventually" get? (b) Represent the amount of cake each person gets as a geometric series: (amount of

first piece) + (amount of second piece) + . . .

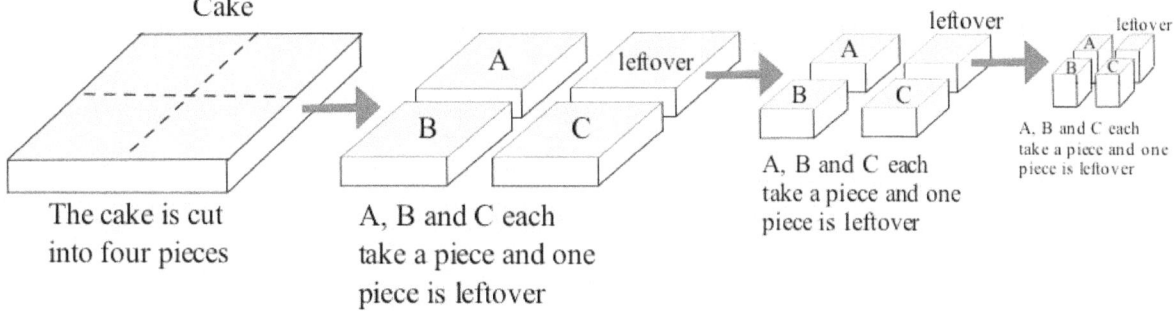

Fig. 2: Sharing a cake evenly among three students

Each series in the previous Example and Practice problems is a Geometric series, a series in which each term is a fixed

multiple of the previous term. Geometric series have the form

$$\sum_{k=0}^{\infty} C \cdot r^k = C + C \cdot r + C \cdot r^2 + C \cdot r^3 + \ldots = C \cdot \sum_{k=0}^{\infty} r^k$$

with $C \neq 0$ and $r \neq 0$ representing fixed numbers. Each term in the series is r times the previous term. Geometric

series are among the most common and easiest series we will encounter. A simple test determines whether a

geometric series converges, and we can even determine the "sum" of the geometric series.

Geometric Series Theorem

The geometric series $\displaystyle\sum_{k=0}^{\infty} r^k = 1 + r + r^2 + r^3 + \ldots$ $\begin{cases} \textbf{converges to } \dfrac{1}{1-r} & \textbf{if } |r| < 1 \\[2mm] \textbf{diverges} & \textbf{if } |r| \geq 1 \end{cases}$

Proof: **If $|r| \geq 1$,** then $|r^k|$ approaches 1 or $+\infty$ as k becomes arbitrarily large, so the terms $a_k = r^k$ of

the geometric series do not approach 0. Therefore, by the nth term test for divergence, the series diverges.

If $|r| < 1$, then the terms $a_k = r^k$ of the geometric series approach 0 so the series may or may not converge, and we

need to examine the limit of the partial sums $s_n = 1 + r + r^2 + r^3 + \ldots + r^n$ of the series. For a geometric series, a

clever insight allows us to calculate those partial sums:

$$(1-r) \cdot s_n \quad = (1-r) \cdot (1 + r + r^2 + r^3 + \ldots + r^n)$$

$$= 1 \cdot (1 + r + r^2 + r^3 + \ldots + r^n) - r \cdot (1 + r + r^2 + r^3 + \ldots + r^n)$$

$$= (1 + r + r^2 + r^3 + \ldots + r^n) - (r + r^2 + r^3 + r^4 + \ldots + r^n + r^{n+1})$$

$$= 1 - r^{n+1} \quad .$$

Since $|r| < 1$ we know $r \neq 1$ so we can divide the previous result by $1 - r$ to get

$$s_n = 1 + r + r^2 + r^3 + \ldots + r^n = \frac{1 - r^{n+1}}{1 - r} = \frac{1}{1-r} - \frac{r^{n+1}}{1-r} \quad .$$

This formula for the nth partial sum of a geometric series is sometimes useful, but now we are interested in the limit of s_n as n approaches infinity. Since $|r| < 1$, r^{n+1} approaches 0 as n approaches infinity, so we can conclude that the partial sums $s_n = \frac{1}{1-r} - \frac{r^{n+1}}{1-r}$ approach $\frac{1}{1-r}$ (as "$n \to \infty$").

The geometric series $\sum\limits_{k=0}^{\infty} r^k$ converges to the value $\frac{1}{1-r}$ when $-1 < r < 1$.

Finally, $\sum\limits_{k=0}^{\infty} C \cdot r^k = C \cdot \sum\limits_{k=0}^{\infty} r^k$ so we can easily determine whether or not $\sum\limits_{k=0}^{\infty} C \cdot r^k$ converges and to what number.

Example 2:   How far did the ball in Example 1 travel?

Solution: The distance traveled, $20(1 + \frac{4}{5} + (\frac{4}{5})^2 + (\frac{4}{5})^3 + \ldots)$, is a geometric series with $C = 20$ and $r = $ 4/5. Since $|r| < 1$, the series $1 + \frac{4}{5} + (\frac{4}{5})^2 + (\frac{4}{5})^3 + \ldots$ converges to

$\frac{1}{1-r} = \frac{1}{1 - 4/5} = 5$, so the total distance traveled is

$$20(1 + \frac{4}{5} + (\frac{4}{5})^2 + (\frac{4}{5})^3 + \ldots) = 20(5) = \textbf{100 feet}.$$

Repeating decimal numbers are really geometric series in disguise, and we can use the Geometric Series Theorem to represent the exact value of the sum as a fraction.

Example 3: Represent the repeating decimals $0.\overline{4}$ and $0.\overline{13}$ as geometric series and find their sums.

Solution: $0.\overline{4} = 0.444 \ldots = \frac{4}{10} + \frac{4}{100} + \frac{4}{1000} + \ldots = \frac{4}{10} \cdot (1 + \frac{1}{10} + (\frac{1}{10})^2 + (\frac{1}{10})^3 + \ldots)$

which is a geometric series with $a = 4/10$ and $r = 1/10$. Since $|r| < 1$, the geometric series converges to $\frac{1}{1-r} = \frac{1}{1-1/10} = \frac{10}{9}$, and $0.\overline{4} = \frac{4}{10}(\frac{10}{9}) = \frac{4}{9}$.

Similarly, $0.\overline{13} = 0.131313 \ldots = \frac{13}{100} + \frac{13}{10000} + \frac{13}{1000000} + \ldots$

$$= \frac{13}{100} \cdot (1 + \frac{1}{100} + (\frac{1}{100})^2 + (\frac{1}{100})^3 + \ldots)$$

$$= \frac{13}{100} \cdot (\frac{1}{1-1/100}) = \frac{13}{100}(\frac{100}{99}) = \frac{13}{99} .$$

Practice 2: Represent the repeating decimals $0.\overline{3}$ and $0.\overline{432}$ as geometric series and find their sums.

One reason geometric series are important for us is that some series involving powers of x are geometric series.

Example 4: $\sum_{k=0}^{\infty} 3x^k = 3 + 3x + 3x^2 + \ldots$ and $\sum_{k=0}^{\infty} (2x-5)^k = 1 + (2x-5) + (2x-5)^2 + \ldots$

are geometric series with $r = x$ and $r = 2x - 5$, respectively. Find the values of x for each series so that the series converges.

Solution: A geometric series converges if and only if $|r| < 1$, so the first series converges if and

only if $|x| < 1$, or, equivalently, $-1 < x < 1$. The sum of the first series is $\frac{3}{1-x}$.

In the second series $r = 2x - 5$ so the series converges if and only if $|2x - 5| < 1$. Removing the absolute

value and solving for x, we get $-1 < 2x - 5 < 1$, and (adding 5 to each side and then dividing by 2) $2 < x < 3$.

The second series converges if and only if $2 < x < 3$. The sum of the second series is $\frac{1}{1-(2x-5)}$ or

$\frac{1}{6-2x}$.

Practice 3: The series $\sum_{k=0}^{\infty} (2x)^k$ and $\sum_{k=0}^{\infty} (3x-4)^k$ are geometric series. Find the ratio r

for each series, and find all values of x for each series so that the series converges.

The series in the previous Example and Practice are called "power series" because they involve powers of the

variable x. Later in this chapter we will investigate other power series which are not geometric series

(e.g., $1 + x + x^2/2 + x^3/3 + \ldots$), and we will try to find values of x which guarantee that the series converge.

Harmonic Series: $\displaystyle\sum_{k=1}^{\infty} \frac{1}{k}$

The series $\displaystyle\sum_{k=1}^{\infty} \frac{1}{k}$ is one of the best known and most important **divergent** series. It is called the **harmonic series**

because of its ties to music (Fig. 3).

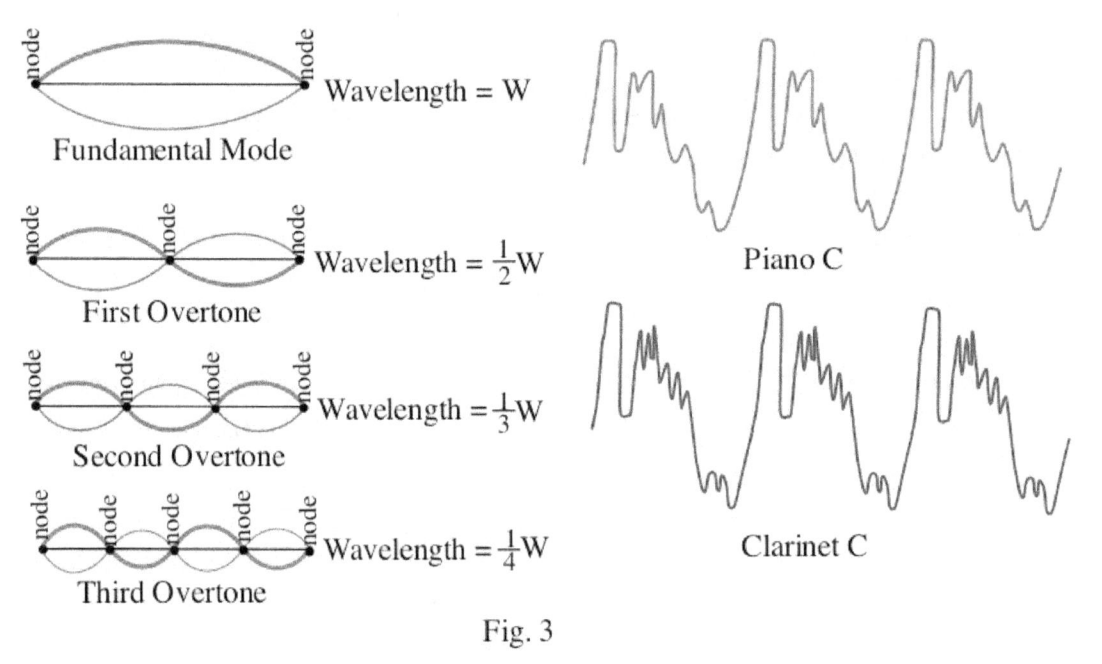

A taut piece of string such as a guitar string or a piano wire can only vibrate so that an integer number of waves are formed. The fundamental mode determines the note being played, and the number and intensity of the harmonics (overtones) present determine the characteristic quality of the sound. Because of these different characteristic qualities, a middle C (264 vibrations per second) played on a piano can be distinguished from a C on a clarinet.

Wavelength = W

Fundamental Mode

Wavelength = $\frac{1}{2}$W

First Overtone

Wavelength = $\frac{1}{3}$W

Second Overtone

Wavelength = $\frac{1}{4}$W

Third Overtone

Piano C

Clarinet C

Fig. 3

If we simply calculate partial sums of the harmonic series, it is not clear that the series diverges — the partial sums s_n grow, but as n becomes large, the values of s_n grow very, very slowly. Fig. 4 shows the values of n needed for the partial sums s_n to finally exceed the integer values 4, 5, 6, 8, 10, and 15. To examine the divergence of the harmonic series, brain power is much more effective than a lot of computing power.

n	s_n
31	4.0224519544
83	5.00206827268
227	6.00436670835
1,674	8.00048557200
12,367	10.00004300827
1,835,421	15.00000378267

Fig. 4

We can show that the harmonic series is divergent by showing that the terms of the harmonic series can be grouped into an infinite number of disjoint "chunks" each of which has a sum larger than 1/2. The series $\sum\limits_{k=1}^{\infty} \frac{1}{2}$ is clearly divergent because the partial sums grow arbitrarily large: by adding enough of the terms together we can make the partial sums, $s_n > n/2$, larger than any predetermined number. Then we can conclude that the partial sums of the harmonic series also approach infinity so the harmonic series diverges.

Theorem: The harmonic series $\sum\limits_{k=1}^{\infty} \frac{1}{k} = 1 + \frac{1}{2} + \frac{1}{3} + \frac{1}{4} + \dots$ **diverges**.

Proof: (This proof is essentially due to Oresme in 1630, twelve years before Newton was born. In 1821 Cauchy included Oresme's proof in a "Course in Analysis" and it became known as Cauchy's argument.)

Let S represent the sum of the harmonic series, $S = 1 + \frac{1}{2} + \frac{1}{3} + \frac{1}{4} + \dots$, and group the terms of the series as indicated by the parentheses:

$$S = 1 + \frac{1}{2} + \left(\frac{1}{3} + \frac{1}{4} \right) + \left(\frac{1}{5} + \frac{1}{6} + \frac{1}{7} + \frac{1}{8} \right) + \left(\frac{1}{9} + \frac{1}{10} + \dots + \frac{1}{16} \right) + \left(\frac{1}{17} + \dots + \frac{1}{32} \right) + \dots$$

↑	↑	↑	↑
2 terms, each greater than or equal to 1/4	4 terms, each greater than or equal to 1/8	8 terms, each greater than or equal to 1/16	16 terms, each greater than or equal to 1/32

Each group in parentheses has a sum greater than 1/2, so

$$S > 1 + \frac{1}{2} + \left(\frac{1}{2} \right) + \left(\frac{1}{2} \right) + \left(\frac{1}{2} \right) + \dots \text{ and}$$

the sequence of partial sums $\{ s_n \}$ does not converge to a finite number. Therefore, the harmonic series diverges.

The harmonic series is an example of a **divergent** series whose terms, $a_k = 1/k$, approach 0. If the terms of a series approach 0, the series may or may not converge — we need to investigate further.

Telescoping Series

Sailors in the seventeenth and eighteenth centuries used telescopes (Fig. 5) which could be extended for viewing and collapsed for storing. Telescoping series get their name because they exhibit a similar "collapsing" property. Telescoping series are rather uncommon. But they are easy to analyze, and it can be useful to recognize them.

Sailor's telescope

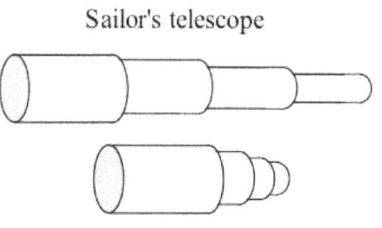

Fig. 5

Example 5: Determine a formula for the partial sum s_n of the series

$$\sum_{k=1}^{\infty} \left[\frac{1}{k} - \frac{1}{k+1} \right] .$$

Then find $\lim_{n \to \infty} s_n$. (Suggestion: It is tempting to algebraically consolidate terms, but the pattern is clearer in this case if you first write out all of the terms.)

Solution: $s_1 = a_1 = 1 - \frac{1}{2}$. In later values of s_n , part of each term cancels part of the next term:

$$s_2 = a_1 + a_2 = \left(1 - \frac{1}{2} \right) + \left(\frac{1}{2} - \frac{1}{3} \right) = 1 - \frac{1}{3}$$

$$s_3 = a_1 + a_2 + a_3 = \left(1 - \frac{1}{2} \right) + \left(\frac{1}{2} - \frac{1}{3} \right) + \left(\frac{1}{3} - \frac{1}{4} \right) = 1 - \frac{1}{4}$$

In general, many of the pieces in each partial sum "collapse" and we are left with a simple form of s_n :

$$s_n = a_1 + a_2 + ... + a_{n-1} + a_n = \left(1 - \frac{1}{2} \right) + \left(\frac{1}{2} - \frac{1}{3} \right) + ... + \left(\frac{1}{n-1} - \frac{1}{n} \right) + \left(\frac{1}{n} - \frac{1}{n+1} \right) = 1 - \frac{1}{n+1}$$

Finally, $\lim_{n \to \infty} s_n = \lim_{n \to \infty} 1 - \frac{1}{n+1} = 1$ so the series converges to 1: $\sum_{k=1}^{\infty} \left[\frac{1}{k} - \frac{1}{k+1} \right] = 1$.

Practice 4:   Find the sum of the series $\sum_{k=3}^{\infty} \left[\sin(\frac{1}{k}) - \sin(\frac{1}{k+1}) \right]$.

PROBLEMS

In problems 1 – 6, rewrite each geometric series using the sigma notation and calculate the value of the sum.

1. $1 + \frac{1}{3} + \frac{1}{9} + \frac{1}{27} + ...$

2. $1 + \frac{2}{3} + \frac{4}{9} + \frac{8}{27} + ...$

3. $\frac{1}{8} + \frac{1}{16} + \frac{1}{32} + \frac{1}{64} + ...$

4. $1 - \frac{1}{2} + \frac{1}{4} - \frac{1}{8} + \frac{1}{16} - ...$

5. $-\frac{2}{3} + \frac{4}{9} - \frac{8}{27} + ...$

6. $1 + \frac{1}{e} + \frac{1}{e^2} + \frac{1}{e^3} + ...$

7.   Rewrite each series in the form of a sum of r^k , and then show that

   (a) $\frac{1}{2} + \frac{1}{4} + \frac{1}{8} + ... = 1$, $\frac{1}{3} + \frac{1}{9} + \frac{1}{27} + ... = \frac{1}{2}$, and (b) for $a > 1$, $\frac{1}{a} + \frac{1}{a^2} + \frac{1}{a^3} + ... = \frac{1}{a-1}$.

8.   A ball is thrown 10 feet straight up into the air, and on each bounce, it rebounds to 60% of its previous height. (a) How far does the ball travel (up and down) during its nth bounce. (b) Use a sum to represent the total distance traveled by the ball. (c) Find the total distance traveled by the ball.

9. An old tennis ball is thrown 20 feet straight up into the air, and on each bounce, it rebounds to 40% of its previous height. (a) How far does the ball travel (up and down) during its nth bounce? (b) Use a sum to represent the total distance traveled by the ball. (c) Find the total distance traveled by the ball.

10. Eighty people are going on an expedition by horseback through desolate country. The people and gear require 90 horses, and additional horses are needed to carry food for the original 90 horses. Each additional horse can carry enough food to feed 3 horses for the trip. How many additional horses are needed? (The original 90 horses will require 30 extra horses to carry their food. The 30 extra horses require 10 more horses to carry their food. etc.)

11. The mathematical diet you are following says you can eat "half of whatever is on the plate," so first you bite off one half of the cake and put the other half back on the plate. Then you pick up the remaining half from the plate (it's "on the plate"), bite off half of that, and return the rest to the plate. And you continue this silly process of picking up the piece from the plate, biting off half, and returning the rest to the plate. (a) Represent the total amount you eat as a series. (b) How much of the cake is **left** after 1 bite, 2 bites, n bites? (c) "Eventually," how much of the cake do you eat?

12. Suppose in Fig. 6 we begin with a square with sides of length 1 (area = 1) and construct another square inside by connecting the midpoints of the sides. Then the new square has area 1/2. If we continue the process of constructing each new square by connecting the midpoints of the sides of the previous square, we get a sequence of squares each of which has 1/2 half the area of the previous square. Find the total area of all of the squares.

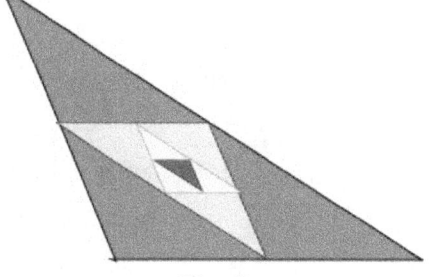

Fig. 6

13. Suppose in Fig. 7 we begin with a triangle with area 1 and construct another triangle inside by connecting the midpoints of the sides. Then the new triangle has area 1/4. Imagine that this construction process is continued and find the total area of all of the triangles.

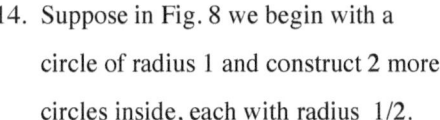

Fig. 7

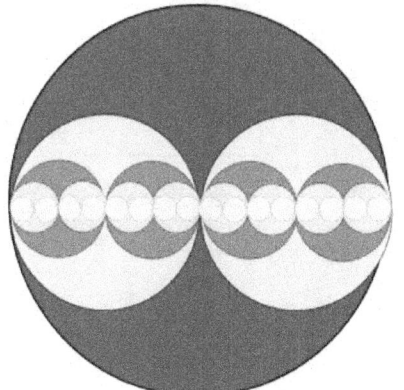

14. Suppose in Fig. 8 we begin with a circle of radius 1 and construct 2 more circles inside, each with radius 1/2. Continue the process of constructing two new circles inside each circle from the previous step and find the total area of the circles.

Fig. 8

15. The construction of the Helga von Koch snowflake begins with an equilateral triangle of area 1 (Fig. 9).

Then each edge is subdivided into three equal lengths, and three equilateral triangles, each with area 1/9, are

built on these "middle thirds" adding a total of 3(1/9) to the original area. The process is repeated: at the next

stage, 3•4 equilateral triangles, each with area 1/81, are built on the new "middle thirds" adding 3•4•(1/81)

more area. (a) Find the total area that results when this process is repeated forever.

number of edges	# triangles added	area of each triangle	total area added
3	0	0	0
3•4	3	1/9	3/9
$3•4^2$	3•4	$1/9^2$	$3•4/9^2 = (3/9)(4/9)$
$3•4^3$	$3•4^2$	$1/9^3$	$3•4^2/9^3 = (3/9)(4^2/9^2)$
$3•4^4$	$3•4^3$	$1/9^4$	$3•4^3/9^4 = (3/9)(4^3/9^3)$

so the total added area of the snowflake is

$$1 + \frac{3}{9} + (\frac{3}{9})(\frac{4}{9}) + (\frac{3}{9})(\frac{4^2}{9^2}) + (\frac{3}{9})(\frac{4^3}{9^3}) + \ldots$$

(b) Express the perimeter of the Koch Snowflake as

a geometric series and find its sum.

(The area is finite, but the perimeter is infinite.)

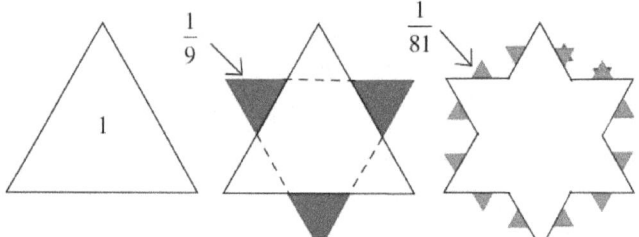

Fig. 9: Helga von Koch Snowflake

16. Harmonic Tower: The base of a tower is a cube

whose edges are each one foot long. On top of it are cubes with edges of length

1/2, 1/3, 1/4, ... (Fig. 10).

(a) Represent the total height of the tower as a series. Is the height finite?

(b) Represent the total surface area of the cubes as a series.

(c) Represent the total volume of the cubes as a series.

(In the next section we will be able to determine if this surface area and

volume are finite or infinite.)

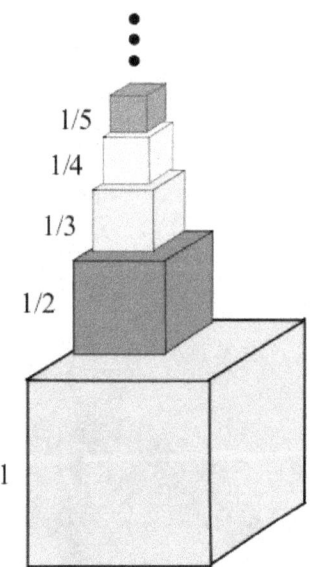

Fig. 10: Harmonic Tower

17. The base of a tower is a sphere whose radius is each one foot long. On top of

each sphere is another sphere with radius one half the radius of the sphere

immediately beneath it (Fig 11). (a) Represent the total height of the

tower as a series and find its sum. (b) Represent the total surface area

of the spheres as a series and find its sum. (c) Represent the total

volume of the spheres as a series and find its sum.

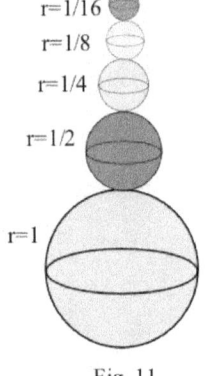

18. Represent the repeating decimals $0.\overline{6}$ and $0.\overline{63}$

as geometric series and find the value of each series as a simple fraction.

19. Represent the repeating decimals $0.\overline{8}$, $0.\overline{9}$, and $0.\overline{285714}$ as geometric

series and find the value of each series as a simple fraction.

Fig. 11

20. Represent the repeating decimals $0.\overline{a}$, $0.\overline{ab}$, and $0.\overline{abc}$ as geometric series and find the value of each series as a simple fraction. What do you think the simple fraction representation is for $0.\overline{abcd}$?

In problems 21 – 32, find all values of x for which each geometric series converges.

21. $\displaystyle\sum_{k=1}^{\infty} (2x + 1)^k$

22. $\displaystyle\sum_{k=1}^{\infty} (3 - x)^k$

23. $\displaystyle\sum_{k=1}^{\infty} (1 - 2x)^k$

24. $\displaystyle\sum_{k=1}^{\infty} 5x^k$

25. $\displaystyle\sum_{k=1}^{\infty} (7x)^k$

26. $\displaystyle\sum_{k=1}^{\infty} (x/3)^k$

27. $1 + \dfrac{x}{2} + \dfrac{x^2}{4} + \dfrac{x^3}{8} + \ldots$

28. $1 + \dfrac{2}{x} + \dfrac{4}{x^2} + \dfrac{8}{x^3} + \ldots$

29. $1 + 2x + 4x^2 + 8x^3 + \ldots$

30. $\displaystyle\sum_{k=1}^{\infty} (2x/3)^k$

31. $\displaystyle\sum_{k=1}^{\infty} \sin^k(x)$

32. $\displaystyle\sum_{k=1}^{\infty} e^{kx} = \sum_{k=1}^{\infty} (e^x)^k$

33. One student thought the formula was $1 + x + x^2 + x^3 + \ldots = \dfrac{1}{1 - x}$. The second student said "That can't be right. If we replace x with 2, then the formula says the sum of the positive numbers $1 + 2 + 4 + 8 + \ldots$ is a negative number $\dfrac{1}{1 - 2} = -1$." Who is right? Why?

34. The Classic Board Problem: If you have identical 1 foot long boards, they can be arranged to hang over the edge of a table. One board can extend 1/2 foot beyond the edge (Fig. 12), two boards can extend 1/2 + 1/4 feet, and, in general, n boards can extend 1/2 + 1/4 + 1/6 + . . . + 1/(2n) feet beyond the edge.

(a) How many boards are needed for an arrangement in which the entire top board is beyond the edge of the table?

(b) How many boards are needed for an arrangement in which the entire top two boards are beyond the edge of the table?

(c) How far can an arrangement extend beyond the edge of the table

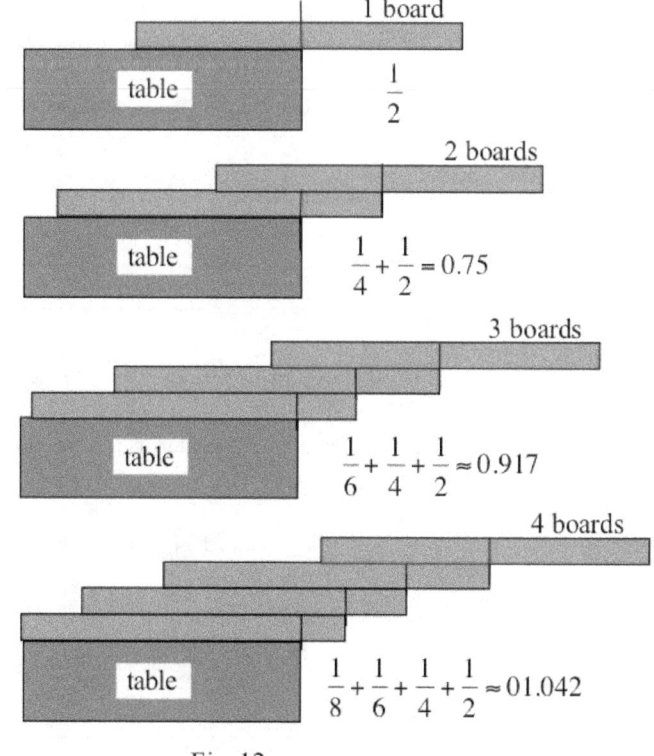

Fig. 12

In problems 35 – 40, calculate the value of the partial sum for n = 4 and n = 5 and find a formula for s_n.

(The patterns may be more obvious if you do not simplify each term.)

35. $\displaystyle\sum_{k=3}^{\infty}\left[\frac{1}{k}-\frac{1}{k+1}\right]$ 36. $\displaystyle\sum_{k=1}^{\infty}\left[\frac{1}{k}-\frac{1}{k+2}\right]$ 37. $\displaystyle\sum_{k=1}^{\infty}[k^3-(k+1)^3]$

38. $\displaystyle\sum_{k=1}^{\infty}\ln\left(\frac{k}{k+1}\right)$ 39. $\displaystyle\sum_{k=3}^{\infty}[f(k)-f(k+1)]$ 40. $\displaystyle\sum_{k=1}^{\infty}[g(k)-g(k+2)]$

In problems 41 – 44, calculate s_4 and s_5 for each series and find the limit of s_n as n approaches infinity.

If the limit is a finite value, it represents the value of the infinite series.

41. $\displaystyle\sum_{k=1}^{\infty}\sin(\tfrac{1}{k})-\sin(\tfrac{1}{k+1})$ 42. $\displaystyle\sum_{k=2}^{\infty}\cos(\tfrac{1}{k})-\cos(\tfrac{1}{k+1})$ 43. $\displaystyle\sum_{k=2}^{\infty}\frac{1}{k^2}-\frac{1}{(k+1)^2}$

44. $\displaystyle\sum_{k=3}^{\infty}\ln(1-\frac{1}{k^2})$ (Suggestion: Rewrite $1-\frac{1}{k^2}$ as $\dfrac{1-\frac{1}{k}}{1-\frac{1}{k+1}}$)

Problems 45 and 46 are outlines of two "proofs by contradiction" that the harmonic series is divergent.
Each proof starts with the assumption that the "sum" of the harmonic series is a finite number, and then an
obviously false conclusion is derived from the assumption. Verify that each step follows from the assumption and
previous steps, and explain why the conclusion is false.

45. Assume that $H = 1 +\frac{1}{2} + \frac{1}{3} + \frac{1}{4} + \frac{1}{5} + \frac{1}{6} + \frac{1}{7} + \dots$ is a finite number, and let

$O = 1 + \frac{1}{3} + \frac{1}{5} + \frac{1}{7} + \dots$ be the sum of the "odd reciprocals," and $E = \frac{1}{2} + \frac{1}{4} + \frac{1}{6} + \frac{1}{8} +\dots$

be the sum of the "even reciprocals." Then

(i) $H = O + E$,   (ii)   each term of O is larger than the corresponding term of E so $O > E$,

and   (iii) $E = \frac{1}{2} + \frac{1}{4} + \frac{1}{6} + \frac{1}{8} +\dots= \frac{1}{2}\{1+\frac{1}{2} + \frac{1}{3} + \frac{1}{4} +\dots\} =\frac{1}{2} H$.

Therefore $H = O + E >\frac{1}{2} H + \frac{1}{2} H = H$ (so "H is strictly bigger than H," a contradiction).

II. Assume that $H = 1 +\frac{1}{2} + \frac{1}{3} + \frac{1}{4} + \frac{1}{5} + \frac{1}{6} + \frac{1}{7} + \dots$ is a finite number, and, starting with the

second term, group the terms into groups of three. Then, using $\frac{1}{n-1} + \frac{1}{n} + \frac{1}{n+1} > 3\cdot\frac{1}{n}$, we have

$H = 1 + (\frac{1}{2} + \frac{1}{3} + \frac{1}{4}) + (\frac{1}{5} + \frac{1}{6} + \frac{1}{7}) + (\frac{1}{8} + \frac{1}{9} + \frac{1}{10}) + \dots$

$\quad > 1 + (\quad 1 \quad) + (\quad \frac{1}{2} \quad) + (\quad \frac{1}{3} \quad) + \dots$

$\quad = 1 + (1+\frac{1}{2} + \frac{1}{3} + \frac{1}{4} +\dots) = 1 + H$. Therefore, $H > 1 + H$ (so "H is bigger than 1 + H").

46. Jacob Bernoulli (1654–1705) was a master of understanding and manipulating series by breaking a difficult series into a sum of easier series. He used that technique to find the sum of the non–geometric series

$$\sum_{k=1}^{\infty} \frac{k}{2^k} = 1/2 + 2/4 + 3/8 + 4/16 + 5/32 + \ldots + k/2^k + \ldots$$ in his book <u>Ars Conjectandi</u> , 1713.

Show that   1/2 +   2/4 +   3/8 +   4/16 +   5/32 +...+ $n/2^n$ + ...   can be written as

	1/2 +	1/4 +	1/8 +	1/16 +	1/32 +...+	$1/2^n$ + ...	(a)
plus		1/4+	1/8 +	1/16 +	1/32 +...+	$1/2^n$ + ...	(b)
plus			1/8 +	1/16 +	1/32 +...+	$1/2^n$ + ...	(c)

plus . . . etc.

Find the values of the geometric series (a), (b), (c), etc. and then find the sum of these values (another geometric series).

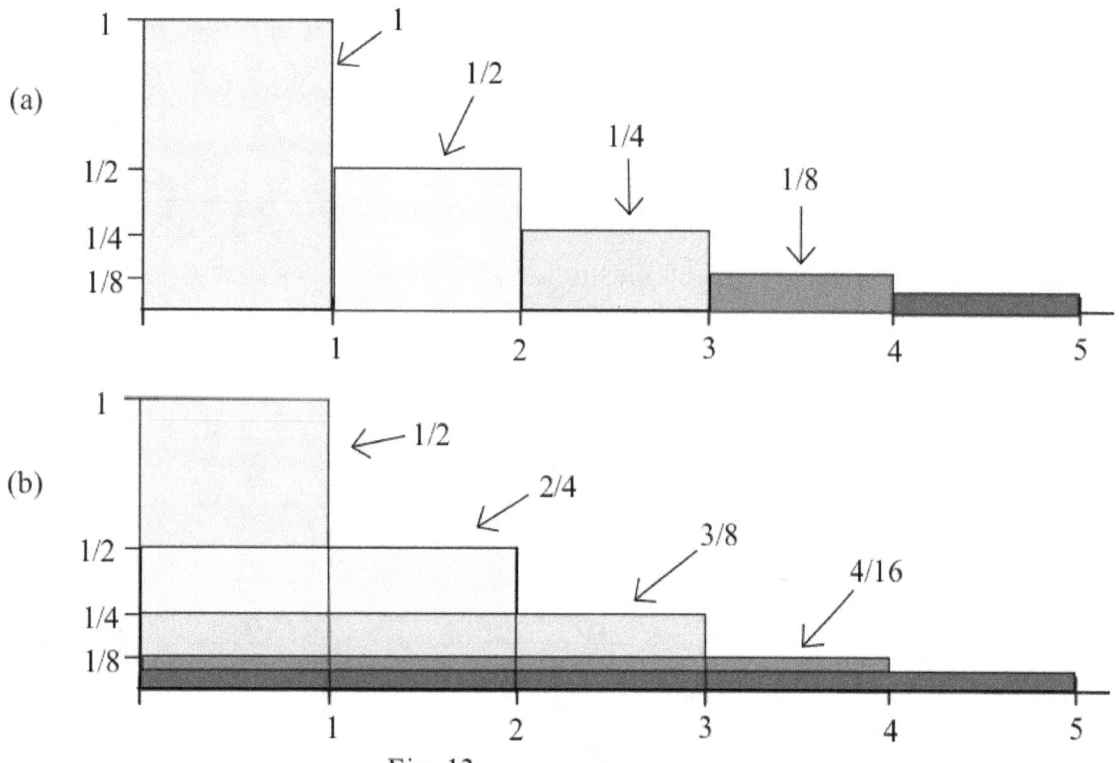

Fig. 13

47. Bernoulli's approach in problem 46 can also be interpreted as a geometric argument for representing the area in Fig. 13 in two different ways. (a) Represent the total area in Fig. 13a as a (geometric) sum of the areas of the side–by–side rectangles, and find the sum of the series. (b) Represent the total area of the stacked rectangles in Fig. 13b as a sum of the areas of the horizontal slices.

Since both series represent the same total area, the values of the series are equal.

48. Use the approach of problem 46 to find a formula for the sum of

(a) the value of $\displaystyle\sum_{k=1}^{\infty} \frac{k}{3^k} = 1/3 + 2/9 + 3/27 + 4/81 + \ldots + k/3^k + \ldots$ and

(b) a formula for the value of $\displaystyle\sum_{k=1}^{\infty} \frac{k}{c^k} = 1/c + 2/c^2 + 3/c^3 + 4/c^4 + \ldots + k/c^k + \ldots$ for $c > 1$.

(answers: (a) $3/4$, (b) $c/(c-1)^2$)

Practice Answers

Practice 1: (a) Since they each get equal shares, and the whole cake is distributed, they each get $1/3$ of the cake. More precisely, after step 1, $1/4$ of the cake remains and $3/4$ was shared. After step 2, $(1/4)^2$ of the cake remains and $1 - (1/4)^2$ was shared. After step n, $(1/4)^n$ of the cake remains and $1 - (1/4)^n$ was shared. So after step n, each student has $(\frac{1}{3})(1 - (1/4)^n)$ of the cake. "Eventually," each student gets (almost) $\frac{1}{3}$ of the cake.

(b) $(\frac{1}{4}) + (\frac{1}{4})^2 + (\frac{1}{4})^3 + \ldots = (\frac{1}{4})\left\{ 1 + (\frac{1}{4}) + (\frac{1}{4})^2 + \ldots \right\}$

Practice 2: $0.\overline{3} = 0.333 \ldots = \frac{3}{10} + \frac{3}{100} + \frac{3}{1000} + \ldots = \frac{3}{10} \cdot (1 + \frac{1}{10} + (\frac{1}{10})^2 + (\frac{1}{10})^3 + \ldots)$

which is a geometric series with $a = 3/10$ and $r = 1/10$. Since $|r| < 1$, the geometric series

converges to $\frac{1}{1-r} = \frac{1}{1-1/10} = \frac{10}{9}$, and $0.\overline{3} = \frac{3}{10}(\frac{10}{9}) = \frac{3}{9} = \frac{1}{3}$.

Similarly, $0.\overline{432} = 0.432432432 \ldots = \frac{432}{1000} + \frac{432}{1000000} + \frac{432}{1000000000} + \ldots$

$= \frac{432}{1000} \cdot (1 + \frac{1}{1000} + (\frac{1}{1000})^2 + (\frac{1}{100})^3 + \ldots)$

$= \frac{432}{1000} \cdot (\frac{1}{1-1/1000}) = \frac{432}{1000}(\frac{1000}{999}) = \frac{432}{999} = \frac{16}{37}$.

Practice 3: $r = 2x$: If $|2x| < 1$, then $-1 < 2x < 1$ so $-1/2 < x < 1/2$.

$\displaystyle\sum_{k=0}^{\infty} (2x)^k$ converges (to $\frac{1}{1-2x}$) when $-1/2 < x < 1/2$.

$r = 3x - 4$: If $|3x - 4| < 1$, then $-1 < 3x - 4 < 1$ so $3 < 3x < 5$ and $1 < x < 5/3$.

$\displaystyle\sum_{k=0}^{\infty} (3x - 4)^k$ converges (to $\frac{1}{1-(3x-4)} = \frac{1}{5-3x}$) when $1 < x < 5/3$.

Practice 4:   Let $s_n = \displaystyle\sum_{k=3}^{n} \sin(\frac{1}{k}) - \sin(\frac{1}{k+1})$

$$= \left\{ \sin(\tfrac{1}{3}) - \sin(\tfrac{1}{4}) \right\} + \left\{ \sin(\tfrac{1}{4}) - \sin(\tfrac{1}{5}) \right\} + \left\{ \sin(\tfrac{1}{5}) - \sin(\tfrac{1}{6}) \right\} + ... + \left\{ \sin(\tfrac{1}{n}) - \sin(\tfrac{1}{n+1}) \right\}$$

$$= \sin(\tfrac{1}{3}) - \sin(\tfrac{1}{n+1}) .$$

Then $\displaystyle\lim_{n\to\infty} s_n = \lim_{n\to\infty} \left\{ \sin(\tfrac{1}{3}) - \sin(\tfrac{1}{n+1}) \right\} = \sin(\tfrac{1}{3})$ so the series converges to $\sin(\tfrac{1}{3})$:

$$\sum_{k=1}^{\infty} \sin(\tfrac{1}{k}) - \sin(\tfrac{1}{k+1}) = \sin(\tfrac{1}{3}) \approx 0.327 .$$

10.3.5   AN INTERLUDE

The previous three sections introduced the topics of sequences and series, discussed the meaning of convergence of series, and examined geometric series in some detail. The ideas, definitions, and results in those sections are fundamental for understanding and working with the material in the rest of this chapter and for later work in theoretical and applied mathematics.

The material in the next several sections is of a different sort — it is more technical and specialized. In order to work effectively with power series we need to know where (for which values of x) the power series converge. And to determine that convergence we need additional methods. In the next several sections we examine several methods for determining where particular series converge or diverge. These methods are called "convergence tests."

- The Integral Test in Section 10.4 says that a series converges if and only if a certain related improper integral is finite. This result lets us change a question about convergence of a series into a question about the convergence of an integral. Sometimes the related integral is easy to evaluate so it is easy to determine the convergence of the series. (Sometimes the related integral is very difficult to evaluate.) The integral test is then used to determine the convergence of P–series, the whole family of series of the form $\sum_{k=1}^{\infty} \dfrac{1}{k^p}$.

- Section 10.5 introduces some methods for determining the convergence of a new series by comparing the new series with some series which we already know converge or diverge. These comparison tests can be very powerful and useful, but their power and usefulness depends on already knowing about the convergence of some particular series to compare against the new series. Typically we will compare new series against two types of series, geometric series and P–series.

- In Section 10.6 we derive a result about the convergence of a series $\sum_{k=1}^{\infty} a_k$ by examining the ratios of successive terms of the series, a_{k+1}/a_k . If this ratio is small enough, then we will be able to conclude that the series converges. If the ratio is large enough, then we will be able to conclude that the series diverges. Unfortunately, sometimes the value of the ratio will not allow us to conclude anything.

- Section 10.7 examines series whose terms alternate in sign, such as the "alternating harmonic series," $1 - 1/2 + 1/3 - 1/4 + \ldots$, and discusses methods to determine whether these alternating series converge.

Each of these sections is rather short and focuses on one or two tests of convergence. As you study the material in each section by itself, you need to be able to use the method discussed in that section. When you finish all of the sections, you also need to be able to decide which convergence test to use.

PROBLEMS

These problems illustrate some of the reasoning that is used in sections 10.4 – 10.7, but they do not assume any information from those sections.

Integrals and sums

1. Which shaded region in Fig. 1 has the larger area, the sum or the integral?

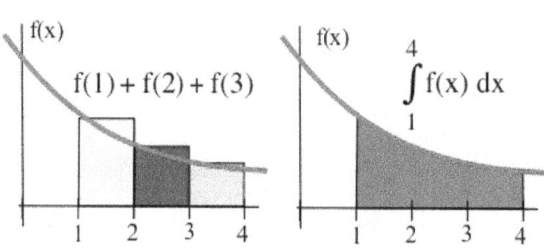

2. Which shaded region in Fig. 2 has the larger area, the sum or the integral?

3. Represent the area of the shaded region in Fig. 3 as an infinite series.

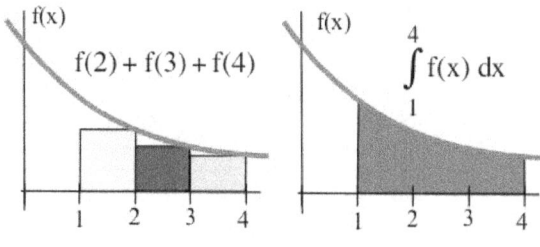

Fig. 1

4. Represent the area of the shaded region in Fig. 4 as an infinite series.

Fig. 2

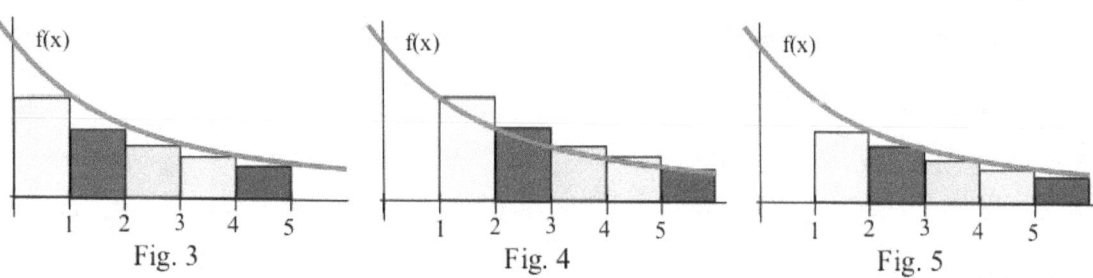

Fig. 3   Fig. 4   Fig. 5

5. Represent the area of the shaded region in Fig. 5 as an infinite series.

6. Which of the following represents the shaded area in Fig. 6?

    (a) f(0) + f(1)   (b) f(1) + f(2)

    (c) f(2) + f(3)   (d) f(3) + f(4)

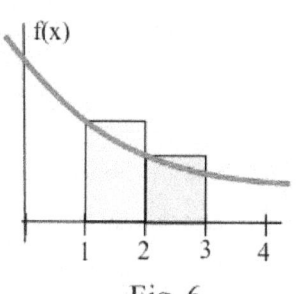

Fig. 6

7. Which of the following represents the shaded area in Fig. 7?

(a) f(0) + f(1)   (b) f(1) + f(2)   (c) f(2) + f(3)   (d) f(3) + f(4)

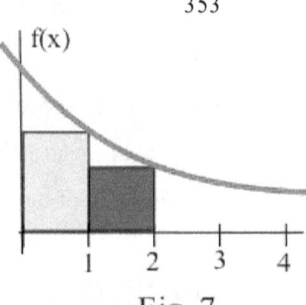

Fig. 7

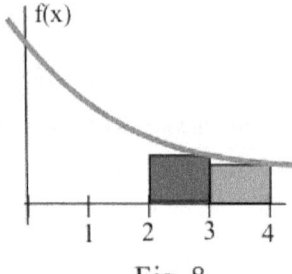

Fig. 8

8. Which of the following represents the shaded area in Fig. 8?

(a)   f(0) + f(1)   (b) f(1) + f(2)   (c) f(2) + f(3)   (d) f(3) + f(4)

9. Which of the following represents the shaded area in Fig. 9?

(a) f(0) + f(1)   (b) f(1) + f(2)   (c) f(2) + f(3)   (d) f(3) + f(4)

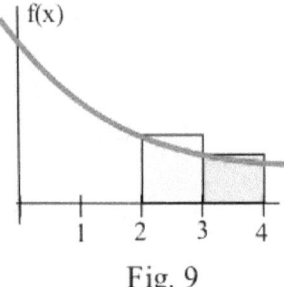

Fig. 9

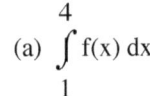

Fig. 10

10. Arrange the following four values in increasing order (Fig. 10):

(a) $\int_{1}^{3} f(x)\, dx$ (b) $\int_{2}^{4} f(x)\, dx$ (c) f(1) + f(2)   (d) f(2) + f(3)

11. Arrange the following four values in increasing order (Fig. 11):

(a) $\int_{1}^{4} f(x)\, dx$ (b) $\int_{2}^{5} f(x)\, dx$

(c) f(1) + f(2) + f(3)   (d) f(2) + f(3) + f(4)

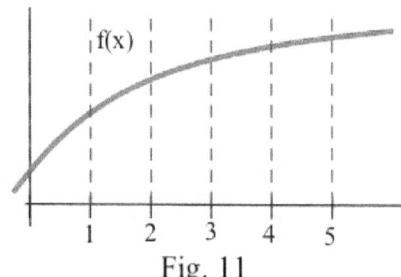

Fig. 11

Comparisons

12. You want to get a summer job operating a type of heavy equipment, and you know there are certain height requirements in order for the operator to fit safely in the cab of the machine. You don't remember what the requirements are, but three of your friends applied. Tom was rejected as too tall. Sam was rejected as too short. Justin got a job. Should you apply for the job if

(a) you are taller than Tom? Why?   (b) you are taller than Sam? Why?

(c) you are shorter than Sam? Why?   (d) you are shorter than Justin? Why?

(e) List the comparisons which indicate that you are the wrong height for the job.

(f) List the comparisons which do not give you enough information about whether you are an acceptable height for the job.

13. You know Wendy did well on the Calculus test and Paula did poorly, but you haven't received your test back yet. If the instructor tells you the following, what can you conclude?

    (a) "You did better than Wendy."   (b) "You did better than Paula."

    (c) "You did worse than Wendy."   (d) "You did worse than Paula."

14. You have recently taken up mountain climbing and are considering a climb of Mt. Baker. You know that Mt. Index is too easy to be challenging, but that Mt. Liberty Bell is too difficult for you. Should you plan a climb of Mt. Baker if an experienced climber friend tells you that

    (a) "Baker is easier than Index."   (b) "Baker is more difficult than Index."

    (c) "Baker is easier than Liberty Bell."   (d) "Baker is more difficult than Liberty Bell."

    (e) Which comparisons indicate that Baker is appropriate: challenging but not too difficult?

    (f) Which comparisons indicate that Baker is not appropriate?

15. As a student you have had Professors Good and Bad for classes, and they each lived up to their names. Now you are considering taking a class from Prof. Unknown whom you don't know. What can you expect if

    (a) a classmate who had Good and Unknown says "Unknown was better than Good" ?

    (b) a classmate who had Good and Unknown says "Good was better than Unknown" ?

    (c) a classmate who had Bad and Unknown says "Unknown was better than Bad" ?

    (d) a classmate who had Bad and Unknown says "Bad was better than Unknown" ?

In problems 16 – 19, all of the series converge. In each pair, which series has the larger sum?

16. $\displaystyle\sum_{k=1}^{\infty} \frac{1}{k^2+1}$, $\displaystyle\sum_{k=1}^{\infty} \frac{1}{k^2}$
           17. $\displaystyle\sum_{k=2}^{\infty} \frac{1}{k^3-5}$, $\displaystyle\sum_{k=2}^{\infty} \frac{1}{k^3}$

18. $\displaystyle\sum_{k=1}^{\infty} \frac{1}{k^2+3k-1}$, $\displaystyle\sum_{k=1}^{\infty} \frac{1}{k^2}$
      19. $\displaystyle\sum_{k=3}^{\infty} \frac{1}{k^2+5k}$, $\displaystyle\sum_{k=3}^{\infty} \frac{1}{k^3+k-1}$

Ratios of successive terms

In problems 20 – 28, a formula is given for each term a_k of a sequence. For each sequence (a) write a formula for a_{k+1} , (b) write the ratio a_{k+1}/a_k , and (c) simplify the ratio a_{k+1}/a_k .

20. $a_k = 3k$ 21. $a_k = k+3$ 22. $a_k = 2k+5$

23. $a_k = 3/k$ 24. $a_k = k^2$ 25. $a_k = 2^k$

26. $a_k = (1/2)^k$ 27. $a_k = x^k$ 28. $a_k = (x-1)^k$

In problems 29–36, state whether the series converges or diverges and calculate the ratio a_{k+1}/a_k .

29. $\sum\limits_{k=1}^{\infty} (\frac{1}{2})^k$

30. $\sum\limits_{k=1}^{\infty} (\frac{1}{5})^k$

31. $\sum\limits_{k=1}^{\infty} 2^k$

32. $\sum\limits_{k=1}^{\infty} (-3)^k$

33. $\sum\limits_{k=1}^{\infty} 4$

34. $\sum\limits_{k=1}^{\infty} (-1)^k$

35. $\sum\limits_{k=1}^{\infty} \frac{1}{k}$

36. $\sum\limits_{k=1}^{\infty} \frac{7}{k}$

Alternating terms

For problems 37 – 40, s_n represents the nth partial sum of the series with terms a_k (e.g., $s_3 = a_1 + a_2 + a_3$).

In each problem, circle the appropriate symbol: "<" or "=" or ">."

37. If $a_5 > 0$, then $s_4 <=> s_5$.

38. If $a_5 = 0$, then $s_4 <=> s_5$.

39. If $a_5 < 0$, then $s_4 <=> s_5$.

40. If $a_{n+1} > 0$ for all n, then $s_n <=> s_{n+1}$ for all n.

41. If $a_{n+1} < 0$ for all n, then $s_n <=> s_{n+1}$ for all n.

42. If $a_4 > 0$ and $a_5 < 0$, then how do $s_3, s_4,$ and s_5 compare?

43. If $a_4 = 0.2$ and $a_5 = -0.1$ and $a_6 = 0.2$, then how do $s_3, s_4,$ and s_5 compare?

44. If $a_4 = -0.3$ and $a_5 = 0.2$ and $a_6 = -0.1$, then how do $s_3, s_4,$ and s_5 compare?

45. If $a_4 = -0.3$ and $a_5 = -0.2$ and $a_6 = 0.1$, then how do $s_3, s_4,$ and s_5 compare?

In problems 46 – 50, the first 8 terms $a_1, a_2, ... , a_8$ of a series are given. Calculate and graph the first 8 partial sums $s_1, s_2, ... , s_8$ of the series and describe the pattern of the graph of the partial sums.

46. $a_1 = 2, a_2 = -1, a_3 = 2, a_4 = -1, a_5 = 2, a_6 = -1, a_7 = 2, a_8 = -1$.

47. $a_1 = 2, a_2 = -1, a_3 = 0.9, a_4 = -0.8, a_5 = 0.7, a_6 = -0.6, a_7 = 0.5, a_8 = -0.4$.

48. $a_1 = 2, a_2 = -1, a_3 = 1, a_4 = -1, a_5 = 1, a_6 = -1, a_7 = 1, a_8 = -1$.

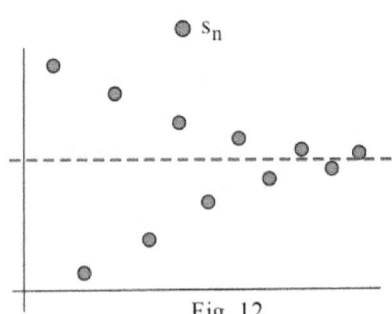

49. $a_1 = -2, a_2 = 1.5, a_3 = -0.8, a_4 = 0.6, a_5 = -0.4, a_6 = 0.2, a_7 = 2, a_8 = -0.1$

50. $a_1 = 5, a_2 = 1, a_3 = -0.6, a_4 = -0.4, a_5 = 0.2, a_6 = 0.1, a_7 = 0.1, a_8 = -0.2$.

51. What condition on the terms a_k guarantees that the graph of 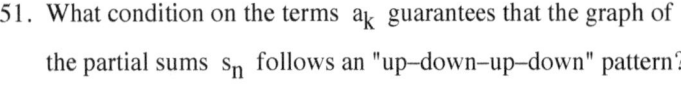 the partial sums s_n follows an "up–down–up–down" pattern?

Fig. 12

52. What condition on the terms a_k guarantees that the graph of
the partial sums s_n forms a "narrowing funnel" pattern in Fig. 12 ?

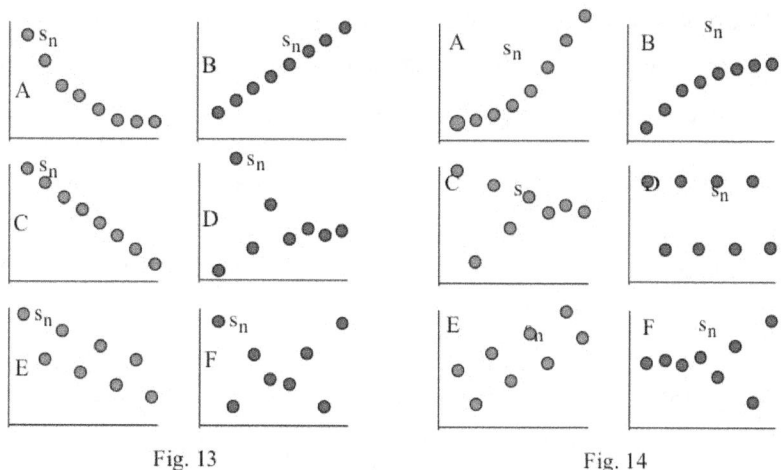

Fig. 13   Fig. 14

53. Fig. 13 shows the graphs of several partial sums s_n .

  (a) For which graphs do the terms a_k alternate in sign?

  (b) For which graphs do the $|a_k|$ decrease?

  (c) For which graphs do the terms a_k alternate in sign and decrease in absolute value?

54. Fig. 14 shows the graphs of several partial sums s_n .

  (a) For which graphs do the terms a_k alternate in sign?

  (b) For which graphs do the $|a_k|$ decrease?

  (c) For which graphs do the terms a_k alternate in sign and decrease in absolute value?

55. The geometric series $\sum\limits_{k=0}^{\infty} \left(-\tfrac{1}{2}\right)^k = 1 - \tfrac{1}{2} + \tfrac{1}{4} - ...$ converges to $\dfrac{1}{1-(-1/2)} = \dfrac{1}{3/2} = \dfrac{2}{3}$.

  Graph the horizontal line $y = \tfrac{2}{3}$ and then graph the partial sums $s_0 , ... , s_8$ of $\sum\limits_{k=0}^{\infty} \left(-\tfrac{1}{2}\right)^k$.

56. The geometric series $\sum\limits_{k=0}^{\infty} (-0.6)^k = 1 - 0.6 + 0.36 - ...$ converges to $\dfrac{1}{1-(-0.6)} = \dfrac{1}{1.6} = 0.625$.

  Graph the horizontal line $y = 0.625$ and then graph the partial sums $s_0 , ... , s_8$ of $\sum\limits_{k=0}^{\infty} (-0.6)^k$.

57. The geometric series $\sum\limits_{k=0}^{\infty} (-2)^k = 1 - 2 + 4 - ...$ diverges ($|r| = |-2| = 2 > 1$). Graph the partial

  sums $s_0 , ... , s_8$ of $\sum\limits_{k=0}^{\infty} (-2)^k$.

10.4   POSITIVE TERM SERIES: INTEGRAL TEST & P–TEST

This section discusses two methods for determining whether some series are convergent. The first, the integral test, says that a given series converges if and only if a related improper integral converges. This lets us trade a question about the convergence of a series for a question about the convergence of an improper integral. The second convergence test, the P–test, says that the convergence of one particular type of series, the sum $\displaystyle\sum_{k=1}^{\infty} \frac{1}{k^p}$, depends only on the value of p. These tests only apply to series whose terms are positive. And, unfortunately, the tests only tell us if the series converge or diverge, but they do not tell us the actual sum of the series.

The Integral Test is the more fundamental and general of the two tests examined in this section, and it is used to prove the P–Test. The P–Test, however, is easier to apply and is likely to be the test you use more often.

Integral Test

A series can be thought of as a sum of areas of rectangles each having a base of one unit (Fig. 1). With this area interpretation of series there is a natural connection between series and integrals and between the convergence of a series and the convergence of an appropriate improper integral.

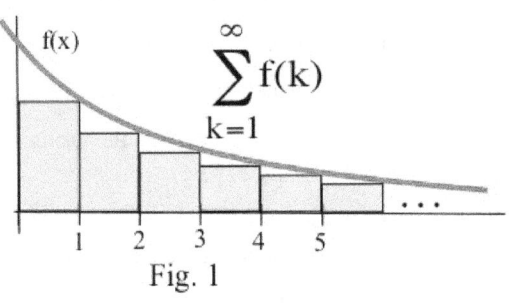

Fig. 1

Example 1:   Suppose the shaded region in Fig. 2a can be painted using 3 gallons of paint. How much paint is needed for the shaded region in Fig. 2b?

Solution: We don't have enough information to determine the exact amount of paint needed for the region in Fig. 2b, but the total of the rectangular areas is smaller than the area in Fig. 2a so less than 3 gallons of paint are needed for the region in Fig. 2b.

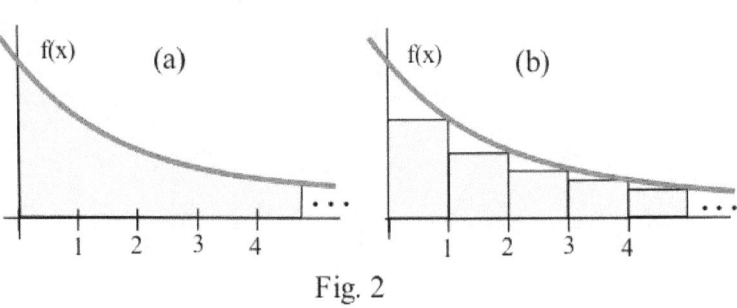

Fig. 2

Practice 1: Suppose the area of the shaded region in Fig. 3a is infinite. What can you say about the total area of the rectangular regions in Fig. 3b?

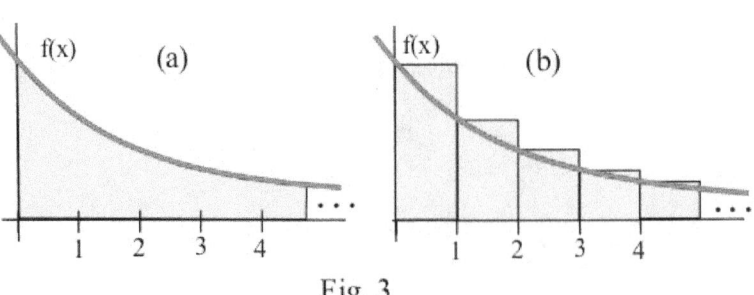

Fig. 3

The geometric reasoning used in Example 1 and Practice 1 can also be used to determine the convergence and divergence of some series.

Example 2:   (a) Which is larger: $\displaystyle\sum_{k=2}^{\infty} \frac{1}{k^2}$ or $\displaystyle\int_{1}^{\infty} \frac{1}{x^2}\, dx$?

(b) Use the result of (a) to show that $\displaystyle\sum_{k=2}^{\infty} \frac{1}{k^2}$ is convergent.

Solution: Fig. 4 illustrates that the area of the rectangles, $\dfrac{1}{2^2} + \dfrac{1}{3^2} + \dfrac{1}{4^2} + \ldots + \dfrac{1}{n^2}$, is less than the area under the graph of the function $f(x) = \dfrac{1}{x^2}$ for $1 \le x \le n$:

$$\sum_{k=2}^{n} \frac{1}{k^2} < \int_{1}^{n} \frac{1}{x^2}\, dx \ \ \text{so} \ \ \sum_{k=2}^{n} \frac{1}{k^2} < \int_{1}^{\infty} \frac{1}{x^2}\, dx = 1 \ \ \text{for every n} \ge 2.$$

Therefore, the partial sums of $\displaystyle\sum_{k=2}^{\infty} \frac{1}{k^2}$ are bounded. Also, each term $a_k = \dfrac{1}{k^2}$ is positive, so the partial sums of $\displaystyle\sum_{k=2}^{\infty} \frac{1}{k^2}$ are monotonically increasing. So, by the Monotonic Theorem of Section

10.1, the sequence of partial sums converges, so the series $\displaystyle\sum_{k=2}^{\infty} \frac{1}{k^2}$ is convergent.

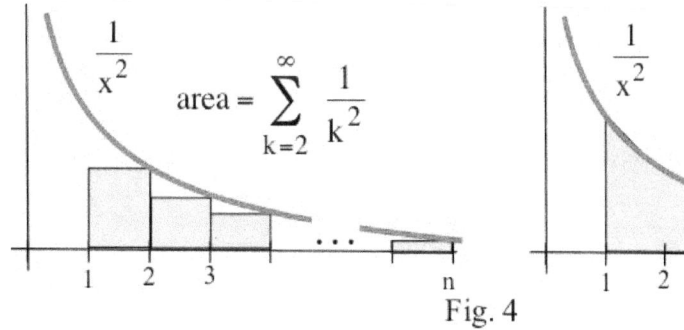 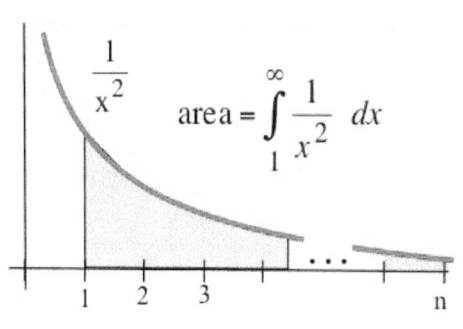

Fig. 4

The reasoning of Example 2 can be extended to the comparison of other series and the appropriate integrals.

Integral Test

Suppose f is a **continuous, positive, decreasing** function on $[1, \infty)$ and $a_k = f(k)$.

The series $\displaystyle\sum_{k=1}^{\infty} a_k$ converges **if and only if** the integral $\displaystyle\int_1^{\infty} f(x)\,dx$ converges.

Equivalently, (a)   if $\displaystyle\int_1^{\infty} f(x)\,dx$ converges, then $\displaystyle\sum_{k=1}^{\infty} a_k$ converges,

and (b)   if $\displaystyle\int_1^{\infty} f(x)\,dx$ diverges, then $\displaystyle\sum_{k=1}^{\infty} a_k$ diverges.

The proof is simply a careful use of the reasoning in the previous Examples.

Proof: Assume that f is a continuous, positive, decreasing function on $[1, \infty)$ and that $a_k = f(k)$.

Part (a): Assume that $\displaystyle\int_1^{\infty} f(x)\,dx$ converges: $\displaystyle\lim_{n\to\infty} \int_1^{n} f(x)\,dx$ is a finite number .

Since each $a_k > 0$, the sequence of partial sums s_n is increasing. If we arrange the rectangles

under the graph of f as in Fig. 5, it is clear that

$$s_n = \sum_{k=1}^{n} a_k = a_1 + \sum_{k=2}^{n} a_k \le a_1 + \int_1^{n} f(x)\,dx \le a_1 + \int_1^{\infty} f(x)\,dx \quad .$$

$\{ s_n \}$ is a bounded, increasing sequence so, by the

Monotone Convergence Theorem of Section 10.1, $\{ s_n \}$

converges and $\displaystyle\sum_{k=1}^{\infty} a_k$ is convergent.

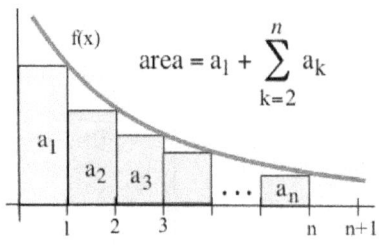

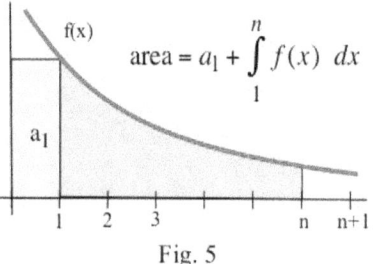

Fig. 5

Part (b): Assume that $\int\limits_{1}^{\infty} f(x)\,dx$ diverges: $\lim\limits_{n\to\infty} \int\limits_{1}^{n} f(x)\,dx = \infty$.

If we arrange the rectangles under the graph of f as in Fig. 6, it is clear that

$$s_n = \sum_{k=1}^{n} a_k \geq \int\limits_{1}^{n+1} f(x)\,dx \quad \text{for all } n,$$

so $\lim\limits_{n\to\infty} s_n \geq \lim\limits_{n\to\infty} \int\limits_{1}^{n} f(x)\,dx = \infty$.

In other words, $\lim\limits_{n\to\infty} s_n = \infty$ and $\sum\limits_{k=1}^{\infty} a_k$ diverges.

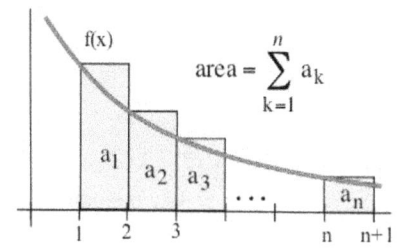

The inequalities in the proof relating the partial sums of the series to the values of integrals are sometimes used to approximate the values of the partial sums of a series:

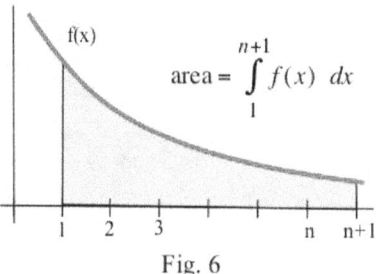

Fig. 6

$$\int\limits_{1}^{n+1} f(x)\,dx \leq \sum_{k=1}^{n} a_k \leq a_1 + \int\limits_{1}^{n} f(x)\,dx .$$

We can use this last set of inequalities with $a_k = \frac{1}{k}$ and n = 1,000 to conclude that

$$\int\limits_{1}^{10^3+1} \frac{1}{x}\,dx \leq \sum_{k=1}^{10^3} \frac{1}{k} \leq 1 + \int\limits_{1}^{10^3} \frac{1}{x}\,dx \quad \text{so} \quad 7.48646986155 \leq \sum_{k=1}^{10^3} \frac{1}{k} \leq 8.48547086055 .$$

(If n = 1,000,000, then the same type of reasoning shows that the partial sum of 1/k from k = 1 to $k = 10^6$ is between 13.815511 and 14.815510 .)

If the series does not start with k = 1, a Corollary of the Integral Test can be used.

Corollary: If   f satisfies the hypotheses of the Integral Test on $[N, \infty)$ and $a_k = f(k)$,

then $\sum\limits_{k=N}^{\infty} a_k$ and $\int\limits_{N}^{\infty} f(x)\,dx$ both converge or both diverge.

Example 3:   Use the Integral Test to determine whether (a) $\sum\limits_{k=1}^{\infty} \frac{1}{k^3}$ and (b) $\sum\limits_{k=2}^{\infty} \frac{1}{k \cdot \ln(k)}$ converge.

Solution: (a) If $f(x) = \dfrac{1}{x^3}$, then $a_k = f(k)$ and f is continuous, positive and decreasing on $[1, \infty)$.

$$\text{Then } \int_1^\infty \frac{1}{x^3}\, dx = \lim_{n\to\infty} \int_1^n \frac{1}{x^3}\, dx = \lim_{n\to\infty} \left(-\frac{1}{2x^2} \right) \Big|_1^n$$

$$= \lim_{n\to\infty} \left(-\frac{1}{2n^2} \right) - \left(-\frac{1}{2\cdot 1^2} \right) = \frac{1}{2} .$$

The integral $\displaystyle\int_1^\infty \frac{1}{x^3}\, dx$ converges so the series $\displaystyle\sum_{k=1}^\infty \frac{1}{k^3}$ converges.

(b) If $f(x) = \dfrac{1}{x\cdot\ln(x)}$, then $a_k = f(k)$ and f is continuous, positive and decreasing on $[2, \infty)$.

$$\text{Then } \int_2^\infty \frac{1}{x\cdot\ln(x)}\, dx = \lim_{n\to\infty} \int_2^n \frac{1}{x\cdot\ln(x)}\, dx = \lim_{n\to\infty} \ln(\ln(x)) \Big|_2^n$$

$$= \lim_{n\to\infty} \ln(\ln(n)) - \ln(\ln(2)) = \infty .$$

The integral $\displaystyle\int_2^\infty \frac{1}{x\cdot\ln(x)}\, dx$ diverges so the series $\displaystyle\sum_{k=2}^\infty \frac{1}{k\cdot\ln(k)}$ diverges.

Practice 2:   Use the Integral Test to determine whether (a) $\displaystyle\sum_{k=4}^\infty \frac{1}{\sqrt{k}}$ and (b) $\displaystyle\sum_{k=1}^\infty e^{-k}$ converge.

Note: The Integral Test does not give the value of the sum, it only answers the question of whether the series converges or diverges. Typically the value of the improper integral is not equal to the sum of the series.

P–Test for Convergence of $\displaystyle\sum_{k=1}^\infty \frac{1}{k^p}$

The P–Test is very easy to use. And it answers the convergence question for a whole family of series.

P–Test

The series $\displaystyle\sum_{k=1}^\infty \frac{1}{k^p}$ $\begin{cases} \text{converges} & \text{if } p > 1 \\ \text{diverges} & \text{if } p \le 1 . \end{cases}$

Proof: If **p = 1**, then $\sum_{k=1}^{\infty} \frac{1}{k^p} = \sum_{k=1}^{\infty} \frac{1}{k}$, the harmonic series, which we already know diverges (by

Section 10.3 or, using the Integral Test, since $\int_{1}^{\infty} \frac{1}{x} \, dx$ diverges to infinity.)

The proof for $p \neq 1$ is a straightforward application of the Integral Test on $f(x) = 1/x^p$.

If $p \neq 1$, then $\int_{1}^{\infty} \frac{1}{x^p} \, dx = \int_{1}^{\infty} x^{-p} \, dx = \lim_{A \to \infty} \int_{1}^{A} x^{-p} \, dx$

$$= \lim_{A \to \infty} \left(\frac{1}{1-p} \right) \cdot x^{1-p} \Big|_{1}^{A}$$

$$= \lim_{A \to \infty} \left(\frac{1}{1-p} \right) \cdot A^{1-p} - \left(\frac{1}{1-p} \right) \cdot 1$$

As we examine the limit of A^{1-p}, there are two cases to consider: $p < 1$ and $p > 1$.

If **p < 1**, then $1 - p > 0$ so A^{1-p} approaches infinity as A approaches infinity. Then

$\int_{1}^{\infty} \frac{1}{x^p} \, dx$ diverges, so, by the Integral Test, $\sum_{k=1}^{\infty} \frac{1}{k^p}$ diverges.

If **p > 1**, then $p - 1 > 0$ and $A^{1-p} = \frac{1}{A^{p-1}}$ approaches 0 as A approaches infinity. Then

$\int_{1}^{\infty} \frac{1}{x^p} \, dx$ converges, so, by the Integral Test, $\sum_{k=1}^{\infty} \frac{1}{k^p}$ converges.

Example 3:   Use the P–Test to determine whether (a) $\sum_{k=1}^{\infty} \frac{1}{k^2}$ and (b) $\sum_{k=4}^{\infty} \frac{1}{\sqrt{k}}$ converge.

Solution: The convergence of both series have already been determined using the Integral Test, but the

P–Test is much easier to apply.

(a) $p = 2 > 1$ so $\sum_{k=1}^{\infty} \frac{1}{k^2}$ converges.   (b) $p = 1/2 < 1$ so $\sum_{k=4}^{\infty} \frac{1}{\sqrt{k}}$ diverges.

The P–Test is very easy to use (Is the exponent $p > 1$ or is $p \leq 1$?), and it is also very useful. In the next

section we will compare new series with series whose convergence we already know, and most often this

comparison is with some P–series whose convergence we know about from the P–Test.

Note: **The P–Test does not give the value of the sum, it only answers the question of whether the**

series converges of diverges.

PROBLEMS

In problems 1 – 15 show that the function determined by the terms of the given series satisfies the hypotheses of the Integral Test, and then use the Integral Test to determine whether the series converges or diverges.

1. $\displaystyle\sum_{k=1}^{\infty} \frac{1}{2k+5}$

2. $\displaystyle\sum_{k=1}^{\infty} \frac{1}{(2k+5)^2}$

3. $\displaystyle\sum_{k=1}^{\infty} \frac{1}{(2k+5)^{3/2}}$

4. $\displaystyle\sum_{k=1}^{\infty} \frac{\ln(k)}{k}$

5. $\displaystyle\sum_{k=2}^{\infty} \frac{1}{k\cdot(\ln(k))^2}$

6. $\displaystyle\sum_{k=1}^{\infty} \frac{1}{k^2}\cdot\sin\left(\frac{1}{k}\right)$

7. $\displaystyle\sum_{k=1}^{\infty} \frac{1}{k^2+1}$

8. $\displaystyle\sum_{k=1}^{\infty} \frac{1}{k^2+100}$

9. $\displaystyle\sum_{k=1}^{\infty} \left\{\frac{1}{k}-\frac{1}{k+3}\right\}$

10. $\displaystyle\sum_{k=1}^{\infty} \left\{\frac{1}{k}-\frac{1}{k+1}\right\}$

11. $\displaystyle\sum_{k=1}^{\infty} \frac{1}{k(k+5)}$

12. $\displaystyle\sum_{k=2}^{\infty} \frac{1}{k^2-1}$

13. $\displaystyle\sum_{k=1}^{\infty} k\cdot e^{-(k^2)}$

14. $\displaystyle\sum_{k=1}^{\infty} k^2\cdot e^{-(k^3)}$

15. $\displaystyle\sum_{k=1}^{\infty} \frac{1}{\sqrt{6k+10}}$

For problems 16 – 20, (a) use the P–Test to determine whether the given series converges, and then (b) use the Integral Test to verify your convergence conclusion of part (a).

16. $\displaystyle\sum_{k=1}^{\infty} \frac{1}{k^2}$

17. $\displaystyle\sum_{k=1}^{\infty} \frac{1}{k^3}$

18. $\displaystyle\sum_{k=2}^{\infty} \frac{1}{k}$

19. $\displaystyle\sum_{k=2}^{\infty} \frac{1}{\sqrt{k}}$

20. $\displaystyle\sum_{k=3}^{\infty} \frac{1}{k^{2/3}}$

21. $\displaystyle\sum_{k=3}^{\infty} \frac{1}{k^{3/2}}$

In the proof of the Integral Test, we derived an inequality bounding the values of the partial sums $s_n = \displaystyle\sum_{k=1}^{n} a_k$

between the values of two integrals: $\displaystyle\int_1^{n+1} f(x)\,dx \leq \sum_{k=1}^{n} a_k \leq a_1 + \int_1^{n} f(x)\,dx$. For problems 22 – 27, use this inequality to determine bounds on the values of $s_{10}, s_{100},$ and $s_{1,000,000}$ for the given series.

22. $\displaystyle\sum_{k=1}^{\infty} \frac{1}{k^2}$ (Note: The exact value of $\displaystyle\sum_{k=1}^{\infty}\frac{1}{k^2}$ is $\frac{\pi^2}{6}$ but it beyond our means to prove that in this course.)

23. $\displaystyle\sum_{k=1}^{\infty} \frac{1}{k^3}$

24. $\displaystyle\sum_{k=1}^{\infty} \frac{1}{k}$

25. $\displaystyle\sum_{k=1}^{\infty} \frac{1}{k+1000}$

26. $\displaystyle\sum_{k=1}^{\infty} \frac{1}{k^2+1}$

27. $\displaystyle\sum_{k=1}^{\infty} \frac{1}{k^2+100}$

28. Euler's Constant: Define $g_1 = 1 - \ln(1) = 1$,

$g_2 = (1 + \frac{1}{2}) - \ln(2) \approx 0.806853$,

$g_3 = (1 + \frac{1}{2} + \frac{1}{3}) - \ln(3) \approx 0.734721$, and, in general,

$g_n = (1 + \frac{1}{2} + \frac{1}{3} + \frac{1}{4} + ... + \frac{1}{n}) - \ln(n)$.

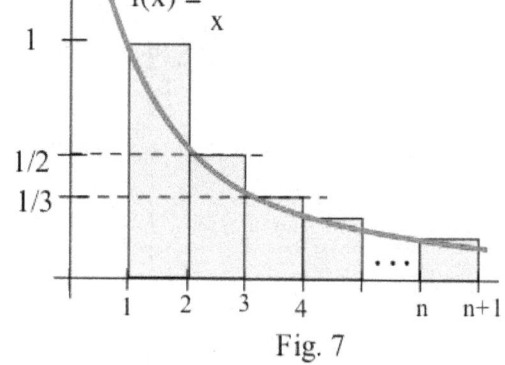

Fig. 7

(a) Make several copies of Fig. 7, and shade the
      regions represented by g_2, g_3, g_4, and g_n.

(b) Provide a geometric argument that $g_n > 0$ for

      all $n \geq 1$.

(c) Provide a geometric argument that $\{ g_n \}$ is monotonically decreasing: $g_{n+1} < g_n$ for all $n \geq 1$.

(d) Conclude from parts (b) and (c) and the Monotone Convergence Theorem (Section 10.1) that $\{ g_n \}$

      converges.

(Note: The value to which $\{ g_n \}$ converges is denoted by "γ," the lower case Greek letter gamma, and is

        called Euler's constant. It is not even known if γ is a rational number. $\gamma \approx 0.5772157 \cdots$.)

29. (a)   Show that the integral $\displaystyle\int_{2}^{\infty} \frac{1}{x \cdot (\ln x)^q} \, dx$ converges for $q>1$ and diverges for $q \leq 1$.

    (b)   Use the result of part (a) to state a "Q test" for $\displaystyle\sum_{k=2}^{\infty} \frac{1}{k \cdot (\ln k)^q}$.

In problems 30 – 33, use the result of Problem 29 to determine whether the given series converge.

30. $\displaystyle\sum_{k=2}^{\infty} \frac{1}{k \cdot \ln(k)}$

31. $\displaystyle\sum_{k=2}^{\infty} \frac{1}{k \cdot (\ln k)^3}$

32. $\displaystyle\sum_{k=2}^{\infty} \frac{1}{k \cdot \sqrt{\ln k}}$

33. $\displaystyle\sum_{k=2}^{\infty} \frac{1}{k \cdot \ln(k^3)}$

Practice Answers

Practice 1:   { total area of rectangular pieces } > { area under the curve in Fig. 3a } so if the shaded

area in Fig. 3a is infinite, then the shaded area in Fig. 3b is also infinite.

Practice 2:   (a) Let $f(x) = \dfrac{1}{\sqrt{x}}$. Then $a_k = f(k)$ and f is continuous, positive and decreasing on $[1, \infty)$.

Then $\displaystyle\int_{4}^{\infty} \dfrac{4}{\sqrt{x}}\, dx = \lim_{n\to\infty} \int_{1}^{n} \dfrac{1}{\sqrt{x}}\, dx = \lim_{n\to\infty} 2\, x^{1/2}\Big|_{4}^{n}$

$\qquad\qquad\qquad\qquad\qquad = \lim_{n\to\infty} 2\sqrt{n} - 2\sqrt{4} = \infty$.

The integral $\displaystyle\int_{4}^{\infty} \dfrac{1}{\sqrt{x}}\, dx$ diverges so the series $\displaystyle\sum_{k=4}^{\infty} \dfrac{1}{\sqrt{k}}$ diverges.

(Note: It will be easier to determine that this series diverges by using the P–Test which

occurs right after this Practice problem in the text.)

(b) Let $f(x) = e^{-x}$. Then $e^{-k} = a_k = f(k)$ and f is continuous, positive and decreasing on $[1, \infty)$.

Then $\displaystyle\int_{1}^{\infty} e^{-x}\, dx = \lim_{n\to\infty} \int_{1}^{n} e^{-x}\, dx = \lim_{n\to\infty} -e^{-x}\Big|_{1}^{n} = \lim_{n\to\infty} \left(-e^{-n}\right) - \left(-e^{-1}\right)$

$\qquad\qquad\qquad\qquad\qquad\qquad = \lim_{n\to\infty} \left(-\dfrac{1}{e^{n}}\right) - \left(-\dfrac{1}{e}\right) = \dfrac{1}{e} \approx 0.368$.

The integral $\displaystyle\int_{1}^{\infty} e^{-x}\, dx$ converges so the series $\displaystyle\sum_{k=1}^{\infty} e^{-k}$ converges.

10.5   POSITIVE TERM SERIES: COMPARISON TESTS

This section discusses how to determine whether some series converge or diverge by comparing them with other series which we already know converge or diverge. In the basic Comparison Test we compare the two series term by term. In the more powerful Limit Comparison Test, we compare limits of ratios of the terms of the two series. Finally, we focus on the parts of the terms of a series that determine whether the series converges or diverges.

Comparison Test

Informally, if the individual terms of our series are smaller than the corresponding terms of a known convergent series, then our series converges. If our series is larger, term by term, than a known divergent series then our series diverges. If the individual terms of our series are larger than the corresponding terms of a convergent series or smaller than the corresponding terms of a divergent series, then our series may converge or diverge — the Comparison Test does not tell us.

Comparison Test

Suppose we want to determine whether $\displaystyle\sum_{k=1}^{\infty} a_k$ converges or diverges.

(a) If there is a convergent series $\displaystyle\sum_{k=1}^{\infty} c_k$ with $0 < a_k \le c_k$ for all k, then $\displaystyle\sum_{k=1}^{\infty} a_k$ converges.

(b) If there is a divergent series $\displaystyle\sum_{k=1}^{\infty} d_k$ with $a_k \ge d_k > 0$ for all k, then $\displaystyle\sum_{k=1}^{\infty} a_k$ diverges.

Proof: Since all of the terms of the $a_k, c_k,$ and d_k series are positive, their sequences of partial sums are all monotonic increasing. The proof compares the partial sums of the various series.

(a) Suppose that $0 < a_k \le c_k$ for all k and that $\displaystyle\sum_{k=1}^{\infty} c_k$ converges. Since $\displaystyle\sum_{k=1}^{\infty} c_k$ converges,

then the partial sums $t_n = \displaystyle\sum_{k=1}^{n} c_k$ approach a finite limit: $\displaystyle\lim_{n\to\infty} t_n = L$.

For each n, $s_n \le t_n$ (why?) so $\displaystyle\lim_{n\to\infty} s_n \le \lim_{n\to\infty} t_n = L$, and the sequence $\{ s_n \}$ is bounded by L. Finally, by the Monotone Convergence Theorem, we can conclude that $\{ s_n \}$

converges and that the series $\displaystyle\sum_{k=1}^{\infty} a_k$ converges.

(b) Suppose that $a_k \geq d_k > 0$ for all k and that $\displaystyle\sum_{k=1}^{\infty} d_k$ diverges. Since $\displaystyle\sum_{k=1}^{\infty} d_k$ diverges,

then the partial sums $u_n = \displaystyle\sum_{k=1}^{n} d_k$ approach infinity: $\displaystyle\lim_{n \to \infty} u_n = \infty$. Then

$$\lim_{n \to \infty} s_n \geq \lim_{n \to \infty} u_n = \infty$$

so the sequence of partial sums $\{ s_n \}$ diverges and the series $\displaystyle\sum_{k=1}^{\infty} a_k$ diverges.

The Comparison Test requires that we select and compare our series against a series whose convergence or

divergence is known, and that choice requires that we know a collection of series that converge and some

that diverge. Typically, we pick a p–series or a geometric series to compare with our series, but this choice

requires some experience and practice.

Example 1:   Use the Comparison Test to determine the convergence or divergence of

(a) $\displaystyle\sum_{k=1}^{\infty} \frac{1}{k^2 + 3}$ and (b) $\displaystyle\sum_{k=1}^{\infty} \frac{k+1}{k^2}$.

Solution: For these two series it is useful to compare with p–series for appropriate values of p.

(a) For all k, $\dfrac{1}{k^2 + 3} < \dfrac{1}{k^2}$, and $\displaystyle\sum_{k=1}^{\infty} \frac{1}{k^2}$ converges (P–Test, p = 2)

so $\displaystyle\sum_{k=1}^{\infty} \frac{1}{k^2 + 3}$ converges.

(b) For all k, $\dfrac{k+1}{k^2} = \dfrac{1}{k} + \dfrac{1}{k^2} > \dfrac{1}{k}$.

Since $\displaystyle\sum_{k=1}^{\infty} \frac{1}{k}$ diverges (P–Test, p = 1), we conclude that $\displaystyle\sum_{k=1}^{\infty} \frac{k+1}{k^2}$ diverges.

Practice 1:   Use the Comparison Test to determine the convergence or divergence of

(a) $\displaystyle\sum_{k=3}^{\infty} \frac{1}{\sqrt{k} - 2}$ and (b) $\displaystyle\sum_{k=1}^{\infty} \frac{1}{2^k + 7}$.

Example 2:   A student has shown algebraically that $\frac{1}{k^2} < \frac{1}{k^2-1} < \frac{1}{k}$ for all $k \geq 2$. From this information and the Comparison Test, what can the student conclude about the convergence of the series $\sum\limits_{k=2}^{\infty} \frac{1}{k^2-1}$?

Solution: Nothing. The Comparison Test only gives a definitive answer if our series is smaller than a convergent series or if our series is larger than a divergent series. In this example, our series is larger than a convergent series, $\sum \frac{1}{k^2}$, and is smaller than a divergent series, $\sum \frac{1}{k}$, so we can not conclude anything about the convergence of $\sum\limits_{k=2}^{\infty} \frac{1}{k^2-1}$.

However, we can show that if $k \geq 2$ then $\frac{1}{k^2-1} < \frac{2}{k^2}$. Since $\sum\limits_{k=2}^{\infty} \frac{2}{k^2}$ converges (P–Test), we can conclude that $\sum\limits_{k=2}^{\infty} \frac{1}{k^2-1}$ converges. Next in this section we present a variation on the Comparison Test that allows us to quickly conclude that $\sum\limits_{k=2}^{\infty} \frac{1}{k^2-1}$ converges.

Limit Comparison Test

The exact value of the sum of a series depends on every part of the terms of the series, but if we are only asking about convergence or divergence, some parts of the terms can be safely ignored. For example, the three series with terms $1/k^2$, $1/(k^2+1)$, and $1/(k^2-1)$ converge to different values,

$$\sum\limits_{k=2}^{\infty} \frac{1}{k^2} \approx 0.645 , \qquad \sum\limits_{k=2}^{\infty} \frac{1}{k^2+1} \approx 0.577 , \qquad \sum\limits_{k=2}^{\infty} \frac{1}{k^2-1} = 0.750 ,$$

but they all do converge. The "+ 1" and "– 1" in the denominators affect the value of the final sum, but they do not affect whether that sum is finite or infinite. When k is a large number, the values of $1/(k^2+1)$ and $1/(k^2-1)$ are both very close to the value of $1/k^2$, and the convergence or divergence of the series $\sum\limits_{k=2}^{\infty} \frac{1}{k^2+1}$ and $\sum\limits_{k=2}^{\infty} \frac{1}{k^2-1}$ can be predicted from the convergence or divergence of the series $\sum\limits_{k=2}^{\infty} \frac{1}{k^2}$.

The Limit Comparison Test states these ideas precisely.

Limit Comparison Test

Suppose $a_k > 0$ for all k, and we want to determine whether $\displaystyle\sum_{k=1}^{\infty} a_k$ converges or diverges.

If   there is a series $\displaystyle\sum_{k=1}^{\infty} b_k$ so that $\displaystyle\lim_{k\to\infty} \frac{a_k}{b_k} = L$, a **positive, finite** value,

then $\displaystyle\sum_{k=1}^{\infty} a_k$ and $\displaystyle\sum_{k=1}^{\infty} b_k$ both converge or both diverge.

Idea for a proof: The key idea is that if $\displaystyle\lim_{k\to\infty} \frac{a_k}{b_k} = L$ is a **positive, finite** value, then, when n is

very large, $\frac{a_k}{b_k} \approx L$ so $a_k \approx L \cdot b_k$ and $\displaystyle\sum_{k=N}^{\infty} a_k \approx L \cdot \sum_{k=N}^{\infty} b_k$. If one of these series converges, then so

does the other. If one of these series diverges, then so does the other. When n is a relatively small

number, the a_k and b_k values may not have a ratio close to L, but the first "few" values of a series do not

affect the convergence or divergence of the series. A proof of the Limit Comparison Test is given in an

Appendix after the Practice Answers.

Example 3:   Put $a_k = \frac{1}{k^2}$, $b_k = \frac{1}{k^2 + 1}$, and $c_k = \frac{1}{k^2 - 1}$ and show that $\displaystyle\lim_{k\to\infty} \frac{a_k}{b_k} = 1$

and $\displaystyle\lim_{k\to\infty} \frac{a_k}{c_k} = 1$. Since $\displaystyle\sum_{k=2}^{\infty} \frac{1}{k^2}$ converges (P–Test, p = 2) we can conclude

that $\displaystyle\sum_{k=2}^{\infty} \frac{1}{k^2 + 1}$ and $\displaystyle\sum_{k=2}^{\infty} \frac{1}{k^2 - 1}$ both converge too.

Solution: $\dfrac{a_k}{b_k} = \dfrac{1/k^2}{1/(k^2 + 1)} = \dfrac{k^2 + 1}{k^2} = 1 + \dfrac{1}{k^2} \longrightarrow 1$ so L = 1 is positive and finite.

Similarly, $\dfrac{a_k}{c_k} = \dfrac{1/k^2}{1/(k^2 - 1)} = \dfrac{k^2 - 1}{k^2} = 1 - \dfrac{1}{k^2} \longrightarrow 1$ so L = 1 is positive and finite.

Practice 2:   (a)   Find a p–series to "limit–compare" with $\sum\limits_{k=2}^{\infty} \dfrac{k^2 + 5k}{k^3 + k^2 + 7}$.

(Suggestion: put $a_k = \dfrac{k^2 + 5k}{k^3 + k^2 + 7}$ and find a value of p so that $b_k = \dfrac{1}{k^p}$ and

$\lim\limits_{k \to \infty} \dfrac{a_k}{b_k} = L$, a positive, finite value.) Does $\sum\limits_{k=2}^{\infty} \dfrac{k^2 + 5k}{k^3 + k^2 + 7}$ converge?

(b) Find a p–series to compare with $\sum\limits_{k=3}^{\infty} \dfrac{5}{\sqrt{k^4 - 11}}$. Does $\sum\limits_{k=3}^{\infty} \dfrac{5}{\sqrt{k^4 - 11}}$ converge ?

The Limit Comparison Test is particularly useful because it allows us to ignore some parts of the terms that cause algebraic difficulties but that have no effect on the convergence of the series.

Using "Dominant Terms"

To use the Limit Comparison Test we need to pick a new series to compare with our given series. One effective way to pick the new series is to form the new series using the largest power of the variable (dominant term) from the numerator and the largest power of the variable (dominant term) from the denominator. The "dominant term" series consists of $\dfrac{\text{dominant term in the numerator}}{\text{dominant term in the denominator}}$. Then the Limit Comparison Test allows us to conclude that the original series converges if and only if the "dominant term" series converges.

Example 4:   For each of the given series, form a new series consisting of the dominant terms from the numerator and the denominator. Does the series of dominant terms converge?

(a) $\sum\limits_{k=3}^{\infty} \dfrac{5k^2 - 3k + 2}{17 + 2k^4}$ (b) $\sum\limits_{k=1}^{\infty} \dfrac{1 + 4k}{\sqrt{k^3 + 5k}}$ (c) $\sum\limits_{k=1}^{\infty} \dfrac{k^{23} + 1}{5k^{10} + k^{26} + 3}$.

Solution: (a)   The dominant terms of the numerator and denominator are $5k^2$ and $2k^4$, respectively, so the

"dominant term" series is $\sum\limits_{k=3}^{\infty} \dfrac{5k^2}{2k^4} = \dfrac{5}{2} \sum\limits_{k=3}^{\infty} \dfrac{1}{k^2}$ which converges (P–Test, $p = 2$).

(b) The dominant terms of the numerator and denominator are $4k$ and $k^{3/2}$, respectively, so the

"dominant term" series is $\sum\limits_{k=1}^{\infty} \dfrac{4k}{k^{3/2}} = 4 \sum\limits_{k=3}^{\infty} \dfrac{1}{k^{1/2}}$ which diverges (P–Test, $p = 1/2$).

(c) The dominant terms of the numerator and denominator are k^{23} and k^{26} , respectively, so the

"dominant term" series is $\sum\limits_{k=1}^{\infty} \dfrac{k^{23}}{k^{26}} = \sum\limits_{k=1}^{\infty} \dfrac{1}{k^3}$ which converges (P–Test, $p = 3$).

Using the Limit Comparison Test to compare the given series with the "dominant term" series, we can conclude that the given series (a) and (c) converge and that the given series (b) diverges.

Practice 3:   For each of the given series, form a new series consisting of the dominant terms from the numerator and the denominator. Does the series of dominant terms converge? Do the given series converge?

(a) $\displaystyle\sum_{k=1}^{\infty} \frac{3k^4 - 5k + 2}{1 + 17k^2 + 9k^5}$
(b) $\displaystyle\sum_{k=1}^{\infty} \frac{\sqrt{1+9k}}{k^2 + 5k - 2}$
(c) $\displaystyle\sum_{k=1}^{\infty} \frac{k^{25} + 1}{5k^{10} + k^{26} + 3}$.

Experienced users of series commonly use "dominant terms" to make quick and accurate judgments about the convergence or divergence of a series. With practice, so can you.

PROBLEMS

In problems 1 – 12 use the Comparison Test to determine whether the given series converge or diverge.

1. $\displaystyle\sum_{k=1}^{\infty} \frac{\cos^2(k)}{k^2}$
2. $\displaystyle\sum_{k=1}^{\infty} \frac{3}{k^3 + 7}$
3. $\displaystyle\sum_{n=3}^{\infty} \frac{5}{n-1}$

4. $\displaystyle\sum_{n=1}^{\infty} \frac{2 + \sin(n)}{n^3}$
5. $\displaystyle\sum_{j=1}^{\infty} \frac{3 + \cos(j)}{j}$
6. $\displaystyle\sum_{j=1}^{\infty} \frac{\arctan(j)}{j^{3/2}}$

7. $\displaystyle\sum_{k=1}^{\infty} \frac{\ln(k)}{k}$
8. $\displaystyle\sum_{k=1}^{\infty} \frac{k-1}{k \cdot 1.5^k}$
9. $\displaystyle\sum_{k=1}^{\infty} \frac{k+9}{k \cdot 2^k}$

10. $\displaystyle\sum_{n=1}^{\infty} \frac{n^3 + 7}{n^4 - 1}$
11. $\displaystyle\sum_{n=1}^{\infty} \frac{1}{1+2+3+...+(n-1)+n}$
12. $\displaystyle\sum_{k=1}^{\infty} \frac{1}{k!} = \sum_{k=1}^{\infty} \frac{1}{1 \cdot 2 \cdot 3 \cdot ... \cdot (k-1) \cdot k}$

In problems 13 – 21 use the Limit Comparison Test (or the Nth Term Test) to determine whether the given series converge or diverge.

13. $\displaystyle\sum_{k=3}^{\infty} \frac{k+1}{k^2 + 4}$
14. $\displaystyle\sum_{j=1}^{\infty} \frac{7}{\sqrt{j^3 + 3}}$
15. $\displaystyle\sum_{w=1}^{\infty} \frac{5}{w+1}$

16. $\displaystyle\sum_{n=1}^{\infty} \frac{7n^3 - 4n + 3}{3n^4 + 7n^3 + 9}$
17. $\displaystyle\sum_{k=1}^{\infty} \frac{k^3}{(1+k^2)^3}$
18. $\displaystyle\sum_{k=1}^{\infty} \left(\frac{\arctan(k)}{k} \right)^2$

19. $\displaystyle\sum_{n=1}^{\infty} \left(\frac{5 - \frac{1}{n}}{n} \right)^3$
20. $\displaystyle\sum_{w=1}^{\infty} \left(1 + \frac{1}{w} \right)^w$
21. $\displaystyle\sum_{j=1}^{\infty} \left(1 - \frac{1}{j} \right)^j$

In problems 22 – 30 use "dominant term" series to determine whether the given series converge or diverge.

22. $\displaystyle\sum_{n=3}^{\infty} \frac{n+100}{n^2-4}$

23. $\displaystyle\sum_{k=1}^{\infty} \frac{7k}{\sqrt{k^3+5}}$

24. $\displaystyle\sum_{k=1}^{\infty} \frac{5}{k+1}$

25. $\displaystyle\sum_{j=1}^{\infty} \frac{j^3-4j+3}{2j^4+7j^6+9}$

26. $\displaystyle\sum_{n=1}^{\infty} \frac{5n^3+7n^2+9}{\left(1+n^3\right)^2}$

27. $\displaystyle\sum_{n=1}^{\infty} \left(\frac{\arctan(3n)}{2n}\right)^2$

28. $\displaystyle\sum_{k=1}^{\infty} \left(\frac{3-\frac{1}{k}}{k}\right)^2$

29. $\displaystyle\sum_{j=1}^{\infty} \frac{\sqrt{j^3+4j^2}}{j^2+3j-2}$

30. $\displaystyle\sum_{k=1}^{\infty} \frac{k+9}{k\cdot 2^k}$

Putting it all together

In problems 31 – 51 use any of the methods from this or previous sections to determine whether the given series converge or diverge. Give reasons for your answers.

31. $\displaystyle\sum_{n=2}^{\infty} \frac{n^2+10}{n^3-2}$

32. $\displaystyle\sum_{k=1}^{\infty} \frac{3k}{\sqrt{k^5+7}}$

33. $\displaystyle\sum_{k=1}^{\infty} \frac{3}{2k+1}$

34. $\displaystyle\sum_{j=1}^{\infty} \frac{j^2-j+1}{3j^4+2j^2+1}$

35. $\displaystyle\sum_{n=1}^{\infty} \frac{2n^3+n^2+6}{\left(3+n^2\right)^2}$

36. $\displaystyle\sum_{n=1}^{\infty} \left(\frac{\arctan(2n)}{3n}\right)^3$

37. $\displaystyle\sum_{k=1}^{\infty} \left(\frac{1-\frac{2}{k}}{k}\right)^3$

38. $\displaystyle\sum_{j=1}^{\infty} \frac{\sqrt{j^2+4j}}{j^3-2}$

39. $\displaystyle\sum_{k=1}^{\infty} \frac{k+5}{k\cdot 3^k}$

40. $\displaystyle\sum_{n=1}^{\infty} \frac{1+\sin(n)}{n^2+4}$

41. $\displaystyle\sum_{k=1}^{\infty} \frac{k+2}{\sqrt{k^2+1}}$

42. $\displaystyle\sum_{k=1}^{\infty} \frac{\sin(k\pi)}{k+1}$

43. $\displaystyle\sum_{j=1}^{\infty} \frac{3}{e^j+j}$

44. $\displaystyle\sum_{n=1}^{\infty} \frac{(2+3n)^2+9}{\left(1+n^3\right)^2}$

45. $\displaystyle\sum_{n=1}^{\infty} \left(\frac{\tan(3)}{2+n}\right)^2$

46. $\displaystyle\sum_{n=1}^{\infty} \sin^2\left(\frac{1}{n}\right)$

47. $\displaystyle\sum_{n=1}^{\infty} \sin^3\left(\frac{1}{n}\right)$

48. $\displaystyle\sum_{n=1}^{\infty} \cos^2\left(\frac{1}{n}\right)$

49. $\displaystyle\sum_{j=1}^{\infty} \cos^3\left(\frac{1}{j}\right)$

50. $\displaystyle\sum_{n=1}^{\infty} \tan^2\left(\frac{1}{n}\right)$

51. $\displaystyle\sum_{n=1}^{\infty} \left(1-\frac{2}{n}\right)^n$

Review for Positive Term Series: Converge or Diverge

State whether the given series converge or diverge and give reasons for your answer. You may need any of the methods discussed so far as well as some ingenuity.

R1. $\displaystyle\sum_{n=3}^{\infty} \frac{5}{3^n}$

R2. $\displaystyle\sum_{j=3}^{\infty} \frac{5+\cos(j^3)}{j^2}$

R3. $\displaystyle\sum_{w=1}^{\infty} \frac{2}{3+\sin(w^3)}$

R4. $\displaystyle\sum_{n=1}^{\infty} \frac{5}{(1/3)^n}$

R5. $\displaystyle\sum_{k=1}^{\infty} e^{-k}$

R6. $\displaystyle\sum_{w=1}^{\infty} \sin^2\!\left(\frac{1}{w}\right)$ (Hint: for $0\le x\le 1$, $\sin(x)\le x$)

R7. $\displaystyle\sum_{k=1}^{\infty} \sin\!\left(\frac{1}{k^3}\right)$ (see the R6 hint)

R8. $\displaystyle\sum_{j=1}^{\infty} \cos^2\!\left(\frac{1}{j}\right)$

R9. $\displaystyle\sum_{n=3}^{\infty} \frac{5+\cos(n^2)}{n^3}$

R10. $\displaystyle\sum_{k=1}^{\infty} \frac{1}{k\cdot(3+\ln(k))}$

R11. $\displaystyle\sum_{j=1}^{\infty} \frac{1}{j\cdot(3+\ln(j))^2}$

R12. $\displaystyle\sum_{n=1}^{\infty} \frac{4}{n\cdot \arctan(n)}$

R13. $\displaystyle\sum_{n=3}^{\infty} \frac{4\cdot \arctan(n)}{n}$

R14. $\displaystyle\sum_{k=1}^{\infty} \frac{\ln(k)}{k^3}$

R15. $\displaystyle\sum_{k=1}^{\infty} \frac{\ln(k)}{k^2}$

R16. $\displaystyle\sum_{j=1}^{\infty} \left(\frac{j}{2j+3}\right)^j$

R17. $\displaystyle\sum_{n=1}^{\infty} \frac{1+n}{1+n^2}$

R18. $\displaystyle\sum_{n=1}^{\infty} \left[\sin(n) - \sin(n+1)\right]$

R19. $\displaystyle\sum_{k=1}^{\infty} \sqrt{\frac{k^3+5}{k^5+3}}$

R20. $\displaystyle\sum_{k=1}^{\infty} \frac{1}{k^k}$

R21. $\displaystyle\sum_{n=1}^{\infty} n^{1/n}$

Practice Answers

Practice 1:   (a) For $k > 3$, $0 < k-2 < k$ so $0 < \sqrt{k-2} < \sqrt{k}$ and $\dfrac{1}{\sqrt{k-2}} > \dfrac{1}{\sqrt{k}} = \dfrac{1}{k^{1/2}}$.

$\displaystyle\sum_{k=3}^{\infty} \frac{1}{k^{1/2}}$ diverges (P–test, $p = 1/2$) so $\displaystyle\sum_{k=3}^{\infty} \frac{1}{\sqrt{k-2}}$ diverges.

(b) For $k > 1$, $2^k + 7 > 2^k > 0$ so $\dfrac{1}{2^k+7} < \dfrac{1}{2^k}$.

$\displaystyle\sum_{k=3}^{\infty} \frac{1}{2^k} = \sum_{k=3}^{\infty} \left(\frac{1}{2}\right)^k$ which is a convergent geometric series ($r = 1/2$) so $\displaystyle\sum_{k=3}^{\infty} \frac{1}{2^k+7}$ converges.

Practice 2:   (a) $a_k = \dfrac{k^2 + 5k}{k^3 + k^2 + 7}$. Put $b_k = \dfrac{1}{k^1}$. Then

$$\frac{a_k}{b_k} \ = \ \frac{\dfrac{k^2 + 5k}{k^3 + k^2 + 7}}{\dfrac{1}{k^1}} \ = \left(\frac{k^1}{1}\right) \frac{k^2 + 5k}{k^3 + k^2 + 7} \ = \ \frac{k^3 + 5k^2}{k^3 + k^2 + 7} \ = \ \frac{k^3\left(1 + \dfrac{5}{k}\right)}{k^3\left(1 + \dfrac{1}{k} + \dfrac{7}{k^3}\right)} \ \longrightarrow 1$$

so L = 1 is positive and finite, and $\displaystyle\sum_{k=1}^{\infty} a_k$ and $\displaystyle\sum_{k=1}^{\infty} b_k$ both converge or both diverge.

Since we know $\displaystyle\sum_{k=1}^{\infty} b_k \ = \ \sum_{k=1}^{\infty} \frac{1}{k}$ diverges (P–test, p=1 or as the Harmonic series),

we can conclude that $\displaystyle\sum_{k=1}^{\infty} a_k \ = \ \sum_{k=1}^{\infty} \frac{k^2 + 5k}{k^3 + k^2 + 7}$ diverges

(b) $a_k = \dfrac{5}{\sqrt{k^4 - 11}}$. Put $b_k = \dfrac{1}{\sqrt{k^4}} \ = \ \dfrac{1}{k^2}$. Then

$$\frac{a_k}{b_k} \ = \ \frac{\dfrac{5}{\sqrt{k^4 - 11}}}{\dfrac{1}{k^2}} \ = \ \frac{k^2}{1} \cdot \frac{5}{\sqrt{k^4 - 11}} \ = \ \frac{5}{1} \sqrt{\frac{k^4}{k^4 - 11}} \ \longrightarrow \frac{5}{1}\sqrt{1} \ = \ 5$$

so L = 5 is positive and finite, and $\displaystyle\sum_{k=1}^{\infty} a_k$ and $\displaystyle\sum_{k=1}^{\infty} b_k$ both converge or both diverge.

Since we know $\displaystyle\sum_{k-1}^{\infty} b_k \ = \ \sum_{k=1}^{\infty} \frac{1}{k^2}$ converges (P–test, p=2) , we can conclude that

$$\sum_{k=1}^{\infty} a_k \ = \ \sum_{k=1}^{\infty} \frac{5}{\sqrt{k^4 - 11}} \text{ converges.}$$

Practice 3:   (a) $\displaystyle\sum_{k=1}^{\infty} \frac{3k^4}{9k^5} \ = \ \frac{1}{3} \sum_{k=1}^{\infty} \frac{1}{k}$ which diverges (P–test, p = 1) so $\displaystyle\sum_{k=1}^{\infty} \frac{3k^4 - 5k + 2}{1 + 17k^2 + 9k^5}$ diverges.

(b) $\displaystyle\sum_{k=1}^{\infty} \frac{\sqrt{9k}}{k^2} \ = \ 3\sum_{k=1}^{\infty} \frac{k^{1/2}}{k^2} \ = \ 3\sum_{k=1}^{\infty} \frac{1}{k^{3/2}}$ which converges (P–test, p = 3/2)

so $\displaystyle\sum_{k=1}^{\infty} \frac{\sqrt{1 + 9k}}{k^2 + 5k - 2}$ converges.

(c) $\displaystyle\sum_{k=1}^{\infty} \frac{k^{25}}{k^{26}} = \sum_{k=1}^{\infty} \frac{1}{k}$ which diverges (P–test, p = 1) so $\displaystyle\sum_{k=1}^{\infty} \frac{k^{25}+1}{5k^{10}+k^{26}+3}$ diverges.

Appendix: Proof of the Limit Comparison Test

(a)   Suppose $\displaystyle\sum_{k=1}^{\infty} b_k$ converges and that $\displaystyle\lim_{k\to\infty} \frac{a_k}{b_k} = L$, a **positive, finite** value.

Since $\displaystyle\lim_{k\to\infty} \frac{a_k}{b_k} = L$, there is a value N so that $\frac{a_k}{b_k} < L + 1$ when $k \geq N$. Then

$a_k < b_k{\cdot}(L + 1)$ when $k \geq N$, and $\displaystyle\sum_{k=N}^{\infty} a_k < (L+1){\cdot} \sum_{k=N}^{\infty} b_k$. Since $\displaystyle\sum_{k=N}^{\infty} b_k$

converges, we can conclude that $\displaystyle\sum_{k=N}^{\infty} a_k$ converges so $\displaystyle\sum_{k=1}^{\infty} a_k$ also converges.

(b)   Suppose $\displaystyle\sum_{k=1}^{\infty} b_k$ diverges and that $\displaystyle\lim_{k\to\infty} \frac{a_k}{b_k} = L$, a **positive, finite** value.

Since $\displaystyle\lim_{k\to\infty} \frac{a_k}{b_k} = L$, there is a value N so that $\frac{a_k}{b_k} > L/2 > 0$ when $k \geq N$. Then

$a_k > \frac{L}{2}{\cdot}b_k$ when $k \geq N$, and $\displaystyle\sum_{k=N}^{\infty} a_k \geq \frac{L}{2}{\cdot} \sum_{k=N}^{\infty} b_k$. Since $\displaystyle\sum_{k=N}^{\infty} b_k$ diverges, we

can conclude that $\displaystyle\sum_{k=N}^{\infty} a_k$ diverges so $\displaystyle\sum_{k=1}^{\infty} a_k$ also diverges.

10.6   ALTERNATING SERIES

In the last two sections we considered tests for the convergence of series whose terms were all positive. In this section we examine series whose terms change signs in a special way: they alternate between positive and negative. And we present a very easy–to–use test to determine if these alternating series converge.

An **alternating series** is a series whose terms alternate between positive and negative. For example, the following are alternating series:

$$\text{(1)} \qquad 1 - \frac{1}{2} + \frac{1}{3} - \frac{1}{4} + \frac{1}{5} - \ldots + (-1)^{k+1}\frac{1}{k} + \ldots = \sum_{k=1}^{\infty} (-1)^{k+1}\frac{1}{k} \quad \text{(alternating harmonic series)}$$

$$\text{(2)} \qquad -\frac{1}{3} + \frac{2}{4} - \frac{3}{5} + \frac{4}{6} - \frac{5}{7} + \ldots + (-1)^{k}\frac{k}{k+2} + \ldots = \sum_{k=1}^{\infty} (-1)^{k}\frac{k}{k+2}$$

$$\text{(3)} \qquad -\frac{1}{\sqrt{3}} + \frac{1}{\sqrt{5}} - \frac{1}{\sqrt{7}} + \frac{1}{\sqrt{9}} + \frac{1}{\sqrt{11}} - \ldots + (-1)^{k}\frac{1}{\sqrt{2k+1}} + \ldots = \sum_{k=1}^{\infty} (-1)^{k}\frac{1}{\sqrt{2k+1}} \;.$$

Figures 1, 2 and 3 show graphs and tables of values of several partial sums s_n for each of these series. As we move from left to right in each graph (as n increases), the partial sums alternately get larger and smaller, a common pattern for the partial sums of alternating series. The same pattern appears in the tables.

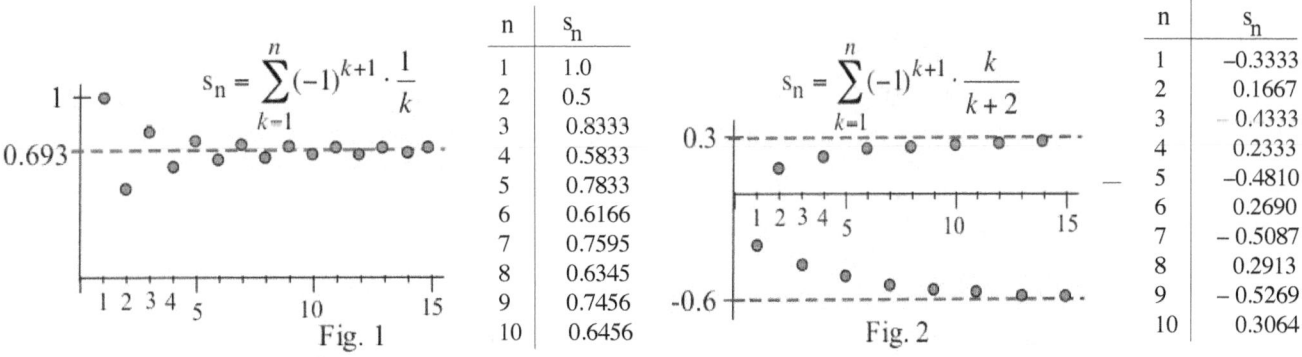

n	s_n
1	1.0
2	0.5
3	0.8333
4	0.5833
5	0.7833
6	0.6166
7	0.7595
8	0.6345
9	0.7456
10	0.6456

$$s_n = \sum_{k=1}^{n} (-1)^{k+1}\cdot\frac{1}{k}$$

Fig. 1

n	s_n
1	-0.3333
2	0.1667
3	-0.4333
4	0.2333
5	-0.4810
6	0.2690
7	-0.5087
8	0.2913
9	-0.5269
10	0.3064

$$s_n = \sum_{k=1}^{n} (-1)^{k+1}\cdot\frac{k}{k+2}$$

Fig. 2

$$s_n = \sum_{k=1}^{n} (-1)^{k}\cdot\frac{1}{\sqrt{2k+1}}$$

n	s_n
1	-0.5774
2	-0.1301
3	-0.5081
4	-0.1748
50	-0.2828
51	-0.3813
1000	-0.3211
1001	-0.3435

Fig. 3

Alternating Series Test

The following result provides a very easy way to determine that some alternating series converge. It says that if the absolute values of the terms decrease monotonically to 0 then the series converges. This is the main result for alternating series.

Alternating Series Test

If   the numbers a_n satisfy the three conditions

      (i) $a_n > 0$ for all n (each a_n is positive)

      (ii) $a_n > a_{n+1}$ (the terms a_n are monotonically decreasing)

      (iii) $\lim\limits_{n \to \infty} a_n = 0$

then   the **alternating series** $\sum\limits_{n=1}^{\infty} (-1)^{n+1} a_n$ converges.

Proof: In order to show that the alternating series converges, we need to show that the sequence of partial sums approaches a finite limit, and we do so in this case by showing that the sequence of even partial sums $\{ s_2, s_4, s_6, \ldots \}$ and the sequence of odd partial sums $\{ s_1, s_3, s_5, \ldots \}$ each approach the same value.

Even partial sums:

$$s_2 = a_1 - a_2 > 0 \qquad\qquad\qquad\qquad\qquad \text{since } a_1 > a_2$$

$$s_4 = a_1 - a_2 + a_3 - a_4 = s_2 + (a_3 - a_4) > s_2 \qquad \text{since } a_3 > a_4$$

$$s_6 = a_1 - a_2 + a_3 - a_4 + a_5 - a_6 = s_4 + (a_5 - a_6) > s_4 \qquad \text{since } a_5 > a_6.$$

In general, the sequence of even partial sums is positive and increasing

$$s_{2n+2} = s_{2n} + (a_{2n+1} - a_{2n+2}) > s_{2n} > 0 \qquad\qquad \text{since } a_{2n+1} > a_{2n+2}.$$

Also,

$$s_{2n} = a_1 - a_2 + a_3 - a_4 + a_5 - \ldots - a_{2n-2} + a_{2n-1} - a_{2n}$$

$$= a_1 - (a_2 - a_3) - (a_4 - a_5) \ldots - (a_{2n-2} - a_{2n-1}) - a_{2n} < a_1.$$

so the sequence of even partial sums is bounded above by a_1.

Since the sequence $\{ s_2, s_4, s_6, \ldots \}$ of even partial sums is increasing and bounded, the

Monotone Convergence Theorem of Section 10.1 tells us that sequence of even partial sums

converges to some finite limit: $\lim\limits_{n \to \infty} s_{2n} = L$.

Odd partial sums:

$$s_{2n+1} = s_{2n} + a_{2n+1} \text{ so } \lim_{n \to \infty} s_{2n+1} = \lim_{n \to \infty} s_{2n} + \lim_{n \to \infty} a_{2n+1} = L + 0 = L .$$

Since the sequence of even partial sums and the sequence of odd partial sums both approach the same

limit L, we can conclude that the limit of the sequence of partial sums is L and that the series

$$\sum_{n=1}^{\infty} (-1)^{n+1} a_n \text{ converges to L: } \sum_{n=1}^{\infty} (-1)^{n+1} a_n = L.$$

Example 1:   Show that each of the three alternating series satisfies the three conditions in the hypothesis

of the Alternating Series Test. Then we can conclude that each of them converges.

(a) $1 - \dfrac{1}{2} + \dfrac{1}{3} - \dfrac{1}{4} + \ldots = \sum_{n=1}^{\infty} (-1)^{n+1} \dfrac{1}{n}$

(b) $\dfrac{3}{1} - \dfrac{3}{\sqrt{2}} + \dfrac{3}{\sqrt{3}} - \dfrac{3}{\sqrt{4}} + \ldots = \sum_{n=1}^{\infty} (-1)^{n+1} \dfrac{3}{\sqrt{n}}$

(c) $\dfrac{7}{2 \cdot \ln(2)} - \dfrac{7}{3 \cdot \ln(3)} + \dfrac{7}{4 \cdot \ln(4)} - \ldots = \sum_{n=2}^{\infty} (-1)^{n} \dfrac{7}{n \cdot \ln(n)}$.

Solution: (a) This series is called the **alternating harmonic series**. (i) $a_n = \dfrac{1}{n} > 0$ for

all positive n. (ii) Since $n < n+1$, then $\dfrac{1}{n} > \dfrac{1}{n+1}$ and $a_n > a_{n+1}$.

(iii) $\lim_{n \to \infty} a_n = \lim_{n \to \infty} \dfrac{1}{n} = 0$. Therefore, the alternating harmonic series converges.

Fig. 1 shows some partial sums for the alternating harmonic series.

(b) $a_n = \dfrac{3}{\sqrt{n}} > 0$. Since $n < n+1$, we have $\sqrt{n} < \sqrt{n+1}$ and $\dfrac{3}{\sqrt{n}} > \dfrac{3}{\sqrt{n+1}}$.

$\lim_{n \to \infty} a_n = \lim_{n \to \infty} \dfrac{3}{\sqrt{n}} = 0$. The series $\dfrac{3}{1} + \dfrac{3}{\sqrt{2}} + \dfrac{3}{\sqrt{3}} + \dfrac{3}{\sqrt{4}} + \ldots$ converges.

(c) $a_n = \dfrac{7}{n \cdot \ln(n)} > 0$ for n ≥ 2. Since $n < n+1$ and $\ln(n) < \ln(n+1)$, we have

$n \cdot \ln(n) < (n+1) \cdot \ln(n+1)$ and $\dfrac{7}{n \cdot \ln(n)} > \dfrac{7}{(n+1) \cdot \ln(n+1)}$.

$\lim_{n \to \infty} a_n = \lim_{n \to \infty} \dfrac{7}{n \cdot \ln(n)} = 0$. The series $\sum_{n=2}^{\infty} (-1)^{n} \dfrac{7}{n \cdot \ln(n)}$ converges.

Practice 1:   Show that these two alternating series satisfy the three conditions in the hypothesis of the

Alternating Series Test. Then we can conclude that each of them converges.

$$\text{(a)}\quad 1 - \frac{1}{4} + \frac{1}{9} - \frac{1}{16} + \ldots = \sum_{n=1}^{\infty} (-1)^{n+1} \frac{1}{n^2}$$

$$\text{(b)}\quad \frac{3}{\ln(2)} - \frac{3}{\ln(3)} + \frac{3}{\ln(4)} - \frac{3}{\ln(5)} + \ldots = \sum_{n=2}^{\infty} (-1)^{n} \frac{3}{\ln(n)}$$

Example 2:   Does $\sum_{n=1}^{\infty} (-1)^{n+1} \frac{n}{n+2}$ converge?

Solution: $a_n = \frac{n}{n+2} > 0$, but $\lim_{n \to \infty} a_n = \lim_{n \to \infty} \frac{n}{n+2} = 1 \neq 0$. Since the terms do not approach 0, we

can conclude from the Nth Term Test For Divergence (Section 10.2) that the series diverges.

Fig. 2 shows some of the partial sums for this series. You should notice that the even and the odd

partial sums in Fig. 2 are approaching two different values.

Practice 2:   (a) Does $\sum_{n=1}^{\infty} (-1)^{n+1} n$ converge? (b) Does $\sum_{n=1}^{\infty} (-1)^{n+1} \frac{1}{\sqrt{2n+1}}$ converge?

Example Of A Divergent Alternating Series

If the terms of a series, any series, do not approach 0, then the series must diverge (N^{th} Term Test For

Divergence). If the terms do approach 0 the series may converge or may diverge. There are divergent

alternating series whose terms approach 0 (but the approach to 0 is not monotonic). For example,

$\frac{3}{2} - \frac{1}{2} + \frac{3}{4} - \frac{1}{4} + \frac{3}{6} - \frac{1}{6} + \frac{3}{8} - \frac{1}{8} + \ldots$ is an alternating series whose terms approach 0, but the series diverges.

The even partial sums of our new series are

$$s_2 = \frac{3}{2} - \frac{1}{2} = 1,$$

$$s_4 = \frac{3}{2} - \frac{1}{2} + \frac{3}{4} - \frac{1}{4} = (\frac{3}{2} - \frac{1}{2}) + (\frac{3}{4} - \frac{1}{4}) = 1 + \frac{1}{2},$$

$$s_6 = \frac{3}{2} - \frac{1}{2} + \frac{3}{4} - \frac{1}{4} + \frac{3}{6} - \frac{1}{6} = (\frac{3}{2} - \frac{1}{2}) + (\frac{3}{4} - \frac{1}{4}) + (\frac{3}{6} - \frac{1}{6}) = 1 + \frac{1}{2} + \frac{1}{3},$$

and $s_{2n} = 1 + \frac{1}{2} + \frac{1}{3} + \ldots + \frac{1}{n}$.

You should recognize that these partial sums are the partial sums of the harmonic series, a divergent series,

so the partial sums of our new series diverge and our new series is divergent. If the terms of an alternating

series approach 0, but not monotonically, then the Alternating Series Test does not apply, and the series

may converge or it may diverge.

Approximating the Sum of an Alternating Series

If we know that a series converges and if we add the first "many" terms together, then we expect that the resulting partial sum is close to the value S obtained by adding all of the terms together. Generally, however, we do not know how close the partial sum is to S. The situation with many alternating series is much nicer. The next result says that for some alternating series (those that satisfy the three conditions in the next box), the difference between S and the nth partial sum of the alternating series, $|S - s_n|$, is less than the magnitude of the next term in the series, a_{n+1}.

Estimation Bound for Alternating Series

If   S is the sum of an alternating series $\displaystyle\sum_{n=1}^{\infty} (-1)^{n+1} a_n$

that satisfies the three conditions

    (i) $a_n > 0$ for all n   (each a_n is positive),

    (ii) $a_n > a_{n+1}$ (the terms are monotonically decreasing),

and (iii) $\displaystyle\lim_{n\to\infty} a_n = 0$ (the terms approach 0),

then   the nth partial sum s_n is within a_{n+1} of the sum S: $s_n - a_{n+1} < S < s_n + a_{n+1}$

and | approximation "error" using s_n as an estimate of S $|$ = $|S - s_n| < a_{n+1}$.

Note: This Estimation Bound only applies to alternating series. It is tempting, but wrong, to use it with other types of series.

Geometric idea behind the Estimation Bound:

If we have an alternating series that satisfies the hypothesis of the Estimation Bound, then the graph of the sequence $\{s_n\}$ of partial sums is "trumpet–shaped" or "funnel–shaped" (Fig. 4). The partial sums are alternately above and below the value S, and they "squeeze" in on the value S. Since the distance from s_n to S is less than the distance between the successive terms s_n and s_{n+1} (Fig. 5), then

$|S - s_n| < |s_n - s_{n+1}| = a_{n+1}$.

Proof of the Estimation Bound for Alternating Series:

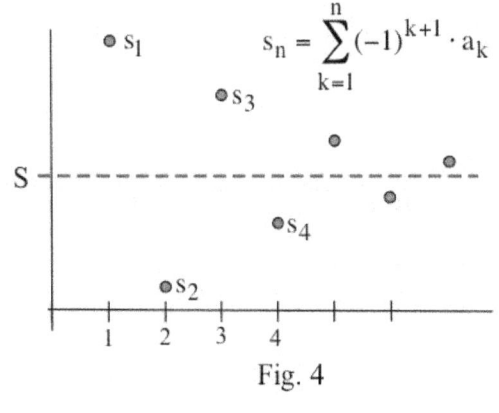

Fig. 4

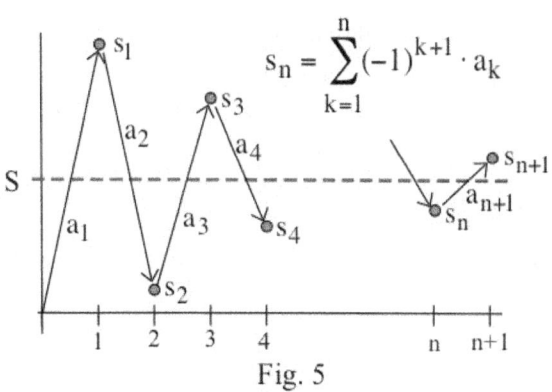

Fig. 5

$$S - s_n = (\, a_1 - a_2 + a_3 - a_4 + \ldots + (-1)^{n+1} a_n + (-1)^{n+2} a_{n+1} \ldots) - (\, a_1 - a_2 + a_3 - a_4 + \ldots + (-1)^{n+1} a_n \,)$$

$$= (-1)^{n+2} a_{n+1} + (-1)^{n+3} a_{n+2} + (-1)^{n+4} a_{n+3} + \ldots$$

$$= (-1)^{n+2} (\, a_{n+1} - a_{n+2} + a_{n+3} - \ldots \,) \, .$$

Then $0 \le |\, S - s_n \,| \quad = |\, (-1)^{n+2} (\, a_{n+1} - a_{n+2} + a_{n+3} - a_{n+4} + a_{n+5} \ldots) \,|$

$$= a_{n+1} - a_{n+2} + a_{n+3} - a_{n+4} + a_{n+5} \ldots \qquad \text{since } |(-1)^{n+2}| = 1 \text{ and the rest is positive}$$

$$= a_{n+1} - (a_{n+2} - a_{n+3}) - (a_{n+4} - a_{n+5}) - \ldots \; < a_{n+1} \quad .$$

The Estimation Bound is typically used in two different ways. Sometimes we know the value of n, and we want to know how close s_n is to S. Sometimes we know how close we want s_n to be to S, and we want to find a value of n to ensure that level of closeness. The next two Examples illustrate these two different uses of the Estimation Bound.

Example 3:   How close is $\displaystyle\sum_{n=1}^{4} (-1)^{n+1} \frac{1}{n^2} = 1 - \frac{1}{4} + \frac{1}{9} - \frac{1}{16} \approx 0.79861$ to the sum $\displaystyle\sum_{n=1}^{\infty} (-1)^{n+1} \frac{1}{n^2}$?

Solution: $a_n = \dfrac{1}{n^2}$ and $s_4 = \displaystyle\sum_{n=1}^{4} (-1)^{n+1} \frac{1}{n^2} = 1 - \frac{1}{4} + \frac{1}{9} - \frac{1}{16} \approx 0.79861$ so, by the Estimation Bound, we

can conclude that $|\, S - s_4 \,| < a_5 = \dfrac{1}{25} = 0.04 : 0.79861$ is within 0.04 of the exact value S. Then

0.79861 − 0.04 < S < 0.79861 + 0.04 and 0.7**5**861 < S < 0.8**3**861 .

Similarly, $s_9 = 1 - \dfrac{1}{4} + \dfrac{1}{9} - \dfrac{1}{16} + \ldots + \dfrac{1}{81} \approx 0.82796$ is within $a_{10} = \dfrac{1}{100} = 0.01$ of the exact

value of S, and $s_{99} \approx 0.822517$ is within $a_{100} = \dfrac{1}{100^2} = 0.0001$ of the exact value of S.

Then 0.822517 − 0.0001 < S < 0.822517 + 0.0001 and 0.822**4**17 < S < 0.822**6**17 .

Practice 3:   Evaluate s_4 and s_9 for the alternating series $\displaystyle\sum_{n=1}^{\infty} (-1)^{n+1} \frac{1}{n^3}$ and determine bounds

for $|\, S - s_4 \,|$ and $|\, S - s_9 \,|$.

Example 4:   Find the number of terms needed so that s_n is within 0.001 of the exact

value of $\displaystyle\sum_{n=1}^{\infty} (-1)^{n+1} \frac{1}{n!}$ and evaluate s_n.

Solution: We know $|\, S - s_n \,| < a_{n+1}$ so we want to find n so that $a_{n+1} \le 0.001 = \dfrac{1}{1000}$. With a little

numerical experimentation on a calculator, we see that 6! = 720 and 1/720 is not small enough, but

$7! = 5,040 > 1,000$ so $\frac{1}{7!} = \frac{1}{5040} \approx 0.000198 < 0.001$. Since $n+1 = 7$, $s_6 \approx 0.631944$ is the first

partial sum guaranteed to be within 0.001 of S. In fact, s_6 is within $\frac{1}{5040} \approx 0.000198$ of S, so

$0.631746 < S < 0.632142$.

Practice 4:   Find the number of terms needed so s_n is within 0.001 of the value of $\displaystyle\sum_{n=1}^{\infty} (-1)^{n+1} \frac{1}{n^3 + 5}$

and evaluate s_n.

The Estimation Bound guarantees that s_n is **within** a_{n+1} of S. In fact, s_n is often much closer than a_{n+1}
to S. The value a_{n+1} is an **upper bound** on how far s_n can be from S.

Note 1:   The first **finite** number of terms do not affect the convergence or divergence of a series (they do effect
the sum S) so we can use the Alternating Series Test and the Estimation Bound if the terms of a series
"eventually" satisfy the conditions of the hypotheses of these results. By "eventually" we mean there is
a value M so that for $n > M$ the series is an alternating series.

Note 2:   If a series has some positive terms and some negative terms and if those terms do NOT "eventually" alternate
signs, then we can NOT use the Alternating Series Test – it simply does not apply to such series.

PROBLEMS

In problems 1 – 6 you are given the values of the first four **terms** of a series. (a) Calculate and graph the
first four partial sums for each series. (b) Which of the series are not alternating series?

1. $1, -0.8, 0.6, -0.4$ 2. $-1, 1.5, -0.7, 1$ 3. $-1, 2, -3, 4$

4. $2, -1, -0.5, 0.3$ 5. $-1, -0.6, 0.4, 0.2$ 6. $2, -1, 0.5, -0.3$

In problems 7 – 12 you are given the values of the first five **partial sums** of a series. Which of the series
are not alternating series. Why?

7. $2, 1, 3, 2, 4$ 8. $2, 1, 1.8, 1.4, 1.6$ 9. $2, 3, 2.1, 2.9, 2.8$

10. $-3, -1, -2.5, -1.5, -2$ 11. $-1, 1, -0.8, -0.6, -0.4$ 12. $-2.3, -1.6, -1.4, -1.8, -1.7$

13. Fig. 6 shows the graphs of the partial sums of three series.
Which is/are not the partial sums of alternating series? Why?

14. Fig. 7 shows the graphs of the partial sums of three series.
Which is/are not the partial sums of alternating series? Why?

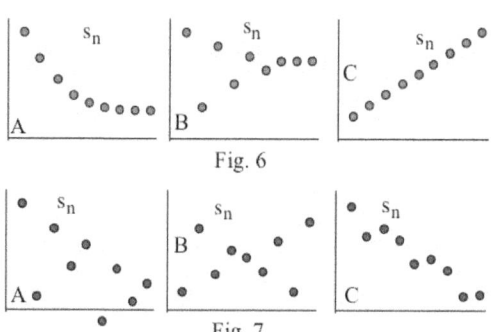
Fig. 6

Fig. 7

15. Fig. 8 shows the graphs of the partial sums of three series. Which is/are not the partial sums of alternating series? Why?

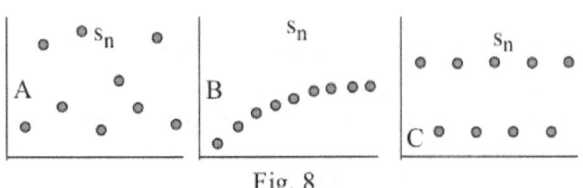

Fig. 8

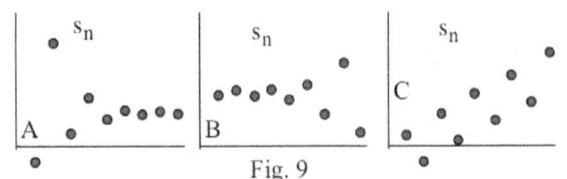

Fig. 9

16. Fig. 9 shows the graphs of the partial sums of three series. Which is/are not the partial sums of alternating series? Why?

In problems 17 – 31 determine whether the given series converge or diverge.

17. $\displaystyle\sum_{n=1}^{\infty} (-1)^{n+1} \frac{1}{n+5}$

18. $\displaystyle\sum_{n=2}^{\infty} (-1)^{n} \frac{1}{\ln(n)}$

19. $\displaystyle\sum_{n=1}^{\infty} (-1)^{n+1} \frac{n}{n^2+3}$

20. $\displaystyle\sum_{n=5}^{\infty} (-0.99)^{n+1}$

21. $\displaystyle\sum_{n=1}^{\infty} (-1)^{n} \frac{|n+3|}{|n+7|}$

22. $\displaystyle\sum_{n=1}^{\infty} (-1)^{n+1} \sin(\tfrac{1}{n})$

23. $\displaystyle\sum_{n=4}^{\infty} \frac{\cos(n\pi)}{n}$

24. $\displaystyle\sum_{n=3}^{\infty} \frac{\sin(n\pi)}{n}$

25. $\displaystyle\sum_{n=1}^{\infty} (-1)^{n+1} \frac{\ln(n)}{n}$

26. $\displaystyle\sum_{n=3}^{\infty} (-1)^{n+1} \frac{\ln(n)}{\ln(n^3)}$

27. $\displaystyle\sum_{n=2}^{\infty} (-1)^{n+1} \frac{\ln(n^3)}{\ln(n^{10})}$

28. $\displaystyle\sum_{n=1}^{\infty} (-1)^{n+1} \frac{5}{\sqrt{n+7}}$

29. $\displaystyle\sum_{n=1}^{\infty} \frac{(-2)^{n+1}}{1+(-3)^{n}}$

30. $\displaystyle\sum_{n=3}^{\infty} (-1)^{n+1} \frac{(-2)^{n+1}}{1+3^{n}}$

31. $\displaystyle\sum_{n=1}^{\infty} \cos(n\pi)\cdot\sin(\pi/n)$

In problems 32 – 40, (a) calculate s_4 for each series and determine an upper bound for how far s_4 is from the exact value S of the infinite series. Then (b) use s_4 find lower and upper bounds on the value of S so that $\{\text{lower bound}\} < S < \{\text{upper bound}\}$.

32. $\displaystyle\sum_{n=1}^{\infty} (-1)^{n+1} \frac{1}{n+6}$

33. $\displaystyle\sum_{n=1}^{\infty} (-1)^{n} \frac{1}{\ln(n+1)}$

34. $\displaystyle\sum_{n=1}^{\infty} (-1)^{n+1} \frac{2}{\sqrt{n+21}}$

35. $\displaystyle\sum_{n=1}^{\infty} (-0.8)^{n+1}$

36. $\displaystyle\sum_{n=1}^{\infty} (-\tfrac{1}{3})^{n}$

37. $\displaystyle\sum_{n=1}^{\infty} (-1)^{n+1} \sin(\tfrac{1}{n})$

38. $\displaystyle\sum_{n=1}^{\infty} (-1)^{n+1} \frac{1}{n^2}$

39. $\displaystyle\sum_{n=1}^{\infty} (-1)^{n} \frac{1}{n^3}$

40. $\displaystyle\sum_{n=1}^{\infty} \frac{\cos(n\pi)}{n+\ln(n)}$

In problems 41 – 50 find the number of terms needed to guarantee that s_n is within the specified

distance D of the exact value S of the sum of the given series.

41. $\sum_{n=1}^{\infty} \frac{(-1)^{n+1}}{n+6}$, D = 0.01   42. $\sum_{n=1}^{\infty} \frac{(-1)^n}{\ln(n+1)}$, D = 0.01   43. $\sum_{n=1}^{\infty} \frac{(-1)^n}{\sqrt{n}}$, D = 0.01

44. $\sum_{n=1}^{\infty} (-0.8)^{n+1}$, D = 0.003   45. $\sum_{n=1}^{\infty} (-\frac{1}{3})^n$, D = 0.002   46. $\sum_{n=1}^{\infty} (-1)^{n+1} \sin(\frac{1}{n})$, D = 0.06

47. $\sum_{n=1}^{\infty} (-1)^{n+1} \frac{1}{n^2}$, D = 0.001   48. $\sum_{n=1}^{\infty} (-1)^n \frac{1}{n^3}$, D = 0.0001   49. $\sum_{n=1}^{\infty} \frac{\cos(n\pi)}{n + \ln(n)}$, D = 0.04

Problems 50 – 60 ask you to use the series S(x), C(x), and E(x) given below. For each problem,
(a) substitute the given value for x in the series, (b) evaluate s_3 , the sum of the first three terms
of the series, (c) determine an upper bound on the error | actual sum – s_3 | = | S – s_3 | .

(The partial sums of S(x), C(x), and E(x) approximate sin(x), cos(x), and e^x, respectively.)

$$S(x) = x - \frac{x^3}{2\cdot3} + \frac{x^5}{2\cdot3\cdot4\cdot5} - \frac{x^7}{2\cdot3\cdot4\cdot5\cdot6\cdot7} + \ldots + (-1)^n \frac{x^{2n+1}}{(2n+1)!} + \ldots$$

$$C(x) = 1 - \frac{x^2}{2} + \frac{x^4}{2\cdot3\cdot4} - \frac{x^6}{2\cdot3\cdot4\cdot5\cdot6} + \ldots + (-1)^n \frac{x^{2n}}{(2n)!} + \ldots$$

$$E(x) = 1 + x + \frac{x^2}{2} + \frac{x^3}{2\cdot3} + \frac{x^4}{2\cdot3\cdot4} + \ldots + \frac{x^n}{n!} + \ldots$$

50. x = 0.5 in S(x)   51. x = 0.3 in S(x)   52. x = 1 in S(x)

53. x = 0.1 in S(x)   54. x = –0.3 in S(x)   55. x = 1 in C(x)

56. x = 0.5 in C(x)   57. x = –0.2 in C(x)   58. x = –0.3 in C(x)

59. x = –1 in E(x)   60. x = –0.5 in E(x)   61. x = –0.2 in E(x)

Practice Answers

Practice 1:   (a) (i) $a_n = \frac{1}{n^2} > 0$ for all n. (ii) $n^2 < (n+1)^2$ so $\frac{1}{n^2} > \frac{1}{(n+1)^2}$ and $a_n > a_{n+1}$.

(iii) $\lim_{n\to\infty} a_n = \lim_{n\to\infty} \frac{1}{n^2} = 0$. The series $\sum_{n=1}^{\infty} (-1)^{n+1} \frac{1}{n^2}$ converges.

(b) (i) $a_n = \frac{3}{\ln(n)} > 0$ for all n≥2. (ii) $\ln(n) < \ln(n+1)$ so $\frac{3}{\ln(n)} > \frac{3}{\ln(n+1)}$ and $a_n > a_{n+1}$.

(iii) $\lim_{n\to\infty} a_n = \lim_{n\to\infty} \frac{3}{\ln(n)} = 0$. The series $\sum_{n=1}^{\infty} (-1)^n \frac{3}{\ln(n)}$ converges.

Practice 2:   (a) $|(-1)^{n+1} n| = |n| \longrightarrow \infty \neq 0$ so the terms $(-1)^{n+1} n$ do NOT approach 0, and the

series diverges by the nth Term Test for Divergence (Section 10.2).

      (b)   (i) $a_n = \dfrac{1}{\sqrt{2n+1}} > 0$ for all $n \geq 1$.

         (ii) $2n+1 < 2(n+1) + 1$ so $\sqrt{2n+1} < \sqrt{2(n+1)+1}$ and $\dfrac{1}{\sqrt{2n+1}} > \dfrac{1}{\sqrt{2(n+1)+1}}$ so $a_n > a_{n+1}$.

         (iii) $\displaystyle\lim_{n\to\infty} a_n = \lim_{n\to\infty} \dfrac{1}{\sqrt{2n+1}} = 0$. Therefore, by the Alternating Series Test,

           the series $\displaystyle\sum_{n=1}^{\infty} (-1)^{n+1} \dfrac{1}{\sqrt{2n+1}}$ converges.

Practice 3: $\displaystyle\sum_{n=1}^{\infty} (-1)^{n+1} \dfrac{1}{n^3}$: $s_4 = 1 - \dfrac{1}{8} + \dfrac{1}{27} - \dfrac{1}{64} \approx 0.896412$.

      $|S - s_4| < |a_5| = |(-1)^6 \dfrac{1}{5^3}| = \dfrac{1}{125} = 0.008$ so $s_4 - 0.008 < S < s_4 + 0.008$

      and $0.888412 < S < 0.904412$.

      $s_9 \approx 0.9021164$. $|S - s_9| < |a_{10}| = |(-1)^{11} \dfrac{1}{10^3}| = \dfrac{1}{1000} = 0.001$ so

      $s_9 - 0.001 < S < s_9 + 0.001$ and $0.901116 < S < 0.903116$.

Practice 4:   We need to find a value for n so $|a_{n+1}| < 0.001$.

      $|a_{n+1}| = |(-1)^{n+2} \dfrac{1}{(n+1)^3 + 5}| = \dfrac{1}{(n+1)^3 + 5}$

n	$\lvert a_{n+1} \rvert$
1	$\dfrac{1}{2^3 + 5} = \dfrac{1}{13} \approx 0.0769$
2	$\dfrac{1}{3^3 + 5} = \dfrac{1}{32} \approx 0.03125$
8	$\dfrac{1}{9^3 + 5} = \dfrac{1}{734} \approx 0.00136$
9	$\dfrac{1}{10^3 + 5} = \dfrac{1}{1005} \approx 0.000995 < 0.001$

Since $|a_{10}| < 0.001$ we can be sure that s_9 is within 0.001 of S.

$s_9 = \displaystyle\sum_{n=1}^{9} (-1)^{n+1} \dfrac{1}{n^3 + 5} = \dfrac{1}{6} - \dfrac{1}{13} + \dfrac{1}{32} - \dots + \dfrac{1}{1005} \approx 0.112156571145$ so

$0.111156571145 < S < 0.113156571145$.

10.7   ABSOLUTE CONVERGENCE and the RATIO TEST

The series we examined in the previous sections all behaved very regularly with regard to the signs of the terms: the signs of the terms were either all the same or they alternated between $+$ and $-$. However, there are series whose signs do not behave in such regular ways, and in this section we examine some techniques for determining whether those series converge or diverge.

Two Examples

Suppose we have the two series whose terms have magnitudes $\frac{1}{n}$ and $\frac{1}{n^2}$, but we don't know whether each term is positive or negative:

(a) $\quad \sum\limits_{n=1}^{\infty} * \frac{1}{n} = (*1) + (* \frac{1}{2}) + (* \frac{1}{3}) + (* \frac{1}{4}) + ... + (* \frac{1}{n}) + ...$ where each $*$ is either $+$ or $-$.

(b) $\quad \sum\limits_{n=1}^{\infty} * \frac{1}{n^2} = (*1) + (* \frac{1}{4}) + (* \frac{1}{9}) + (* \frac{1}{16}) + ... + (* \frac{1}{n^2}) + ...$ where each $*$ is either $+$ or $-$.

These two series behave very differently depending on how we replace the each "$*$" with $+$ or $-$ signs.

Series (a): $\sum\limits_{n=1}^{\infty} * \frac{1}{n}$

If we **always** replace "$*$" with a $+$, then we have the harmonic series which diverges.

If we **always** replace "$*$" with a $-$, then we have the -1 times the harmonic series which diverges.

If we **alternate** replacing "$*$" with $+$ and $-$, then we have the alternating harmonic series which converges.

The answer we have to give to the question "Does series (a) converge?" is "It depends on how the signs of the terms are chosen."

Series (b): $\sum\limits_{n=1}^{\infty} * \frac{1}{n^2}$

If we always replace "$*$" with a $+$, then we have a series which converges (by the P–Test with

$p = 2$). This is the largest value series (b) can have.

If we always replace "$*$" with a $-$, then we have -1 times a convergent series so the series

converges. This is the smallest value series (b) can have.

If we replace the "$*$" with $+$ or $-$ in some other way, then the result is some number between the largest and

smallest possible values (both of which are finite numbers), and the result must be some finite number.

The answer to the question "Does series (b) converge?" is "Yes, no matter how the $*$ are replaced with

$+$ or $-$." The value of the sum is affected by how the signs are chosen, but the series does converge.

Series (a) and (b) illustrate the distinction we want to examine in this section. Series (a) is an example of a **conditionally convergent** series since the convergence depends on how the "∗" are replaced. Series (b) is an example of an **absolutely convergent** series since it does not matter how the "∗" are replaced. Series (b) is convergent even when we are adding all positive terms or all negative terms.

The following definitions make the distinctions precise.

Definitions

Definition: A series $\sum\limits_{n=1}^{\infty} a_n$ is **absolutely convergent** if $\sum\limits_{n=1}^{\infty} |a_n|$ converges.

Series (b) is absolutely convergent.

Definition: A series $\sum\limits_{n=1}^{\infty} a_n$ is **conditionally convergent** if it is convergent but

not absolutely convergent (i.e., if $\sum\limits_{n=1}^{\infty} a_n$ converges, but $\sum\limits_{n=1}^{\infty} |a_n|$ diverges).

The alternating harmonic series $\sum\limits_{n=1}^{\infty} (-1)^{n+1}\dfrac{1}{n}$ is conditionally convergent because

$\sum\limits_{n=1}^{\infty} (-1)^{n+1}\dfrac{1}{n}$ converges but $\sum\limits_{n=1}^{\infty} \left| (-1)^{n+1}\dfrac{1}{n} \right| = \sum\limits_{n=1}^{\infty} \dfrac{1}{n}$ diverges.

Example 1:   Determine whether these series are absolutely convergent, conditionally convergent, or divergent.

(a) $\sum\limits_{n=1}^{\infty} (-1)^{n+1}\dfrac{1}{\sqrt{n}}$ (b) $\sum\limits_{k=1}^{\infty} \dfrac{\sin(k)}{k^2}$ (c) $\sum\limits_{j=1}^{\infty} (-1)^{j+1}\dfrac{j^2}{j+1}$

Solution: (a) $\sum\limits_{n=1}^{\infty} \left| (-1)^{n+1}\dfrac{1}{\sqrt{n}} \right| = \sum\limits_{n=1}^{\infty} \dfrac{1}{\sqrt{n}}$ which diverges by the P–Test with $p = 1/2 < 1$

so the series is not absolutely convergent. $\dfrac{1}{\sqrt{n}} \longrightarrow 0$ monotonically so the

alternating series $\sum\limits_{n=1}^{\infty} (-1)^{n+1}\dfrac{1}{\sqrt{n}}$ converges by the Alternating Series Test.

The series $\sum\limits_{n=1}^{\infty} (-1)^{n+1} \dfrac{1}{\sqrt{n}}$ is conditionally convergent.

(b) $\sum\limits_{k=1}^{\infty} \left| \dfrac{\sin(k)}{k^2} \right| \leq \sum\limits_{k=1}^{\infty} \dfrac{1}{k^2}$ which converges by the P–Test,

so $\sum\limits_{k=1}^{\infty} \dfrac{\sin(k)}{k^2}$ is absolutely convergent.

(c) $\left| (-1)^{j+1} \dfrac{j^2}{j+1} \right| = j - 1 + \dfrac{1}{j+1} \longrightarrow \infty \neq 0$ so $\sum\limits_{j=1}^{\infty} (-1)^{j+1} \dfrac{j^2}{j+1}$ diverges.

Practice 1:   Determine whether these series are absolutely convergent, conditionally convergent, or divergent.

(a) $\sum\limits_{n=2}^{\infty} (-1)^n \dfrac{5}{\ln(n)}$ (b) $\sum\limits_{k=1}^{\infty} \dfrac{\cos(\pi k)}{k^2}$

An important result about absolutely convergent series is that they are also convergent.

Absolute Convergence Theorem

Every absolutely convergent series is convergent: if $\sum\limits_{n=1}^{\infty} | a_n |$ converges, then $\sum\limits_{n=1}^{\infty} a_n$ converges.

Since the series $\sum\limits_{n=1}^{\infty} \left| * \dfrac{1}{n^2} \right| = \sum\limits_{n=1}^{\infty} \dfrac{1}{n^2}$ is absolutely convergent, the series $\sum\limits_{n=1}^{\infty} * \dfrac{1}{n^2}$ converges no matter how the "*" is replaced with + and – signs.

Proof of the Absolute Convergence Theorem:

If $a_n \geq 0$ then $a_n = | a_n |$, and if $a_n < 0$ then $a_n = -| a_n |$. In either case we have $-| a_n | \leq a_n \leq | a_n |$.

Adding $| a_n |$ to each piece of $-| a_n | \leq a_n \leq | a_n |$, we have $0 \leq | a_n | + a_n \leq 2 | a_n |$ for all n.

Let $b_n = | a_n | + a_n \geq 0$ for all n.

Since $b_n \geq 0$ and $b_n \leq 2 | a_n |$, the terms of a convergent

series, then, by the Comparison Test we know that $\sum\limits_{n=1}^{\infty} b_n$ converges.

Finally, $a_n = (| a_n | + a_n) - | a_n | = b_n - | a_n |$ so the series $\sum\limits_{n=1}^{\infty} a_n = \sum\limits_{n=1}^{\infty} b_n - \sum\limits_{n=1}^{\infty} | a_n |$

is the difference of two convergent series, and we can conclude that the series $\sum\limits_{n=1}^{\infty} a_n$ converges.

The following corollary (the contrapositive form of the Absolute Convergence Theorem) is sometimes useful for showing that a series is not absolutely convergent.

Corollary:

If a series is not convergent, then it is not absolutely convergent:

$$\text{if } \sum_{n=1}^{\infty} a_n \text{ diverges, then } \sum_{n=1}^{\infty} |a_n| \text{ diverges.}$$

The Ratio Test

The following test is useful for determining whether a given series is absolutely convergent, and it will be used often in Section 10.8 when we want to determine where a power series converges. It says that we can be certain that a given series is absolutely convergent, if the limit of the ratios of successive terms has a value less than 1. The Ratio Test is very important and will be used very often in the next sections on power series.

The Ratio Test

Suppose $\lim_{n \to \infty} \left| \dfrac{a_{n+1}}{a_n} \right| = L$.

(a) If $L < 1$, then the series $\sum_{n=1}^{\infty} a_n$ is absolutely convergent (and also convergent).

(b) If $L > 1$, then the series $\sum_{n=1}^{\infty} a_n$ is divergent.

(c) If $L = 1$, then the series $\sum_{n=1}^{\infty} a_n$ may converge or may diverge (the Ratio Test does not help).

A proof of the Ratio Test is rather long and is included in an Appendix after the Practice Answers for this section. Part (a) is proved by showing that if $L < 1$, then the series is, term–by–term, less than a convergent geometric series. Part (b) is proved by showing that if $L > 1$, then the terms of the series do not approach 0 so the series diverges. Part (c) is proved by giving two series, one convergent and one divergent, that both have $L = 1$.

One powerful aspect of the Ratio Test is that it is very "mechanical" — we simply calculate a particular limit and then we (often) have a conclusion about the convergence or divergence of the series.

Example 2:   Use the Ratio Test to determine if these series are absolutely convergent:

(a) $\displaystyle\sum_{n=1}^{\infty} \frac{2^n\, n}{5^n}$ (b) $\displaystyle\sum_{n=1}^{\infty} \frac{n^2}{n!}$.

Solution: (a) $a_n = \dfrac{2^n\, n}{5^n}$ so $a_{n+1} = \dfrac{2^{n+1}\,(n+1)}{5^{n+1}}$. Then

$$\left|\frac{a_{n+1}}{a_n}\right| = \left|\frac{\dfrac{2^{n+1}\,(n+1)}{5^{n+1}}}{\dfrac{2^n\, n}{5^n}}\right| = \left|\frac{2^{n+1}}{2^n}\,\frac{5^n}{5^{n+1}}\,\frac{n+1}{n}\right| = \left|\frac{2}{5}\,\frac{n+1}{n}\right| \longrightarrow \frac{2}{5} < 1$$

so $\displaystyle\sum_{n=1}^{\infty} \frac{2^n\, n}{5^n}$ is absolutely convergent.

(b) $a_n = \dfrac{n^2}{n!}$ so $a_{n+1} = \dfrac{(n+1)^2}{(n+1)!}$. Then

$$\left|\frac{a_{n+1}}{a_n}\right| = \left|\frac{\dfrac{(n+1)^2}{(n+1)!}}{\dfrac{n^2}{n!}}\right| = \left|\frac{n!}{(n+1)!}\,\frac{(n+1)^2}{n^2}\right| = \left|\frac{1}{n+1}\left(\frac{n+1}{n}\right)^2\right| \longrightarrow 0 < 1$$

so $\displaystyle\sum_{n=1}^{\infty} \frac{n^2}{n!}$ is absolutely convergent.

Practice 2:   Use the Ratio Test to determine if these series are absolutely convergent:

(a) $\displaystyle\sum_{n=1}^{\infty} (-1)^{n+1}\frac{e^n}{n!}$ (b) $\displaystyle\sum_{n=1}^{\infty} \frac{n^5}{3^n}$.

The Ratio Test is very useful for determining values of a variable which guarantee the absolute

convergence (and convergence) of a series. This is a method we will use often in the rest of this chapter.

Example 3:   For which values of x is the series $\displaystyle\sum_{n=1}^{\infty} \frac{(x-3)^n}{n}$ absolutely convergent?

Solution: $a_n = \dfrac{(x-3)^n}{n}$ so $a_{n+1} = \dfrac{(x-3)^{n+1}}{n+1}$. Then

$$\left|\frac{a_{n+1}}{a_n}\right| = \left|\frac{\dfrac{(x-3)^{n+1}}{n+1}}{\dfrac{(x-3)^n}{n}}\right| = \left|\frac{(x-3)^{n+1}}{(x-3)^n}\,\frac{n}{n+1}\right| \rightarrow \;\; |x-3| = L.$$

Now we simply need to solve the inequality $|x-3| < 1$ $(L < 1$) for x.

If $|x-3| < 1$, then $-1 < x - 3 < 1$ so **$2 < x < 4$** .

If $2 < x < 4$, then $L = |x-3| < 1$ so $\displaystyle\sum_{n=1}^{\infty} \frac{(x-3)^n}{n}$ is absolutely convergent (and convergent).

If $x < 2$ or $x > 4$ then $L > 1$ so the series diverges. (What happens at the "endpoints," x=2 and x=4, where L=1?)

Practice 3:   For which values of x is the series $\displaystyle\sum_{n=1}^{\infty} \frac{(x-5)^n}{n^2}$ absolutely convergent?

Note: If the terms of a series contain **factorials** or things raised to the **nth power**, it is usually a good idea to use the Ratio Test as the first test you try.

Rearrangements

Absolutely convergent series share an important property with finite sums — no matter what order we add the numbers, the sum is always the same. Stated another way, the sum of an absolutely convergent series is always the same value, even if the terms are rearranged.

Conditionally convergent series do not have this property — the order in which we add the terms does matter. If the terms of a conditionally convergent series are rearranged or reordered, the sum after the rearrangement may be different than the sum before the rearrangement. A rather amazing fact is that the terms of a conditionally convergent series can be rearranged to obtain any sum we want.

We illustrate this strange result by showing that we can rearrange the alternating harmonic series, a series conditionally convergent to approximately 0.69, so that the sum of the rearranged series is 2 rather than 0.69..

$$\sum_{n=1}^{\infty} (-1)^{n+1} \frac{1}{n} = 1 - \frac{1}{2} + \frac{1}{3} - \frac{1}{4} + \frac{1}{5} - \frac{1}{6} + \dots \text{ is conditionally convergent to approximately } 0.69 .$$

First we note that the sum of the positive terms alone is a divergent series:

$$1 + \frac{1}{3} + \frac{1}{5} + \frac{1}{7} + \dots = \sum_{n=1}^{\infty} \frac{1}{2n-1} \text{ which diverges by the Limit Comparison Test,}$$

so the partial sums of the positive terms eventually exceed any number we pick.

Similarly, the sum of the negative terms alone is also a divergent series:

$$-\frac{1}{2} - \frac{1}{4} - \frac{1}{6} - \frac{1}{8} + \dots = -\frac{1}{2} \sum_{n=1}^{\infty} \frac{1}{n} \text{ which diverges,}$$

so the partial sums of the negative terms eventually become as large negatively as we want.

Finally, we pick a target number we want for the sum (in this illustration, we picked a target of 2).

Then the following clever strategy tells us how to chose the order of the terms, the rearrangement, so the sum of the rearranged series is the target number, 2 :

(1) select the positive terms, in the order they appear in the original series, until the partial sum exceeds our target number, then

(2) select the negative terms, in the order they appear in the original series, until the partial sum falls below our target number, and

(3) keep repeating steps (1) and (2) with the previously unused terms of the original series.

In order to rearrange the terms of the alternating harmonic series so the sum is 2, we pick positive terms, in order, until the partial sum **is larger than** 2. This requires the first 8 positive terms:

$s_1 = 1$,

$s_2 = 1 + \frac{1}{3} \approx 1.3333,\ldots$

$s_7 = 1 + \frac{1}{3} + \frac{1}{5} + \frac{1}{7} + \frac{1}{9} + \frac{1}{11} + \frac{1}{13} \approx 1.955133755$

$s_8 = 1 + \frac{1}{3} + \frac{1}{5} + \frac{1}{7} + \frac{1}{9} + \frac{1}{11} + \frac{1}{13} + \frac{1}{15} \approx 2.021800422$.

Then we pick negative terms, in order, until the partial sum **is less than** two. This only requires 1 negative term:

$s_9 = 1 + \frac{1}{3} + \frac{1}{5} + \frac{1}{7} + \frac{1}{9} + \frac{1}{11} + \frac{1}{13} + \frac{1}{15} - \frac{1}{2} \approx 1.521800422$.

Then we pick more unused positive terms until the partial sum exceeds 2. This requires many more positive terms:

$s_{10} = 1 + \frac{1}{3} + \frac{1}{5} + \frac{1}{7} + \frac{1}{9} + \frac{1}{11} + \frac{1}{13} + \frac{1}{15} - \frac{1}{2} + \frac{1}{17} \approx 1.580623951$

$s_{11} = 1 + \frac{1}{3} + \frac{1}{5} + \frac{1}{7} + \frac{1}{9} + \frac{1}{11} + \frac{1}{13} + \frac{1}{15} - \frac{1}{2} + \frac{1}{17} + \frac{1}{19} \approx 1.63325553$

$s_{12} = s_{11} + \frac{1}{21} \approx 1.680874578,\ s_{13} = s_{12} + \frac{1}{23} \approx 1.724352839$

$s_{14} \approx 1.764352839,\ s_{15} \approx 1.801389876,\ s_{16} \approx 1.835872634,\ s_{17} \approx 1.868130699$

$s_{18} \approx 1.898433729,\ s_{19} \approx 1.927005158,\ s_{20} \approx 1.954032185,\ s_{21} \approx 1.97967321,$

$s_{22} \approx s_{21} + \frac{1}{41} \approx 2.004063454$.

Then we pick more previously unused negative terms, in order, until the partial sum is less than two. Again only 1 negative term is required:

$s_{23} = s_{22} - \frac{1}{4} \approx 1.754063454$.

As we continue to repeat this process, we "eventually" use all of the terms of the original conditionally convergent series and the partial sums of the new "rearranged" series get, and stay, arbitrarily close to the target number, 2. The same method can be used to rearrange the terms of the alternating harmonic series (or any conditionally convergent series) to sum to $0.3, 3, 30$ or any positive target number we want. How do you think the strategy needs to be changed to rearrange a conditionally convergent series to sum to a **negative** target number?

PROBLEMS

In problems 1 – 30 determine whether the given series Converge Absolutely, Converge Conditionally, or Diverge and give reasons for your conclusions.

1. $\sum_{n=1}^{\infty} (-1)^n \frac{1}{n+2}$ 2. $\sum_{n=1}^{\infty} (-1)^{n+1} \frac{1}{\sqrt{n}}$ 3. $\sum_{n=1}^{\infty} (-1)^{n+1} \frac{5}{n^3}$

4. $\displaystyle\sum_{n=1}^{\infty} (-1)^n \frac{1}{2 + \ln(n)}$

5. $\displaystyle\sum_{n=1}^{\infty} (-0.5)^n$

6. $\displaystyle\sum_{n=1}^{\infty} (-0.5)^{-n}$

7. $\displaystyle\sum_{n=1}^{\infty} (-1)^n \frac{1}{n^2}$

8. $\displaystyle\sum_{n=1}^{\infty} (-1)^{n+1} \frac{1}{3 + n^2}$

9. $\displaystyle\sum_{n=1}^{\infty} (-1)^{n+1} \frac{\ln(n)}{n}$

10. $\displaystyle\sum_{n=1}^{\infty} (-1)^{n+1} \frac{\ln(n)}{n^2}$

11. $\displaystyle\sum_{n=1}^{\infty} (-1)^n \frac{1}{n + \ln(n)}$

12. $\displaystyle\sum_{n=1}^{\infty} (-1)^{n+1} \frac{5}{\sqrt{n + 7}}$

13. $\displaystyle\sum_{n=1}^{\infty} (-1)^n \sin(\frac{1}{n})$ (recall that if $0 < x < 1$, then $0 < \sin(x) < x$)

14. $\displaystyle\sum_{n=1}^{\infty} (-1)^n \sin(\frac{1}{n^2})$

15. $\displaystyle\sum_{n=1}^{\infty} (-1)^n \sqrt{n}\, \sin(\frac{1}{n^2})$

16. $\displaystyle\sum_{n=1}^{\infty} \frac{\cos(\pi n)}{n}$

17. $\displaystyle\sum_{n=2}^{\infty} (-1)^{n+1} \frac{\ln(n)}{\ln(n^3)}$

18. $\displaystyle\sum_{n=1}^{\infty} (-1)^{n+1} \frac{n^2 + 7}{n^2 + 10}$

19. $\displaystyle\sum_{n=1}^{\infty} (-1)^{n+1} \frac{n^2 + 7}{n^3 + 10}$

20. $\displaystyle\sum_{n=1}^{\infty} (-1)^{n+1} \frac{n^2 + 7}{n^4 + 10}$

21. $\displaystyle\sum_{n=1}^{\infty} (-1)^{n+1} \frac{(n^2 + 7)^2}{n^2 + 10}$

22. $\displaystyle\sum_{n=1}^{\infty} \frac{\sin(n)}{n^2}$

23. $\displaystyle\sum_{n=1}^{\infty} \frac{\sin(\pi n)}{n}$

24. $\displaystyle\sum_{n=1}^{\infty} (-n)^{-n}$

25. $\displaystyle\sum_{n=1}^{\infty} (-1)^{n+1} \frac{1 + \sqrt{3n}}{n + 2}$

26. $\displaystyle\sum_{n=1}^{\infty} \frac{(-2)^n}{n^2}$

27. $\displaystyle\sum_{n=1}^{\infty} (-1)^{n+1} \frac{(-3)^n}{n^3}$

28. $\displaystyle\sum_{n=2}^{\infty} (-1)^{n+1} \left(\frac{\ln(n)}{\ln(n^5)} \right)^2$

29. $\displaystyle\sum_{n=1}^{\infty} \frac{(-2)^n}{n \cdot 3^n}$

30. $\displaystyle\sum_{n=1}^{\infty} \frac{3^n}{n^3}$

The Ratio Test is commonly used with series that contain factorials, and factorials are also going to become more common in the next few sections. Problems 31 – 40 ask you to simplify factorial expressions in order to get ready for that usage.

(Definitions: $0! = 1$, $1! = 1$, $2! = 1 \cdot 2 = 2$, $3! = 1 \cdot 2 \cdot 3 = 6$, $4! = 24$, and, in general, $n! = 1 \cdot 2 \cdot 3 \cdot \ldots \cdot (n-1) \cdot n$)

31. Show that $\dfrac{n!}{(n+1)!} = \dfrac{1}{n+1}$

32. Show that $\dfrac{n!}{(n+2)!} = \dfrac{1}{(n+1)(n+2)}$

33. Show that $\dfrac{n!}{(n+3)!} = \dfrac{1}{(n+1)(n+2)(n+3)}$

34. Show that $\dfrac{(n+1)!}{(n+2)!} = \dfrac{1}{n+2}$

35. Show that $\dfrac{(n-1)!}{(n+1)!} = \dfrac{1}{(n)(n+1)}$

36. Show that $\dfrac{2(n!)}{(2n)!} = \dfrac{2}{(n+1) \cdot (n+2) \cdot \ldots \cdot (2n)}$

37. Show that $\dfrac{(2n)!}{(2n+1)!} = \dfrac{1}{2n+1}$

38. Show that $\dfrac{(2n)!}{(2(n+1))!} = \dfrac{1}{(2n+1)(2n+2)}$

39. Show that $\dfrac{n^n}{n!} = \dfrac{n}{1} \cdot \dfrac{n}{2} \cdot \dfrac{n}{3} \cdot ... \cdot \dfrac{n}{n-1} \cdot \dfrac{n}{n}$

40. For $n > 0$, which is larger, $7!$ or $\dfrac{(n+7)!}{n!}$?

In problems 41 – 56, (a) determine the value of $L = \lim\limits_{n\to\infty} \left| \dfrac{a_{n+1}}{a_n} \right|$ for the given series, and

(b) state the conclusion of the Ratio Test as it applies to the series. (c) If the Ratio Test is inconclusive,

use some other method to determine if the given series converges or diverges.

41. $\displaystyle\sum_{n=1}^{\infty} \dfrac{1}{n}$

42. $\displaystyle\sum_{n=1}^{\infty} \dfrac{1}{n^2}$

43. $\displaystyle\sum_{n=2}^{\infty} \dfrac{1}{n^3}$

44. $\displaystyle\sum_{n=2}^{\infty} \dfrac{1}{\sqrt{n}}$

45. $\displaystyle\sum_{n=1}^{\infty} \left(\dfrac{1}{2}\right)^n$

46. $\displaystyle\sum_{n=1}^{\infty} \left(\dfrac{1}{3}\right)^n$

47. $\displaystyle\sum_{n=5}^{\infty} 1^n$

48. $\displaystyle\sum_{n=1}^{\infty} (-2)^n$

49. $\displaystyle\sum_{n=1}^{\infty} \dfrac{1}{n!}$

50. $\displaystyle\sum_{n=1}^{\infty} \dfrac{5}{n!}$

51. $\displaystyle\sum_{n=1}^{\infty} \dfrac{2^n}{n!}$

52. $\displaystyle\sum_{n=1}^{\infty} \dfrac{5^n}{n!}$

53. $\displaystyle\sum_{n=1}^{\infty} \left(\dfrac{1}{2}\right)^{3n}$

54. $\displaystyle\sum_{n=1}^{\infty} \left(\dfrac{1}{3}\right)^{2n}$

55. $\displaystyle\sum_{n=5}^{\infty} (0.9)^{2n+1}$

56. $\displaystyle\sum_{n=5}^{\infty} (-0.8)^{2n+1}$

For each series in problems 57 – 74 find the values of x for which the value of $L = \lim\limits_{n\to\infty} \left| \dfrac{a_{n+1}}{a_n} \right| < 1$.

From the Ratio Test we can conclude that each series converges absolutely (and thus converges) for those

values of x.

57. $\displaystyle\sum_{n=5}^{\infty} (x-5)^n$

58. $\displaystyle\sum_{n=1}^{\infty} \dfrac{(x-5)^n}{n}$

59. $\displaystyle\sum_{n=1}^{\infty} \dfrac{(x-5)^n}{n^2}$

60. $\displaystyle\sum_{n=1}^{\infty} \dfrac{(x-2)^n}{n^2}$

61. $\displaystyle\sum_{n=1}^{\infty} \dfrac{(x-2)^n}{n!}$

62. $\displaystyle\sum_{n=1}^{\infty} \dfrac{(x-10)^n}{n!}$

63. $\displaystyle\sum_{n=1}^{\infty} \dfrac{(2x-12)^n}{n^2}$

64. $\displaystyle\sum_{n=1}^{\infty} \dfrac{(4x-12)^n}{n^2}$

65. $\displaystyle\sum_{n=1}^{\infty} \dfrac{(6x-12)^n}{n!}$

66. $\displaystyle\sum_{n=1}^{\infty} (x-3)^{2n}$

67. $\displaystyle\sum_{n=2}^{\infty} \frac{(x+1)^{2n}}{n}$

68. $\displaystyle\sum_{n=2}^{\infty} \frac{(x+2)^{2n+1}}{n^2}$

69. $\displaystyle\sum_{n=1}^{\infty} \frac{(x-5)^{3n+1}}{n^2}$

70. $\displaystyle\sum_{n=1}^{\infty} \frac{(x+4)^{2n+1}}{n!}$

71. $\displaystyle\sum_{n=1}^{\infty} \frac{(x+3)^{2n-1}}{(n+1)!}$

72. $S(x) = \displaystyle\sum_{n=0}^{\infty} (-1)^n \frac{x^{2n+1}}{(2n+1)!}$

73. $C(x) = \displaystyle\sum_{n=0}^{\infty} (-1)^n \frac{x^{2n}}{(2n)!}$

74. $E(x) = \displaystyle\sum_{n=0}^{\infty} \frac{x^n}{n!}$

Rearrangements

In problems 75 – 80 use the strategy in the illustrative example to find the first 15 terms of a rearrangement of the given conditionally convergent series so that the rearranged series converges to the given target number.

75. $\displaystyle\sum_{n=1}^{\infty} (-1)^n \frac{1}{n}$, target number = 0.3 .

76. $\displaystyle\sum_{n=1}^{\infty} (-1)^{n+1} \frac{1}{n}$, target number = 0.7 .

77. $\displaystyle\sum_{n=1}^{\infty} (-1)^{n+1} \frac{1}{n}$, target number = 1 .

78. $\displaystyle\sum_{n=1}^{\infty} (-1)^{n+1} \frac{1}{\sqrt{n}}$, target number = 0.4 .

79. $\displaystyle\sum_{n=1}^{\infty} (-1)^{n+1} \frac{1}{\sqrt{n}}$, target number = 1 .

80. $\displaystyle\sum_{n=1}^{\infty} (-1)^{n+1} \frac{1}{n}$, target number = −1 .

Practice Answers

Practice 1:   (a) $\displaystyle\sum_{n=2}^{\infty} (-1)^n \frac{5}{\ln(n)}$ converges by the Alternating Series Test

$\displaystyle\sum_{n=2}^{\infty} \left| (-1)^n \frac{5}{\ln(n)} \right| = \sum_{n=2}^{\infty} \frac{5}{\ln(n)}$ which diverges by comparison with the divergent series $\displaystyle\sum_{n=2}^{\infty} \frac{5}{n}$.

Therefore, $\displaystyle\sum_{n=2}^{\infty} (-1)^n \frac{5}{\ln(n)}$ is **conditionally convergent** .

(b) $\displaystyle\sum_{k=1}^{\infty} \frac{\cos(\pi k)}{k^2} = \frac{-1}{1^2} + \frac{1}{2^2} + \frac{-1}{3^2} + ...$

$\displaystyle\sum_{k=1}^{\infty} \left| \frac{\cos(\pi k)}{k^2} \right| = \sum_{k=1}^{\infty} \frac{1}{k^2}$ which is convergent (P–test, p = 2) so

$\displaystyle\sum_{k=1}^{\infty} \frac{\cos(\pi k)}{k^2}$ is **absolutely convergent** (and convergent).

Practice 2:   (a) $a_n = (-1)^{n+1} \dfrac{e^n}{n!}$ so $a_{n+1} = (-1)^{n+2} \dfrac{e^{n+1}}{(n+1)!}$. Then

$$\left| \frac{a_{n+1}}{a_n} \right| = \left| \frac{(-1)^{n+2} \frac{e^{n+1}}{(n+1)!}}{(-1)^{n+1} \frac{e^n}{n!}} \right| = \left| \frac{e^{n+1}}{e^n} \frac{n!}{(n+1)!} \right| = \left| \frac{e}{n+1} \right| \longrightarrow 0 < 1$$

so $\displaystyle\sum_{n=1}^{\infty} (-1)^{n+1} \frac{e^n}{n!}$ is absolutely convergent.

(b) $a_n = \dfrac{n^5}{3^n}$ so $a_{n+1} = \dfrac{(n+1)^5}{3^{n+1}}$. Then

$$\left| \frac{a_{n+1}}{a_n} \right| = \left| \frac{\frac{(n+1)^5}{3^{n+1}}}{\frac{n^5}{3^n}} \right| = \left| \frac{3^n}{3^{n+1}} \frac{(n+1)^5}{n^5} \right| = \left| \frac{1}{3} \left(\frac{n+1}{n} \right)^5 \right| \longrightarrow \frac{1}{3} < 1$$

so $\displaystyle\sum_{n=1}^{\infty} \frac{n^5}{3^n}$ is absolutely convergent.

Practice 3: $a_n = \dfrac{(x-5)^n}{n^2}$ so $a_{n+1} = \dfrac{(x-5)^{n+1}}{(n+1)^2}$. Then

$$\left| \frac{a_{n+1}}{a_n} \right| = \left| \frac{\frac{(x-5)^{n+1}}{(n+1)^2}}{\frac{(x-5)^n}{n^2}} \right| = \left| \frac{(x-5)^{n+1}}{(x-5)^n} \frac{n^2}{(n+1)^2} \right| \rightarrow \quad |x - 5| = L.$$

Now we simply need to solve the inequality $|x - 5| < 1$ $(L < 1)$ for x.

If $|x - 5| < 1$, then $-1 < x - 5 < 1$ so **4 < x < 6** .

If $4 < x < 6$, then $L = |x - 5| < 1$ so $\displaystyle\sum_{n=1}^{\infty} \frac{(x-5)^n}{n^2}$ is absolutely convergent (and convergent).

Also, if $x < 4$ or $x > 6$ then $L > 1$ so $\displaystyle\sum_{n=1}^{\infty} \frac{(x-5)^n}{n^2}$ diverges.

"Endpoints": If $x = 4$, then $\displaystyle\sum_{n=1}^{\infty} \frac{(x-5)^n}{n^2} = \sum_{n=1}^{\infty} \frac{(-1)^n}{n^2}$ which converges.

If $x=6$, then $\displaystyle\sum_{n=1}^{\infty} \frac{(x-5)^n}{n^2} = \sum_{n=1}^{\infty} \frac{1}{n^2}$ which converges.

Appendix: A Proof of the Ratio Test

The Ratio Test has three parts, (a), (b), and (c), and each part requires a separate proof.

(a) $L<1 \Rightarrow$ the series is absolutely convergent. The basic pattern of the proof for this part is to show that the given series is, term–by–term, less than a convergent geometric series. Then we conclude by the Comparison Test that the given series converges.

Suppose $\lim\limits_{n\to\infty} \left| \dfrac{a_{n+1}}{a_n} \right| = L < 1$. Then there is a number r between

L and 1 so that the ratios $\left| \dfrac{a_{n+1}}{a_n} \right|$ are eventually (for all n > some N) less than r :

for all $n > N$, $\left| \dfrac{a_{n+1}}{a_n} \right| < r < 1$.

Then $|a_{N+1}| < r\,|a_N|$, $|a_{N+2}| < r\,|a_{N+1}| < r^2\,|a_N|$, $|a_{N+3}| < r\,|a_{N+2}| < r^3\,|a_N|$,

and, in general, $|a_{N+k}| < r^k\,|a_N|$.

So $|a_N| + |a_{N+1}| + |a_{N+2}| + |a_{N+3}| + |a_{N+4}| + ...$

$< |a_N| + r\,|a_N| + r^2\,|a_N| + r^3\,|a_N| + r^4\,|a_N| + ...$

$= |a_N|\cdot\{\, 1 + r + r^2 + r^3 + r^4 + ... \,\}$

$= |a_N|\cdot \dfrac{1}{1-r}$, since the powers of r form a convergent geometric series,

and the series $\sum\limits_{n=N}^{\infty} |a_n|$ is convergent by the Comparison Test.

Finally, we can conclude that the series $\sum\limits_{n=1}^{\infty} |a_n| = \sum\limits_{n=1}^{N-1} |a_n| + \sum\limits_{n=N}^{\infty} |a_n|$ is convergent, since

it is the sum of two convergent series. $\sum\limits_{n=1}^{\infty} a_n$ is absolutely convergent.

(b) $L>1 \Rightarrow$ the series is divergent. The basic idea in this part is to show that the terms of the given series do not approach 0. Then we can conclude by the Nth Term Test that the given series diverges.

Suppose $\lim\limits_{n\to\infty} \left| \dfrac{a_{n+1}}{a_n} \right| = L > 1$. Then the ratios $\left| \dfrac{a_{n+1}}{a_n} \right|$ are eventually (for all n > some N)

larger than 1 so $|a_{N+1}| > |a_N|$, $|a_{N+2}| > |a_{N+1}| > |a_N|$, $|a_{N+3}| > |a_{N+2}| > |a_N|$,

and, for all $k > N$, $|a_k| > |a_N|$.

Thus $\lim\limits_{n\to\infty} |a_n| \geq |a_N| \neq 0$ and $\lim\limits_{n\to\infty} a_n \neq 0$, so by the Nth Term Test for Divergence

(Section 10.2) we can conclude that the series $\sum\limits_{n=1}^{\infty} a_n$ is divergent.

(c) $L=1 \implies$ nothing. Part (c) can be verified by giving two series, one convergent and one divergent, for which $L = 1$.

If $a_n = \dfrac{1}{n^2}$, then $\displaystyle\sum_{n=1}^{\infty} a_n = \sum_{n=1}^{\infty} \dfrac{1}{n^2}$ is convergent by the P–Test with p=2, and

$$\lim_{n\to\infty} \left| \frac{a_{n+1}}{a_n} \right| = \lim_{n\to\infty} \left| \frac{1/(n+1)^2}{1/n^2} \right| = \lim_{n\to\infty} \left(\frac{n}{n+1} \right)^2 = 1.$$

If $a_n = \dfrac{1}{n}$, then $\displaystyle\sum_{n=1}^{\infty} a_n = \sum_{n=1}^{\infty} \dfrac{1}{n}$ is the divergent harmonic series, and

$$\lim_{n\to\infty} \left| \frac{a_{n+1}}{a_n} \right| = \lim_{n\to\infty} \left| \frac{1/(n+1)}{1/n} \right| = \lim_{n\to\infty} \frac{n}{n+1} = 1.$$

Since $L = 1$ for each of these series, one convergent and one divergent, knowing that $L = 1$ for

a series does not let us conclude that the series converges or that it diverges.

10.8   POWER SERIES: $\displaystyle\sum_{n=0}^{\infty} a_n x^n$ and $\displaystyle\sum_{n=0}^{\infty} a_n(x-c)^n$

So far most of the series we have examined have consisted of numbers (numerical series), but the most important series contain powers of a variable, and they define functions of that variable.

Definition of Power Series

A **power series** is an expression of the form $\displaystyle\sum_{n=0}^{\infty} a_n x^n = a_0 + a_1 x + a_2 x^2 + a_3 x^3 + \ldots + a_n x^n + \ldots$

where $a_0, a_1, a_2, a_3, \ldots$ are constants, called the coefficients of the series, and x is a variable.

(Note: For $n = 0$ we use the convention for power series that $x^0 = 1$ even when $x = 0$. This convention simply makes it easier for us to represent the series using the summation notation.)

For each value of the variable, the power series is simply a numerical series that may converge or diverge. If the power series does converge, the value of the function is the sum of the series, and the domain of the function is the set of x values for which the series converges. Power series are particularly important in mathematics and applications because many important functions such as $\sin(x)$, $\cos(x)$, e^x, and $\ln(x)$ can be represented and approximated by power series.

The following are examples of power series:

$$f(x) = 1 + x + x^2 + x^3 + x^4 + \ldots = \sum_{n=0}^{\infty} x^n \quad (a_n = 1 \text{ for all n})$$

$$g(x) = 1 + x + \frac{x^2}{2!} + \frac{x^3}{3!} + \frac{x^4}{4!} + \ldots = \sum_{n=0}^{\infty} \frac{x^n}{n!} \quad (a_n = \frac{1}{n!} \text{ for all n and with the definition that } 0! = 1)$$

$$h(x) = x - \frac{x^3}{3!} + \frac{x^5}{5!} - \frac{x^7}{7!} + \ldots = \sum_{n=0}^{\infty} (-1)^n \cdot \frac{x^{2n+1}}{(2n+1)!} \quad (a_n = \frac{(-1)^n}{(2n+1)!} \text{ for all n.})$$

Power series look like long (very long) polynomials, and in many ways they behave like polynomials.

This section focuses on what a power series is and on determining where a given power series converges. Section 10.9 looks at the arithmetic (sums, differences, products) and calculus (derivative and integrals) of power series. Section 10.10 examines how to represent particular functions such as $\sin(x)$ and e^x and others as power series.

Finding Where a Power Series Converges

The power series $f(x) = \sum\limits_{n=0}^{\infty} a_n x^n$ always converges at $x = 0$: $f(0) = \sum\limits_{n=0}^{\infty} a_n(0)^n = a_0$. To find which

other values of x make a power series converge, we could try x values one–by–one, but that is very

inefficient and time consuming. Instead, the Ratio Test allows us to get answers for lots of x values all at

once. (Since we are using a_n to represent the coefficient of the nth term $a_n x^n$, we let $c_n = a_n x^n$

represent the nth term and we use the ratio $\dfrac{c_{n+1}}{c_n}$.)

Example 1: Find all of the values of x for which the power series $\sum\limits_{n=0}^{\infty} (2n+1)\cdot x^n$ converges.

Solution: $c_n = (2n+1)\cdot x^n$ so $c_{n+1} = (2(n+1)+1)\cdot x^{n+1} = (2n+3)\cdot x^{n+1}$. Then, using the Ratio Test,

$$\left| \frac{c_{n+1}}{c_n} \right| = \left| \frac{(2n+3)\cdot x^{n+1}}{(2n+1)\cdot x^n} \right| = \left| \frac{2n+3}{2n+1} \cdot \frac{x^{n+1}}{x^n} \right| = \left| \frac{2n+3}{2n+1} \cdot x \right| \rightarrow |x| = L .$$

From the Ratio Test we know the series converges if $L < 1$: if $|x| < 1$ or, in other words, $-1 < x < 1$.

We also know the series diverges if $L > 1$ so the series diverges if $|x| > 1$: if $x > 1$ or $x < -1$.

Finally, we need to check the two remaining values of x: the **endpoints** $x = -1$ and $x = 1$.

When $x = 1$, $\sum\limits_{n=0}^{\infty} (2n+1)\cdot x^n = \sum\limits_{n=0}^{\infty} (2n+1)\cdot 1^n = \sum\limits_{n=0}^{\infty} (2n+1)$ which diverges since the

terms do not approach 0. When $x = -1$, $\sum\limits_{n=0}^{\infty} (2n+1)\cdot x^n = \sum\limits_{n=0}^{\infty} (2n+1)\cdot (-1)^n$ which also

diverges because the terms do not approach 0.

The power series $\sum\limits_{n=0}^{\infty} (2n+1)\cdot x^n$ converges if and only if $-1 < x < 1$. In other words, the series

converges when x is in the interval $(-1, 1)$ and it diverges when x is outside the interval $(-1, 1)$.

Example 2: Find all of the values of x for which the power series $\sum\limits_{n=1}^{\infty} \dfrac{x^n}{n\cdot 3^n}$ converges.

Solution: $c_n = \dfrac{x^n}{n\cdot 3^n}$ so $c_{n+1} = \dfrac{x^{n+1}}{(n+1)\cdot 3^{n+1}}$. Using the Ratio Test,

$$\left| \frac{c_{n+1}}{c_n} \right| = \left| \frac{\dfrac{x^{n+1}}{(n+1)\cdot 3^{n+1}}}{\dfrac{x^n}{n\cdot 3^n}} \right| = \left| \frac{n}{n+1} \cdot \frac{3^n}{3^{n+1}} \cdot \frac{x^{n+1}}{x^n} \right| = \left| \frac{n}{n+1} \cdot \frac{1}{3} \cdot x \right| \rightarrow \left| \frac{x}{3} \right| = L .$$

Solving $|\frac{x}{3}| < 1$, we have $-1 < \frac{x}{3} < 1$ or $-3 < x < 3$. The series converges for $-3 < x < 3$ and it

diverges for $x < -3$ and for $x > 3$.

Finally, we need to check the two remaining values of x: the **endpoints** $x = -3$ and $x = 3$.

When $x = -3$, **Error! Not a valid embedded object.** $= \sum_{n=1}^{\infty} \frac{(-3)^n}{n \cdot 3^n} = \sum_{n=1}^{\infty} \frac{(-1)^n}{n}$ which

converges by the

Alternating Series Test.

When $x = 3$, $\sum_{n=1}^{\infty} \frac{x^n}{n \cdot 3^n} = \sum_{n=1}^{\infty} \frac{(3)^n}{n \cdot 3^n} = \sum_{n=1}^{\infty} \frac{1}{n}$, the harmonic series, which diverges.

In summary, the power series $\sum_{n=1}^{\infty} \frac{x^n}{n \cdot 3^n}$ converges if $-3 \leq x < 3$, if x is in the interval $[-3, 3)$.

Note: The Ratio Test is very powerful for determining where a power series converges: put $c_n = a_n x^n$,

calculate the limit of the ratio $\left| \frac{c_{n+1}}{c_n} \right|$, and then solve the resulting absolute value inequality for x.

Typically, we also need to check the endpoints of the interval by replacing x with the two endpoint

values and then determining if the resulting numerical series converge or diverge at these endpoints.

The Ratio Test does not help with the endpoints.

Practice 1:   Find all of the values of x for which the series $\sum_{n=1}^{\infty} \frac{5^n \cdot x^n}{n}$ converges.

Interval of Convergence, Radius of Convergence

In each of the previous examples, the values of x for which the power series converge form an interval.

The next theorem says that always happens.

Interval of Convergence Theorem for Power Series

The values of x for which the power series $\sum_{n=0}^{\infty} a_n x^n$ converges form an interval.

(i)   If this power series converges for $x = c$, then the series converges for all satisfying $|x| < |c|$.

(ii)   If this power series diverges for $x = d$, then the series diverges for all satisfying $|x| > |d|$.

A proof of the Interval of Convergence Theorem is given after the Practice Answers.

Meaning of the Interval of Convergence Theorem: If a power series $\displaystyle\sum_{n=0}^{\infty} a_n x^n$ converges for a value x = c,

then the series also converges for all values of x closer to the origin than c. If the power series diverges

for a value x = d, then the power series diverges for all values of x farther from the origin than d.

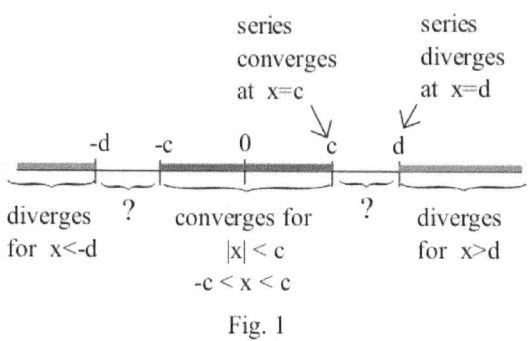

series
converges
at x=c

series
diverges
at x=d

-d   -c   0   c   d

diverges   ?   converges for   ?   diverges
for x<-d $|x| < c$ for x>d
                  -c < x < c

Fig. 1

This guarantees that the values where the power
series converges form an interval, from –|c| to |c|.
This Theorem does **not** tell us about the
convergence of the power series at the endpoints of
the interval — we need to check those two points
individually. This Theorem also does **not** tell us
about the convergence of the power series for
values of x with |c| < |x| < |d|. See Fig. 1.

Example 3:   Suppose we know that a power series $\displaystyle\sum_{n=0}^{\infty} a_n x^n$ converges at x = 4 and diverges at x = 9.

What can we conclude (converge or diverge or not enough information) about the series

when x = 2, –3, –4, 5, –6, 8, –9, 10, –11?

Solution: We know the power series converges at x = 4 (c = 4) so we can conclude that the series

converges for x = 2 and x = –3 since |2|<|4| and |–3|<|4|.

We know the power series diverges at x = 9 (d = 9) so we can conclude that the series diverges

for x = 10 and x = –11 since |10|>|9| and |–11|>|9|.

The remaining values of x (–4, 5, 6, 8, and 9) do not satisfy |x| < 4 or |x| > 9 so the

series may converge or may diverge — we don't

have enough information.

Fig. 2 shows the regions where
convergence of this power series is
guaranteed, where divergence is
guaranteed, and where we don't have
enough information.

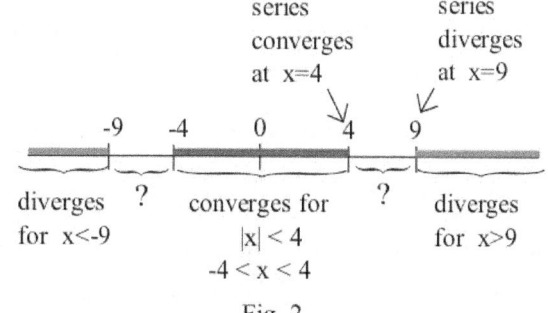

series
converges
at x=4

series
diverges
at x=9

-9   -4   0   4   9

diverges   ?   converges for   ?   diverges
for x<-9 $|x| < 4$ for x>9
                  -4 < x < 4

Fig. 2

Practice 2:   Suppose we know that a power series $\sum\limits_{n=0}^{\infty} a_n x^n$ converges at $x = 3$ and diverges at $x = -7$.

What can we say about the convergence of the series for $x = -1, 2, -3, 4, -6, 7, -8$, and 17?

Sketch the regions of known convergence and known divergence.

Definition: The **interval of convergence** of a power series $\sum\limits_{n=0}^{\infty} a_n x^n$ is the interval of

values of x for which the series converges.

From Example 1 we know the interval of convergence of $\sum\limits_{n=0}^{\infty} (2n+1) \cdot x^n$ is $(-1, 1)$.

From Example 2 we know the interval of convergence of $\sum\limits_{n=1}^{\infty} \dfrac{x^n}{n \cdot 3^n}$ is $[-3, 3)$.

Definition:

For each power series $\sum\limits_{n=0}^{\infty} a_n x^n$ there is a

number R, called the **radius of convergence**, so that

the series

converges for $|x| < R$ and diverges when $|x| > R$.

(The series may converge or may diverge at $|x| = R$.)

(Fig. 3)

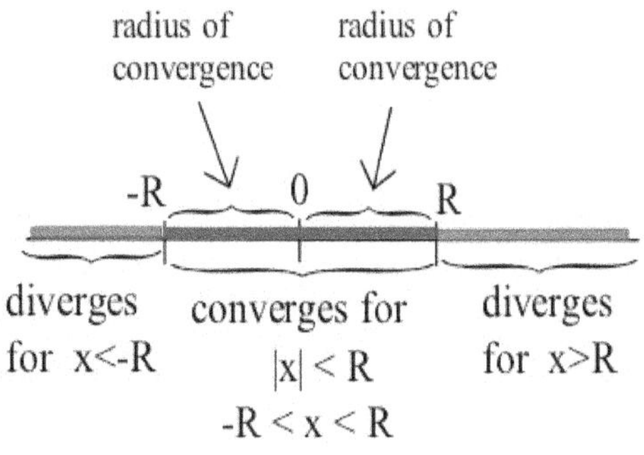

Fig. 3

The radius of convergence is half of the length of the interval of convergence.

Example 4:   What is the radius of convergence of each series in Examples 1 and 2?

Solution: The power series $\sum\limits_{n=0}^{\infty} (2n+1) \cdot x^n$ converges if $-1 < x < 1$ so $R = 1$.

The power series $\sum\limits_{n=1}^{\infty} \dfrac{x^n}{n \cdot 3^n}$ converges if $-3 \leq x < 3$, so $R = 3$.

The convergence or divergence of a power series at the endpoints of the interval of convergence does not affect the value of the radius of convergence R, and the value of R does not tell us about the convergence of the power series at the endpoints of the interval of convergence, at $x = R$ and $x = -R$.

Practice 3:   What is the radius of convergence of the series in Practice 1?

Summary

For a power series $\sum_{n=0}^{\infty} a_n x^n$ exactly one of these three situations occurs:

(i)   the series converges only for $x = 0$. (Then we say the radius of convergence is 0.)

(ii)   the series converges for all x with $|x| < R$ and diverges for all x with $|x| > R$. (Then we say the radius of convergence is R.)

(iii) the series converges for all values of x. (Then we say the radius of convergence is infinite.)

The following list shows the intervals and radii of convergence for several power series. Four of the series in the list have the same radius of convergence, $R = 1$, but slightly different intervals of convergence. (The $\sum$ below simply means a series whose starting index is some finite value of n and whose upper index is ∞.)

Series	Radius of Convergence	Interval of Convergence	Series converges for x Values in the Shaded Intervals
$\sum n!\cdot x^n$	$R = 0$	{ 0 }, a single point	
$\sum x^n$	$R = 1$	$(-1, 1)$	
$\sum \dfrac{x^n}{n}$	$R = 1$	$[-1, 1)$	
$\sum (-1)^n \dfrac{x^n}{n}$	$R = 1$	$(-1, 1]$	
$\sum \dfrac{x^n}{n^2}$	$R = 1$	$[-1, 1]$	
$\sum \dfrac{x^n}{2^n}$	$R = 2$	$(-2, 2)$	
$\sum \dfrac{x^n}{n!}$	$R = \infty$	$(-\infty, \infty)$	

Power Series Centered at c

Sometimes it is useful to "shift" a power series. These shifted power series contain powers of "x – c" instead of powers of "x," but many of the properties we have examined still hold.

Definition: A **power series centered at c** is a series of the form

$$\sum_{n=0}^{\infty} a_n(x-c)^n = a_0 + a_1(x-c) + a_2(x-c)^2 + a_3(x-c)^3 + \ldots + a_n(x-c)^n + \ldots$$

where $a_0, a_1, a_2, a_3, \ldots$ are constants, called the coefficients of the series, and x is a variable.

Note: As usual for power series, for $n = 0$ we use the convention that $(x-c)^0 = 1$, even when $x = c$.

A power series centered at c always converges for $x = c$, and the interval of convergence is an interval **centered at c**. The radius of convergence is half the length of the interval of convergence (Fig. 4).

If a power series centered at c converges for a value of x, then the series converges for all values closer to c. If a power series centered at c diverges for a value of x, then the series diverges for all values farther from c.

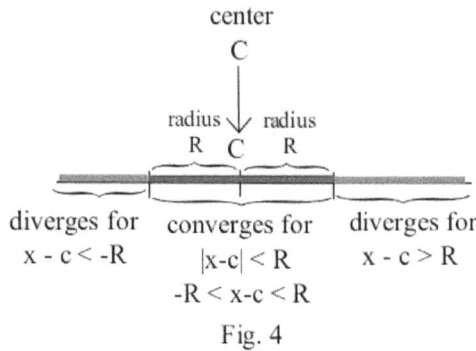

Fig. 4

Example 5:   Suppose we know that a power series $\sum_{n=0}^{\infty} a_n(x-4)^n$ converges at $x = 6$ and diverges at $x = 0$. What can we conclude (converge or diverge or not enough information) about the series when $x = 3, 9, -1, 2,$ and 7?

Solution: We know the power series converges at $x = 6$ so we can conclude that the series converges for values of x **closer to 4** than $|6 - 4| = 2$ units: the series converges at $x = 3$.

We know the power series diverges at $x = 0$ so we can conclude that the series diverges for values of x **farther from 4** than $|0 - 4| = 4$ units: the series diverges at $x = 9$ and -1.

The remaining values of x (2 and 7) do not satisfy $|x - 4| < 2$ or $|x - 4| > 4$ so the series may converge or may diverge.

Fig. 5 shows the regions where convergence of this power series is guaranteed, where divergence is guaranteed, and where we don't have enough information.

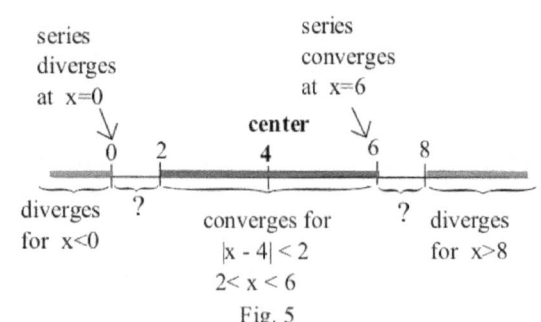

Fig. 5

Practice 4:   Suppose we know that a power series $\sum\limits_{n=0}^{\infty} a_n(x + 5)^n$ converges at $x = -1$ and diverges

at $x = 1$. What can we conclude about the series when $x = -2, -9, 0, -11$, and 3?

The Ratio Test is still our primary tool for finding an interval of convergence of a power series, even if the

power series is centered at c rather than at 0.

Example 6:   Find the interval and radius of convergence of $\sum\limits_{n=1}^{\infty} \dfrac{(x - 5)^n}{n \cdot 2^n}$.

Solution: $c_n = \dfrac{(x-5)^n}{n \cdot 2^n}$ so $c_{n+1} = \dfrac{(x-5)^{n+1}}{(n+1) \cdot 2^{n+1}}$. The ratio $\left| \dfrac{c_{n+1}}{c_n} \right|$ for the Ratio Test appears to

be messy, but if we group the similar pieces algebraically, the ratio simplifies nicely:

$$\left| \frac{c_{n+1}}{c_n} \right| = \left| \frac{\dfrac{(x-5)^{n+1}}{(n+1) \cdot 2^{n+1}}}{\dfrac{(x-5)^n}{n \cdot 2^n}} \right| = \left| \frac{n}{n+1} \cdot \frac{2^n}{2^{n+1}} \cdot \frac{(x-5)^{n+1}}{(x-5)^n} \right| = \left| \frac{n}{n+1} \cdot \frac{1}{2} \cdot (x-5) \right| \rightarrow \left| \frac{(x-5)}{2} \right| = L .$$

Solving $\left| \dfrac{(x-5)}{2} \right| < 1$, we have $-1 < \dfrac{(x-5)}{2} < 1$ so $-2 < x-5 < 2$ and $3 < x < 7$. The series

converges for $3 < x < 7$, and it diverges for $x < 3$ and for $x > 7$.

Finally, we need to check the two remaining values of x: the **endpoints** $x = 3$ and $x = 7$.

When $x = 3$, $\sum\limits_{n=1}^{\infty} \dfrac{(x-5)^n}{n \cdot 2^n} = \sum\limits_{n=1}^{\infty} \dfrac{(-2)^n}{n \cdot 2^n} = \sum\limits_{n-1}^{\infty} \dfrac{(-1)^n}{n}$ which converges by the

Alternating Series Test.

When $x = 7$, $\sum\limits_{n=1}^{\infty} \dfrac{(x-5)^n}{n \cdot 2^n} = \sum\limits_{n=1}^{\infty} \dfrac{(2)^n}{n \cdot 2^n} = \sum\limits_{n=1}^{\infty} \dfrac{1}{n}$, the harmonic series, which diverges.

The power series $\sum\limits_{n=1}^{\infty} \dfrac{(x-5)^n}{n \cdot 2^n}$ converges if $3 \le x < 7$. The interval of convergence is $[3, 7)$, and the

radius of convergence is $R = \dfrac{1}{2}$ (length of the interval of convergence) $= \dfrac{1}{2}(7 - 3) = 2$.

Practice 5:   Find the interval and radius of convergence of $\sum\limits_{n=0}^{\infty} \dfrac{n \cdot (x - 3)^n}{5^n}$.

A power series looks like a very long polynomial. However, a regular polynomial with a finite number of terms is defined at every value of x, but a power series may diverge for many values of the variable x. As we continue to work with power series we need to be alert to where the power series converges (and behaves like a finite polynomial) and where the power series diverges. We need to know the interval of convergence of the power series, and, typically, we use the Ratio Test to find that interval.

PROBLEMS

In problems 1 – 24, (a) find all values of x for which each given power series converges, and (b) graph the interval of convergence for the series on a number line.

1. $\sum\limits_{n=1}^{\infty} x^n$

2. $\sum\limits_{n=1}^{\infty} (x-3)^n$

3. $\sum\limits_{n=1}^{\infty} (x+2)^n$

4. $\sum\limits_{n=1}^{\infty} (x+5)^n$

5. $\sum\limits_{n=3}^{\infty} \frac{x^n}{n}$

6. $\sum\limits_{n=3}^{\infty} \frac{(x-2)^n}{n}$

7. $\sum\limits_{n=1}^{\infty} \frac{(x+3)^n}{n}$

8. $\sum\limits_{n=1}^{\infty} \frac{(x-5)^n}{n^2}$

9. $\sum\limits_{n=1}^{\infty} \frac{(x-7)^{2n+1}}{n^2}$

10. $\sum\limits_{n=1}^{\infty} \frac{(x+1)^{2n}}{n^3}$

11. $\sum\limits_{n=2}^{\infty} (2x)^n$

12. $\sum\limits_{n=2}^{\infty} (5x)^n$

13. $\sum\limits_{n=1}^{\infty} (\frac{x}{3})^{2n+1}$

14. $\sum\limits_{n=1}^{\infty} n(\frac{x}{4})^{2n+1}$

15. $\sum\limits_{n=1}^{\infty} (2x-6)^n$

16. $\sum\limits_{n=1}^{\infty} (3x+1)^n$

17. $\sum\limits_{n=0}^{\infty} \frac{x^n}{n!}$

18. $\sum\limits_{n=0}^{\infty} \frac{(x-5)^n}{n!}$

19. $\sum\limits_{n=1}^{\infty} \frac{n^2 \cdot x^n}{3^n}$

20. $\sum\limits_{n=1}^{\infty} \frac{n^5 \cdot x^n}{3^n}$

21. $\sum\limits_{n=1}^{\infty} n! \cdot x^n$

22. $\sum\limits_{n=1}^{\infty} n! \cdot (x+2)^n$

23. $\sum\limits_{n=3}^{\infty} n! \cdot (x-7)^n$

24. $\sum\limits_{n=0}^{\infty} \frac{3^n \cdot x^n}{n!}$

In problems 25 – 34, the letters "a" and "b" represent positive constants. Find all values of x for which each given power series converges.

25. $\sum\limits_{n=1}^{\infty} (x-a)^n$

26. $\sum\limits_{n=1}^{\infty} (x+b)^n$

27. $\sum\limits_{n=1}^{\infty} \frac{(x-a)^n}{n}$

28. $\sum\limits_{n=1}^{\infty} \frac{(x-a)^n}{n^2}$

29. $\sum\limits_{n=1}^{\infty} (ax)^n$

30. $\sum\limits_{n=1}^{\infty} (\frac{x}{a})^n$

31. $\sum\limits_{n=1}^{\infty} (ax-b)^n$

32. $\sum\limits_{n=1}^{\infty} (ax+b)^n$

33. A friend claimed that the interval of convergence for a power series of the form $\sum\limits_{n=1}^{\infty} a_n \cdot (x-4)^n$ is the

    interval $(1, 9)$. Without checking your friend's work, how can you be certain that your friend is wrong?

34. Which of the following intervals are possible intervals of convergence for the power series in problem 33?

    $(2, 6), (0, 4), x = 0, [1, 7], (-1, 9], x = 4, [3, 5), [-4, 4), x = 3$.

35. Which of the following intervals are possible intervals of convergence for $\sum\limits_{n=1}^{\infty} a_n \cdot (x-7)^n$?

    $(3, 10), (5, 9), x = 0, [1, 13], (-1, 15], x = 4, [3, 11), [0, 14), x = 7$.

36. Fill in each blank with a number so the resulting interval could be the interval of convergence for the power

    series $\sum\limits_{n=1}^{\infty} a_n \cdot (x-3)^n$: $(0, \_\_\_), (\_\_\_, 7), [1, \_\_\_], (\_\_\_, 15], [\_\_\_, 11), [0, \_\_\_), x = \_\_\_$.

37. Fill in each blank with a number so the resulting interval could be the interval of convergence for the power

    series $\sum\limits_{n=1}^{\infty} a_n \cdot (x-1)^n$: $(0, \_\_\_), (\_\_\_, 7), [1, \_\_\_], (\_\_\_, 5], [\_\_\_, 11), [0, \_\_\_), x = \_\_\_$.

In problems 38 – 45, use the patterns you noticed in the earlier problems and examples to build a power series
with the given intervals of convergence. (There are many possible correct answers — find one.)

38. $(-5, 5)$ 39. $[-3, 3)$ 40. $[-2, 2]$ 41. $(0, 6)$

42. $[0, 8)$ 43. $[2, 8)$ 44. $[3, 7]$ 45. $x = 3$

In problems 46 – 59, find the interval of convergence for each series. For x in the interval of convergence, find
the sum of the series as a function of x. (Hint: You know how to find the sum of a geometric series.)

46. $\sum\limits_{n=0}^{\infty} x^n$ 47. $\sum\limits_{n=0}^{\infty} (x-3)^n$ 48. $\sum\limits_{n=0}^{\infty} (2x)^n$ 49. $\sum\limits_{n=0}^{\infty} (3x)^n$

50. $\sum\limits_{n=0}^{\infty} x^{2n}$ 51. $\sum\limits_{n=0}^{\infty} x^{3n}$ 52. $\sum\limits_{n=0}^{\infty} (\frac{x-6}{2})^n$ 53. $\sum\limits_{n=0}^{\infty} (\frac{x-6}{5})^n$

54. $\sum\limits_{n=0}^{\infty} (\frac{x}{2})^n$ 55. $\sum\limits_{n=0}^{\infty} (\frac{x}{5})^n$ 56. $\sum\limits_{n=0}^{\infty} (\frac{x}{3})^{2n}$ 57. $\sum\limits_{n=0}^{\infty} (\frac{x}{2})^{3n}$

58. $\sum\limits_{n=0}^{\infty} (\frac{1}{2}\sin(x))^n$ 59. $\sum\limits_{n=0}^{\infty} (\frac{1}{3}\cos(x))^n$

Practice Answers

Practice 1: $c_n = \dfrac{5^n \cdot x^n}{n}$ so $c_{n+1} = \dfrac{5^{n+1} \cdot x^{n+1}}{n+1}$. Then, using the Ratio Test,

$$\left| \frac{c_{n+1}}{c_n} \right| = \left| \frac{\dfrac{5^{n+1} x^{n+1}}{n+1}}{\dfrac{5^n x^n}{n}} \right| = \left| \frac{n}{n+1} \cdot \frac{5^{n+1}}{5^n} \cdot \frac{x^{n+1}}{x^n} \right| = \left| \frac{n}{n+1} \cdot 5 \cdot x \right| \rightarrow |5x| = L .$$

Solving $|5x| < 1$, we have $-1 < 5x < 1$ or $-1/5 < x < 1/5$. The series converges for $-1/5 < x < 1/5$
and it diverges for $x < -1/5$ and for $x > 1/5$.

Finally, we need to check the two remaining values of x: the **endpoints** $x = -1/5$ and $x = 1/5$.

When $x = -1/5$, $\displaystyle\sum_{n=1}^{\infty} \frac{5^n x^n}{n} = \sum_{n=1}^{\infty} \frac{5^n(-1/5)^n}{n} = \sum_{k=1}^{\infty} \frac{(-1)^n}{n}$ which converges by the

Alternating Series Test.

When $x = 1/5$, $\displaystyle\sum_{n=1}^{\infty} \frac{5^n x^n}{n} = \sum_{n=1}^{\infty} \frac{5^n(1/5)^n}{n} = \sum_{k=1}^{\infty} \frac{1}{n}$, the harmonic series, which diverges.

In summary, the power series $\displaystyle\sum_{n=1}^{\infty} \frac{5^n x^n}{n}$ converges if $-1/5 \le x < 1/5$, if x is in the interval $[-1/5, 1/5)$.

Practice 2:   The series converges at $x = -1$ and $x = 2$.

The convergence is unknown at

$x = -3$ (endpoint?), 4, –6, and 7 (endpoint?)

The series diverges at $x = -8$ and $x = 17$.

The regions of known convergence and known

divergence are shown in Fig. 6.

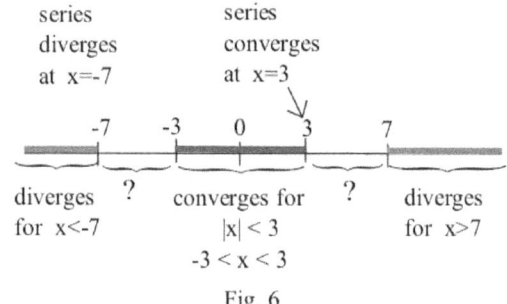

Fig. 6

Practice 3:   The radius of convergence is **1/5** .

Practice 4:   The series converges at $x = -2$.

The convergence is unknown at

$x = -9$ (endpoint?), 0, and –11 (endpoint?)

The series diverges at $x = 3$.

The regions of known convergence and known

divergence are shown in Fig. 7.

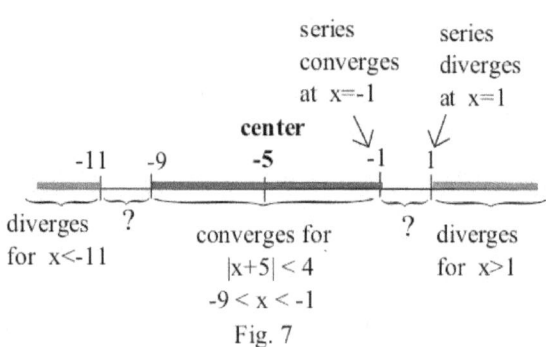

Fig. 7

Practice 5: $c_n = \dfrac{n \cdot (x-3)^n}{5^n}$ so $c_{n+1} = \dfrac{(n+1) \cdot (x-3)^{n+1}}{5^{n+1}}$. Then, using the Ratio Test,

$$\left| \frac{c_{n+1}}{c_n} \right| = \left| \frac{\dfrac{(n+1) \cdot (x-3)^{n+1}}{5^{n+1}}}{\dfrac{n \cdot (x-3)^n}{5^n}} \right| = \left| \frac{n+1}{n} \cdot \frac{5^n}{5^{n+1}} \cdot \frac{(x-3)^{n+1}}{(x-3)^n} \right| = \left| \frac{n+1}{n} \cdot \frac{1}{5} \cdot (x-3) \right| \to \left| \frac{x-3}{5} \right| = L .$$

Solving $\left| \frac{x-3}{5} \right| < 1$, we have $-1 < \frac{x-3}{5} < 1$ so $-5 < x-3 < 8$. The series converges for $-2 < x < 8$

and it diverges for $x < -2$ and for $x > 8$. The radius of convergence is 5.

To find the interval of convergence we still need to check the endpoints $x = -2$ and $x = 8$:

When $x = -2$, $\displaystyle\sum_{n=0}^{\infty} \frac{n \cdot (x-3)^n}{5^n} = \sum_{n=1}^{\infty} \frac{n \cdot (-5)^n}{5^n} = \sum_{n=1}^{\infty} n \cdot (-1)^n$ which diverges by the nth Term Test.

When $x = 8$, $\displaystyle\sum_{n=0}^{\infty} \frac{n \cdot (x-3)^n}{5^n} = \sum_{n=1}^{\infty} \frac{n \cdot (5)^n}{5^n} = \sum_{n=1}^{\infty} n$ which diverges by the nth Term Test.

The interval of convergence is $-2 < x < 8$ or $(-2, 8)$.

Appendix: Proof of the Interval of Convergence Theorem

(i)   Suppose the power series $\displaystyle\sum_{n=0}^{\infty} a_n x^n$ converges at $x = c$: $\displaystyle\sum_{n=0}^{\infty} a_n c^n$ converges.

Then the terms of the series $a_n c^n$ must approach 0, $\displaystyle\lim_{n \to \infty} a_n c^n = 0$, so there is a number N so that

if $n \geq N$, then $|a_n c^n| < 1$ and $|a_n| < \dfrac{1}{|c^n|}$ for all $n > N$.

If $|x| < |c|$, then $\displaystyle\sum_{n=0}^{\infty} |a_n x^n| = \left\{ |a_0| + |a_1 x| + \ldots + |a_{N-1} x^{N-1}| \right\} + \left\{ |a_N x^N| + |a_{N+1} x^{N+1}| + \ldots \right\}$.

The first piece, $|a_0| + |a_1 x| + \ldots + |a_{N-1} x^{N-1}|$, consists of a finite number of terms so it is a finite number.

The second piece, $|a_N x^N| + |a_{N+1} x^{N+1}| + \ldots$ is less than a convergent geometric series:

$$|a_N x^N| + |a_{N+1} x^{N+1}| + |a_{N+2} x^{N+2}| + |a_{N+3} x^{N+3}| + \ldots$$

$$< \frac{1}{|c^N|} |x^N| + \frac{1}{|c^{N+1}|} |x^{N+1}| + \frac{1}{|c^{N+2}|} |x^{N+2}| + \frac{1}{|c^{N+3}|} |x^{N+3}| + \ldots$$

$$= \left| \frac{x}{c} \right|^N + \left| \frac{x}{c} \right|^{N+1} + \left| \frac{x}{c} \right|^{N+2} + \left| \frac{x}{c} \right|^{N+3} + \ldots \text{ which converges since } |x| < |c| \text{ and } \left| \frac{x}{c} \right| < 1.$$

If $|x| < |c|$, then $\displaystyle\sum_{n=0}^{\infty} |a_n x^n|$ converges so $\displaystyle\sum_{n=0}^{\infty} a_n x^n$ converges.

(ii) This part follows from part (i) of the Theorem. Suppose that the power series $\sum\limits_{n=0}^{\infty} a_n x^n$ diverges

at $x = d$: $\sum\limits_{n=0}^{\infty} a_n d^n$ diverges. If the series converges for some x_0 with $|x_0| > |d|$, we can put $c = x_0$ and

conclude from part (i) that the series must converge at $x = d$ because $|c| = |x_0| > |d|$. This contradicts the

fact that the series diverges at $x = d$, so the series cannot converge for any x_0 with $|x_0| > |d|$.

If $\sum\limits_{n=0}^{\infty} a_n d^n$ diverges and $|x| > |d|$, then the power series $\sum\limits_{n=0}^{\infty} a_n x^n$ diverges.

10.9 REPRESENTING FUNCTIONS AS POWER SERIES

Power series define functions, but how are these power series functions related to functions we know about such as $\sin(x)$, $\cos(x)$, e^x, and $\ln(x)$? How can we represent common functions as power series, and why would we want to do so? The next two sections provide partial answers to these questions. In this section we start with a function defined by a geometric series and show how we can obtain power series representations for several related functions. And we look at a few ways in which power series representations of functions are used. The next section examines a more general method for obtaining power series representations for functions.

The foundation for the examples in this section is a power series whose sum we know. The power series $\sum\limits_{n=0}^{\infty} x^n$ is also a geometric series, with the common ratio $r = x$, and, for $|x| < 1$, we know the sum of the series is $\dfrac{1}{1-x}$.

Geometric Series Formula

For $|x| < 1$, $\quad \sum\limits_{n=0}^{\infty} x^n = 1 + x + x^2 + x^3 + x^4 + ... = \dfrac{1}{1-x}$.

One simple but powerful method of obtaining power series for related functions is to replace each "x" with a function of x.

Substitution in Power Series

Suppose $f(x)$ is defined by a power series

$$f(x) = \sum\limits_{n=0}^{\infty} a_n x^n = a_0 + a_1 x + a_2 x^2 + a_3 x^3 + ... + a_n x^n + ...$$

that converges for $-R < x < R$.

Then $f(x^p) = \sum\limits_{n=0}^{\infty} a_n \{ x^p \}^n = a_0 + a_1 x^p + a_2 \{ x^p \}^2 + a_3 \{ x^p \}^3 + ... + a_n \{ x^p \}^n + ...$

$$= a_0 + a_1 x^p + a_2 x^{2p} + a_3 x^{3p} + ... + a_n x^{np} + ...$$

converges for $-R < x^p < R$, and

$$f(x-c) = \sum\limits_{n=0}^{\infty} a_n (x-c)^n = a_0 + a_1(x-c) + a_2(x-c)^2 + a_3(x-c)^3 + ... + a_n(x-c)^n + ...$$

converges for $-R < x - c < R$.

We can use this substitution method to obtain power series for some functions related to the Geometric Series Formula $\frac{1}{1-x}$.

Example 1:   Find power series for (a) $\frac{1}{1-x^2}$, (b) $\frac{1}{1+x}$, and (c) $\frac{x}{1-x}$.

Solution:

(a) Substituting "x^2" for "x" in the Geometric Series Formula, we get

$$\frac{1}{1-x^2} = 1 + \{x^2\} + \{x^2\}^2 + \{x^2\}^3 + \{x^2\}^4 + \ldots = \sum_{n=0}^{\infty} (x^2)^n$$

$$= 1 + x^2 + x^4 + x^6 + x^8 + \ldots = \sum_{n=0}^{\infty} x^{2n} \quad \text{for } -1 < x < 1.$$

(b) Substituting "$-x$" for "x" in the Geometric Series Formula,

$$\frac{1}{1+x} = \frac{1}{1-(-x)} = 1 + \{-x\} + \{-x\}^2 + \{-x\}^3 + \{-x\}^4 + \ldots = \sum_{n=0}^{\infty} (-x)^n$$

$$= 1 - x + x^2 - x^3 + x^4 + \ldots = \sum_{n=0}^{\infty} (-1)^n x^n \quad \text{for } -1 < x < 1.$$

(c) We need to recognize that $\frac{x}{1-x}$ is a product, $\frac{x}{1-x} = x \cdot \frac{1}{1-x}$. Then

$$x \cdot \frac{1}{1-x} = x \{ 1 + x + x^2 + x^3 + x^4 + \ldots \}$$

$$= x + x^2 + x^3 + x^4 + x^5 + \ldots = \sum_{n=0}^{\infty} x^{n+1} \quad \text{or, equivalently,} \quad \sum_{n=1}^{\infty} x^n \quad \text{for } -1 < x < 1.$$

Practice 1:   Find power series for (a) $\frac{1}{1-x^3}$, (b) $\frac{1}{1+x^2}$, and (c) $\frac{5x}{1+x}$.

One of the features of polynomials that makes them very easy to differentiate and integrate is that we can differentiate and integrate them term–by–term. The same result is true for power series.

Term–by–Term Differentiation and Integration of Power Series

Suppose f(x) is defined by a power series

$$f(x) = \sum_{n=0}^{\infty} a_n x^n = a_0 + a_1 x + a_2 x^2 + a_3 x^3 + \ldots + a_n x^n + \ldots$$

that converges for $-R < x < R$.

Then,

(a) the derivative of f is given by the power series obtained by term–by–term differentiation of f:

$$f'(x) = \sum_{n=1}^{\infty} n \cdot a_n\, x^{n-1} = a_1 + 2a_2 x + 3a_3 x^2 + 4a_4 x^3 + \ldots + n \cdot a_n x^{n-1} + \ldots$$

(b) an antiderivative of f is given by the power series obtained by term–by–term integration of f:

$$\int f(x)\, dx = C + \sum_{n=0}^{\infty} a_n \frac{x^{n+1}}{n+1} = C + a_0 x + a_1 \frac{x^2}{2} + a_2 \frac{x^3}{3} + a_3 \frac{x^4}{4} + \ldots + a_n \frac{x^{n+1}}{n+1} + \ldots$$

The power series for the derivative and antiderivative of f converge for $-R < x < R$.

(The power series for f, f ' and the antiderivative of f may differ in whether or not they converge at the **endpoints** of the interval of convergence, but they all converge for $-R < x < R$.)

The proof of this result is rather long and technical and is omitted.

Like the previous substitution method, term–by–term differentiation and integration can be used to obtain power series for some functions related to the Geometric Series Formula $\frac{1}{1-x}$.

Example 2:   Find power series for (a) $\ln(1-x)$, and (b) $\arctan(x)$.

Solution: These two are more challenging than the previous examples because we need to recognize that these two functions are integrals of functions whose power series we already know.

(a) $\ln(1-x) = \int \frac{-1}{1-x}\, dx = -\int \{ 1 + x + x^2 + x^3 + x^4 + \ldots \}\, dx$

$$= -\{ x + \frac{x^2}{2} + \frac{x^3}{3} + \frac{x^4}{4} + \ldots \} + C = C - \sum_{n=1}^{\infty} \frac{x^n}{n} \ .$$

We can find the value of C by using the fact that $\ln(1) = 0$:

for x = 0, $0 = \ln(1 - 0) = C - \sum_{n=1}^{\infty} \frac{0^n}{n}$ = C so C = 0 and

$$\ln(1 - x) = -\left\{ x + \frac{x^2}{2} + \frac{x^3}{3} + \frac{x^4}{4} + ... \right\} = - \sum_{n=1}^{\infty} \frac{x^n}{n} \quad \text{for } -1 \le x < 1.$$

(b) $\arctan(x) = \int \frac{1}{1 + x^2}\, dx = \int \left\{ 1 - x^2 + x^4 - x^6 + x^8 - .... \right\}\, dx$

$$= C + \left\{ x - \frac{x^3}{3} + \frac{x^5}{5} - \frac{x^7}{7} + \frac{x^9}{9} - ... \right\} = C + \sum_{n=0}^{\infty} (-1)^n \frac{x^{2n+1}}{2n+1} \ .$$

We can find the value of C by using the fact that arctan(0) = 0:

for x = 0, $0 = \arctan(0) = C + \sum_{n=0}^{\infty} (-1)^n \frac{0^{2n+1}}{2n+1}$ = C so C = 0 and

$$\arctan(x) = x - \frac{x^3}{3} + \frac{x^5}{5} - \frac{x^7}{7} + \frac{x^9}{9} - ... = \sum_{n=0}^{\infty} (-1)^n \frac{x^{2n+1}}{2n+1} \quad \text{for } -1 \le x \le 1.$$

Practice 2:   Find a power series for ln(1 + x) .

Power series can also be used to help us evaluate definite integrals.

Example 3:   Use the power series for arctan(x) to represent the definite integral $\int_{0}^{0.5} \arctan(x)\, dx$

as a numerical series. Then approximate the value of the integral by calculating the
sum of the first four terms of the numerical series.

Solution: $\int_{0}^{0.5} \arctan(x)\, dx = \int_{0}^{0.5} \left\{ x - \frac{x^3}{3} + \frac{x^5}{5} - \frac{x^7}{7} + \frac{x^9}{9} - ... \right\}\, dx$

$$= \frac{x^2}{2} - \frac{x^4}{4 \cdot 3} + \frac{x^6}{6 \cdot 5} - \frac{x^8}{8 \cdot 7} + \frac{x^{10}}{10 \cdot 9} - ... \ \Bigg|_{0}^{0.5}$$

$$= \left\{ \frac{1}{2}(0.5)^2 - \frac{1}{12}(0.5)^4 + \frac{1}{30}(0.5)^6 - \frac{1}{56}(0.5)^8 + \frac{1}{90}(0.5)^{10} - ... \right\} - \left\{ 0 \right\} .$$

The sum of the first four terms is approximately 0.120243 . Since the numerical series is an
alternating series, we know that the fourth partial sum, 0.120243 , is within the value of the next term,
$\frac{1}{90}(0.5)^{10} \approx 0.000011$, of the exact value of the sum. The exact value of the definite integral
is between 0.120243 − 0.000011 = 0.120232 and 0.120243 + 0.000011 = 0.120254 .

Practice 3:   Use the power series for $x^2 \cdot \ln(1+x)$ to represent the definite integral

$$\int_0^{0.2} x^2 \cdot \ln(1+x)\, dx$$ as a numerical series. Then approximate the value of the integral by

calculating the sum of the first three terms of the numerical series.

All of the power series used in this section have followed from the Geometric Series Formula, and

their main purpose here was to illustrate some uses of substitution and term–by–term differentiation and

integration to obtain power series for related functions. Many functions, however, are not related to a

geometric series, and the next section discusses a method for representing them using power series. Once

we can represent these new functions using power series, we can then use substitution and term–by–term

differentiation and integration to obtain power series for functions related to them.

The following table collects some of the power series representations we have obtained in this section.

$$\textbf{Table of Series Based on}\quad \frac{1}{1-x} \;=\; \sum_{n=0}^{\infty} x^n$$

$$\frac{1}{1-x} = 1 + x + x^2 + x^3 + x^4 + \ldots = \sum_{n=0}^{\infty} x^n \text{ with interval of convergence } (-1, 1).$$

$$\frac{1}{1+x} = 1 - x + x^2 - x^3 + x^4 + \ldots = \sum_{n=0}^{\infty} (-1)^n x^n = \frac{1}{1-(-x)}$$

$$\frac{1}{1-x^2} = 1 + x^2 + x^4 + x^6 + x^8 + \ldots = \sum_{n=0}^{\infty} x^{2n} = \frac{1}{1-(x^2)}$$

$$\frac{1}{1+x^2} = 1 - x^2 + x^4 - x^6 + x^8 - \ldots = \sum_{n=0}^{\infty} (-1)^n x^{2n} = \frac{1}{1-(-x^2)}$$

$$\frac{1}{1-x^3} = 1 + x^3 + x^6 + x^9 + x^{12} + \ldots = \sum_{n=0}^{\infty} x^{3n} = \frac{1}{1-(x^3)}$$

$$\ln(1-x) = -\left\{ x + \frac{x^2}{2} + \frac{x^3}{3} + \frac{x^4}{4} + \ldots \right\} = -\sum_{n=1}^{\infty} \frac{x^n}{n} = \int \frac{-1}{1-x}\, dx$$

$$\ln(1+x) = x - \frac{x^2}{2} + \frac{x^3}{3} - \frac{x^4}{4} + \ldots = \sum_{n=1}^{\infty} (-1)^{n+1} \frac{x^n}{n} = \int \frac{1}{1+x}\, dx$$

$$\arctan(x) = x - \frac{x^3}{3} + \frac{x^5}{5} - \frac{x^7}{7} + \frac{x^9}{9} - \ldots = \sum_{n=0}^{\infty} (-1)^n \frac{x^{2n+1}}{2n+1} = \int \frac{1}{1+x^2}\, dx$$

$$\frac{1}{(1-x)^2} = 1 + 2x + 3x^2 + 4x^3 + 5x^4 + \ldots = \sum_{n=1}^{\infty} n \cdot x^{n-1} = \mathbf{D}\left(\frac{1}{1-x}\right)$$

$$\frac{1}{(1+x)^2} = 1 - 2x + 3x^2 - 4x^3 + 5x^4 - \ldots = \sum_{n=1}^{\infty} (-1)^n\, n \cdot x^{n-1} = \mathbf{D}\left(\frac{-1}{1+x}\right)$$

Fig. 1 shows the graphs of arctan(x) and the first few polynomials $x, x - \frac{x^3}{3}$, $x - \frac{x^3}{3} + \frac{x^5}{5}$ that

approximate arctan(x). This type of approximation will be discussed a little in Section 10.10 and a lot in

Section 10.11.

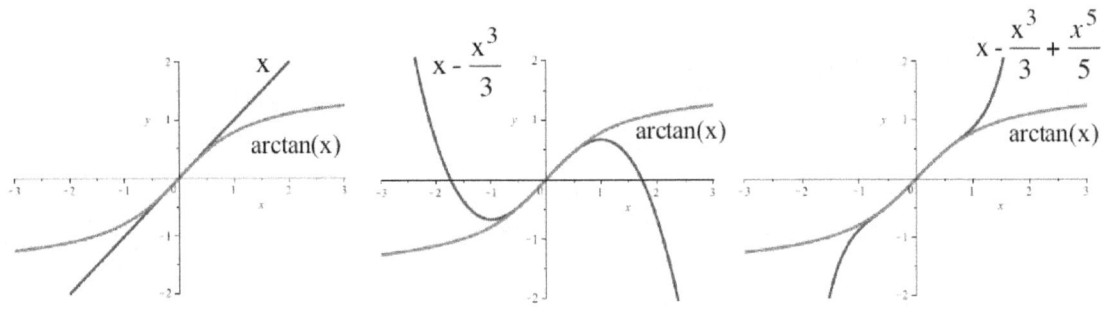

Fig. 1

PROBLEMS

In problems 1 – 14, use the substitution method and a known power series to find power series for the

given functions.

1. $\dfrac{1}{1-x^4}$

2. $\dfrac{1}{1-x^5}$

3. $\dfrac{1}{1+x^4}$

4. $\dfrac{1}{1+x^5}$

5. $\dfrac{1}{5+x} = \dfrac{1}{5} \cdot \dfrac{1}{1+(x/5)}$

6. $\dfrac{1}{3-x} = \dfrac{1}{3} \cdot \dfrac{1}{1-(x/3)}$

7. $\dfrac{x^2}{1+x^3}$

8. $\dfrac{x}{1+x^4}$

9. $\ln(1+x^2)$

10. $\ln(1+x^3)$

11. $\arctan(x^2)$

12. $\arctan(x^3)$

13. $\dfrac{1}{(1-x^2)^2}$

14. $\dfrac{1}{(1+x^2)^2}$

In problems 15–21, represent each definite integral as a numerical series. Calculate the sum of the first three terms for each series.

15. $\displaystyle\int_0^{0.5} \frac{1}{1-x^3}\ dx$

16. $\displaystyle\int_0^{0.5} \frac{1}{1+x^3}\ dx$

17. $\displaystyle\int_0^{0.6} \ln(1+x)\ dx$

18. $\displaystyle\int_0^{0.5} x^2 \cdot \arctan(x)\ dx$

19. $\displaystyle\int_0^{0.3} \frac{1}{(1-x)^2}\ dx$

20. $\displaystyle\int_0^{0.7} \frac{x^3}{(1-x)^2}\ dx$

In problems 21 – 27, use the Table of Series to help represent each function as a power series. Then calculate each limit.

21. $\displaystyle\lim_{x\to 0} \frac{\arctan(x)}{x}$

22. $\displaystyle\lim_{x\to 0} \frac{\ln(1-x)}{2x}$

23. $\displaystyle\lim_{x\to 0} \frac{\ln(1+x)}{2x}$

24. $\displaystyle\lim_{x\to 0} \frac{\arctan(x^2)}{x}$

25. $\displaystyle\lim_{x\to 0} \frac{\ln(1-x^2)}{3x}$

26. $\displaystyle\lim_{x\to 0} \frac{\ln(1+x^2)}{3x}$

In problems 27 – 32, use the Table of Series or the substitution method to determine a power series for each function and then determine the interval of convergence of each power series.

27. $\dfrac{1}{1+x}$

28. $\dfrac{1}{1-x^2}$

29. $\ln(1-x)$

30. $\ln(1+x)$

31. $\arctan(x)$

32. $\arctan(x^2)$

The series given below for $\sin(x)$ and e^x are derived in Section 10.10. Use these given series and the methods of this section to answer problems 33 – 42.

$$\sin(x) = x - \frac{x^3}{3!} + \frac{x^5}{5!} - \frac{x^7}{7!} + \ldots = \sum_{k=0}^{\infty} (-1)^k \frac{x^{2k+1}}{(2k+1)!}\ ,\quad e^x = 1 + x + \frac{x^2}{2!} + \frac{x^3}{3!} + \frac{x^4}{4!} + \ldots = \sum_{k=0}^{\infty} \frac{x^k}{k!}$$

33. Find a power series for $\sin(x^2)$

34. Find a power series for $\sin(2x)$

35. Find a power series for $e^{(-x^2)}$

36. Find a power series for $e^{(2x)}$

37. Find a power series for $\cos(x)$

38. Find a power series for $\cos(x^2)$

39. Represent the integral as a numerical series: $\displaystyle\int_0^1 \sin(x^2)\ dx$

40. Represent the integral as a numerical series: $\displaystyle\int_0^1 e^{(-x^2)}\ dx$

41. $\displaystyle\lim_{x\to 0} \frac{\sin(x)}{x}$

42. $\displaystyle\lim_{x\to 0} \frac{\sin(x)-x}{x^3}$

Practice Answers

Practice 1:   Geometric Series Formula: $\dfrac{1}{1-x} = \sum\limits_{n=0}^{\infty} x^n$ for $|x| < 1$.

(a) Replacing "x" with "x^3" we have $\dfrac{1}{1-x^3} = \sum\limits_{n=0}^{\infty} x^{3n}$ for $|x^3| < 1$ or $|x| < 1$.

(b) Replacing "x" with "$-x^2$" we have $\dfrac{1}{1-(-x^2)} = \dfrac{1}{1+x^2} = \sum\limits_{n=0}^{\infty} (-x^2)^n = \sum\limits_{n=0}^{\infty} (-1)^n x^{2n}$

   for $|-x| < 1$ or $|x| < 1$.

(c) Using the result of part (b), $\dfrac{5x}{1+x} = 5x \dfrac{1}{1+x} = 5x \cdot \sum\limits_{n=0}^{\infty} (-1)^n x^n = 5 \cdot \sum\limits_{n=0}^{\infty} (-1)^n x^{n+1}$

   for $|-x| < 1$ or $|x| < 1$.

Practice 2: $\ln(1+x) = \displaystyle\int \dfrac{1}{1+x}\, dx = \int \sum\limits_{n=0}^{\infty} (-1)^n x^n\, dx = \int 1 - x + x^2 - x^3 + x^4 - \ldots\, dx$

$= x - \dfrac{x^2}{2} + \dfrac{x^3}{3} - \dfrac{x^4}{4} + \dfrac{x^5}{5} - \ldots + C$. Putting $x = 0$, we get $C = 0$. Then

$\ln(1+x) = \sum\limits_{n=0}^{\infty} (-1)^n \dfrac{x^{n+1}}{n+1}$ or, equivalently, $\sum\limits_{n=1}^{\infty} (-1)^{n+1} \dfrac{x^n}{n}$

Practice 3:   Using the result of Practice 2,

$x^2 \cdot \ln(1+x) = x^2 \cdot \sum\limits_{n=1}^{\infty} (-1)^{n+1} \dfrac{x^n}{n} = \sum\limits_{n=1}^{\infty} (-1)^{n+1} \dfrac{x^{n+2}}{n}$. Then

$\displaystyle\int x^2 \cdot \ln(1+x)\, dx = \int \sum\limits_{n=1}^{\infty} (-1)^{n+1} \dfrac{x^{n+2}}{n}\, dx = \sum\limits_{n=1}^{\infty} \dfrac{(-1)^{n+1}}{n} \dfrac{x^{n+3}}{n+3} \Bigg|_0^{0.2}$

$= \left\{ \dfrac{1}{4}(0.2)^4 - \dfrac{1}{10}(0.2)^5 + \dfrac{1}{18}(0.2)^6 - \dfrac{1}{28}(0.2)^7 + \ldots \right\} - \left\{ 0 \right\}$.

$s_3 = \dfrac{1}{4}(0.2)^4 - \dfrac{1}{10}(0.2)^5 + \dfrac{1}{18}(0.2)^6 \approx 0.0003716$ with an error less than $|a_4| = \dfrac{1}{28}(0.2)^7 \approx 4.57 \cdot 10^{-7}$

10.10   TAYLOR AND MACLAURIN SERIES

This section discusses a method for representing a variety of functions as power series, and power series representations are derived for $\sin(x), \cos(x), e^x$, and several functions related to them. These power series are used to evaluate the functions and limits and to approximate definite integrals.

We start with an examination of how to determine the formula for a polynomial from information about the polynomial when $x = 0$, and then this process is extended to determine a series representation for a function from information about the function when $x = 0$.

Polynomials

Polynomials are among the easiest functions to work with, and they have a variety of "nice" properties including the following:

> The values of $P(x)$ and its derivatives at $x = 0$ completely determine the formula for $P(x)$.

If the values of $P(x)$ and all of its derivatives at $x = 0$ are known, then we can use those values to find a formula for $P(x)$.

Example 1:   Suppose $P(x)$ is a cubic polynomial with $P(0) = 7, P'(0) = 5, P''(0) = 16$, and
$P'''(0) = 18$. (Since $P(x)$ is a cubic, its higher derivatives are all 0.) Find a formula for $P(x)$.

Solution: Since $P(x)$ is a cubic polynomial, then $P(x) = a_0 + a_1 x + a_2 x^2 + a_3 x^3$ for some numbers
a_0, a_1, a_2, and a_3. We want to find the values of those numbers, and we can do so by
substituting 0 for x in the expressions for P, P', P'', and P''' and using the given information.

$$7 = P(0) = a_0 + a_1 \cdot 0 + a_2 \cdot 0^2 + a_3 \cdot 0^3 = a_0 \text{ so } a_0 = 7.$$

$$P'(x) = a_1 + 2a_2 x + 3a_3 x^2.\ 5 = P'(0) = a_1 + 2a_2 \cdot 0 + 3a_3 \cdot 0^2 = a_1 \text{ so } a_1 = 5 .$$

$$P''(x) = 2a_2 + 6a_3 x.\ 16 = P''(0) = 2a_2 + 6a_3 \cdot 0 = 2a_2 \text{ so } a_2 = 16/2 = 8 .$$

$$P'''(x) = 6a_3.\ 18 = P''(0) = 6a_3 \text{ so } a_3 = 18/6 = 3 .$$

$P(x) = 7 + 5x + 8x^2 + 3x^3$. You should verify that this cubic polynomial and its derivatives have the values specified in the problem.

Practice 1:   Suppose $P(x)$ is a 4th degree polynomial with $P(0) = -3, P'(0) = 4, P''(0) = 10$,
$P'''(0) = 12$, and $P^{(4)}(0) = 24$. (Since $P(x)$ is a 4th degree polynomial, the higher
derivatives are all 0.) Find a formula for $P(x)$.

For polynomials, the n_{th} derivative evaluated at $x = 0$ is $P^{(n)}(0) = (n)(n-1)(n-2)...(2)(1) a_n = n! \cdot a_n$, so the coefficient a_n of the nth term of the polynomial is $a_n = P^{(n)}(0)/n!$.

Series

In many important ways power series behave like polynomials, very big polynomials, and this is one of those ways. The next result says that if a function can be represented by a power series, then the coefficients of the power series just depend on the values of the derivatives of the function evaluated at 0.

Maclaurin Series for f(x)

If   a function $f(x)$ has a power series representation $f(x) = \sum_{n=0}^{\infty} a_n x^n$ for $|x| < R$

then   the coefficients are given by $a_n = \dfrac{f^{(n)}(0)}{n!}$.

The **Maclaurin Series** for $f(x)$ is

$$f(x) = \sum_{n=0}^{\infty} \frac{f^{(n)}(0)}{n!} x^n = f(0) + f'(0) \cdot x + \frac{f''(0)}{2!} \cdot x^2 + \frac{f'''(0)}{3!} \cdot x^3 + ... + \frac{f^{(n)}(0)}{n!} \cdot x^n + ...$$

Proof: Suppose $f(x) = \sum_{n=0}^{\infty} a_n x^n = a_0 + a_1 x + a_2 x^2 + a_3 x^3 + ... + a_n x^n + ...$ for $|x| < R$.

Then $f(0) = a_0 + a_1 \cdot 0 + a_2 \cdot 0^2 + a_3 \cdot 0^3 + ... + a_n \cdot 0^n + ... = a_0$ so $a_0 = f(0) = \dfrac{f^{(0)}(0)}{0!}$.

(We are using the conventions that $f^{(0)}(x) = f(x)$ and that $0! = 1$.)

$f'(x) = a_1 + 2a_2 x + 3a_3 x^2 + ... + na_n x^{n-1} + ...$

so $f'(0) = a_1 + 2a_2 \cdot 0 + 3a_3 \cdot 0^2 + ... + na_n \cdot 0^{n-1} + ... = a_1$ and $a_1 = \dfrac{f'(0)}{1!}$.

$f''(x) = 2a_2 + 2 \cdot 3a_3 x + ... + (n-1) \cdot n \cdot a_n x^{n-2} + ...$

so $f''(0) = 2a_2 + 2 \cdot 3a_3 \cdot 0 + ... + (n-1) \cdot n \cdot a_n \cdot 0^{n-2} + ... = 2a_2$ and $a_2 = \dfrac{f''(0)}{2} = \dfrac{f''(0)}{2!}$.

$f'''(x) = 2 \cdot 3a_3 + ... + (n-2) \cdot (n-1) \cdot n \cdot a_n x^{n-3} + ...$

so $f'''(0) = 2 \cdot 3a_3 + ... + (n-2) \cdot (n-1) \cdot n \cdot a_n \cdot 0^{n-3} + ... = 2 \cdot 3a_3$ and $a_3 = \dfrac{f'''(0)}{2 \cdot 3} = \dfrac{f'''(0)}{3!}$.

In general, $f^{(n)}(x) = 1 \cdot 2 \cdot 3 \cdot ... \cdot (n-1) \cdot n \, a_n + \{$ terms still containing powers of x $\}$

so $f^{(n)}(0) = 1 \cdot 2 \cdot 3 \cdot ... \cdot (n-1) \cdot n \, a_n + \{ \, 0 \, \} = n! \, a_n$ and $a_n = \dfrac{f^{(n)}(0)}{n!}$.

A similar result, and proof, is also true for a "shifted" power series, a power series centered at some value c. Such shifted series are called Taylor series.

Taylor Series for $f(x)$ centered at c

If   a function $f(x)$ has a power series representation $f(x) = \sum_{n=0}^{\infty} a_n (x - c)^n$ for $|x - c| < R$

then   the coefficients are given by $a_n = \dfrac{f^{(n)}(c)}{n!}$.

The **Taylor Series** for $f(x)$ at c is

$$f(x) = \sum_{n=0}^{\infty} \frac{f^{(n)}(c)}{n!} (x-c)^n = f(c) + f'(c) \cdot (x-c) + \frac{f''(c)}{2!} \cdot (x-c)^2 + \frac{f'''(c)}{3!} \cdot (x-c)^3 + \dots + \frac{f^{(n)}(c)}{n!} \cdot (x-c)^n + \dots$$

The proof is very similar to the proof for Maclaurin series and is not included here.

A Maclaurin series is a Taylor series centered at $c = 0$, and Maclaurin series are a special case of Taylor series.

Note: These statements for Maclaurin series and Taylor series do not say that every function is or can be written as a power series. However, if a function is a power series, then its coefficients must follow the given pattern. Fortunately, most of the important functions such as $\sin(x)$, $\cos(x)$, e^x, and $\ln(x)$ can be written as power series.

You should notice that the first term of the Taylor series is simply the value of the function f at the point $x - c$: it provides the best constant function approximation of f near $x = c$. The sum of the first two terms of the Taylor series pattern for a function, $f(c) + f'(c) \cdot (x-c)$, is the formula for the tangent line to f at $x = c$ and is the linear approximation of $f(x)$ near $x = c$ that we first examined in Chapter 2. The Taylor series formula extends these approximations to higher degree polynomials, and the partial sums of the Taylor series provide higher degree polynomial approximations of f near $x = c$.

Taylor series and Maclaurin series were first discovered by the Scottish mathematician/astronomer James Gregory (1638–1675), but the results were not published until after his death. The English mathematician Brook Taylor (1685–1731) independently rediscovered these results and included them in a book in 1715. The Scottish mathematician/engineer Colin Maclaurin (1698–1746) quoted Taylor's work in his *Treatise on Fluxions* published in 1742. Maclaurin's book was widely read, and the Taylor series centered at $c = 0$ became known as Maclaurin series.

Example 2:   Find the Maclaurin series for $f(x) = \sin(x)$ and the radius of convergence of the series.

Solution: $f(x) = \sin(x)$ so $f(0) = \sin(0)$ and $a_0 = f(0) = 0$.

$f'(x) = \cos(x)$ so $f'(0) = \cos(0) = 1$ and $a_1 = f'(0) = 1$.

$f''(x) = -\sin(x)$ so $f''(0) = -\sin(0) = 0$ and $a_2 = \dfrac{f''(0)}{2!} = 0$.

$f'''(x) = -\cos(x)$ so $f'''(0) = -\cos(0) = -1$ and $a_3 = \dfrac{f'''(0)}{3!} = \dfrac{-1}{3!}$.

$f^{(4)}(x) = \sin(x)$ and the pattern repeats:

$a_4 = 0, a_5 = \dfrac{1}{5!}$, $a_6 = 0$, $a_7 = \dfrac{-1}{7!}$, $a_8 = 0$, $a_9 = \dfrac{1}{9!}$, ...

$$\sin(x) = x - \frac{x^3}{3!} + \frac{x^5}{5!} - \frac{x^7}{7!} + \frac{x^9}{9!} - \frac{x^{11}}{11!} + \ldots = \sum_{n=0}^{\infty} (-1)^n \frac{x^{2n+1}}{(2n+1)!} .$$

Notice that the Maclaurin series for $\sin(x)$ alternates and contains only odd powers of x.

We use the Ratio Test to find the radius of convergence. Let $c_n = (-1)^n \dfrac{x^{2n+1}}{(2n+1)!}$. Then

$$c_{n+1} = (-1)^{n+1} \frac{x^{2(n+1)+1}}{(2(n+1)+1)!} = (-1)^{n+1} \frac{x^{2n+3}}{(2n+3)!} \quad \text{so}$$

$$\left| \frac{c_{n+1}}{c_n} \right| = \left| \frac{(-1)^{n+1} \dfrac{x^{2n+3}}{(2n+3)!}}{(-1)^n \dfrac{x^{2n+1}}{(2n+1)!}} \right| = \left| \frac{x^{2n+3}}{x^{2n+1}} \frac{(2n+1)!}{(2n+3)!} \right| = \left| x^2 \frac{1}{(2n+2)(2n+3)} \right| \to 0 < 1$$

for every value of x.

The radius of convergence is $R = \infty$, and the interval of convergence is $(-\infty, \infty)$.

The Maclaurin series for $\sin(x)$ converges for every value of x.

Fig. 1 shows the graphs of $\sin(x)$ and the first few approximating polynomials $x, x - \dfrac{x^3}{3!}$, and

$x - \dfrac{x^3}{3!} + \dfrac{x^5}{5!}$ for $-\pi \leq x \leq \pi$.

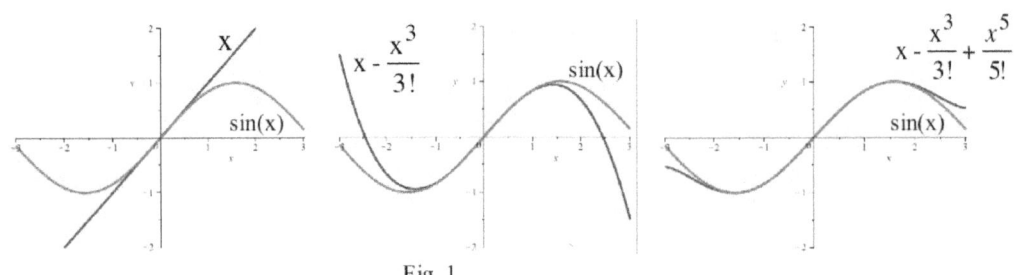

Fig. 1

By focusing our attention near x = 0, Fig. 1 shows the "goodness" of the Taylor polynomial fit to the function f(x) = sin(x). However, Fig. 2 shows that if x is not close to 0 then the values of the Taylor polynomials are far from the values of f(x) = sin(x). Typically the Taylor polynomials of a function are closest to the function when x is close to the number at which the series was centered, the value of c.

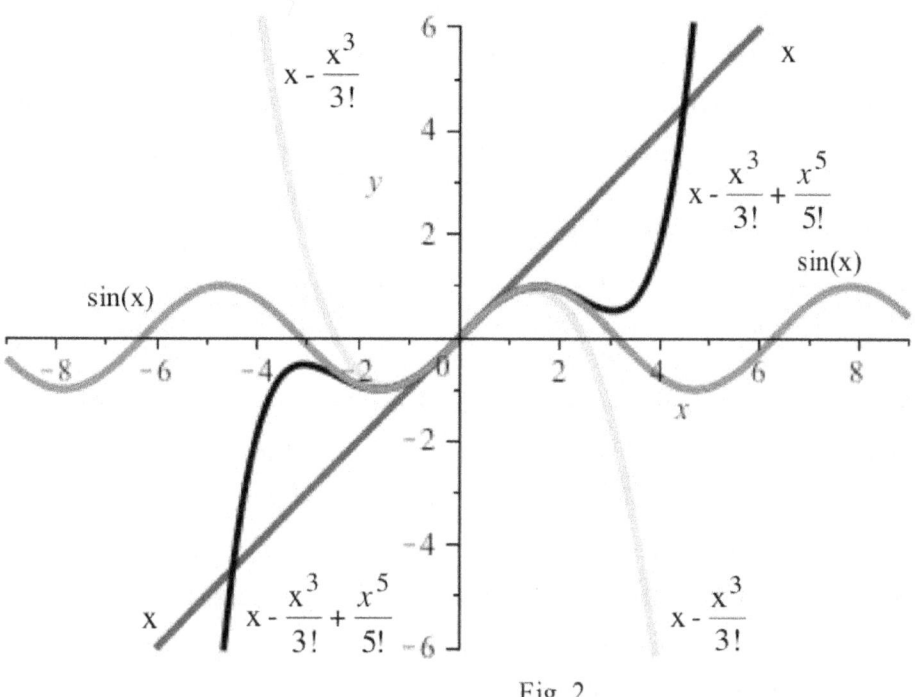

Fig. 2

Practice 2:   Find the Maclaurin series for f(x) = cos(x) and the radius of convergence of the series. (Suggestion: Use the Term–by–Term Differentiation result of Section 10.9 .)

Once we have power series representations for sin(x) and cos(x), we can use the methods of Section 10.9 and the known series to approximate values of sine and cosine, determine power series representations of related functions, calculate limits, and approximate definite integrals.

Example 3:   Use the series $\sin(x) = x - \frac{x^3}{3!} + \frac{x^5}{5!} - \frac{x^7}{7!} + \frac{x^9}{9!} - \frac{x^{11}}{11!} + \dots$ to represent sin(0.5) as a numerical series. Approximate the value of sin(0.5) by calculating the partial sum of the first three non–zero terms and give a bound on the "error" between this approximation and the exact value of sin(0.5).

Solution: $\sin(0.5) = (0.5) - \frac{(0.5)^3}{3!} + \frac{(0.5)^5}{5!} - \frac{(0.5)^7}{7!} + \frac{(0.5)^9}{9!} - \frac{(0.5)^{11}}{11!} + \dots$.

$\sin(0.5) \approx (0.5) - \frac{(0.5)^3}{3!} + \frac{(0.5)^5}{5!} = \frac{1}{2} - \frac{1}{48} + \frac{1}{3840} \approx 0.479427083333$.

Since this is an alternating series, the difference between the approximation and the exact value is

less than the next term in the alternating series: "error" $< \dfrac{(0.5)^7}{7!} = \dfrac{1}{645120} \approx 0.00000155$.

If we use the sum of the first four nonzero terms to approximate the value of sin(0.5), then the

"error" of the approximation is less than $\dfrac{(0.5)^9}{9!} = \dfrac{1}{185794560} \approx 5.4 \times 10^{-9}$.

We were able to obtain a bound for the error in the approximation of sin(0.5) because we were dealing

with an alternating series, a type of series for which we have an error bound. However, many power

series are not alternating series. In Section 10.11 we discuss a general error bound for Taylor series.

Practice 3:   Use the sum of the first two nonzero terms of the Maclaurin series for cos(x) to

approximate the value of cos(0.2). Give a bound on the "error" between this

approximation and the exact value of cos(0.2).

Calculator Note: When you press the buttons on a calculator to evaluate sin(0.5) or cos(0.2), the

calculator does not look up the answer in a table. Instead, the calculator is programmed with series

representations for sine and cosine and other functions, and it calculates a partial sum of the

appropriate series to obtain a numerical answer. It adds enough terms so that the 8 or 9 digits shown

on the display are (usually) correct. In Section 10.11 we examine these methods in more detail and

consider how to determine the number of terms needed in the partial sum to achieve the desired

number of accurate digits in the answer.

Example 4:   Represent $\sin(x^3)$ and $\int \sin(x^3)\, dx$ as power series. Then use the first three nonzero

terms to approximate the value of $\displaystyle\int_0^1 \sin(x^3)\, dx$ and obtain a bound on the "error" of

this approximation.

Solution: $\sin(x) = x - \dfrac{x^3}{3!} + \dfrac{x^5}{5!} - \dfrac{x^7}{7!} + \dfrac{x^9}{9!} - \dfrac{x^{11}}{11!} + \ldots$ so

$$\sin(x^3) = (x^3) - \dfrac{(x^3)^3}{3!} + \dfrac{(x^3)^5}{5!} - \dfrac{(x^3)^7}{7!} \ldots = x^3 - \dfrac{x^9}{3!} + \dfrac{x^{15}}{5!} - \dfrac{x^{21}}{7!} \ldots .$$

$$\int_0^1 \sin(x^3)\, dx = \int_0^1 \left\{ x^3 - \dfrac{x^9}{3!} + \dfrac{x^{15}}{5!} - \dfrac{x^{21}}{7!} \ldots \right\} dx = \dfrac{x^4}{4} - \dfrac{x^{10}}{10 \cdot 3!} + \dfrac{x^{16}}{16 \cdot 5!} - \dfrac{x^{22}}{22 \cdot 7!} + \ldots \Bigg|_0^1$$

$$= \left\{ \dfrac{1}{4} - \dfrac{1}{10 \cdot 3!} + \dfrac{1}{16 \cdot 5!} - \dfrac{1}{22 \cdot 7!} + \ldots \right\} - \{ 0 \} .$$

$$\frac{1}{4} - \frac{1}{10 \cdot 3!} + \frac{1}{16 \cdot 5!} \approx 0.2338542 \text{ and } \frac{1}{22 \cdot 7!} = \frac{1}{110880} \approx 0.0000090 \text{ so}$$

$$\int_0^1 \sin(x^3) \ dx \approx 0.2338542 \text{ and this approximation is within } 0.0000090 \text{ of the exact value.}$$

If we took just one more term, $\displaystyle\int_0^1 \sin(x^3) \ dx \approx \frac{1}{4} - \frac{1}{10 \cdot 3!} + \frac{1}{16 \cdot 5!} - \frac{1}{22 \cdot 7!} \approx 0.233845515$

is within $\dfrac{1}{28 \cdot 9!} \approx 0.000000098$ of the exact value of the integral.

Practice 4:   Represent $x \cdot \cos(x^3)$ and $\displaystyle\int x \cdot \cos(x^3) \ dx$ as power series. Then use the first two

nonzero terms to approximate the value of $\displaystyle\int_0^{1/2} x \cdot \cos(x^3) \ dx$ and obtain a bound on the

"error" of this approximation.

Graphically

Each partial sum of the series $\sin(x) = x - \dfrac{x^3}{3!} + \dfrac{x^5}{5!} - \dfrac{x^7}{7!} + \dfrac{x^9}{9!} - \dfrac{x^{11}}{11!} + \ldots$ contains a finite number

of terms and is simply a polynomial:

$$P_1(x) = x$$

$$P_3(x) = x - \frac{x^3}{3!} = x - \frac{1}{6} x^3$$

$$P_5(x) = x - \frac{x^3}{3!} + \frac{x^5}{5!} = x - \frac{1}{6} x^3 + \frac{1}{120} x^5 , \ \ldots .$$

Fig. 1 showed the graphs of $\sin(x)$ and $P_1(x), P_3(x)$, and $P_5(x)$. As you saw, all of these polynomials

are "good" approximations of $\sin(x)$ when x is very close to 0. The higher degree polynomials $P_n(x)$

provide "good" approximations of $\sin(x)$ over larger intervals.

Power Series for e^x

Example 5: Find the Maclaurin series for $f(x) = e^x$ and the radius of convergence of the series.

Solution: This is a very important series.

$f(x) = e^x$ so $f(0) = e^0 = 1$ and $a_0 = f(0) = 1$.

$f'(x) = e^x$ so $f'(0) = e^0 = 1$ and $a_1 = f'(0) = 1$.

$f''(x) = e^x$ so $f''(0) = e^0 = 1$ and $a_2 = \dfrac{f''(0)}{2!} = \dfrac{1}{2!}$.

For every value of n, $f^{(n)}(x) = e^x$ so $f^{(n)}(0) = e^0 = 1$ and $a_n = \dfrac{f''(0)}{n!} = \dfrac{1}{n!}$. Then

$$e^x = 1 + x + \frac{x^2}{2!} + \frac{x^3}{3!} + \frac{x^4}{4!} + \frac{x^5}{5!} + \dots = \sum_{n=0}^{\infty} \frac{x^n}{n!} \ .$$

We can use the Ratio Test to find the radius of convergence. $c_n = \dfrac{x^n}{n!}$ so $c_{n+1} = \dfrac{x^{n+1}}{(n+1)!}$.

$$\left| \frac{c_{n+1}}{c_n} \right| = \left| \frac{\dfrac{x^{n+1}}{(n+1)!}}{\dfrac{x^n}{n!}} \right| = \left| \frac{x^{n+1}}{x^n} \frac{n!}{(n+1)!} \right| = \left| x \cdot \frac{1}{n+1} \right| \ \to\ 0 < 1 \text{ for every value of } x.$$

The radius of convergence is $R = \infty$, and the interval of convergence is $(-\infty, \infty)$. The Maclaurin series for e^x converges for every value of x.

Practice 5: Evaluate the partial sums of the first six terms of the numerical series for $e = e^1$ and $\dfrac{1}{\sqrt{e}} = e^{-1/2}$ and compare these partial sums with the values your calculator gives.

(Note: The numerical series for e^1 is not an alternating series so we do not have a bound for the approximation yet. We will in the next section.)

Fig. 3 shows the graphs of e^x and the approximating polynomials $1 + x$, $1 + x + \dfrac{x^2}{2!}$, and $1 + x + \dfrac{x^2}{2!} + \dfrac{x^3}{3!}$ for values of near 0.

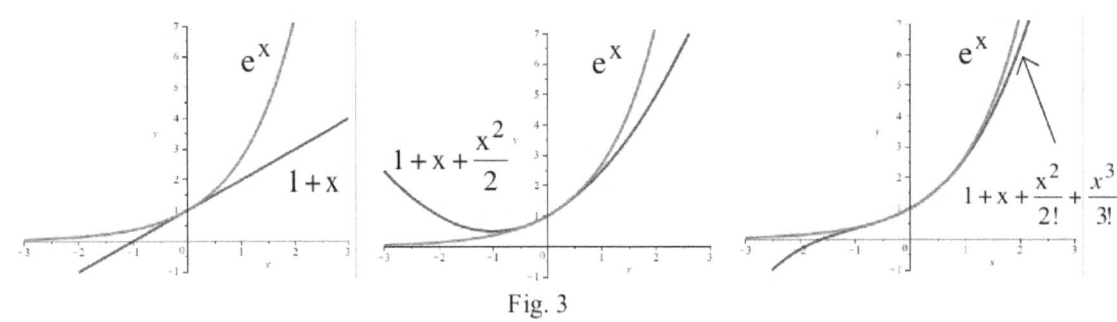

Fig. 3

The following series converge for all values of x in the interval I:

$$\sin(x) = x - \frac{x^3}{3!} + \frac{x^5}{5!} - \frac{x^7}{7!} + \frac{x^9}{9!} - \frac{x^{11}}{11!} + \ldots = \sum_{n=0}^{\infty} (-1)^n \frac{x^{2n+1}}{(2n+1)!} \qquad I = (-\infty, \infty).$$

$$\cos(x) = 1 - \frac{x^2}{2!} + \frac{x^4}{4!} - \frac{x^6}{6!} + \frac{x^8}{8!} - \frac{x^{10}}{10!} + \ldots = \sum_{n=0}^{\infty} (-1)^n \frac{x^{2n}}{(2n)!} \qquad I = (-\infty, \infty).$$

$$e^x = 1 + x + \frac{x^2}{2!} + \frac{x^3}{3!} + \frac{x^4}{4!} + \frac{x^5}{5!} + \ldots = \sum_{n=0}^{\infty} \frac{x^n}{n!} \qquad I = (-\infty, \infty).$$

$$\ln(x) = (x-1) - \frac{1}{2}(x-1)^2 + \frac{1}{3}(x-1)^3 - \frac{1}{4}(x-1)^4 + \ldots = \sum_{n=1}^{\infty} (-1)^{n+1} \frac{(x-1)^n}{n} \qquad I = (0, 2].$$

Typically, these series converge very quickly to the value of the functions when x is close to 0 (or when x is close to 1 for ln(x)), but the convergence can be rather slow when x is far from 0. For example, the first 2 terms of the Taylor series for sine approximate sin(0.1) correctly to 6 decimal places, but 11 terms are needed to approximate sin(5) with the same accuracy.

Multiplying Power Series

We can add and subtract power series term–by–term, and we have already multiplied power series by single terms such as x and x^2 , but occasionally it is useful to multiply a power series by another power series. The method for multiplying series is the same method we use to multiply a polynomial by another polynomial, but it becomes very tedious to get more than the first few terms of the resulting product.

Example 6:   Find the first 5 nonzero terms of

$$\frac{1}{1-x} \cdot \sin(x) = (1 + x + x^2 + x^3 + \ldots) \cdot \left(x - \frac{x^3}{6} + \frac{x^5}{120} - \ldots\right).$$

Solution: 1 $+ x$ $+ x^2$ $+ x^3$ $+ x^4$ $+ x^5$ $+ ...$

times x $-\dfrac{x^3}{6}$ $+\dfrac{x^5}{120}$ $- ...$

---

x $+ x^2$ $+ x^3$ $+ x^4$ $+ x^5$ $+ ....$ (from multiplying by x)

$-\dfrac{1}{6} x^3$ $-\dfrac{1}{6} x^4$ $-\dfrac{1}{6} x^5$ $+ ...$ (from multiplying by $-x^3/6$)

$+\dfrac{1}{120} x^5$ $+ ...$ (from multiplying by $x^5/120$)

$=$ x $+ x^2$ $+\dfrac{5}{6} x^3$ $+\dfrac{5}{6} x^4$ $+\dfrac{101}{120} x^5$ $+ ....$ (from adding previous terms)

Practice 6:   Find the first 3 nonzero terms of

$$e^x \cdot \sin(x) = \left(1 + x + \frac{x^2}{2} + \frac{x^3}{6} + ... \right) \cdot \left(x - \frac{x^3}{6} + \frac{x^5}{120} - ... \right) .$$

It is also possible to divide one power series by another power series using a procedure similar to "long division" of a polynomial by a polynomial, but we will not discuss that algorithm.

PROBLEMS

1. Find a formula for a polynomial P of degree 2 such that $P(0) = 7$, $P'(0) = -5$, and $P''(0) = 8$.

2. Find a formula for a polynomial P of degree 3 such that $P(0) = -1$, $P'(0) = 2$, $P''(0) = -5$, and $P'''(0) = 12$.

3. Find a formula for a polynomial P of degree 2 such that $P(3) = -2$, $P'(3) = 5$, and $P''(3) = 4$.

4. Find a formula for a polynomial P of degree 2 such that $P(1) = -2$, $P'(1) = 5$, and $P''(1) = 4$.

In problems 5 – 8, calculate the first several terms of the Maclaurin series for the given functions and compare with the series representations we found in Section 10.9 . (The series should be the same.)

5. $\ln(1 + x)$ to the x^4 term   6. $\ln(1 - x)$ to the x^4 term

7. $\arctan(x)$ to the x^3 term   8. $1/(1 - x)$ to the x^4 term

In problems 9 – 12, calculate the first several terms of the Maclaurin series for the given functions.

9. $\cos(x)$ to the x^6 term   10. $\tan(x)$ to the x^5 term

11. $\sec(x)$ to the x^4 term   12. e^{3x} to the x^4 term

In problems 13 - 18, calculate the first several terms of the Taylor series for the given functions at the given point c.

13. $\ln(x)$ for $c = 1$ 14. $\sin(x)$ for $c = \pi$

15. $\sin(x)$ for $c = \pi/2$ 16. $\sqrt{x}$ for $c = 1$

17. $\sqrt{x}$ for $c = 9$

In problems 18 – 21, use the first three nonzero terms of the Maclaurin series for each function to approximate the numerical values. Then compare the Maclaurin series approximation with the value your calculator gives.

18. $\sin(0.1)$, $\sin(0.2)$, $\sin(0.5)$, $\sin(1)$, and $\sin(2)$

19. $\cos(0.1)$, $\cos(0.2)$, $\cos(0.5)$, $\cos(1)$, and $\cos(2)$

20. $\ln(1.1)$, $\ln(1.2)$, $\ln(1.3)$, $\ln(2)$, and $\ln(3)$

21. $\arctan(0.1)$, $\arctan(0.2)$, $\arctan(0.5)$, $\arctan(1)$, and $\arctan(2)$

In problems 22 – 27, calculate the first three nonzero terms of the power series for each of the integrals.

22. $\int \cos(x^2)\,dx$ and $\int \cos(x^3)\,dx$ 23. $\int \sin(x^2)\,dx$ and $\int \sin(x^3)\,dx$

24. $\int e^{(x^2)}\,dx$ and $\int e^{(x^3)}\,dx$ 25. $\int e^{(-x^2)}\,dx$ and $\int e^{(-x^3)}\,dx$

26. $\int \ln(x)\,dx$ and $\int x\ln(x)\,dx$ 27. $\int x\sin(x)\,dx$ and $\int x^2\sin(x)\,dx$

In problems 28 – 35, use the series representation of these functions to calculate the limits.

28. $\displaystyle\lim_{x\to 0} \frac{1 - \cos(x)}{x}$ 29. $\displaystyle\lim_{x\to 0} \frac{1 - \cos(x)}{x^2}$ 30. $\displaystyle\lim_{x\to 0} \frac{\ln(x)}{x - 1}$

31. $\displaystyle\lim_{x\to 0} \frac{1 - e^x}{x}$ 32. $\displaystyle\lim_{x\to 0} \frac{1 + x - e^x}{x^2}$ 33. $\displaystyle\lim_{x\to 0} \frac{\sin(x)}{x}$

34. $\displaystyle\lim_{x\to 0} \frac{x - \sin(x)}{x^3}$ 35. $\displaystyle\lim_{x\to 0} \frac{x - \dfrac{x^3}{6} - \sin(x)}{x^5}$

36. Use the series for e^x and e^{-x} to write a series representation for $\cosh(x) = \frac{1}{2}(e^x + e^{-x})$.

   (The function "cosh" is called the hyperbolic cosine function.)

37. Use the series for e^x and e^{-x} to write a series representation for $\sinh(x) = \frac{1}{2}(e^x - e^{-x})$.

   (The function "sinh" is called the hyperbolic sine function.)

38. Show that **D**(series for cosh(x) in problem 36) is the series for sinh(x) in problem 37.

39. Show that **D**(series for sinh(x) in problem 37) is the series for cosh(x) in problem 36.

Euler's Formula

So far we have only discussed series with real numbers, but sometimes it is useful to replace the variable with complex numbers. The next problems ask you to make such a substitution and then to derive and use one of the most famous formulas in mathematics, Euler's formula. (Recall that $i = \sqrt{-1}$ is called the complex unit and that its powers follow the pattern $i^2 = -1$, $i^3 = (i^2)(i) = -i$, $i^4 = (i^2)(i^2) = 1$, $i^5 = (i^4)(i) = i$, ...)

40. Start with the series $e^x = 1 + x + \dfrac{x^2}{2!} + \dfrac{x^3}{3!} + \dfrac{x^4}{4!} + \dfrac{x^5}{5!} + \dfrac{x^6}{6!} + \dfrac{x^7}{7!} + \dfrac{x^8}{8!} + \ldots$

    (a) Substitute "ix" for "x" and write a series for e^{ix}.

    (b) In part (a) simplify each power of i and write a simplified series for e^{ix}. (e.g., $(ix)^3/3!$ simplifies to $i^3 x^3/3! = -i \cdot x^3/3!$)

    (c) Sort the terms in the series in part (b) into those terms that do not contain i and those terms that do contain i. Then rewrite the series for e^{ix} in the form
    $e^{ix} = \{$ terms that did not contain i $\} + i \cdot \{$ terms that did contain i $\}$.

    (d) You should recognize the sum in each bracket in part (c) as the series for an elementary function (hint: think trigonometry). Rewrite the pattern in part (c) as
    $e^{ix} = \{$ function $\} + i \cdot \{$ another function $\}$.

41. The answer you should have gotten in problem 40d, $e^{ix} = \cos(x) + i \cdot \sin(x)$, is called Euler's formula. Use Euler's formula to calculate the values of $e^{i(\pi/2)}$ and $e^{\pi i}$.

42. Use Euler's formula to show that $e^{\pi i} + 1 = 0$. This is one of the most remarkable formulas in mathematics because it connects five of the most fundamental constants (the additive identity 0, the multiplicative identity 1, the complex unit i, and the two most commonly used irrational numbers π and e in a simple but non-obvious way.

Binomial Series

You have probably seen the pattern for expanding $(1 + x)^n$

where n is a nonnegative integer:

$(1 + x)^0 = 1$

$(1 + x)^1 = 1 + x$

$(1 + x)^2 = 1 + 2x + x^2$

$(1 + x)^3 = 1 + 3x + 3x^2 + x^3$

$(1 + x)^4 = 1 + 4x + 6x^2 + 4x^3 + x^4$

$(1 + x)^5 = 1 + 5x + 10x^2 + 10x^3 + 5x^4 + x^5$

Row	Pascal's Triangle
0 · · · · · · · · · · · · · 1
1 · · · · · · · · · · 1   1
2 · · · · · · · · · 1   2   1
3 · · · · · · · · 1   3   3   1
4 · · · · · · 1   4   6   4   1
5 · · · · · 1   5   10   10   5   1
6 · · · 1   6   15   20   15   6   1

Each number in Pascal's Triangle is the sum of the two numbers closest to it in the row immediately above it.

Fig. 4

either using Pascal's triangle (Fig. 4) or using the binomial coefficients, written $\binom{n}{k}$ and defined as

$$\binom{n}{0} = 1 \quad \text{and} \quad \binom{n}{k} = \frac{n(n-1)(n-2) \cdots (n-k+1)}{k!} = \frac{n!}{k! \, (n-k)!} \quad .$$

43. Calculate the binomial coefficients $\binom{3}{0}$, $\binom{3}{1}$, $\binom{3}{2}$, and $\binom{3}{3}$ and verify that

    (i)   they agree with the entries in the 3rd row of Pascal's triangle

    (ii) they agree with the coefficients of the terms of $(1 + x)^3$.

44. Calculate the binomial coefficients $\binom{4}{0}$, $\binom{4}{1}$, $\binom{4}{2}$, $\binom{4}{3}$, and $\binom{4}{4}$ and verify that

    (i)   they agree with the entries in the 4th row of Pascal's triangle

    (ii) they agree with the coefficients of the terms of $(1 + x)^4$.

Using binomial coefficients, the pattern for nonnegative integer powers of $(1 + x)$ can be described in a very compact way:

$$(1 + x)^n = \sum_{k=0}^{n} \binom{n}{k} x^k .$$

When n is a positive integer, $(1 + x)^n$ expands to be a polynomial of degree n.

But what happens when n is a negative integer or perhaps not even an integer? This was a question that Newton himself investigated, and it led him to a general pattern, called the Binomial Series Theorem, for $(1 + x)^m$ when m is any real number. And now you can do it, too.

45. Let $f(x) = (1 + x)^{5/2}$ and determine the first 5 terms of the Maclaurin series for $f(x)$.

46. Let $f(x) = (1 + x)^{-3/2}$ and determine the first 5 terms of the Maclaurin series for $f(x)$.

47. Let $f(x) = (1 + x)^m$ and determine the first 4 terms of the Maclaurin series for $f(x)$. This is the start of the derivation of the Binomial Series Theorem given below.

Binomial Series Theorem

If   m is any real number and $|x| < 1$

then $(1 + x)^m = 1 + mx + \dfrac{m(m-1)}{2} x^2 + \dfrac{m(m-1)(m-2)}{3!} x^3 + ... = \displaystyle\sum_{k=0}^{\infty} \binom{m}{k} x^k$

where $\binom{m}{k} = \dfrac{m(m-1)(m-2) \cdots (m-k+1)}{k!}$ (for $k \geq 1$) and $\binom{m}{0} = 1$.

---

48. Use the Ratio Test to show that $\displaystyle\sum_{k=0}^{\infty} \binom{m}{k} x^k$ converges for $|x| < 1$.

Practice Answers

Practice 1: $P(x) = a_0 + a_1x + a_2x^2 + a_3x^3 + a_4x^4$ with $P(0) = -3, P'(0) = 4, P''(0) = 10$,

$P'''(0) = 12$, and $P^{(4)}(0) = 24$.

$-3 = P(0) = a_0 + a_1{\cdot}0 + a_2{\cdot}0 + a_3{\cdot}0 + a_4{\cdot}0 = a_0$ so $a_0 = -3$

$P'(x) = a_1 + 2a_2x + 3a_3x^2 + 4a_4x^3$

$4 = P'(0) = a_1 + 2a_2{\cdot}0 + 3a_3{\cdot}0 + 4a_4{\cdot}0 = a_1$ so $a_1 = 4$

$P''(x) = 2a_2 + 6a_3x + 12a_4x^2$

$10 = P''(0) = 2a_2 + 6a_3{\cdot}0 + 12a_4{\cdot}0 = 2a_2$ so $a_2 = 10/2 = 5$

$P'''(x) = 6a_3 + 24a_4x$

$12 = P'''(0) = 6a_3 + 24a_4{\cdot}0 = 6a_3$ so $a_3 = 12/6 = 2$

$P^{(4)}(x) = 24a_4$

$24 = P^{(4)}(0) = 24a_4$ so $a_4 = 24/24 = 1$.

Then $P(x) = -3 + 4x + 5x^2 + 2x^3 + 1x^4$.

Practice 2: $\cos(x) = \mathbf{D}(\sin(x)) = \mathbf{D}\left(x - \frac{x^3}{3!} + \frac{x^5}{5!} - \frac{x^7}{7!} + \frac{x^9}{9!} - \frac{x^{11}}{11!} + \ldots\right)$

$$= 1 - 3\cdot\frac{x^2}{3!} + 5\cdot\frac{x^4}{5!} - 7\cdot\frac{x^6}{7!} + 9\cdot\frac{x^8}{9!} - 11\cdot\frac{x^{10}}{11!} + \ldots$$

$$= 1 - \frac{x^2}{2!} + \frac{x^4}{4!} - \frac{x^6}{6!} + \frac{x^8}{8!} - \frac{x^{10}}{10!} + \ldots = \sum_{n=0}^{\infty}(-1)^n\frac{x^{2n}}{(2n)!}$$

Practice 3: $\cos(x) = 1 - \frac{x^2}{2!} + \frac{x^4}{4!} - \frac{x^6}{6!} + \frac{x^8}{8!} - \frac{x^{10}}{10!} + \ldots$

Using the first two nonzero terms, $\cos(0.2) \approx 1 - \frac{(0.2)^2}{2!} = 1 - \frac{0.04}{2} = 0.98$.

Since $\cos(0.2) = 1 - \frac{(0.2)^2}{2!} + \frac{(0.2)^4}{4!} - \frac{(0.2)^6}{6!} + \ldots$ is a convergent alternating series, the error

is less than the absolute value of the next term.

Then $\cos(0.2) \approx 1 - \frac{(0.2)^2}{2!} = 0.98$ with an error less than $\left|\frac{(0.2)^4}{4!}\right| = \frac{0.0016}{24} \approx 0.000067$:

$|\cos(0.2) - 0.98| < 0.000067$. (In fact, $\cos(0.2) \approx 0.9800665778$.)

Practice 4: $\cos(x) = 1 - \frac{x^2}{2!} + \frac{x^4}{4!} - \frac{x^6}{6!} + \frac{x^8}{8!} - \frac{x^{10}}{10!} + \ldots$

$\cos(x^3) = 1 - \frac{(x^3)^2}{2!} + \frac{(x^3)^4}{4!} - \frac{(x^3)^6}{6!} + \frac{(x^3)^8}{8!} - \ldots = 1 - \frac{x^6}{2!} + \frac{x^{12}}{4!} - \frac{x^{18}}{6!} + \frac{x^{24}}{8!} - \frac{x^{30}}{10!} + \ldots$

$x\cdot\cos(x^3) = x - \frac{x^7}{2!} + \frac{x^{13}}{4!} - \frac{x^{19}}{6!} + \frac{x^{25}}{8!} - \frac{x^{31}}{10!} + \ldots$

$\int x\cdot\cos(x^3)\,dx = \frac{x^2}{2} - \frac{1}{8}\cdot\frac{x^8}{2!} + \frac{1}{14}\cdot\frac{x^{14}}{4!} - \frac{1}{20}\cdot\frac{x^{20}}{6!} + \frac{1}{26}\cdot\frac{x^{26}}{8!} - \frac{1}{32}\cdot\frac{x^{32}}{10!} + \ldots + C$

$\int_{0}^{1/2} x\cdot\cos(x^3)\,dx \approx \left.\frac{x^2}{2} - \frac{1}{8}\cdot\frac{x^8}{2!}\right|_{0}^{1/2} = \left(\frac{(0.5)^2}{2} - \frac{(0.5)^8}{2!8}\right) - (0) \approx 0.124755859$

with $|\text{error}| \le \frac{(0.5)^{14}}{4!14} \approx 1.82\cdot10^{-7} = 0.000000182$.

Practice 5: $e^x = 1 + x + \frac{x^2}{2!} + \frac{x^3}{3!} + \frac{x^4}{4!} + \frac{x^5}{5!} + \ldots$. Using the first six terms,

$e^1 \approx 1 + 1 + \frac{1}{2!} + \frac{1}{3!} + \frac{1}{4!} + \frac{1}{5!} \approx 2.71666666666$ (My calculator gives $e^1 \approx 2.718281828$)

$e^{-1/2} \approx 1 + (-1/2) + \frac{(-1/2)^2}{2!} + \frac{(-1/2)^3}{3!} + \frac{(-1/2)^4}{4!} + \frac{(-1/2)^5}{5!} \approx 0.6065104167$

(My calculator gives $e^{-1/2} \approx 0.6065306597$).

Practice 6: $1 +$ $x +$ $\dfrac{x^2}{2} +$ $\dfrac{x^3}{6} +$ ... $(= e^x)$

times $x -$ $\dfrac{x^3}{6} +$ $\dfrac{x^5}{120} -$ ... $(= \sin(x))$

$\rule{450px}{1px}$

                 $x +$ $x^2 +$ $\dfrac{x^3}{2} +$ $\dfrac{x^4}{6} +$ ...   (from multiplying by x)

                         $-\dfrac{x^3}{6} -$ $\dfrac{x^4}{6} -$ $\dfrac{x^5}{12} -$ ...   (from multiplying by $-x^3/6$)

                                 $\dfrac{x^5}{120} +$ ...   (from multiplying by $x^5/120$)

$\rule{400px}{1px}$

product is $x +$ $x^2 +$ $\dfrac{x^3}{3} +$ $0 -$ $\dfrac{-9x^5}{120} +$ ...   (from adding the previous terms)

The sum of the first three nonzero terms is $e^x \cdot \sin(x) = x + x^2 + \dfrac{x^3}{3} + ... $.

10.11   APPROXIMATION USING TAYLOR POLYNOMIALS

The previous two sections focused on obtaining power series representations for functions, finding their intervals of convergence, and using those power series to approximate values of functions, limits, and integrals. In the cases where the power series resulted in an alternating numerical series, we were also able to use the Estimation Bound for Alternating Series (Section 10.6) to get a bound on the "error:"

"error" = | {exact value} – {partial sum approximation} | < | next term in the series | .

If the power series did not result in an alternating numerical series, we did not have a bound on the size of the error of the approximation.

In this section we introduce Taylor Polynomials (partial sums of the Taylor Series) and obtain a bound on the approximation error, the value | { exact value of f(x) } – { Taylor Polynomial approximation of f(x) } | . The bound we get is valid even if the Taylor series is not an alternating series, and the pattern for the error bound looks very much like **the next term in the series**, the first unused term in the partial sum of the Taylor series. In mathematics, this error bound is important for determining which functions are approximated by their Taylor series. In computer and calculator applications, the error bound is important to designers to ensure that their machines calculate enough digits of functions such as e^x and $\sin(x)$ for various values of x. In general, knowing this error bound can help us work efficiently by allowing us to use only the number of terms we really need.

We also examine graphically how well the Taylor Polynomials of f(x) approximate f(x)

Taylor Polynomials

If we add a finite number of terms of a power series, the result is a polynomial.

Definition

For a function f, the **n^{th} degree Taylor Polynomial** (centered at c), written $\mathbf{P_n(x)}$, is the partial sum of the terms up to the nth degree of the Taylor Series for f:

$$P_n(x) \quad = \sum_{k=0}^{n} \frac{f^{(k)}(c)}{k!} (x-c)^k$$

$$= f(c) + f'(c)(x-c) + \frac{f''(c)}{2!}(x-c)^2 + \frac{f'''(c)}{3!}(x-c)^3 + \frac{f^{(4)}(c)}{4!}(x-c)^4 + ... + \frac{f^{(n)}(c)}{n!}(x-c)^n$$

Example 1:   Write the first four Taylor Polynomials , $P_0(x)$ to $P_3(x)$, centered at 0 for e^x, and then

graph them for $-1 < x < 1$.

Solution: The Maclaurin series for e^x is $e^x = 1 + x + \dfrac{x^2}{2!} + \dfrac{x^3}{3!} + \dfrac{x^4}{4!} + \dfrac{x^5}{5!} + ... = \displaystyle\sum_{n=0}^{\infty} \dfrac{x^n}{n!}$ so

$$P_0(x) = 1, \quad P_1(x) = 1 + x, \quad P_2(x) = 1 + x + \dfrac{x^2}{2}, \quad \text{and} \quad P_3(x) = 1 + x + \dfrac{x^2}{2} + \dfrac{x^3}{6} \ .$$

The graphs of e^x and $P_1(x), P_2(x),$ and $P_3(x)$ are shown in Fig. 1.

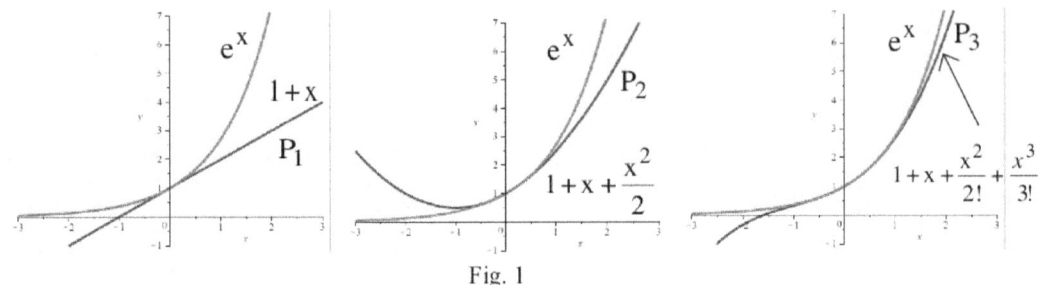

Fig. 1

Notice that $P_0(x)$ and e^x agree in value when $x = 0$,

$P_1(x)$, e^x, and their first derivatives agree in value when $x = 0$,

$P_2(x)$, e^x, their first derivatives, and their second derivatives agree in value when $x = 0$.

Practice 1:   Write the Taylor Polynomials $P_0(x), P_2(x),$ and $P_4(x)$ centered at 0 for $\cos(x)$, and then

graph them for $-\pi < x < \pi$. Write the Taylor Polynomials $P_1(x)$ and $P_3(x)$.

When we center the Taylor Polynomial at $x = c \neq 0$, the Taylor Polynomials approximate the function and

its derivatives well for x close to c.

Example 2:   Write the Taylor Polynomials $P_0(x), P_2(x),$ and $P_4(x)$ **centered at $3\pi/2$** for $\sin(x)$, and

then graph them for $2 < x < 8$.

Solution: The Taylor series, centered at $3\pi/2$, for $\sin(x)$ is

$$\sin(x) = -1 + \dfrac{1}{2!}(x - 3\pi/2)^2 - \dfrac{1}{4!}(x - 3\pi/2)^4 + \dfrac{1}{6!}(x - 3\pi/2)^6 + ... = \displaystyle\sum_{n=0}^{\infty} (-1)^{n+1} \dfrac{1}{(2n)!}(x - 3\pi/2)^{2n} \ .$$

Then $P_0(x) = -1$, $P_2(x) = -1 + \dfrac{1}{2}(x - 3\pi/2)^2$, and $P_4(x) = -1 + \dfrac{1}{2}(x - 3\pi/2)^2 - \dfrac{1}{24}(x - 3\pi/2)^4$.

The graphs of $\sin(x), P_0(x), P_2(x),$ and $P_4(x)$ are shown in Fig. 2.

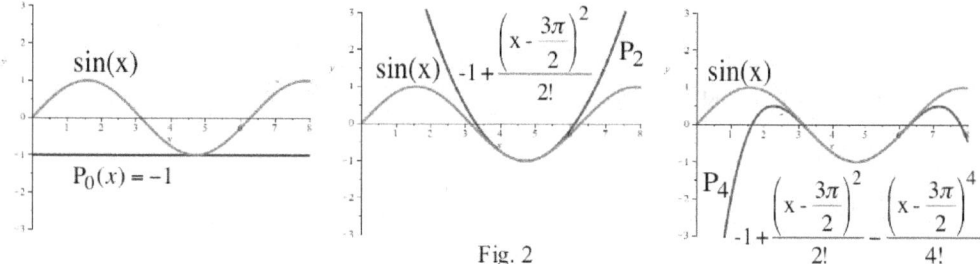

Fig. 2

Practice 2:   Write the Taylor Polynomials $P_0(x), P_1(x),$ and $P_3(x)$ **centered at $\pi/2$** for $\cos(x)$, and

then graph them for $-1 < x < 4$.

The Remainder

Approximation formulas such as the Taylor Polynomials are useful by themselves, but in many applied

situations we want to know how good the approximation is or how many terms of a series are required to

obtain a needed level of accuracy. If 2 terms of a series give you the needed level of accuracy for your

application, it is a waste of time and money to use 100 terms. On the other hand, sometimes even 100

terms may not give the accuracy you need. Fortunately, it is possible to obtain a guarantee on how close a

particular Taylor Polynomial approximation is to the exact value. Then we can work efficiently and use the

number of terms that we need. The next theorem gives a pattern for the amount of "error" in our Taylor

Polynomial approximation and can be used to obtain a bound on the size of the "error."

Taylor's Formula with Remainder

If   f has n+1 derivatives in an interval I containing c, and x is in I,

then   there is a number **z** , strictly <u>between</u> c and x, so that

$$f(x) = P_n(x) + R_n(x) \qquad \text{where} \quad R_n(x) = \frac{f^{(n+1)}(\mathbf{z})}{(n+1)!}(x-c)^{n+1} \; .$$

This says that $f(x)$ is equal to the nth degree Taylor Polynomial plus a Remainder, and the Remainder

$R_n(x)$ has the form given in the theorem. Notice that the pattern for R_n looks like the pattern for the

$(n+1)^{st}$ term of the Taylor series for $f(x)$ except that it contains $f^{(n+1)}(\mathbf{z})$ instead of $f^{(n+1)}(\mathbf{c})$. This

particular pattern for $R_n(x)$ is called the Lagrange form of the remainder, and is named for the French–

Italian mathematician and astronomer Joseph Lagrange (1736–1813).

The main idea of the proof of the Taylor's Formula with Remainder is straightforward, but the technical details are rather complicated. The main idea and the technical details are given in the Appendix. The pattern for the remainder, $\frac{f^{(n+1)}(z)}{(n+1)!}(x-c)^{n+1}$, contains three pieces, $(n+1)!$, $(x-c)^{n+1}$, and $f^{(n+1)}(z)$ for some z between x and c. The Taylor Remainder Formula is typically used in two ways:

In one type of use, the Taylor Polynomial is given so x, c, and n are known, and we can evaluate $(n+1)!$ and $(x-c)^{n+1}$ exactly. That leaves the piece $f^{(n+1)}(z)$ for some z between x and c. If we can find a bound for the value of $|f^{(n+1)}(z)|$ for all z between x and c, then we can put it together with the values of $(n+1)!$ and $(x-c)^{n+1}$ to obtain a bound for the remainder term $R_n(x)$.

In the other common usage, the amount of acceptable "error" is given, so x, c, and $R_n(x)$ are known, and we need to find a value of n that guarantees the required accuracy.

Corollary: A Bound for the Remainder $R_n(x)$

If f has $n+1$ derivatives in an interval I containing c, and x is in I, and
$|f^{(n+1)}(z)| \leq M$ for **all** z between x and c,

then   "error" $= |f(x) - P_n(x)| = |R_n(x)| \leq M \cdot \frac{|x-c|^{n+1}}{(n+1)!}$.

Example 3:   We plan to approximate the values of e^x with $P_3(x) = 1 + x + \frac{x^2}{2} + \frac{x^3}{6}$. Find a bound for the "error" of the approximation, $R_3(x)$, if x is in the interval

(a) $[-1, 1]$, (b) $[-3, 2]$ and (c) $[-0.2, 0.3]$.

Solution: $f(x) = e^x$, $c = 0$ (a Maclaurin series), $n = 3$, and $f^{(n+1)}(x) = f^{(4)}(x) = e^x$.

(a) For x in the interval $[-1, 1]$:
$|(x-c)^{n+1}| = |x^4| \leq |1^4| = 1$. $(n+1)! = 4! = 24$.

For x in $[-1, 1]$, $|f^{(n+1)}(x)| = |e^x| \leq e^1$. A "crude" but "easy to use" bound for e^1 is
$e^1 < (3)^1 = 3 = M$. (A more precise bound is $e^1 < (2.72)^1 < 2.72$.)

Then $|R_3(x)| < M \cdot \frac{|x-c|^{n+1}}{(n+1)!} < 3 \cdot \frac{1}{24} = 0.125$.

For all $-1 < x < 1$, $P_3(x) = 1 + x + \frac{x^2}{2} + \frac{x^3}{6}$ is within 0.125 of e^x .

(b) For x in the interval $[-3, 2]$: $|(x-c)^{n+1}| = |x^4| \leq |(-3)^4| = 81$ and $(n+1)! = 4! = 24$.

For x in $[-3, 2]$, $|f^{(n+1)}(x)| = |e^x| \leq e^2$. A "crude" but "easy to use" bound for e^2 is

$$e^2 < (3)^2 = 9 = M. \quad \text{(A more precise bound is } e^2 < (2.72)^2 < 7.4 \text{ .)}$$

Then $|R_3(x)| < M \cdot \dfrac{|x-c|^{n+1}}{(n+1)!} < 9 \cdot \dfrac{81}{24} = 30.375$. Obviously we cannot have

much confidence in our use of $P_3(x)$ to approximate e^x on the interval $[-3, 2]$.

(c) For x in the interval $[-0.2, 0.3]$: $|(x-c)^{n+1}| = |x^4| \le |0.3^4| = 0.0081$.

$(n+1)! = 4! = 24$.

For x in $[-0.2, 0.3]$, $|f^{(n+1)}(x)| = |e^x| \le e^{0.3}$. A bound for $e^{0.3}$ is

$e^{0.3} < (2.72)^{0.3} < 1.4 = M$ — obtained using a calculator .

Then $|R_3(x)| < M \cdot \dfrac{|x-c|^{n+1}}{(n+1)!} < 1.4 \cdot \dfrac{0.0081}{24} = 0.0004725$.

For all $-0.2 < x < 0.3$, $P_3(x) = 1 + x + \dfrac{x^2}{2} + \dfrac{x^3}{6}$ is within 0.0004725 of e^x .

When the interval is small, we can be confident that $P_3(x)$ provides a good approximation of e^x, but

as the interval grows, so does our bound on the remainder. To guarantee a good approximation on a

larger interval, we typically need $(n+1)!$ to be larger so we need to use a higher degree Taylor

Polynomial $P_n(x)$.

Practice 3:   Find a value of n to guarantee that $P_n(x)$ is within 0.001 of e^x for x in the interval $[-3, 2]$.

Example 4:   We want to approximate the values of $f(x) = \sin(x)$ on the interval $[-\pi/2, \pi/2]$ with an

error less that 10^{-10}. How many terms of the Maclaurin series for $\sin(x)$ are needed?

Solution: For every value of n, $|f^{(n+1)}(x)|$ is $|\sin(x)|$ or $|\cos(x)|$ so $M = 1$ in the Bound for the

Remainder. Then "error" $= |R_n(x)| < 1 \cdot \dfrac{|x-0|^{n+1}}{(n+1)!} \le \dfrac{(\pi/2)^{n+1}}{(n+1)!}$, and we need to find a value of n so

that $\dfrac{(\pi/2)^{n+1}}{(n+1)!}$ is less than 10^{-10}. A bit of numerical experimentation on a calculator shows that

$$\dfrac{(\pi/2)^{14}}{14!} \approx 6.39 \times 10^{-9} , \quad \dfrac{(\pi/2)^{15}}{15!} \approx 6.69 \times 10^{-10} , \text{ and } \dfrac{(\pi/2)^{16}}{16!} \approx 6.57 \times 10^{-11}$$

so we can take n = 15: $P_{15}(x) = x - \dfrac{x^3}{3!} + \dfrac{x^5}{5!} - \dfrac{x^7}{7!} + \dfrac{x^9}{9!} - \dfrac{x^{11}}{11!} + \dfrac{x^{13}}{13!} - \dfrac{x^{15}}{15!}$.

If $-\pi/2 \le x \le \pi/2$, then $|P_{15}(x) - \sin(x)| < 10^{-10}$.

Practice 4:   How many terms of the Maclaurin series for e^x are needed to approximate e^x to within

10^{-10} for $0 \le x \le 1$?

Calculator Notes

Imagine that you are in charge of designing or selecting an algorithm for a calculator to use when the SIN button is pushed. (Smartest move: find a mathematician who knows about "numerical analysis" and the design and implementation of algorithms.) You know that if the value of x is relatively close to 0, then SIN(x) can be quickly approximated to 10 digits (the size of the display of the calculator) by using a "few" terms of the Taylor series for sin(x): if $-1.57 \le x \le 1.57$, then

$$x - \frac{x^3}{3!} + \frac{x^5}{5!} - \frac{x^7}{7!} + \frac{x^9}{9!} - \frac{x^{11}}{11!} + \frac{x^{13}}{13!} - \frac{x^{15}}{15!}$$

$$= x\left(1 - \frac{x^2}{2\cdot3}\left(1 - \frac{x^2}{4\cdot5}\left(1 - \frac{x^2}{6\cdot7}\left(1 - \frac{x^2}{8\cdot9}\left(1 - \frac{x^2}{10\cdot11}\left(1 - \frac{x^2}{12\cdot13}\left(1 - \frac{x^2}{14\cdot15} \right) \right) \right) \right) \right) \right) \right)$$

gives the value of sin(x) with 10 digits of accuracy.

(The second pattern looks more complicated, but is usually preferred because it uses fewer multiplications and avoids very large values such as x^{15} and 15!) But you also want your algorithm to give 10 digits of accuracy even when x is larger, say 10 or 101.7. Rather than computing <u>many</u> more terms of the Maclaurin series for sine, some algorithms simply shift the problem closer to 0. First they use the fact that sin(x) = sin($x - 2\pi$) to keep shifting the problem until the argument is in the interval $[0, 2\pi]$:

sin(10) = sin(10 − 2π) = sin(3.71681469)

sin(101.7) = sin(101.7 − 2π) = sin(101.7 − 4π) = ... = sin(101.7 − 32π) = sin(1.169035085).

Once the argument is between 0 and 2π, additional trigonometric facts are used:

if the new value of x is larger than π, use sin(x) = −sin($x - \pi$) to replace "x" with "x − π" (and keep track of the change in sign of the answer). The new x value is in the interval $[0, \pi]$.

Finally, we can shift the problem into the interval $[0, \pi/2]$:

if the new value of x is larger than $\pi/2$, use sin(x) = sin($\pi - x$) to replace "x" with "π − x."

This new x value is in the interval $[0, \pi/2] \approx [0, 1.57]$ and the 7 terms of the sine series shown above are sufficient to approximate sin(x) with 10 digits of accuracy.

There are, however, major problems when calculators encounter the sine or exponential of a very large number. Since calculators only store the leading finite number of digits of a number (usually 10 or 12 digits), the calculator can not distinguish large numbers that differ past that leading number of stored digits: one calculator correctly says that $(10^{12}+1) - 10^{12} = 1$, but it incorrectly reports that $(10^{13}+1) - 10^{13} = 0$. Since it calculates "$10^{13}+1 = 10^{13}$", it also would falsely report the same values for sin($10^{13}+1$) and sin(10^{13}). In fact, the people who programmed this particular type of calculator recognized the problem, and the calculator gives an error message if it is asked to calculate sin(10^{11}). This particular calculator reports $e^{230} \approx 7.7 \times 10^{99}$. It reports an error for e^{231} since the largest number it can display is 9.9×10^{99} and e^{231} exceeds that value. What happens on your calculator?

PROBLEMS

In problems 1 – 10, calculate the Taylor polynomials P_0, P_1, P_2, P_3, and P_4 for the given function

centered at the given value of c. Then graph the function and the Taylor polynomials on the given interval.

1. $f(x) = \sin(x)$, c = 0, [–2, 4]   2. $f(x) = \cos(x)$, c = 0, [–2, 4]

3. $f(x) = \ln(x)$, c = 1, [0.1, 3]   4. $f(x) = \arctan(x)$, c = 0, [–3, 3]

5. $f(x) = \sqrt{x}$, c = 1, [0, 3]   6. $f(x) = \sqrt{x}$, c = 9, [0, 20]

7. $f(x) = (1 + x)^{-1/2}$, c = 0, [–2, 3]   8. $f(x) = e^{2x}$, c = 0, [–2, 4]

9. $f(x) = \sin(x)$, c = $\pi/2$, [–1, 5]   10. $f(x) = \sin(x)$, c = π, [–1, 5]

In problems 11 – 18, a function f(x) and a value of n are given. Determine a formula for $R_n(x)$ and

find a bound for $|R_n(x)|$ on the given interval. This bound for $|R_n(x)|$ is our "guaranteed accuracy" for

P_n to approximate f(x) on the given interval. (Use c = 0.)

11. $f(x) = \sin(x)$, n = 5, [$-\pi/2$, $\pi/2$]   12. $f(x) = \sin(x)$, n = 9, [$-\pi/2$, $\pi/2$]

13. $f(x) = \sin(x)$, n = 5, [$-\pi$, π]   14. $f(x) = \sin(x)$, n = 9, [$-\pi$, π]

15. $f(x) = \cos(x)$, n = 10, [–1, 2]   16. $f(x) = \cos(x)$, n = 10, [–1, 5]

17. $f(x) = e^x$, n = 6, [–1, 2]   18. $f(x) = e^x$, n = 10, [–1, 3]

In problems 19 – 24, determine how many terms of the Taylor series for f(x) are needed to approximate f

to within the specified error on the given interval. (For each function use the center c = 0.)

19. $f(x) = \sin(x)$ within 0.001 on [1, 1]   20. $f(x) = \sin(x)$ within 0.001 on [–3, 3]

21. $f(x) = \sin(x)$ within 0.00001 on [–1.6, 1.6]   22. $f(x) = \cos(x)$ within 0.001 on [–2, 2]

23. $f(x) = e^x$ within 0.001 on [0, 2]   24. $f(x) = e^x$ within 0.001 on [–1, 4]

Series Approximations of π

The following problems illustrate some of the ways series have been used to obtain very precise

approximations of π. Several of these methods use the series for arctan(x),

$$\arctan(x) = x - \frac{x^3}{3} + \frac{x^5}{5} - \frac{x^7}{7} + \frac{x^9}{9} - \dots = \sum_{n=0}^{\infty} (-1)^n \frac{x^{2n+1}}{2n+1},$$

which converges rapidly if $|x|$ is close to zero.

Method I: $\tan(\frac{\pi}{4}) = 1$ so $\frac{\pi}{4} = \arctan(1) = 1 - \frac{1}{3} + \frac{1}{5} - \frac{1}{7} + \frac{1}{9} - \dots = \sum_{n=0}^{\infty} (-1)^n \frac{1}{2n+1}$ and

$$\pi = 4\left\{ 1 - \frac{1}{3} + \frac{1}{5} - \frac{1}{7} + \frac{1}{9} - \dots \right\}.$$

25. (a) Approximate π as $4\left\{ 1 - \frac{1}{3} + \frac{1}{5} - \frac{1}{7} + \frac{1}{9} \right\}$ and compare this value with the value your

   calculator gives for π.

   (b) The series for arctan(1) is an alternating series so we have an "easy" error bound. Use the error bound

   for an alternating series to find a bound for the error if 50 terms of the arctan(1) series are used.

   (c) Using the error bound for an alternating series, how many terms of the arctan(1) series are needed to

   guarantee that the series approximation of π is within 0.0001 of the exact value of π?

   (The arctan(1) series converges so slowly that it is not used to approximate π.)

Method II: $\tan(a + b) = \frac{\tan(a) + \tan(b)}{1 - \tan(a)\tan(b)}$ so $\tan\left(\arctan\left(\frac{1}{2} \right) + \arctan\left(\frac{1}{3} \right) \right) = \frac{\frac{1}{2} + \frac{1}{3}}{1 - \frac{1}{2} \cdot \frac{1}{3}} = 1$. Then

   $\frac{\pi}{4} = \arctan(1) = \arctan\left(\frac{1}{2} \right) + \arctan\left(\frac{1}{3} \right)$, and the series for $\arctan\left(\frac{1}{2} \right)$ and $\arctan\left(\frac{1}{3} \right)$

   converge much more rapidly than the series for arctan(1).

26. (a) Approximate π as

   $4\left\{ \text{(sum of the first 4 terms of the } \arctan\left(\frac{1}{2} \right) \text{ series)} + \text{(sum of the first 4 terms of the } \arctan\left(\frac{1}{3} \right) \text{ series)} \right\}$.

   Then compare this value with the value your calculator gives for π.

   (b) The series for $\arctan\left(\frac{1}{2} \right)$ and $\arctan\left(\frac{1}{3} \right)$ are each alternating series. Use the error bound for an

   alternating series to find a bound for the error if 10 terms of each series are used.

   (c) How many terms of each series are needed to guarantee that the series approximation of π is within

   0.0001 of the exact value of π?

Other Methods: We will not justify these methods, but they converge to π more rapidly than the first

          two methods.

   A: $\frac{\pi}{4} = 4 \cdot \arctan\left(\frac{1}{5} \right) - \arctan\left(\frac{1}{239} \right)$ (due to Machin in 1706)

   B: $\pi = 48 \cdot \arctan\left(\frac{1}{18} \right) + 32 \cdot \arctan\left(\frac{1}{57} \right) - 20 \cdot \arctan\left(\frac{1}{239} \right)$

27. (a) Use the first 3 terms of each series in formula A to approximate π. How much does it differ from the

   value your calculator gives you?

   (b) Why does formula A converge more rapidly (using fewer terms) than methods I and II?

28. (a) Use the first 3 terms of each series in formula B to approximate π. How much does it differ from the

   value your calculator gives you?

   (b) Why does formula B converge more rapidly (using fewer terms) than Methods I and II and formula A?

Practice Answers

Practice 1: $\cos(x) = 1 - \dfrac{x^2}{2!} + \dfrac{x^4}{4!} - \dfrac{x^6}{6!} + \dfrac{x^8}{8!} - \dfrac{x^{10}}{10!} + \ldots = \displaystyle\sum_{n=0}^{\infty} (-1)^n \dfrac{x^{2n}}{(2n)!}$ so

$$P_0(x) = 1, P_2(x) = 1 - \dfrac{x^2}{2}\ , \text{ and } P_4(x) = 1 - \dfrac{x^2}{2} + \dfrac{x^4}{24}\ . \text{ Their graphs are shown in Fig. 3.}$$

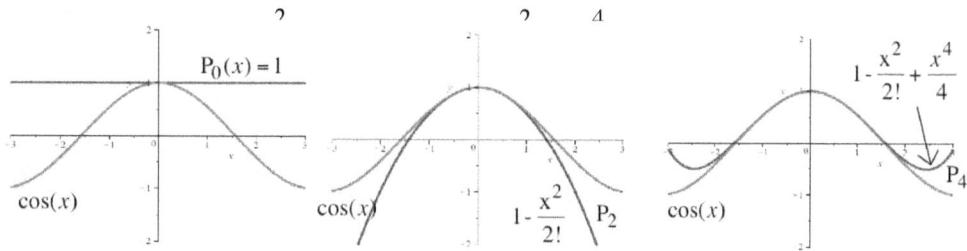

Fig. 3

Practice 2:

$$\cos(x) = -(x - \pi/2) + \dfrac{1}{3!}(x - \pi/2)^3 - \dfrac{1}{5!}(x - \pi/2)^5 + \dfrac{1}{7!}(x - \pi/2)^7 - \ldots$$

$$= \displaystyle\sum_{n=0}^{\infty} (-1)^{n+1} \dfrac{1}{(2n+1)!} (x - \pi/2)^{2n+1}\ .$$

Then $P_0(x) = 0$, $P_1(x) = -(x - \pi/2)$, and $P_3(x) = -(x - \pi/2) + \dfrac{1}{6}(x - \pi/2)^3$. The graphs of
$\cos(x), P_1(x),$ and $P_3(x)$ are shown in Fig. 4.

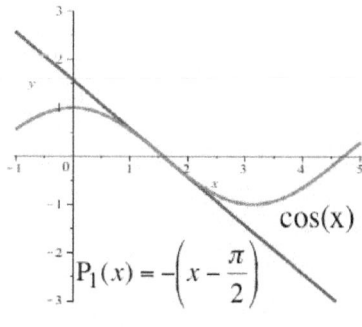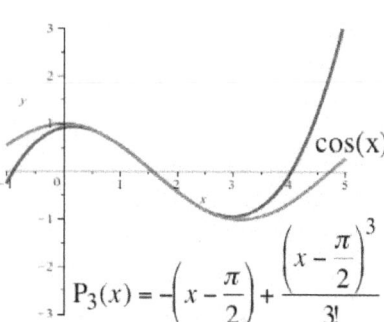

Fig. 4

Practice 3:   For x in the interval $[-3, 2]$, $|(x-c)^{n+1}| = |x^3| \le |(-3)^3| = 27$.

For x in $[-3, 2]$, $|f^{(n+1)}(x)| = |e^x| \le e^2$. A "crude" bound for e^2 is $e^2 < (3)^2 = 9 = M$.

Then $|R_n(x)| < M \cdot \dfrac{(x-c)^{n+1}}{(n+1)!} < 9 \cdot \dfrac{27}{n!}$, and we want a value of n so $9 \cdot \dfrac{27}{n!} \le 0.001$:

we want $n! \ge \dfrac{(9)(27)}{0.001} = 243{,}000$. Using a calculator, we see that $8! = 40{,}320$ is not

large enough, but $9! = 362{,}880 > 243{,}000$ so we can use $n = 9$.

For x in the interval $[-3, 2]$, $P_9(x) = 1 + x + \dfrac{x^2}{2!} + \dfrac{x^3}{3!} + \dfrac{x^4}{4!} + \dfrac{x^5}{5!} + \dfrac{x^6}{6!} + \dfrac{x^7}{7!} + \dfrac{x^8}{8!} + \dfrac{x^9}{9!}$ is

within 0.001 of e^x.

Practice 4: $R_n(x) \le 10^{-10}$. $f(x) = e^x$, $c = 0$, and for every n, $f^{(n+1)}(x) = e^x$. For $0 \le x \le 1$,

$|f^{(n+1)}(x)| = |e^x| \le e < 2.72 = M$. We want to find a value for n so

$M \cdot \dfrac{(x-c)^{n+1}}{(n+1)!} = 2.72 \cdot \dfrac{(1-0)^{n+1}}{(n+1)!} < 10^{-10}$. Some numerical experimentation on a calculator

shows that $2.72 \cdot \dfrac{1}{15!} \approx 1.58 \ 10^{-9}$ and $2.72 \cdot \dfrac{1}{16!} \approx 9.9 \ 10^{-11}$ so we can take $n = 15$.

For $0 \le x \le 1$, $P_{15}(x) = 1 + x + \dfrac{x^2}{2!} + \dfrac{x^3}{3!} + \dfrac{x^4}{4!} + \ldots + \dfrac{x^{15}}{15!}$ is within 10^{-10} of e^x.

Appendix: Idea and details of a Proof of Taylor's Formula with Remainder

Main idea of the proof:

We define a new differentiable function $g(t)$ and show that $g(x) = 0$ and $g(c) = 0$.

Then, by Rolle's Theorem, we can conclude that there is a number z, between x and c, so that $g'(z) = 0$.
Finally, we set $g'(z) = 0$ and algebraically obtain the given formula for $R_n(x)$.

Let $R_n(x) = f(x) - P_n(x)$ be the difference between $f(x)$ and the nth Taylor polynomial for $P_n(x)$.

Define a differentiable function $g(t)$ to be

$$g(t) = f(x) - \left\{ f(t) + f'(t)(x-t) + \frac{f''(t)}{2!}(x-t)^2 + \frac{f'''(t)}{3!}(x-t)^3 + \ldots + \frac{f^{(n)}(t)}{n!}(x-t)^n \right\} - R_n(x)\frac{(x-t)^{n+1}}{(x-c)^{n+1}} \ .$$

This may seem to be a strange way to define a function, but it turns out to have the properties we need:

$$g(x) = f(x) - \left\{ f(x) + 0 + 0 + 0 + \ldots + 0 \right\} - R_n(x)\frac{0}{(x-c)^{n+1}} = f(x) - f(x) = 0 \text{, and}$$

$$g(c) = f(x) - \left\{ f(c) + f'(c)(x-c) + \frac{f''(c)}{2!}(x-c)^2 + \frac{f'''(c)}{3!}(x-c)^3 + \frac{f^{(4)}(c)}{4!}(x-c)^4 + \frac{f^{(n)}(c)}{n!}(x-c)^n \right\}$$

$$- R_n(x)\frac{(x-c)^{n+1}}{(x-c)^{n+1}}$$

$$= f(x) - P_n(x) - R_n(x) = 0 \text{ since } R_n(x) = f(x) - P_n(x).$$

Then, by Rolle's Theorem, there is a number z, strictly between x and c, so $g'(z) = 0$.

Notice that g is defined to be a function of t so we treat x and c as constants and differentiate with respect to t. The key pattern is that when we differentiate a term such as $\dfrac{f'''(t)}{3!} \cdot (x-t)^3$ with respect to t, we need to use the product rule. The resulting derivative has two terms:

$$\frac{d}{dt}\left\{ \frac{f'''(t)}{3!}(x-t)^3 \right\} = \frac{f'''(t)}{3!}\frac{d}{dt}(x-t)^3 + (x-t)^3 \cdot \frac{d}{dt}\frac{f'''(t)}{3!}$$

$$= \frac{f'''(t)}{3!}(3)(x-t)^2(-1) + \frac{f^{(4)}(t)}{3!}(x-t)^3 = -\frac{f'''(t)}{2!}(x-t)^2 + \frac{f^{(4)}(t)}{3!}(x-t)^3 .$$

When we differentiate $g(t)$ with respect to t, we get a complicated pattern, but most of the terms cancel:

$$g'(t) = \frac{d}{dt}g(t) = 0 - \Bigg\{ \ f'(t)$$

$$-f'(t) + f''(t)(x-t)$$

$$-f''(t)(x-t) + \frac{f'''(t)}{2!}(x-t)^2$$

$$-\frac{f'''(t)}{2!}(x-t)^2 + \frac{f^{(4)}(t)}{3!}(x-t)^3$$

$$- \ \ldots \ - \frac{f^{(n)}(t)}{(n-1!)}(x-t)^{n-1} + \frac{f^{(n+1)}(t)}{n!}(x-t)^n \Bigg\} - R_n(x)\cdot(n+1)\cdot\frac{(x-t)^n(-1)}{(x-c)^{n+1}}$$

$$= -\frac{f^{(n+1)}(t)}{n!}(x-t)^n + R_n(x)\cdot(n+1)\cdot\frac{(x-t)^n}{(x-c)^{n+1}}$$

$$= (x-t)^n\left\{ R_n(x)\cdot(n+1)\cdot\frac{1}{(x-c)^{n+1}} - \frac{f^{(n+1)}(t)}{n!} \right\} .$$

By Rolle's Theorem, there is a value z, between x and c, for the variable t so $g'(z) = 0$. Then

$$(x-z)^n\left\{ R_n(x)\cdot(n+1)\cdot\frac{1}{(x-c)^{n+1}} - \frac{f^{(n+1)}(z)}{n!} \right\} = 0 .$$

z is strictly between x and c so $z \neq x$ and we can divide each side by $(x-z)^n$ to get

$$R_n(x)\cdot(n+1)\cdot\frac{1}{(x-c)^{n+1}} - \frac{f^{(n+1)}(z)}{n!} = 0 .$$

Finally, $R_n(x)\cdot(n+1)\cdot\dfrac{1}{(x-c)^{n+1}} = \dfrac{f^{(n+1)}(z)}{n!}$ so $R_n(x) = \dfrac{f^{(n+1)}(z)}{n!(n+1)} \cdot (x-c)^{n+1} = \dfrac{f^{(n+1)}(z)}{(n+1)!}(x-c)^{n+1}$,

the result we wanted to prove.

Chapter 10: Odd Answers

Section 10.0

1. (a) $32, 64$ (b) 2^5 (c) 2^n 3. (a) $-1, +1$ (b) $-1 = (-1)^5$ (c) $(-1)^n$

5. (a) $120, 720$ (b) $5!$ or $5 \cdot 24$ (c) $n!$ or $n \cdot a_{n-1}$ 7. $1, 3/2, 11/6, 25/12, 137/60, 147/60$

9. $1, 1/2, 3/4, 5/8, 11/16, 21/32$ 11. $1, 0, 1, 0, 1, 0$

13. (a) $g(5) = -1, g(6) = +1$ (b) see the figure for $g(x)$.

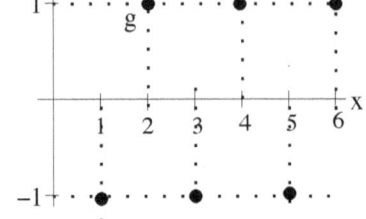

15. (a) $t(5) = 1 - 1/2 + 1/4 - 1/8 + 1/16 - 1/32 = 21/32$,

   $t(6) = 43/64$. The graph of $t(x)$ is shown.

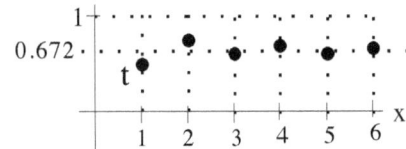

17. (a) $P(x) = 1 - \dfrac{x^2}{2}$ (b) Graphs

x	P(x)	cos(x)	| P(x) − cos(x) |
0	1.0	1.0	0
0.1	0.995	0.99500	0
0.2	0.98	0.98006	0.00006
0.3	0.955	0.95533	0.00033
1.0	0.5	0.54030	0.04030
2.0	−1.0	−0.41615	0.58385

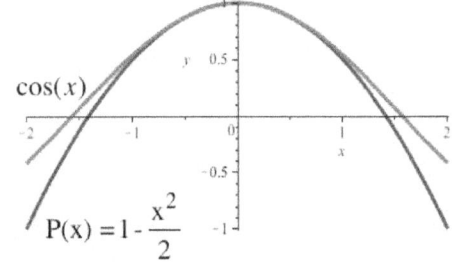

   (c) $P(x) = 1 - \dfrac{x^2}{2} + \dfrac{x^4}{24}$

   Graphs

x	P(x)	cos(x)	| P(x) − cos(x) |
0	1.0	1.0	0
0.1	0.99500	0.99500	0
0.2	0.98006	0.98006	0
0.3	0.95533	0.95533	0
1.0	0.54167	0.54030	0.00137
2.0	−0.33333	−0.41615	0.08282

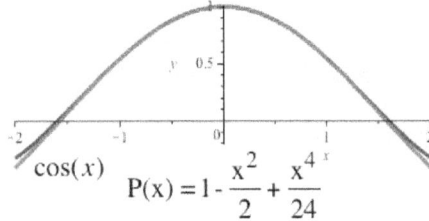

19. $P(x) = Ax + B.$ $5 = P(0) = A \cdot 0 + B = B$ so $B = 5.$ $3 = P'(0) = A.$ $P(x) = 3x + 5.$

21. $P(x) = Ax + B.$ $4 = P(0) = A \cdot 0 + B = B$ so $B = 4.$ $-1 = P'(0) = A.$ $P(x) = -1x + 4.$

23. $P(x) = 0x + 4 = 4.$ 25. $P(x) = Ax + B.$ $P(0) = B.$ $P'(0) = A.$

27. $P(x) = Ax^2 + Bx + C.$ $-2 = P(0) = A \cdot 0 + B \cdot 0 + C = C.$ $7 = P'(0) = 2A \cdot 0 + B = B.$

   $6 = P''(0) = 2A$ so $A = 6/2 = 3.$ $P(x) = 3x^2 + 7x - 2.$

29. $P(x) = Ax^2 + Bx + C.$ $8 = P(0) = A \cdot 0 + B \cdot 0 + C = C.$ $5 = P'(0) = 2A \cdot 0 + B = B.$

   $10 = P''(0) = 2A$ so $A = 10/2 = 5.$ $P(x) = 5x^2 + 5x + 8.$

31. $P(x) = Ax^2 + Bx + C$. $-3 = P(0) = A \cdot 0 + B \cdot 0 + C = C$. $-2 = P'(0) = 2A \cdot 0 + B = B$.

    $4 = P''(0) = 2A$ so $A = 4/2 = 2$. $P(x) = 2x^2 - 2x - 3$.

33. $P(x) = Ax^3 + Bx^2 + Cx + D$. $5 = P(0) = A \cdot 0 + B \cdot 0 + C \cdot 0 + D = D$. $3 = P'(0) = 3A \cdot 0 + 2B \cdot 0 + C = C$.

    $4 = 6A0 + 2B = 2B$ so $B = 4/2 = 2$. $6 = P''(0) = 6A$ so $A = 6/6 = 1$.

    $P(x) = 1x^3 + 2x^2 + 3x + 5$.

35. $P(x) = Ax^3 + Bx^2 + Cx + D$. $4 = P(0) = A \cdot 0 + B \cdot 0 + C \cdot 0 + D = D$. $-1 = P'(0) = 3A \cdot 0 + 2B \cdot 0 + C = C$.

    $-2 = 6A0 + 2B = 2B$ so $B = -2/2 = -1$. $-12 = P''(0) = 6A$ so $A = -12/6 = -2$.

    $P(x) = -2x^3 - 1x^2 - 1x + 4$.

37. $P(x) = Ax^3 + Bx^2 + Cx + D$. $4 = P(0) = A \cdot 0 + B \cdot 0 + C \cdot 0 + D = D$. $0 = P'(0) = 3A \cdot 0 + 2B \cdot 0 + C = C$.

    $-4 = 6A0 + 2B = 2B$ so $B = -4/2 = -2$. $36 = P''(0) = 6A$ so $A = 36/6 = 6$.

    $P(x) = 6x^3 - 2x^2 + 0x + 4 = 6x^3 - 2x^2 + 4$.

39. $A = P'''(0)/6$, $B = P''(0)/2$, $C = P'(0)$, and $D = P(0)$.

Section 10.1

1. $\dfrac{1}{n^2}$ 3. $\dfrac{n-1}{n} = 1 - \dfrac{1}{n}$ 5. $\dfrac{n}{2^n}$

7. $\{-1, 0, 1/3, 1/2, 3/5, 2/3, ...\}$ Graph is shown.   9. $\{1, 2/3, 3/5, 4/7, 5/9, 6/11, ...\}$ Graph is shown.

11. $\{2, 3\frac{1}{2}, 2\frac{2}{3}, 3\frac{1}{4}, 2\frac{4}{5}, 3\frac{1}{6}, ...\}$ Graph is shown.   13. $\{0, \frac{1}{2}, -\frac{2}{3}, \frac{3}{4}, -\frac{4}{5}, \frac{5}{6}, ...\}$ Graph is shown.

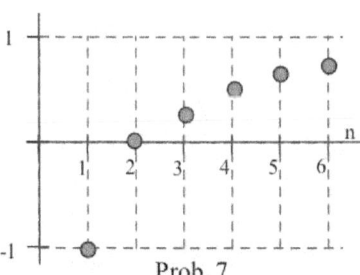
Prob. 7

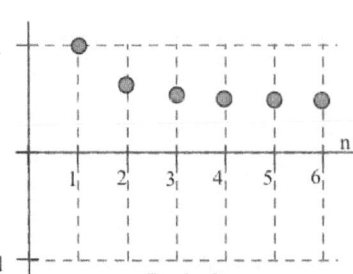
Prob. 9

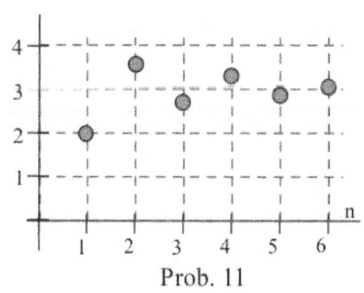
Prob. 11

15. $\{1, \frac{1}{2}, \frac{1}{3!}, \frac{1}{4!}, \frac{1}{5!}, \frac{1}{6!}, ...\} = \{1, \frac{1}{2}, \frac{1}{6}, \frac{1}{24}, \frac{1}{120}, \frac{1}{720}, ...\}$ The graph is shown below.

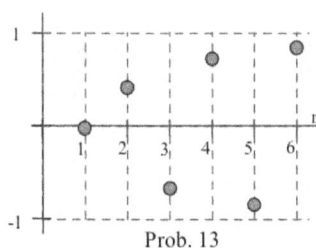
Prob. 13

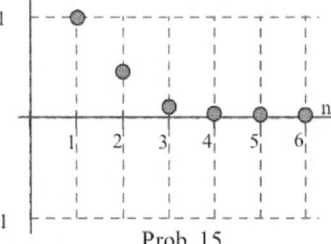
Prob. 15

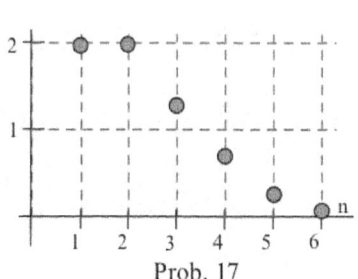
Prob. 17

17. $\{ \frac{2^1}{1!}, \frac{2^2}{2!}, \frac{2^3}{3!}, \frac{2^4}{4!}, \frac{2^5}{5!}, \frac{2^6}{6!}, ... \}$ The graph is shown.

19. $a_1 = 2, a_2 = -2, a_3 = 2, a_4 = -2, a_5 = 2, a_6 = -2, a_7 = 2, a_8 = -2, a_9 = 2, a_{10} = -2$

21. $\{$ $\sin(2\pi/3), \sin(4\pi/3), \sin(6\pi/3), \sin(8\pi/3), \sin(10\pi/3), \sin(12\pi/3), \sin(14\pi/3), \sin(16\pi/3),$
    $\sin(18\pi/3), \sin(20\pi/3), ... \}$

23. $c_1 = 1, c_2 = 3, c_3 = 6, c_4 = 10, c_5 = 15, c_6 = 21, c_7 = 28, c_8 = 36, c_9 = 45, c_{10} = 55$

25. $\{ a_n \}$ appears to converge. $\{ b_n \}$ does not appear to converge.

27. $\{ e_n \}$ does not appear to converge. $\{ f_n \}$ appears to converge.

29. $\{ 1 - \frac{2}{n} \}$ converges to 1.   31. $\{ \frac{n^2}{n+1} \}$ grows arbitrarily large and diverges.

33. $\{ \frac{n}{2n-1} \}$ converges to $\frac{1}{2}$ 35. $\{ \ln(3 + \frac{7}{n}) \}$ converges to $\ln(3) \approx 1.099$.

37. $\{ 4 + (-1)^n \}$ alternates in value between 3 and 5 and does not appraoch a single number.
    The sequence diverges.

39. $\{ \frac{1}{n!} \}$ converges to 0.

41. $\{ (1 - \frac{1}{n})^n \}$ converges to $e^{-1} = \frac{1}{e}$. (See Section 3.7, Example 7.)

43. $\{ \frac{(n+2)(n-5)}{n^2} \} = \{ \frac{n^2 - 3n - 10}{n^2} \} = \{ 1 - \frac{3}{n} - \frac{10}{n^2} \}$ converges to 1.

45. Take $N = \sqrt{\frac{3}{\varepsilon}}$. If $n > N = \sqrt{\frac{3}{\varepsilon}}$ then $n^2 > \frac{3}{\varepsilon}$ and $\varepsilon > \frac{3}{n^2} = \left| \frac{3}{n^2} - 0 \right|$.

47. Take $N = \frac{1}{\varepsilon}$. If $n > N = \frac{1}{\varepsilon}$ then $\varepsilon > \frac{1}{n} = |(3 - \frac{1}{n}) - 3| = | \frac{3n-1}{n} - 3 |$.

49. $\{ \frac{1}{n^{th} \text{ prime}} \}$ is a subsequence of $\{ \frac{1}{n^{th} \text{ integer}} \} = \{ \frac{1}{n} \}$ which converges to 0, so we can
    conclude that $\{ \frac{1}{n^{th} \text{ prime}} \}$ converges to 0.

51. $\{ (-2)^n (\frac{1}{3})^n \} = \{ (-1)^n (\frac{2}{3})^n \}$.
    If n is even, $\{ (-1)^n (\frac{2}{3})^n \} = \{ (\frac{2}{3})^n \}$ which converges to 0. If n is odd,
    $\{ (-1)^n (\frac{2}{3})^n \} = \{ -(\frac{2}{3})^n \}$ which also converges to 0. Since "n even" and "n odd" account
    for all of the positive integers, we can conclude that $\{ (-2)^n (\frac{1}{3})^n \}$ converges to 0.

53. $\left\{ (1 + \frac{5}{n^2})^{(n^2)} \right\}$ is a subsequence of $\left\{ (1 + \frac{5}{n})^n \right\}$ which converges to e^5 (Section 3.7, Example 7)

so $\left\{ (1 + \frac{5}{n^2})^{(n^2)} \right\}$ also converges to e^5.

55. $a_n = 7 - \frac{2}{n}$ so $a_{n+1} = 7 - \frac{2}{n+1}$. $a_{n+1} - a_n = (7 - \frac{2}{n+1}) - (7 - \frac{2}{n}) = \frac{2}{n} - \frac{2}{n+1} = \frac{2}{n(n+1)} > 0$ for all $n \geq 1$.

Therefore, $a_{n+1} > a_n = $ and $\{ a_n \}$ is monotonically increasing.

57. $a_n = 2^n$ so $a_{n+1} = 2^{n+1}$. $a_{n+1} - a_n = 2^{n+1} - 2^n = 2^n 2 - 2^n = 2^n(2 - 1) = 2^n > 0$ for all $n \geq 1$.

Therefore, $a_{n+1} > a_n = $ and $\{ a_n \}$ is monotonically increasing.

59. $a_n = \frac{n+1}{n!}$ so $a_{n+1} = \frac{(n+1)+1}{(n+1)!} = \frac{n+2}{(n+1)!}$. Then

$$\frac{a_{n+1}}{a_n} = \frac{\frac{n+2}{(n+1)!}}{\frac{n+1}{n!}} = \frac{n+2}{n+1} \frac{n!}{(n+1)!} = \frac{n+2}{n+1} \frac{1 \cdot 2 \cdot 3 \cdot \ldots \cdot n}{1 \cdot 2 \cdot 3 \cdot \ldots \cdot n \cdot (n+1)} = \frac{n+2}{n+1} \frac{1}{n+1} < 1 \text{ for all } n \geq 1$$

so $a_{n+1} < a_n$ for all $n \geq 1$ and $\{ a_n \}$ is monotonically decreasing.

61. $a_n = (\frac{5}{4})^n$ so $a_{n+1} = (\frac{5}{4})^{n+1}$. Then

$$\frac{a_{n+1}}{a_n} = \frac{(\frac{5}{4})^{n+1}}{(\frac{5}{4})^n} = \frac{5}{4} > 1 \text{ for all } n \text{ so } a_{n+1} > a_n \text{ for all } n > 0 \text{ and } \{ a_n \} \text{ is montonically increasing.}$$

63. $a_n = \frac{n}{e^n}$ so $a_{n+1} = \frac{n+1}{e^{n+1}}$. Then

$$\frac{a_{n+1}}{a_n} = \frac{\frac{n+1}{e^{n+1}}}{\frac{n}{e^n}} = \frac{n+1}{n} \frac{e^n}{e^{n+1}} = \frac{n+1}{n} \frac{1}{e} < 1 \text{ for } n > 1 \text{ (reason: } e > 2 \text{ so } n \cdot e > 2n > n + 1 \text{ so } \frac{n+1}{n \cdot e} < 1).$$

So $a_{n+1} < a_n$ for all $n > 0$ and $\{ a_n \}$ is monotonically decreasing.

65. Let $f(x) = 5 - \frac{3}{x}$. Then $f'(x) = \frac{3}{x^2} > 0$ for all x so $f(x)$ is increasing. From that we can conclude that

$a_n = f(n)$ is monotonically increasing.

67. Let $f(x) = \cos(\frac{1}{x})$. Then $f'(x) = -\sin(\frac{1}{x}) \cdot (\frac{-1}{x^2}) = \frac{1}{x^2} \cdot \sin(\frac{1}{x}) > 0$ for all $x \geq 1$. From that we

can conclude that $a_n = f(n)$ is monotonically increasing.

69. This is similar to problem 59. The ratio method works nicely.

71. One method is to examine $a_{n+1} - a_n = (1 - \frac{1}{2^{n+1}}) - (1 - \frac{1}{2^n}) = \frac{1}{2^n} - \frac{1}{2^{n+1}} = \frac{1}{2^n} - \frac{1}{2 \cdot 2^n} = \frac{1}{2 \cdot 2^n} > 0$

for all n so $a_{n+1} > a_n$ for all n and $\{a_n\}$ is monotonically increasing.

The ratio and derivative methods also work.

$(\mathbf{D}(1 - \frac{1}{2^x}) = \mathbf{D}(1 - 2^{-x}) = 0 - 2^{-x} \cdot \ln(2) \cdot \mathbf{D}(-x) = 2^{-x} \ln(2) = \frac{\ln(2)}{2^x} > 0$ for x > 0.)

73. This is similar to problem 63. The ratio method works nicely.

75. N = 4: $a_1 = 4, a_2 = \frac{1}{2}(4 + \frac{4}{4}) = \frac{5}{2} = 2.5$, $a_3 = \frac{1}{2}(2.5 + \frac{4}{2.5}) = 2.05$, $a_n = \frac{1}{2}(2.05 + \frac{4}{2.05}) \approx 2.00061$.

N = 9: $a_1 = 9, a_2 = \frac{1}{2}(9 + \frac{9}{9}) = \frac{10}{2} = 5$, $a_3 = \frac{1}{2}(5 + \frac{9}{5}) = 3.2$, $a_n = \frac{1}{2}(3.2 + \frac{9}{3.2}) = 3.00625$.

N = 5: $a_1 = 5, a_2 = \frac{1}{2}(5 + \frac{5}{5}) = \frac{6}{2} = 3$. $a_3 = \frac{1}{2}(3 + \frac{5}{3}) \approx 2.333$, $a_n = \frac{1}{2}(2.333 + \frac{5}{2.333}) \approx 2.238$.

77. (a) p = 0.02 , and we want to solve $0.01 = \frac{0.02}{0.02k + 1}$ for k. Then $0.02k + 1 = \frac{0.02}{0.01} = 2$ so

$0.02k = 1$ and $k = \frac{1}{0.02} = 50$ generations.

(b) We want to solve $\frac{1}{2} p = \frac{p}{kp + 1}$ for k in terms of p. $kp + 1 = \frac{p}{0.5p} = 2$ so $kp = 1$ and

$k = \frac{1}{p}$ generations.

79. (a) The first "few" grains can be anywhere on the x–axis.

(b) After a "lot of grains" have been placed, there will be a large pile of sand close to 3 on the x–axis.

81. $-1 \le \sin(n) \le 1$ for all integers n.

(a) The first few grains will be scattered between −1 and +1 on the x–axis.

(b) After a "lot of grains" have been placed, the sand will be scattered "uniformly" along the interval from −1 to +1. (See part (c).)

(c) This argument is rather sophisticated, but the result is interesting: no two grains ever end up on the same point.

We assume that two grains do end up on the same point, and then derive a contradiction. From this we conclude that our original assumption (two grains on one point) was false.

Assume that two grains do end up on the same point so $a_m = a_n$ for distinct integers m and n.

Then $\sin(m) = \sin(n)$ so $0 = \sin(m) - \sin(n) = 2 \cdot \sin(\frac{m-n}{2}) \cdot \cos(\frac{m+n}{2})$ and either $\sin(\frac{m-n}{2}) = 0$ or $\cos(\frac{m+n}{2}) = 0$. If $\sin(\frac{m-n}{2}) = 0$, then $\frac{m-n}{2} = \pi K$ for some integer K and $\pi = \frac{m-n}{2K}$ where m, n, and K are integers. Then π is a rational number, a contradiction of the fact that π is irrational.

If $\cos(\frac{m+n}{2}) = 0$, then $\frac{m+n}{2} = \frac{\pi}{2} + K\pi = \pi(\frac{1}{2} + K)$ for some integer K so $\pi = \frac{m+n}{1 + 2K}$, a rational number. This again contradicts the irrationality of π, so out original assumption (two grains on the same point) was false.

Section 10.2

1. $\displaystyle\sum_{k=1}^{\infty} \frac{1}{k}$

3. $\displaystyle\sum_{k=1}^{\infty} \frac{2}{3k}$

5. $\displaystyle\sum_{k=1}^{\infty} \left(-\frac{1}{2}\right)^k$ or $\displaystyle\sum_{k=1}^{\infty} (-1)^k \cdot \frac{1}{2^k}$

7. The graph is given.

9. The graph is given.

11. The graph is given.

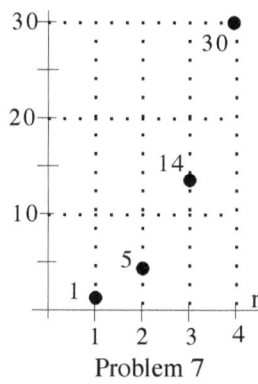
Problem 7

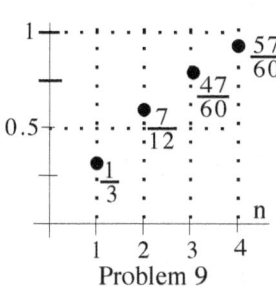
Problem 9

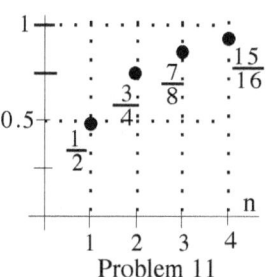
Problem 11

13. $a_1 = 3$, $a_2 = -1$, $a_3 = 2$, $a_4 = 1$

15. $a_1 = 4$, $a_2 = 0.5$, $a_3 = -0.2$, $a_4 = 0.5$

17. $a_1 = 1$, $a_2 = 0.1$, $a_3 = 0.01$, $a_4 = 0.001$

19. $\displaystyle\sum_{k=1}^{\infty} \frac{8}{10^k}$

21. $\displaystyle\sum_{k=1}^{\infty} \frac{5}{10^k}$

23. $\displaystyle\sum_{k=1}^{\infty} \frac{a}{10^k}$

25. $\displaystyle\sum_{k=1}^{\infty} \frac{17}{100^k}$

27. $\displaystyle\sum_{k=1}^{\infty} \frac{7}{100^k}$

29. $\displaystyle\sum_{k=1}^{\infty} \frac{abc}{1000^k}$

31. $\displaystyle\sum_{k=0}^{\infty} 30 \cdot (0.8)^k$

33. $80, 64, 51.2, 100 \cdot (0.8)^n$

35. $(1/4)^n \to 0$ so the series may converge or may diverge. (Later we will see that it converges.)

37. $(4/3)^n \to \infty \neq 0$ so the series diverges.

39. $\dfrac{\sin(n)}{n} \to 0$ so the series may converge or may diverge.

41. $\cos(1/n) \to \cos(0) = 1 \neq 0$ so the series diverges.

43. $\dfrac{n^2 - 20}{n^5 + 4} \to 0$ so the series may converge or may diverge. (Later we will see that it converges.)

Section 10.3

1. $\displaystyle\sum_{k=0}^{\infty} \left(\tfrac{1}{3}\right)^k = \dfrac{1}{1-\left(\tfrac{1}{3}\right)} = \dfrac{3}{2}$

2. $\displaystyle\sum_{k=0}^{\infty} \left(\tfrac{2}{3}\right)^k = \dfrac{1}{1-\left(\tfrac{2}{3}\right)} = 3$

3. $\dfrac{1}{8} \cdot \displaystyle\sum_{k=0}^{\infty} \left(\tfrac{1}{2}\right)^k = \dfrac{1}{8} \cdot \dfrac{1}{1-\left(\tfrac{1}{2}\right)} = \dfrac{1}{4}$

5. $-\dfrac{2}{3} \cdot \displaystyle\sum_{k=0}^{\infty} \left(-\tfrac{2}{3}\right)^k = -\dfrac{2}{3} \cdot \dfrac{1}{1-\left(-\tfrac{2}{3}\right)} = -\dfrac{2}{5}$

7. (a) $\dfrac{1}{2} \cdot \displaystyle\sum_{k=0}^{\infty} \left(\tfrac{1}{2}\right)^k = \dfrac{1}{2} \cdot \dfrac{1}{1-\left(\tfrac{1}{2}\right)} = \dfrac{1}{2} \cdot \dfrac{2}{1} = 1$, $\dfrac{1}{3} \cdot \displaystyle\sum_{k=0}^{\infty} \left(\tfrac{1}{3}\right)^k = \dfrac{1}{3} \cdot \dfrac{1}{1-\left(\tfrac{1}{3}\right)} = \dfrac{1}{3} \cdot \dfrac{3}{2} = \dfrac{1}{2}$

   (b) $\dfrac{1}{a} \cdot \displaystyle\sum_{k=0}^{\infty} \left(\tfrac{1}{a}\right)^k = \dfrac{1}{a} \cdot \dfrac{1}{1-\left(\tfrac{1}{a}\right)} = \dfrac{1}{a} \cdot \dfrac{a}{a-1} = \dfrac{1}{a-1}$

9. (a) $40 \cdot (0.4)^{n-1}$

   (b) $40 \cdot \displaystyle\sum_{k=0}^{\infty} \left(0.4\right)^k$

   (c) $40 \cdot \dfrac{1}{1-0.4} = \dfrac{40}{0.6} = 66\dfrac{2}{3}$ ft.

11. (a) $\displaystyle\sum_{k=1}^{\infty} \left(\tfrac{1}{2}\right)^k$

   (b) $\dfrac{1}{2}$, $\dfrac{1}{4}$, $\left(\tfrac{1}{2}\right)^n$

   (c) All of the cake.

13. $1 + \dfrac{1}{4} + \left(\tfrac{1}{4}\right)^2 + \left(\tfrac{1}{4}\right)^3 + \ldots = \displaystyle\sum_{k=0}^{\infty} \left(\tfrac{1}{4}\right)^k = \dfrac{4}{3}$.

15. (a) Area $= 1 + \dfrac{3}{9} + \dfrac{3}{9} \cdot \dfrac{4}{9} + \dfrac{3}{9}\left(\tfrac{4}{9}\right)^2 + \dfrac{3}{9}\left(\tfrac{4}{9}\right)^3 + \ldots = 1 + \dfrac{3}{9}\left\{ 1 + \dfrac{4}{9} + \left(\tfrac{4}{9}\right)^2 + \left(\tfrac{4}{9}\right)^3 + \ldots \right\}$

$$= 1 + \dfrac{3}{9}\left\{ \dfrac{1}{1-(4/9)} \right\} = 1 + \dfrac{3}{5} = \dfrac{8}{5} = 1.6 .$$

   (b) Let L be the length of the original triangle ($L = 3\sqrt{\dfrac{4}{\sqrt{3}}}$) and P_n be the perimeter at

   the nth step. Then $P_0 = 3L$. $P_1 = 3 \cdot 4 \cdot \left(\dfrac{L}{3}\right) = 4L$,

$$P_2 = 3 \cdot 4^2 \cdot \left(\dfrac{L}{3^2}\right) = 3L\left(\dfrac{4}{3}\right)^2$$

$$P_3 = 3 \cdot 4^3 \cdot \left(\dfrac{L}{3^3}\right) = 3L\left(\dfrac{4}{3}\right)^3$$

$$P_4 = 3 \cdot 4^4 \cdot \left(\dfrac{L}{3^4}\right) = 3L\left(\dfrac{4}{3}\right)^4 , \text{ and, in general,}$$

$$P_n = 3 \cdot 4^n \cdot \left(\dfrac{L}{3^n}\right) = 3L\left(\dfrac{4}{3}\right)^n .$$

   Since $\dfrac{4}{3} > 1$, the sequence of terms $3L\left(\dfrac{4}{3}\right)^n$ grows without bound, and the perimeter

   "approaches infinity."

17. (a) Height $= 2 + 2(\frac{1}{2}) + 2(\frac{1}{4}) + 2(\frac{1}{8}) + \ldots = 2\left\{ 1 + \frac{1}{2} + (\frac{1}{2})^2 + (\frac{1}{2})^3 + \ldots \right\} = 2 \cdot \frac{1}{1 - \frac{1}{2}} = 4.$

   (b) Surface area $= 4\pi(1)^2 + 4\pi(\frac{1}{2})^2 + 4\pi(\frac{1}{4})^2 + 4\pi(\frac{1}{8})^2 + \ldots$

$$= 4\pi\left\{ 1 + \frac{1}{4} + (\frac{1}{4})^2 + (\frac{1}{4})^3 + \ldots \right\} = 4\pi \cdot \frac{1}{1 - \frac{1}{4}} = \frac{16\pi}{3} \approx 16.755 .$$

   (c) Volume $= \frac{4\pi}{3}(1)^3 + \frac{4\pi}{3}(\frac{1}{2})^3 + \frac{4\pi}{3}(\frac{1}{4})^3 + \frac{4\pi}{3}(\frac{1}{8})^3 + \ldots$

$$= \frac{4\pi}{3}\left\{ 1 + \frac{1}{8} + (\frac{1}{8})^2 + (\frac{1}{8})^3 + \ldots \right\} = \frac{4\pi}{3} \cdot \frac{1}{1 - \frac{1}{8}} = \frac{32\pi}{21} \approx 4.787 .$$

19. $0.8888\ldots = \frac{8}{10} + \frac{8}{10^2} + \frac{8}{10^3} + \ldots = \frac{8}{10}\left\{ 1 + \frac{1}{10} + (\frac{1}{10})^2 + (\frac{1}{10})^3 + \ldots \right\} = \frac{8}{10}\left\{ \frac{10}{9} \right\} = \frac{8}{9} .$

   $0.9999\ldots = \frac{9}{10} + \frac{9}{10^2} + \frac{9}{10^3} + \ldots = \frac{9}{10}\left\{ 1 + \frac{1}{10} + (\frac{1}{10})^2 + (\frac{1}{10})^3 + \ldots \right\} = \frac{9}{10}\left\{ \frac{10}{9} \right\} = 1 .$

   $0.285714\ldots = \frac{285714}{1000000} + \frac{285714}{1000000^2} + \frac{285714}{1000000^3} + \ldots$

$$= \frac{285714}{1000000}\left\{ 1 + \frac{1}{1000000} + (\frac{1}{1000000})^2 + (\frac{1}{1000000})^3 + \ldots \right\}$$

$$= \frac{285714}{1000000}\left\{ \frac{1000000}{999999} \right\} = \frac{285714}{999999} .$$

21. Series converges for $|2x + 1| < 1$: $-1 < x < 0$.   23. Series converges for $|1 - 2x| < 1$: $0 < x < 1$.

25. Series converges for $|7x| < 1$: $-\frac{1}{7} < x < \frac{1}{7}$.   27. Series converges for $|\frac{x}{2}| < 1$: $-2 < x < 2$.

29. Series converges for $|2x| < 1$: $-\frac{1}{2} < x < \frac{1}{2}$.

31. Series converges for $|\sin(x)| < 1$: for all $x \neq \frac{\pi}{2} \pm N\pi$ for integer values of N.

33. The formula is correct if $|x| < 1$. The value $x = 2$ does not satisfy the condition $|x| < 1$, so the formula does not apply.

35. $s_4 = (\frac{1}{3} - \frac{1}{4}) + (\frac{1}{4} - \frac{1}{5}) = \frac{1}{3} - \frac{1}{5}$, $s_5 = \frac{1}{3} - \frac{1}{6}$, $s_n = \frac{1}{3} - \frac{1}{n+1} \to \frac{1}{3}$

37. $s_3 = (1^3 - 2^3) + (2^3 - 3^3) + (3^3 - 4^3) = 1 - 4^3$, $s_4 = 1 - 5^3$, $s_n = 1 - (n+1)^3 \to -\infty$.

39. $s_4 = (f(3) - f(4)) + (f(4) - f(5)) = f(3) - f(5)$, $s_5 = f(3) - f(6)$, $s_n = f(3) - f(n+1)$

41. $s_4 = \sin(1) - \sin(\frac{1}{5}) \approx 0.643$, $s_5 = \sin(1) - \sin(\frac{1}{6}) \approx 0.676$, $s_n = \sin(1) - \sin(\frac{1}{n+1}) \to \sin(1) \approx 0.841$.

43. $s_4 = (\frac{1}{2^2} - \frac{1}{3^2}) + (\frac{1}{3^2} - \frac{1}{4^2}) + (\frac{1}{4^2} - \frac{1}{5^2}) = \frac{1}{4} - \frac{1}{25}$, $s_5 = \frac{1}{4} - \frac{1}{36}$, $s_n = \frac{1}{4} - \frac{1}{(n+1)^2} \to \frac{1}{4}$

45. & 47. On your own.

Section 10.3.5

Integrals and sums

1. The sum.

2. The integral.

3. $\displaystyle\sum_{k=1}^{\infty} f(k)$

4. $\displaystyle\sum_{k=1}^{\infty} f(k)$

5. $\displaystyle\sum_{k=2}^{\infty} f(k)$

6. (b) $f(1) + f(2)$

7. (b) $f(1) + f(2)$

8. (d) $f(3) + f(4)$

9. (c) $f(2) + f(3)$

10. (b), (d), (a), (c) : $\displaystyle\int_{2}^{4} f(x)\,dx \;<\; f(2) + f(3) \;<\; \int_{1}^{3} f(x)\,dx \;<\; f(1) + f(2)$

11. (c), (a), (d), (b) : $f(1) + f(2) + f(3) \;<\; \displaystyle\int_{1}^{4} f(x)\,dx \;<\; f(2) + f(3) + f(4) \;<\; \int_{2}^{5} f(x)\,dx$

Comparisons

12. (a)　No. You are definitely too tall.　　(b)　Apply. You **may** meet the requirements.

　　(c)　No. Definitely too short.　　(d)　Apply. You **may** meet the requirements.

　　(e)　You do not meet the requirements if you are {shorter than Sam} or {taller than Tom}.

　　(f)　You do not have enough information if you are {shorter than Tom} or {taller than Sam}.

13. (a)　You did well (better than Wendy).　　(b)　You may have done well or poorly.

　　(c)　You may have done well or poorly.　　(d)　You did poorly (worse than Paula).

14. (a)　Baker is too easy (easier than Index).　　(b)　Baker may be right for you.

　　(c)　Baker may be right for you.　　(d)　Baker is too hard for you (harder than Liberty Bell).

　　(e)　Baker may be a good climb for you if it is harder than Index and easier than Liberty Bell.

　　(f)　Baker is too easy if Baker is easier than Index. Baker is too hard if Baker is harder than Liberty Bell.

15. (a)　Expect Unknown to be (very) good.　　(b)　Unknown is still unknown.

　　(c)　Unknown is still unknown.　　(d)　Unknown is bad.

16. $\displaystyle\sum_{k=1}^{\infty} \frac{1}{k^2 + 1} \;<\; \sum_{k=1}^{\infty} \frac{1}{k^2}$

17. $\displaystyle\sum_{k=2}^{\infty} \frac{1}{k^3 - 5} \;>\; \sum_{k=2}^{\infty} \frac{1}{k^3}$

18. $\displaystyle\sum_{k=1}^{\infty} \frac{1}{k^2 + 3k - 1} \;<\; \sum_{k=1}^{\infty} \frac{1}{k^2}$

19. $\displaystyle\sum_{k=3}^{\infty} \frac{1}{k^2 + 5k} \;>\; \sum_{k=3}^{\infty} \frac{1}{k^3 + k - 1}$

Ratios of successive terms

20. $a_k = 3k$, $a_{k+1} = 3(k+1)$, $\dfrac{a_{k+1}}{a_k} = \dfrac{k+1}{k}$.

21. $a_k = k + 3$, $a_{k+1} = (k+1) + 3$, $\dfrac{a_{k+1}}{a_k} = \dfrac{k+4}{k+3}$.

22. $a_k = 2k + 5$, $a_{k+1} = 2(k+1) + 5$, $\dfrac{a_{k+1}}{a_k} = \dfrac{2k+7}{2k+5}$.

23. $a_k = 3/k$, $a_{k+1} = \dfrac{3}{k+1}$, $\dfrac{a_{k+1}}{a_k} = \dfrac{\frac{3}{k+1}}{\frac{3}{k}} = \dfrac{k}{k+1}$.

24. $a_k = k^2$, $a_{k+1} = (k+1)^2$, $\dfrac{a_{k+1}}{a_k} = \dfrac{(k+1)^2}{k^2}$.   25. $a_k = 2^k$, $a_{k+1} = 2^{k+1}$, $\dfrac{a_{k+1}}{a_k} = \dfrac{2^{k+1}}{2^k} = 2$.

26. $a_k = (\,1/2\,)^k$, $a_{k+1} = (\,1/2\,)^{k+1}$, $\dfrac{a_{k+1}}{a_k} = \dfrac{(1/2)^{k+1}}{(1/2)^k} = \dfrac{1}{2}$.

27. $a_k = x^k$, $a_{k+1} = x^{k+1}$, $\dfrac{a_{k+1}}{a_k} = \dfrac{x^{k+1}}{x^k} = x$.

28. $a_k = (\,x-1\,)^k$, $a_{k+1} = (x-1)^{k+1}$, $\dfrac{a_{k+1}}{a_k} = \dfrac{(x-1)^{k+1}}{(x-1)^k} = x - 1$.

29. $\displaystyle\sum_{k=1}^{\infty} (\tfrac{1}{2})^k$ is a geometric series with $r = 1/2$ so the series converges. $\dfrac{a_{k+1}}{a_k} = \dfrac{(1/2)^{k+1}}{(1/2)^k} = \dfrac{1}{2}$.

30. $\displaystyle\sum_{k=1}^{\infty} (\tfrac{1}{5})^k$ is a geometric series with $r = 1/5$ so the series converges. $\dfrac{a_{k+1}}{a_k} = \dfrac{(1/5)^{k+1}}{(1/5)^k} = \dfrac{1}{5}$.

31. $\displaystyle\sum_{k=1}^{\infty} 2^k$ is a geometric series with $r = 2$ so the series diverges. $\dfrac{a_{k+1}}{a_k} = \dfrac{2^{k+1}}{2^k} = 2$.

32. $\displaystyle\sum_{k=1}^{\infty} (\,-3\,)^k$ is a geometric series with $r = -3$ so the series diverges. $\dfrac{a_{k+1}}{a_k} = \dfrac{(-3)^{k+1}}{(-3)^k} = -3$.

33. $\displaystyle\sum_{k=1}^{\infty} 4 = 4 + 4 + 4 + ...$ diverges by the Nth Term Test for Divergence since $a_n = 4$ for all n, and

$a_n = 4$ does not approach 0. $\dfrac{a_{k+1}}{a_k} = \dfrac{4}{4} = 1$.

34. $\displaystyle\sum_{k=1}^{\infty} (\,-1\,)^k$ diverges by the Nth Term Test for Divergence since a_n does not approach 0. (It also is a

geometric series with $r = -1$ and $|\,r\,| = 1$.) $\dfrac{a_{k+1}}{a_k} = \dfrac{(-1)^{k+1}}{(-1)^k} = -1$.

35. $\displaystyle\sum_{k=1}^{\infty} \dfrac{1}{k}$ is the harmonic series which diverges. $\dfrac{a_{k+1}}{a_k} = \dfrac{\frac{1}{k+1}}{\frac{1}{k}} = \dfrac{k}{k+1}$.

36. $\displaystyle\sum_{k=1}^{\infty} \dfrac{7}{k} = 7 \cdot \displaystyle\sum_{k=1}^{\infty} \dfrac{1}{k}$ is the divergent harmonic series. $\dfrac{a_{k+1}}{a_k} = \dfrac{\frac{7}{k+1}}{\frac{7}{k}} = \dfrac{k}{k+1}$.

Alternating terms

37. If $a_5 > 0$, then $s_4 < s_5$.

38. If $a_5 = 0$, then $s_4 = s_5$.

39. If $a_5 < 0$, then $s_4 > s_5$.

40. If $a_{n+1} > 0$ for all n, then $s_n < s_{n+1}$ for all n.

41. If $a_{n+1} < 0$ for all n, then $s_n > s_{n+1}$ for all n.

42. If $a_4 > 0$ and $a_5 < 0$, then $s_3 < s_4$ and $s_4 > s_5$.

43. If $a_4 = 0.2$ and $a_5 = -0.1$ and $a_6 = 0.2$, then $s_3 < s_5 < s_4$.

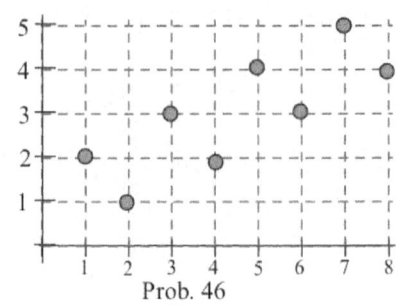
Prob. 46

44. If $a_4 = -0.3$ and $a_5 = 0.2$ and $a_6 = -0.1$, then $s_3 > s_5 > s_4$.

45. If $a_4 = -0.3$ and $a_5 = -0.2$ and $a_6 = 0.1$, then $s_3 > s_4 > s_5$.

46. $s_1 = 2, s_2 = 1, s_3 = 3, s_4 = 2, s_5 = 4, s_6 = 3, s_7 = 5, s_8 = 4$.

47. $s_1 = 2, s_2 = 1, s_3 = 1.9, s_4 = 1.1, s_5 = 1.8, s_6 = 1.2$,

    $s_7 = 1.7, s_8 = 1.3$. The graph is given.

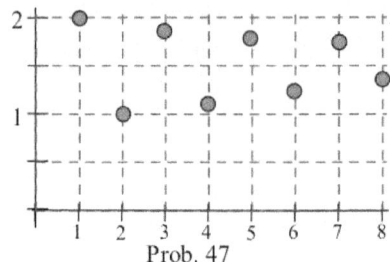
Prob. 47

48. $s_1 = 2, s_2 = 1, s_3 = 2, s_4 = 1, s_5 = 2, s_6 = 1, s_7 = 2, s_8 = 1$.

    The graph is given.

49. $s_1 = -2, s_2 = -0.5, s_3 = -1.3, s_4 = -0.7, s_5 = -1.1, s_6 = -0.9$,

    $s_7 = 1.1, s_8 = 1.0$. The graph is given.

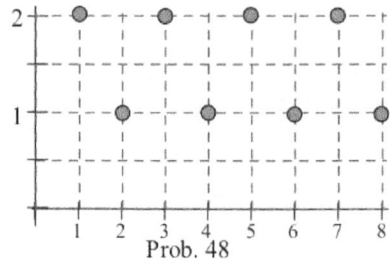
Prob. 48

50. $s_1 = 5, s_2 = 6, s_3 = 5.4, s_4 = 5.0, s_5 = 5.2, s_6 = 5.3$,

    $s_7 = 5.4, s_8 = 5.2$. The graph is given.

51. If the a_k alternate in sign, then the graph of the partial sums

    s_n follows an "up–down–up–down" pattern?

52. If the a_k alternate in sign and decrease in magnitude (the $|a_k|$

    is decreasing), then the graph of the partial sums s_n forms

    a "narrowing funnel" pattern.

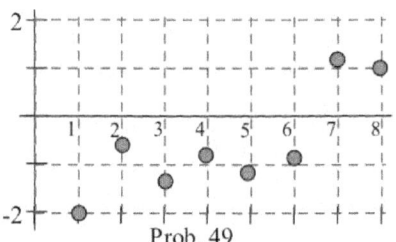
Prob. 49

53. (a) The terms a_k alternate in sign for the graphs D, E, and F.

    (b) The $|a_k|$ decrease for the graphs A and D.

    (c) The terms a_k alternate in sign and decrease in absolute

        value for the graph D.

54. (a) The terms a_k alternate in sign for the graphs C, D, E,

        and F in Fig. 14.

    (b) The $|a_k|$ decrease for the graphs B and C.

    (c) The terms a_k alternate in sign and decrease in absolute

        value for the graph C.

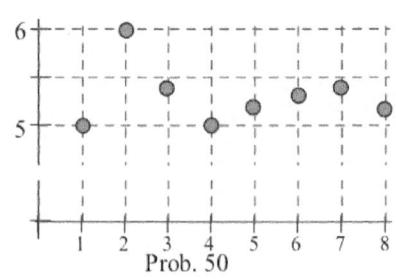
Prob. 50

55. The graph of the first eight partial sums of

$$\sum_{k=0}^{\infty} \left(-\frac{1}{2}\right)^k = 1 - \frac{1}{2} + \frac{1}{4} - \ldots = \frac{2}{3} \quad \text{is given.}$$

Notice the "narrowing funnel" shape of the graph.

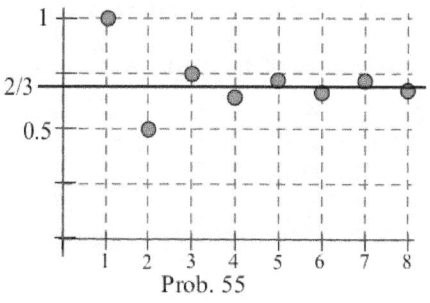
Prob. 55

56. The graph of the first eight partial sums of

$$\sum_{k=0}^{\infty} \left(-0.6\right)^k = 1 - 0.6 + 0.36 - \ldots = 0.625$$

is given.

Notice the "narrowing funnel" shape of the graph.

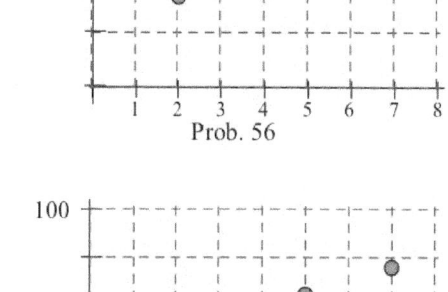
Prob. 56

57. The graph is given of the first eight partial sums of the

divergent series

$$\sum_{k=0}^{\infty} \left(-2\right)^k = 1 - 2 + 4 - \ldots$$

This is a divergent series (N^{th} Term Test for Divergence).

Notice the "widening funnel" shape of the graph.

100

-100

Prob. 57

Section 10.4 (Odd numbered problem solutions followed by even numbered problem answers.)

1. $\displaystyle\int_1^{\infty} \frac{1}{2x+5}\, dx = \frac{1}{2}\,\ln|2x+5|\,\Big|_1^A = \frac{1}{2}\,\ln|2A+5| - \frac{1}{2}\,\ln|7| \to \infty$ (as $A \to \infty$) so $\displaystyle\sum_{k=1}^{\infty} \frac{1}{2k+5}$ diverges.

3. $\displaystyle\int_1^{\infty} (2x+5)^{-3/2}\, dx = -(2x+5)^{-1/2}\,\Big|_1^A = \frac{-1}{\sqrt{2A+5}} - \frac{-1}{\sqrt{7}} \to \frac{1}{\sqrt{7}}$ (as $A \varnothing \infty$) so $\displaystyle\sum_{k=1}^{\infty} \frac{1}{(2k+5)^{3/2}}$

converges.

5. $\displaystyle\int \frac{1}{x \cdot (\ln(x))^2}\, dx = \frac{-1}{\ln(x)} + C$ (using a u–substitution with $u = \ln(x)$ and $du = \frac{1}{x}\, dx$) so

$\displaystyle\int_2^{\infty} \frac{1}{x \cdot (\ln(x))^2}\, dx = \frac{-1}{\ln(x)}\,\Big|_2^A = \frac{-1}{\ln(A)} - \frac{-1}{\ln(2)} \to \frac{1}{\ln(2)}$ (as $A \varnothing \infty$) so $\displaystyle\sum_{k=2}^{\infty} \frac{1}{k \cdot (\ln(k))^2}$ converges.

7. $\displaystyle\int_1^{\infty} \frac{1}{x^2+1}\, dx = \arctan(x)\,\Big|_1^A = \arctan(A) - \arctan(1) \to \frac{\pi}{2} - \frac{\pi}{4}$ (as $A \varnothing \infty$) so $\displaystyle\sum_{k=1}^{\infty} \frac{1}{k^2+1}$ converges.

9. This is a telescoping series:

$$\sum_{k=1}^{\infty} \left\{ \tfrac{1}{k} - \tfrac{1}{k+3} \right\} = \left\{ 1 - \tfrac{1}{4} \right\} + \left\{ \tfrac{1}{2} - \tfrac{1}{5} \right\} + \left\{ \tfrac{1}{3} - \tfrac{1}{6} \right\} + \left\{ \tfrac{1}{4} - \tfrac{1}{7} \right\} + \left\{ \tfrac{1}{5} - \tfrac{1}{8} \right\} + \ldots \to 1 + \tfrac{1}{2} + \tfrac{1}{3} .$$

The Integral Test also works:

$$\int_{1}^{\infty} \tfrac{1}{x} - \tfrac{1}{x+3} \, dx = \ln|x| - \ln|x+3| \, \Big|_{1}^{A} = \{ \ln|A| - \ln|A+3| \} - \{ \ln(1) - \ln(4) \}$$

$$= \ln\left| \tfrac{4A}{A+3} \right| \to \ln(4) \ (\text{as } A \to \infty) \ \text{ so } \ \sum_{k=1}^{\infty} \left\{ \tfrac{1}{k} - \tfrac{1}{k+3} \right\} \text{ converges.}$$

(Notice that the "telescoping series" method gives the value of the series, but the Integral Test only tells us that the series converges. For this series, the "telescoping series" method is both easier and more precise.)

11. $\displaystyle\int_{1}^{\infty} \frac{1}{x(x+5)} \, dx = \int_{1}^{\infty} \frac{1}{5} \left\{ \tfrac{1}{x} - \tfrac{1}{x+5} \right\} dx$ (using the method of Partial Fraction Decomposition)

$$= \tfrac{1}{5} \left\{ \ln|x| - \ln|x+5| \right\} = \tfrac{1}{5} \left\{ \ln\left| \tfrac{x}{x+5} \right| \right\} \Big|_{1}^{A} = \tfrac{1}{5} \left\{ \ln\left| \tfrac{A}{A+5} \right| - \ln\left| \tfrac{1}{6} \right| \right\} \to \tfrac{1}{5} \left\{ \ln(1) - \ln\left(\tfrac{1}{6} \right) \right\}$$

(as $A \to \infty$) so $\displaystyle\sum_{k=1}^{\infty} \frac{1}{k \cdot (k+5)}$ converges.

13. $\displaystyle\int_{1}^{\infty} x \cdot e^{-(x^2)} \, dx = \frac{-1}{2} e^{-(x^2)} \, \Big|_{1}^{A} = \left(\tfrac{-1}{2} e^{-(A^2)} \right) - \left(\tfrac{-1}{2} e^{-1} \right) = \frac{1}{2e} - \frac{1}{2e^{(A^2)}} \to \frac{1}{2e}$ (as $A \varnothing \infty$) so

$\displaystyle\sum_{k=1}^{\infty} k \cdot e^{-(k^2)}$ converges.

15. $\displaystyle\int \frac{1}{\sqrt{6x+10}} \, dx = \frac{1}{3}\sqrt{6x+10} + C$ (using a u–substitution with $u = 6x + 10$ and $du = 6 \, dx$). Then

$$\int_{1}^{\infty} \frac{1}{\sqrt{6x+10}} \, dx = \frac{1}{3}\sqrt{6x+10} \, \Big|_{1}^{A} = \frac{1}{3}\sqrt{6A+10} - \frac{1}{3}\sqrt{16} \to \infty \ (\text{as } A \to \infty)$$

so $\displaystyle\sum_{k=1}^{\infty} \frac{1}{\sqrt{6k+10}}$ diverges.

17. $p = 3 > 1$ so $\displaystyle\sum_{k=1}^{\infty} \frac{1}{k^3}$ converges.   19. $p = 1/2 < 1$ so $\displaystyle\sum_{k=2}^{\infty} \frac{1}{\sqrt{k}}$ diverges.

21. $p = 3/2 > 1$ so $\displaystyle\sum_{k=3}^{\infty} \frac{1}{k^{3/2}}$ converges.

23. $\displaystyle\sum_{k=1}^{\infty}\frac{1}{k^3}$: $\displaystyle\int_1^{11}\frac{1}{x^3}\,dx = 0.4958677$, $1+\displaystyle\int_1^{10}\frac{1}{x^3}\,dx = 1.495$ so $0.4958677 < s_{10} < 1.495$.

$\displaystyle\int_1^{101}\frac{1}{x^3}\,dx = 0.499951$, $1+\displaystyle\int_1^{100}\frac{1}{x^3}\,dx = 1.49995$ so $0.499951 < s_{100} < 1.49995$.

$\displaystyle\int_1^{1000001}\frac{1}{x^3}\,dx = 0.5000000$, $1+\displaystyle\int_1^{1000000}\frac{1}{x^3}\,dx = 1.5000000$ so $0.5000000 < s_{1000000} < 1.5000000$.

25. $\displaystyle\sum_{k=1}^{\infty}\frac{1}{k+1000}$: $\displaystyle\int_1^{11}\frac{1}{x+1000}\,dx = 0.0099404$, $\dfrac{1}{1001}+\displaystyle\int_1^{10}\frac{1}{x+1000}\,dx = 0.0099498$

so $\mathbf{0.0099404} < s_{10} < \mathbf{0.0099498}$. (This is a very precise estimate of s_{10} .)

$\displaystyle\int_1^{101}\frac{1}{x+1000}\,dx = 0.09522$, $\dfrac{1}{1001}+\displaystyle\int_1^{100}\frac{1}{x+1000}\,dx = 0.0953$ so $\mathbf{0.09522} < s_{100} < \mathbf{0.0953}$.

$\displaystyle\int_1^{1000001}\frac{1}{x+1000}\,dx = 6.90776$, $\dfrac{1}{1001}+\displaystyle\int_1^{1000000}\frac{1}{x+1000}\,dx = 6.90875$ so $\mathbf{6.90}776 < s_{1000000} < \mathbf{6.90}875$.

27. $\displaystyle\sum_{k=1}^{\infty}\frac{1}{k^2+100}$: $\displaystyle\int_1^{11}\frac{1}{x^2+100}\,dx = \dfrac{1}{10}\arctan\left(\dfrac{x}{10}\right)\Big|_1^{11} = 0.0733$, $\dfrac{1}{101}+\displaystyle\int_1^{10}\frac{1}{x^2+100}\,dx = 0.0783$

so $\mathbf{0.073} < s_{10} < \mathbf{0.078}$. Also, $\mathbf{0.137} < s_{100} < \mathbf{0.147}$ and $\mathbf{0.1471} < s_{1000000} < \mathbf{0.157}$.

29. For $\mathbf{q \neq 1}$, let $u = \ln(x)$ and $du = \dfrac{1}{x}\,dx$.

Then $\displaystyle\int \frac{1}{x\cdot(\ln(x))^q}\,dx = \int \frac{1}{(u)^q}\,du = \frac{1}{1-q}\,u^{-q+1} = \frac{1}{1-q}\,(\ln(x))^{1-q} + C.$

Then $\displaystyle\int_2^{\infty} \frac{1}{x\cdot(\ln(x))^q}\,dx = \frac{1}{1-q}\,(\ln(x))^{1-q}\Big|_2^A = \frac{1}{1-q}\,(\ln(A))^{1-q} - \frac{1}{1-q}\,(\ln(2))^{1-q}.$

If $q < 1$, then $\dfrac{1}{1-q}\,(\ln(A))^{1-q} - \dfrac{1}{1-q}\,(\ln(2))^{1-q} \to \infty$ (as $A \to \infty$) so $\displaystyle\sum_{k=2}^{\infty}\frac{1}{k\cdot(\ln k)^q}$ diverges.

If $q > 1$, then $\dfrac{1}{1-q}\,(\ln(A))^{1-q} - \dfrac{1}{1-q}\,(\ln(2))^{1-q} \to -\dfrac{1}{1-q}\,(\ln(2))^{1-q}$ (as $A \to \infty$)

so $\displaystyle\sum_{k=2}^{\infty}\frac{1}{k\cdot(\ln k)^q}$ converges.

If $\mathbf{q = 1}$, then

$$\int_{2}^{\infty} \frac{1}{x \cdot (\ln(x))^q} \, dx = \int_{2}^{\infty} \frac{1}{x \cdot \ln(x)} \, dx = \ln|\ln(x)| \Big|_{2}^{A} = \ln|\ln(A)| - \ln|\ln(2)| \rightarrow \infty \ (\text{as } A \rightarrow \infty)$$

so $\displaystyle\sum_{k=2}^{\infty} \frac{1}{k \cdot (\ln k)}$ diverges.

"Q–Test:" $\displaystyle\sum_{k=2}^{\infty} \frac{1}{k \cdot (\ln k)^q}$ $\begin{cases} \text{diverges} & \text{if } q \leq 1 \\ \text{converges} & \text{if } q > 1 \end{cases}$

31. $q = 3 > 1$ so $\displaystyle\sum_{k=2}^{\infty} \frac{1}{k \cdot (\ln k)^3}$ converges.

33. $\displaystyle\sum_{k=2}^{\infty} \frac{1}{k \cdot \ln (k^3)} = \sum_{k=2}^{\infty} \frac{1}{3k \cdot \ln (k)} = \frac{1}{3} \sum_{k=2}^{\infty} \frac{1}{k \cdot \ln (k)}$ which diverges ($q = 1$).

Section 10.4 Some Even Answers

2. Converge	4. Diverge	6. Converge	8. Converge	10. Converge
12. Converge	14. Converge	16. Converge	18. Diverge	20. Diverge
30. Diverge	32. Diverge			

Section 10.5 (Odd numbered problem solutions followed by even numbered problem answers.)

1. $\displaystyle\sum_{k=1}^{\infty} \frac{\cos^2(k)}{k^2} \leq \sum_{k=1}^{\infty} \frac{1}{k^2}$ which converges by the P–Test (p=2) so $\displaystyle\sum_{k=1}^{\infty} \frac{\cos^2(k)}{k^2}$ converges.

3. $\displaystyle\sum_{n=3}^{\infty} \frac{5}{n-1} > 5 \sum_{n=3}^{\infty} \frac{1}{n}$ which is the harmonic series and is divergent, so $\displaystyle\sum_{n=3}^{\infty} \frac{5}{n-1}$ diverges.

5. $-1 \leq \cos(x) \leq 1$ so $2 \leq 3 + \cos(x) \leq 4$. Then $\displaystyle\sum_{j=1}^{\infty} \frac{3 + \cos(j)}{j} > 2 \sum_{j=1}^{\infty} \frac{1}{j}$ which is the harmonic series   and is

divergent, so $\displaystyle\sum_{j=1}^{\infty} \frac{3 + \cos(j)}{j}$ diverges.

7. $\displaystyle\sum_{k=1}^{\infty} \frac{\ln(k)}{k} > \sum_{k=1}^{\infty} \frac{1}{k}$ which is the harmonic series and is divergent, so $\displaystyle\sum_{k=1}^{\infty} \frac{\ln(k)}{k}$ diverges.

9. $\displaystyle\sum_{k=1}^{\infty} \frac{k+9}{k \cdot 2^k} = \sum_{k=1}^{\infty} \frac{k+9}{k} \cdot \frac{1}{2^k} < 10 \sum_{k=1}^{\infty} \frac{1}{2^k}$ which is a convergent geometric series ($r = \frac{1}{2}$)

so $\displaystyle\sum_{k=1}^{\infty} \frac{k+9}{k \cdot 2^k}$ converges.

11. $\displaystyle\sum_{n=1}^{\infty} \frac{1}{1+2+3+\ldots+(n-1)+n} = \sum_{n=1}^{\infty} \frac{1}{\frac{n(n+1)}{2}} = \sum_{n=1}^{\infty} \frac{2}{n(n+1)}) < 2 \sum_{n=1}^{\infty} \frac{1}{n^2}$ which converges by

the P–Test ($p = 2$) so $\displaystyle\sum_{n=1}^{\infty} \frac{1}{1+2+3+\ldots+(n-1)+n}$ converges.

13. Let $a_k = \dfrac{k+1}{k^2+4}$ and $b_k = \dfrac{1}{k}$. Then $\dfrac{a_k}{b_k} = \dfrac{k^2+k}{k^2+4} \to 1$ and $\displaystyle\sum_{k=3}^{\infty} \frac{1}{k}$ diverges so $\displaystyle\sum_{k=3}^{\infty} \frac{k+1}{k^2+4}$ diverges.

15. Let $a_w = \dfrac{5}{w+1}$ and $b_w = \dfrac{5}{w}$. Then $\dfrac{a_w}{b_w} = \dfrac{w}{w+1} \to 1$ and $\displaystyle\sum_{w=1}^{\infty} \frac{1}{w}$ diverges so $\displaystyle\sum_{w=1}^{\infty} \frac{5}{w+1}$ diverges.

17. $\displaystyle\sum_{k=1}^{\infty} \frac{k^3}{(1+k^2)^3} = \sum_{k=1}^{\infty} \left(\frac{k}{1+k^2}\right)^3$. Let $a_k = \left(\dfrac{k}{1+k^2}\right)^3$ and $b_k = \dfrac{1}{k^3}$.

Then $\dfrac{a_k}{b_k} = \left(\dfrac{k}{1+k^2}\right)^3 \cdot \dfrac{k^3}{1} = \left(\dfrac{k^2}{1+k^2}\right)^3 \to (1)^3 = 1$ and $\displaystyle\sum_{k=1}^{\infty} \frac{1}{k^3}$ converges by the

P–Test so $\displaystyle\sum_{k=1}^{\infty} \frac{k^3}{(1+k^2)^3}$ converges.

19. $\displaystyle\sum_{n=1}^{\infty} \left(\frac{5-\frac{1}{n}}{n}\right)^3 = \sum_{n=1}^{\infty} \left(\frac{5}{n}-\frac{1}{n^2}\right)^3 = \sum_{n=1}^{\infty} \left(\frac{5n-1}{n^2}\right)^3$. Let $a_n = \left(\dfrac{5n-1}{n^2}\right)^3$ and $b_n = \dfrac{1}{n^3}$.

Then $\dfrac{a_n}{b_n} = \left(\dfrac{5n-1}{n^2}\right)^3 \dfrac{n^3}{1} = \left(\dfrac{5n-1}{n^2} \cdot \dfrac{n}{1}\right)^3 = \left(\dfrac{5n-1}{n}\right)^3 \to 5^3 = 125$ (positive and finite).

$\displaystyle\sum_{k=1}^{\infty} \frac{1}{k^3}$ converges by the P–Test so $\displaystyle\sum_{n=1}^{\infty} \left(\frac{5-\frac{1}{n}}{n}\right)^3$ converges.

21. $\displaystyle\sum_{j=1}^{\infty} \left(1 - \frac{1}{j}\right)^j$ diverges by the Nth Term Test for Divergence since $a_j = \left(1 - \dfrac{1}{j}\right)^j \to e^{-1} \neq 0$ as $j \to \infty$.

We could use the Limit Comparison Test by taking $b_j = e^{-1}$ and showing that $\dfrac{a_j}{b_j} \to 1$, but the

N^{th} Term Test for Divergence is more direct for this series.

23. $\displaystyle\sum_{k=1}^{\infty} \frac{7k}{\sqrt{k^3+5}}$: The dominant term series is $\displaystyle\sum_{k=1}^{\infty} \frac{k}{k^{3/2}} = \sum_{k=1}^{\infty} \frac{1}{k^{1/2}}$ which diverges by the P–Test ($p=1/2$).

25. $\sum\limits_{j=1}^{\infty} \dfrac{j^3 - 4j + 3}{2j^4 + 7j^6 + 9}$: The dominant term series is $\sum\limits_{j=1}^{\infty} \dfrac{j^3}{j^6} = \sum\limits_{j=1}^{\infty} \dfrac{1}{j^3}$ which converges by the P–Test (p=3).

27. $\sum\limits_{n=1}^{\infty} \left(\dfrac{\arctan(3n)}{2n} \right)^2$: The dominant term series is $\sum\limits_{n=1}^{\infty} \left(\dfrac{\pi/2}{2n} \right)^2 = \dfrac{\pi^2}{16} \sum\limits_{n=1}^{\infty} \dfrac{1}{n^2}$ which converges

by the P–Test (p=2).

29. $\sum\limits_{j=1}^{\infty} \dfrac{\sqrt{j^3 + 4j^2}}{j^2 + 3j - 2}$: The dominant term series is $\sum\limits_{j=1}^{\infty} \dfrac{j^{3/2}}{j^2} = \sum\limits_{j=1}^{\infty} \dfrac{1}{j^{1/2}}$ which diverges by the P–Test.

31. $\sum\limits_{n=2}^{\infty} \dfrac{n^2 + 10}{n^3 - 2}$ diverges using dominant terms and the P–Test (p=1).

33. $\sum\limits_{k=1}^{\infty} \dfrac{3}{2k + 1}$ diverges using dominant terms and the P–Test (p=1).

35. $\sum\limits_{n=1}^{\infty} \dfrac{2n^3 + n^2 + 5}{(3 + n^2)^2}$ diverges using dominant terms and the P–Test (p=1).

37. $\sum\limits_{k=1}^{\infty} \left(\dfrac{1 - \frac{2}{k}}{k} \right)^3 = \sum\limits_{k=1}^{\infty} \left(\dfrac{k - 2}{k^2} \right)^3$ converges using dominant terms $\left(\dfrac{k}{k^2} \right)^3 = \left(\dfrac{1}{k} \right)^3$ and the P–Test (p=3).

39. $\sum\limits_{k=1}^{\infty} \dfrac{k + 5}{k \cdot 3^k}$ converges by using dominant terms $\dfrac{k}{k \cdot 3^k} = \dfrac{1}{3^k} = \left(\dfrac{1}{3} \right)^k$ and the Geometric Series Test (r = 1/3).

41. $\sum\limits_{k=1}^{\infty} \dfrac{k + 2}{\sqrt{k^2 + 1}}$ diverges using dominant terms and the Nth Term Test for Divergence.

43. $\sum\limits_{j=1}^{\infty} \dfrac{3}{e^j + j}$ converges using dominant terms $\dfrac{1}{e^j} = \left(\dfrac{1}{e} \right)^j$ and the Geometric Series Test (r = 1/e).

45. $\sum\limits_{n=1}^{\infty} \left(\dfrac{\tan(3)}{2 + n} \right)^2$ converges by using dominant terms $\dfrac{1}{n^2}$ and the P–Test (p=2).

47. $\sum\limits_{k=1}^{\infty} \sin^3 \left(\dfrac{1}{n} \right)$ converges by comparison with the convergent series $\sum\limits_{n=1}^{\infty} \dfrac{1}{n^3}$

49. $\sum\limits_{j=1}^{\infty} \cos^3(\frac{1}{n})$ diverges using the Nth Term Test for Divergence: the terms approach $1 \neq 0$.

51. $\sum\limits_{n=1}^{\infty} (1 - \frac{2}{n})^n$ diverges using the Nth Term Test for Divergence: the terms approach $\frac{1}{e^2} \neq 0$.

Section 10.5 Some Even Answers

2. Converges   4. Converges   6. Converges   8. Converges   10. Diverges

12. Converges   14. Converges   16. Diverges   18. Converges   20. Diverges

22. Diverges   24. Diverges   26. Converges   28. Converges   30. Converges

32. Converges   34. Converges   36. Converges   38. Converges   40. Converges

42. Converges   44. Converges   46. Converges   48. Diverges   50. Converges

You still need to supply reasons for each answer given below.

R1. Converge   R2. Converge   R3. Diverge   R4. Diverge   R5. Converge

R6. Converge   R7. Converge   R8. Diverge   R9. Converge   R10. Diverge

R11. Converge   R12. Diverge   R13. Diverge   R14. Converge   R15. Converge

R16. Converge   R17. Diverge   R18. Diverge   R19. Diverge   R20. Converge

R21. Diverge

Section 10.6 (Odd numbered problem solutions followed by some even numbered problem answers.)

1. $1, 0.2, 0.8, 0.4$ Alternating (so far). The graph of s_n is shown.

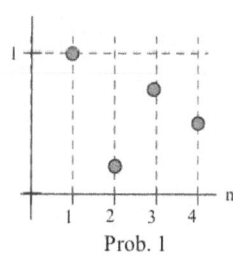
Prob. 1

3. $-1, 1, -2, 2$ Alternating (so far). The graph of s_n is shown.

5. $-1, -1.6, -1.2, -1$ Not alternating. The graph of s_n is shown.

7. Alternating: $a_1 = 2, a_2 = -1, a_3 = 2, a_4 = -1, a_5 = 2$

9. Not alternating: $a_1 = 2, a_2 = 1, a_3 = -0.9, a_4 = 0.8, a_5 = -0.1$

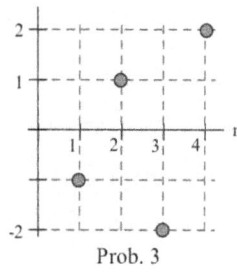
Prob. 3

11. Not alternating: $a_1 = -1, a_2 = 2, a_3 = -1.2, a_4 = 0.2, a_5 = 0.2$

13. Graphs A and C are not the graphs of partial sums of alternating series.

15. Graph B is not the graph of an alternating series.

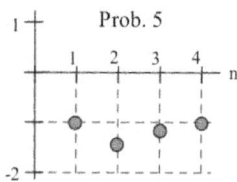
Prob. 5

17. Converges.   19. Converges.   21. Diverges.   23. Converges.

25. Converges.   27. Diverges.   29. Converges.   31. Converges.

33. $s_4 = -\dfrac{1}{\ln(2)} + \dfrac{1}{\ln(3)} - \dfrac{1}{\ln(4)} + \dfrac{1}{\ln(5)} \approx -0.63247$.

   $|s_4 - S| < \dfrac{1}{\ln(6)} \approx 0.55811$. $-1.19058 < S < -0.07436$

35. $s_4 = (-0.8)^2 + (-0.8)^3 + (-0.8)^4 + (-0.8)^5 \approx 0.20992$. $|s_4 - S| < (0.8)^6 \approx 0.26214$.

   $-0.05222 < S < 0.47206$. (For your information: $s_{10} \approx 0.317378$, $s_{50} \approx 0.355550$, $s_{100} \approx 0.355556$)

37. $s_4 = \sin(1) - \sin(\tfrac{1}{2}) + \sin(\tfrac{1}{3}) - \sin(\tfrac{1}{4}) \approx 0.441836$. $|s_4 - S| < \sin(\tfrac{1}{5}) \approx 0.198669$.

   $0.243167 < S < 0.640505$. ($s_{10} \approx 0.503356$, $s_{50} \approx 0.0.540897$, $s_{100} \approx 0.5458219$)

39. $s_4 = -1 + \dfrac{1}{8} - \dfrac{1}{27} + \dfrac{1}{64} \approx -0.896412$. $|s_4 - S| < \dfrac{1}{125} = 0.008$. $-0.904412 < S < -0.888412$.

   ($s_{10} \approx -0.901116$, $s_{50} \approx -0.901539$, $s_{100} \approx -0.901542$)

41. $|\dfrac{1}{(n+1)+6}| \le 0.01$ so $n = 93$ works. Use s_{93} .

43. $|\dfrac{1}{\sqrt{n+1}}| \le \dfrac{1}{100}$ so $\sqrt{n+1} \ge 100$ and $n = 10{,}000 - 1$ works. Use s_{9999} .

45. $|\dfrac{1}{3^{n+1}}| \le \dfrac{2}{1000}$ so $3^{n+1} \ge 500$ and $n = 5$ works. Use s_5 .

47. $|\dfrac{1}{(n+1)^2}| \le \dfrac{1}{1000}$ so $(n+1)^2 \ge 1000$ and $n = 31$ works. Use s_{31} .

49. $|\dfrac{1}{(n+1) + \ln(n+1)}| \le \dfrac{4}{100}$ so $(n+1) + \ln(n+1) \ge 25$. Some "calculator experimentation" shows

   that $n = 21$ works. $(20+1) + \ln(20+1) \approx 24.04$ so $n = 20$ is too small. $(21+1) + \ln(21+1) \approx 25.09$

   so $n = 21$ works. Use s_{21} .

51. (a) $S(0.3) = x - \dfrac{(0.3)^3}{2 \cdot 3} + \dfrac{(0.3)^5}{2 \cdot 3 \cdot 4 \cdot 5} - \dfrac{(0.3)^7}{2 \cdot 3 \cdot 4 \cdot 5 \cdot 6 \cdot 7} + \ldots + (-1)^n \dfrac{(0.3)^{2n+1}}{(2n+1)!} + \ldots$

   (b) $s_3 = (0.3) - \dfrac{(0.3)^3}{2 \cdot 3} + \dfrac{(0.3)^5}{2 \cdot 3 \cdot 4 \cdot 5} \approx 0.29552025$.

   (c) $|S - s_3| \le \dfrac{(0.3)^7}{2 \cdot 3 \cdot 4 \cdot 5 \cdot 6 \cdot 7} \approx 0.000000043$. ($\sin(0.3) \approx 0.295520206661$)

53. (a) $S(0.1) = x - \dfrac{(0.1)^3}{2 \cdot 3} + \dfrac{(0.1)^5}{2 \cdot 3 \cdot 4 \cdot 5} - \dfrac{(0.1)^7}{2 \cdot 3 \cdot 4 \cdot 5 \cdot 6 \cdot 7} + \ldots + (-1)^n \dfrac{(0.1)^{2n+1}}{(2n+1)!} + \ldots$

   (b) $s_3 = (0.1) - \dfrac{(0.1)^3}{2 \cdot 3} + \dfrac{(0.1)^5}{2 \cdot 3 \cdot 4 \cdot 5} \approx 0.099833416667$.

   (c) $|S - s_3| \le \dfrac{(0.1)^7}{2 \cdot 3 \cdot 4 \cdot 5 \cdot 6 \cdot 7} \approx 1.98 \cdot 10^{-11}$ .. ($\sin(0.1) \approx 0.099833416647$)

55. (a) $C(1) = 1 - \dfrac{1^2}{2} + \dfrac{1^4}{2\cdot3\cdot4} - \dfrac{1^6}{2\cdot3\cdot4\cdot5\cdot6} + ... + (-1)^n \dfrac{1^{2n}}{(2n)!} + ... = 1 - \dfrac{1}{2} + \dfrac{1}{24} - \dfrac{1}{720} + ... + (-1)^n \dfrac{1^{2n}}{(2n)!} + ...$

   (b) $s_3 = 1 - \dfrac{1^2}{2} + \dfrac{1^4}{2\cdot3\cdot4} \approx 0.5416667$.

   (c) $| S - s_3 | \le \dfrac{(1)^7}{2\cdot3\cdot4\cdot5\cdot6\cdot7} \approx 0.0013889 ..$ ($\cos(1) \approx 0.5403023$)

57. (a) $C(-0.2) = 1 - \dfrac{(-0.2)^2}{2} + \dfrac{(-0.2)^4}{2\cdot3\cdot4} - \dfrac{(-0.2)^6}{2\cdot3\cdot4\cdot5\cdot6} + ... + (-1)^n \dfrac{1^{2n}}{(2n)!} + ...$

   (b) $s_3 = 1 - \dfrac{(-0.2)^2}{2} + \dfrac{(-0.2)^4}{2\cdot3\cdot4} \approx 0.98006666667$.

   (c) $| S - s_3 | \le \dfrac{(-0.2)^6}{2\cdot3\cdot4\cdot5\cdot6} \approx 8.9 \cdot 10^{-8} ....$ ($\cos(-0.2) \approx 0.980066577841$)

59. (a) $E(-1) = 1 + (-1) + \dfrac{(-1)^2}{2} + \dfrac{(-1)^3}{2\cdot3} + \dfrac{(-1)^4}{2\cdot3\cdot4} + ... + \dfrac{(-1)^n}{n!} + ... = 1 - 1 + \dfrac{1}{2} - \dfrac{1}{6} + \dfrac{1}{24} - \dfrac{1}{120} + ...$

   (b) $s_3 = 1 - 1 + \dfrac{1}{2} = 0.5$.

   (c) $| S - s_3 | \le \dfrac{1}{6} \approx 0.16667 ....$ ($e^{-1} \approx 0.36787944$)

61. (a) $E(-0.2) = 1 + (-0.2) + \dfrac{(-0.2)^2}{2} + \dfrac{(-0.2)^3}{2\cdot3} + \dfrac{(-0.2)^4}{2\cdot3\cdot4} + ... + \dfrac{(-1)^n}{n!} + ...$

   (b) $s_3 = 1 + (-0.2) + \dfrac{(-0.2)^2}{2} = 0.82$ (c) $| S - s_3 | \le \dfrac{(0.2)^3}{6} \approx 0.0013333 .... $ ($e^{-0.2} \approx 0.8187307$)

Section 10.6 Some Even Answers

2. Alternating (so far)   4. Not alternating   6. Alternating (so far)

8. Alternating (so far)   10. Alternating (so far)   12. Not alternating

14. B and C .   16. A, B, and C   18. Converges.   20. Converges.

22. Converges.   24. Converges (to 0).   26. Diverges.   28. Converges.

30. Converges.   42. $n \approx 2.7 \cdot 10^{43}$ 44. $n = 26$ works.   46. $n = 16$ works.

48. $n = 21$ works.

Section 10.7 (Odd numbered problem solutions followed by some even numbered problem answers.)

1. Conditionally convergent   3. Absolutely convergent   5. Absolutely convergent

7. Absolutely convergent   9. Conditionally convergent   11. Conditionally convergent

13. Conditionally convergent   15. Absolutely convergent   17. Divergent

19. Conditionally convergent   21. Divergent   23. Absolutely convergent

25. Conditionally convergent   27. Divergent   29. Absolutely convergent

31. $\dfrac{n!}{(n+1)!} = \dfrac{1 \cdot 2 \cdot 3 \ldots \cdot n}{1 \cdot 2 \cdot 3 \ldots \cdot n \cdot (n+1)} = \dfrac{1}{n+1}$ 33. $\dfrac{n!}{(n+3)!} = \dfrac{1 \cdot 2 \cdot 3 \ldots \cdot n}{1 \cdot 2 \cdot 3 \ldots \cdot n \cdot (n+1) \cdot (n+2) \cdot (n+3)} = \dfrac{1}{(n+1)(n+2)(n+3)}$

35. $\dfrac{(n-1)!}{(n+1)!} = \dfrac{1 \cdot 2 \cdot 3 \ldots \cdot (n-1)}{1 \cdot 2 \cdot 3 \ldots \cdot (n-1) \cdot (n) \cdot (n+1)} = \dfrac{1}{(n)(n+1)}$

37. $\dfrac{(2n!)}{(2n+1)!} = \dfrac{1 \cdot 2 \cdot 3 \ldots \cdot n \cdot (n+1) \cdot \ldots \cdot (2n)}{1 \cdot 2 \cdot 3 \ldots \cdot n \cdot (n+1) \cdot \ldots \cdot (2n) \cdot (2n+1)} = \dfrac{1}{2n+1}$ 39. $\dfrac{n^n}{n!} = \dfrac{n \cdot n \cdot n \cdot n \cdot \ldots \cdot n \cdot n}{1 \cdot 2 \cdot 3 \cdot 4 \cdot \ldots \cdot (n-1) \cdot n}$

41. Ratio $= \left| \dfrac{n}{n+1} \right| \to 1 = L.$ Diverges   (Harmonic series)

43. Ratio $= \left| \left(\dfrac{n}{n+1} \right)^3 \right| \to 1 = L.$ Converges by the P–Test (p = 3).

45. Ratio $= \left| \dfrac{1}{2} \right| \to \dfrac{1}{2} = L.$ Converges   (Geometric series with r = 1/2).

47. Ratio $= |1| \to 1 = L.$ Diverges by the Nth Term Test for Divergence.

49. Ratio $= \left| \dfrac{n!}{(n+1)!} \right| = \left| \dfrac{1}{n+1} \right| \to 0 = L.$ Converges by the Ratio Test.

51. Ratio $= \left| \dfrac{2^{n+1}}{(n+1)!} \dfrac{n!}{2^n} \right| = \left| \dfrac{2}{n+1} \right| \to 0 = L.$ Converges by the Ratio Test.

53. Ratio $= \left| \dfrac{(1/2)^{3(n+1)}}{(1/2)^{3n}} \right| = \left| \left(\dfrac{1}{2} \right)^3 \right| = \dfrac{1}{8} \to \dfrac{1}{8} = L.$ Converges by the Ratio Test.

55. Ratio $= \left| \dfrac{(0.9)^{2(n+1)+1}}{(0.9)^{2n+1}} \right| = \left| (0.9)^2 \right| = 0.81 \to 0.81 = L.$ Converges by the Ratio Test.

57. Ratio $= \left| \dfrac{(x-5)^{n+1}}{(x-5)^n} \right| = |x-5| \to |x-5| = L.$ Series converges absolutely if and only if $|x-5| < 1$: $4 < x < 6$.

59. Ratio $= \left| \dfrac{(x-5)^{n+1}}{(n+1)^2} \dfrac{n^2}{(x-5)^n} \right| = \left| (x-5) \cdot \left(\dfrac{n}{n+1} \right)^2 \right| \to |x-5| = L.$ Series converges absolutely if and

only if $|x-5| < 1$: $4 < x < 6$.

61. Ratio $= \left| \dfrac{(x-2)^{n+1}}{(n+1)!} \dfrac{n!}{(x-2)n} \right| = \left| (x-2) \cdot \dfrac{1}{n+1} \right| \to 0 = L$ for all values of x so the series converges

absolutely for all values of x.

63. Ratio $= \left| \dfrac{(2x-12)^{n+1}}{(n+1)^2} \dfrac{n^2}{(2x-12)^n} \right| = \left| (2x-12) \cdot \left(\dfrac{n}{n+1} \right)^2 \right| \to |2x-12| = L.$ Series converges absolutely

if and only if $|2x-12| < 1$: $\dfrac{11}{2} < x < \dfrac{13}{2}$.

65. Ratio = $\left| \dfrac{(6x-12)^{n+1}}{(n+1)!} \dfrac{n!}{(6x-12)^n} \right| = \left| (6x-12)\cdot \dfrac{1}{n+1} \right| \to 0 = L$ for all values of x so the series

converges absolutely for all values of x.

67. Ratio = $\left| \dfrac{(x+1)^{2(n+1)}}{n+1} \dfrac{n}{(x+1)^{2n}} \right| = \left| (x+1)^2\cdot \dfrac{n}{n+1} \right| \to (x+1)^2 = L$ for all values of x. Series converges

absolutely if and only if $(x+1)^2 < 1$: $-2 < x < 0$.

69. Ratio = $\left| \dfrac{(x-5)^{3(n+1)+1}}{(n+1)^2} \dfrac{n^2}{(x-5)^{3n+1}} \right| = \left| (x-5)^3\cdot \left(\dfrac{n}{n+1} \right)^2 \right| \to \left| (x-5)^3 \right| = L$ for all values of x. Series

converges absolutely if and only if $\left| (x-5)^3 \right| < 1$: $4 < x < 6$.

71. Ratio = $\left| \dfrac{(x+3)^{2(n+1)-1}}{((n+1)+1)!} \dfrac{(n+1)!}{(x+3)^{2n-1}} \right| = \left| (x+3)^2\cdot \dfrac{1}{n+2} \right| \to 0 = L$ for all values of x so the series

converges absolutely for all values of x.

73. Ratio = $\left| \dfrac{x^{2(n+1)}}{(2(n+1))!} \dfrac{(2n)!}{x^{2n}} \right| = \left| x^2 \cdot \dfrac{1}{(2n+1)(2n+2)} \right| \to 0 = L$ for all values of x so the series

converges absolutely for all values of x.

Section 10.7 Some Even Answers

2. Conditionally convergent   4. Conditionally convergent   6. Divergent

8. Absolutely convergent   10. Absolutely convergent   12. Conditionally convergent

14. Absolutely convergent   16. Conditionally convergent   18. Divergent

20. Absolutely convergent   22. Absolutely convergent   24. Absolutely convergent

26. Divergent   28. Divergent   30. Divergent

42. L = 1. Convergent by P–Test.   44. L = 1. Divergent by P–Test.

46. L = 1/3. Convergent Geo. series.   48. Divergent by N^{th} Term Test.

50. L=0. Convergent by Ratio Test.   52. L = 0. Convergent by Ratio Test.

54. L = $(1/3)^2$. Convergent by Ratio Test.   56. L = 0.64. Convergent by Ratio Test.

58. Absolutely convergent for $4 < x < 6$.   60. Absolutely convergent for $1 < x < 3$.

62. Absolutely convergent for all x.   64. Absolutely convergent for $11/4 < x < 13/4$.

66. Absolutely convergent for $2 < x < 4$.   68. Absolutely convergent for $-3 < x < -1$.

70. Absolutely convergent for all x.   72. Absolutely convergent for all x.

74. Absolutely convergent for all x.

Section 10.8 (Odd numbered problem solutions)

1.   Ratio test: $\left|\dfrac{a_{n+1}}{a_n}\right| = \left|\dfrac{x^{n+1}}{x^n}\right| = |x| \to |x| = L.$ $|x| < 1$ if and only if $-1 < x < 1.$

Endpoints: if $x = -1$ or $x = 1$, then the terms do not approach 0 so the series diverges.

Interval of convergence: $-1 < x < 1.$ (You can provide the graph of this interval.)

3.   Ratio test: $\left|\dfrac{a_{n+1}}{a_n}\right| = \left|\dfrac{(x+2)^{n+1}}{(x+2)^n}\right| = |x+2| \to |x+2| = L.$ $|x+2| < 1$ if and only if

$-1 < x+2 < 1$ so $-3 < x < -1.$

Endpoints: if $x = -3$ or $x = -1$, then the terms do not approach 0 so the series diverges.

Interval of convergence: $-3 < x < -1.$ (You can provide the graph of this interval.)

5.   Ratio test: $\left|\dfrac{a_{n+1}}{a_n}\right| = \left|\dfrac{x^{n+1}/(n+1)}{x^n/n}\right| = \left|x \cdot \dfrac{n}{n+1}\right| \to |x| = L.$ $|x| < 1$ if and only if $-1 < x < 1.$

Endpoints: if $x = -1$, then $\displaystyle\sum_{n=1}^{\infty} \dfrac{x^n}{n} = \sum_{n=1}^{\infty} \dfrac{(-1)^n}{n}$ which converges by the Alternating Series Test

if $x = 1$, then $\displaystyle\sum_{n=1}^{\infty} \dfrac{x^n}{n} = \sum_{n=1}^{\infty} \dfrac{(1)^n}{n}$ which diverges (harmonic series)

Interval of convergence: $-1 \le x < 1.$ (You can provide the graph of this interval.)

7.   Ratio test: $\left|\dfrac{a_{n+1}}{a_n}\right| = \left|\dfrac{(x+3)^{n+1}/(n+1)}{(x+3)^n/n}\right| = \left|(x+3) \cdot \dfrac{n}{n+1}\right| \to |x+3| = L.$ $|x+3| < 1$ if and only if

$-1 < x+3 < 1$ or $-4 < x < -2.$

Endpoints: if $x = -4$, then $\displaystyle\sum_{n=1}^{\infty} \dfrac{(x+3)^n}{n} = \sum_{n=1}^{\infty} \dfrac{(-1)^n}{n}$ which converges by the Alternating Series Test

if $x = -2$, then $\displaystyle\sum_{n=1}^{\infty} \dfrac{(x+3)^n}{n} = \sum_{n=1}^{\infty} \dfrac{(1)^n}{n}$ which diverges (harmonic series)

Interval of convergence: $-4 \le x < -2.$ (You can provide the graph of this interval.)

9.   Ratio test: $\left|\dfrac{a_{n+1}}{a_n}\right| = \left|\dfrac{(x-7)^{2(n+1)+1}/(n+1)^2}{(x-7)^{2n+1}/n^2}\right| = \left|(x-7)^2 \cdot \left(\dfrac{n}{n+1}\right)^2\right| \to (x-7)^2 = L.$ $(x-7)^2 < 1$ if

and only if $-1 < x-7 < 1$ or $6 < x < 8.$

Endpoints: if $x = 6$, then $\displaystyle\sum_{n=1}^{\infty} \dfrac{(x-7)^{2n+1}}{n^2} = \sum_{n=1}^{\infty} \dfrac{(-1)^{2n+1}}{n^2} = \sum_{n=1}^{\infty} \dfrac{-1}{n^2}$ which converges by the P–Test

if $x = 8$, then $\displaystyle\sum_{n=1}^{\infty} \dfrac{(x-7)^{2n+1}}{n^2} = \sum_{n=1}^{\infty} \dfrac{(1)^{2n+1}}{n^2}$ which converges by the P–Test (p=2).

Interval of convergence: $6 \le x \le 8.$ (You can provide the graph of this interval.)

11. Ratio = $|2x| \rightarrow |2x| = L$. $|2x| < 1$ if and only if $-1 < 2x < 1$ or $-1/2 < x < 1/2$.

Endpoints: if $x = -1/2$, then the series $= \sum\limits_{n=1}^{\infty}(2 \cdot (\frac{-1}{2}))^n = \sum\limits_{n=1}^{\infty}(-1)^n$ which diverges

if $x = 1/2$, then the series $= \sum\limits_{n=1}^{\infty}(2 \cdot (\frac{1}{2}))^n = \sum\limits_{n=1}^{\infty}(1)^n$ which diverges

Interval of convergence: $-1/2 < x < 1/2$. (You can provide the graph of this interval.)

13. Ratio = $|(\frac{x}{3})^2| \rightarrow |\frac{x^2}{9}| = L$. $|\frac{x^2}{9}| < 1$ if and only if $-1 < \frac{x^2}{9} < 1$ or $-3 < x < 3$.

The series diverges at both endpoints, $x = -3$ and $x = 3$, so the interval of convergence is $-3 < x < 3$.

15. Ratio = $|2x - 6| \rightarrow |2x - 6| = L$. $|2x - 6| < 1$ if and only if $-1 < 2x - 6 < 1$ or $5/2 < x < 7/2$.

The series diverges at both endpoints, $x = 5/2$ and $x = 7/2$, so the interval of convergence is $5/2 < x < 7/2$.

17. Ratio = $|\frac{n!}{(n+1)!} \frac{x^{n+1}}{x^n}| = |\frac{1}{n+1} \cdot x| \rightarrow 0 = L$, and $L < 1$ for all x so the interval of convergence is

the entire real number line.

19. Ratio = $|\frac{(n+1)^2}{n^2} \frac{3^n}{3^{n+1}} \frac{x^{n+1}}{x^n}| = |(\frac{n+1}{n})^2 \cdot \frac{x}{3}| \rightarrow |\frac{x}{3}| = L$. $|\frac{x}{3}| < 1$ if and only if $-1 < \frac{x}{3} < 1$.

The series diverges at both endpoints, $x = -3$ and $x = 3$, so the interval of convergence is $-3 < x < 3$.

21. Ratio = $|\frac{(n+1)!}{n!} \frac{x^{n+1}}{x^n}| = |(n+1) \cdot x| \rightarrow \infty > 1$ if $x \neq 0$ and $|(n+1) \cdot x| \rightarrow 0 < 1$ if $x = 0$.

The series diverges for $x \neq 0$, and the series converges (and is boring) when $x = 0$. The "interval" of

convergence is a single point: $\{0\}$.

23. Ratio = $|\frac{(n+1)!}{n!} \frac{(x-7)^{n+1}}{(x-7)^n}| = |(n+1) \cdot (x-7)| \rightarrow \infty > 1$ if $x \neq 7$ and $|(n+1) \cdot (x-7)| \rightarrow 0 < 1$ if $x = 7$.

The series diverges for $x \neq 7$, and the series converges (and is boring) when $x = 7$. The "interval" of

convergence is a single point: $\{7\}$.

25. Ratio = $|\frac{(x-a)^{n+1}}{(x-a)^n}| = |x - a| \rightarrow |x - a| = L$. $|x - a| < 1$ if and only if $a - 1 < x < a + 1$.

The series diverges at both endpoints, $x = a-1$ and $x = a+1$, so the interval of convergence is $a-1 < x < a+1$.

27. Ratio = $|\frac{(x-a)^{n+1}/(n+1)}{(x-a)^n/n}| = |\frac{n}{n+1}(x-a)| \rightarrow |x - a| = L$. $|x - a| < 1$ if and only if $a - 1 < x < a + 1$.

Endpoints: if $x = a-1$, then $\sum\limits_{n=1}^{\infty}\frac{(x-a)^n}{n} = \sum\limits_{n=1}^{\infty}\frac{(-1)^n}{n}$ which converges by the Alternating Series Test

if $x = a+1$, then $\sum\limits_{n=1}^{\infty}\frac{(x-a)^n}{n} = \sum\limits_{n=1}^{\infty}\frac{(1)^n}{n}$ which diverges (harmonic series)

Interval of convergence: $a-1 \le x < a+1$.

29. Ratio = $\left| \dfrac{(ax)^{n+1}}{(ax)^n} \right| = |ax| \rightarrow |ax| = L.$ $|ax| < 1$ if and only if $\dfrac{-1}{a} < x < \dfrac{1}{a}$.

The series diverges at both endpoints, $x = -1/a$ and $x = 1/a$, so the interval of convergence is $\dfrac{-1}{a} < x < \dfrac{1}{a}$.

31. Ratio = $\left| \dfrac{(ax-b)^{n+1}}{(ax-b)^n} \right| = |ax - b| \rightarrow |ax - b| = L.$ $|ax - b| < 1$ if and only if $-1 < ax - b \leq 1$ or

$\dfrac{b-1}{a} < x < \dfrac{b+1}{a}$.

The series diverges at both endpoints so the interval of convergence is $\dfrac{b-1}{a} < x < \dfrac{b+1}{a}$.

33. We can be certain the friend is wrong because the interval of convergence must be symmetric about the point

$x = 4$, and the friend's interval, $1 < x < 9$, is not symmetric about $x = 4$.

35. $(5, 9), [1, 13], (-1, 15], [3, 11), [0, 14), x = 7$ are all possible intervals of convegence for the series.

37. $(0, \mathbf{2}), (-5, 7), [1, \mathbf{1}], (-3, 5], [-9, 11], [0, \mathbf{2}), x = \mathbf{1}$.

Note: There are many possible correct answers for problems 38 – 45.

39. $\displaystyle\sum_{n=1}^{\infty} \dfrac{1}{n} \left(\dfrac{x}{3} \right)^n$

41. $\displaystyle\sum_{n=1}^{\infty} \dfrac{(x-3)^n}{3^n} = \sum_{n=1}^{\infty} \left(\dfrac{x-3}{3} \right)^n$

43. $\displaystyle\sum_{n=1}^{\infty} \dfrac{1}{n} \left(\dfrac{x-5}{3} \right)^n$

45. $\displaystyle\sum_{n=1}^{\infty} n! \cdot (x-3)^n$

47. Interval of convergence: $2 < x < 4$. $\displaystyle\sum = \dfrac{1}{1-(x-3)} = \dfrac{1}{4-x}$.

49. Interval of convergence: $\dfrac{-1}{3} < x < \dfrac{1}{3}$. $\displaystyle\sum = \dfrac{1}{1-3x}$.

51. Interval of convergence: $-1 < x < 1$. $\displaystyle\sum = \dfrac{1}{1-x^3}$

53. Interval of convergence: $1 < x < 11$. $\displaystyle\sum = \dfrac{1}{1-\frac{x-6}{5}} = \dfrac{5}{11-x}$.

55. Interval of convergence: $-5 < x < 5$. $\displaystyle\sum = \dfrac{1}{1-\frac{x}{5}} = \dfrac{5}{5-x}$.

57. Interval of convergence: $-2 < x < 2$. $\displaystyle\sum = \dfrac{1}{1-(x/2)^3} = \dfrac{8}{8-x^3}$.

59. $\left| \dfrac{1}{3} \cos(x) \right| < 1$ **for all x**, so for all x the sum $\displaystyle\sum = \dfrac{1}{1-\frac{1}{3}\cos(x)} = \dfrac{3}{3-\cos(x)}$.

Section 10.9 (Odd numbered problem solutions)

1. $\dfrac{1}{1-x^4} = 1 + x^4 + x^8 + x^{12} + x^{16} + \ldots = \displaystyle\sum_{n=0}^{\infty} x^{4n}$

3. $\dfrac{1}{1+x^4} = 1 - x^4 + x^8 - x^{12} + x^{16} + \ldots = \displaystyle\sum_{n=0}^{\infty} (-1)^n x^{4n}$

5. $\dfrac{1}{5+x} = \dfrac{1}{5} \cdot \dfrac{1}{1+(x/5)} = \dfrac{1}{5}\left\{ 1 - \dfrac{x}{5} + \left(\dfrac{x}{5}\right)^2 - \left(\dfrac{x}{5}\right)^3 + \left(\dfrac{x}{5}\right)^4 + \ldots \right\} = \dfrac{1}{5}\displaystyle\sum_{n=0}^{\infty} (-1)^n \left(\dfrac{x}{5}\right)^n$

7. $\dfrac{x^2}{1+x^3} = x^2\left\{ \dfrac{1}{1+x^3} \right\} = x^2\left\{ 1 - x^3 + x^6 - x^9 + x^{12} + \ldots \right\} = x^2 \displaystyle\sum_{n=0}^{\infty}(-1)^n x^{3n} = \displaystyle\sum_{n=0}^{\infty}(-1)^n x^{3n+2}$

9. $\ln(1+x^2) = x^2 - \dfrac{x^4}{2} + \dfrac{x^6}{3} - \dfrac{x^8}{4} + \ldots = \displaystyle\sum_{n=1}^{\infty}(-1)^{n+1}\dfrac{x^{2n}}{n}$

11. $\arctan(x^2) = x^2 - \dfrac{x^6}{3} + \dfrac{x^{10}}{5} - \dfrac{x^{14}}{7} + \dfrac{x^{18}}{9} - \ldots = \displaystyle\sum_{n=0}^{\infty}(-1)^n\dfrac{x^{2(2n+1)}}{2n+1} = \displaystyle\sum_{n=0}^{\infty}(-1)^n\dfrac{x^{4n+2}}{2n+1}$

13. $\dfrac{1}{(1-x^2)^2} = 1 + 2x^2 + 3x^4 + 4x^6 + 5x^8 + \ldots = \displaystyle\sum_{n=1}^{\infty} n \cdot x^{2(n-1)} = \displaystyle\sum_{n=1}^{\infty} n \cdot x^{2n-2}$

15. $\displaystyle\int_0^{0.5} \dfrac{1}{1-x^3}\, dx \approx \int_0^{0.5} 1 + x^3 + x^6 + x^9 + \ldots\, dx = x + \dfrac{x^4}{4} + \dfrac{x^7}{7} + \ldots \Big|_0^{0.5} \approx 0.516741$

17. $\displaystyle\int_0^{0.6} \ln(1+x)\, dx \approx \int_0^{0.6} x - \dfrac{x^2}{2} + \dfrac{x^3}{3} - \ldots\, dx = \dfrac{x^2}{2} - \dfrac{x^3}{2\cdot 3} + \dfrac{x^4}{3\cdot 4} - \ldots \Big|_0^{0.6} = 0.1548$

19. $\displaystyle\int_0^{0.3} \dfrac{1}{(1-x)^2}\, dx \approx \int_0^{0.3} 1 + 2x + 3x^2 + \ldots\, dx = x + x^2 + x^3 + \ldots \Big|_0^{0.3} = 0.417$

21. $\displaystyle\lim_{x\to 0} \dfrac{\arctan(x)}{x} = \lim_{x\to 0} \dfrac{1}{x}\left\{ x - \dfrac{x^3}{3} + \dfrac{x^5}{5} - \dfrac{x^7}{7} + \dfrac{x^9}{9} - \ldots \right\} = \lim_{x\to 0} 1 - \dfrac{x^2}{3} + \dfrac{x^4}{5} - \ldots = \mathbf{1}$

23. $\displaystyle\lim_{x\to 0} \dfrac{\ln(1+x)}{2x} = \lim_{x\to 0} \dfrac{1}{2x}\left\{ x - \dfrac{x^2}{2} + \dfrac{x^3}{3} - \dfrac{x^4}{4} + \ldots \right\} = \lim_{x\to 0} \dfrac{1}{2} - \dfrac{x}{4} + \dfrac{x^2}{6} - \dfrac{x^3}{8} + \ldots = \mathbf{\dfrac{1}{2}}$

25. $\displaystyle\lim_{x\to 0} \dfrac{\ln(1-x^2)}{3x} = \lim_{x\to 0} \dfrac{1}{3x}\left\{ -x^2 - \dfrac{x^4}{2} - \dfrac{x^6}{3} - \dfrac{x^8}{4} - \ldots \right\} = \lim_{x\to 0} -\dfrac{x}{3} - \dfrac{x^3}{6} - \dfrac{x^5}{9} - \ldots = \mathbf{0}$

27. $\frac{1}{1+x} = \sum\limits_{n=0}^{\infty} (-1)^n x^n$. Using the Ratio Test, the ratio = $\left| \frac{x^{n+1}}{x^n} \right| = |x| \to |x| = L$. $|x| < 1$ if and

only if $-1 < x < 1$.

The series diverges at the endpoints $x = -1$ and $x = 1$ so the interval of convergence is $-1 < x < 1$.

29. $\ln(1-x) = -\sum\limits_{n=1}^{\infty} \frac{x^n}{n}$. Using the Ratio Test, the ratio = $\left| \frac{x^{n+1}}{x^n} \frac{n}{n+1} \right| = \left| x \cdot \frac{n}{n+1} \right| \to |x| = L$.

$|x| < 1$ if and only if $-1 < x < 1$. The series converges when $x = -1$ and diverges when $x = 1$ so the interval of convergence is $-1 \le x < 1$.

31. $\arctan(x) = \sum\limits_{n=0}^{\infty} (-1)^n \frac{x^{2n+1}}{2n+1}$.

Using the Ratio Test, the ratio = $\left| \frac{x^{2(n+1)+1}}{x^{2n+1}} \frac{2n+1}{2(n+1)+1} \right| = \left| x^2 \cdot \frac{2n+1}{2n+3} \right| \to |x^2| = L$. $|x^2| < 1$

if and only if $-1 < x^2 < 1$ so $-1 < x < 1$. The series converges when $x = -1$ and when $x = 1$ so the interval of convergence is $-1 \le x \le 1$.

33. $\sin(x^2) = (x^2) - \frac{(x^2)^3}{3!} + \frac{(x^2)^5}{5!} - \frac{(x^2)^7}{7!} + ... = x^2 - \frac{x^6}{3!} + \frac{x^{10}}{5!} - \frac{x^{14}}{7!} + ...$

$= \sum\limits_{k=0}^{\infty} (-1)^k \frac{(x^2)^{2k+1}}{(2k+1)!} = \sum\limits_{k=0}^{\infty} (-1)^k \frac{x^{4k+2}}{(2k+1)!}$

35. $e^{(-x^2)} = 1 + (-x^2) + \frac{(-x^2)^2}{2!} + \frac{(-x^2)^3}{3!} + \frac{(-x^2)^4}{4!} + ... = 1 - x^2 + \frac{x^4}{2!} - \frac{x^6}{3!} + \frac{x^8}{4!} + ... = \sum\limits_{k=0}^{\infty} (-1)^k \frac{x^{2k}}{k!}$

37. $\cos(x) = \mathbf{D}(\sin(x)) = \mathbf{D}\left(x - \frac{x^3}{3!} + \frac{x^5}{5!} - \frac{x^7}{7!} + ... \right) = 1 - \frac{x^2}{2!} + \frac{x^4}{4!} - \frac{x^6}{6!} + ... = \sum\limits_{k=0}^{\infty} (-1)^k \frac{x^{2k}}{(2k)!}$

39. Using the result of Problem 33,

$$\int\limits_0^1 \sin(x^2) \, dx = \int\limits_0^1 x^2 - \frac{x^6}{3!} + \frac{x^{10}}{5!} - \frac{x^{14}}{7!} + ... \, dx = \frac{x^3}{3} - \frac{x^7}{3! \cdot 7} + \frac{x^{11}}{5! \cdot 11} - \frac{x^{15}}{7! \cdot 15} + ... \bigg|_0^1$$

$$= \left\{ \frac{1}{3} - \frac{1}{3! \cdot 7} + \frac{1}{5! \cdot 11} - \frac{1}{7! \cdot 15} + ... \right\} - \{ 0 \}.$$

41. $\lim\limits_{x \to 0} \frac{\sin(x)}{x} = \lim\limits_{x \to 0} \frac{x - \frac{x^3}{3!} + \frac{x^5}{5!} - \frac{x^7}{7!} + ...}{x} = \lim\limits_{x \to 0} 1 - \frac{x^2}{3!} + \frac{x^4}{5!} - \frac{x^6}{7!} + ... = 1 + (\text{all 0s}) = 1.$

Section 10.10 (Odd numbered problem solutions)

1. $P(x) = 4x^2 - 5x + 7$ 2. $P(x) = 2x^3 - \frac{5}{2}x^2 + 2x - 1$ 3. $P(x) = 2(x-3)^2 + 5(x-3) - 2$

5. $\ln(1 + x) = x - \frac{x^2}{2} + \frac{x^3}{3} - \frac{x^4}{4} + \ldots$ 7. $\arctan(x) = x - \frac{x^3}{3} + \ldots$

9. $\cos(x) = 1 - \frac{x^2}{2!} + \frac{x^4}{4!} - \frac{x^6}{6!} + \ldots$

11. $\sec(x) = 1 + \frac{x^2}{2!} + \frac{5x^4}{4!} + \ldots$ (The higher derivatives of sec(x) get "messy.")

13. Around $c = 1$, $\ln(x) = (x-1) - \frac{(x-1)^2}{2} + \frac{(x-1)^3}{3} - \frac{(x-1)^4}{4} + \ldots$

15. Around $c = \pi/2$, $\sin(x) = 1 - \frac{(x - \pi/2)^2}{2!} + \frac{(x - \pi/2)^4}{4!} - \frac{(x - \pi/2)^6}{6!} + \ldots$

17. Around $c = 9$, $\sqrt{x} = 3 + \frac{1}{2 \cdot 3}(x-9) - \frac{1}{4 \cdot 3^3}\frac{(x-9)^2}{2!} + \frac{3}{8 \cdot 3^5}\frac{(x-9)^3}{3!} - \frac{15}{16 \cdot 3^7}\frac{(x-9)^4}{4!} + \ldots$

19. Using the first three nonzero terms for cos(x),

$P(x) = 1 - \frac{x^2}{2!} + \frac{x^4}{4!}$. See the Cosine Table.

x	cos(x)	P(x)
0.1	0.995004165	0.995004167
0.2	0.98006657	0.98006666
0.5	0.87758	0.87604
1	0.54030	0.54167
2	–0.4161	–0.3333

21. Using the first three nonzero terms for arctan(x),

$P(x) = x - \frac{x^3}{3} + \frac{x^5}{5}$. See the Arctan Table.

x	arctan(x)	P(x)
0.1	0.09966865	0.09966867
0.2	0.197396	0.197397
0.5	0.4636	0.4646
1	0.7854	0.8667
2	1.1071	5.7333

23. $\int \sin(x^2)\, dx = \int x^2 - \frac{x^6}{6} + \frac{x^{10}}{120}\, dx$

$= \frac{x^3}{3} - \frac{x^7}{42} + \frac{x^{11}}{1320} + C$

$\int \sin(x^3)\, dx = \int x^3 - \frac{x^9}{6} + \frac{x^{15}}{120}\, dx$

$= \frac{x^4}{4} - \frac{x^{10}}{60} + \frac{x^{16}}{1920} + C$

25. $\int e^{(-x^2)}\, dx = \int 1 + (-x^2) + \frac{(-x^2)^2}{2}\, dx = \int 1$

$- x^2 + \frac{x^4}{2}\, dx = x - \frac{x^3}{3} + \frac{x^5}{10} + C$

$\int e^{(-x^3)}\, dx = \int 1 + (-x^3) + \frac{(-x^3)^2}{2}\, dx = \int 1 - x^3 + \frac{x^6}{2}\, dx = x - \frac{x^4}{4} + \frac{x^7}{14} + C$

27. $\int x \sin(x)\, dx = \int x \left\{ x - \frac{x^3}{6} + \frac{x^5}{120} \right\} dx = \int x^2 - \frac{x^4}{6} + \frac{x^6}{120}\, dx = \frac{x^3}{3} - \frac{x^5}{30} + \frac{x^7}{840} + C$

$\int x^2 \sin(x)\, dx = \int x^2 \left\{ x - \frac{x^3}{6} + \frac{x^5}{120} \right\} dx = \int x^3 - \frac{x^5}{6} + \frac{x^7}{120}\, dx = \frac{x^4}{4} - \frac{x^6}{36} + \frac{x^8}{960} + C$

29. $\displaystyle \lim_{x \to 0} \frac{1 - \cos(x)}{x^2} = \lim_{x \to 0} \frac{1 - \left\{ 1 - \frac{x^2}{2!} + \frac{x^4}{4!} - \frac{x^6}{6!} + \dots \right\}}{x^2}$

$\displaystyle = \lim_{x \to 0} \frac{\frac{x^2}{2!} - \frac{x^4}{4!} + \frac{x^6}{6!} - \dots}{x^2} = \lim_{x \to 0} \frac{1}{2!} - \frac{x^2}{4!} + \frac{x^4}{6!} - \dots = \mathbf{\frac{1}{2}}$.

31. $\displaystyle \lim_{x \to 0} \frac{1 - e^x}{x} = \lim_{x \to 0} \frac{1 - \left\{ 1 + x + \frac{x^2}{2!} + \frac{x^3}{3!} + \dots \right\}}{x}$

$\displaystyle = \lim_{x \to 0} \frac{-x - \frac{x^2}{2!} - \frac{x^3}{3!} - \dots}{x} = \lim_{x \to 0} -1 - \frac{x}{2!} - \frac{x^2}{3!} - \dots = \mathbf{-1}$

33. $\displaystyle \lim_{x \to 0} \frac{\sin(x)}{x} = \lim_{x \to 0} \frac{x - \frac{x^3}{3!} + \frac{x^5}{5!} - \dots}{x} = \lim_{x \to 0} 1 - \frac{x^2}{3!} + \frac{x^4}{5!} - \dots = 1$

35. $\displaystyle \lim_{x \to 0} \frac{x - \frac{x^3}{6} - \sin(x)}{x^5} = \lim_{x \to 0} \frac{x - \frac{x^3}{6} - \left\{ x - \frac{x^3}{3!} + \frac{x^5}{5!} - \frac{x^7}{7!} \dots \right\}}{x^5}$

$\displaystyle = \lim_{x \to 0} \frac{-\frac{x^5}{5!} + \frac{x^7}{7!} \dots}{x^5} = \lim_{x \to 0} -\frac{1}{5!} + \frac{x^2}{7!} \dots = -\mathbf{\frac{1}{120}}$

37. $\sinh(x) = \frac{1}{2}(e^x - e^{-x}) = \frac{1}{2}\left(\left\{ 1 + x + \frac{x^2}{2!} + \frac{x^3}{3!} + \frac{x^4}{4!} + \dots \right\} - \left\{ 1 - x + \frac{x^2}{2!} - \frac{x^3}{3!} + \frac{x^4}{4!} + \dots \right\} \right)$

$= \frac{1}{2}\left(2x + 2\frac{x^3}{3!} + 2\frac{x^5}{5!} + 2\frac{x^7}{7!} + \dots \right) = x + \frac{x^3}{3!} + \frac{x^5}{5!} + \frac{x^7}{7!} + \dots$

39. $\mathbf{D}\left(x + \frac{x^3}{3!} + \frac{x^5}{5!} + \frac{x^7}{7!} + \dots \right) = 1 + 3\frac{x^2}{3!} + 5\frac{x^4}{5!} + 7\frac{x^6}{7!} + \dots = 1 + \frac{x^2}{2!} + \frac{x^4}{4!} + \frac{x^6}{6!} + \dots$

41. $e^{ix} = \cos(x) + i \cdot \sin(x)$. When $x = \pi/2$, then $e^{ix} = e^{i(\pi/2)} = \cos(\pi/2) + i \cdot \sin(\pi/2) = i$.

If $x = \pi$, then $e^{ix} = e^{\pi i} = \cos(\pi) + i \cdot \sin(\pi) = -1$. This result is often restated in the form

$e^{\pi i} + 1 = 0$, a formula that relates the 5 most common constants in all of mathematics, $e, \pi, i, 1$, and 0.

43. $\binom{3}{0} = 1$ (by the definition), $\binom{3}{1} = \dfrac{3!}{1! \cdot 2!} = \dfrac{3 \cdot 2 \cdot 1}{(1) \cdot (2 \cdot 1)} = 3$,

$\binom{3}{2} = \dfrac{3!}{2! \cdot 1!} = \dfrac{3 \cdot 2 \cdot 1}{(2 \cdot 1) \cdot (1)} = 3$, and $\binom{3}{3} = \dfrac{3!}{0! \cdot 3!} = \dfrac{3 \cdot 2 \cdot 1}{(1) \cdot (3 \cdot 2 \cdot 1)} = 1$.

45. The Maclaurin series for $(1 + x)^{5/2}$ is

$$1 + \frac{5}{2} x + \left(\frac{5}{2}\right)\left(\frac{3}{2}\right) \frac{x^2}{2!} + \left(\frac{5}{2}\right)\left(\frac{3}{2}\right)\left(\frac{1}{2}\right) \frac{x^3}{3!} + \left(\frac{5}{2}\right)\left(\frac{3}{2}\right)\left(\frac{1}{2}\right)\left(\frac{-1}{2}\right) \frac{x^4}{4!} + \left(\frac{5}{2}\right)\left(\frac{3}{2}\right)\left(\frac{1}{2}\right)\left(\frac{-1}{2}\right)\left(\frac{-3}{2}\right) \frac{x^5}{5!} + \ldots$$

47. If $f(x) = (1 + x)^m$, then $f(0) = 1$. $f'(x) = m(1 + x)^{m-1}$ so $f'(0) = m$.

$f''(x) = m(m-1)(1 + x)^{m-2}$ so $f''(0) = m(m-1)$. $f'''(x) = m(m-1)(m-2)(1 + x)^{m-3}$

so $f'''(0) = m(m-1)(m-2)$. $f^{(4)}(x) = m(m-1)(m-2)(m-3\_(1 + x)^{m-4}$ so $f'''(0) = m(m-1)(m-2)(m-3)$.

And so on. Then the Maclaurin series for $(1 + x)^m$ is

$$1 + mx + m(m-1) \frac{x^2}{2!} + m(m-1)(m-2) \frac{x^3}{3!} + m(m-1)(m-2)(m-3) \frac{x^4}{4!} + \ldots$$

Section 10.11 (Odd numbered problem solutions)

1. $f(x) = \sin(x), c = 0, [-2, 4]$: $f'(x) = \cos(x), f''(x) = -\sin(x), f'''(x) = -\cos(x), f^{(iv)}(x) = \sin(x)$ so

$a_0 = 0, a_1 = 1, a_2 = 0, a_3 = -1/6, a_4 = 0$. Then $P_0(x) = 0$, $P_1(x) = 0 + x = x$,

$P_2(x) = 0 + x + 0 = x$, $P_3(x) = 0 + x + 0 - x^3/6 = x - x^3/6$,

and $P_4(x) = 0 + x + 0 - x^3/6 + 0 = x - x^3/6$.

3. $f(x) = \ln(x), c = 1, [0.1, 3]$: $f'(x) = 1/x$, $f''(x) = -1/x^2$, $f'''(x) = 2/x^3$, $f^{(iv)}(x) = -6/x^4$ so

(using $c = 1$) $a_0 = 0$, $a_1 = 1$, $a_2 = -1/2$, $a_3 = 1/3$, $a_4 = -1/4$. Then $P_0(x) = 0$, $P_1(x) = (x - 1)$,

$P_2(x) = (x - 1) - \dfrac{1}{2}(x - 1)^2$, $P_3(x) = (x - 1) - \dfrac{1}{2}(x - 1)^2 + \dfrac{1}{3}(x - 1)^3$, and

$P_4(x) = (x - 1) - \dfrac{1}{2}(x - 1)^2 + \dfrac{1}{3}(x - 1)^3 - \dfrac{1}{4}(x - 1)^4$.

5. $f(x) = \sqrt{x} = x^{1/2}, c = 1, [1, 3]$: $f'(x) = \dfrac{1}{2} x^{-1/2}$, $f''(x) = \dfrac{-1}{4} x^{-3/2}$, $f'''(x) = \dfrac{3}{8} x^{-5/2}$, and

$f^{(iv)}(x) = \dfrac{-15}{16} x^{-7/2}$ so $a_0 = 1$, $a_1 = \dfrac{1}{2}$, $a_2 = -\dfrac{1}{8}$, $a_3 = \dfrac{1}{16}$, $a_4 = \dfrac{-5}{128}$. Then $P_0(x) = 1$,

$P_1(x) = 1 + \dfrac{1}{2}(x - 1)$, $P_2(x) = 1 + \dfrac{1}{2}(x - 1) - \dfrac{1}{8}(x - 1)^2$,

$P_3(x) = 1 + \dfrac{1}{2}(x - 1) - \dfrac{1}{8}(x - 1)^2 + \dfrac{1}{16}(x - 1)^3$, and

$P_4(x) = 1 + \dfrac{1}{2}(x - 1) - \dfrac{1}{8}(x - 1)^2 + \dfrac{1}{16}(x - 1)^3 - \dfrac{5}{128}(x - 1)^4$.

7. $f(x) = (1 + x)^{-1/2}$, $c = 0$, $[-2, 3]$: $f'(x) = \frac{-1}{2}(1 + x)^{-3/2}$, $f''(x) = \frac{3}{4}(1 + x)^{-5/2}$,

$f'''(x) = \frac{-15}{8}(1 + x)^{-7/2}$, $f^{(iv)}(x) = \frac{105}{16}(1 + x)^{-9/2}$ so $a_0 = 1$, $a_1 = \frac{-1}{2}$, $a_2 = \frac{3}{8}$, $a_3 = \frac{-5}{16}$,

and $a_4 = \frac{35}{128}$. Then $P_0(x) = 1$, $P_1(x) = 1 - \frac{1}{2}x$, $P_2(x) = 1 - \frac{1}{2}x + \frac{3}{8}x^2$,

$P_3(x) = 1 - \frac{1}{2}x + \frac{3}{8}x^2 - \frac{5}{16}x^3$, and $P_4(x) = 1 - \frac{1}{2}x + \frac{3}{8}x^2 - \frac{5}{16}x^3 + \frac{35}{128}x^4$.

9. $f(x) = \sin(x)$, $c = \pi/2$, $[-1, 5]$: $f'(x) = \cos(x)$, $f''(x) = -\sin(x)$, $f'''(x) = -\cos(x)$, $f^{(iv)}(x) = \sin(x)$ so

(using $c = \pi/2$) $a_0 = 1$, $a_1 = 0$, $a_2 = -1/2$, $a_3 = 0$, $a_4 = 1/4!$. Then $P_0(x) = 1$, $P_1(x) = 1$,

$P_2(x) = 1 - \frac{1}{2}(x - \frac{\pi}{2})^2$, $P_3(x) = 1 - \frac{1}{2}(x - \frac{\pi}{2})^2$, and $P_4(x) = 1 - \frac{1}{2}(x - \frac{\pi}{2})^2 + \frac{1}{4!}(x - \frac{\pi}{2})^4$.

11. $f(x) = \sin(x)$, $c = 0$, $n = 5$, $-\pi/2 \le x \le \pi/2$: $R_5(x) = \frac{f^{(6)}(z)}{6!}(x - c)^6 = \frac{f^{(6)}(z)}{720}x^6$ for some z between

x and 0.

$|R_5(x)| \le \frac{|-\sin(z)|}{720}|x|^6 \le \frac{1}{720}(\frac{\pi}{2})^6 \approx 0.02086$.

13.   same as problem 11 except $-\pi \le x \le \pi$: $R_5(x) = \frac{f^{(6)}(z)}{6!}(x - c)^6 = \frac{f^{(6)}(z)}{720}x^6$ for some

z between x and 0. $|R_5(x)| \le \frac{|-\sin(z)|}{720}|x|^6 \le \frac{1}{720}(\pi)^6 \approx 1.335$.

15. $f(x) = \cos(x)$, $c = 0$, $n = 10$, $-1 \le x \le 2$: $R_{10}(x) = \frac{f^{(11)}(z)}{11!}(x - c)^{11} = \frac{\sin(z)}{11!}x^{11}$ for some z

between x and 0. $|R_{10}(x)| = \frac{|\sin(z)|}{11!}x^{11} \le \frac{1}{11!}(2)^{11} = \frac{2048}{39916800} \approx 0.000051$.

Since the only nonzero terms of the power series for $f(x) = \cos(x)$ (about $c = 0$) are the even powers of

x, P_{10} is the same as P_{11} so we could use the error term for P_{11} instead of the one for P_{10}. The

advantage of using the P_{11} error term is that we have a larger factorial in the denominator of R_{11} and

get a smaller bound on the error of the approximation.

Taking $n = 11$, $-1 \le x \le 2$: $R_{11}(x) = \frac{f^{(12)}(z)}{12!}(x - c)^{12} = \frac{|\cos(z)|}{12!}x^{12}$ for some z between x

and 0. $|R_{11}(x)| = \frac{|\cos(z)|}{12!}x^{12} \le \frac{1}{12!}(2)^{12} = \frac{4096}{479001600} \approx 0.00000855$.

17. $f(x) = e^x$, $c = 0$, $n = 6$, $-1 \le x \le 2$: $R_6(x) = \frac{f^{(7)}(z)}{7!}(x - c)^7 = \frac{f^{(7)}(z)}{5040}x^7$ for some z between x and

0. Using the estimate $e < 3$, $|R_6(x)| = \frac{|e^z|}{7!}|x|^7 < \frac{3^2}{7!}(2)^7 = \frac{1152}{5040} \approx 0.2286$.

19. $f(x) = \sin(x), c = 0, [-1, 1], E = \text{"error"} = 0.001$. The derivatives of $\sin(x)$ are $\pm\sin(x)$ or $\pm\cos(x)$ and all of them have maximum absolute value less than or equal to 1.

$|R_n(x)| = \dfrac{|f^{(n+1)}(z)|}{(n+1!)} |x - c|^{n+1} \leq \dfrac{1}{(n+1)!} |1|^{n+1} = \dfrac{1}{(n+1)!}$ so we want to find a value of n

that makes $\dfrac{1}{(n+1)!} \leq 0.001$. It is difficult to solve factorial equations algebraically, but a

"calculator investigation" shows that $\dfrac{1}{(5+1)!} = \dfrac{1}{720} \approx 0.00139 > 0.001$, and

$\dfrac{1}{(6+1)!} = \dfrac{1}{5040} \approx 0.000198 < 0.001$ so we should **use n = 6**. That means we need to use the 3

terms involving x, x^3 and x^5.

21. $f(x) = \sin(x), c = 0, [-1.6, 1.6], E = \text{"error"} = 0.00001$. The derivatives of $\sin(x)$ are $\pm\sin(x)$ or $\pm\cos(x)$ and all of them have maximum absolute value less than or equal to 1.

$|R_n(x)| = \dfrac{|f^{(n+1)}(z)|}{(n+1!)} |x - c|^{n+1} \leq \dfrac{1}{(n+1)!} |1.6|^{n+1} = \dfrac{1}{(n+1)!}(1.6)^{n+1}$ so we want to find a

value of n that makes $\dfrac{1}{(n+1)!}(1.6)^{n+1} \leq 0.00001$. Some calculator investigation shows that

$\dfrac{1}{(9+1)!}(1.6)^{9+1} \approx 0.00003 > 0.00001$ and

$\dfrac{1}{(10+1)!}(1.6)^{10+1} \approx 0.0000044 < 0.00001$ so we should **use n = 10**. That means we need to use

the 5 terms involving x, x^3, x^5, x^7 and x^9.

23. $f(x) = e^x, c = 0, [0, 2], E = \text{"error"} = 0.001$. The derivatives of e^x are all e^x.

$|R_n(x)| = \dfrac{|f^{(n+1)}(z)|}{(n+1!)} |x - c|^{n+1} \leq \dfrac{|e^z|}{(n+1!)} |x|^{n+1} \leq \dfrac{3^2}{(n+1)!}(2)^{n+1}$ (using $e < 3$).

Some calulator investigation shows $\dfrac{3^2}{(9+1)!}(2)^{9+1} \approx 0.0025 > 0.001$ and

$\dfrac{3^2}{(10+1)!}(2)^{10+1} \approx 0.00046 < 0.001$ so we should **use n = 10**. (Using a better upper bound for

the value of e such as 2.75 or 2.72 does not change the conclusion: use n = 10.)

25. (a) $4\{1 - \dfrac{1}{3} + \dfrac{1}{5} - \dfrac{1}{7} + \dfrac{1}{9}\} \approx 4\{0.83492063\} = 3.33968252$ compared with $\pi \approx 3.14159265$

from a calculator.

(b) Let $A = 4\{1 - \dfrac{1}{3} + \dfrac{1}{5} - \dfrac{1}{7} + ... - \dfrac{1}{99}\}$. Then $|A - \pi| < |51^{st} \text{ term}| = \dfrac{4}{101} \approx 0.0396$.

(c) Let $B = 4\{1 - \dfrac{1}{3} + \dfrac{1}{5} - \dfrac{1}{7} + ... \pm \dfrac{1}{2n - 1}\}$.

Then $|B - \pi| < |\text{ next term }| = \dfrac{4}{2(n+1) - 1} = \dfrac{4}{2n + 1}$. We want $\dfrac{4}{2n + 1} \leq 0.0001$ so

$n \geq 1999.5$. Take **n = 2000 terms** to get the precision we want.

27.   (a) A: Put $C = 16 \cdot \arctan(\frac{1}{5}) - 4 \cdot \arctan(\frac{1}{239})$

$$= 16\{ \frac{1}{5} - (\frac{1}{3})(\frac{1}{5})^3 + (\frac{1}{5})(\frac{1}{5})^5 \} - 4\{ \frac{1}{239} - (\frac{1}{3})(\frac{1}{239})^3 + (\frac{1}{5})(\frac{1}{239})^5 \}$$

$$\approx 16\{ 0.197397333333 \} - 4\{ 0.004184076002 \} = 3.14162102879 .$$

$(C - \pi \approx 0.0000284 .)$

(b) Formula A converges more rapidly than Methods I and II because we are using smaller values

of x, x = 1/5 and x = 1/239, and the powers of these smaller values of x approach 0 more

quickly than the values of x, x = 1 and x = 1/2 and x = 1/3, used in Methods I and II.

11.0   MOVING BEYOND TWO DIMENSIONS

So far our study of calculus has taken place almost exclusively in the xy–plane, a 2–dimensional space. The functions we worked with typically had the form y = f(x) so the graphs of these functions could be drawn in the xy–plane. And we have considered limits, derivatives, integrals, and their applications in two dimensions. However, we live in a three (or more) dimensional space, and some ideas and applications require that we move beyond two dimensions.

This chapter marks the start of our move into three dimensions and the mathematics of higher dimensions. The next several chapters extend the ideas and techniques of limits, derivatives, rates of change, maximums and minimums, and integrals beyond two dimensions. The work you have already done in two dimensions is an absolutely vital foundation for these extensions. As we work beyond two dimensions you should be alert for the the parts of the ideas and techniques that extend very easily (many of them) and those that require more extensive changes.

> Section 11.1 introduces vectors and some of the vocabulary, techniques and applications of
> vectors in the plane. This section still takes place in two dimensions, but the ideas are
> important for our move into higher dimensional spaces.

> Section 11.2 introduces the three–dimensional rectangular coordinate system, visualization in
> three dimensions, and measuring distances between points in three dimensions.

> Section 11.3 extends the basic vector ideas, techniques and applications to three dimensions.

> Sections 11.4 and 11.5 introduce two important types of multiplication, the dot product and the
> cross product, for vectors in 3–dimensional space and examines what they measure and
> some of their applications.

> Section 11.6 considers the simplest objects, lines and planes, in 3–dimensional space.

> Section 11.7 introduces surfaces described by second–degree equations and catalogs the possible
> shapes they can have.

Chapter 11 is the first step in our move beyond two dimensions. It contains the fundamental geometry of points and vectors in three dimensions. The concepts and techniques of this chapter are important and useful by themselves, and they are a necessary foundation for the study of calculus in three and more dimensions in Chapter 12 and beyond.

11.1   VECTORS IN THE PLANE

Measurements of some quantities such as mass, speed, temperature, and height can be given by a single

number, but a single number is not enough to describe measurements of some quantities in the plane such

as displacement or velocity. Displacement or velocity not only tell us how much or how fast something has

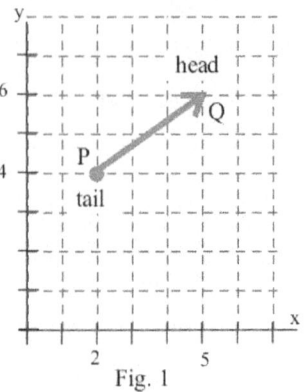

Fig. 1

moved but also tell the direction of that movement. For quantities that have

both length (magnitude) and direction, we use vectors.

A **vector** is a quantity that has both a magnitude and a direction, and vectors

are represented geometrically as directed line segments (arrows). The vector

V given by the directed line segment from the starting point P = (2,4) to

the point Q = (5,6) is shown in Fig. 1. The starting point is called the **tail**

of the vector and the ending point is called the **head** of the vector.

Geometrically, two vectors are equal if they have the same length and point

in the same direction. Fig. 2 shows a number of vectors that are equal to

vector **V** in Fig. 1. The equality of vectors in the plane depends only on

their lengths and directions. Equality of vectors does not depend on their

locations in the plane.

A vector in the plane can be represented algebraically as an ordered pair of

numbers measuring the horizontal and vertical displacement of the endpoint of

the vector from the beginning point of the vector. The numbers in the ordered

pair are called the **components** of the vector. Vector **V** in Fig. 1 can be

represented as $\mathbf{V} = \langle\, 5{-}2\, , 6{-}4\,\rangle = \langle\, 3, 2\,\rangle$, an ordered pair of numbers

enclosed by "bent" brackets. All of the vectors in Fig. 2 are also represented

algebraically by $\langle\, 3, 2\,\rangle$.

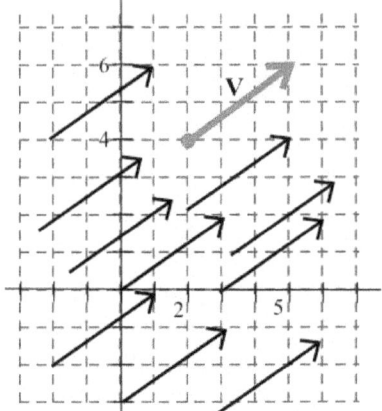

All of these directed line segments
represent the vector V
Fig. 2

Notation: In our work with vectors, it is important to recognize when we are describing a number or a

point or a vector. To help keep those distinctions clear, we use different notations for

numbers, points and vectors:

a number: regular lower case letter: $a, b, x, y, x_1, y_1, \dots$
A number is called a scalar quantity or simply a **scalar**.

a point: regular upper case letter: A, B, ...
ordered pair of numbers enclosed by ( ) : $(2, 3), (a, b), (x_1, y_1), \dots$

a vector: **bold** upper case letter: **A, B, U, V,** ...
ordered pair of numbers enclosed by $\langle\ \rangle$: $\langle\, 2, 3\,\rangle, \langle\, a, b\,\rangle, \langle\, x_1, y_1\,\rangle, \dots$
a letter with an arrow over it: $\vec{A}, \vec{B}, \vec{U}, \vec{V}, \dots$

Definition: Equality of Vectors

Geometrically, two vectors are equal if their lengths are equal and their directions are the same.

Algebraically, two vectors are equal if their respective components are equal:

if $U = \langle a, b \rangle$ and $V = \langle x, y \rangle$, then $U = V$ if and only if $a = x$ and $b = y$.

Example 1:   Vectors U and V are given in Fig. 3.

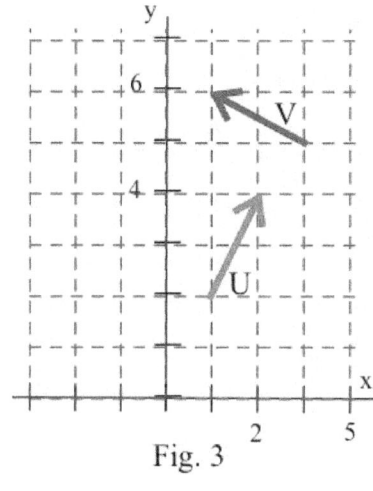

Fig. 3

(a) Represent U and V using the $\langle \ \rangle$ notation.

(b) Sketch U and V as line segments starting at the point $(0,0)$.

(c) Sketch U and V as line segments starting at the point $(-1,5)$.

(d) If (x,y) is the starting point, what is the ending point of the line segment representing the vector U ?

Solution:

(a) The components of a vector are the displacements from the starting to the ending points so $U = \langle 2-1, 4-2 \rangle = \langle 1, 2 \rangle$ and $V = \langle 1-3, 6-5 \rangle = \langle -2, 1 \rangle$.

(b) If U starts at $(0,0)$, then the ending point of the line segment is $(0+1, 0+2) = (1,2)$.

The ending point of V is $(0-2, 0+1) = (-2, 1)$. See Fig. 4.

(c) The ending point of the line segment for U is $(-1+1, 5+2) = (0,7)$. For V, the ending point is $(-1-2, 5+1) = (-3, 6)$. See Fig. 4.

(d) If (x,y) is the starting point for $U = \langle 1, 2 \rangle$ then the ending point is $(x+1, y+2)$.

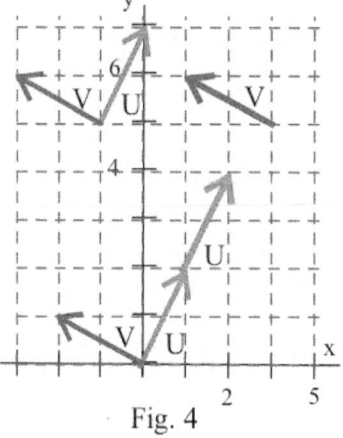

Fig. 4

Practice 1: $A = \langle 3, 4 \rangle$ and $W = \langle -2, 3 \rangle$.

(a) Represent A and W as line segments beginning at the point $(0,0)$.

(b) Represent A and W as line segments beginning at the point $(2,-4)$.

(c) If A and W begin at the point (p, q), at which points do they end?

(d) How long is a line segment representing vector A? W?

(e) What is the slope of a line segment representing vector A? W?

(f) Find a vector whose line segment representation is perpendicular to A. To W.

The magnitude of a vector **V**, written |**V**| , is the length of the line segment representing the vector. That length is the distance between the starting point and the ending point of the line segment. The magnitude can be calculated by using the distance formula.

The **magnitude** or **length** of a vector $\mathbf{V} = \langle a, b \rangle$ is $|\mathbf{V}| = \sqrt{a^2 + b^2}$.

The only vector in the plane with magnitude 0 is $\langle 0, 0 \rangle$, called the **zero vector** and written **0** or $\vec{0}$. The zero vector is a line segment of length 0 (a point), and it has no specific direction or slope.

Adding Vectors

If two people are pushing a box in the same direction along a line, one with a force of 30 pounds and the other with a force of 40 pounds (Fig. 5), then the result of their efforts is equivalent to a single force of 70 pounds along the same line. However, if the people are pushing in different directions (Fig. 6), the problem of finding the result of their combined effort is slightly more difficult. Vector addition provides a simple solution.

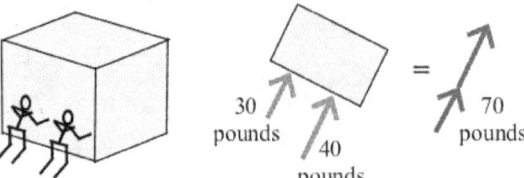

Fig. 5

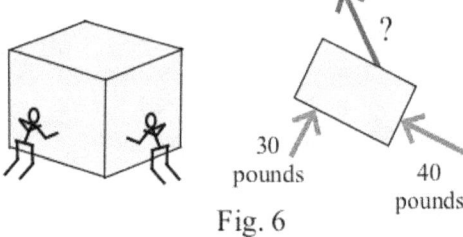

Fig. 6

Definition: Vector Addition

If $\mathbf{A} = \langle a_1, a_2 \rangle$ and $\mathbf{B} = \langle b_1, b_2 \rangle$, then $\mathbf{A} + \mathbf{B} = \langle a_1 + b_1, a_2 + b_2 \rangle$.

The result of applying two forces, represented by the vectors **A** and **B**, is equivalent to the single force represented by the vector **A** + **B**. In Fig. 6, the effort of person A can be represented by the vector $\mathbf{A} = \langle 30, 0 \rangle$ and the effort of person B by the vector $\mathbf{B} = \langle 0, 40 \rangle$. Their combined effort is equivalent to a single force vector $\mathbf{C} = \mathbf{A} + \mathbf{B} = \langle 30, 40 \rangle$. (Since |**C**| = 50 pounds, if the two people cooperated and pushed in the direction of **C**, they could achieve the same result by each exerting 10 pounds less force.)

Example 2:   Let $A = \langle 3, 5 \rangle$ and $B = \langle -1, 4 \rangle$. (a) Graph A and B each starting at the origin.

(b) Calculate $C = A + B$ and graph it, starting at the origin.

(c) Calculate the magnitudes of A, B, and C.

(d) Find a vector V so $A + V = \langle 4, 2 \rangle$.

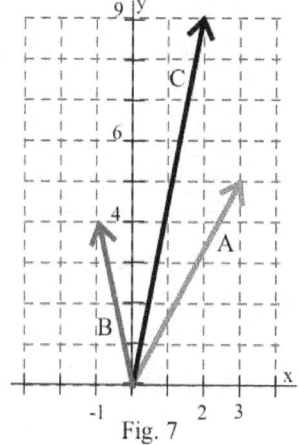

Fig. 7

Solution: (a) The graphs of A and B are shown in Fig. 7.

(b) $C = \langle 3, 5 \rangle + \langle -1, 4 \rangle = \langle 2, 9 \rangle$. The graph of C is

shown in Fig. 7.

(c) $|A| = \sqrt{3^2 + 5^2} = \sqrt{34} \approx 5.8$, $|B| = \sqrt{17} \approx 4.1$, and

$|C| = \sqrt{85} \approx 9.2$.

(d) Let $V = \langle x, y \rangle$. Then $A + V = \langle 3+x, 5+y \rangle = \langle 4, 2 \rangle$

so $3+x = 4$ and $x = 1$. Also, $5+y = 2$ so $y = -3$ and

$V = \langle 1, -3 \rangle$

Practice 2:   Let $A = \langle -2, 5 \rangle$ and $B = \langle 7, -4 \rangle$. (a) Graph A and B each

starting at the origin.

(b) Calculate $C = A + B$ and graph it, starting at the origin.

(c) Find and graph a vector V so $V + C = 0$.

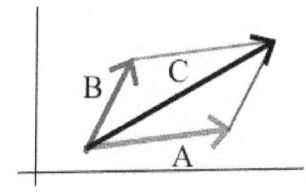

Fig. 8

The parallelogram method and the head–to–tail method are two commonly used methods for adding

vectors graphically.

The parallelogram method (Fig. 8):

i)   arrange the vectors A and B to have a common starting point

ii) use the two given vectors to complete a parallelogram

iii) draw a vector C from the common starting point of the two original

vectors to the opposite corner of the parallelogram: $C = A + B$

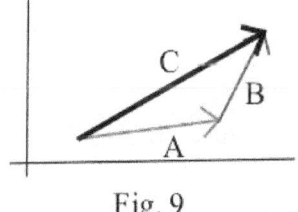

Fig. 9

The head–to–tail method (Fig. 9):

i)   position the tail of vector B at the head of the vector A

ii) draw a vector C from the tail of A to the head of B: $C = A + B$

The head–to–tail method is particularly useful when we need to add

several vectors together. We can simply string them along head–to–tail,

head–to–tail, ... and finally draw their sum as a directed line segment

from the tail of the first vector to the head of the last vector (Fig. 10).

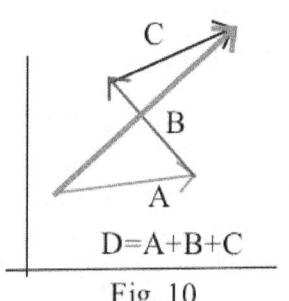

D=A+B+C

Fig. 10

Practice 3: Draw the vectors **U** = **A** + **C** and **V** = **A** + **B** + **C** for **A**, **B**, and **C** given in Fig. 11.

A
B
C
Fig. 11

Scalar Multiplication

We can multiply a vector by a number (a scalar) by multiplying each component by that number.

> **Definition**: Scalar Multiplication
>
> If k is a scalar and $\mathbf{A} = \langle a_1, a_2 \rangle$ is a vector,
>
> then $k\mathbf{A} = \langle ka_1, ka_2 \rangle$, a vector.

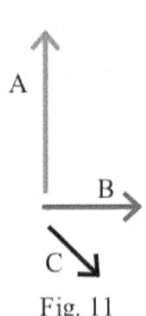

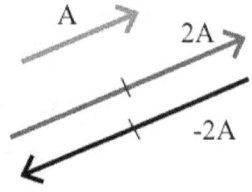

A
2A
-2A
Fig. 12

Multiplying by a scalar k gives a vector that is |k| times as long as the original vector. If k is positive, then **A** and k**A** have the same direction. If k is negative, then **A** and k**A** point in opposite directions (Fig. 12).

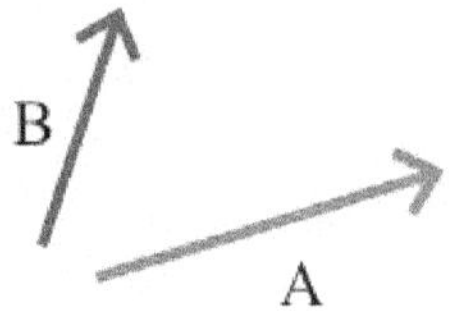

B
A
Fig. 13

Example 3: Vectors **A** and **B** are shown in Fig. 13.
Graph and label **C** = 2**A**, $\mathbf{D} = \frac{1}{2}\mathbf{B}$, **E** = –2**B**, **F** = **B** + 2**A**, and **G** = **A** + (–1)**A** .

Solution: The vectors are shown in Fig. 14.

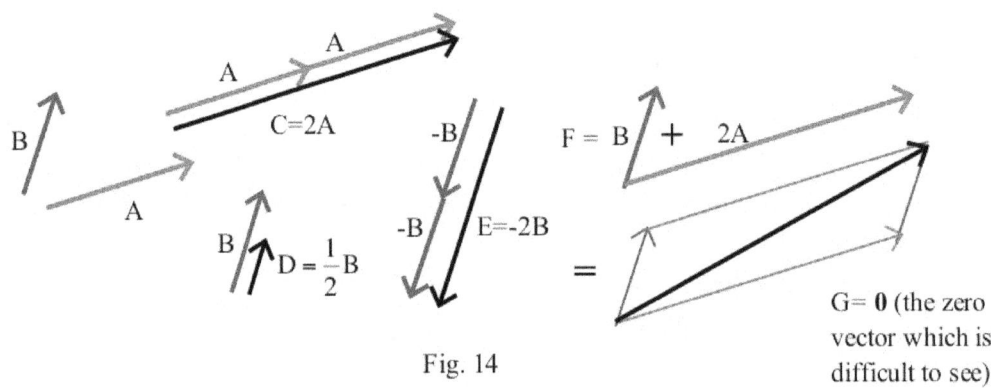

B
A
A
A
C=2A
B
D = 1/2 B
-B
-B
E=-2B
F = B + 2A
=
G= **0** (the zero vector which is difficult to see)
Fig. 14

Practice 4: Vectors **U** and **V** are shown in Fig. 15.
Graph and label **A** = 2**U**, $\mathbf{B} = -\frac{1}{2}\mathbf{U}$, **C** = (–1)**V**, and **D** = **U** + (–1)**V**.

U
V
Fig. 15

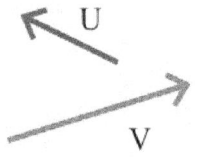

Subtracting Vectors

We can also subtract vectors algebraically and graphically.

Definition: Vector Subtraction

If $\mathbf{A} = \langle a_1, a_2 \rangle$ and $\mathbf{B} = \langle b_1, b_2 \rangle$, then $\mathbf{A} - \mathbf{B} = \mathbf{A} + (-1)\mathbf{B} = \langle a_1 - b_1, a_2 - b_2 \rangle$, a vector.

Graphically, we can construct the line segment representing the vector

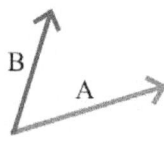
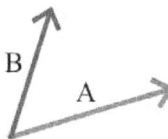

$\mathbf{A} - \mathbf{B}$ either using the head–to–tail method (Fig. 16) to add $\mathbf{A}$ and $-\mathbf{B}$, or by moving $\mathbf{A}$ and $\mathbf{B}$ so they have a common starting point (Fig. 17) and then drawing the line segment from the head of $\mathbf{B}$ to the head of $\mathbf{A}$.

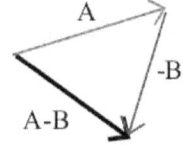

Practice 5:   Vectors $\mathbf{A}, \mathbf{B}$ and $\mathbf{C}$ are shown in Fig. 18.

Graph $\mathbf{U} = \mathbf{A} - \mathbf{B}$ and $\mathbf{V} = \mathbf{C} - \mathbf{A}$.

Fig. 16

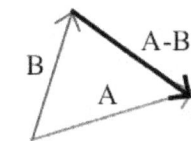

Fig. 17

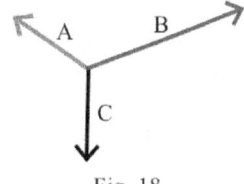

Fig. 18

Algebraic Properties of Vector Operations

Some of the properties of vectors are given below.

If $\mathbf{A}, \mathbf{B}$, and $\mathbf{C}$ are vectors in the plane and x and y are scalars, then

1. $\mathbf{A} + \mathbf{B} = \mathbf{B} + \mathbf{A}$

2. $(\mathbf{A} + \mathbf{B}) + \mathbf{C} = \mathbf{A} + (\mathbf{B} + \mathbf{C})$

3. $\mathbf{A} + \mathbf{0} = \mathbf{A}$ ($\mathbf{0} = \langle 0,0 \rangle$ is called the additive identity vector)

4. $\mathbf{A} + (-1)\mathbf{A} = \mathbf{0}$ ($-\mathbf{A} = (-1)\mathbf{A}$ is called the additive inverse vector of A)

5. $x(\mathbf{A} + \mathbf{B}) = x\mathbf{A} + x\mathbf{B}$

6. $(x + y)\mathbf{A} = x\mathbf{A} + y\mathbf{A}$

These and additional properties of vectors are easily verified using the definitions of vector equality, vector addition and scalar multiplication.

Unit Vector, Direction of a Vector, Standard Basis Vectors

Definitions:

A **unit vector** is a vector whose length is 1.

The **direction** of a nonzero vector $\mathbf{A}$ is the unit vector $\frac{1}{|\mathbf{A}|}\,\mathbf{A} = \frac{\mathbf{A}}{|\mathbf{A}|}$.

The **standard basis vectors** in the plane are $\mathbf{i} = \langle\, 1, 0\, \rangle$
and $\mathbf{j} = \langle\, 0, 1\, \rangle$. (Fig. 19)

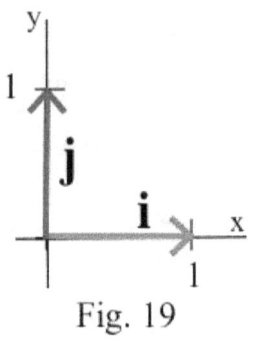

Fig. 19

Every nonzero vector $\mathbf{A}$ is the product of its magnitude and its direction: $\mathbf{A} = |\mathbf{A}|\,\frac{\mathbf{A}}{|\mathbf{A}|}$.

The standard basis vectors are unit vectors.

Every vector in the plane can be written as a sum of scalar multiples of these two basis vectors:

if $\mathbf{A} = \langle\, a_1, a_2\, \rangle$, then $\mathbf{A} = \langle\, a_1, 0\, \rangle + \langle\, 0, a_2\, \rangle = a_1 \mathbf{i} + a_2 \mathbf{j}$.

Example 4:   Let $\mathbf{A} = 5\mathbf{i} + 12\mathbf{j}$ and $\mathbf{B} = 3\mathbf{i} - 4\mathbf{j}$.

(a) Determine the directions of $\mathbf{A}, \mathbf{B}$, and $\mathbf{C} = \mathbf{A} - 3\mathbf{j}$.

(b) Write $3\mathbf{A} + 2\mathbf{B}$ and $4\mathbf{A} - 5\mathbf{B}$ in terms of the standard basis vectors.

Solution: (a) $|\mathbf{A}| = \sqrt{25 + 144}\ = \sqrt{169}\ = 13$ so the direction of $\mathbf{A}$ is

$\frac{1}{13}\,\langle\, 5, 12\, \rangle = \langle\, \frac{5}{13}, \frac{12}{13}\, \rangle = \frac{5}{13}\,\mathbf{i} + \frac{12}{13}\,\mathbf{j}$. The direction of $\mathbf{B}$ is $\frac{3}{5}\,\mathbf{i} - \frac{4}{5}\,\mathbf{j}$.

$|\mathbf{C}| = |\, 5\mathbf{i} + 9\mathbf{j}\, | = \sqrt{106}$ so the direction of $\mathbf{C}$ is $\frac{5}{\sqrt{106}}\,\mathbf{i} + \frac{9}{\sqrt{106}}\,\mathbf{j}$.

(b) $3\mathbf{A} + 2\mathbf{B} = 3(5\mathbf{i} + 12\mathbf{j}) + 2(3\mathbf{i} - 4\mathbf{j}) = 15\mathbf{i} + 36\mathbf{j} + 6\mathbf{i} - 8\mathbf{j} = 21\mathbf{i} + 28\mathbf{j}$.

$4\mathbf{A} - 5\mathbf{B} = 4(5\mathbf{i} + 12\mathbf{j}) - 5(3\mathbf{i} - 4\mathbf{j}) = 20\mathbf{i} + 48\mathbf{j} - 15\mathbf{i} + 20\mathbf{j}\ = 5\mathbf{i} + 68\mathbf{j}$.

Practice 6:   Let $\mathbf{U} = 7\mathbf{i} + 24\mathbf{j}$ and $\mathbf{V} = 15\mathbf{i} - 8\mathbf{j}$.

Determine the directions of $\mathbf{U}, \mathbf{V}$, and $\mathbf{W} = \mathbf{U} + 3\mathbf{i}$.

Often in applications a vector is described in terms of a magnitude at an angle to some line. In those situations we typically need to use trigonometry to determine the components of the vector.

Example 5:   Suppose you are pulling on the rope with a force of 50 pounds at an angle of 25° to the horizontal ground (Fig. 20). What are the components of this force vector parallel and perpendicular to the ground?

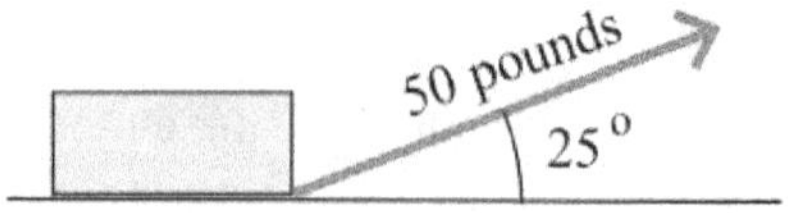

Fig. 20

Solution: The horizontal component (parallel to the

ground) is 50 cos(25°) ≈ 45.3 pounds.

The vertical component (perpendicular to the ground) is

50 sin(25°) ≈ 21.1 pounds.

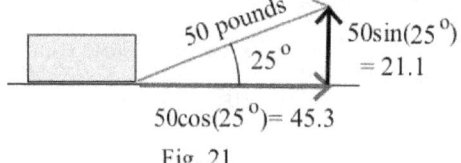

Fig. 21

A force of approximately 45.3 pounds operates to pull the box along the ground (Fig. 21), and a

force of approximately 21.1 pounds is operating to lift the box.

The following result from trigonometry is used to find the components of vectors in the plane.

> If   a vector **V** with magnitude |**V**| makes an angle of θ with a horizontal line,
>
> then   **V** = |**V**| cos(θ) **i** + |**V**|sin(θ) **j** = ⟨ |**V**| cos(θ), |**V**|sin(θ) ⟩ .

Practice 7:   A horizontal force of 50 pounds is required to move the box in Fig. 22. If you can pull on the

rope with a total force of 70 pounds, what is the largest angle that the

rope can make with the ground and still move the box?

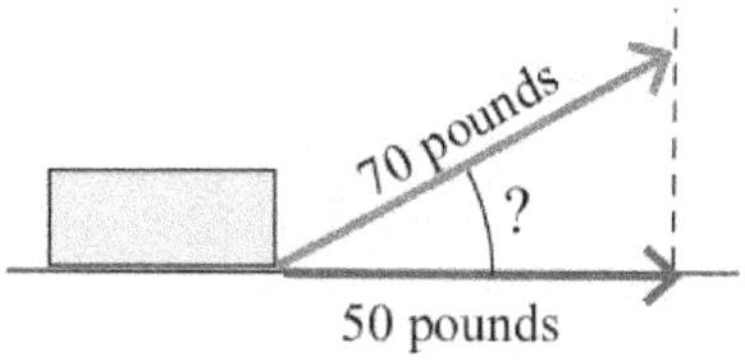

Fig. 22

Additional Applications of Vectors in the Plane

The following applications are more complicated than the previous ones, but they begin to illustrate the

range and power of vector methods for solving applied problems. In general, vector methods allow us to

work separately with the horizontal and vertical components of a problem, and then put the

results together into a complete answer.

Example 6:   In water with no current, your boat can travel at 20 knots (nautical

miles per hour). Suppose your boat is on the ocean and you want to

follow a course to travel due north, but the water current is **W** = –6**i** – 8**j**

(Fig. 23). At what angle θ , east of due north, should

you steer your boat so the resulting course **R** is due north?

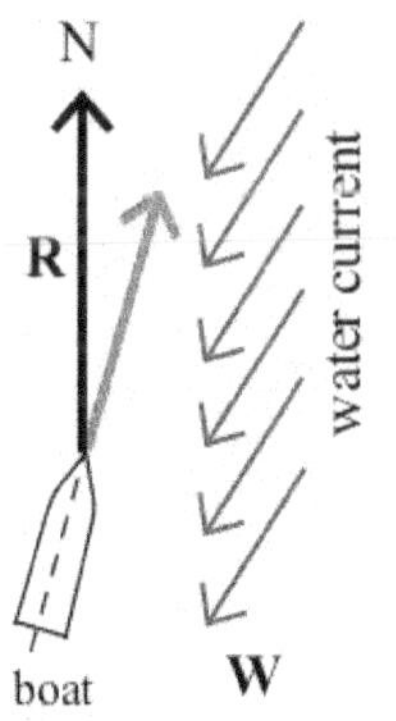

Fig. 23

Solution: First, we should notice that since **R** points due north, the **i**

component **R** is 0: **R** = ⟨ 0, s ⟩. We also need to recognize that **V**   makes an

angle of 90° – θ with the horizontal so   **V** = ⟨ 20 cos(90° – θ) , 20 sin(90° – θ) ⟩ . Finally

V + **W** = **R** so ⟨ 20 cos(90° – θ) – 6, 20 sin(90° – θ) – 8 ⟩ = ⟨ 0, s ⟩. Equating the first components

of this vector equation, we have 20 cos(90° – θ) – 6 = 0 so cos(90° – θ) = 6/20 = 0.3 ,

90° – θ ≈ 72.5° , and θ ≈ 17.5° . You should steer your boat approximately 17.5° east of due north

in order to maintain a course taking you due north. Your speed along this course is

|**R**| = s = 20 sin(90° – θ) – 8 ≈ 20 sin(72.5°) – 8 ≈ 11.1 knots.

Practice 8:   With the same boat and water current as in Example 6, at what angle θ , east of due north, should you steer your boat so the resulting course **R** is due east? What is your resulting speed due east?

Example 7:   A video camera weighing 15 pounds is going to be suspended by two wires from the ceiling of a room as shown in Fig. 24. What is the resulting tension in each wire? (The tension in a wire is the magnitude of the force vector.)

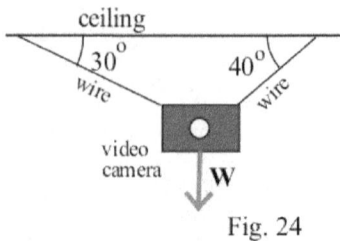

Fig. 24

Solution: The force vector of the camera is straight down so

$\mathbf{W} = \langle 0, -15 \rangle$. Let vector **A** be the force vector for the left ($30°$) wire, and vector **B** be the force vector for the right ($40°$) vector. Vector **A** has magnitude |**A**| and can be represented as $\langle -|\mathbf{A}|\cos(30°), |\mathbf{A}| \sin(30°) \rangle$. Similarly, $\mathbf{B} = \langle |\mathbf{B}|\cos(40°), |\mathbf{B}|\sin(40°) \rangle$. Since the system is in equilibrium, the sum of the force vectors is 0 so

$$\mathbf{0} = \mathbf{A} + \mathbf{B} + \mathbf{W} = \langle -|\mathbf{A}|\cos(30°) + |\mathbf{B}|\cos(40°) + 0, |\mathbf{A}| \sin(30°) + |\mathbf{B}|\sin(40°) - 15 \rangle .$$

From the components of the vector equation we have two equations,

$$0 = -|\mathbf{A}|\cos(30°) + |\mathbf{B}|\cos(40°) + 0 \ \text{ and } \ 0 = |\mathbf{A}| \sin(30°) + |\mathbf{B}|\sin(40°) - 15 ,$$

that we want to solve for the tensions |**A**| and |**B**| .

From the first, we get $|\mathbf{A}|\cos(30°) = |\mathbf{B}|\cos(40°)$ so $|\mathbf{B}| = |\mathbf{A}|\dfrac{\cos(30°)}{\cos(40°)}$.

Substituting this value for |**B**| into the second equation we have

$$0 = |\mathbf{A}| \sin(30°) + |\mathbf{A}| \frac{\cos(30°)}{\cos(40°)} \sin(40°) - 15 \ = \ |\mathbf{A}| \left\{ \sin(30°) + \cos(30°)\tan(40°) \right\} - 15$$

so $|\mathbf{A}| \ = \ \dfrac{15}{\sin(30°) + \cos(30°)\tan(40°)} \ \approx \ 12.2$ pounds. Putting this value back into

$|\mathbf{B}| = |\mathbf{A}|\dfrac{\cos(30°)}{\cos(40°)}$, we get $|\mathbf{B}| = (12.2)\dfrac{\cos(30°)}{\cos(40°)} \ \approx \ 13.9$ pounds .

Practice 9:   What are the tensions in the wires if the angles are changed to 35° and 50° ?

PROBLEMS

In problems 1 – 4, vectors **U** and **V** are given graphically. Sketch the vectors 3**U**, –2**V** , **U** + **V** , and **U** – **V** .

1. **U** and **V** are given in Fig. 25.

2. **U** and **V** are given in Fig. 26.

3. **U** and **V** are given in Fig. 27.

4. **U** and **V** are given in Fig. 28.

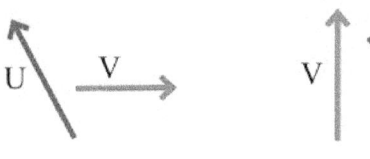

Fig. 25

Fig. 26

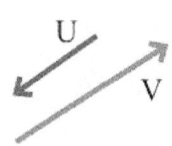

Fig. 27

Fig. 28

In problems 5 – 10, vectors **U** and **V** are given.

(a) Sketch the vectors **U**, **V**, 3**U** , –2**V** , **U** + **V** , and **U** – **V** .

(b) Calculate | **U** | and | **V** | and find the directions of **U** and **V**.

(c) Find the slopes of the line segments representing **U** and **V** and their angles with the x–axis.

5. $\mathbf{U} = \langle 1, 4 \rangle$ and $\mathbf{V} = \langle 3, 2 \rangle$

6. $\mathbf{U} = \langle 2, 5 \rangle$ and $\mathbf{V} = \langle 6, 1 \rangle$

7. $\mathbf{U} = \langle -2, 5 \rangle$ and $\mathbf{V} = \langle 3, -7 \rangle$

8. $\mathbf{U} = \langle -3, 4 \rangle$ and $\mathbf{V} = \langle 1, -5 \rangle$

9. $\mathbf{U} = \langle -4, -3 \rangle$ and $\mathbf{V} = \langle 3, -4 \rangle$

10. $\mathbf{U} = \langle -5, 2 \rangle$ and $\mathbf{V} = \langle -2, -5 \rangle$

In problems 11 – 14, vectors **A, B,** and **C** are given. Calculate **U** = **A** + **B** – **C** and **V** = **A** – **B** + **C**.

11. **A, B,** and **C** in Fig. 29.

12. **A, B,** and **C** in Fig. 30.

13. $\mathbf{A} = \langle 1, 4 \rangle , \mathbf{B} = \langle 3, 1 \rangle , \mathbf{C} = \langle 5, 2 \rangle$

14. $\mathbf{A} = \langle -1, 5 \rangle , \mathbf{B} = \langle 2, 0 \rangle , \mathbf{C} = \langle 6, -2 \rangle$

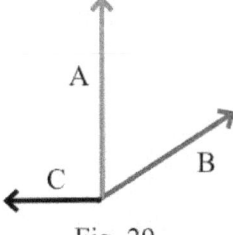

Fig. 29

In problems 15 – 20, find a vector **V** with the given properties.

15. | **V** | = 3 and direction $0.6\mathbf{i} + 0.8\mathbf{j}$

16. | **V** | = 2 and direction $0.6\mathbf{i} - 0.8\mathbf{j}$

17. | **V** | = 5 and **V** makes an angle of 35° with the positive x–axis.

18. | **V** | = 2 and **V** makes an angle of 150° with the positive x–axis.

19. | **V** | = 7 and **V** has slope 3.

20. | **V** | = 7 and **V** has slope –2.

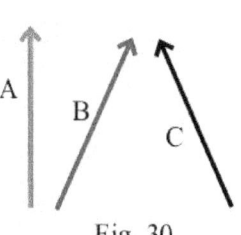

Fig. 30

In problems 21 – 26, find the direction of each function at the given point. That is, find a unit vector, there are two, parallel to the tangent line to the curve at the given point.

21. $f(x) = x^2 + 3x - 2$ at $(1, 2)$ 22. $f(x) = \sin(3x) + e^x$ at $(0, 1)$

23. $f(x) = \ln(x^2 + 1)$ at $(0, 0)$ 24. $f(x) = \dfrac{2 + \sin(x)}{1 + x}$ at $(0, 2)$

25. $f(x) = \cos^2(3x)$ at $(\pi, 1)$ 26. $f(x) = \arctan(x)$ at $(0, 0)$

In problems 27 – 34, a vector **U** is shown or given as $\mathbf{U} = \langle a, b \rangle$.

(a) Sketch the "shadow vector" (Fig. 31) of **U** on the x–axis. (This is the
        "projection of **U**" onto the x–axis.)

(b) Sketch the "shadow vector" of **U** on the y–axis. (This is the "projection of **U**"
        onto the y–axis.)

(c) Represent each of these "shadow vectors" in the form $a\mathbf{i} + b\mathbf{j}$.

27. **U** is given in Fig. 32.   28. **U** is given in Fig. 33.

29. **U** is given in Fig. 34.   30. **U** is given in Fig. 35.

"shadow vector" of
U on the x-axis
Fig. 31

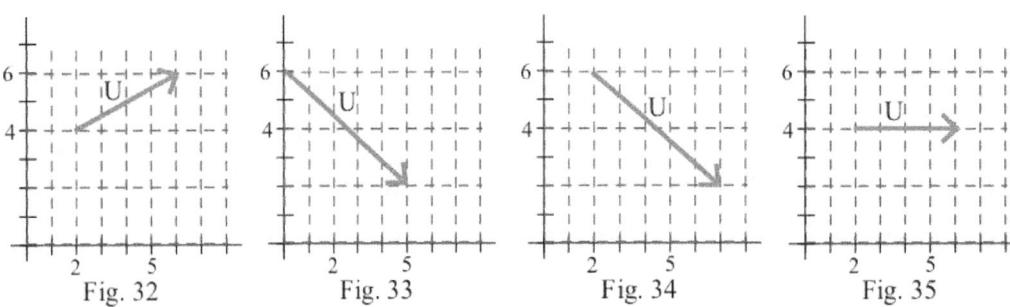

Fig. 32   Fig. 33   Fig. 34   Fig. 35

31. $\mathbf{U} = \langle 1, 4 \rangle$ 32. $\mathbf{U} = \langle -2, 3 \rangle$ 33. $\mathbf{U} = \langle 5, -2 \rangle$ 34. $\mathbf{U} = \langle -1, -3 \rangle$

In problems 35 – 38, vectors **A** and **B** are given. Find a vector **C** so that $\mathbf{A} + \mathbf{B} + \mathbf{C} = \mathbf{0}$.

35. **A** and **B** are shown in Fig. 36.   36. **A** and **B** are shown in Fig. 37.

37. $\mathbf{A} = \langle 7, 4 \rangle$ and $\mathbf{B} = \langle -3, 2 \rangle$ 38. $\mathbf{A} = \langle -5, -3 \rangle$ and $\mathbf{B} = \langle 2, -4 \rangle$

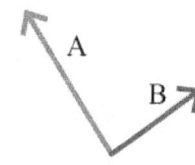

Fig. 36

39. Suppose you are pulling on the rope with a force of **60 pounds** at an angle of **65°** to
        the horizontal ground. What are the components of this force vector parallel and
        perpendicular to the ground?

40. Suppose you are pulling on the rope with a force of **100 pounds** at an angle of **15°** to
        the horizontal ground. What are the components of this force vector parallel and
        perpendicular to the ground?

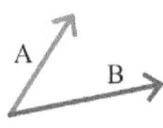

Fig. 37

41. Two ropes are attached to the bumper of a car. Rope A is pulled with a force of 50 pounds at an angle of 45° to the horizontal ground, and rope B is pulled with a force of 70 pounds at an angle of 35° to the horizontal ground. The same effect can be produced by a single rope pulling with what force and at what angle to the ground?

42. Two ropes are attached to the bumper of a car. Rope A is pulled with a force of 60 pounds at an angle of 30° to the horizontal ground, and rope B is pulled with a force of 80 pounds at an angle of 15° to the horizontal ground. The same effect can be produced by a single rope pulling with what force and at what angle to the ground?

43. Rope A is pulled with a force of 100 pounds at an angle of 30° to the horizontal ground, rope B is pulled with a force of 90 pounds at an angle of 25° to the horizontal, and rope C is pulled with a force of 80 pounds at an angle of 15° to the horizontal. The same effect can be produced pulling on a single rope with what force and at what angle to the ground?

44. A plane is flying due east at 200 miles per hour when it encounters a wind $\mathbf{W} = 30\mathbf{i} + 40\mathbf{j}$.

    (a) What is the path of the plane in this wind if the pilot keeps it pointed due east?

    (b) What direction should the pilot point the plane in order to fly due east?

45. A boat is moving due north at 18 miles per hour when it encounters a current $\mathbf{C} = -3\mathbf{i} + 4\mathbf{j}$.

    (a) What is the path of the boat in this current if the boater keeps it pointed due north?

    (b) What direction should the boater steer in order to go due north?

46. Suppose you leave home and hike 10 miles due north, then 8 miles in the direction 40° east of north, and then 6 miles due east. (a) How far are you from home? (b) What direction should you hike in order to return home?

47. A 60 foot rope is attached to the tops of two poles that are 50 feet apart, and a 100 pound person is going hand–over–hand from one end of the rope to the other (Fig. 38).

    (a) What is the tension in each part of the rope when the person is half way?

    (b) What is the tension in the rope when the person is 10

       feet (horizontally) from the start?

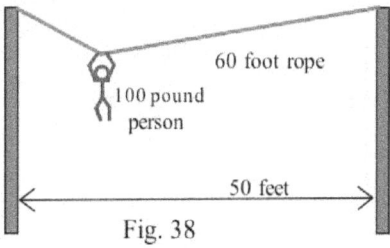

Fig. 38

Practice Answers

Practice 1:   (a) and (b) See Fig. 39.

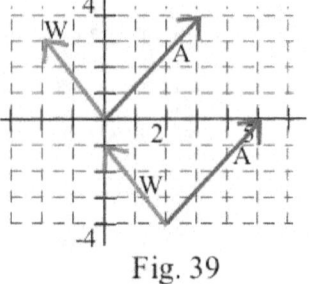

(c) **A** ends at the point $(p+3, q+4)$. **W** ends at the point $(p-2, q+3)$.

(d) We can use the Pythagorean distance formula to determine the length
of the hypotenuse.

For **A**, length $= \sqrt{3^2 + 4^2} = \sqrt{25} = 5$.

For **W**, length $= \sqrt{(-2)^2 + 3^2} = \sqrt{13} \approx 3.6$.

Fig. 39

(e) Slope = rise/run. For **A**, slope = 4/3. For **W**, slope = 3/(–2).

(f) Two lines are perpendicular if one slope is the negative reciprocal of the other. If **B** is
perpendicular to **A**, then the slope of **B** is –1/(slope of **A**) = –3/4. The vector $\mathbf{B} = \langle 4, -3 \rangle$ is
one vector that has the slope we want. $\mathbf{B} = \langle -4, 3 \rangle$, and $\mathbf{B} = \langle 2, -3/2 \rangle$ are two other vectors with the same slope, and there
are lots of others. $\mathbf{U} = \langle 3, 2 \rangle$ has slope 2/3 and is
perpendicular to **W**.

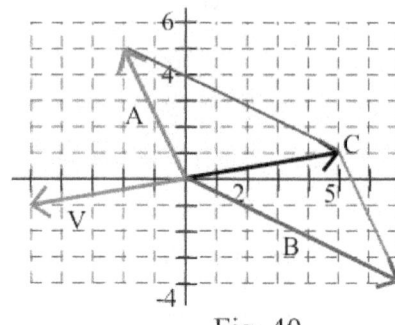

Practice 2:   (a)   **A** and **B** are shown in Fig. 40.

(b) $\mathbf{C} = \langle -2, 5 \rangle + \langle 7, -4 \rangle = \langle 5, 1 \rangle$. The graph of **C** is shown
in Fig. 40.

(c) $\mathbf{V} = \langle -5, -1 \rangle$. The graph of **V** is shown in Fig. 40.

Fig. 40

Practice 3:   See Fig. 41.

Practice 4:   See Fig. 42.

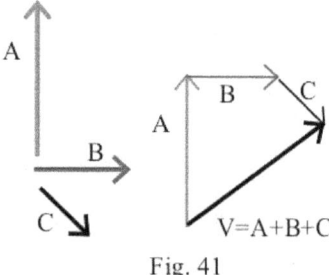

Fig. 41

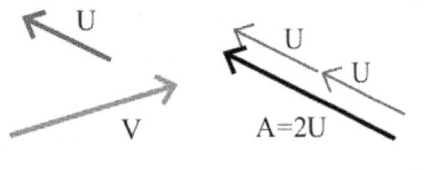 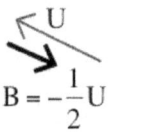 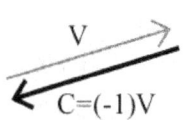 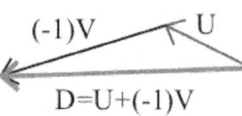

Fig. 42

Practice 5:   See Fig. 43.

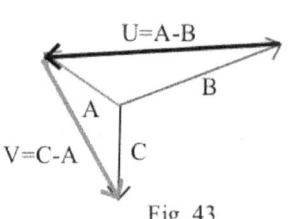

Fig. 43

Practice 6: $|U| = \sqrt{(7)^2 + (24)^2} = \sqrt{625} = 25$, so the direction of **U** is

$$\frac{U}{|U|} = \frac{7\mathbf{i} + 24\mathbf{j}}{25} = \frac{7}{25}\mathbf{i} + \frac{24}{25}\mathbf{j} .$$

$|V| = \sqrt{(15)^2 + (-8)^2} = \sqrt{289} = 17$, so the direction of **V** is $\dfrac{V}{|V|} = \dfrac{15\mathbf{i} - 8\mathbf{j}}{17} = \dfrac{15}{17}\mathbf{i} - \dfrac{8}{17}\mathbf{j}$.

$\mathbf{W} = \mathbf{U} + 3\mathbf{i} = (7\mathbf{i} + 24\mathbf{j}) + 3\mathbf{i} = 10\mathbf{i} + 24\mathbf{j}$, so $|W| \sqrt{(10)^2 + (24)^2} = \sqrt{676} = 26$.

The direction of **W** is $\dfrac{10\mathbf{i} + 24\mathbf{j}}{26} = \dfrac{5}{13}\mathbf{i} + \dfrac{12}{13}\mathbf{j}$.

Practice 7: $|V| = 70$ pounds, and you want the horizontal component, $|V|\cos(\theta)$, to be 50 pounds, so $70\cos(\theta) = 50$ and $\theta = \arccos(5/7) \approx 44.4°$.

Practice 8: $\mathbf{R} = \langle s, 0 \rangle$. **V** makes an angle of $90° - \theta$ with the horizontal so

$\mathbf{V} = \langle 20\cos(90° - \theta), 20\sin(90° - \theta) \rangle$.

Also, $\mathbf{V} + \mathbf{W} = \mathbf{R}$ so $\langle 20\cos(90° - \theta) - 6, 20\sin(90° - \theta) - 8 \rangle = \langle s, 0 \rangle$.

Equating the second components of this vector equation, we have

$20\sin(90° - \theta) - 8 = 0$ so $\sin(90° - \theta) = 8/20 = 0.8$, $90° - \theta \approx 23.6°$, and $\theta \approx 66.4°$.

You should steer your boat approximately 66.4° east of due north in order to maintain a course

taking you due north. Your speed due east is

$|R| = s = 20\cos(90° - \theta) - 6 \approx 20\cos(90° - 66.4°) - 6 = 12.3$ knots.

Practice 9:   The method of solution is the same as Example 7.

$\mathbf{W} = \langle 0, -15 \rangle$, $\mathbf{A} = \langle -|A|\cos(35°), |A|\sin(35°) \rangle$, $\mathbf{B} = \langle |B|\cos(50°), |B|\sin(50°) \rangle$, and

$\mathbf{0} = \mathbf{A} + \mathbf{B} + \mathbf{W} = \langle -|A|\cos(35°) + |B|\cos(50°) + 0, |A|\sin(35°) + |B|\sin(50°) - 15 \rangle$.

Imitating the algebraic steps of Example 7, we get

$|A| = \dfrac{15}{\sin(35°) + \cos(35°)\tan(50°)} \approx 9.68$ pounds and

$|B| = |A|\,\dfrac{\cos(35°)}{\cos(50°)} \approx 12.33$ pounds.

11.2   RECTANGULAR COORDINATES IN THREE DIMENSIONS

In this section we move into 3–dimensional space. First we examine the 3–dimensional rectangular coordinate system, how to locate points in three dimensions, distance between points in three dimensions, and the graphs of some simple 3–dimensional objects. Then, as we did in two dimensions, we discuss vectors in three dimensions, the basic properties and techniques with 3–dimensional vectors, and some of their applications. The extension of the algebraic representations and techniques from 2 to three dimensions is straightforward, but it usually takes practice to visualize 3–dimensional objects and to sketch them on a 2–dimensional piece of paper.

3–Dimensional Rectangular Coordinate System

In the 2–dimensional rectangular coordinate system we have two coordinate axes that meet at right angles at the origin (Fig. 1), and it takes two numbers, an ordered pair (x, y),

to specify the rectangular coordinate location of a point in the plane (2 dimensions). Each ordered pair (x, y) specifies the location of exactly one point, and the location of each point is given by exactly one ordered pair (x, y). The x and y values are the coordinates of the point (x, y).

The situation in three dimensions is very similar. In the 3–dimensional rectangular coordinate system we have three coordinate axes that meet at right angles (Fig. 2), and three numbers, an ordered triple (x, y, z), are needed to specify the location of a point. Each ordered triple (x, y, z) specifies the location of exactly one point, and the location of each point is given by exactly one ordered triple (x, y, z). The x, y and z values are the coordinates of the point (x, y, z). Fig. 3 shows the location of the point $(4, 2, 3)$.

Right–hand orientation of the coordinate axes (Fig. 4): Imagine your right hand in front hand in front of you with the palm toward your face, your thumb pointing up, you index finger straight out, and your next finger toward your face (and the two bottom fingers bent into the palm (Fig. 4). Then, in the right hand coordinate system, your thumb points along the positive z-axis, your index finger along the positive x-axis, and the other finger along the positive y-axis. Other orientations of the axes are possible and valid (with appropriate labeling), but the right–hand system is the most common orientation and is the one we will generally use.

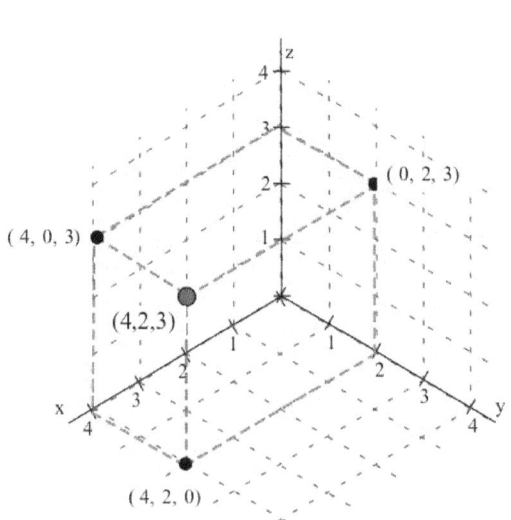

Fig. 3: Locating the point (4, 2, 3)

The three coordinate axes determine three planes (Fig.5): the xy–plane consisting of all points with z–coordinate 0, the xz–plane consisting of all points with y–coordinate 0, and the yz–plane with x–coordinate 0. These three planes then divide the 3–dimensional space into 8 pieces called **octants**. The only octant we shall refer to by name is the **first octant** which is the octant determined by the positive x, y, and z–axes (Fig. 6).

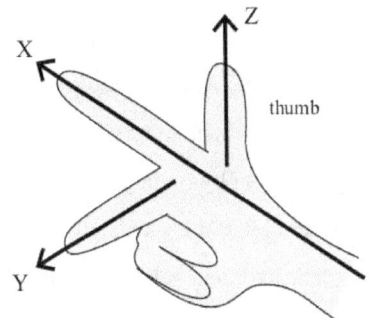

Fig. 4: Right-hand coordinate system

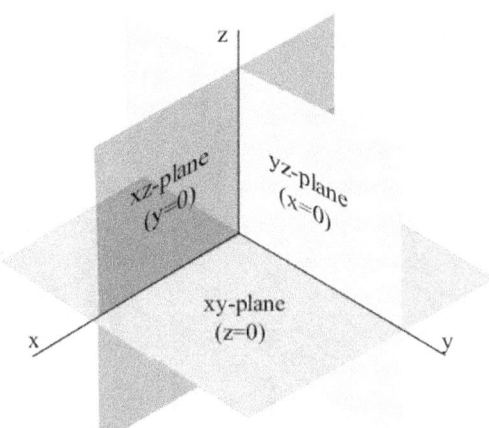

Fig. 5: Coordinate Planes

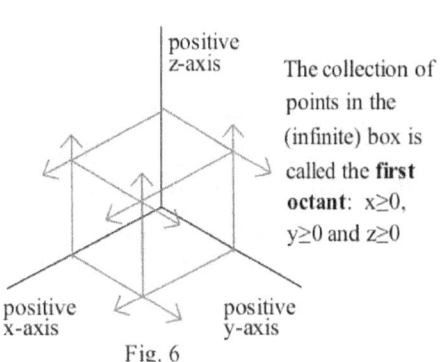

The collection of points in the (infinite) box is called the **first octant**: $x \geq 0$, $y \geq 0$ and $z \geq 0$

Fig. 6

Visualization in three dimensions: Some people have difficulty visualizing points and other objects in three dimensions, and it may be useful for you to spend a few minutes to create a small model of the 3–dimensional axis system for your desk. One model consists of a corner of a box (or room) as in Fig. 7: the floor is the xy–plane; the wall with the window is the xz–plane; and the wall with the door is the yz–plane. Another simple model uses a small Styrofoam ball and three pencils (Fig. 8): just stick the pencils into the ball as in Fig. 8, label each pencil as the appropriate axis, and mark a few units along each axis (pencil). By referring to such a model for your early work in three dimensions, it becomes easier to visualize the locations of points and the shapes of other objects later.

This visualization can be very helpful.

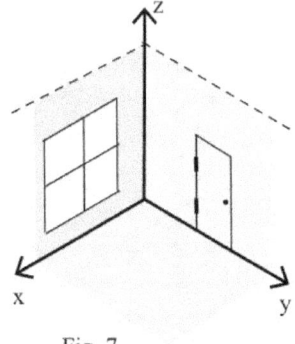

Fig. 7

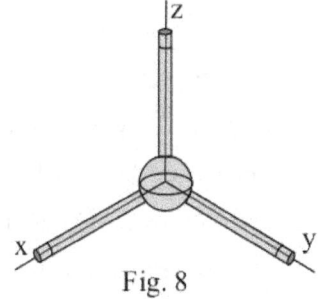

Fig. 8

Each ordered triple (x, y, z) specifies the location of a single point, and this location point can be plotted by locating the point $(x, y, 0)$ on the xy–plane and then going up z units (Fig. 9). (We could also get to the same (x, y, z) point by finding the point $(x, 0, z)$ on the xz–plane and then going y units parallel to the y–axis, or by finding $(0, y, z)$ on the yz–plane and then going x units parallel to the x–axis.)

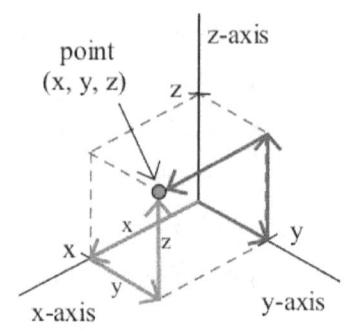

Fig. 9

Example 1:   Plot the locations of the points $P = (0, 3, 4), Q = (2, 0, 4),$

$R = (1, 4, 0), \ S = (3, 2, 1), \ \text{and} \ T(-1, 2, 1) \ .$

Solution: The points are shown in Fig. 10.

Practice 1:   Plot and label the locations of the points

$A = (0, -2, 3), B = (1, 0, -5), C = (-1, 3, 0), \ \text{and}$

$D = (1, -2, 3)$ on the coordinate system in Fig. 11.

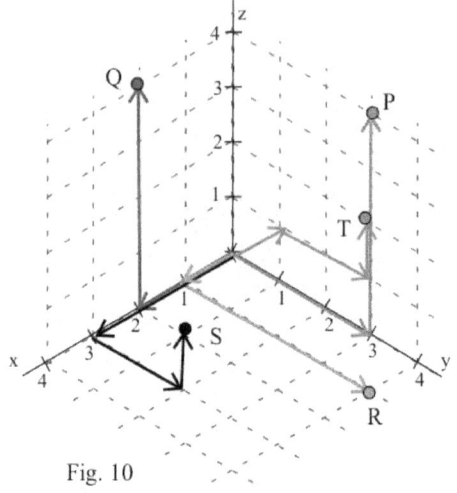

Fig. 10

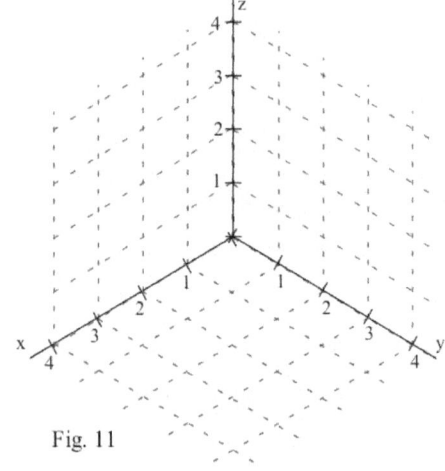

Fig. 11

Practice 2:   The opposite corners of a rectangular box are at $(0, 1, 2)$ and $(2, 4, 3)$. Sketch the box and find its volume.

Once we can locate points, we can begin to consider the graphs of various collections of points. By the graph of "z = 2" we mean the collection of all points (x, y, z) which have the form "$(x, y, 2)$". Since no condition is imposed on the x and y variables, they take all possible values. The graph of $z = 2$ (Fig. 12) is a plane parallel to the xy–plane and 2 units above the xy–plane. Similarly, the graph of $y = 3$ is a plane parallel to the xz–plane (Fig. 13a), and $x = 4$ is a plane parallel to the yz–plane (Fig. 13b). (Note: The planes have been drawn as rectangles, but they actually extend infinitely far.)

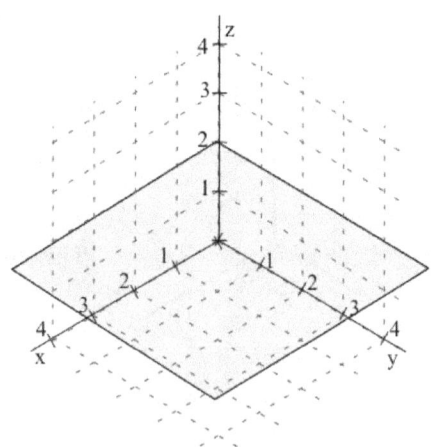

Fig. 12: Plane z=2

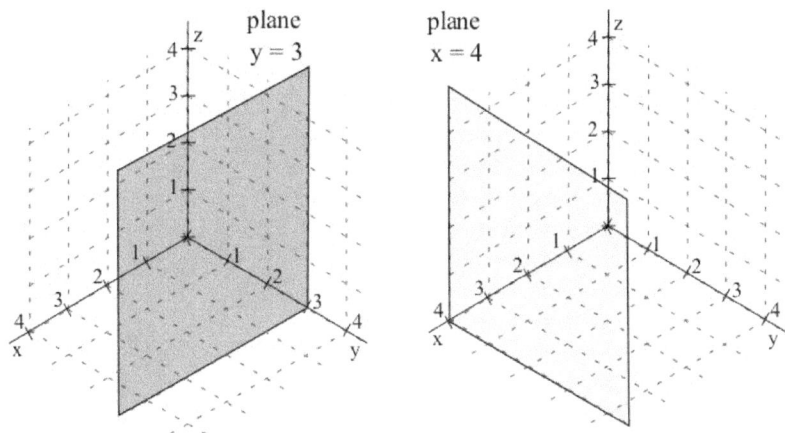

Fig. 13: Planes y = 3 and x = 4

Practice 3:   Graph the planes (a) x = 2, (b) y = −1, and
(c) z = 3 in Fig. 14. Give the coordinates of
the point that lies on all three planes.

Example 2:   Graph the set of points (x, y, z) such that
x = 2 and y = 3.

Solution: The points that satisfy the conditions all have the form
(2, 3, z), and, since no restriction has been placed on the z–
variable, z takes all values. The result is the line (Fig. 15)
through the point (2, 3, 0) and parallel to the z–axis.

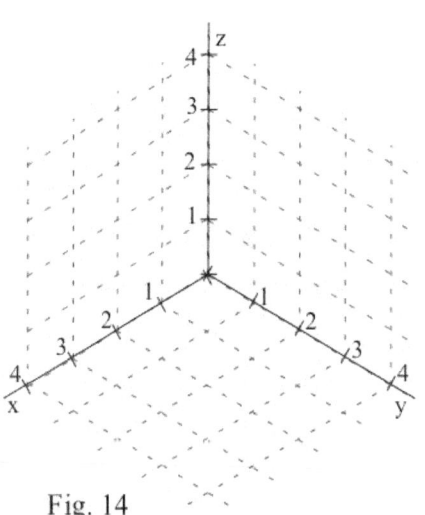

Fig. 14

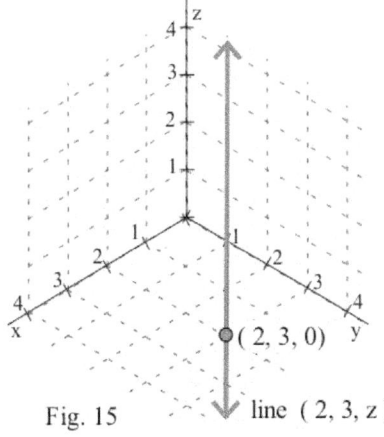

Fig. 15   line (2, 3, z)

Practice 4:   On Fig. 15, graph the points that have the form
(a) (x, 1, 4) and (b) (2, y, −1).

In Section 11.5 we will examine planes and lines that are not
parallel to any of the coordinate planes or axes.

Example 3:   Graph the set of points (x, y, z) such that
$x^2 + z^2 = 1$.

Solution: In the xz–plane (y = 0), the graph of $x^2 + z^2 = 1$ is a circle centered at the origin and with

radius 1 (Fig. 16a). Since no restriction has been placed on the y–variable, y takes all values. The

result is the cylinder in Figs. 16b and 16c, a circle moved parallel to the y–axis.

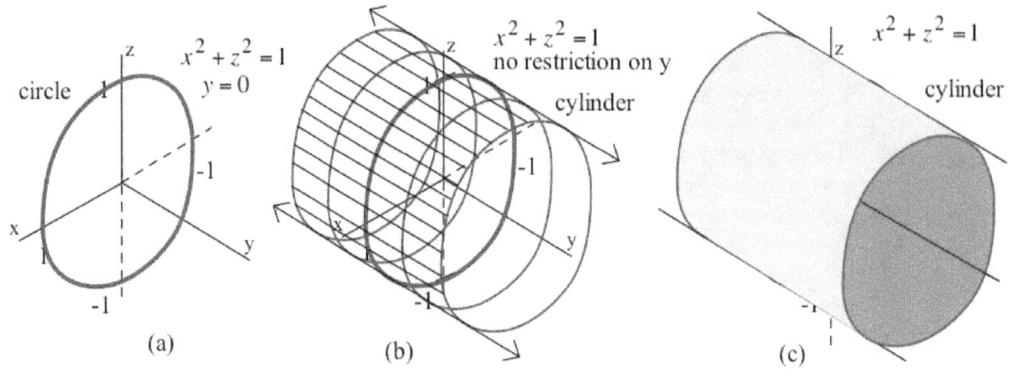

Fig. 16

Practice 5:   Graph the set of points (x, y, z) such that $y^2 + z^2 = 4$. (Suggestion: First graph

$y^2 + z^2 = 4$ in the yz–plane (x = 0) and then extend the result as x takes on all values.)

Distance Between Points

In two dimensions we can think of the distance between points as the length

of the hypothenuse of a right triangle (Fig. 17), and that leads to the Pythagorean

formula: distance = $\sqrt{\Delta x^2 + \Delta y^2}$. In three dimensions

we can also think of the distance between points as the length of the

hypothenuse of a right triangle (Fig. 18), but in this situation the calculations appear

more complicated. Fortunately, they are straightforward:

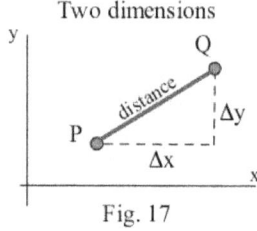

Two dimensions

Fig. 17

$$\text{distance}^2 = \text{base}^2 + \text{height}^2 = \left(\sqrt{\Delta x^2 + \Delta y^2} \right)^2 + \Delta z^2$$
$$= \Delta x^2 + \Delta y^2 + \Delta z^2 \quad \text{so distance} = \sqrt{\Delta x^2 + \Delta y^2 + \Delta z^2} \quad .$$

Three dimensions

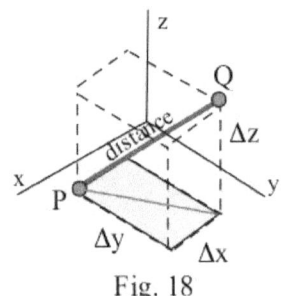

Fig. 18

If $P = (x_1, y_1, z_1)$ and $Q = (x_2, y_2, z_2)$ are points in space,

then   the distance between P and Q is

distance = $\sqrt{\Delta x^2 + \Delta y^2 + \Delta z^2}$

$= \sqrt{(x_2 - x_1)^2 + (y_2 - y_1)^2 + (z_2 - z_1)^2}$.

The 3–dimensional pattern is very similar to the 2–dimensional pattern with the additional piece Δz^2 .

Example 4:   Find the distances between all of the pairs of the given points. Do any three of these points

form a right triangle? Do any three of these points lie on a straight line?

Points: $A = (1, 2, 3), B = (7, 5, -3), C = (8, 7, -1), D = (11, 13, 5)$.

Solution: $Dist(A, B) = \sqrt{6^2 + 3^2 + (-6)^2}$ $= \sqrt{36 + 9 + 36}$ $= \sqrt{81}$ $= 9$. Similarly,

$Dist(A,C) = \sqrt{90}$, $Dist(A, D) = 15$, $Dist(B, C) = 3$, $Dist(B, D) = 12$, and $Dist(C, D) = 9$.

$\{ Dist(A, B) \}^2 + \{ Dist(B, D) \}^2 = \{ Dist(A, D) \}^2$ so the points A, B, and D form a right triangle

with the right angle at point B. Also, the points A, B,

and C form a right triangle with the right angle at point

B since $9^2 + 3^2 = (\sqrt{90})^2$

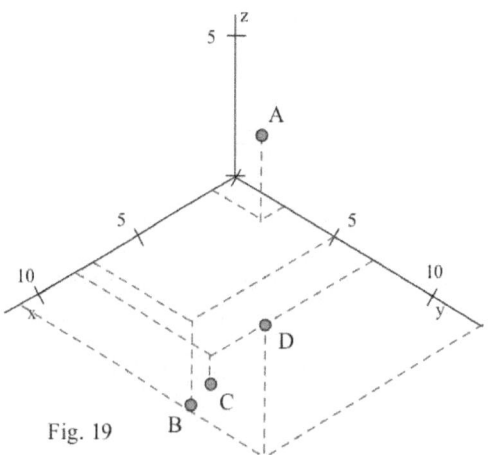

$Dist(B, C) + Dist(C, D) = Dist(B, D)$ so the points

B, C, and D line on a straight line. The points are shown in

Fig. 19. In three dimensions it is often difficult to determine

the size of an angle from a graph or to determine whether

points are collinear.

Fig. 19

Practice 6:   Find the distances between all of the pairs of the

points $A = (3, 1, 2), B = (9, 7, 5), C = (9, 7, 9)$. Which two of these points are closest together?

Which two are farthest apart? Do these three points form a right triangle?

In two dimensions, the set of points at a fixed distance from a given point

is a circle, and we used the distance formula to determine equations

describing circles: the circle with center $(2, 3)$ and radius 5 (Fig. 20) is

given by $(x-2)^2 + (y-3)^2 = 5^2$ or $x^2 + y^2 - 4x - 6y = 12$.

The same ideas work for spheres in three dimensions.

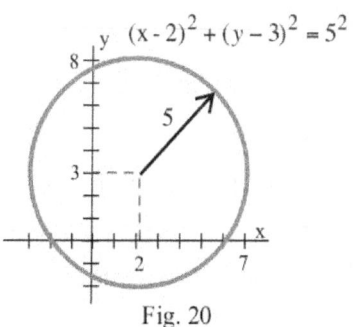

Fig. 20

Spheres: The set of points (x, y, z) at a fixed distance r from a point (a, b, c) is a

sphere (Fig. 21) with center (a, b, c) and radius r.

The sphere is given by the equation $(x-a)^2 + (y-b)^2 + (z-c)^2 = r^2$.

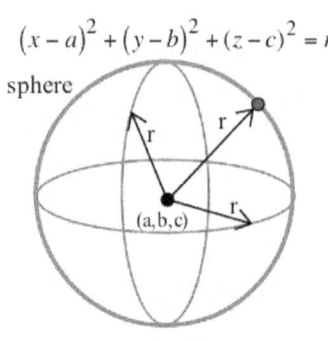

$$(x-a)^2 + (y-b)^2 + (z-c)^2 = r^2$$

sphere

Fig. 21

Example 5:   Write the equations of the following two spheres: (A) center $(2, -3, 4)$ and radius 3, and (B) center $(4, 3, -5)$ and radius 4. What is the minimum distance between a point on A and a point on B? What is the maximum distance between a point on A and a point on B?

Solution: (A) $(x-2)^2 + (y+3)^2 + (z-4)^2 = 3^2$.

(B) $(x-4)^2 + (y-3)^2 + (z+5)^2 = 4^2$.

The distance between the centers is
$$\sqrt{(4-2)^2 + (3+3)^2 + (-5-4)^2} = \sqrt{121} = 11, \text{ so}$$

the minimum distance between points on the spheres (Fig. 22) is

$$11 - (\text{one radius}) - (\text{other radius}) = 11 - 3 - 4 = 4.$$

The maximum distance between points on the spheres is $11 + 3 + 4 = 18$.

Practice 7:   Write the equations of the following two spheres:

(A) center $(1, -5, 3)$ and radius 10, and (B) center $(7, -7, 0)$ and radius 2. What is the minimum distance between a point on A and a point on B? What is the maximum distance between a point on A and a point on B?

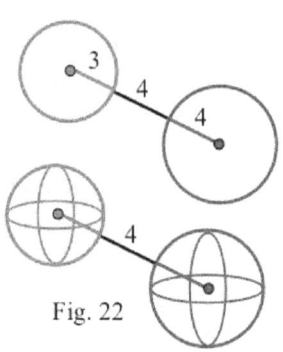

Fig. 22

Beyond Three Dimensions

At first it may seem strange that there is anything beyond three dimensions, but fields as different as physics, statistics, psychology, economics and genetics routinely work in higher dimensional spaces. In three dimensions we use an ordered 3–tuple (x, y, z) to represent and locate a point, but there is no logical or mathematical reason to stop at three. Physicists talk about "space–time space," a 4–dimensional space where a point is represented by a 4–tuple (x, y, z, t) with x, y, and z representing a location and t represents time. This is very handy for describing complex motions, and the "distance" between two "space–time" points tells how far apart they are in (3–dimensional) distance and time. "String theorists," trying to model the early behavior and development of the universe, work in 10–dimensional space and use 10–tuples to represent points in that space.

On a more down to earth scale, any object described by 5 separate measures (numbers) can be thought of as a point in 5–dimensional space. If a pollster asks students 5 questions, then one student's responses can be represented as an ordered 5–tuple (a, b, c, d, e) and can be thought of as a point in five dimensional space. The collection of responses from an entire class of students is a cloud of points in 5–space, and the center of mass of that cloud (the point formed as the mean of all of the individual points) is often used as a group response. Psychologists and counselors sometimes use a personality profile that rates people on four independent scales (IE, SN, TF, JP). The "personality type" of each person can be represented as an ordered 4–tuple, a point in 4–dimensional "personality–type space." If the distance between two people is small in "personality–type space," then they have similar "personality types." Some matchmaking services ask clients a number of questions (each question is a dimension in this "matching space") and then try to find a match a small distance away.

Many biologists must deal with huge amounts of data, and often this data is represented as ordered n– tuples, points in n–dimensional space. In the book The History and Geography of Human Genes (1994), the authors summarize more than 75,000 allele frequencies in nearly 7,000 human populations in the form of maps. "To construct one of the maps, eighty–two genes were examined in many populations throughout the world. Each population was represented on a computer grid as a point in eighty–two dimensional space, with its position along each dimensional axis representing the frequency of one of the alleles in question." (Natural History, 6/94, p. 84)

Geometrically, it is difficult to work in more than three dimensions, but length/distance calculations are still easy.

Definitions for n dimensions:

A point in n–dimensional space is an ordered n–tuple $(a_1, a_2, a_3, \ldots , a_n)$.

If $A = (a_1, a_2, a_3, \ldots , a_n)$ and $B = (b_1, b_2, b_3, \ldots , b_n)$ are points in n–dimensional space,

then   the distance between A and B is

$$\text{distance} = \sqrt{(b_1-a_1)^2 + (b_2-a_2)^2 + (b_3-a_3)^2 + \ldots + (b_n-a_n)^2} \ .$$

Example 6:   Find the distance between the points $P = (1, 2, -3, 5, 6)$ and $Q = (5, -1, 4, 0, 7)$.

Solution: Distance $= \sqrt{(5-1)^2+(-1-2)^2+(4--3)^2+(0-5)^2+(7-6)^2} \quad = \sqrt{16+9+49+25+1} \quad = \sqrt{100} = 10$.

Practice 8:   Write an equation for the 5–dimensional sphere with radius 8 and center $(3, 5, 0, -2, 4)$.

PROBLEMS

In problems 1 – 4, plot the given points.

1. $A = (0,3,4)$, $B = (1,4,0)$, $C = (1,3,4)$, $D = (1,4,2)$

2. $E = (4,3,0)$, $F = (3,0,1)$, $G = (0,4,1)$, $H = (3,3,1)$

3. $P = (2,3,-4)$, $Q = (1,-2,3)$, $R = (4,-1,-2)$, $S = (-2,1,3)$

4. $T = (-2,3,-4)$, $U = (2,0,-3)$, $V = (-2,0,0)$, $W = (-3,-1,-2)$

In problems 5 – 8, plot the lines.

5. $(3, y, 2)$ and $(1, 4, z)$

6. $(x, 3, 1)$ and $(2, 4, z)$

7. $(x, -2, 3)$ and $(-1, y, 4)$

8. $(3, y, -2)$ and $(-2, 4, z)$

In problems 9 – 12, three collinear points are given. Plot the points and the draw a line through them.

9. $(4,0,0)$, $(5,2,1)$, and $(6,4,2)$

10. $(1,2,3)$, $(3,4,4)$, and $(5,6,5)$

11. $(3,0,2)$, $(3,2,3)$, and $(3,6,5)$

12. $(-1,3,4)$, $(2,3,2)$, and $(5,3,0)$

In problems 13 – 16, calculate the distances between the given points and determine if any three of them are collinear. (Note: P, Q, R are collinear if $\text{dist}(P,Q) + \text{dist}(Q,R) = \text{dist}(P,R)$)

13. $A = (5,3,4)$, $B = (3,4,4)$, $C = (2,2,3)$, $D = (1,6,4)$

14. $A = (6,2,1)$, $B = (3,2,1)$, $C = (3,2,5)$, $D = (1, -4,2)$

15. $A = (3,4,2)$, $B = (-1,6,-2)$, $C = (5,3,4)$, $D = (2,2,3)$

16. $A = (-1,5,0)$, $B = (1,3,2)$, $C = (5,-1,3)$, $D = (3,1,2)$

In problems 17 – 20 you are given three corners of a box whose sides are parallel to the xy, xz, and yz planes. Find the other five corners and calculate the volume of the box.

17. $(1,2,1)$, $(4,2,1)$, and $(1,4,3)$

18. $(5,0,2)$, $(1,0,5)$, and $(1,5,2)$

19. $(4,5,0)$, $(1,4,3)$, and $(1,5,3)$

20. $(4,0,1)$, $(0,3,1)$, and $(0,0,5)$

In problems 21 – 24, graph the given planes.

21. $y = 1$ and $z = 2$

22. $x = 4$ and $y = 2$

23. $x = 1$ and $y = 0$

24. $x = 2$ and $z = 0$

In problems 25 – 28, the center and radius of a sphere are given. Find an equation for the sphere.

25. Center = $(4, 3, 5)$, radius = 3

26. Center = $(0, 3, 6)$, radius = 2

27. Center = $(5, 1, 0)$, radius = 5

28. Center = $(1, 2, 3)$, radius = 4

In problems 29 – 32, the equation of a sphere is given. Find the center and radius of the sphere.

29. $(x-3)^2 + (y+4)^2 + (z-1)^2 = 16$ 30. $(x+2)^2 + y^2 + (z-4)^2 = 25$

31. $x^2 + y^2 + z^2 - 4x - 6y - 8z = 71$ 32. $x^2 + y^2 + z^2 + 6x - 4y = 12$

Problems 33 – 36 name all of the shapes that are possible for the intersection of the two given shapes in three dimensions.

33. A line and a plane   34. Two planes

35. A plane and a sphere   36. Two spheres

In problems 37 – 44, sketch the graphs of each collection of points. Name the shape of each graph.

37. All (x, y, z) such that (a) $x^2 + y^2 = 4$ and $z = 0$, (b) $x^2 + y^2 = 4$ and $z = 2$.

38. All (x, y, z) such that (a) $x^2 + z^2 = 4$ and $y = 0$, (b) $x^2 + z^2 = 4$ and $y = 1$.

39. All (x, y, z) such that $x^2 + y^2 = 4$ and no restriction on z.

40. All (x, y, z) such that $x^2 + z^2 = 4$ and no restriction on y.

41. All (x,y,z) such that (a) $y = \sin(x)$ and $z = 0$, (b) $y = \sin(x)$ and $z = 1$,

    (c) $y = \sin(x)$ and no restriction on z.

42. All (x,y,z) such that (a) $z = x^2$ and $y = 0$, (b) $z = x^2$ and $y = 2$,

    (c) $z = x^2$ and no restriction on y.

43. All (x,y,z) such that (a) $z = 3 - y$ and $x = 0$, (b) $z = 3 \quad y$ and $x = 2$,

    (c) $z = 3 - y$ and no restriction on x.

44. All (x,y,z) such that (a) $z = 3 - x$ and $y = 0$, (b) $z = 3 - x$ and $y = 2$,

    (c) $z = 3 - x$ and no restriction on y.

The volume of a sphere with radius r is $\frac{4}{3} \pi r^3$. Use that formula to help determine the volumes of the following parts of spheres in problems 45 and 46.

45. All (x,y,z) such that (a) $x^2 + y^2 + z^2 \leq 4$ and $z \geq 0$, (b) $x^2 + y^2 + z^2 \leq 4$, $z \geq 0$, and $y \geq 0$,

    (c) $x^2 + y^2 + z^2 \leq 4$, $z \geq 0$, $y \geq 0$, and $x \geq 0$.

46. All (x,y,z) such that (a) $x^2 + y^2 + z^2 \leq 9$ and $x \geq 0$, (b) $x^2 + y^2 + z^2 \leq 9$, $x \geq 0$, and $z \geq 0$,

    (c) $x^2 + y^2 + z^2 \leq 9$, $x \geq 0$, $z \geq 0$, and $y \geq 0$.

"Shadow" Problems

The following "shadow" problems assume that we have an object in the first octant. Then light rays parallel to the x–axis cast a shadow of the object on the yz–plane (Fig. 23). Similarly, light rays parallel to the y–axis cast a shadow of the object on the xz–plane, and rays parallel to the z–axis cast a shadow on the xy–plane. (The point of these and many of the previous problems is to get you thinking and visualizing in three dimensions.)

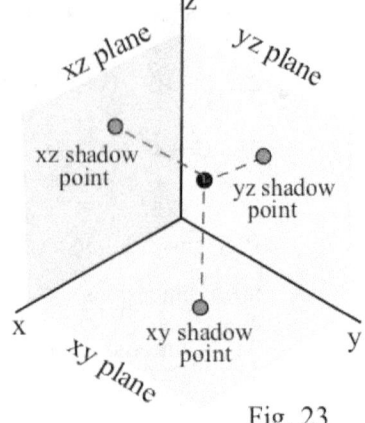

Fig. 23

S1.   Give the coordinates of the shadow points of the point $(1,2,3)$ on each of the coordinate planes.

S2.   Give the coordinates of the shadow points of the point $(4,1,2)$ on each of the coordinate planes.

S3.   Give the coordinates of the shadow points of the point (a,b,c) on each of the coordinate planes.

S4.   A line segment in the first octant begins at the point $(4,2,1)$ and ends at $(1,3,3)$. Where do the shadows of the line segment begin and end on each of the coordinate planes? Are the shadows of the line segment also line segments?

S5.   A line segment begins at the point $(1,2,4)$ and ends at $(1,4,3)$. Where do the shadows of the line segment begin and end on each of the coordinate planes? Are the shadows of the line segment also line segments?

S6.   A line segment begins at the point (a,b,c) and ends at (p,q,r). Where do the shadows of the line segment begin and end on each of the coordinate planes? Are the shadows of the line segment also line segments?

S7.   The three points $(0,0,0), (4,0,3)$, and $(4,0,2)$ are the vertices of a triangle in the first octant. Describe the shadow of this triangle on each of the coordinate planes. Are the shadows always triangles?

S8.   The three points $(1,2,3), (4,3,1)$, and $(2,3,4)$ are the vertices of a triangle in the first octant. Describe the shadow of this triangle on each of the coordinate planes. Are the shadows always triangles?

S9.   The three points $(a,b,c), (p,q,r)$, and (x,y,z) are the vertices of a triangle in the first octant. Describe the shadow of this triangle on each of the coordinate planes. Are the shadows always triangles?

S10.   A line segment in the first octant is 10 inches long. (a) What is the shortest shadow it can have on a coordinate plane? (b) What is the longest shadow it can have on a coordinate plane?

S11.   A triangle in the first octant has an area of 12 square inches. (a) What is the smallest area its shadow can have on a coordinate plane? (b) What is the largest area?

S12.   Design a solid 3–dimensional object whose shadow on one coordinate plane is a square, on another coordinate plane a circle, and on the third coordinate plane a triangle.

Practice Answers

Practice 1:   The points are plotted in Fig. 24 .

Practice 2:   The box is shown in Fig. 25.

$\Delta x = 2$ = width, $\Delta y = 3$ = length,

and $\Delta z = 1$ = height, so

volume = $(2)(3)(1) = 6$ cubic units.

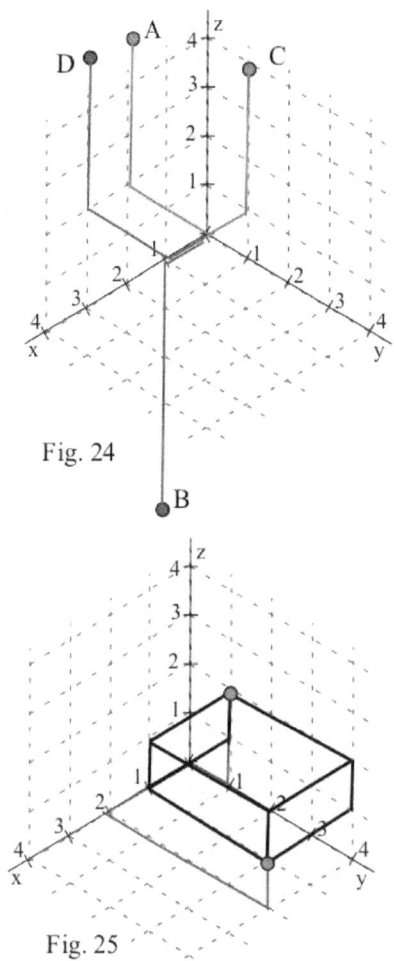

Fig. 24

Practice 3:   The planes are shown in Fig. 26(a).

Each pair of planes intersects along a line, shown as a dark

lines in Fig. 26(b), and the three lines intersect at the point

$(2, -1, 3)$. This is the only point that lies on all three of the

planes.

Fig. 25

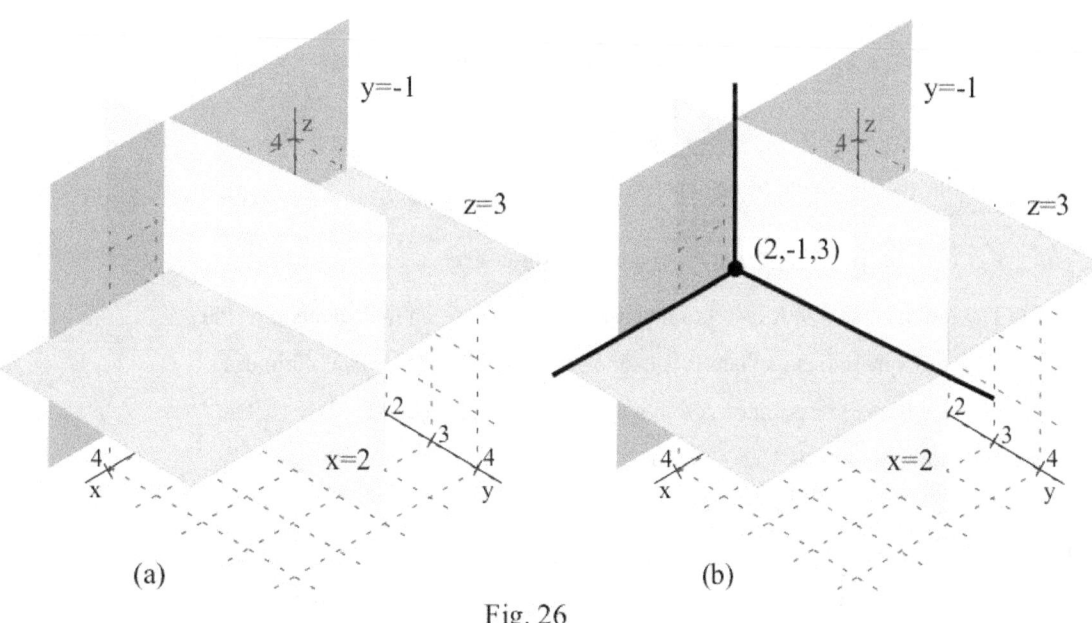

(a)

(b)

Fig. 26

Practice 4:   The points that satisfy $(x, 1, 4)$ are shown in Fig.

27. The collection of these points form a line. One way to

sketch the graph of the line is to first plot the point where the

line crosses one of the coordinate planes, $(0, 1, 4)$ in this case,

and then sketch a line through that point and parallel to the

appropriate axis.

The line of points that satisfy $(2, y, -1)$ are also shown in Fig.

27, as well as the point $(2, 0, -1)$ where the line intersects the

xz–plane.

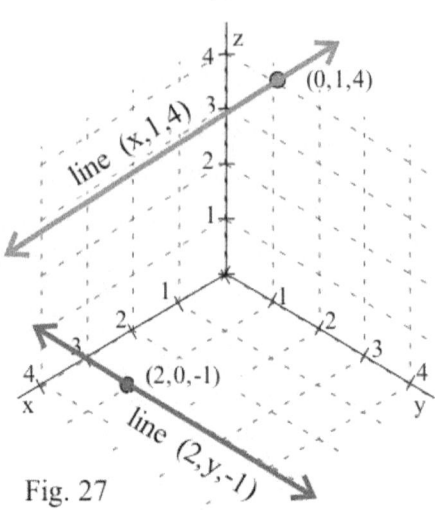

Fig. 27

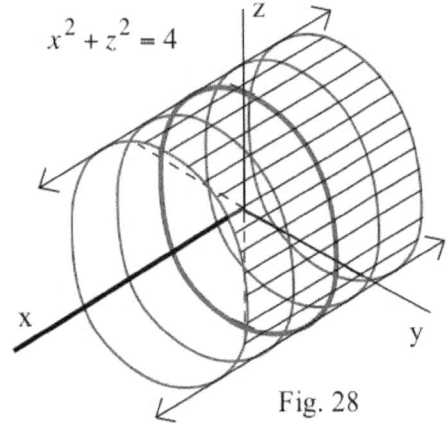

$x^2 + z^2 = 4$

Fig. 28

Practice 5:   The graph, a cylinder with radius 2 around the

x–axis, is shown in Fig. 28. The dark circle is the graph

of points satisfying $y^2 + z^2 = 4$ **and** $x = 0$.

Practice 6:

$$\text{Dist}(A,B) = \sqrt{6^2 + 6^2 + 3^2} = 9,$$

$$\text{Dist}(A,C) = \sqrt{6^2 + 6^2 + 7^2} = 11, \text{ and}$$

$\text{Dist}(B,C) = \sqrt{0^2 + 0^2 + 4^2} = 4$. B and C are closest. A and C are farthest apart.

$4^2 + 9^2 \neq 11^2$ so the points do not form a right triangle.

Practice 7:   (A) $(x-1)^2 + (y+5)^2 + (z-3)^2 = 10^2$.   (B) $(x-7)^2 + (y+7)^2 + (z-0)^2 = 2^2$.

The distance between the centers is

$$\sqrt{(7-1)^2 + (-7+5)^2 + (0-3)^2} = \sqrt{49} = 7.$$

The radius of sphere A is larger than the distance between the centers plus the

radius of sphere B so sphere B is inside sphere A (Fig. 29). The minimum

distance between a point on A and a point on B is $10 - (5 + 2 + 2) = 1$. The

maximum distance is $10 + (5 + 2 + 2) = 19$.

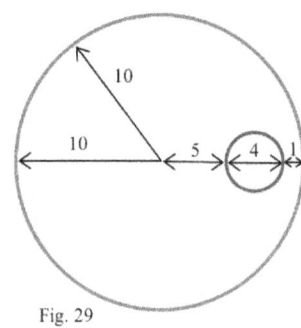

Fig. 29

Practice 8:   A point $P = (v,w,x,y,z)$ is on the sphere if and only if the distance from P to the center

$(3, 5, 0, -2, 4)$ is 8. Using the distance formula, and squaring each side, we have

$$(v-3)^2 + (w-5)^2 + (x-0)^2 + (y+2)^2 + (z-4)^2 = 8^2 .$$

11.3   VECTORS IN THREE DIMENSIONS

Once you understand the 3–dimensional coordinate system, 3–dimensional vectors are a straightforward
extension of vectors in two dimensions. Vectors in three dimensions are more difficult to visualize and sketch,
but all of the 2–dimensional algebraic techniques extend very naturally, with just one more component.

A vector in any setting is a **quantity that has both a direction and a**
magnitude, and in three dimensions vectors can be represented
geometrically as directed line segments, (arrows). The vector **V**
given by the line segment from the starting point (tail) $P = (1, 2, 3)$ to
the ending point (head) $Q = (4, 1, 4)$ is shown in Fig. 1. The vector
V from P to Q is represented algebraically by the ordered triple
enclosed in "bent" brackets: $\mathbf{V} = \langle 3, -1, 1 \rangle$ with each component
representing the displacement from P to Q. (We continue to reserve
the "round brackets" ( ) to represent points.) Fig. 1 also shows
several other geometric representations of vector **V**.

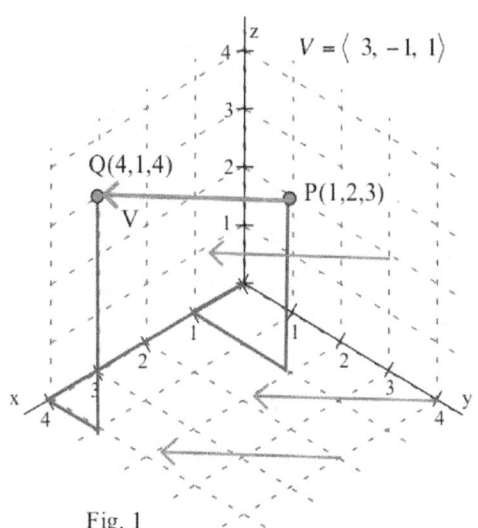

Fig. 1

Definitions and Properties

The following definitions, arithmetic operations, and properties of
vectors in 3–dimensional space are straightforward generalizations from two dimensions.

Definition: Equality of Vectors

Geometrically, two vectors are equal if their lengths are equal and their directions are the same.

Algebraically, two vectors are equal if their respective components are equal:

if $\mathbf{U} = \langle a, b, c \rangle$ and $\mathbf{V} = \langle x, y, z \rangle$,

then $\mathbf{U} = \mathbf{V}$ if and only if $a = x$, $b = y$, and $c = z$.

The definitions of scalar multiplication, vector addition and vector subtraction in three dimensions are
similar to the definitions in two dimensions, but each vector has one more component.

Definitions: Vector Arithmetic

If $\mathbf{A} = \langle a_1, a_2, a_3 \rangle$ and $\mathbf{B} = \langle b_1, b_2, b_3 \rangle$ are vectors and k is a scalar, then

then $k\mathbf{A} = \langle ka_1, ka_2, ka_3 \rangle$

$\mathbf{A} + \mathbf{B} = \langle a_1 + b_1, a_2 + b_2, a_3 + b_3 \rangle$

$\mathbf{A} - \mathbf{B} = \mathbf{A} + (-1)\mathbf{B} = \langle a_1 - b_1, a_2 - b_2, a_3 - b_3 \rangle$.

Example 1: For $\mathbf{A} = \langle\, 3, -4, 2\,\rangle$ and $\mathbf{B} = \langle\, 5, 1, -3\,\rangle$, calculate $\mathbf{C} = 3\mathbf{B}$, $\mathbf{D} = 2\mathbf{A} + 3\mathbf{B}$, and

$\mathbf{E} = 5\mathbf{A} - 2\mathbf{B}$.

Solution: $\mathbf{C} = 3\mathbf{B} = 3\langle\, 5, 1, -3\,\rangle = \langle\, 15, 3, -9\,\rangle$.

$\mathbf{D} = 2\mathbf{A} + 3\mathbf{B} = 2\langle\, 3, -4, 2\,\rangle + 3\langle\, 5, 1, -3\,\rangle = \langle\, 6{+}15, -8{+}3, 4{-}9\,\rangle = \langle\, 21, -5, -5\,\rangle$.

$\mathbf{E} = 5\mathbf{A} - 2\mathbf{B} = 5\langle\, 3, -4, 2\,\rangle - 2\langle\, 5, 1, -3\,\rangle = \langle\, 5, -22, 16\,\rangle$.

Practice 1:   For $\mathbf{A} = \langle\, 5, -4, 1\,\rangle$ and $\mathbf{B} = \langle\, 2, -3, 4\,\rangle$, calculate

$\mathbf{C} = 5\mathbf{A}$, $\mathbf{D} = 3\mathbf{A} + 4\mathbf{B}$, and $\mathbf{E} = 2\mathbf{B} - 3\mathbf{A}$.

Each of the given vector arithmetic operations also has a geometric

interpretation:

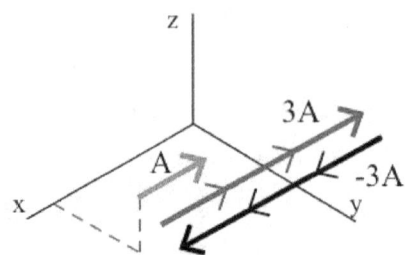

Fig. 2

Multiplying by a scalar k results in a vector that is $|k|$ times as

long as the original vector, $|\,k\mathbf{A}\,| = |k|\,|\mathbf{A}|$. If k is positive, then

$\mathbf{A}$ and $k\mathbf{A}$ have the same direction. If k is negative, then $\mathbf{A}$ and $k\mathbf{A}$ point in opposite

directions (Fig. 2)

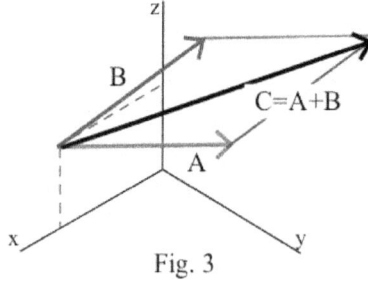

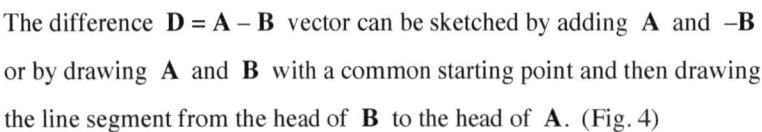

The sum $\mathbf{C} = \mathbf{A} + \mathbf{B}$ vector can be found geometrically by using the

parallelogram or head–to–tail methods described in Section 11.1. (Fig. 3)

Fig. 3

The difference $\mathbf{D} = \mathbf{A} - \mathbf{B}$ vector can be sketched by adding $\mathbf{A}$ and $-\mathbf{B}$

or by drawing $\mathbf{A}$ and $\mathbf{B}$ with a common starting point and then drawing

the line segment from the head of $\mathbf{B}$ to the head of $\mathbf{A}$. (Fig. 4)

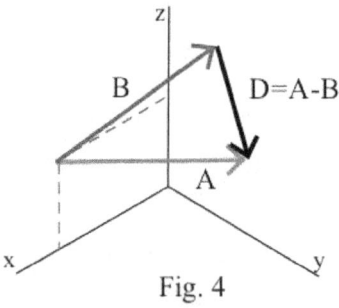

Fig. 4

Because it is more difficult to make precise drawings and measurements in three dimensions, the geometric

methods are seldom used to perform vector arithmetic in three dimensions. These geometric interpretations

are still very powerful and are important for understanding the meaning of various arithmetic operations

and for understanding how certain algorithms are developed.

Visualizing vector arithmetic in three dimensions: If you took the

time in Section 11.2 to build a small model of a 3–dimensional

coordinate system, you can use it now to see and handle some

3–dimensional vectors. Sharpened pencils or skewer sticks

make good physical "vectors."

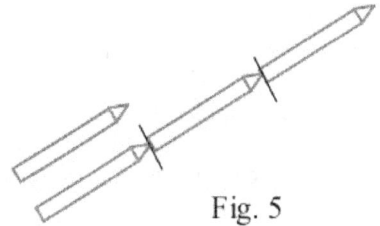

Fig. 5

Visualizing the vectors **A** and 3**A** is easy — just tape

together a short pencil and one three times as long

(Fig. 5). How could you modify this arrangement to

illustrate **A** and –3**A**?

Addition and subtraction are more difficult, but the

following vectors fit together nicely:

$\mathbf{C} = \mathbf{A} + \mathbf{B}$ with $\mathbf{A} = \langle\, 2, 3, 6 \,\rangle$ and

$\mathbf{B} = \langle\, 4, 0, 0 \,\rangle$ (Fig. 6), and $\mathbf{D} = \mathbf{A} - \mathbf{B}$ with

$\mathbf{A} = \langle\, 4, 7, 4 \,\rangle$ and $\mathbf{B} = \langle\, 4, 2, 4 \,\rangle$.

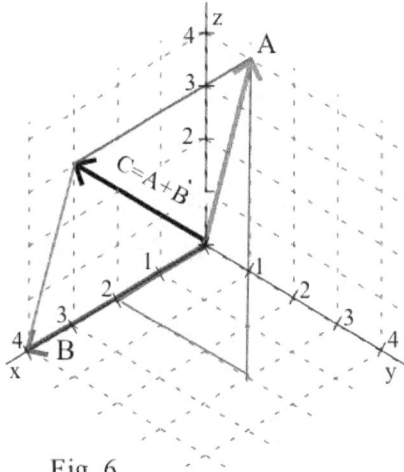

Fig. 6

The length of a vector in three dimensions follows directly from the formula for the distance between

points in 3–dimensional space.

The **magnitude** or **length** of a vector $\mathbf{V} = \langle\, a, b, c \,\rangle$ is $|\mathbf{V}| = \sqrt{a^2 + b^2 + c^2}$.

Since the components $a, b,$ and c of the vector $\mathbf{V} = \langle\, a, b, c \,\rangle$ represent the displacements of the ending

point from the starting point in the $x, y,$ and z directions, we can represent $\mathbf{V}$ as a line segment from the

point $(\, 0, 0, 0 \,)$ to the point $(\, a, b, c \,)$. Then the length of $\mathbf{V}$ is the distance from the point $(\, 0, 0, 0 \,)$ to the

point $(\, a, b, c \,)$: length = { distance from $(\, 0, 0, 0 \,)$ to $(\, a, b, c \,)$ } = $\sqrt{a^2 + b^2 + c^2}$.

The only vector in 3–dimensional space with magnitude 0 is the zero vector $\mathbf{0} = \langle\, 0, 0, 0 \,\rangle$. The zero

vector has no specific direction.

Example 2:   Determine the lengths of $\mathbf{A} = \langle\, 2, 8, 16 \,\rangle$, $\mathbf{B} = \langle\, -4, 8, 8 \,\rangle$, $\mathbf{C}$ = {vector represented by the

line segment from (1,2,3) to (7,–1,9) } , $\mathbf{D} = \mathbf{A} - \mathbf{B}$, and $\mathbf{E} = \mathbf{B} + \mathbf{C}$.

Solution: $|\mathbf{A}| = \sqrt{2^2 + 8^2 + 16^2} = \sqrt{324} = 18$. $|\mathbf{B}| = \sqrt{(-4)^2 + 8^2 + 8^2} = \sqrt{144} = 12$.

$\mathbf{C} = \langle\, 7\text{–}1, -1\text{–}2, 9\text{–}3 \,\rangle = \langle\, 6, -3, 6 \,\rangle$ so $|\mathbf{C}| = \sqrt{81} = 9$.

$\mathbf{D} = \mathbf{A} - \mathbf{B} = \langle\, 2\text{–}(-4), 8\text{–}8, 16\text{–}8 \,\rangle = \langle\, 6, 0, 8 \,\rangle$ so $|\mathbf{D}| = \sqrt{100} = 10$.

$\mathbf{E} = \mathbf{B} + \mathbf{C} = \langle\, -4\text{+}6, 8\text{+}(-3), 8\text{+}6 \,\rangle = \langle\, 2, 5, 14 \,\rangle$ so $|\mathbf{E}| = \sqrt{225} = 15$.

Practice 2:   Determine the lengths of $A = \langle 2, 3, 6 \rangle$, $B = \langle 2, 1, 2 \rangle$, and $C = A - 2B$.

Definitions:

The **direction** of a nonzero vector **A** is the unit vector

$$U = \frac{1}{|A|} A = \frac{A}{|A|}.$$

The **standard basis vectors** in the plane are

$$i = \langle 1, 0, 0 \rangle, j = \langle 0, 1, 0 \rangle, \text{ and } k = \langle 0, 0, 1 \rangle. \text{ (Fig. 7)}$$

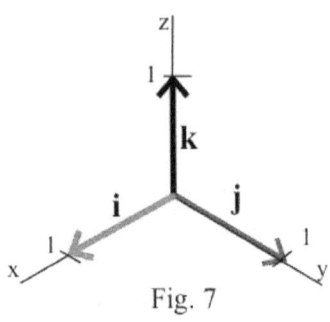

Fig. 7

Example 3:   Determine the lengths and directions of $A = 6i + 2j + 9k$,

$$B = 6i + 3j - 6k, C = 3A + 2B, \text{ and } D = B - 2A.$$

Solution: $|A| = \sqrt{6^2 + 2^2 + 9^2} = \sqrt{121} = 11$. Direction of **A** is $\frac{A}{|A|} = \frac{6}{11}i + \frac{2}{11}j + \frac{9}{11}k$.

$|B| = \sqrt{6^2 + 3^2 + (-6)^2} = \sqrt{81} = 9$. Direction of **B** is $\frac{B}{|B|} = \frac{2}{3}i + \frac{1}{3}j - \frac{2}{3}k$.

$C = 3A + 2B = 30i + 12j + 15k$. $|C| = \sqrt{1269} \approx 35.6$.

The direction of **C** is $\frac{C}{|C|} = \frac{30}{\sqrt{1269}}i + \frac{12}{\sqrt{1269}}j + \frac{15}{\sqrt{1269}}k \approx 0.84i + 0.34j + 0.42k$.

$D = B - 2A = -6i - 1j - 24k$. $|D| = \sqrt{613} \approx 24.8$.

The direction of **D** is $\frac{D}{|D|} = \frac{-6}{\sqrt{613}}i - \frac{1}{\sqrt{613}}j - \frac{24}{\sqrt{613}}k \approx 0.24i - 0.04j - 0.97k$.

Practice 3:   Determine the lengths and directions of $A = 3i + 2j - 6k$, $B = 6j - 8k$,

$$C = A + 3B, \text{ and } D = 2B - 3A.$$

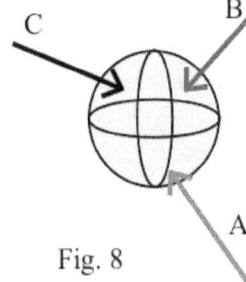

Fig. 8

Example 4: Three players are pushing on a ball, but the ball is not moving (Fig. 8).

Player A is pushing with a force of 45 pounds in the direction

$\frac{1}{9}i - \frac{4}{9}j + \frac{8}{9}k$, and player B is pushing with a force of 60 pounds in the

direction $\frac{4}{6}i + \frac{4}{6}j + \frac{2}{6}k$. How hard and in what direction is player C pushing?

Solution: Since we know the magnitude and direction of the force vectors for players A and B, we can

determine the force vector for each player:

$A = \{ \text{magnitude} \} \{ \text{direction} \} = 45 \{ \frac{1}{9}i - \frac{4}{9}j + \frac{8}{9}k \} = 5i - 20j + 40k$, and

$B = \{ \text{magnitude} \} \{ \text{direction} \} = 60 \{ \frac{4}{6}i + \frac{4}{6}j + \frac{2}{6}k \} = 40i + 40j + 20k$.

Pushing together, their force vector is $A + B = 45i + 20j + 60k$. Since the ball is not moving,

the force vectors of A, B, and C equal the zero vector, and we can solve for C's force vector:

$$C = -(A + B) = -45i - 20j - 60k .$$

C's force is $|C| = \sqrt{(-45)^2 + (-20)^2 + (-60)^2} = \sqrt{6025} \approx 77.6$ pounds. C is pushing in

the direction $\dfrac{C}{|C|} = \dfrac{-45}{\sqrt{6025}}i - \dfrac{20}{\sqrt{6025}}j - \dfrac{60}{\sqrt{6025}}k \approx -0.60i - 0.26j - 0.77k .$

At this point, you should find the arithmetic of vectors in three dimensions is not much different or harder than in two dimensions. Angles, however, are another story, one we consider in Section 11.4. Fortunately, there is a straightforward process for determining the angle between two 3–dimensional vectors and it is useful in a variety of applications.

Beyond Three Dimensions

Just as we can represent points in 4, 5, or n–dimensional space, we can also work with n–dimensional vectors, $\langle a_1, a_2, a_3, ... a_n \rangle$. Even though it is no longer easy (or possible?) to work geometrically with these vectors, the arithmetic operations of scalar multiplication, vector addition, vector subtraction, and length are still defined component–by–component and are still easy.

Example 5:   Write the vector **V** for the directed line segment from $P = (1,2,-3,5,6)$ to

$Q = (5,-1,4,0,7)$, and find the length and direction of **V**.

Solution: $V = \langle 5-1, -1-2, 4--3, 0-5, 7-6 \rangle = \langle 4, -3, 7, -5, 1 \rangle .$

$|V| = \sqrt{4^2 + (-3)^2 + 7^2 + (-5)^2 + 1^2} = \sqrt{100} = 10 .$

Direction of **V** is $\dfrac{V}{|V|} = \langle 0.4, -0.3, 0.7, -0.5, 0.1 \rangle .$

Practice 4:   Calculate the lengths of $A = \langle 4, 2, -5, 1, 0 \rangle$, $B = \langle 6, 0, 2, -3, 6 \rangle$, and $C = 2A - 3B .$

We can even determine formulas for some collections of points in higher dimensions.

Example 6:   Find a formula for the set of points (w,x,y,z) in 4–dimensional space that are at a distance of 5

units from the point (5,3,–2,1). (This set is "4–sphere" with radius 5 and center (5,3,–2,1).)

Solution: We want the distance from (w,x,y,z) to (5,3,–2,1), $\sqrt{(w-5)^2 + (x-3)^2 + (y+2)^2 + (z-1)^2}$,

to be 5 so

$$\sqrt{(w-5)^2 + (x-3)^2 + (y+2)^2 + (z-1)^2} = 5 \text{ and}$$

$$(w-5)^2 + (x-3)^2 + (y+2)^2 + (z-1)^2 = 25 .$$

PROBLEMS

In problems 1 – 4 the vectors **A** and **B** are shown. Sketch and label **C** = 2**A**, **D** = –**B**, and **E** = **A** – **B**.

1.   See Fig. 9.   2.   See Fig. 10.   3.   See Fig. 11.   4.   See Fig. 12.

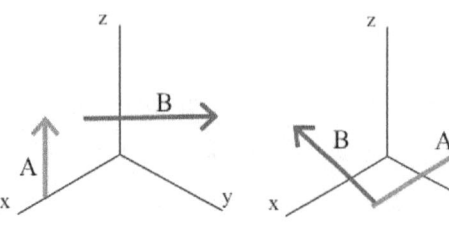
Fig. 9   Fig. 10

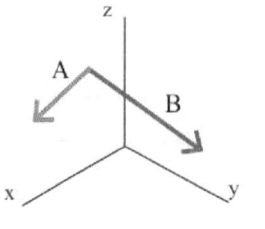
Fig. 11

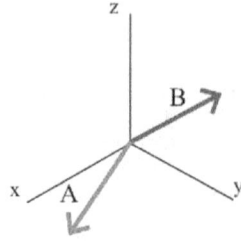
Fig. 12

In problems 5 – 12, vectors **U** and **V** are given. Calculate **W** = 2**U** + **V**, |**U**|, |**V**|, and |**W**|, and the directions of **U**, **V**, and **W**.

5. $U = \langle 2, 3, 6 \rangle$, $V = \langle 2, -9, 6 \rangle$ 6. $U = \langle 6, 3, 6 \rangle$, $V = \langle 2, 4, 4 \rangle$

7. $U = \langle 5, 2, 14 \rangle$, $V = \langle 4, -7, 4 \rangle$ 8. $U = \langle 8, 4, 1 \rangle$, $V = \langle 4, 4, -2 \rangle$

9. $U = 9\mathbf{i} + 6\mathbf{j} + 2\mathbf{k}$, $V = 3\mathbf{i} + 6\mathbf{j} - 6\mathbf{k}$ 10. $U = 24\mathbf{i} + 2\mathbf{j} + 24\mathbf{k}$, $V = 10\mathbf{i} - 25\mathbf{j} + 2\mathbf{k}$

11. $U = 10\mathbf{i} + 11\mathbf{j} + 2\mathbf{k}$, $V = 6\mathbf{i} + 3\mathbf{j} - 6\mathbf{k}$ 12. $U = 8\mathbf{i} + 1\mathbf{j} + 4\mathbf{k}$, $V = 3\mathbf{i} + 12\mathbf{j} + 4\mathbf{k}$

13. $A = \langle 3, 6, -2 \rangle$, $B = \langle 5, 0, -4 \rangle$. Find **C** so $A + B + C = 0$.

14. $A = \langle 1, -3, 7 \rangle$, $B = \langle -5, 2, -3 \rangle$. Find **C** so $A + B + C = 0$.

15. $A = \langle 9, \pi, -3 \rangle$, $B = \langle e, 7, 4 \rangle$, and $C = \langle -6, 2, 0 \rangle$. Find **D** so $A + B + C + D = 0$.

16. $A = \langle -4, 3, 1 \rangle$, $B = \langle 5, 5, -\pi \rangle$, and $C = \langle 0, -2, 1 \rangle$. Find **D** so $A + B + C + D = 0$.

17. Which of the following vectors has the smallest magnitude and which has the largest:
    $A = \langle 6, 9, 2, 3 \rangle$, $B = \langle -8, 0, 6, 0 \rangle$, $C = \langle -6, 3, 3, 6 \rangle$, and $D = \langle 9, 1, 8, -5 \rangle$?

18. Which of the following vectors has the smallest magnitude and which has the largest:
    $A = \langle 11, 2, -10, 1 \rangle$, $B = \langle -5, 14, 2, 6 \rangle$, $C = \langle 2, 8, 16, -5 \rangle$, and $D = \langle 12, 3, 7, 3 \rangle$?

19. $A = \langle 2, 4, 3 \rangle$. Sketch **A** and find the "shadows" of **A** on the coordinate planes (e.g., on the xy, xz, and yz planes).

20. $B = \langle 4, 1, 2 \rangle$. Sketch **B** and find the "shadows" of **B** on the coordinate planes (e.g., on the xy, xz, and yz planes).

21. $C = \langle 5, 2, 3 \rangle$. Sketch C and find the "shadows" of C on the coordinate planes.

22. $D = \langle 3, 4, 0 \rangle$. Sketch D and find the "shadows" of D on the coordinate planes.

23. $A = \langle 1, 0, 0 \rangle$. Sketch A and find three nonparallel vectors that are perpendicular to A. How many vectors are perpendicular to A?

24. $B = \langle 0, 0, 1 \rangle$. Sketch B and find three nonparallel vectors that are perpendicular to B. How many vectors are perpendicular to B?

25. $C = \langle 1, 2, 0 \rangle$. Sketch C and find two nonparallel vectors that are perpendicular to C.

26. $C = \langle 0, 2, 3 \rangle$. Sketch C and find two nonparallel vectors that are perpendicular to C.

In problems 27 – 30, you are asked to sketch smooth curves that go through given points with given directions. Later, in Section 12.1, we will discuss how to find parametric equations for curves that satisfy conditions of this type.

27. Sketch a smooth curve that goes through the point (0,0,1) with direction $\langle 1, 0, 0 \rangle$ and then bends and goes through the point (0,1,0) with direction $\langle 0, 1, 0 \rangle$.

28. Sketch a smooth curve that goes through the point (0,0,1) with direction $\langle 0.8, 0, 0.6 \rangle$ and then bends and goes through the point (0,2,0) with direction $\langle 0, 0, 1 \rangle$.

29. Sketch a smooth curve that goes through the point (2,0,0) with direction $\langle 0, 0, 1 \rangle$ and then bends and goes through the point (0,1,2) with direction $\langle -1, 0, 0 \rangle$.

30. Sketch a smooth curve that goes through the point (4, 1, 2) with direction $\langle 0, 0, -1 \rangle$ and then bends and goes through the point (0,0,0) with direction $\langle 0, -1, 0 \rangle$.

31. Find a linear equation for $x(t)$ so $x(0) = 3$ and $x(1) = 7$, for $y(t)$ so $y(0) = 5$ and $y(1) = 4$, and for $z(t)$ so $z(0) = 1$ and $z(1) = 1$.

32. Find a linear equation for $x(t)$ so $x(0) = 1$ and $x(1) = 5$, for $y(t)$ so $y(0) = 2$ and $y(1) = 0$, and for $z(t)$ so $z(0) = 3$ and $z(1) = 5$.

33. Find a linear equation for $x(t)$ so $x(0) = 2$ and $x(1) = 5$, for $y(t)$ so $y(0) = 3$ and $y(1) = 3$, and for $z(t)$ so $z(0) = 6$ and $z(1) = 1$.

34. Find a linear equation for $x(t)$ so $x(0) = 0$ and $x(1) = -3$, for $y(t)$ so $y(0) = -2$ and $y(1) = 1$, and for $z(t)$ so $z(0) = 4$ and $z(1) = 1$.

Practice Answers

Practice 1: $C = 5A = 5\langle\, 5, -4, 1\,\rangle = \langle\, 25, -20, 5\,\rangle$.

$D = 3A + 4B = 3\langle\, 5, -4, 1\,\rangle + 4\langle\, 2, -3, 4\,\rangle = \langle\, 15+8, -12-12, 3+16\,\rangle = \langle\, 23, -24, 19\,\rangle$.

$E = 2B - 3A = 2\langle\, 2, -3, 4\,\rangle - 3\langle\, 5, -4, 1\,\rangle = \langle\, 4-15, -6+12, 8-3\,\rangle = \langle\, -11, 6, 5\,\rangle$.

Practice 2: $|A| = \sqrt{2^2 + 3^2 + 6^2} = \sqrt{49} = 7$. $|B| = \sqrt{2^2 + 1^2 + 2^2} = \sqrt{9} = 3$.

$C = A - 2B = \langle\, 2, 3, 6\,\rangle - 2\langle\, 2, 1, 2\,\rangle = \langle\, -2, 1, 2\,\rangle$ so $|C| = \sqrt{9} = 3$.

Practice 3: $|A| = \sqrt{3^2 + 2^2 + (-6)^2} = \sqrt{49} = 7$. Direction of A is $\dfrac{A}{|A|} = \dfrac{3}{7}\mathbf{i} + \dfrac{2}{7}\mathbf{j} - \dfrac{6}{7}\mathbf{k}$.

$|B| = \sqrt{0^2 + 6^2 + (-8)^2} = \sqrt{100} = 10$. Direction of B is $\dfrac{B}{|B|} = \dfrac{6}{10}\mathbf{j} - \dfrac{8}{10}\mathbf{k}$.

$C = A + 3B = 3\mathbf{i} + 20\mathbf{j} - 30\mathbf{k}$. $|C| = \sqrt{1309} \approx 36.18$.

Direction of C is $\dfrac{C}{|C|} = \dfrac{3}{\sqrt{1309}}\mathbf{i} + \dfrac{20}{\sqrt{1309}}\mathbf{j} - \dfrac{30}{\sqrt{1309}}\mathbf{k} \approx 0.08\,\mathbf{i} + 0.55\,\mathbf{j} - 0.83\,\mathbf{k}$.

$D = 2B - 3A = -9\mathbf{i} + 6\mathbf{j} + 2\mathbf{k}$. $|D| = \sqrt{121} = 11$.

Direction of D is $\dfrac{D}{|D|} = \dfrac{-9}{11}\mathbf{i} + \dfrac{6}{11}\mathbf{j} + \dfrac{2}{11}\mathbf{k} \approx -0.82\mathbf{i} + 0.55\mathbf{j} + 0.18\mathbf{k}$.

Practice 4: $|A| = \sqrt{4^2 + 2^2 + (-5)^2 + 1^2 + 0} = \sqrt{46} \approx 6.8$.

$|B| = \sqrt{6^2 + 0^2 + 2^2 + (-3)^2 + 6^2} = \sqrt{85} \approx 9.2$.

$C = 2A - 3B = 2\langle\, 4, 2, -5, 1, 0\,\rangle - 3\langle\, 6, 0, 2, -3, 6\,\rangle = \langle\, -10, 4, -16, 11, -18\,\rangle$.

$|C| = \sqrt{(-10)^2 + 4^2 + (-16)^2 + 11^2 + (-18)^2} = \sqrt{817} \approx 28.6$.

11.4   DOT PRODUCT

In the previous sections we looked at the meaning of vectors in two and three dimensions, but the only operations we used were addition and subtraction of vectors and multiplication by a scalar. Some of the applications of 2–dimensional vectors used the angles that the vectors made with the coordinate axes and with each other, but, so far, in three dimensions we have not used angles. This section addresses both of those situations. It introduces a way to multiply two vectors, in two and three dimensions, called the dot product, and this dot product provides us with a relatively easy way to determine angles between vectors. Section 11.5 introduces a different method of multiplying two vectors, the cross product, in three dimensions that has other useful applications.

Since we will soon have three different types of multiplications for a vector (scalar, dot, and cross), it is important that you distinguish among them and call each multiplication operation by its full name.

Definition: Dot Product

Two dimensions: The **dot product** of $\mathbf{A} = \langle\, a_1, a_2\, \rangle$ and $\mathbf{B} = \langle\, b_1, b_2\, \rangle$

is $\mathbf{A} \cdot \mathbf{B} = a_1 b_1 + a_2 b_2$.

Three dimensions: The **dot product** of $\mathbf{A} = \langle\, a_1, a_2, a_3\, \rangle$ and $\mathbf{B} = \langle\, b_1, b_2, b_3\, \rangle$

is $\mathbf{A} \cdot \mathbf{B} = a_1 b_1 + a_2 b_2 + a_3 b_3$.

Both vectors in the dot product must have the same number of components, and the result of the dot product $\mathbf{U} \cdot \mathbf{V}$ is a scalar.

Example 1:   For $\mathbf{A} = \langle\, 4, 1, 8\, \rangle$ and $\mathbf{B} = \langle\, 2, -4, 4\, \rangle$, calculate $\mathbf{A} \cdot \mathbf{B}$, $\mathbf{A} \cdot \mathbf{A}$, $\mathbf{B} \cdot \mathbf{B}$, and

$(\mathbf{A}-\mathbf{B}) \cdot (\mathbf{A}+2\mathbf{B})$.

Solution: $\mathbf{A} \cdot \mathbf{B} = \langle\, 4, 1, 8\, \rangle \cdot \langle\, 2, -4, 4\, \rangle = (4)(2) + (1)(-4) + (8)(4) = 8 - 4 + 32 = 36$.

$\mathbf{A} \cdot \mathbf{A} = \langle\, 4, 1, 8\, \rangle \cdot \langle\, 4, 1, 8\, \rangle = (4)(4) + (1)(1) + (8)(8) = 81$.

$\mathbf{B} \cdot \mathbf{B} = \langle\, 2, -4, 4\, \rangle \cdot \langle\, 2, -4, 4\, \rangle = (2)(2) + (-4)(-4) + (4)(4) = 36$.

You should notice that $\mathbf{A} \cdot \mathbf{A} = |\mathbf{A}|^2$ and $\mathbf{B} \cdot \mathbf{B} = |\mathbf{B}|^2$.

Finally, $\mathbf{A}-\mathbf{B} = \langle\, 2, 5, 4\, \rangle$ and $\mathbf{A} + 2\mathbf{B} = \langle\, 8, -7, 16\, \rangle$

so $(\mathbf{A}-\mathbf{B}) \cdot (\mathbf{A}+2\mathbf{B}) = \langle\, 2, 5, 4\, \rangle \cdot \langle\, 8, -7, 16\, \rangle = (2)(8) + (5)(-7) + (4)(16) = 45$.

Practice 1:   For $\mathbf{U} = \langle\, 2, 6, -3\, \rangle$ and $\mathbf{V} = \langle\, -1, 2, 2\, \rangle$, calculate $\mathbf{U} \cdot \mathbf{V}$, $\mathbf{U} \cdot \mathbf{U}$, $\mathbf{V} \cdot \mathbf{V}$, $\mathbf{U} \cdot (\mathbf{U}+\mathbf{V})$,

and $\mathbf{U} \cdot \mathbf{U} + \mathbf{U} \cdot \mathbf{V}$. Does $\mathbf{U} \cdot \mathbf{U} = |\mathbf{U}|^2$? Does $\mathbf{U} \cdot \mathbf{V} = \mathbf{V} \cdot \mathbf{U}$?

As you might have noticed in Example 1 and Practice 1, the dot product seems to have some of the
properties of ordinary multiplication of numbers.

Properties of the Dot Product:　　(1)　$A \cdot A = |A|^2$

(2)　$A \cdot B = B \cdot A$

(3)　$k(A \cdot B) = (kA) \cdot B = A \cdot (kB)$

(4)　$A \cdot (B + C) = A \cdot B + A \cdot C$

All of these properties can be proved using the definition of the dot product.

Proof of (1):   If $A = \langle a_1, a_2, a_3 \rangle$ then $A \cdot A = (a_1)^2 + (a_2)^2 + (a_3)^2$ and

$$|A|^2 = (\sqrt{(a_1)^2 + (a_2)^2 + (a_3)^2} \;)^2 = (a_1)^2 + (a_2)^2 + (a_3)^2 \text{ so } A \cdot A = |A|^2 \;.$$

Proof of (3): $k(A \cdot B) = k(a_1 b_1 + a_2 b_2 + a_3 b_3) = ka_1 b_1 + ka_2 b_2 + ka_3 b_3 \;.$

$(kA) \cdot B = \langle ka_1, ka_2, ka_3 \rangle \cdot \langle b_1, b_2, b_3 \rangle = ka_1 b_1 + ka_2 b_2 + ka_3 b_3 \;. \text{ And}$

$A \cdot (kB) = \langle a_1, a_2, a_3 \rangle \cdot \langle kb_1, kb_2, kb_3 \rangle = a_1(kb_1) + a_2(kb_2) + a_3(kb_3)$

$= ka_1 b_1 + ka_2 b_2 + ka_3 b_3 \text{ so } k(A \cdot B) = (kA) \cdot B = A \cdot (kB)$

Practice 2:   Prove Property (2) for 3–dimensional vectors.

The next result about dot products is very important, and much of the usefulness of dot products follows from
it. It enables us to easily determine the angle between two vectors in two or three (or more) dimensions.

Angle Property of Dot Products

$A \cdot B = |A| \, |B| \cos(\theta)$ where θ is the angle between A and B . Equivalently,

if A and B are nonzero vectors, then the angle θ between A and B satisfies $\cos(\theta) = \dfrac{A \cdot B}{|A| \, |B|}$.

Proof of the Angle Property: The proof uses the Law of Cosines and
several of the properties of the dot product.

The vectors A, B and $A–B$ can be arranged to form a triangle
(Fig. 1) with the angle θ between A and B. Applying the Law of
Cosines to this triangle, we have

$$| A - B |^2 = |A|^2 + |B|^2 - 2|A||B| \cos(\theta) \;.$$

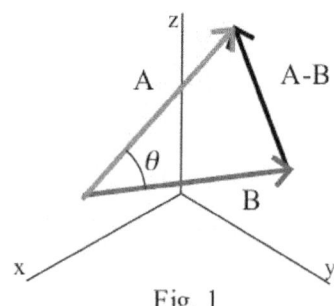

Fig. 1

We can also use the properties of the dot product to expand $| \mathbf{A} - \mathbf{B} |^2$

in a different way: $| \mathbf{A} - \mathbf{B} |^2 = (\mathbf{A} - \mathbf{B})\bullet(\mathbf{A} - \mathbf{B})$

$$= \mathbf{A}\bullet\mathbf{A} - \mathbf{A}\bullet\mathbf{B} - \mathbf{B}\bullet\mathbf{A} + \mathbf{B}\bullet\mathbf{B} = |\mathbf{A}|^2 - 2\mathbf{A}\bullet\mathbf{B} + |\mathbf{B}|^2.$$

From these two representations for $| \mathbf{A} - \mathbf{B} |^2$ we have that

$$|\mathbf{A}|^2 - 2\mathbf{A}\bullet\mathbf{B} + |\mathbf{B}|^2 = |\mathbf{A}|^2 + |\mathbf{B}|^2 - 2|\mathbf{A}||\mathbf{B}| \cos(\theta) \quad \text{so}$$

$$- 2\mathbf{A}\bullet\mathbf{B} = - 2|\mathbf{A}||\mathbf{B}| \cos(\theta) \quad \text{and} \quad \mathbf{A}\bullet\mathbf{B} = |\mathbf{A}| |\mathbf{B}| \cos(\theta) .$$

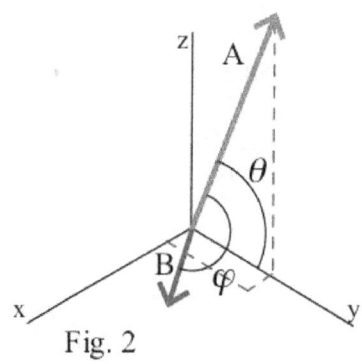

Fig. 2

Example 2:   Let $\mathbf{A} = \langle 2, 5, 14 \rangle$ and $\mathbf{B} = \langle 2, 1, -2 \rangle$. Find the angles (a) between $\mathbf{A}$ and $\mathbf{B}$ and (b)

between $\mathbf{A}$ and the positive y–axis (Fig. 2).

Solution: (a) $|\mathbf{A}| = \sqrt{4+25+196} = 15$, $|\mathbf{B}| = \sqrt{4+1+4} = 3$, $\mathbf{A}\bullet\mathbf{B} = 4+5-28 = -19$, and

$$\cos(\varphi) = \frac{\mathbf{A}\bullet\mathbf{B}}{|\mathbf{A}| |\mathbf{B}|} = \frac{-19}{(15)(3)} \approx -0.4222 \text{ so } \varphi \approx 2.01 \text{ or about } 115.0°.$$

(b)   The basis vector $\mathbf{j} = \langle 0, 1, 0 \rangle$ points along the positive y–axis so the angle between

$\mathbf{A}$ and the positive y–axis is the same as the angle between $\mathbf{A}$ and $\mathbf{j}$.

$|\mathbf{A}| = 15, |\mathbf{j}| = 1$, and $\mathbf{A}\bullet\mathbf{j} = (2)(0) + (5)(1) + (14)(0) = 5$ so

$$\cos(\theta) = \frac{\mathbf{A}\bullet\mathbf{j}}{|\mathbf{A}| |\mathbf{j}|} = \frac{5}{(15)(1)} \approx 0.333 \text{ so } \theta \approx 1.23 \text{ or about } 70.5°.$$

Practice 3:   Let $\mathbf{A} = \langle 2, -6, 3 \rangle$, $\mathbf{B} = \langle 4, 8, -1 \rangle$ and $\mathbf{C} = \langle 3, 0, -4 \rangle$ and

determine the angles between the vectors (a) $\mathbf{A}$ and $\mathbf{B}$, (b) $\mathbf{A}$ and

$\mathbf{C}$, (c) $\mathbf{B}$ and the negative x–axis, and (d) $\mathbf{C}$ and the positive y–axis.

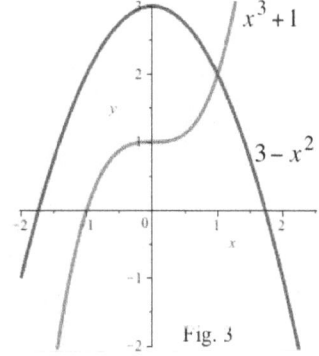

Example 3:   Find the angle of intersection of the graphs of $f(x) = x^3 + 1$ and

$g(x) = 3 - x^2$ at the point $(1, 2)$. (The angle of intersection is the angle

between tangent vectors to the graphs at the point.) Fig. 3.

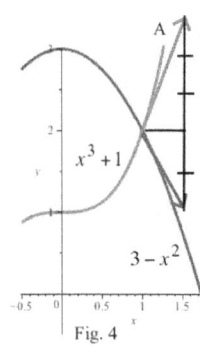

Fig. 4

Solution: $f '(x) = 3x^2$, so the slope of f at (1,2) is $f '(1) = 3(1)^2 = 3 : \dfrac{\text{rise}}{\text{run}} = \dfrac{3}{1}$

and a vector, $\mathbf{A} = 1\mathbf{i} + 3\mathbf{j}$, with this slope is shown in Fig. 4. Similarly, $g '(x) = -2x$,

so the slope of g at (1,2) is $g '(1) = -2(1) = -2: \dfrac{\text{rise}}{\text{run}} = \dfrac{-2}{1}$ and vector $\mathbf{B} = 1\mathbf{i} - 2\mathbf{j}$

has the same slope. Then $|\mathbf{A}| = \sqrt{10}$, $|\mathbf{B}| = \sqrt{5}$, and $\mathbf{A}\bullet\mathbf{B} = -5$, so

$$\cos(\theta) = \frac{-5}{\sqrt{10}\sqrt{5}} \approx -0.707 . \text{ Then } \theta = \arccos(-0.707) \approx 2.356 \text{ (or } 135^{\circ} \text{).}$$

Note: If $y = f(x)$, then a tangent vector to the graph of f at the point $(x_0, f(x_0))$ is $\mathbf{T} = \langle 1, f '(x_0) \rangle$.

If a curve is given parametrically by $x(t)$ and $y(t)$,

then a tangent vector to the curve when $t = t_0$ is $\mathbf{T} = x '(t_0)\mathbf{i} + y '(t_0)\mathbf{j} = \langle x '(t_0), y '(t_0) \rangle$.

Some books define the dot product $\mathbf{A} \cdot \mathbf{B}$ as $\mathbf{A} \cdot \mathbf{B} = |\mathbf{A}| \, |\mathbf{B}| \cos(\theta)$ and then derive the definition we gave, $\mathbf{A} \cdot \mathbf{B} = a_1 b_1 + a_2 b_2 + a_3 b_3$, as a property. Either way, the pattern $\mathbf{A} \cdot \mathbf{B} = a_1 b_1 + a_2 b_2 + a_3 b_3$ is typically used to compute the dot product, and the angle property $\mathbf{A} \cdot \mathbf{B} = |\mathbf{A}| \, |\mathbf{B}| \cos(\theta)$ is typically used to help us see what the dot product measures and to help us derive and simplify some vector algorithms.

Criteria for A and B to be Perpendicular

Let $\mathbf{A}$ and $\mathbf{B}$ be nonzero vectors. $\mathbf{A}$ and $\mathbf{B}$ are perpendicular if and only if $\mathbf{A} \cdot \mathbf{B} = 0$.

Proof: If $\mathbf{A}$ and $\mathbf{B}$ are perpendicular, then $\theta = \pm\, \pi/2$ so $\mathbf{A} \cdot \mathbf{B} = |\mathbf{A}| \, |\mathbf{B}| \cos(\theta) = |\mathbf{A}| \, |\mathbf{B}| \, (0) = 0$.

If $0 = \mathbf{A} \cdot \mathbf{B} = |\mathbf{A}| \, |\mathbf{B}| \cos(\theta)$, then $\cos(\theta) = 0$ so $\theta = \pm\, \pi/2 \, (\pm\, 2\pi n)$ and $\mathbf{A}$ and $\mathbf{B}$ are perpendicular.

Example 4:   (a) Find a vector perpendicular to $\mathbf{V} = \langle\, 1, 2, 3 \,\rangle$.

(b) Find a vector perpendicular to the line $4x + 3y = 24$.

Solution: (a) We need a nonzero vector $\mathbf{N}$ so that $\mathbf{N} \cdot \mathbf{V} = 0$. If $\mathbf{N} = \langle\, a, b, c \,\rangle$, then we want to find values

for $a, b,$ and c so that $a + 2b + 3c = 0$, and there are lots of values for $a, b,$ and c that work: $\mathbf{N} = \langle\, 0, -3, 2 \,\rangle, \langle\, 3, 0, -1 \,\rangle, \langle\, 2, -1, 0 \,\rangle, \langle\, 1, 1, -1 \,\rangle, \langle\, 1, -2, 1 \,\rangle$ and lots of others all give $\mathbf{N} \cdot \mathbf{V} = 0$. Fig. 5 illustrates why a single vector in three dimensions can have perpendicular vectors that point in an infinite number of different directions.

(b) The points $P = (\, 6, 0)$ and $Q = (\, 0, 8)$ both lie on the line, so the vector $\mathbf{V}$ from P to Q is parallel to the line, and $\mathbf{V} = \langle\, 0{-}6, 8{-}0 \,\rangle = \langle\, -6, 8 \,\rangle$.

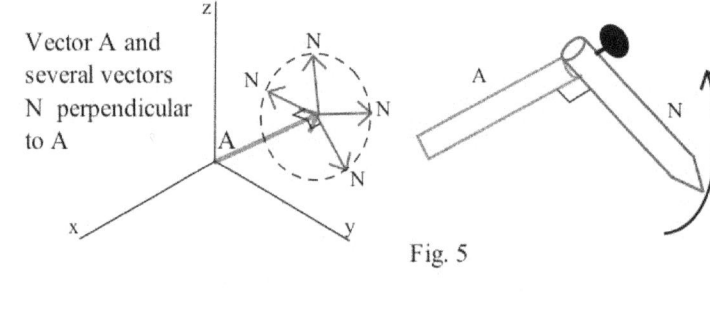

Vector A and several vectors N perpendicular to A

Fig. 5

If $\mathbf{N}$ is perpendicular to $\mathbf{V}$, then $\mathbf{N}$ is perpendicular to the line, and $\mathbf{N} = \langle\, 4, 3 \,\rangle$ is perpendicular to $\mathbf{V}$ since $\mathbf{N} \cdot \mathbf{V} = 0$. The vector $\mathbf{N} = \langle\, 4, 3 \,\rangle$ is perpendicular to the line $4x + 3y = 24$. Every scalar multiple $k\mathbf{N}$ with $k \neq 0$ is also perpendicular to the line.

Practice 4:   Find a vector $\mathbf{N}$ perpendicular to the line $-5x + 2y = 30$. Is $\mathbf{N} = \langle\, -5, 2 \,\rangle$ perpendicular to $-5x + 2y = 30$?

You might have noticed a pattern in the vectors perpendicular to the lines in Example 4 and Practice 4. The vector $\langle\, 4, 3 \,\rangle$ is perpendicular to the line $4x + 3y = 24$, and the vector $\langle\, -5, 2 \,\rangle$ is perpendicular to the line $-5x + 2y = 30$. The next result says this pattern is not an accident.

Finding a Vector Perpendicular to a Line

The vector $\mathbf{N} = a\mathbf{i} + b\mathbf{j}$ is perpendicular to the line $ax + by = c$.

Problem **66** asks you to prove this result.

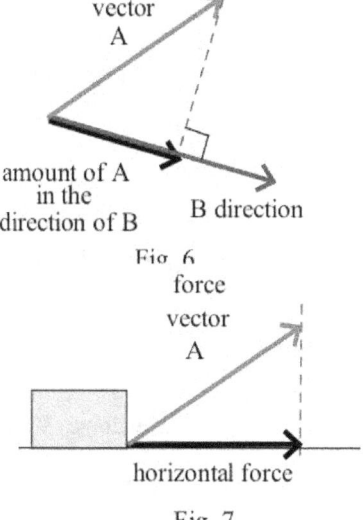

force
vector
A

amount of A
in the
direction of B   B direction

Fig. 6

Projection of a Vector onto a Vector

The length of a force vector tells us the amount of force in the direction of the
vector, but sometimes we want to know the size of the force in another direction
(Fig. 6). One of the examples in two dimensions (Fig. 7) involved finding the

force
vector
A

horizontal force

Fig. 7

amount of "horizontal force" obtained when we pulled on a box at an angle to the
horizontal. Similar questions can also be asked if one of the directions is not
horizontal (Fig. 8) and in three or more dimensions (Fig. 9). Since we now have a
method for determining the angle between two vectors in three dimensions, the

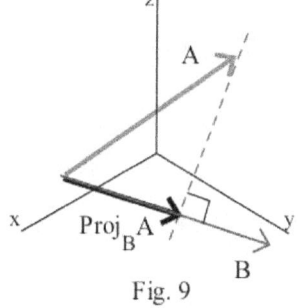

Fig. 9

solutions are relatively straightforward. The vector
representing the amount of a vector **A** in the direction of
a vector **B** is called the "projection of **A** onto **B**" and is denoted as **Proj$_B$ A**.

Visualizing the projection of A onto B : Fig. 10 shows
several geometric examples of the projection of a vector **A**
onto a vector **B**. We arrange **A** and **B** to have the same starting point, draw a
(dotted) line through the head of **A** and perpendicular to **B**, and form the
projection of **A** onto **B** as the vector from the starting point of **B** to the
point where the dotted line intersects **B** (or an extension of **B**). The projection of **A** onto **B**
is a vector along the line of **B** — the direction of the projection of **A** onto **B** is either the
same direction as **B**, **B**/|**B**|, or opposite the direction of **B**, –**B**/|**B**| .

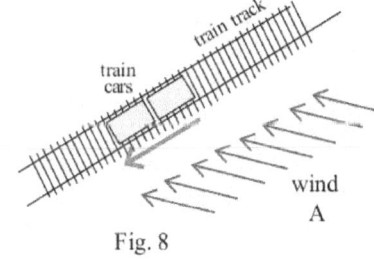

Fig. 8

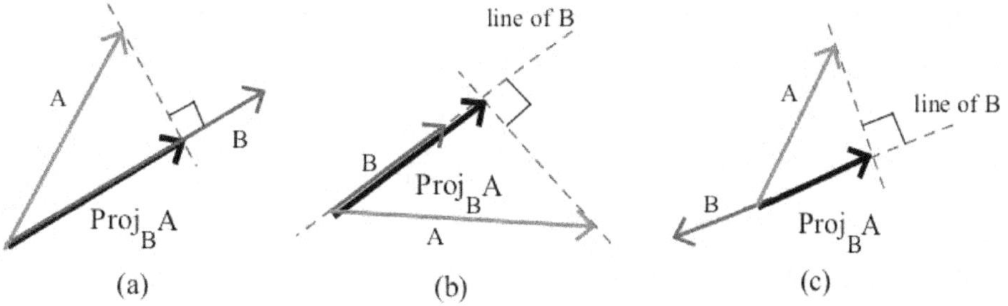

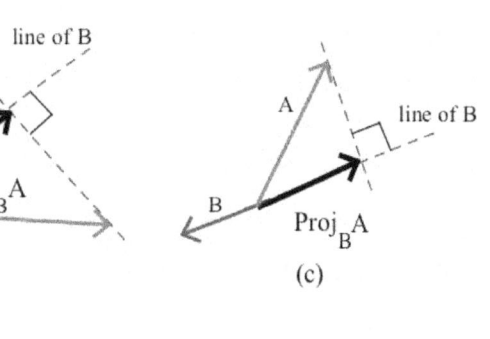

(a)   (b)   (c)

Fig. 10

Fig. 11 shows some examples of the
projection of **B** onto **A**, **Proj$_A$ B** .

The resulting projection vector is
always along the line of **A** .

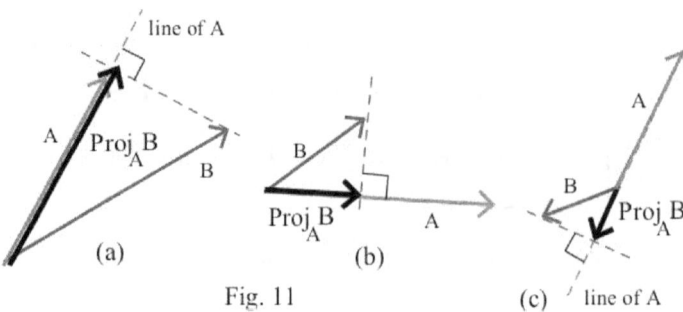

(a)   (b)

Fig. 11

(c)

Once we understand the geometric meaning of
"the projection of **A** onto **B**," trigonometry
enables us to determine the vector projection of
A onto **B** : **Proj$_B$ A** (Fig. 12), and its magnitude,
| **Proj$_B$ A** | , called the scalar projection.

$$\left| Proj_B A \right| = \left| A \right| \cos(\theta)$$

Direction of $Proj_B A$ is $\dfrac{B}{|B|}$

Fig. 12

Definitions: Vector Projection, Scalar Projection

The **vector projection of A onto B** is the

(magnitude of the projection) times (the direction of **B**): **Proj$_B$ A** $= \left(\ |A| \cos(\theta) \ \right) \left(\dfrac{B}{|B|} \right)$

The **scalar projection of A onto B** is $|A| \cos(\theta)$ where θ is the angle between **A** and **B** .

Note: "Projection of **A** onto **B**" usually means Vector Projection.

We can use properties of the dot product to simplify the calculation of projections.

Since θ is the angle between **A** and B, then $\cos(\theta) = \dfrac{A \cdot B}{|A||B|}$ so the scalar projection of **A** onto **B** is

$|A| \cos(\theta) = |A| \dfrac{A \cdot B}{|A||B|} = \dfrac{A \cdot B}{|B|}$. Putting this result into the definition of the vector projection of **A**

onto **B**, we get

$$\left(\ |A| \cos(\theta) \ \right)\left(\frac{B}{|B|} \right) = \frac{A \cdot B}{|B|} \left(\frac{B}{|B|} \right) = \frac{A \cdot B}{|B|^2} \ B \ .$$

Calculating Scalar and Vector Projections

Vector projection of **A** onto **B** is **Proj$_B$ A** $= \left(\dfrac{A \cdot B}{|B|^2} \right) B$ (a vector).

Scalar projection of **A** onto **B** $= \dfrac{A \cdot B}{|B|}$ (a scalar) $=$ magnitude of **Proj$_B$ A** .

Example 5:   For $\mathbf{A} = \langle\, 6, -2, 3\,\rangle$ and $\mathbf{B} = \langle\, 4, 8, -1\,\rangle$, calculate scalar and vector projections of

(a) $\mathbf{A}$ onto $\mathbf{B}$, (b) $\mathbf{B}$ onto $\mathbf{A}$, (c) $\mathbf{A}$ onto the positive x–axis,

(d) $\mathbf{A}$ onto the positive y–axis, and (e) $\mathbf{A}$ onto the positive z–axis.

Solution: $|\mathbf{A}| = 7$, $|\mathbf{B}| = 9$, and $\mathbf{A} \bullet \mathbf{B} = 24 - 16 - 3 = 5$.

(a) The scalar projection of $\mathbf{A}$ onto $\mathbf{B}$ is $\dfrac{\mathbf{A} \bullet \mathbf{B}}{|\mathbf{B}|} = \dfrac{5}{9}$.

The vector projection of $\mathbf{A}$ onto $\mathbf{B}$ is $\mathbf{Proj}_{\mathbf{B}}\,\mathbf{A} = \left(\dfrac{\mathbf{A} \bullet \mathbf{B}}{|\mathbf{B}|^2} \right) \mathbf{B} = \left(\dfrac{5}{81} \right)\langle\, 4, 8, -1\,\rangle = \langle\, \dfrac{20}{81}, \dfrac{40}{81}, \dfrac{-5}{81}\,\rangle$.

(b) Scalar projection of $\mathbf{B}$ onto $\mathbf{A}$ is $\dfrac{\mathbf{A} \bullet \mathbf{B}}{|\mathbf{A}|} = \dfrac{5}{7}$. Vector projection is $\left(\dfrac{\mathbf{A} \bullet \mathbf{B}}{|\mathbf{A}|^2} \right) \mathbf{A} = \langle\, \dfrac{30}{49}, \dfrac{-10}{49}, \dfrac{15}{49}\,\rangle$.

(c) $\mathbf{i}$ has the same direction as the positive x–axis. Scalar projection of $\mathbf{A}$ onto $\mathbf{i}$ is $\dfrac{\mathbf{A} \bullet \mathbf{i}}{|\mathbf{i}|} = 6$.

Vector projection of $\mathbf{A}$ onto $\mathbf{i}$ is $\mathbf{Proj}_{\mathbf{i}}\,\mathbf{A} = \left(\dfrac{\mathbf{A} \bullet \mathbf{i}}{|\mathbf{i}|^2} \right) \mathbf{i} = \langle\, 6, 0, 0\,\rangle$.

(d) and (e) The scalar projections of $\mathbf{A}$ onto $\mathbf{j}$ and $\mathbf{k}$ are –2 and 3, respectively.

The vector projections of $\mathbf{A}$ onto $\mathbf{j}$ and $\mathbf{k}$ are $\langle\, 0, -2, 0\,\rangle$ and $\langle\, 0, 0, 3\,\rangle$, respectively.

Practice 5:   For $\mathbf{U} = \langle\, 9, -2, 6\,\rangle$ and $\mathbf{V} = \langle\, 1, 2, -2\,\rangle$, calculate the scalar and vector projections of

(a) $\mathbf{U}$ onto $\mathbf{V}$, (b) $\mathbf{V}$ onto $\mathbf{U} + \mathbf{V}$, and (c) $\mathbf{V}$ onto the positive y–axis.

Applications

Projections are useful in a number of situations. The two examples given here illustrate only two of a variety of those uses. In the first example below, we use the geometric meaning of projection to derive a formula for the distance from a point to a line. In the second example, we illustrate how projections can be used to calculate work.

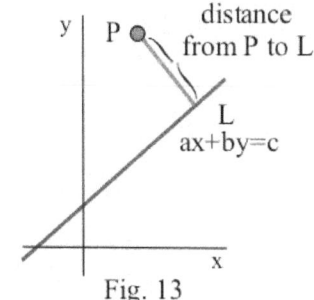

Fig. 13

Example 6:   Find the distance from the point $P = (\, p, q\,)$ to a line $ax + by = c$. (Fig. 13)

Solution: Using vectors and projections, finding this distance can be broken into several simple steps (Fig. 14). The algorithm:

(1)   Find a vector $\mathbf{N}$ perpendicular to the line: $\mathbf{N} = \langle\, a, b\,\rangle$.(Fig. 14a)

(2)   Find a point Q on the line and call it $(\, x_0, y_0\,)$.   (Fig. 14b)

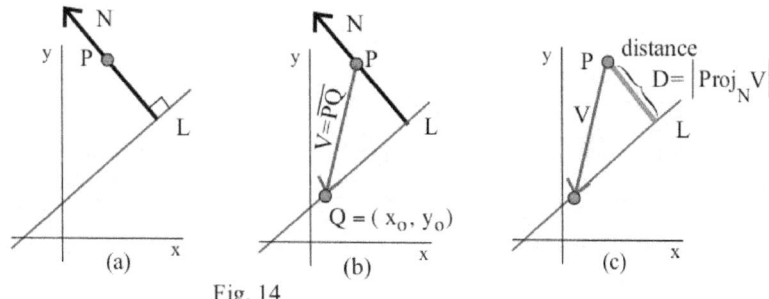

Fig. 14

(3)   Form the vector $\mathbf{V}$ from P to Q: $\mathbf{V} = \langle\, p - x_0, q - y_0\,\rangle$.

(4)   Find the absolute value of the scalar projection of $\mathbf{V}$ onto $\mathbf{N}$: $\left|\, \dfrac{\mathbf{V}\cdot\mathbf{N}}{|\mathbf{N}|}\,\right|$ (Fig. 14c)

The formula: The distance from P to the line is this absolute value of the scalar projection:

$$\text{distance} = \left|\ \frac{\mathbf{V}\cdot\mathbf{N}}{|\mathbf{N}|}\ \right| = \left|\ \frac{a(p - x_0) + b(q - y_0)}{\sqrt{a^2 + b^2}}\ \right|$$

$$= \frac{|\,ap + bq - (ax_0 + by_0)\,|}{\sqrt{a^2 + b^2}} = \frac{|\,ap + bq - c\,|}{\sqrt{a^2 + b^2}} \quad \text{since } ax_0 + by_0 = c.$$

Practice 6:   Step through the algorithm in Example 6 to find the distance from the point $P = (\,2, 7\,)$ to
the line $3x - 4y = 20$.

The projection of a force vector onto a vector with a different direction tells us the amount of the force in
that other direction, a useful result to know for solving work problems.

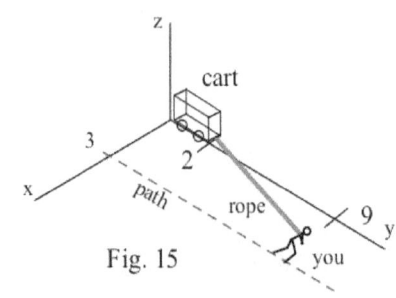

Fig. 15

Example 7:   A cart moves on a track located on the y–axis, and you are
pulling on a rope with a force of 70 pounds. Find the amount
of work that you do in moving the cart from the point $P = (0, 2, 0)$
to the point $Q = (0, 9, 0)$ if you have the rope over your shoulder
4 feet above the ground and walk 12 feet in front of the cart along a
path 3 feet to the side of the y–axis (Fig. 15).

Solution: Work = (distance moved)(force in the direction of the movement)

First, we can convert the given information into a vector form: the force vector has a

magnitude of 70 pounds in the direction $\langle\, \frac{3}{13}, \frac{12}{13}, \frac{4}{13}\,\rangle$ so the force vector

is $\mathbf{F} = \langle\, \frac{210}{13}, \frac{840}{13}, \frac{280}{13}\,\rangle$. The movement of the cart is in the direction of the vector from P to Q,

$\mathbf{D} = Q - P = \langle\, 0, 7, 0\,\rangle$, so the amount of work is

work   = (distance moved)(magnitude of the force in the direction of the movement)

= (length of PQ)(scalar projection of $\mathbf{F}$ onto direction of PQ)

$= (\,|\mathbf{D}|\,)(\dfrac{\mathbf{F}\cdot\mathbf{V}}{|\mathbf{V}|}\,) = \mathbf{F}\cdot\mathbf{V} = \langle\, \frac{210}{13}, \frac{840}{13}, \frac{280}{13}\,\rangle\cdot\langle\, 0, 7, 0\,\rangle = \dfrac{5880}{13} \approx 452.3$ foot–pounds.

Work

If   a constant force vector $\mathbf{F}$ moves an object from a point P to a point Q,

then   the amount of work done is   Work $= \mathbf{F}\cdot\mathbf{D}$ where $\mathbf{D}$ is the displacement vector from P to Q.

Practice 7:   In Example 7, determine the amount of work done if you walk **6** feet to the side of the

y–axis. (All of the other distances are the same as in Example 7.)

In Chapter 13, we discuss how to determine the amount of work done if the force is variable or if the object

moves along a curved path.

Beyond Three Dimensions

If **A** and **B** have the same number of components, then we can define their dot product, the angle between

them, and the projection of one onto the other with the same patterns as for two and three dimensions.

Definitions:

If $\mathbf{A} = \langle a_1, a_2, ..., a_n \rangle$ and $\mathbf{B} = \langle b_1, b_2, ..., b_n \rangle$ are nonzero vectors in n–dimensional space,

then $\mathbf{A} \bullet \mathbf{B} = a_1 b_1 + a_2 b_2 + ... + a_n b_n$,

the angle θ between **A** and **B** satisfies $\cos(\theta) = \dfrac{\mathbf{A} \bullet \mathbf{B}}{|\mathbf{A}||\mathbf{B}|}$, and

the vector projection of A onto B is $\mathbf{Proj_B\ A} = \left(\dfrac{\mathbf{A} \bullet \mathbf{B}}{|\mathbf{B}|^2} \right) \mathbf{B}$.

Now, even though we may not visualize 4 or 5–dimensional vectors, we can calculate the dot product and the

angle between two vectors.

Practice 8:   The psychological profiles for you and a friend were $\mathbf{Y} = \langle 5, 1, -7, 5 \rangle$ and

$\mathbf{F} = \langle 6, 10, 8, 5 \rangle$ for the four personality categories measured by the profile. Should you say you

and your friend are "very alike" ($\theta < 30°$), "somewhat alike" ($30° \leq \theta < 60°$), "different"

($60° \leq \theta \leq 120°$), "somewhat opposite" ($120° < \theta \leq 150°$). or "very opposite" ($150° < \theta \leq 180°$)?

What about you and and another friend with the profile $\mathbf{A} = \langle 10, -4, -10, 3 \rangle$?

This section has involved very little "calculus," but the ideas of dot products and projections of vectors are

very powerful and useful, and they will be used often as we develop "vector calculus" in later chapters.

PROBLEMS

1. $\mathbf{A} = \langle 1, 2, 3 \rangle$, $\mathbf{B} = \langle -2, 4, -1 \rangle$. Calculate $\mathbf{A} \bullet \mathbf{B}, \mathbf{B} \bullet \mathbf{A}, \mathbf{A} \bullet \mathbf{A}, \mathbf{A} \bullet (\mathbf{B} + \mathbf{A})$, and $(2\mathbf{A} + 3\mathbf{B}) \bullet (\mathbf{A} - 2\mathbf{B})$.

2. $\mathbf{A} = \langle 6, -1, 2 \rangle$, $\mathbf{B} = \langle 2, 4, -3 \rangle$. Calculate $\mathbf{A} \bullet \mathbf{B}, \mathbf{B} \bullet \mathbf{A}, \mathbf{A} \bullet \mathbf{A}, \mathbf{A} \bullet (\mathbf{B} + \mathbf{A})$, and $(2\mathbf{A} + 3\mathbf{B}) \bullet (\mathbf{A} - 2\mathbf{B})$.

3. $U = \langle 6, -1, 2 \rangle$, $V = \langle 2, 4, -3 \rangle$. Calculate $U \cdot V, U \cdot U, U \cdot i, U \cdot j, U \cdot k$, and $(V + i) \cdot U$.

4. $U = \langle -3, 3, 2 \rangle$, $V = \langle 2, 4, -3 \rangle$. Calculate $U \cdot V, U \cdot U, V \cdot i, V \cdot j, V \cdot k, (V + k) \cdot U$.

5. $S = 2i - 4j + k$, $T = 3i + j - 5k$, $U = i + 3j + 2k$. Calculate $S \cdot T, T \cdot U, T \cdot T$,

     $(S + T) \cdot (S - T)$, and $(S \cdot T)U$.

6. $S = i - 3j + 2k$, $T = 5i + 3j - 2k$, $U = 2i + 4j + 2k$. Calculate $S \cdot T, S \cdot U, S \cdot S$,

     $(T + U) \cdot (T - U)$, and $S(T \cdot U)$.

In problems 7 – 18, calculate the angle between the given vectors. Also calculate the angles between the first vector and each of the coordinate axes.

7. $A = \langle 1, 2, 3 \rangle$, $B = \langle -2, 4, -1 \rangle$.   8. $A = \langle 6, -1, 2 \rangle$, $B = \langle 2, 4, -3 \rangle$.

9. $U = \langle 6, -1, 2 \rangle$, $V = \langle 2, 4, -3 \rangle$.   10. $U = \langle -3, 3, 2 \rangle$, $V = \langle 2, 4, -3 \rangle$.

11. $S = 2i - 4j + k$, $T = 3i + j - 5k$.   12. $S = i - 3j + 2k$, $T = 5i + 3j - 2k$.

13. $A = 2i - 3j + 5k$, $B = -5i + 0j + 2k$.   14. $A = 4i - 3j + 2k$, $B = 3i + 2j - 3k$.

15. $A = \langle 5, -2, 0 \rangle$, $B = \langle -3, 4, 0 \rangle$.   16. $A = \langle 5, 0, 0 \rangle$, $B = \langle 0, 4, -3 \rangle$.

17. $U = \langle 1, 0, 3 \rangle$, $V = \langle -2, 0, 1 \rangle$.   18. $U = \langle 0, 1, 2 \rangle$, $V = \langle 2, 4, 0 \rangle$.

In problems 19 – 28, determine the angle of intersection of the graphs of the given functions at the given point (i.e., determine the angle between the vectors tangent to the functions at the given point).

19. $f(x) = x^2 + 3x - 2$, $g(x) = 3x - 1$ at $(1, 2)$ 20. $f(x) = x^2 + 3x - 2$, $g(x) = 3 - x^2$ at $(1, 2)$

21. $f(x) = e^x$, $g(x) = \cos(x)$ at $(0, 1)$ 22. $f(x) = \sin(x)$, $g(x) = \cos(x)$ at $(\pi/4, \frac{\sqrt{2}}{2})$

23. $f(x) = 1 + \arctan(3x)$, $g(x) = \ln(e + x)$ at $(0, 1)$ 24. $f(x) = \sin(x^2)$, $g(x) = 1 - \cos(x^2)$ at $(0, 0)$

25. f: $x = 3t$, $y = t^2 + t - 1$; g: $x = 2t + 1$, $y = 2t - 1$ at the point when $t = 1$

26. f: $x = \sin(t)$, $y = \cos(t)$; g: $x = t^3$, $y = t^2 + 1$ at the point when $t = 0$

27. f: $x = e^t + 3$, $y = 2 + \cos(5t)$; g: $x = t^2 + 3t + 4$, $y = 3 + \ln(1 - t)$ at the point when $t = 0$

28. f: $x = 2 + \cos(5t)$, $y = e^t + 3$; g: $x = 3 + \ln(1 - t)$, $y = t^2 + 3t + 4$ at the point when $t = 0$

In problems 29 – 48, find a vector N that is perpendicular to the given vector or line. (Typically there are several correct answers.)

29. $A = \langle 1, -2, 0 \rangle$ 30. $B = \langle -5, 0, 3 \rangle$ 31. $C = \langle 7, 3 \rangle$

32. $D = \langle 7, -3 \rangle$ 33. $E = \langle 2, -1, 3 \rangle$ 34. $S = \langle 1, 2, 5 \rangle$

35. $T = 3i + j - 5k$

36. $U = i + 3j + 2k$

37. $V = 2i - 4j + k$

38. $W = -6i - 4j + k$

39. $A = 2i - 3j$

40. $B = 3i + 2j$

41. $C = 3i - 2j$

42. $x + y = 6$

43. $3x + 2y = 6$

44. $5x - 3y = 30$

45. $x - 4y = 8$

46. $5x + y = 10$

47. $y = 3$

48. $x = 2$

In problems 49 – 52 sketch $\mathbf{Proj_B\ A}$.

49. **A** and **B** in Fig. 16.   50. **A** and **B** in Fig. 17.   51. **A** and **B** in Fig. 18.   52. **A** and **B** in Fig. 19.

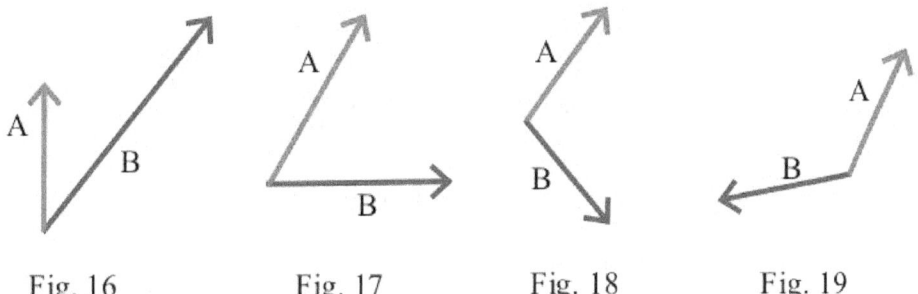

Fig. 16   Fig. 17   Fig. 18   Fig. 19

In problems 53 – 56 sketch $\mathbf{Proj_A\ B}$.

53. **A** and **B** in Fig. 16.   54. **A** and **B** in Fig. 17.   55. **A** and **B** in Fig. 18.   56. **A** and **B** in Fig. 19.

In problems 57 – 63, calculate $\mathbf{Proj_B\ A}$ and $\mathbf{Proj_A\ B}$.

57. $A = \langle 1, -2, 0 \rangle$, $B = \langle 5, 0, 3 \rangle$

58. $A = \langle 1, 2, 3 \rangle$, $B = \langle -2, 4, -1 \rangle$

59. $A = 2i - 3j + 5k$, $B = -5i + 0j + 2k$

60. $A = 4i - 3j + 2k$, $B = 3i + 2j - 3k$

61. $A = \langle 5, 0, 0 \rangle$, $B = \langle 0, 4, -3 \rangle$

62. $A = \langle 2, -1, 3 \rangle$, $B = \langle 1, 2, 5 \rangle$

63. $A = \langle 1, -2, 3 \rangle$, $B = j$

64. Suppose **A** and **B** have the same length. Which has the larger magnitude: $\mathbf{Proj_B\ A}$ or $\mathbf{Proj_A\ B}$? Justify your answer.

65. Suppose $|A| = 3\ |B|$. Which has the larger magnitude: $\mathbf{Proj_B\ A}$ or $\mathbf{Proj_A\ B}$? Justify your answer.

66. Prove that the vector $N = ai + bj$ is perpendicular to the line $ax + by = c$.

(Suggestion: Pick any two points $P = (x_0, y_0)$ and $Q = (x_1, y_1)$ on the line. Then the vector **V** with

starting point P and ending point Q has the same direction as the line, and

$V = \langle x_1 - x_0, y_1 - y0 \rangle$. Now show that **N** is perpendicular to **V** .)

In problems 67 – 72, calculate the distance from the given point to the given line using (a) the algorithm

of Example 6, and (b) the formula found in Example 6.

67. $(1, 3), y = 7 - x$ 68. $(-2, 1), 3x - 2y = 6$ 69. $(5, 3), 3y = 3x + 7$

70. $(-3, -4), y = 3x + 2$ 71. $(0, 0), 4x + 3y = 7$ 72. $(0, 0), ax + by = c$

73. Fig. 20 shows the position of a road and a house. How close is the road to

     the house (minimum distance)?

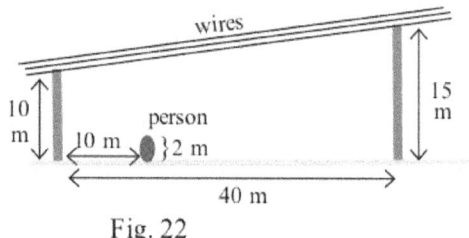

Fig. 20

                         74. In Fig. 21, how close is the wire to the magnet?

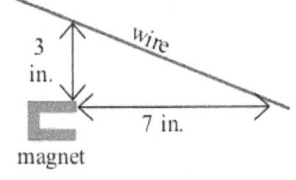

Fig. 21

                         75. A person is standing directly below the electrical

                         transmission wires in Fig. 22. Assuming the wires are so taut that they follow a

                         straight line, how close do they come to the person's head?

76. The **direction cosines** of a vector $\mathbf{A} = a_1\mathbf{i} + a_2\mathbf{j} + a_3\mathbf{k}$ are the

     cosines of the angles $\mathbf{A}$ makes with each of the coordinate

     vectors $\mathbf{i}, \mathbf{j}$, and $\mathbf{k}$: if $\mathbf{A}$ makes angles θ_x , θ_y , and θ_z with

     the x, y, and z axes, respectively, then the direction cosines of $\mathbf{A}$

     are $\cos(\theta_x)$, $\cos(\theta_y)$, and $\cos(\theta_z)$.

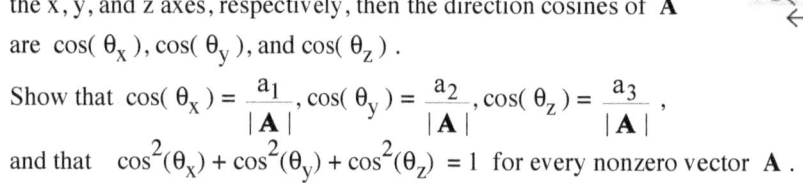

Fig. 22

     Show that $\cos(\theta_x) = \dfrac{a_1}{|\mathbf{A}|}$, $\cos(\theta_y) = \dfrac{a_2}{|\mathbf{A}|}$, $\cos(\theta_z) = \dfrac{a_3}{|\mathbf{A}|}$,

     and that $\cos^2(\theta_x) + \cos^2(\theta_y) + \cos^2(\theta_z) = 1$ for every nonzero vector $\mathbf{A}$.

77. A car moves on a track located on the y–axis, and you are pulling on a rope with a force of 50 pounds. Find

     the amount of work that you do in moving the cart from the origin to the point $Q = (0, 10, 0)$ if you have the

     rope over your shoulder 4 feet above the ground and walk 10 feet in front of the cart along a path 5 feet to the

     side of the y–axis.

78. Redo Problem 77 assuming that you are now pulling on the rope with a force of 100 pounds.

79. A wind blowing parallel to the y–axis exerts a force of 8 pounds on a kite. How much work does the

     wind do in moving the kite in a straight line from the point $(20, 30, 40)$ to the point $(50, 90, 150)$.

80. Redo problem 79 assuming that the wind is blowing parallel to the x–axis.

81. How much work does the person in Fig. 23 do moving the

     box 20 feet along the ground?

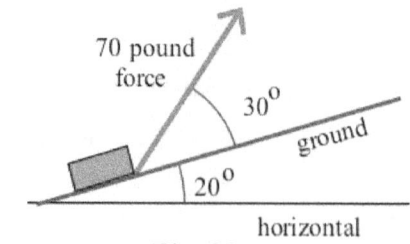

Fig. 23

Beyond Three Dimensions

83. $\mathbf{A} = \langle\, 1, 2, 3, 4\, \rangle$, $\mathbf{B} = \langle\, -2, 4, -1, 3\, \rangle$. Calculate $\mathbf{A \bullet B}$, $\mathbf{B \bullet A}$, and the angle between $\mathbf{A}$ and $\mathbf{B}$.

84. $\mathbf{A} = \langle\, -3, 4, 5, -1\, \rangle$, $\mathbf{B} = \langle\, -2, 4, -2, -4\, \rangle$. Calculate $\mathbf{A \bullet B}$, $\mathbf{B \bullet A}$, and the angle between $\mathbf{A}$ and $\mathbf{B}$.

In problems 85 – 88, a particular personality profile assigns a number between –1 and + 1 to each person on each of five personality characteristics. Using the categories ("very alike," "somewhat alike," , etc.) of Practice 8, determine the correct category for the pair of personality profiles given in each problem.

85. $\mathbf{A} = \langle\, 1, -0.2, 0.3, -0.6, 0.4\, \rangle$, $\mathbf{B} = \langle\, -0.5, 0.3, 0.3, -0.1, 0.2\, \rangle$

86. $\mathbf{A} = \langle\, 0.2, -0.3, -0.3, -0.7, 0.6\, \rangle$, $\mathbf{B} = \langle\, 0.4, -0.2, -0.4, -0.5, 0.8\, \rangle$

87. $\mathbf{A} = \langle\, 0.1, 0.2, 0.3, 0.4, 0.5\, \rangle$, $\mathbf{B} = \langle\, 0.3, 0.4, 0.5, 0.6, 0.7\, \rangle$

88. $\mathbf{A} = \langle\, 0.8, 0.3, -0.5, 0.2, 0.6\, \rangle$, $\mathbf{B} = \langle\, -0.2, -0.4, -0.6, -0.4, 0.1\, \rangle$

89. Prove the Parallelogram Law for vectors: $|\,\mathbf{A} + \mathbf{B}\,|^2 + |\,\mathbf{A} - \mathbf{B}\,|^2 \; 2|\,\mathbf{A}\,|^2 + 2|\,\mathbf{B}\,|^2$

Practice Answers

Practice 1: $\mathbf{U \bullet V} = (2)(-1) + (6)(2) + (-3)(2) = 4.$ $\mathbf{U \bullet U} = (2)(2) + (6)(6) + (-3)(-3) = 49.$

$\mathbf{V \bullet V} = (-1)(-1) + (2)(2) + (2)(2) = 9.$

$\mathbf{U \bullet (U+V)} = \langle\, 2, 6, -3\, \rangle \langle\, 1, 8, -1\, \rangle = 53.$ $\mathbf{U \bullet U} + \mathbf{U \bullet V} = 49 + 4 = 53.$

$|\mathbf{U}|^2 = (\sqrt{2^2 + 6^2 + (-3)^2}\;)2 = 2^2 + 6^2 + (-3)^2 = 49 = \mathbf{U \bullet U}$.

$\mathbf{V \bullet U} = (-1)(2) + (2)(6) + (2)(-3) = 4$ so $\mathbf{U \bullet V} = \mathbf{V \bullet U}$.

Practice 2: $\mathbf{A \bullet B} = a_1 b_1 + a_2 b_2 + a_3 b_3 = b_1 a_1 + b_2 a_2 + b_3 a_3 = \mathbf{B \bullet A}$.

Practice 3: $|\mathbf{A}| = \sqrt{4+36+9} = 7,$ $|\mathbf{B}| = \sqrt{16+64+1} = 9,$ and $|\mathbf{C}| = \sqrt{9+0+16} = 5$.

(a) $\cos(\theta) = \dfrac{\mathbf{A \bullet B}}{|\mathbf{A}|\,|\mathbf{B}|} = \dfrac{8-48-3}{(7)(9)} = \dfrac{-43}{63} \approx -0.683$ so $\theta \approx 2.32$ (radians) or $133.04°$.

(b) $\cos(\theta) = \dfrac{\mathbf{A \bullet C}}{|\mathbf{A}|\,|\mathbf{C}|} = \dfrac{6+0-12}{(7)(5)} = \dfrac{-6}{35} \approx -0.171$ so $\theta \approx 1.74$ or $99.87°$.

(c) $\cos(\theta) = \dfrac{\mathbf{B \bullet (-i)}}{|\mathbf{B}|\,|-i|} = \dfrac{-4+0+0}{(9)(1)} = \dfrac{-4}{9} \approx -0.444$ so $\theta \approx 2.03$ or $116.36°$.

(d) $\cos(\theta) = \dfrac{\mathbf{C \bullet j}}{|\mathbf{C}|\,|j|} = \dfrac{0+0+0}{(5)(1)} = 0$ so $\theta = \pi/2$ or $90°$. $\mathbf{C}$ and $\mathbf{j}$ are perpendicular.

Practice 4:   First we need to find two points on the line and then use those two points to form a vector

parallel to the line: $P = (-6, 0)$ and $Q = (0, 15)$ are on the line, and

$\mathbf{V} = \langle\, 0-(-6),\, 15-0 \,\rangle = \langle\, 6, 15 \,\rangle$ is parallel to the line. $\mathbf{N \cdot V} = \langle\, -5, 2 \,\rangle \cdot \langle\, 6, 15 \,\rangle = -30 + 30 = 0$ so

$\mathbf{N} = \langle\, -5, 2 \,\rangle$ is perpendicular to $-5x + 2y = 30$. Every scalar multiple $k\mathbf{N}$ with $k \neq 0$ is also

perpendicular to the line.

Practice 5:   (a) $|\mathbf{U}| = 11, |\mathbf{V}| = 3, \mathbf{U \cdot V} = -7$. scalar projection of $\mathbf{U}$ onto $\mathbf{V}$ is $\dfrac{\mathbf{U \cdot V}}{|\mathbf{V}|} = \dfrac{-7}{3}$.

Vector projection of $\mathbf{U}$ onto $\mathbf{V}$ is $\left(\dfrac{\mathbf{U \cdot V}}{|\mathbf{V}|^2} \right) \mathbf{V} = \langle\, \dfrac{-7}{9}, \dfrac{-14}{9}, \dfrac{14}{9} \,\rangle$.

(b) $\mathbf{U + V} = \langle\, 10, 0, 4 \,\rangle$, $|\mathbf{U + V}| = \sqrt{116}$, $\mathbf{V \cdot (U + V)} = 10+0-8 = 2$.

Scalar projection of $\mathbf{V}$ onto $\mathbf{U + V}$ is $\dfrac{\mathbf{V \cdot (U + V)}}{|\mathbf{U + V}|} = \dfrac{2}{\sqrt{116}} \approx 0.19$.

Vector projection of $\mathbf{V}$ onto $\mathbf{U + V}$ is $\left(\dfrac{\mathbf{V \cdot (U + V)}}{|\mathbf{U + V}|^2} \right) (\mathbf{U + V}) = \langle\, \dfrac{5}{29}, 0, \dfrac{2}{29} \,\rangle$.

(c) Scalar projection of $\mathbf{V}$ onto the positive y–axis is $\dfrac{\mathbf{V \cdot j}}{|\mathbf{j}|} = 2$.

Vector projection of $\mathbf{V}$ onto $\mathbf{j}$ is $\left(\dfrac{\mathbf{V \cdot j}}{|\mathbf{j}|^2} \right) \mathbf{j} = \langle\, 0, 2, 0 \,\rangle$.

Practice 6:   (1) $\mathbf{N} = \langle\, 3, -4 \,\rangle$. Take $Q = (\, 0, -5 \,)$ (any other point on the line also works -- try one).

(3) $\mathbf{V}$ is the vector from P to Q: $\mathbf{V} = \langle\, -2, -12 \,\rangle$.

(4) scalar projection of $\mathbf{V}$ onto $\mathbf{N}$ is $\dfrac{\mathbf{V \cdot N}}{|\mathbf{N}|} = \dfrac{-6 + 48}{\sqrt{9 + 16}} = \dfrac{42}{5} = 8.4$, the distance from

the point to the line.

We get the same answer, but perhaps less understanding, using the formula:

$$\dfrac{|\, ap + bq - c \,|}{\sqrt{a^2 + b^2}} = \dfrac{|\,(3)(2) + (-4)(7) - 20\,|}{\sqrt{9 + 16}} = \dfrac{|\,6 - 28 - 20\,|}{5} = \dfrac{42}{5} .$$

Practice 7:   The direction of the force vector is $\langle\, 6, 12, 4 \,\rangle / |\langle\, 6, 12, 4 \,\rangle| = \langle\, \dfrac{6}{14}, \dfrac{12}{14}, \dfrac{4}{14} \,\rangle$ so

$\mathbf{F} = 70 \langle\, \dfrac{6}{14}, \dfrac{12}{14}, \dfrac{4}{14} \,\rangle = \langle\, 30, 60, 20 \,\rangle$. Then the work done is

Work $= \mathbf{F \cdot D} = \langle\, 30, 60, 20 \,\rangle \cdot \langle\, 0, 7, 0 \,\rangle = 0 + (60 \text{ pounds})(7 \text{ feet}) + 0 = 420 \text{ foot–pounds}$.

Practice 8: $|\mathbf{Y}| = \sqrt{5^2 + 1^2 + (-7)^2 + 5^2} = \sqrt{100} = 10$, $|\mathbf{F}| = 15$, and $|\mathbf{A}| = 15$.

For $\mathbf{Y}$ and $\mathbf{F}$, $\cos(\,\theta\,) = \dfrac{\mathbf{Y \cdot F}}{|\mathbf{Y}||\mathbf{F}|} = \dfrac{30+10-56+25}{(10)(15)} = \dfrac{9}{150}$ so $\theta \approx 86.6°$: "different."

For $\mathbf{Y}$ and $\mathbf{A}$, $\cos(\,\theta\,) = \dfrac{\mathbf{Y \cdot A}}{|\mathbf{Y}||\mathbf{A}|} = \dfrac{50-4+70+15}{(10)(15)} = \dfrac{131}{150}$ so $\theta \approx 29.2°$: "very alike."

11.5   CROSS PRODUCT

This section is the final one about the arithmetic of vectors, and it introduces a second type of vector–vector multiplication called the cross product. The material in this section and the previous sections is the foundation for the next several chapters on the calculus of vector–valued functions and functions of several variables, and all of the vector arithmetic is used extensively.

The dot product of two vectors results in a scalar, a number related to the magnitudes of the two original vectors and to the angle between them, and the dot product is defined for two vectors in 2–dimensional, 3–dimension, and higher dimensional spaces. The cross product of a vector and a vector differs from the dot product in several significant ways: the cross product is only defined for two vectors in 3–dimensional space, and the cross product of two vectors is a vector. At first, the definition of the cross product given below may seem strange, but the resulting vector has some very useful properties as well as some unusual ones. The torque wrench described next illustrates some of the properties we get with the cross product.

Torque wrench: As you pull down on the torque wrench in Fig. 1, a force
is applied to the bolt that twists it into the wall. This "twisting" force is
the result of two vectors, the length and a direction of the wrench **A** and
the magnitude and direction of the pulling force **B** . To model this
"twisting into the wall" result of **A** and **B**, we want a vector **C** that
points into the wall and depends on the magnitudes of **A** and **B** as well
as on the angle between the wrench and the direction of the pull.

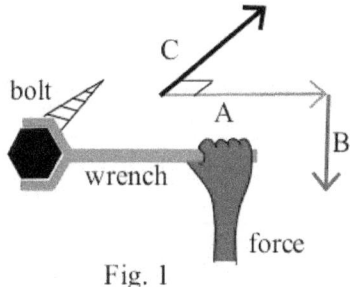

Fig. 1

If the pull is downward (Fig. 1), we want **C** to point
into the page.

If the pull is upward (Fig. 2), we want **C** to point out of the page.

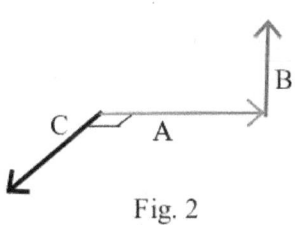

Fig. 2

If the angle between **A** and **B** is close to $\pm 90^{\circ}$ (Fig. 3), we want the magnitude
of **C** to be large.

If the angle between **A** and **B** is small (Fig. 4), we want the magnitude
of **C** to be small.

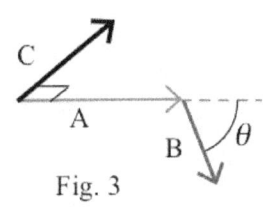

Fig. 3

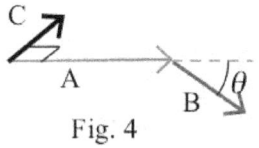

Fig. 4

There is a way to combine the vectors **A** and **B** to produce a vector **C** with the properties that model the torque wrench. This vector **C**, called the cross product of **A** and **B**, also turns out to be very useful when we discuss planes through given points and tangent planes to surfaces. It is not at all obvious from the definition given below for the cross product that the cross product has a relationship to torque wrenches, planes, or anything else of interest or use to us, but it does and we will investigate those applied and geometric properties. Try not to be repelled by the unusual definition — it leads to some lovely results.

Definition of the **Cross Product (Vector Product):**

For $\mathbf{A} = \langle a_1, a_2, a_3 \rangle$ and $\mathbf{B} = \langle b_1, b_2, b_3 \rangle$, the **cross product** of $\mathbf{A}$ and $\mathbf{B}$ is

$$\mathbf{A} \times \mathbf{B} = (a_2 b_3 - a_3 b_2)\mathbf{i} + (a_3 b_1 - a_1 b_3)\mathbf{j} + (a_1 b_2 - a_2 b_1)\mathbf{k}, \text{ a vector.}$$

The symbols "$\mathbf{A} \times \mathbf{B}$" are read "$\mathbf{A}$ cross $\mathbf{B}$."

Many people find it difficult to remember a complicated formula like the definition of the cross product.

Fortunately, there is an easy way to do so, but it requires a digression into the calculation of determinants.

Determinants

Determinants appear in a number of areas of mathematics, and you may have already seen them as part of Cramer's Rule for solving systems of equations. Here we only need them as a device for making it easier to remember and calculate and use the cross product.

Definition of the 2x2 ("two by two") Determinant: The determinant $\begin{vmatrix} a & b \\ c & d \end{vmatrix} = ad - bc$.

Some people prefer to remember the visual pattern in Fig. 5.

Example 1:   Evaluate the determinants: $\begin{vmatrix} 1 & 4 \\ 3 & 5 \end{vmatrix}$,

$\begin{vmatrix} x & y \\ -2 & 3 \end{vmatrix}$, and $\begin{vmatrix} \mathbf{i} & \mathbf{j} \\ 0 & 4 \end{vmatrix}$.

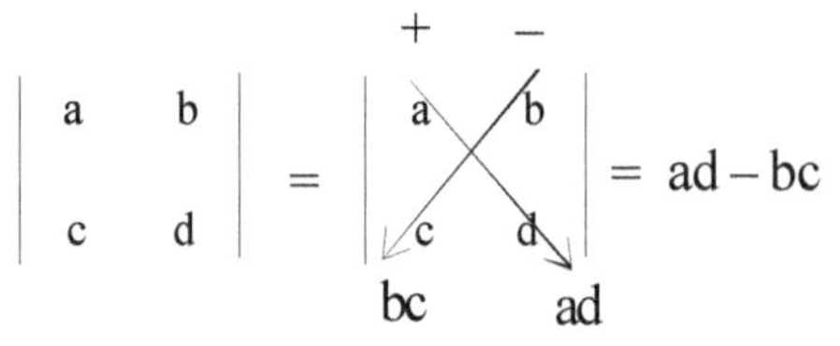

Solution: $\begin{vmatrix} 1 & 4 \\ 3 & 5 \end{vmatrix} = (1)(5) - (4)(3) = 5 - 12 = -7.$

$\begin{vmatrix} x & y \\ -2 & 3 \end{vmatrix} = (x)(3) - (y)(-2) = 3x + 2y.$

$$\begin{vmatrix} a & b \\ c & d \end{vmatrix} = \begin{vmatrix} a & b \\ c & d \end{vmatrix} = ad - bc$$

Fig. 5: Pattern for a 2x2 determinant

$\begin{vmatrix} \mathbf{i} & \mathbf{j} \\ 0 & 4 \end{vmatrix} = (\mathbf{i})(4) - (\mathbf{j})(0) = 4\mathbf{i}.$

Practice 1:   Evaluate the determinants: $\begin{vmatrix} -3 & 4 \\ 5 & 6 \end{vmatrix}$, $\begin{vmatrix} x & y \\ 0 & -3 \end{vmatrix}$, and $\begin{vmatrix} \mathbf{i} & \mathbf{j} \\ -4 & 3 \end{vmatrix}$.

A 3x3 determinant can be defined in terms of several 2x2 determinants.

Definition of the 3x3 Determinant: $\begin{vmatrix} a_1 & a_2 & a_3 \\ b_1 & b_2 & b_3 \\ c_1 & c_2 & c_3 \end{vmatrix} = a_1 \begin{vmatrix} b_2 & b_3 \\ c_2 & c_3 \end{vmatrix} - a_2 \begin{vmatrix} b_1 & b_3 \\ c_1 & c_3 \end{vmatrix} + a_3 \begin{vmatrix} b_1 & b_2 \\ c_1 & c_2 \end{vmatrix}.$

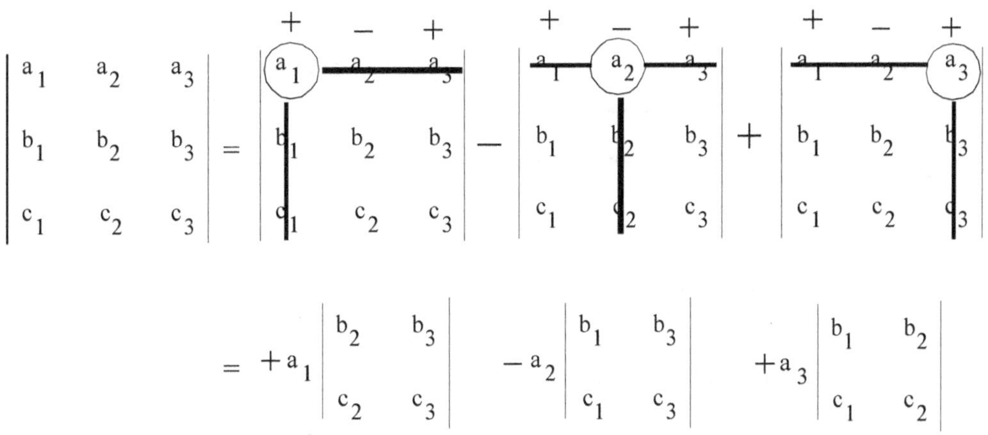

$$\begin{vmatrix} a_1 & a_2 & a_3 \\ b_1 & b_2 & b_3 \\ c_1 & c_2 & c_3 \end{vmatrix} = \begin{vmatrix} a_1 & a_2 & a_3 \\ b_1 & b_2 & b_3 \\ c_1 & c_2 & c_3 \end{vmatrix} - \begin{vmatrix} a_1 & a_2 & a_3 \\ b_1 & b_2 & b_3 \\ c_1 & c_2 & c_3 \end{vmatrix} + \begin{vmatrix} a_1 & a_2 & a_3 \\ b_1 & b_2 & b_3 \\ c_1 & c_2 & c_3 \end{vmatrix}$$

$$= +a_1 \begin{vmatrix} b_2 & b_3 \\ c_2 & c_3 \end{vmatrix} - a_2 \begin{vmatrix} b_1 & b_3 \\ c_1 & c_3 \end{vmatrix} + a_3 \begin{vmatrix} b_1 & b_2 \\ c_1 & c_2 \end{vmatrix}$$

Fig. 6: Pattern for a 3x3 determinant

Many people prefer to remember the visual pattern in Fig. 6.

In the 3x3 definition, the first 2x2 determinant, $\begin{vmatrix} b_2 & b_3 \\ c_2 & c_3 \end{vmatrix}$, is the part of the original 3x3 table after the first row and the first column have been removed, the row and column containing a_1 . The second 2x2 determinant $\begin{vmatrix} b_1 & b_3 \\ c_1 & c_3 \end{vmatrix}$ is what remains of the original table after the first row and the second column have been removed, the row and column containing a_2 . The third 2x2 determinant $\begin{vmatrix} b_1 & b_2 \\ c_1 & c_2 \end{vmatrix}$ is what remains of the original table after the first row and the third column have been removed, the row and column containing a_3 . (Note: The leading signs attached to the three terms alternate: $+ - +$.)

Example 2:   Evaluate the determinants $\begin{vmatrix} 2 & 3 & 5 \\ 0 & -4 & 1 \\ -3 & 4 & 0 \end{vmatrix}$ and $\begin{vmatrix} i & j & k \\ 3 & 4 & 1 \\ 0 & -2 & 5 \end{vmatrix}$.

Solution: $\begin{vmatrix} 2 & 3 & 5 \\ 0 & -4 & 1 \\ -3 & 4 & 0 \end{vmatrix} = (2)\begin{vmatrix} -4 & 1 \\ 4 & 0 \end{vmatrix} - (3)\begin{vmatrix} 0 & 1 \\ -3 & 0 \end{vmatrix} + (5)\begin{vmatrix} 0 & -4 \\ -3 & 4 \end{vmatrix} = (2)(-4) - (3)(3) + (5)(-12) = -77.$

$\begin{vmatrix} i & j & k \\ 3 & 4 & 1 \\ 0 & -2 & 5 \end{vmatrix} = i\begin{vmatrix} 4 & 1 \\ -2 & 5 \end{vmatrix} - j\begin{vmatrix} 3 & 1 \\ 0 & 5 \end{vmatrix} + k\begin{vmatrix} 3 & 4 \\ 0 & -2 \end{vmatrix} = +22i - 15j + (-6)k .$

Practice 2:   Evaluate the determinants $\begin{vmatrix} 3 & 5 & 0 \\ 1 & 4 & -1 \\ -2 & 0 & 6 \end{vmatrix}$ and $\begin{vmatrix} i & j & k \\ 2 & -1 & 3 \\ 4 & 0 & 5 \end{vmatrix}$.

The original definition of the cross product can now be rewritten using the determinant notation.

Determinant Form of the Cross Product Definition:

$$\text{For } \mathbf{A} = \langle a_1, a_2, a_3 \rangle \text{ and } \mathbf{B} = \langle b_1, b_2, b_3 \rangle, \ \mathbf{A} \times \mathbf{B} = \begin{vmatrix} \mathbf{i} & \mathbf{j} & \mathbf{k} \\ a_1 & a_2 & a_3 \\ b_1 & b_2 & b_3 \end{vmatrix}.$$

The second determinant in Example 2 represents the cross product $\mathbf{A} \times \mathbf{B}$ for $\mathbf{A} = \langle 3, 4, 1 \rangle$ and $\mathbf{B} = \langle 0, -2, 5 \rangle$, and the second determinant in Practice 2 is the cross product $\mathbf{A} \times \mathbf{B}$ for $\mathbf{A} = \langle 2, -1, 3 \rangle$ and $\mathbf{B} = \langle 4, 0, 5 \rangle$.

The cross products of various pairs of the basis vectors $\mathbf{i}, \mathbf{j},$ and $\mathbf{k}$ are relatively easy to evaluate, and they begin to illustrate some of the properties of cross products.

Example 3:   Use the determinant form of the definition of the cross product to evaluate

(a) $\mathbf{i} \times \mathbf{j}$, (b) $\mathbf{j} \times \mathbf{i}$, and (c) $\mathbf{i} \times \mathbf{i}$.

Solution: $\mathbf{i} \times \mathbf{j} = \begin{vmatrix} \mathbf{i} & \mathbf{j} & \mathbf{k} \\ 1 & 0 & 0 \\ 0 & 1 & 0 \end{vmatrix} = \mathbf{i} \begin{vmatrix} 0 & 0 \\ 1 & 0 \end{vmatrix} - \mathbf{j} \begin{vmatrix} 1 & 0 \\ 0 & 0 \end{vmatrix} + \mathbf{k} \begin{vmatrix} 1 & 0 \\ 0 & 1 \end{vmatrix} = 0\mathbf{i} + 0\mathbf{j} + 1\mathbf{k} = \mathbf{k}$.

$\mathbf{j} \times \mathbf{i} = \begin{vmatrix} \mathbf{i} & \mathbf{j} & \mathbf{k} \\ 0 & 1 & 0 \\ 1 & 0 & 0 \end{vmatrix} = \mathbf{i} \begin{vmatrix} 1 & 0 \\ 0 & 0 \end{vmatrix} - \mathbf{j} \begin{vmatrix} 0 & 0 \\ 1 & 0 \end{vmatrix} + \mathbf{k} \begin{vmatrix} 0 & 1 \\ 1 & 0 \end{vmatrix} = 0\mathbf{i} + 0\mathbf{j} + (-1)\mathbf{k} = -\mathbf{k}$.

$\mathbf{i} \times \mathbf{i} = \begin{vmatrix} \mathbf{i} & \mathbf{j} & \mathbf{k} \\ 1 & 0 & 0 \\ 1 & 0 & 0 \end{vmatrix} = \mathbf{i} \begin{vmatrix} 0 & 0 \\ 0 & 0 \end{vmatrix} - \mathbf{j} \begin{vmatrix} 1 & 0 \\ 1 & 0 \end{vmatrix} + \mathbf{k} \begin{vmatrix} 1 & 0 \\ 1 & 0 \end{vmatrix} = 0\mathbf{i} + 0\mathbf{j} + 0\mathbf{k} = \mathbf{0}$.

You should note that $\mathbf{i} \times \mathbf{j} = \mathbf{k}$ is perpendicular to the vectors $\mathbf{i}$ and $\mathbf{j}$ and the xy–plane. Similarly, $\mathbf{j} \times \mathbf{i} = -\mathbf{k}$ is also perpendicular to $\mathbf{i}$ and $\mathbf{j}$ and the xy–plane.

Practice 3:   Use the determinant form of the definition of the cross product to evaluate

(a) $\mathbf{j} \times \mathbf{k}$, (b) $\mathbf{k} \times \mathbf{j}$, and (c) $\mathbf{j} \times \mathbf{j}$.

The cross products of pairs of basis vectors follow a simple pattern given below and in Fig. 7.

$\mathbf{i} \times \mathbf{j} = \mathbf{k}$ $\mathbf{j} \times \mathbf{i} = -\mathbf{k}$ $\mathbf{i} \times \mathbf{i} = 0$

$\mathbf{j} \times \mathbf{k} = \mathbf{i}$ $\mathbf{k} \times \mathbf{j} = -\mathbf{i}$ $\mathbf{j} \times \mathbf{j} = 0$

Pattern for the Cross Product of the basis vectors

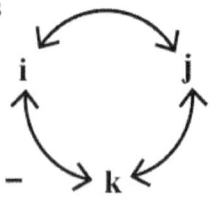

clockwise is +

counterclockwise is –

Fig. 7

$$\mathbf{k} \times \mathbf{i} = \mathbf{j} \qquad\qquad \mathbf{i} \times \mathbf{k} = -\mathbf{j} \qquad\qquad \mathbf{k} \times \mathbf{k} = 0$$

Example 4:   Evaluate $\mathbf{A} \times \mathbf{B}$ for $A = \langle\, 2, -3, 0 \,\rangle$ and $B = \langle\, 3, 1, -4 \,\rangle$.

Solution: $$\mathbf{A} \times \mathbf{B} = \begin{vmatrix} \mathbf{i} & \mathbf{j} & \mathbf{k} \\ 2 & -3 & 0 \\ 3 & 1 & -4 \end{vmatrix} = \mathbf{i}\begin{vmatrix} -3 & 0 \\ 1 & -4 \end{vmatrix} - \mathbf{j}\begin{vmatrix} 2 & 0 \\ 3 & -4 \end{vmatrix} + \mathbf{k}\begin{vmatrix} 2 & -3 \\ 3 & 1 \end{vmatrix} = 12\mathbf{i} + 8\mathbf{j} + 11\mathbf{k}.$$

Practice 4:   Evaluate $\mathbf{A} \times \mathbf{B}$ for $A = \langle\, 3, 0, -5 \,\rangle$ and $B = \langle\, -2, 4, 1 \,\rangle$.

Properties of the Cross Product:

(a) $0 \times \mathbf{A} = \mathbf{A} \times 0 = 0$

(b) $\mathbf{A} \times \mathbf{A} = 0$

(c) $\mathbf{A} \times \mathbf{B} = -\mathbf{B} \times \mathbf{A}$

(d) $k(\mathbf{A} \times \mathbf{B}) = k\mathbf{A} \times \mathbf{B} = \mathbf{A} \times k\mathbf{B}$

(e) $\mathbf{A} \times (\mathbf{B} + \mathbf{C}) = (\mathbf{A} \times \mathbf{B}) + (\mathbf{A} \times \mathbf{C})$

The proofs of all of these properties are straightforward applications of the definition of the cross product as is illustrated below for part (c). Proofs of (a) and (b) are given in the Appendix after the Practice Answers, and the proofs of (d) and (e) are left as exercises.

Proof of (c): If $\mathbf{A} = \langle\, a_1, a_2, a_3 \,\rangle$ and $\mathbf{B} = \langle\, b_1, b_2, b_3 \,\rangle$, then

$$\mathbf{B} \times \mathbf{A} = \begin{vmatrix} \mathbf{i} & \mathbf{j} & \mathbf{k} \\ b_1 & b_2 & b_3 \\ a_1 & a_2 & a_3 \end{vmatrix} = \mathbf{i}\begin{vmatrix} b_2 & b_3 \\ a_2 & a_3 \end{vmatrix} - \mathbf{j}\begin{vmatrix} b_1 & b_3 \\ a_1 & a_3 \end{vmatrix} + \mathbf{k}\begin{vmatrix} b_1 & b_2 \\ a_1 & a_2 \end{vmatrix}$$

$$= \mathbf{i}\,(a_3 b_2 - a_2 b_3) - \mathbf{j}\,(a_3 b_1 - a_1 b_3) + \mathbf{k}\,(a_2 b_1 - a_1 b_2).$$

$$\mathbf{A} \times \mathbf{B} = \begin{vmatrix} \mathbf{i} & \mathbf{j} & \mathbf{k} \\ a_1 & a_2 & a_3 \\ b_1 & b_2 & b_3 \end{vmatrix} = \mathbf{i}\begin{vmatrix} a_2 & a_3 \\ b_2 & b_3 \end{vmatrix} - \mathbf{j}\begin{vmatrix} a_1 & a_3 \\ b_1 & b_3 \end{vmatrix} + \mathbf{k}\begin{vmatrix} a_1 & a_2 \\ b_1 & b_2 \end{vmatrix}$$

$$= \mathbf{i}\,(a_2 b_3 - a_3 b_2) - \mathbf{j}\,(a_1 b_3 - a_3 b_1) + \mathbf{k}\,(a_1 b_2 - a_2 b_1) = -\mathbf{B} \times \mathbf{A}.$$

A vector has both a direction and a magnitude, and the direction and magnitude of the vector $\mathbf{A} \times \mathbf{B}$ each give us useful information: the direction of $\mathbf{A} \times \mathbf{B}$ is perpendicular to $\mathbf{A}$ and $\mathbf{B}$, and the magnitude of $\mathbf{A} \times \mathbf{B}$ is the area of the parallelogram with sides $\mathbf{A}$ and $\mathbf{B}$. These are the two properties of the cross product that get used most often in the later chapters, and they enable us to visualize $\mathbf{A} \times \mathbf{B}$.

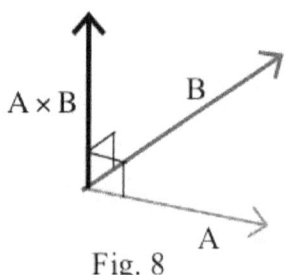

Fig. 8

The **direction** property of **A** x **B**:

$\mathbf{A} \cdot (\mathbf{A} \times \mathbf{B}) = 0$ and $\mathbf{B} \cdot (\mathbf{A} \times \mathbf{B}) = 0$

so **A** x **B** is perpendicular to **A** and to **B** . (Fig. 8)

Proof: $\mathbf{A} \cdot (\mathbf{A} \times \mathbf{B}) = \langle a_1, a_2, a_3 \rangle \cdot \langle a_2 b_3 - a_3 b_2, a_3 b_1 - a_1 b_3, a_1 b_2 - a_2 b_1 \rangle$

$= a_1 a_2 b_3 - a_1 a_3 b_2 + a_2 a_3 b_1 - a_2 a_1 b_3 + a_3 a_1 b_2 - a_3 a_2 b_1 = 0 .$

The proof that $\mathbf{B} \cdot (\mathbf{A} \times \mathbf{B}) = 0$ is similar .

The **magnitude** property of **A** x **B**:

   If   **A** and **B** are nonzero vectors with angle θ

   between them $(0 \le \theta \le \pi)$,

   then $|\mathbf{A} \times \mathbf{B}| = |\mathbf{A}||\mathbf{B}| |\sin(\theta)|$

                 = **area of the parallelogram** formed

                 by the vectors **A** and **B**. (Fig. 9)

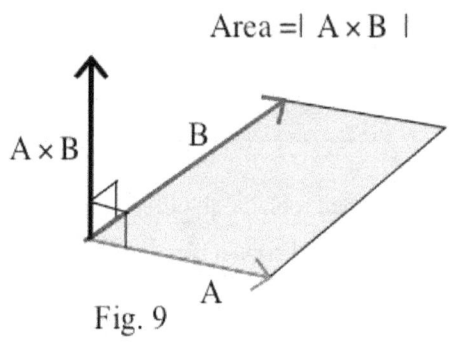

Area $= |\,\mathbf{A} \times \mathbf{B}\,|$

Fig. 9

The proof that $|\mathbf{A} \times \mathbf{B}| = |\mathbf{A}||\mathbf{B}| |\sin(\theta)|$ is algebraically complicated and is given in the Appendix after

the Practice Answers. These properties of the direction and magnitude are sometimes used to define the

cross product, and then the algebraic definition is derived from them.

Corollary to the magnitude property of **A** x **B**:

   The **area of the triangle** formed by the vectors **A** and **B** is $\frac{1}{2} |\mathbf{A} \times \mathbf{B}|$.

Visualizing the cross product: The cross product **A** x **B** is perpendicular to both **A** and **B** and

satisfies a "right hand rule" so we can visualize the direction of **A** x **B** using, of course, your right hand.

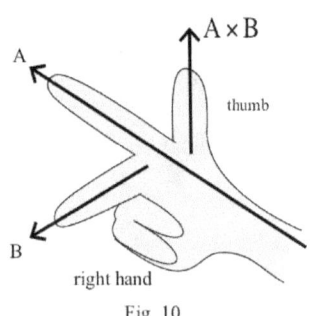

Fig. 10

If **A** and **B** are the indicated fingers (Fig. 10 and Fig. 11))   then your extended

right thumb points in the direction of **A** x **B**.

The relative length of **A** x **B** can be estimated

from the magnitude property of

A x **B** : $|\mathbf{A} \times \mathbf{B}|$ = parallelogram area.

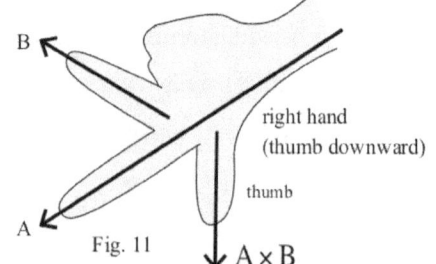

Fig. 11

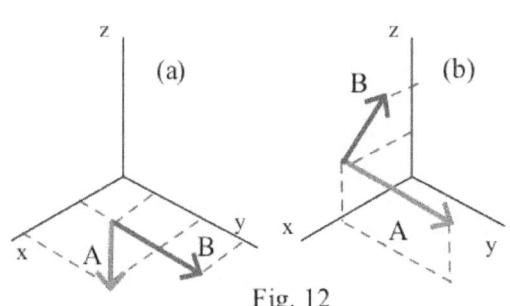

Fig. 12

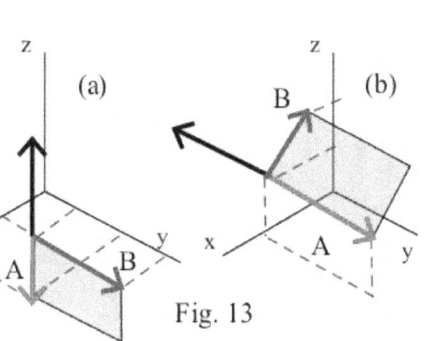

Fig. 13

Example 5:   Fig. 12 shows two pairs of vectors **A** and **B**. For each pair sketch the direction of **A** x **B** . For which pair is | **A** x **B** | larger?

Solution: See Fig. 13. | **A** x **B** | is larger for (b) since the area of their parallelogram is larger in (b).

Practice 5:   Fig. 14 shows two pairs of vectors **A** and **B**. For each pair sketch the direction of **A** x **B** . For which pair is | **A** x **B** | larger?

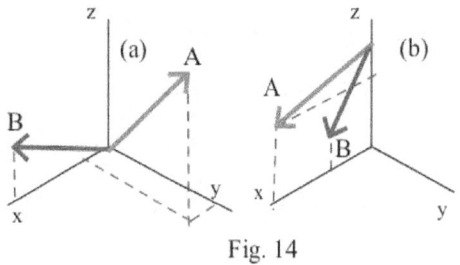

Fig. 14

Torque Wrench Revisited

Now we can use the ideas and properties of the cross product to analyze the original torque wrench problem.

Definition: The **torque vector** produced by a lever arm vector **A** and a force vector **B** is **A** x **B** .

The direction of the torque vector tells us whether the wrench is driving the bolt into the wall or pulling it out of the wall. The magnitude of the torque vector describes the strength of the tendency of the wrench to drive the bolt in or pull it out.

Example 6:   Fig. 15 shows two force vectors acting at the end of a 10 inch wrench. Which vector produces the larger torque?

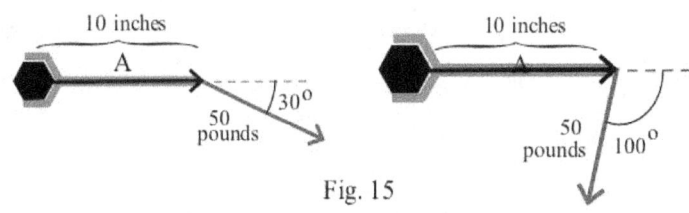

Fig. 15

Solution: |**B**| = 50 pounds with $\theta = 30°$ so

| **A** x **B** | = |**A**||**B**| |sin(θ)|

= (10 inches)(50 pounds) |sin(30°)| $\approx$ 250 pound–inches of force.

|**C**| = 50 pounds with $\theta = 100°$ so

| **A** x **C** | = |**A**||**C**| |sin(θ)| = (10 inches)(50 pounds)| sin(100°) | $\approx$ 492.4 pound–inches of force.

Vector **C** produces the larger torque: the smaller force, used intelligently, produced the larger result.

Practice 6:   In Fig. 16 which force vector

produces the larger torque?

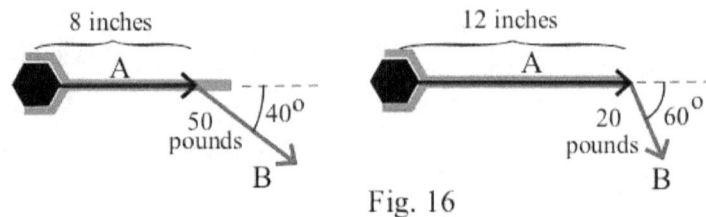

Fig. 16

The Triple Scalar Product A•(B x C)

The combination of the dot and cross products, **A•(B x C)** , for the three

vectors **A, B**, and **C** in 3–dimensional space is called

the triple scalar product because the result of these operations is

a scalar:**A•(B x C)** = **A•**(vector) = scalar. And the magnitude

of this scalar has a nice geometric meaning.

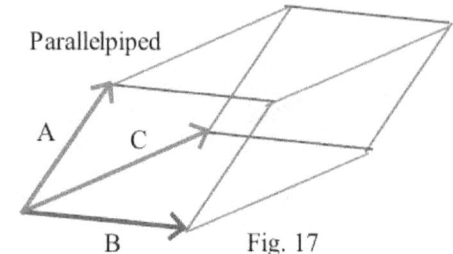

Parallelpiped

Fig. 17

> Geometric meaning of | **A•(B x C)** |
>
> For the vectors **A, B**, and **C** in 3–dimensional space (Fig.17)
>
> | **A•(B x C)** | = volume of the parallelpiped (box) with sides **A, B**, and **C** .

Proof: From the definition of the dot product,

$$\mathbf{A \cdot (B \times C)} = |\mathbf{B \times C}||\mathbf{A}|\cos(\theta) \text{ so}$$

$$|\mathbf{A \cdot (B \times C)}| = |\mathbf{B \times C}||\mathbf{A}||\cos(\theta)| .$$

Volume = (area of the base)(height),

and the area of the base of the box is | **B x C** | .

Since **B x C** is perpendicular to the base, the height h is the

projection of **A** onto **B x C** (Fig. 18):   h = |**A**| | cos(θ) |.

Then Volume = | **B x C**| |**A**| | cos(θ) | which we showed was

equal to | **A•(B x C)** | .

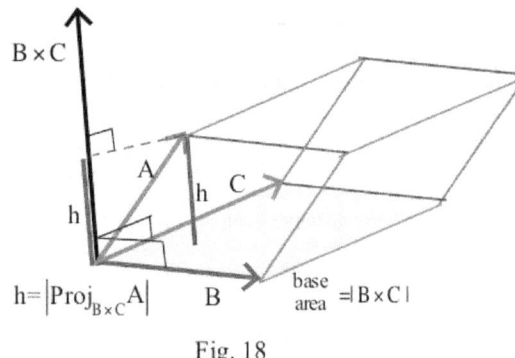

Fig. 18

Problem 46 asks you to show that the triple scalar product can be calculated as a 3x3 determinant.

Beyond Three Dimensions

The objects we examined in previous sections (points, distances, vectors, dot products, angles between vectors,

projections) all had rather nice extensions to more than three dimensions. The cross product is different: the

cross product **A x B** we have defined requires that **A** and **B** be 3–dimensional vectors, and there is no easy

extension to vectors in more than three dimensions that preserves the properties of the cross product.

PROBLEMS

In problems 1 – 12, evaluate the determinants.

1. $\begin{vmatrix} 3 & 4 \\ 2 & 5 \end{vmatrix}$
2. $\begin{vmatrix} 4 & -1 \\ 3 & 1 \end{vmatrix}$
3. $\begin{vmatrix} x & 5 \\ y & 2 \end{vmatrix}$
4. $\begin{vmatrix} 5 & a \\ b & 3 \end{vmatrix}$

5. $\begin{vmatrix} 1 & 0 \\ 0 & 1 \end{vmatrix}$
6. $\begin{vmatrix} 0 & 1 \\ 1 & 0 \end{vmatrix}$
7. $\begin{vmatrix} 1 & 3 & 2 \\ 0 & 5 & 2 \\ 1 & 1 & 0 \end{vmatrix}$
8. $\begin{vmatrix} 2 & 3 & 0 \\ 1 & -3 & 2 \\ -1 & 0 & 4 \end{vmatrix}$

9. $\begin{vmatrix} x & y & z \\ 1 & 2 & 3 \\ 3 & 1 & 2 \end{vmatrix}$
10. $\begin{vmatrix} a & b & c \\ 0 & 3 & 5 \\ 2 & 1 & 3 \end{vmatrix}$
11. $\begin{vmatrix} 2 & 3 & 5 \\ 0 & -4 & 1 \\ -3 & 4 & 0 \end{vmatrix}$
12. $\begin{vmatrix} x & 0 & 0 \\ 0 & 3 & 0 \\ 0 & 0 & 1-x \end{vmatrix}$

In problems 13 – 18, vectors **A** and **B** are given. Calculate (a) **A** x **B** , (b) (**A** x **B**)·**A,**
(c) (**A** x **B**)·**B** , and (d) | **A** x **B** | .

13. $\mathbf{A} = \langle 3,4,5 \rangle$, $\mathbf{B} = \langle -1,2,0 \rangle$ 14. $\mathbf{A} = \langle -2,2,2 \rangle$, $\mathbf{B} = \langle 3,1,2 \rangle$

15. $\mathbf{A} = \langle 1,-3,2 \rangle$, $\mathbf{B} = \langle -2,6,4 \rangle$ 16. $\mathbf{A} = \langle 6,8,-2 \rangle$, $\mathbf{B} = \langle 3,4,-1 \rangle$

17. $\mathbf{A} = 3\mathbf{i} - 1\mathbf{j} + 4\mathbf{k}$, $\mathbf{B} = 1\mathbf{i} - 2\mathbf{j} + 5\mathbf{k}$ 18. $\mathbf{A} = 4\mathbf{i} - 1\mathbf{j} + 0\mathbf{k}$, $\mathbf{B} = 3\mathbf{i} - 2\mathbf{j} + 0\mathbf{k}$

In problems 19 – 22, state whether the result of the given calculation is a vector, a scalar, or is not defined.

19. **A**•(**B** x **C**)   20. **A** x (**B**•**C**)   21. (**A**•**B**) x (**A**•**C**)   22. **A**(**B** x **C**)

23. Prove property (d) of the Properties of the Cross Product: k(**A** x **B**) = k**A** x **B** = **A** x k**B** .

24. Prove property (e) of the Properties of the Cross Product: **A** x (**B** + **C**) = (**A** x **B**) + (**A** x **C**) .

25. Explain geometrically why **A** x **A** = **0** .

26. If |**A**| and |**B**| are fixed, what angle(s) between **A** and **B** maximizes | **A** x **B** |? Why?

In problems 27 – 30, vectors **A** and **B** are given graphically. Sketch **A** x **B** .

27. See Fig. 19.   28. See Fig. 20.   29. See Fig. 21.   30. See Fig. 22.

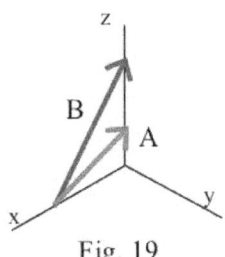

Fig. 19

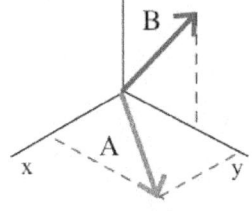

Fig. 20

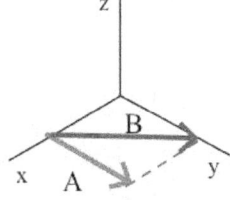

Fig. 21

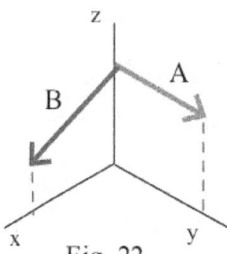

Fig. 22

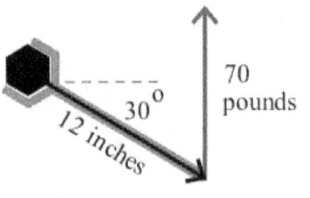

Fig. 23

31. (a) Calculate the torque produced by the
        wrench and force shown in Fig. 23.

    (b) Calculate the torque produced by the
        wrench and force shown in Fig. 24.

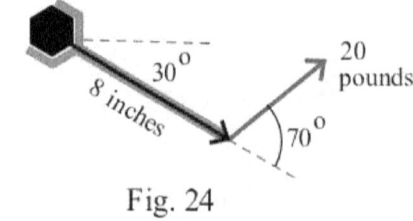

Fig. 24

32. When a tire nut is "frozen" (stuck), a pipe is sometimes put over
    the handle of the tire wrench (Fig. 25). Using the vocabulary and
    ideas of vectors, explain why this is effective.

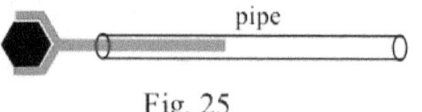

Fig. 25

33. Does the torque on wrench **A** produced by **B** plus the torque
    produced by **C** equal the torque produced by **B** + **C** ? Why or why not?

Areas

34. Sketch the parallelogram with sides **A** = $\langle 5, 1, 0 \rangle$ and **B** = $\langle 2, 4, 0 \rangle$ and find its area. Sketch and
    find the area of the triangle with sides **A** and **B** .

35. Sketch the parallelogram with sides **A** = $\langle 1, 2, 0 \rangle$ and **B** = $\langle 0, 4, 2 \rangle$ and find its area. Sketch and
    find the area of the triangle with sides **A** and **B** .

36. Sketch the triangle with vertices P = (4, 0, 1), Q =(1, 3, 1), and R = (2, 0, 5) and find its area.

37. Sketch the triangle with vertices P = (1, 4, –2), Q =(3, 5, 1), and R = (5, 2, 2) and find its area.

38. Sketch the triangle with vertices P = (a, 0, 0), Q =(0, b, 0), and R = (0, 0, c) and find its area.

Triple Scalar Products

39. Sketch a parallelpiped with edges **A** = $\langle 2, 1, 0 \rangle$, **B** = $\langle -1, 4, 1 \rangle$, **C** = $\langle 1, 1, 2 \rangle$ and find its volume.

40. Sketch a parallelpiped with edges **A** = $\langle 2, 0, 3 \rangle$, **B** = $\langle 0, 4, 5 \rangle$, **C** = $\langle 4, 3, 0 \rangle$ and find its volume.

41. Sketch a parallelpiped with edges **A** = $\langle a, 0, 0 \rangle$, **B** = $\langle 0, b, 0 \rangle$, **C** = $\langle 0, 0, c \rangle$ and find its volume.

Use the result that "the volume of the tetrahedron with edges **A**, **B**, **C** (Fig. 26) is 1/6 the volume of
the parallelpiped with the same edges" to find the areas of the tetrahedrons in problems 42 – 45.

42. Sketch the tetrahedron with vertices
    P = (0,0,0), Q =(3,1,0), R = (0,4,0),
    and S = (0,0,3) and find its volume.

43. Sketch the tetrahedron with vertices P =
    (1,0,2), Q =(3,1,2), R = (0,4,3), and S =
    (0,1,4) and find its volume.

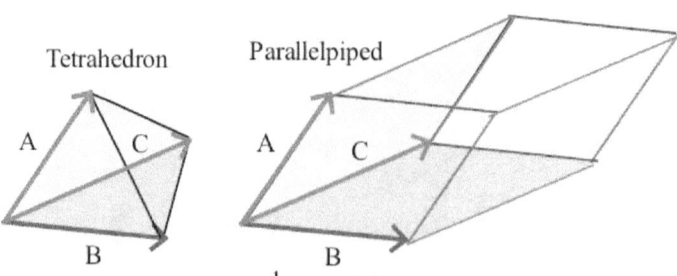

Tetrahedron volume = $\frac{1}{6}$ (Parallelpiped volume)

Fig. 26

44.   Sketch the tetrahedron with vertices $P = (0,0,0), Q = (2,0,0), R = (0,4,0)$, and $S = (0,0,6)$ and find its volume.

45.   Sketch the tetrahedron with vertices $P = (0,0,0), Q = (a,0,0), R = (0,b,0)$, and $S = (0,0,c)$ and find its volume.

46. Show that if $\mathbf{A} = \langle a_1, a_2, a_3 \rangle, \mathbf{B} = \langle b_1, b_2, b_3 \rangle$, and $\mathbf{C} = \langle c_1, c_2, c_3 \rangle$,

$$\text{then } \mathbf{A} \cdot (\mathbf{B} \times \mathbf{C}) = \begin{vmatrix} a_1 & a_2 & a_3 \\ b_1 & b_2 & b_3 \\ c_1 & c_2 & c_3 \end{vmatrix} .$$

Right Tetrahedrons

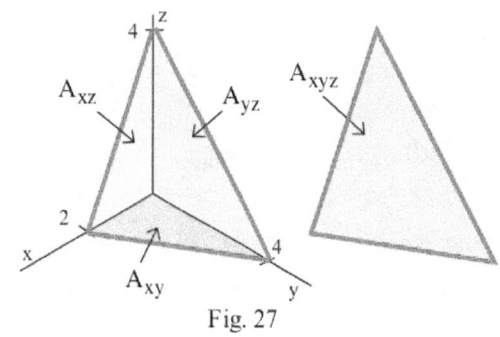

Fig. 27

47. The four points $(0,0,0), (2,0,0), (0,4,0)$, and $(0,0,4)$ form a tetrahedron (Fig. 27) with four triangular faces. Find the areas A_{xy}, A_{xz}, A_{yz}, and A_{xyz} of the four triangular faces.

48. The four points $(0,0,0), (2,0,0), (0,4,0)$, and $(0,0,6)$ form a tetrahedron (Fig. 28) with four triangular faces. Find the areas A_{xy}, A_{xz}, A_{yz}, and A_{xyz} of the four triangular faces.

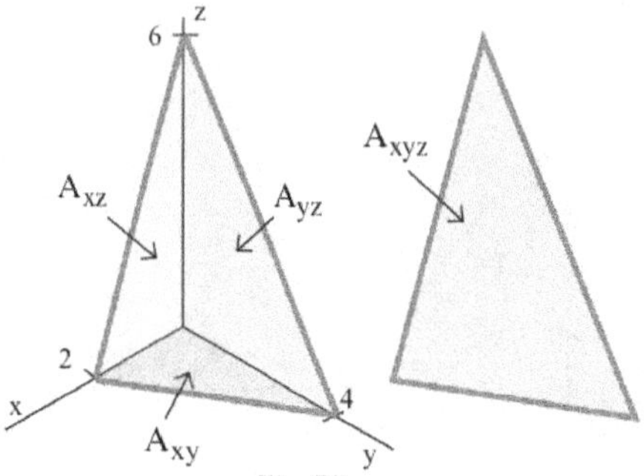

Fig. 28

49. Verify that the answers to problems 47 and 48 satisfy the relationship $(A_{xy})^2 + (A_{xz})^2 + (A_{yz})^2 = (A_{xyz})^2$.

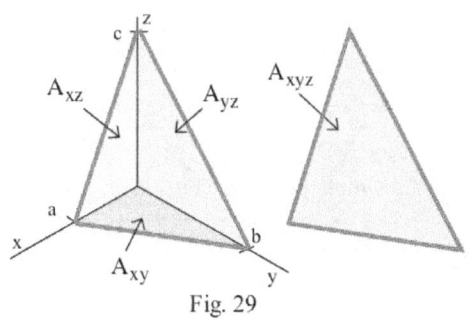

Fig. 29

50. For the right tetrahedron with vertices $(0,0,0), (a,0,0), (0,b,0)$, and $(0,0,c)$, determine the areas of the four triangular faces (Fig. 29) and prove the Pythagorean type result for areas of triangles in a right tetrahedron:

$$(A_{xy})^2 + (A_{xz})^2 + (A_{yz})^2 = (A_{xyz})^2 .$$

51.   The Pythagorean pattern

$$a^2 + b^2 = c^2$$

can be thought of as relating a line segment

C in two dimensions and its "shadows" a

and b on the coordinate axes. Show that

this "shadow" interpretation also holds for

the area of a triangle in three dimensions

and the areas of its "shadows" on the three

coordinate planes (Fig. 30):

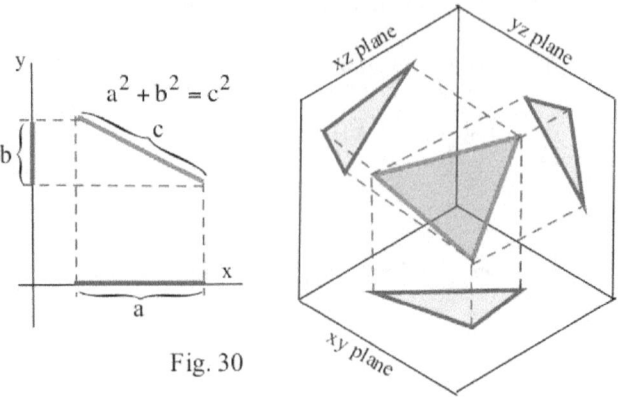

Fig. 30

(area of dark triangle)^2 = (xy shadow area)^2 + (xz shadow area)^2 + (yz shadow area)^2 .

Areas of Regions in the Plane

Among its several uses, the cross product also leads to a simple, easily programmed algorithm for finding

the area of a "simple" (no edges cross) polygon in the plane, and this algorithm is used to approximate the

areas of other regions as well.

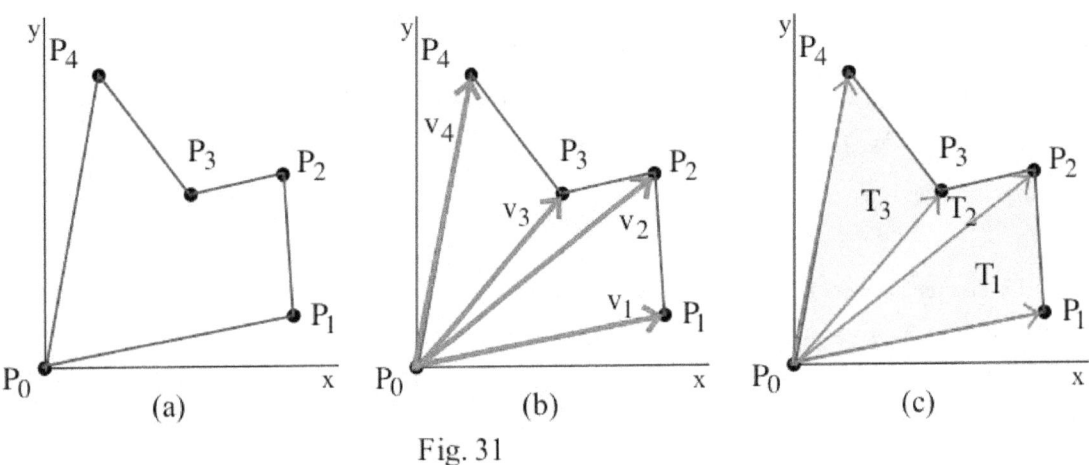

Fig. 31

Suppose $P_0 = (0, 0)$, $P_1 = (x_1, y_1)$, ... , $P_4 = (x_4, y_4)$ are 5 vertices of a simple polygon (Fig. 31a),

with one vertex at the origin and then labeling the others as we travel counterclockwise around the

polygon. Let V_1 be the vector from P_0 to P_1, V_2 from P_0 to P_2, ... (Fig. 31b). Then the area of the

polygon in Fig. 31c is the sum of the 3 triangular areas T_1, T_2, and T_3 and each triangular area can

be found using a cross product: $T_1 = \frac{1}{2} |V_1 \times V_2| = \frac{1}{2}(x_1 y_2 - x_2 y_1)$,

$T_2 = \frac{1}{2} |V_2 \times V_3| = \frac{1}{2}(x_2 y_3 - x_3 y_2)$, and $T_3 = \frac{1}{2} |V_3 \times V_4| = \frac{1}{2}(x_3 y_4 - x_4 y_4)$.

Finally, the total area is the sum

$$\text{Area} \quad = T_1 + T_2 + T_3$$

$$= \frac{1}{2} \left\{ (x_1 y_2 - x_2 y_1) + (x_2 y_3 - x_3 y_2) + (x_3 y_4 - x_4 y_3) \right\}$$

$$= \frac{1}{2} \sum_{k=1}^{n-1} \left(x_k y_{k+1} - x_{k+1} y_k \right) \quad \text{with } n = 4 .$$

The last summation formula works for polygons with at least three vertices. In fact, this algorithm is used by computers to report the area of a region traced by a cursor or stylus: the computer reads the (x,y) location of the cursor several times per second and uses the data and this algorithm to calculate the area of the region (as approximated by a many–sided polygon).

52.   Use the given pattern to find the area of the rectangle with vertices $(0,0), (2,0), (2,3)$, and $(0,3)$. Does the pattern give the area of the rectangle?

53.   Use the given pattern to find the area of the pentagon with vertices $(0,0), (4,1), (5,3), (4,4), (2,4)$, and $(1,3)$.

54.   How can we modify the algorithm to handle the situation in which none of the vertices are at the origin? Show that your modification works for the rectangle with vertices $(1,3), (3,3), (3,6)$, and $(1,6)$.

Note: The cross product satisfies a "right hand rule" so if we go counterclockwise from **U** to **V** (Fig. 32a) then **U** x **V** is positive, and if we go clockwise from **U** to **V** (Fig. 32b) then **U** x **V** is negative.

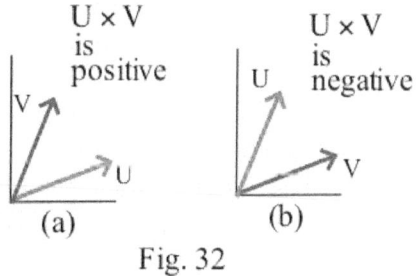

Fig. 32

55.   In Fig. 33a, $\frac{1}{2}$ **V**$_1$ x **V**$_2$ gives the area of T_1 as a positive number; $\frac{1}{2}$ **V**$_2$ x **V**$_3$ gives the area of T_2 as a negative number; and $\frac{1}{2}$ **V**$_3$ x **V**$_4$ gives the area of T_3 as a positive number.

Explain geometrically how these positive and negative numbers "fit together" to give the correct area for the region in Fig. 33b.

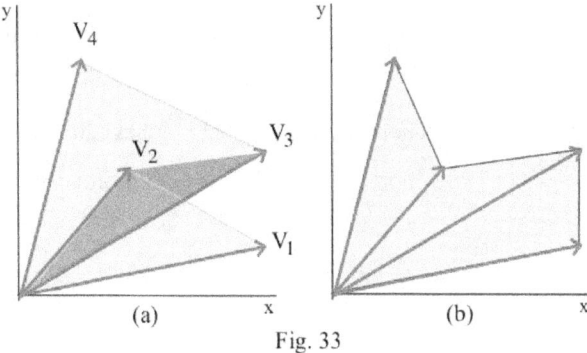

Fig. 33

Practice Answers

Practice 1: $\begin{vmatrix} -3 & 4 \\ 5 & 6 \end{vmatrix} = (-3)(6) - (4)(5) = -38.$ $\begin{vmatrix} x & y \\ 0 & -3 \end{vmatrix} = (x)(-3) - (y)(0) = -3x$.

$\begin{vmatrix} \mathbf{i} & \mathbf{j} \\ -4 & 3 \end{vmatrix} = (\mathbf{i})(3) - (\mathbf{j})(-4) = 3\mathbf{i} + 4\mathbf{j}$.

Practice 2: $\begin{vmatrix} 3 & 5 & 0 \\ 1 & 4 & -1 \\ -2 & 0 & 6 \end{vmatrix} = (3)\begin{vmatrix} 4 & -1 \\ 0 & 6 \end{vmatrix} - (5)\begin{vmatrix} 1 & -1 \\ -2 & 6 \end{vmatrix} + (0)\begin{vmatrix} 1 & 4 \\ -2 & 0 \end{vmatrix} = (3)(24) - (5)(4) + (0)(8) = 52.$

$\begin{vmatrix} \mathbf{i} & \mathbf{j} & \mathbf{k} \\ 2 & -1 & 3 \\ 4 & 0 & 5 \end{vmatrix} = \mathbf{i}\begin{vmatrix} -1 & 3 \\ 0 & 5 \end{vmatrix} - \mathbf{j}\begin{vmatrix} 2 & 3 \\ 4 & 5 \end{vmatrix} + \mathbf{k}\begin{vmatrix} 2 & -1 \\ 4 & 0 \end{vmatrix} = -5\mathbf{i} + 2\mathbf{j} + 4\mathbf{k}$.

Practice 3: $\mathbf{j} \times \mathbf{k} = \begin{vmatrix} \mathbf{i} & \mathbf{j} & \mathbf{k} \\ 0 & 1 & 0 \\ 0 & 0 & 1 \end{vmatrix} = \mathbf{i}\begin{vmatrix} 1 & 0 \\ 0 & 1 \end{vmatrix} - \mathbf{j}\begin{vmatrix} 0 & 0 \\ 0 & 1 \end{vmatrix} + \mathbf{k}\begin{vmatrix} 0 & 1 \\ 0 & 0 \end{vmatrix} = 1\mathbf{i} - 0\mathbf{j} + 0\mathbf{k} = \mathbf{i}$.

$\mathbf{k} \times \mathbf{j} = \begin{vmatrix} \mathbf{i} & \mathbf{j} & \mathbf{k} \\ 0 & 0 & 1 \\ 0 & 1 & 0 \end{vmatrix} = -\mathbf{i}.$ $\mathbf{j} \times \mathbf{j} = \begin{vmatrix} \mathbf{i} & \mathbf{j} & \mathbf{k} \\ 0 & 1 & 0 \\ 0 & 1 & 0 \end{vmatrix} = \mathbf{0}$.

Practice 4: $\mathbf{A} \times \mathbf{B} = \begin{vmatrix} \mathbf{i} & \mathbf{j} & \mathbf{k} \\ 3 & 0 & -5 \\ -2 & 4 & 1 \end{vmatrix} = \mathbf{i}\begin{vmatrix} 0 & -5 \\ 4 & 1 \end{vmatrix} - \mathbf{j}\begin{vmatrix} 3 & -5 \\ -2 & 1 \end{vmatrix} + \mathbf{k}\begin{vmatrix} 3 & 0 \\ -2 & 4 \end{vmatrix} = 20\mathbf{i} + 7\mathbf{j} + 12\mathbf{k}$.

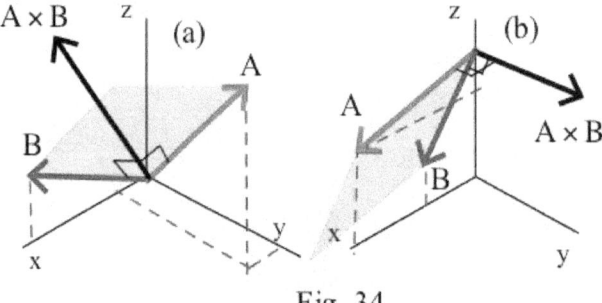

Practice 5:   See Fig. 34. The pair in (a)
produces the larger torque.

Fig. 34

Practice 6:   For **B**,

| torque | = (8 inches)(50 pounds) sin(40°)

≈ 257.1 inch–pounds.

For **C**, | torque | = (12 inches)(20 pounds) sin(60°) ≈ 207.8 inch–pounds.

Force **B** produces the larger torque: sometimes strength is enough.

Appendix: Some Proofs

Proof of (a): $\mathbf{0} \times \mathbf{A} = \mathbf{A} \times \mathbf{0} = \mathbf{0}.$ If $\mathbf{A} = \left\langle a_1, a_2, a_3 \right\rangle$, then

$$\mathbf{0} \times \mathbf{A} = \begin{vmatrix} \mathbf{i} & \mathbf{j} & \mathbf{k} \\ 0 & 0 & 0 \\ a_1 & a_2 & a_3 \end{vmatrix} = \mathbf{i} \begin{vmatrix} 0 & 0 \\ a_2 & a_3 \end{vmatrix} - \mathbf{j} \begin{vmatrix} 0 & 0 \\ a_1 & a_3 \end{vmatrix} + \mathbf{k} \begin{vmatrix} 0 & 0 \\ a_1 & a_2 \end{vmatrix} = 0\mathbf{i} - 0\mathbf{j} + 0\mathbf{k} .$$

The proof that $\mathbf{A} \times \mathbf{0} = \mathbf{0}$ is similar.

Proof of (b): $\mathbf{A} \times \mathbf{A} = \mathbf{0}.$ If $\mathbf{A} = \left\langle a_1, a_2, a_3 \right\rangle$, then

$$\mathbf{A} \times \mathbf{A} = \begin{vmatrix} \mathbf{i} & \mathbf{j} & \mathbf{k} \\ a_1 & a_2 & a_3 \\ a_1 & a_2 & a_3 \end{vmatrix} = \mathbf{i} \begin{vmatrix} a_2 & a_3 \\ a_2 & a_3 \end{vmatrix} - \mathbf{j} \begin{vmatrix} a_1 & a_3 \\ a_1 & a_3 \end{vmatrix} + \mathbf{k} \begin{vmatrix} a_1 & a_2 \\ a_1 & a_2 \end{vmatrix} = 0\mathbf{i} - 0\mathbf{j} + 0\mathbf{k} = \mathbf{0} .$$

Proof that $| \mathbf{A} \times \mathbf{B} | = |\mathbf{A}||\mathbf{B}| |\sin(\theta)|$:

$$\mathbf{A} \times \mathbf{B} = (a_2 b_3 - a_3 b_2)\mathbf{i} + (a_3 b_1 - a_1 b_3)\mathbf{j} + (a_1 b_2 - a_2 b_1)\mathbf{k} \quad \text{so}$$

$$\begin{aligned}
| \mathbf{A} \times \mathbf{B} |^2 &= (\mathbf{A} \times \mathbf{B}) \bullet (\mathbf{A} \times \mathbf{B}) \\
&= (a_2 b_3 - a_3 b_2)^2 + (a_3 b_1 - a_1 b_3)^2 + (a_1 b_2 - a_2 b_1)^2 \\
&= a_2^2 b_3^2 - 2a_2 b_3 a_3 b_2 + a_3^2 b_2^2 + a_3^2 b_1^2 - 2a_3 b_1 a_1 b_3 + a_1^2 b_3^2 + a_1^2 b_2^2 - 2a_1 b_2 a_2 b_1 + a_2^2 b_1^2 \\
&= (a_1^2 + a_2^2 + a_3^2)(b_1^2 + b_2^2 + b_3^2) - (a_1 b_1 + a_2 b_2 + a_3 b_3)^2 \quad \text{(expand \& check)} \\
&= |\mathbf{A}|^2 |\mathbf{B}|^2 - (\mathbf{A} \bullet \mathbf{B})^2 \\
&= |\mathbf{A}|^2 |\mathbf{B}|^2 - (|\mathbf{A}||\mathbf{B}|\cos(\theta))^2 \quad \text{since} \quad \mathbf{A} \bullet \mathbf{B} = |\mathbf{A}||\mathbf{B}|\cos(\theta) \\
&= |\mathbf{A}|^2 |\mathbf{B}|^2 - |\mathbf{A}|^2 |\mathbf{B}|^2 \cos^2(\theta) \\
&= |\mathbf{A}|^2 |\mathbf{B}|^2 \{ 1 - \cos^2(\theta) \} \\
&= |\mathbf{A}|^2 |\mathbf{B}|^2 \sin^2(\theta) .
\end{aligned}$$

Then, taking the square root of each side of $| \mathbf{A} \times \mathbf{B} |^2 = |\mathbf{A}|^2 |\mathbf{B}|^2 \sin^2(\theta)$, we have

$$| \mathbf{A} \times \mathbf{B} | = |\mathbf{A}||\mathbf{B}| | \sin(\theta) | .$$

11.6   LINES AND PLANES IN THREE DIMENSIONS

In two dimensions the graph of a linear equation, either y=ax+b or parametric (x(t),y(t)), is always a straight line, and a point and slope or two points completely determine the line. In three dimensions the graph of a linear equation, either z=ax+by+c or parametric (x(t), y(t), z(t)), has more freedom, and it can be a line or a plane. In this section we examine the equations of lines and planes and their graphs in 3–dimensional space, discuss how to determine their equations from information known about them, and look at ways to determine intersections, distances, and angles in three dimensions. Lines and planes are the simplest graphs in three dimensions, and they are useful for a variety of geometric and algebraic applications. They are also the building blocks we need to find tangent lines to curves and tangent planes to surfaces (Fig. 1) in three dimensions.

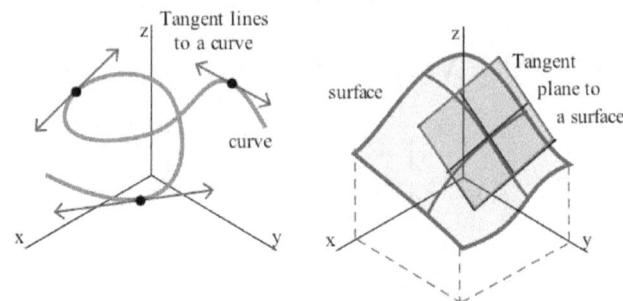

Fig. 1: Tangent lines and planes in three dimensions

Lines in Three Dimensions

Early in calculus we often used the point–slope formula to write the equation of the line through a given point $P = (x_0, y_0)$ with a given slope m: $y - y_0 = m(x - x_0)$. If the point $P = (x_0, y_0)$ is given and the direction of the line is parallel to a given vector $\mathbf{A} = \langle a, b \rangle$, then it is easier to use parametric equations to specify an equation for the line:

a point $Q = (x, y) \neq P$ is on the line if and only if $\dfrac{y - y_0}{x - x_0} = \dfrac{rise}{run} = \dfrac{b}{a} = \dfrac{b \cdot t}{a \cdot t}$ for some $t \neq 0$

so the equations $y - y_0 = b \cdot t$, and $x - x_0 = a \cdot t$ describe points on the line and

$x = x_0 + a \cdot t, y = y_0 + b \cdot t$.

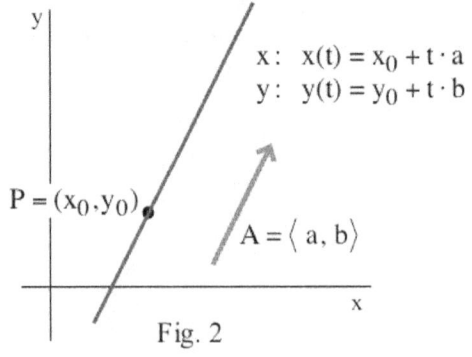

x: $x(t) = x_0 + t \cdot a$
y: $y(t) = y_0 + t \cdot b$

$P = (x_0, y_0)$

$A = \langle a, b \rangle$

Fig. 2

Parametric Equation of a Line in Two Dimensions
Parametric equations for a line through the point $P = (x_0, y_0)$ and parallel to the vector $\mathbf{A} = \langle a, b \rangle$ are $x = x(t) = x_0 + a \cdot t$ and $y = y(t) = y_0 + b \cdot t$. (Fig. 2)

Notice that the coordinates of the point (x_0, y_0) become the constant terms for x(t) and y(t), and the components of the vector $\langle a, b \rangle$ become the coefficients of the variable terms in the parametric equations. This same pattern is true for lines in three (and more) dimensions.

Example 1: Find parametric equations for the lines through the point P = (1,2) that are

(a) parallel to the vector $\mathbf{A} = \langle\, 3, 5 \,\rangle$, and (b) parallel to the vector $\mathbf{B} = \langle\, 6, 10 \,\rangle$.

Then graph the two lines.

Solution: (a) x(t) = 1 + 3t, y(t) = 2 + 5t.   (b)   x(t) = 1 + 6t, y(t) = 2 + 10t .

The graphs of the lines are shown in Fig. 3. In this example, both sets of parametric equations

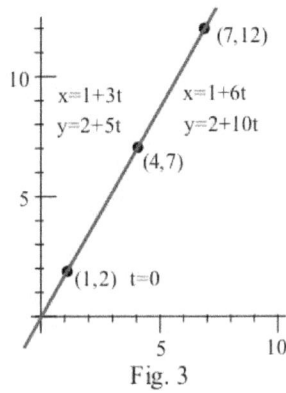

Fig. 3

have the same line when graphed. If we interpret t as time, and (x(t), y(t))

as the location of an object at time t, then the graph is a picture of all of the

points on the path of the object (for all times). The objects in (a) and (b)

have the same path, but they are at different points on the path at time t

(except when t = 0).

Practice 1:   Find parametric equations for the lines through the point

P = (3,–1) that are   (a) parallel to the vector $\mathbf{A} = \langle\, 2, -4 \,\rangle$, and

(b) parallel to the vector $\mathbf{B} = \langle\, 1, 5 \,\rangle$. Then graph the two lines.

The parametric pattern works for lines in three dimensions.

Parametric Equation of a Line in Three Dimensions

An equation of a line through the point $P = (x_0, y_0, z_0)$ and

parallel to the vector $\mathbf{A} = \langle\, a, b, c \,\rangle$ is given by the parametric

equations $x = x(t) = x_0 + at,$

$y = y(t) = y_0 + bt,$

$z = z(t) = z_0 + ct .$ (Fig. 4)

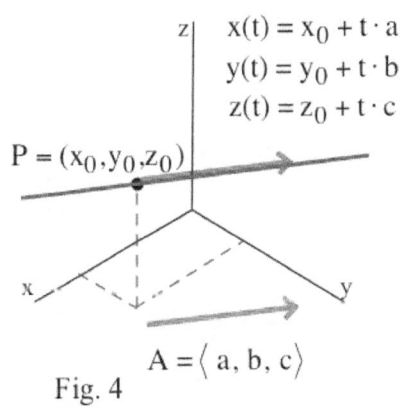

Fig. 4

Proof: To show that $P = (x_0, y_0, z_0)$ is on the line, put t = 0 and evaluate the three parametric equations:

$x = x(0) = x_0 + 0, y = y(0) = y_0 + 0,$ and $z = z(0) = z_0 + 0.$

To show that the line described by the parametric equations has the same direction as **A**, we pick

another point Q on the line and show that the vector from P to Q is parallel to **A**.

Put t = 1, and let $Q = (x(1), y(1), z(1)) = (x_0 + a, y_0 + b, z_0 + c)$. Then the vector

from P to Q is $\mathbf{V} = \langle\, (x_0 + a) - x_0, (y_0 + b) - y_0, (z_0 + c) - z_0 \,\rangle = \langle\, a, b, c \,\rangle = \mathbf{A}$ so

the line described by the parametric equations is parallel to **A** .

Example 2:   Find parametric equations for the line (a) through the point P = (3,0,2) and parallel to the

vector $\mathbf{A} = \langle\, 1, 2, 0 \,\rangle$, and (b) through the two points P = (3,0,2) and Q = (5,–1, 1).

Solution: (a) $x(t) = 3 + 1t$, $y(t) = 0 + 2t$, $z(t) = 2 + 0t$.

(b) We can use the two points to get a direction for the line. The direction of the line is parallel to the vector from P to Q, $\langle 2, -1, -1 \rangle$, so

$$x(t) = 3 + 2t, \; y(t) = 0 - 1t , \; z(t) = 2 - 1t .$$

The graphs of these two lines are shown in Fig. 5.

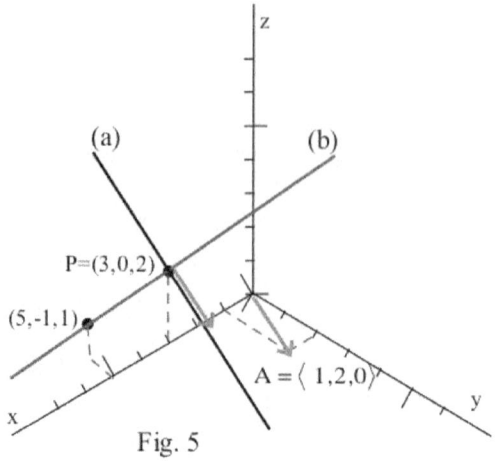

Fig. 5

Practice 2:   Find parametric equations for the line

(a) through the point $P = (2, -1, 0)$ and parallel to the vector $\mathbf{A} = \langle 3, -4, 1 \rangle$, and (b) through the two points $P = (3,0,2)$ and $Q = (2,5,4)$.

Given a point $P = (x_0, y_0, z_0)$ and a vector $\mathbf{A} = \langle a, b, c \rangle$, the point Q is on the line through P in the direction of $\mathbf{A}$ if and only if $Q = (x_0 + at, y_0 + bt, z_0 + ct)$ for some value of t.

Once we are able to write the equations of lines in three dimensions, it is natural to ask about the points and angles of intersection of these lines.

Example 3:   (a)   Find the point of intersection of the lines $K: x_K = 2 + t, y_K = 3 + 2t, z_K = 0 + t$ and $L: x_L = 5 - t, y_L = 1 + 2t, z_L = -1 + t$.

(b)   Find the angle of intersection of the lines K and L.

(c)   Where does line L intersect the xy–plane?

Solution: (a) If we could find a value of t so that $x_K = x_L , y_K = y_L$, and $z_K = z_L$, that would say that the lines not only intersect at the point (x_K, y_K, z_K) , but that the two objects were at that point at the same time. For these parametric equations, there is no value of t such that $x_K = x_L , y_K = y_L$, and $z_K = z_L$, but it is still possible for the lines to intersect, just not at the same "time."

To see if the lines go through a common point, but at different "times," we can change the parameter for the line L to "s" instead of "t," and represent L as $x_L = 5 - s, y_L = 1 + 2s, z_L = -1 + s$. Then the equations $x_K = x_L , y_K = y_L$, and $z_K = z_L$ become

$$\text{x: } 2 + t = 5 - s, \quad \text{y: } 3 + 2t = 1 + 2s, \quad \text{and z: } 0 + t = -1 + s .$$

Solving the first two equations for s and t, we get $s = 2$ and $t = 1$. These values also satisfy the equation for the z–coordinate, $0 + t = -1 + s$, so $x = 2 + t = 3, y = 3 + 2t = 5$, and $z = 0 + t = 1$. The point $(3, 5, 1)$ lies on both lines. (If the values of s and t from the x–component and y–component equations do not satisfy the z–component equation, the lines do not intersect.)

(b) Since the lines intersect, we can find their angle of intersection. Line K is parallel to

  $\mathbf{A} = \langle 1, 2, 1 \rangle$, the coefficients of the t terms, and L is parallel to $\mathbf{B} = \langle -1, 2, 1 \rangle$, and the angle

  between the lines equals the angle between $\mathbf{A}$ and $\mathbf{B}$:

$$\cos(\theta) = \frac{\mathbf{A} \cdot \mathbf{B}}{|\mathbf{A}||\mathbf{B}|} \; = \; \frac{4}{\sqrt{6}\sqrt{6}} \; = \; \frac{2}{3} \; \text{ so } \; \theta \approx 0.84 \;\; \text{(about } 48.2°\text{)}.$$

(c) Every point on the xy–plane has z–coordinate equal to 0, so we can set $-1 + t = 0$ and solve for

  $t = 1$. When t =1, then $x = 5 - (1) = 4$ and $y = 1 + 2(1) = 3$ so line L intersects the xy–plane at

  the point $(4, 3, 0)$.

Practice 3:   If the pairs of lines in (a) or (b) intersect, find the point and angle of intersection.

  (a)   K: x = 1 + t, y = 1 – 2t, z = –3 + 2t and L: x = 8 + 4t, y = –4 + t, z = –5 – 8t .

  (b)   K: x = 1 + t, y = 1 – 2t, z = 2 + 2t and L: x = 8 + 4t, y = –4 + t, z = 3 + t .

  (c)   Where does L: x = 8 + 4t, y = –4 + t, z = 3 + t intersect the yz–plane?

Practice 4:   An arrow is shot from the point (1,2,3) and travels in a straight line in the direction

  $\langle 4, 5, 1 \rangle$. Will the arrow go over a 10 foot high wall built on the xy–plane

  along the line y = 20?   (Fig. 6)

Planes in Three Dimensions

The vectors in a plane point in infinitely many directions
(Fig. 7) so, at first thought, you might think it would be more
difficult to find an equation for a plane than for a line.

Fortunately, however, there is only one
vector (and its scalar multiples) that is
perpendicular to the plane (Fig. 8), and this
"normal" vector makes the task easy.

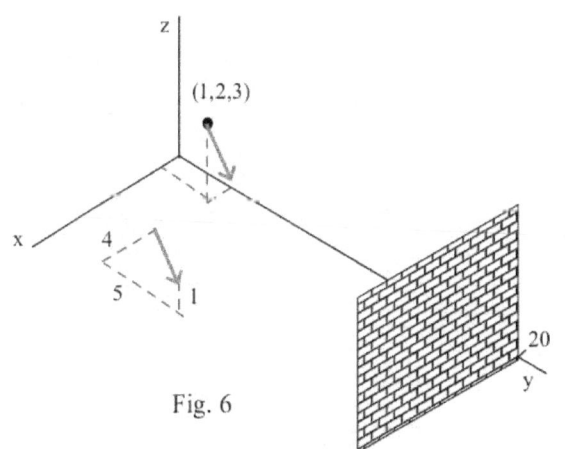

Fig. 6

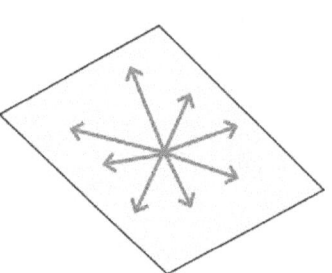

Fig. 7: Vectors in a plane

Suppose $P = (x_0, y_0, z_0)$ is a point on the plane that has
normal vector $\mathbf{N} = \langle a, b, c \rangle$. Let $Q = (x, y, z)$ be
another point. Since $\mathbf{N}$ is perpendicular to every vector on
the plane, the point Q is on the plane if and only if $\mathbf{N}$ is perpendicular to the
vector from P to Q.

That is the idea that leads to an easy equation for the plane.

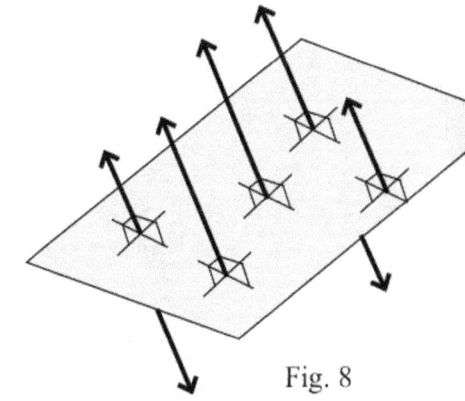

Fig. 8

Let **V** = vector from P to Q = $\langle x - x_0, y - y_0, z - z_0 \rangle$. Then Q is on the plane if and only if **V** is perpendicular to **N**: **V•N** = 0. But **V•N** = $a(x - x_0) + b(y - y_0) + c(z - z_0)$, so the point Q is on the plane if and only if $a(x - x_0) + b(y - y_0) + c(z - z_0) = 0$.

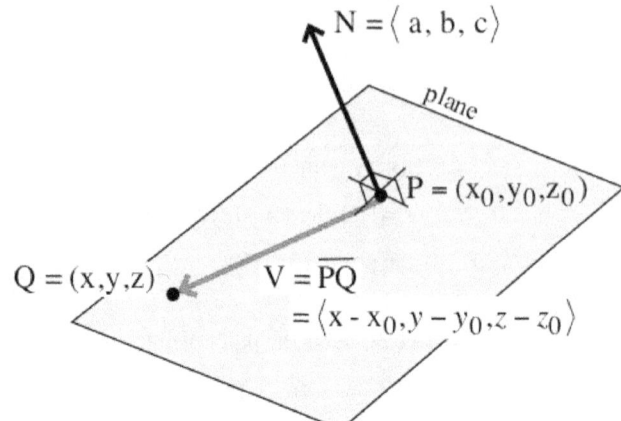

Equation for a Plane in Three Dimensions:

(Point–Normal Form)

An equation for a plane through the point
$P = (x_0, y_0, z_0)$ with normal vector $\mathbf{N} = \langle a, b, c \rangle$
is $a(x - x_0) + b(y - y_0) + c(z - z_0) = 0$. (Fig. 9)

Equation of the plane
$$a(x - x_0) + b(y - y_0) + c(z - z_0) = 0$$
Fig. 9

The Point–Normal form is the fundamental pattern for the equation of a plane, and other information can usually be translated so the Point–Normal form can be used. If we have point P and two vectors **U** and **V** in the plane, then we can use the result that the cross product of two vectors is perpendicular to each of them to find a vector perpendicular to the plane: **N = U x V**. Once we have **N**, we can use the Point–Normal form for the equation of the plane. If we have three points P, Q, and R, we can form the vectors **U** from P to Q and **V** from P to R, calculate the normal vector **N = U x V** and then use the Point–Normal form for the equation for the plane.

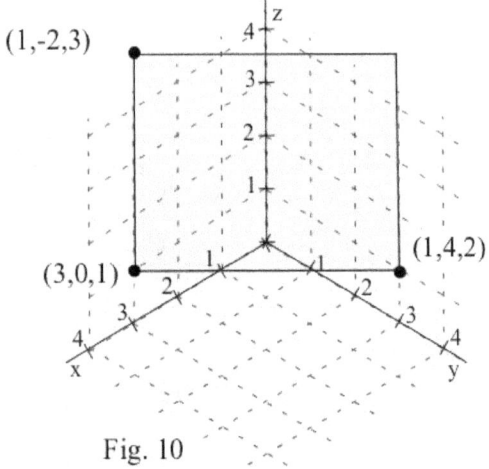

Fig. 10

Example 4:   Find the equation of the plane:

(a) through the point $(1, -2, 3)$ and with normal vector $\mathbf{N} = \langle 5, 3, -4 \rangle$,

(b) through the points $P = (1, -2, 3)$, $Q = (3, 0, 1)$ and $R = (1, 4, 2)$. (Fig. 10)

Solution: (a)   The point (x,y,z) is on the plane if and only if
$5(x-1) + 3(y+2) - 4(z-3) = 0$, or, equivalently,
$5x + 3y - 4z = -13$.

(b) Let **U** = vector from P to Q = $\langle 2, 2, -2 \rangle$ and **V** = vector from P to R = $\langle 0, 6, -1 \rangle$.

Then $\mathbf{N = U \times V} = \begin{vmatrix} \mathbf{i} & \mathbf{j} & \mathbf{k} \\ 2 & 2 & -2 \\ 0 & 6 & -1 \end{vmatrix} = \mathbf{i}\begin{vmatrix} 2 & -2 \\ 6 & -1 \end{vmatrix} - \mathbf{j}\begin{vmatrix} 2 & -2 \\ 0 & -1 \end{vmatrix} + \mathbf{k}\begin{vmatrix} 2 & 2 \\ 0 & 6 \end{vmatrix} = 10\mathbf{i} + 2\mathbf{j} + 12\mathbf{k}$.

The equation of the plane is $10(x-1) + 2(y+2) + 12(z-3) = 0$ or $10x + 2y + 12z = 42$ (or $5x+y+6z=21$).

Practice 5:   Find the equation of the plane:

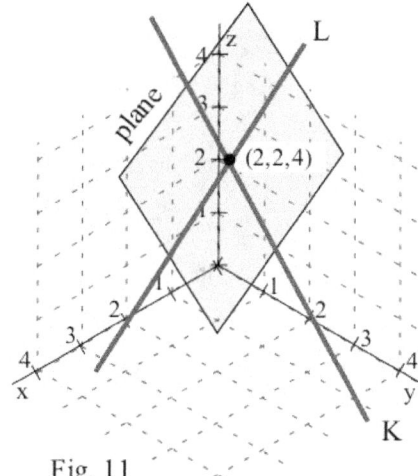

Fig. 11

    (a)   through the point $(4, 1, 0)$ and with normal vector

        $\mathbf{N} = \langle 3, -2, 6 \rangle$,

    (b)   determined by the lines   K: $x = 2 + t, y = 2 + 2t, z = 4 - t$ and

        L: $x = 2 - 3t, y = 2 + t, z = 4 + 4t$. (Fig. 11)

        (The lines intersect at $(2,2,4)$.)

Example 5:   (a)   Find a normal vector to the plane $3x - 2y + 4z = 12$.

    (b)   Where does the plane $3x - 2y + 4z = 12$ intersect each

        coordinate axis?

    (c)   Where does the line $x = 3 + 5t$, $y = -4 + 2t$, $z = 1 - 2t$ intersect the plane $3x - 2y + 4z = 12$?

Solution: (a)   We can get one normal vector to the plane simply by using the coefficients of the

        variables of the equation of the plane: $\mathbf{N} = \langle 3, -2, 4 \rangle$. Any other nonzero vector is normal

        to the plane if and only if it is a nonzero scalar multiple of $\mathbf{N}$.

    (b)   Every point on the x–axis has y=0 and z=0, so we can find where the plane intersects the x–

        axis by setting y and z equal to zero in the plane equation and solving for x:

        $3x - 2(0) + 4(0) = 12$ so $x = 4$. The plane intersects the x–axis at $(4,0,0)$. Similarly, the

        plane intersects the y–axis at $(0,-6,0)$ and the z–axis at $(0,0,3)$.

    (c)   Substitute the parametric patterns for x, y, and z from the line into the equation for the plane

        and then solve for t: $3(3 + 5t) - 2(-4 + 2t) + 4(1 - 2t) = 12$ so

            $9 + 15t + 8 - 4t + 4 - 8t = 12$

            $21 + 3t = 12$ and $t = -3$.

        Then $x = 3 + 5(-3) = -12$, $y = -10$, and $z = 7$. The point $(-12, -10, 7)$ is on the line and on

        the plane.

Practice 6:   (a)   Find a normal vector to the plane $3x + 10y - 4z = 30$.

    (b)   Where does the plane $3x + 10y - 4z = 30$ intersect each coordinate axis?

    (c)   Where does the line $x = 4 + t$, $y = -2 + 2t$, $z = 4 - t$ intersect the plane $3x + 10y + 4z = 30$?

Normal vectors also provide us with a way to determine the angle between a line and a plane and the angle

between two planes.

We can also find where a parametric line intersects a sphere by substituting the $x(t), y(t)$ and $z(t)$ equations

for x, y, and z in the equation of the sphere and then solving for the value of t. The line $x = 3t, y = 12 - 8t$,

$z = 5 + 7t$ intersects the sphere $x^2 + y^2 + z^2 = 13^2$ when t=0 and t=1: at the points $(0, 12, 5)$ and $(3, 4, 12)$.

Angles of Intersection: Line & Plane and Plane & Plane

The angle between a line and a plane (Fig. 12) is

$\pi/2$ – (angle between the line and the normal vector of the plane) .

The angle between two planes (Fig. 13) is the angle between the normal vectors of the planes.

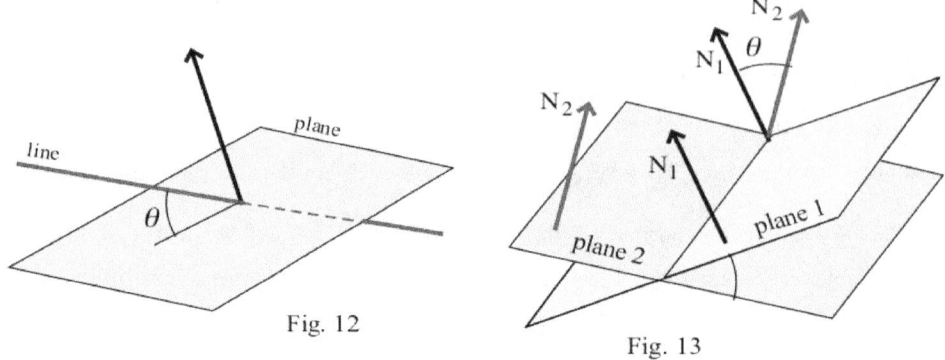

Fig. 12

Fig. 13

If the normal vectors of two planes have different directions, then the planes intersect in a line (the set of points common to both planes is a line), and we can determine parametric equations for that line. One method is to eliminate the y variable from the equations of the planes and write x in terms of z: x = x(z). Then we eliminate the x variable and write y in terms of z: y = y(z) . Finally, we treat the variable z as the parameter "t" in the parametric equations for the line, and we have x = x(t), y = y(t), z = t .

Example 6:   Let P be the plane $12x + 7y – 3z = 43$ and Q be the plane $4x + 7y + 13z = 19$.

(a) Find a parametric equation representation of the line of intersection of the two planes.

(b) Find the angle between the planes.

Solution: (a) This is basically an algebra problem to use the equations of the two planes to solve for two of the variables in terms of the third variable. Then we can treat that third variable as the parameter t and write the equation of the line of intersection in parametric form. We can eliminate y and solve for x in terms of z: (equation for P) – (equation for Q) is $8x – 16z = 24$ so $x = 3 + 2z$. Then we can eliminate x and solve for y in terms of z: (equation for P) – 3(equation for Q) is $–14y – 42z = –14$ so $y = 1 – 3z$. Treating the variable z as our parameter "t" we have the line **x = 3 + 2t, y = 1 – 3t, and z = t** .

As a check, when t = 0 then x = 3 , y = 1 , and z = 0 , and the point (3,1,0) lies on both planes. When t = 1 , then x = 5 , y = –2 , and z = 1 , and the point (5,–2,1) also lies on both planes.

(b) The angle between the planes is the angle between the normal vectors of the planes. $\mathbf{N}_P = \langle\, 12, 7, –3 \,\rangle$, $\mathbf{N}_Q = \langle\, 4, 7, 13 \,\rangle$ so

$$\cos(\theta) = \frac{\mathbf{N}_P \cdot \mathbf{N}_Q}{|\mathbf{N}_P||\mathbf{N}_Q|} = \frac{58}{\sqrt{202}\,\sqrt{234}} \approx 0.267 \text{ so } \theta \approx 1.30 \text{ (about } 74.5°) .$$

Practice 7:   Let R be the plane $2x + 3y - z = 13$ and S be the plane $2x - y + 3z = 1$.

(a) Find the line of intersection of the two planes. (b) Find the angle between the planes.

Sometimes the following alternate method is easier: find two points that lie on the intersection and then write the parametric equations for the line through those two points.

Example 7:   Let P be the plane $7x - y - 11z = 10$ and Q be the plane $9x + y - 5z = 22$.

Find a parametric equation representation of the line of intersection of the two planes.

Solution: Set one variable equal to 0, say z=0, so we then have $7x - y = 10$ and $9x + y = 22$. Adding these two equations together gives $16x = 32$ so x=2 and then y=4 so one point on the intersection of the planes is $A = (2, 4, 0)$. Setting x=0, we have $-y - 11z = 10$ and $y - 5z = 22$. Adding these together gives $-16z = 32$ so z=-2 and then y=12 so another point on the intersection is $B = (0, 12, -2)$. A parametric equation representation of the line through A and B is $x(t) = 2 - 2t, y(t) = 4 + 8t, z(t) = -2t$. (We could have put y=0 and then found the point $C = (3, 0, 1)$. Check that point C satisfies the equations we just found.)

Distance Algorithms

Distances between objects are needed in a variety of applications in three dimensions:

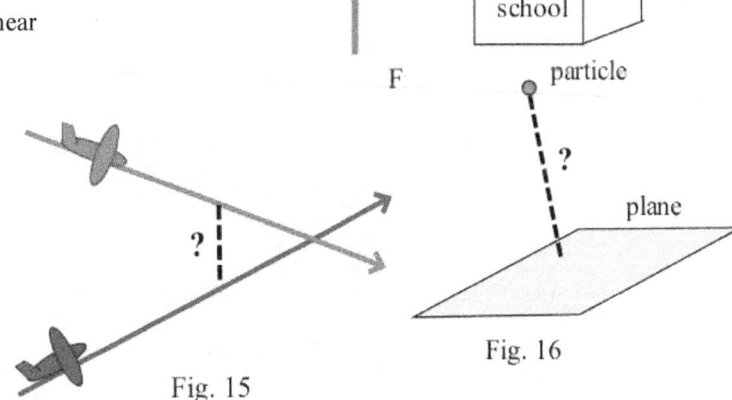

- How close does the power line come to the school (Fig. 14)?

- How close can airplanes on two different linear flight paths come to each other (Fig. 15)?

- How far is the charged particle from the planar plate (Fig. 16)?

Below we give a collection of distance algorithms, geometric constructions, and formulas for the distances between different types of objects. At first this may seem to be a large task, but most of the patterns use the idea of vector projection and follow from thinking about the geometry of the situations.

The distance from a point to a line in two dimensions was discussed in Section 11.3 and is presented again here because it uses the type of reasoning we need for three dimensions and because the resulting formula for two dimensions reappears for some distances in three dimensions.

Two dimensions: Distance from a **point** $P = (x_0, y_0)$ to a **line** $ax + by = c$ (Fig.17).

    (1) Find **N** perpendicular to the line: $\mathbf{N} = \langle a, b \rangle$ works .

    (2) Find a point Q on the line and form the vector **V** from P to Q .

    (3) Distance = | Projection of **V** onto **N** |

$$= \frac{|\mathbf{V \cdot N}|}{|\mathbf{N}|} = \frac{|ax_0 + by_0 - c|}{\sqrt{a^2 + b^2}} \quad .$$

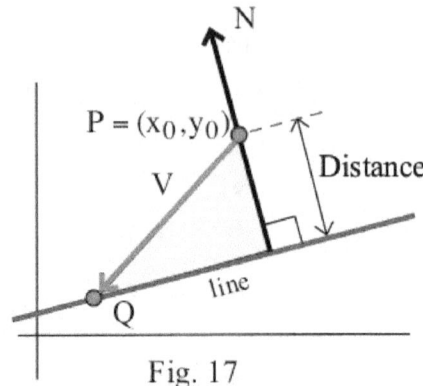

Fig. 17

Three dimensions: Distance from a **point** P to a **plane** with

    normal vector **N** (Fig. 18).

    (1) Find a point Q on the plane and form the vector **V**

        from P to Q .

    (2) Distance = | Projection of **V** onto **N** |

$$= \frac{|\mathbf{V \cdot N}|}{|\mathbf{N}|} \quad .$$

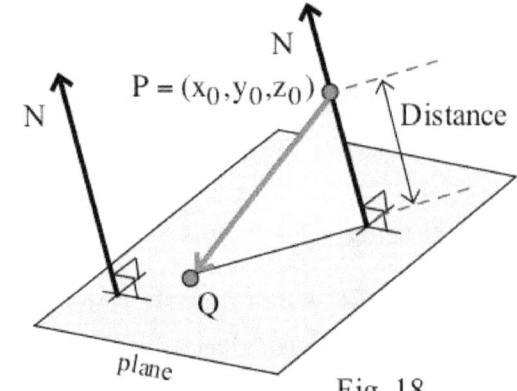

Fig. 18

Three dimensions: Distance from a **line** parallel to

U to a **line** parallel to **V** (**U** not parallel to **V**) (Fig. 19).

    (1) Find a point P on one line and a point Q on the other line .

    (2) Form the vector **W** from P to Q .

    (3) Calculate $\mathbf{N} = \mathbf{U} \times \mathbf{V}$.

    (4) Distance = | projection of **W** onto **N** | $= \frac{|\mathbf{W \cdot N}|}{|\mathbf{N}|}$.

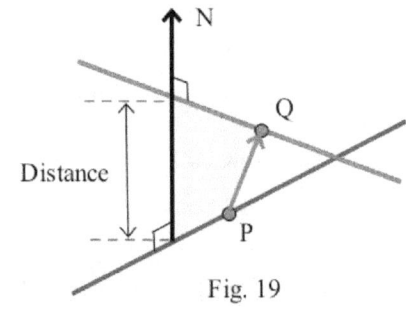

Fig. 19

    (If **U** and **V** are parallel, pick a point on one line and then use the "point to line" formula.

In each of the previous situations, the distances were the lengths of suitable projections and the distance
patterns involved dot products. In the next case, we also need the length of a projection, but then we use
the Pythagorean formula to solve for the distance we want. The simplified result looks similar to the
previous patterns, but has a cross product instead of a dot product.

Three dimensions: Distance from a **point** P to a **line** parallel to U (Fig. 20).

   (1) Find a point Q on the line and form the vector **V** from P to Q .

   (2) Put W = | projection of **V** onto U | = $\dfrac{|\,\mathbf{V}\bullet\mathbf{U}\,|}{|U|}$ = $\dfrac{|V|\,|U|\,|\cos(\theta)|}{|U|}$ = $|V|\,|\cos(\theta)|$.

   (3) Then (distance $)^2$ = $|V|^2 - |W|^2$

$$= |V|^2 - |V|^2 \cos^2(\theta)$$

$$= |V|^2 (1 - \cos^2(\theta))$$

$$= |V|^2 \sin^2(\theta)$$

$$= \frac{|V|^2\,|U|^2\,\sin^2(\theta)}{|U|^2} = \left(\frac{|\,\mathbf{V}\times\mathbf{U}\,|}{|\,U\,|}\right)^2 .$$

   (4) Distance = $\dfrac{|\,\mathbf{V}\times\mathbf{U}\,|}{|\,U\,|}$.

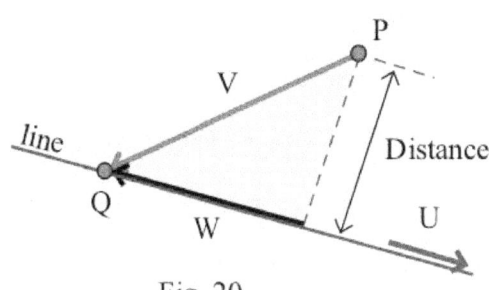

Fig. 20

Example 7:   A power line runs in a straight line from the point A = (200, 0, 100) to the point

                 B = (100, 700, 200). How close does the line come to the corner of a school at P = (300, 400, 0)?

Solution: This problem uses the pattern for the distance from a point to a line in three dimensions.

     P = (300, 400, 0) and **U** = the vector from A to B = $\langle -100, 700, 100 \rangle$ = $100\langle -1, 7, 1 \rangle$:

     $|U| = 100\sqrt{51}$.

     Since A is on the line, let **V** = vector from P to A = $\langle -100, -400, 100 \rangle$ = $100\langle -1, -4, 1 \rangle$.

$$\mathbf{V}\times\mathbf{U} = (100)(100)\begin{vmatrix} \mathbf{i} & \mathbf{j} & \mathbf{k} \\ -1 & -4 & 1 \\ -1 & 7 & 1 \end{vmatrix} = (100)^2(-11\mathbf{i} - 0\mathbf{j} - 11\mathbf{k}) .\ |\,\mathbf{V}\times\mathbf{U}| = 100^2\sqrt{242} .$$

    Finally, distance from power line to school = $\dfrac{|\,\mathbf{V}\times\mathbf{U}\,|}{|\,U\,|}$ = $\dfrac{100^2\sqrt{242}}{100\sqrt{51}}$ ≈ 217.8 feet.

PROBLEMS

In problems 1 – 8, find parametric equations for the lines.

1. The line through the point $(2, -3, 1)$ and parallel to the vector $\langle 3, 4, 2 \rangle$.

2. The line through the point $(0, 5, -2)$ and parallel to the vector $\langle -3, 1, -4 \rangle$.

3. The line through the point $(-2, 1, 4)$ and parallel to the vector $\langle 5, 0, -3 \rangle$.

4. The line through the origin and parallel to the vector $\langle 1, 2, 3 \rangle$.

5. The line through the points $(2, -1, 3)$ and $(3, 4, -2)$.

6. The line through the points $(7, 3, -4)$ and $(5, 0, -2)$.

7. The line through the points $(3, -2, 1)$ and $(3, 4, -1)$.

8.   The line through the origin and $(3, 4, -2)$.

In problems 9 – 12, the equations for a pair of lines are given. Determine whether the lines intersect, and if they do intersect, find the point and angle of intersection.

9.   Line L: $x = 2 + t$, $y = -1 + t$, $z = 3 + 2t$. Line K: $x = 2 - t$, $y = -1 + 2t$, $z = 3 + 4t$.

10.   Line L: $x = 6 + 2t$, $y = 3 - 2t$, $z = -2 + 3t$. Line K: $x = 6 + t$, $y = 3 + 5t$, $z = -2 - 3t$.

11.   Line L: $x = 1 + 3t$, $y = 5 - t$, $z = -2 + 2t$. Line K: $x = 9 + 4t$, $y = 5$, $z = 4 + 3t$.

12.   Line L: $x = 1 + 2t$, $y = 1 + 3t$, $z = 3 + 5t$. Line K: $x = 2 + t$, $y = 3 + t$, $z = 1 + t$.

In problems 13 – 26, find equations for the planes.

13.   The plane through the point $(2, 3, 1)$ and perpendicular to the vector $\langle 5, -2, 4 \rangle$.

14.   The plane through the point $(4, 0, -2)$ and perpendicular to the vector $\langle 3, 1, -5 \rangle$.

15.   The plane through the point $(-3, 5, 6)$ and perpendicular to the vector $\langle 0, 3, 0 \rangle$.

16.   The plane through the origin and perpendicular to the vector $\langle 2, -2, 1 \rangle$.

17.   The plane through the points $(1, 2, 3), (5, 2, 1)$, and $(4, -1, 3)$.

18.   The plane through the points $(3, 5, 3), (-2, 5, 4)$, and $(1, 5, 6)$.

19.   The plane through the points $(-4, 2, 5), (1, 2, 1)$, and $(3, -3, 5)$.

20.   The plane through the origin, $(1, 2, 3)$, and $(4, 5, 6)$.

21.   The plane through the point $(2, -5, 7)$ and parallel to the xy–plane.

22.   The plane through the point $(2, 4, 1)$ and parallel to the yz–plane.

23.   The plane through the point $(4, 2, 3)$ and parallel to the plane $3x - 2y + 5z = 15$.

24.   The plane through the origin and parallel to the plane $2x + 3y - z = 12$.

25.   The plane through the point $(4, 1, 3)$ and perpendicular to the line $x = 2 + 5t$, $y = 1 - 3t$, $z = 2t$.

26.   The plane through the point $(2, 7, 4)$ and perpendicular to the y–axis.

Try to answer problems 27 – 32 without doing any algebra — just think visually.

27.   Where does the plane $x = 0$ intersect the plane $z = 0$?

28.   Where does the plane $y = 2$ intersect the plane $z = 3$?

29.   Where does the plane $3x + 2y + z = 30$ intersect the x–axis?

30.   Where does the plane $3x + 2y + z = 30$ intersect the y–axis?

31.   Where do the three planes $x = 4, y = 2$, and $z = 1$ intersect?

32.   Where does the y–axis intersect the plane $z = 3$?

In problems 33 – 36, two planes are given. Represent their line of intersection using parametric equations and find the angle between the planes.

33. $4x - 2y + 2z = 10$ and $3x - 2y + 3z = 36$.   34. $x + y + 3z = 9$ and $2x + y - 3z = 18$.

35. $5y - z = 10$ and $x + 2y + 4z = 16$.   36. $x = 4$ and $3x - 5y + z = 20$.

In problems 37 – 40, find where the given line intersects the plane and find the angle the line makes with the plane.

37. Line L: $x = 7 + 2t, y = 5 + t, z = 2 + 4t$ and the plane $5x + y - 2z = 20$.

38. Line L: $x = -2 + t, y = 12 - 3t, z = -5 - t$ and the plane $5x + y - 2z = 20$.

39. Line L: $x = 4 + 2t, y = 2 - 3t, z = 7 + t$, and the plane $z = 5$.

40. The x–axis and the plane $4x - 2y + 5z = 12$.

41. Is it possible for three distinct planes to intersect at a single point?

42. Is it possible for three distinct planes to intersect along a line?

43. A bird is flying in a straight line from the feeder (Fig. 21) to the corner of the house. Write an equation for its line of flight.

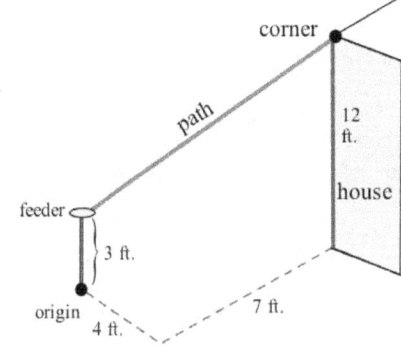

Fig. 21

44. What angle does the plane $3x + 2y + z = 10$ make with

    (a) the xy–plane? (b) the xz–plane?   (c) the yz–plane?

45. What angle does the plane $ax + by + cz = d$ make with each coordinate plane?

46. The four corners of a mirror are located at A(2,2,0), B(2,6,0), C(0,2,3), and D(0,6,3) (Fig. 22). A laser at L(5,0,0) directs a beam of light along the line $x = 5 - t, y = t, z = 0.5t$.

    (a) Write an equation for the plane of the mirror.

    (b) Where does the light beam hit the mirror?

    (c) What is the angle of incidence of the light with the mirror?

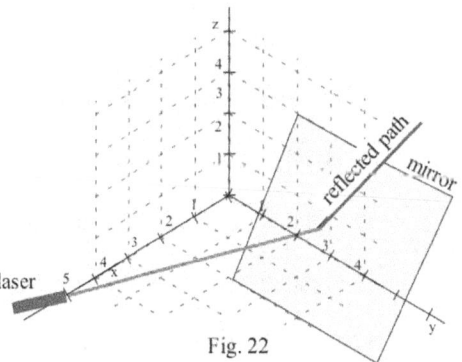

Fig. 22

Many of the following applications can be solved in several ways including the methods of vectors and dot products from section 11.4.

47. When it was built, the Great Pyramid at Giza, Egypt was 481 feet tall and had a square base with length 756 feet on each side.

    (a) Find the angle each side of the pyramid makes with the base.

    (b) Find the angle each side makes with an adjacent side.

    (c) Find the angle each edge makes with the base.

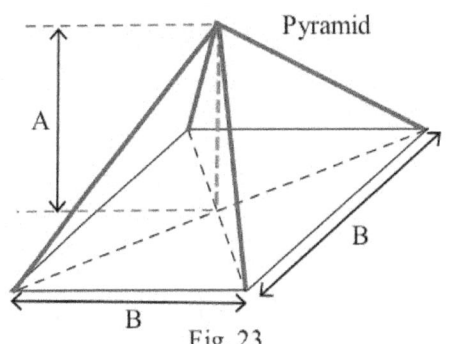

Fig. 23

48. In order to write a program to analyze a general pyramid (Fig. 23) with square base of length B and height A you need to determine the following:

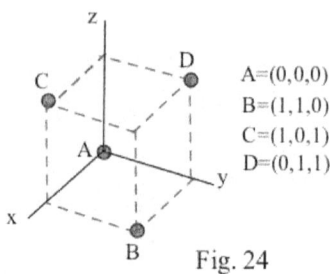

A=(0,0,0)
B=(1,1,0)
C=(1,0,1)
D=(0,1,1)

Fig. 24

   (a) the angle each side makes with the base,

   (b) the area of each side,

   (c) the angle each edge makes with the base,

   (d) the angle each side makes with an adjacent side, and

   (e) the volume of the pyramid.

49. Four molecules are located at the corners of a cube as shown in Fig. 24. The center of the cube is the point O = (0.5, 0.5, 0.5) Show that the angles AOB, AOC, and BOC are equal.

50. The four points (1,0,0), (–0.5,0.866,0), (–0.5,–0.866,0), and (0,0,1.414) are the vertices of an equilateral tetrahedron with center O = (0, 0, 0.354) (Fig. 25), and a molecule is located at each vertex.

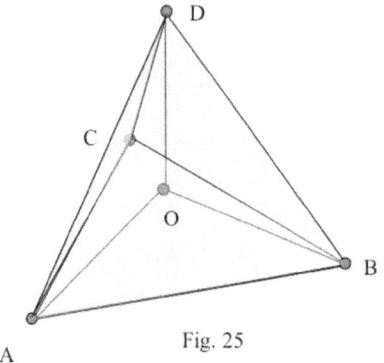

Fig. 25

   (a) Show that the tetrahedron is really equilateral.

   (b) Show that the angles AOB, AOC, and BOC are equal.

   (c) Find the angle between the plane determined by AOB and the plane determined by AOC.

Distances in Three Dimensions

51. Find the distance from the point (4, –1, 2) to the line x = 2 + t, y = 3 + 2t, z = 4t.

52. Find the distance from the point (4, –1, 2) to the line x = 1 – 3t, y = 2 + t, z = 3.

53. Find the distance from the point (2, 3, 1) to the plane 4x + 3y – z = 10.

54. Find the distance from the point (4, –3, 0) to the plane –2x + 5y + 3z = 15.

55. Find the distance (shortest distance) between the line x = 1 + 3t, y = 2 + 5t, z = 1 – t and the line x = 3 – 2t, y = 5 + t, z = 2 + 2t.

56. Find the distance (shortest distance) between the line x = 5 + 2t, y = 2 + t, z = 2 – 3t and the line x = 3 – 2t, y = 1 – 3t, z = 1.

In problems 57 – 62, the parameterized straight–line paths of two objects are given.

(a) Do the objects "crash" (so they are at the same location at the same time)? If so, at what time?

(b) Do the paths of the objects intersect (so the objects are at the same point but at different times)? If so, how close do the objects get to each other?

(c) Do the objects and their paths miss each other? If so, how close do the objects get to each other and how close do their paths get to each other?

57. Object A is at $x = 9 + t, y = 18 + t, z = 25 - 2t$ and object B is at $x = 3 + 2t, y = 30 - t$, and $z = 7 + t$.

58. Object A is at $x = -1 + t, y = -3 + 2t, z = 1 + t$ and object B is at $x = 2, y = t$, and $z = -2 + 2t$.

59. Object A is at $x = 5 - 5t, y = t, z = 5t$ and object B is at $x = 6 - 3t, y = 5 - 2t$, and $z = -3 + 4t$.

60. Object A is at $x = -4 + 5t, y = 3 + 2t, z = 16 - 3t$ and object B is at $x = 12 - t, y = 6 + 3t$, and $z = 3 + 4t$.

61. Object A is at $x = 5 - 2t, y = 0, z = 1$ and object B is at $x = 0, y = -1 + t$, and $z = 0$.

62. Object A is at $x = 1 + 3t, y = 2 + 2t, z = 3 + t$ and object B is at $x = 7 - t, y = 5 + 2t$, and $z = 3 + 3t$.

63. The bearing of an airplane is due north, and it passes directly above the origin at an altitude of 30,000 feet. How close does the airplane come to a balloon 5,000 feet directly above the point $x = 2,000$ and $y = 4,000$? (Assume the earth is "almost flat" in this region.)

64. At time t, airplane A is at $(-3 + t, 0, 1)$ and car B is at $(0, -5 + 2t, 0)$.

(a) How close do they come to each other?

(b) How close do their paths come to each other?

65. At time t, car A is at $(-3 + t, 2 + 2t, 0)$ and airplane B is at $(t, -5 + 2t, t)$.

(a) How close do they come to each other?

(b) How close do their come to each other?

66. Create your own problem, like problems 59 – 64, so the objects on different paths crash at the point $(3, 4, 5)$ at $t = 2$.

67. Create your own problem, like problems 59 – 64, with two objects on different paths so object A goes through the point $(3, 4, 5)$ at $t = 2$ and object B goes through the point $(3, 4, 5)$ at $t = 3$. (Then the paths intersect, but the objects do not crash.)

Practice Answers

Practice 1:   (a) $x(t) = 3 + 2t, y(t) = -1 - 4t$.

(b) $x(t) = 3 + t, y(t) = -1 + 5t$.

The graphs are shown in Fig. 26.

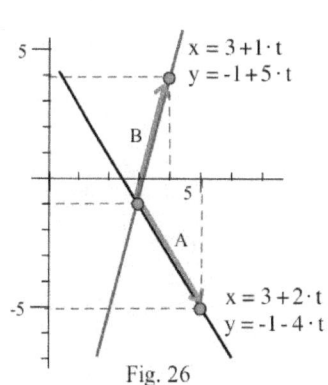

Fig. 26

Practice 2:   (a) $x(t) = 2 + 3t, y(t) = -1 - 4t, z(t) = 0 + t$.

(b) The line is parallel to the vector $\mathbf{B} = \langle -1, 5, 2 \rangle$ from P to Q. Using P as the starting point, a parametric equation of the line is $x(t) = 3 - 1t$,

$y(t) = 0 + 5t, z(t) = 2 + 2t$. (Using Q as the starting point, $x(t) = 2 - t$, $y(t) = 5 + 5t$,

$z(t) = 4 + 2t$.)

Practice 3:   Replacing the t parameter with s for line L and then setting the components equal to

each other, we get: x: $1 + t = 8 + 4s$, y: $1 - 2t = -4 + s$, and z: $-3 + 2t = -5 - 8s$.

Solving the first two equations for t and s, $t = 3$ and $s = -1$. These values for t and s also

satisfy the third equation so $x = 1 + (3) = 4, y = 1 - 2(3) = -5$, and $z = -3 + 2(3) = 3$.

The point $(4, -5, 3)$ lies on both lines.

The lines are parallel to $\mathbf{A} = \langle 1, -2, 2 \rangle$ and $\mathbf{B} = \langle 4, 1, -8 \rangle$ so

$$\cos(\theta) = \frac{\mathbf{A} \cdot \mathbf{B}}{|\mathbf{A}||\mathbf{B}|} = \frac{-14}{(3)(9)} \quad \text{and} \quad \theta \approx 2.12 \quad (\text{about } 121.2°).$$

(b) Replacing the t parameter with s for line L and then setting the components equal to each

other, we get: x: $1 + t = 8 + 4s$, y: $1 - 2t = -4 + s$, and z: $2 + 2t = 3 + s$.

Solving the first two equations (they are the same as in part (a)) for t and s, $t = 3$ and $s = -1$. But

these values do not satisfy the third equation, $2 + 2(3) \neq 3 + (-1)$, so the lines do not intersect.

(c) Every point in the yz–plane has x coordinate 0, so set $8 + 4t = 0$ to get $t = -2$. Then

$y = -4 + (-2) = -6$ and $z = 3 + (-2) = 1$ so the line intersects the yz–plane at $(0, -6, 1)$.

Practice 4:   The parametric equations for the line of travel of the arrow are $x = 1 + 4t, y = 2 + 5t$,

$z = 3 + t$. The arrow reaches the wall when $y = 20$ so $2 + 5t = 20$ and $t = 18/5$. At that time,

$t = 18/5$, the height of the arrow is $z = 3 + t = 3 + (18/5) = 33/5 = 6.6$ feet so the arrow does not

go over the wall.

Practice 5:   (a) The point (x,y,z) is on the plane if and only if $3(x-4) - 2(y-1) + 6(z-0) = 0$, or,

equivalently, $3x - 2y + 6z = 10$.

(b) Taking the directions of the lines K and L from their parametric equations, we know line

K is parallel to $\mathbf{U} = \langle 1, 2, -1 \rangle$ and line L is parallel to $\mathbf{V} = \langle -3, 1, 4 \rangle$. Then

$$\mathbf{N} = \mathbf{U} \times \mathbf{V} = \begin{vmatrix} \mathbf{i} & \mathbf{j} & \mathbf{k} \\ 1 & 2 & -1 \\ -3 & 1 & 4 \end{vmatrix} = \mathbf{i} \begin{vmatrix} 2 & -1 \\ 1 & 4 \end{vmatrix} - \mathbf{j} \begin{vmatrix} 1 & -1 \\ -3 & 4 \end{vmatrix} + \mathbf{k} \begin{vmatrix} 1 & 2 \\ -3 & 1 \end{vmatrix} = 9\mathbf{i} - 1\mathbf{j} + 7\mathbf{k} .$$

The point $(2,2,4)$ is on both lines so it is on the plane. Using the point $(2,2,4)$ and the

normal vector $\mathbf{N} = \langle 9, -1, 7 \rangle$, the equation of the plane is

$9(x-2) - 1(y-2) + 7(z-4) = 0$ or $9x - y + 7z = 44$.

Practice 6:   (a) $\mathbf{N} = \langle\, 3, 10, -4 \,\rangle$ and all nonzero scalar multiples of $\mathbf{N}$ are normal to the plane.

(b)   The plane crosses the x–axis at $(10,0,0)$, the y–axis at $(0,3,0)$, and the z–axis at $(0,0,-7.5)$.

(c)   Solving $3(4 + t) + 10(-2 + 2t) - 4(4 - t) = 30$ for t, we get

$12 + 3t - 20 + 20t - 16 + 4t = 30$ so $-24 + 27t = 30$ and $t = 2$. Then $x = 6, y = 2$,

and $z = 2$. The point $(6,2,2)$ is on the line and on the plane.

Practice 7:   (a) We can eliminate y and solve for x in terms of z: (equation for R) + 3(equation for S) is

$8x + 8z = 16$ so $x = 2 - z$. Then we can eliminate x and solve for y in terms of z:

(equation for R) – (equation for S) is $4y - 4z = 12$ so $y = 3 + z$. Treating the variable z as our

parameter "t" we have the line $\mathbf{x = 2 - t, y = 3 + t, and\ z = t}$.

As a check, when $t = 0$ then $x = 2, y = 3$, and $z = 0$, and the point $(2,3,0)$ lies on both planes.

When $t = 1$, then $x = 1, y = 4$, and $z = 1$, and the point $(1,4,1)$ also lies on both planes.

(b) The angle between the planes is the angle between the normal vectors of the planes.

$\mathbf{N_R} = \langle\, 2, 3, -1 \,\rangle$, $\mathbf{N_S} = \langle\, 2, -1, 3 \,\rangle$ so

$$\cos(\theta) = \frac{\mathbf{N_R \cdot N_S}}{|\mathbf{N_R}||\mathbf{N_S}|} \ = \ \frac{-2}{\sqrt{14}\sqrt{14}} \ = \frac{-1}{7} \ \approx -0.143 \ \text{ so } \ \theta \approx 1.71 \ \ \text{(about } 98.2°) \ .$$

Appendix: Sketching Planes and Conics in the XYZ Coordinate System

Some mathematicians draw horrible sketches of 3–dimensional objects and they still lead productive, happy lives. Some mathematics students are terrible 3D artists and still get A grades. But it really takes very little time and practice to do decent sketches of planes and conic sections in 3D, and it can be satisfying to have the curve you know is an ellipse actually look like an ellipse. This Appendix illustrates some step–by–step ways to draw the basic building blocks for many 3D shapes: the axis system, planes and conic sections.

First Step: Invest in a pencil and a small, inexpensive, clear plastic $30-60-90^o$ triangle with a centimeter scale (Fig. 1). They make it much easier to create good 3D sketches. Many of the directions in this Appendix suggest drawing "help lines" to assist you in putting the various objects in the

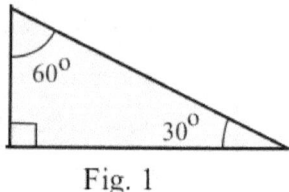

Fig. 1

appropriate positions and orientations. The final drawings often look better if these "help lines" are removed when the sketch is finished, so a pencil is better than a pen. Many of the directions require you to draw lines parallel to the axes and to draw parallelograms, and a clear plastic straightedge (or a triangle) is very useful for this. Finally, the centimeter scale produces small sketches appropriate for personal work.

The following directions are for "by hand" sketches, but they can also help you when using CAD (computer–aided design) and Draw computer programs.

A . The XYZ axis system — everything starts here!

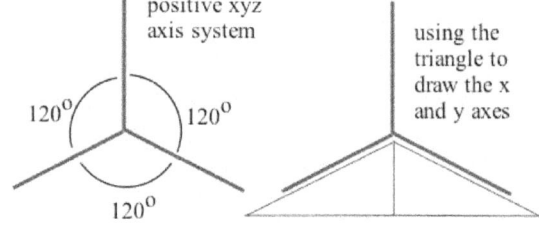

Fig. 2

Including an axis system in a 3D sketch gives the viewer an orientation and a perspective for the location of an object.

(1) Draw a vertical line segment.

(2) Sketch 2 other line segments making 120° angles with the vertical segment. The 30–60–90° triangle is useful here (Fig. 2).

(3) If any of the variables take negative values, you can extend the appropriate axis to the negative values using dashed lines.

(4) If a scale is needed, put "tick" marks along the axes at uniform intervals. (Fig. 3)

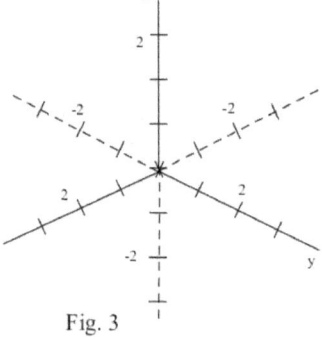

Fig. 3

Since the xyz–coordinate system is needed in so many sketches, you should eventually get used to sketching it without the aid of a ruler. ("Trick:" To get the y-axis, start at the origin, go right 2 units and down one unit, plot a point, and draw the line through this point and the origin. This line is the y-axis. The x-axis is similar: go left two and down one.)

B . The XY, XZ, and YZ coordinate planes. (Actually, rectangular pieces of the coordinate planes.)

The following rectangular pieces of the coordinate planes are useful by themselves, and they are very important aids for drawing curves and conics later.

Keys: Start with an xyz–axis system. Use lines parallel to each axis. The result should be a parallelogram.

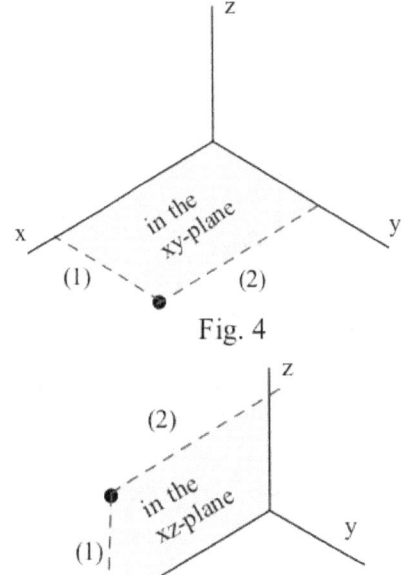

Piece of the XY coordinate plane: (Fig. 4)

(1) Pick a point on the x–axis and draw a line segment to the
    left of and parallel to the y–axis.

(2) From the end of the segment in step (1), draw a line to
    the y–axis that is parallel to the x–axis.

Fig. 4

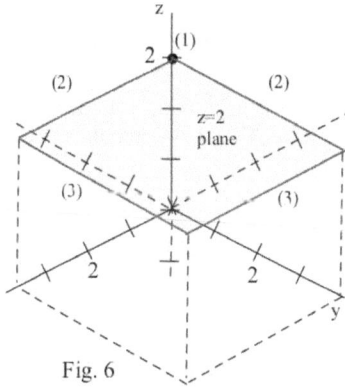

Piece of the XZ coordinate plane: (Fig. 5)

(1) Pick a point on the x–axis and draw a vertical line segment (up and parallel to the z–axis).

(2) From the end of the segment in step (1), draw a line to
    the z–axis that is parallel to the x–axis.

Fig. 5

Plane (piece) parallel to the xy–plane: (Fig. 6)

(1) Locate and label the appropriate point on the z–axis, for example, z = 2.

(2) From the point in step (1), draw line segments parallel to the
    x–axis and the y–axis.

(3) From the ends of the segments in step (2), draw additional lines parallel
    to the y–axis and the x–axis to complete the parallelogram.

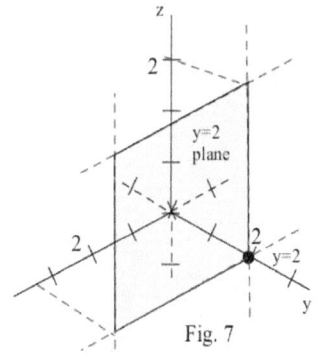

Fig. 6

Plane (piece) parallel to the xz–plane: (Fig. 7)

(1) Locate and label the appropriate point on the y–axis, for example, y = 2.

(2) From the point in step (1), draw line segments parallel to the x–axis
    and the z–axis.

(3) From the ends of the segments in step (2), draw additional lines
    parallel to the z–axis and the x–axis to complete the parallelogram.

Fig. 7

Practice 1: Sketch the planes z = 1 and x = 2 in Fig. 8.

Practice 2: Sketch the planes z = –1 and y = 1 on an XYZ system.

Practice 3: Sketch the rectangular box with opposite corners at

(0, 0, 0) and (4, 3, 2).

Challenge 1: Sketch the plane that contains the points

(0, 0, 3), (4, 0, 3), and (0, 2, 3).

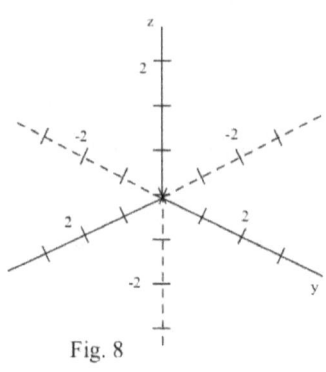

Fig. 8

C . Ellipses

Keys: Plot the vertices and make a parallelogram "frame" for the ellipse. Sketch short tangent segments at the midpoints of the sides of the parallelogram.

Example 1: An ellipse in the xy–plane with vertices at $(3,0,\mathbf{0})$, $(–3,0,\mathbf{0})$, $(0,\mathbf{2},\mathbf{0})$, and $(0,–\mathbf{2},\mathbf{0})$. (Fig. 9)
(1) Plot the vertices in the xy–plane (z=0).
(2) Sketch a parallelogram "frame" by drawing lines through (3,0) and (–3,0) that are parallel to the y–axis, and lines through (0,2) and (0,–2) that are parallel to the x–axis.
(3) At each vertex sketch a short "tangent segment" that lies on the side of the parallelogram frame.
(4) Finish the sketch by smoothly connecting the "tangent segments."

If your sketch requires several conics, it is useful to erase all or most of the parallelogram frame.

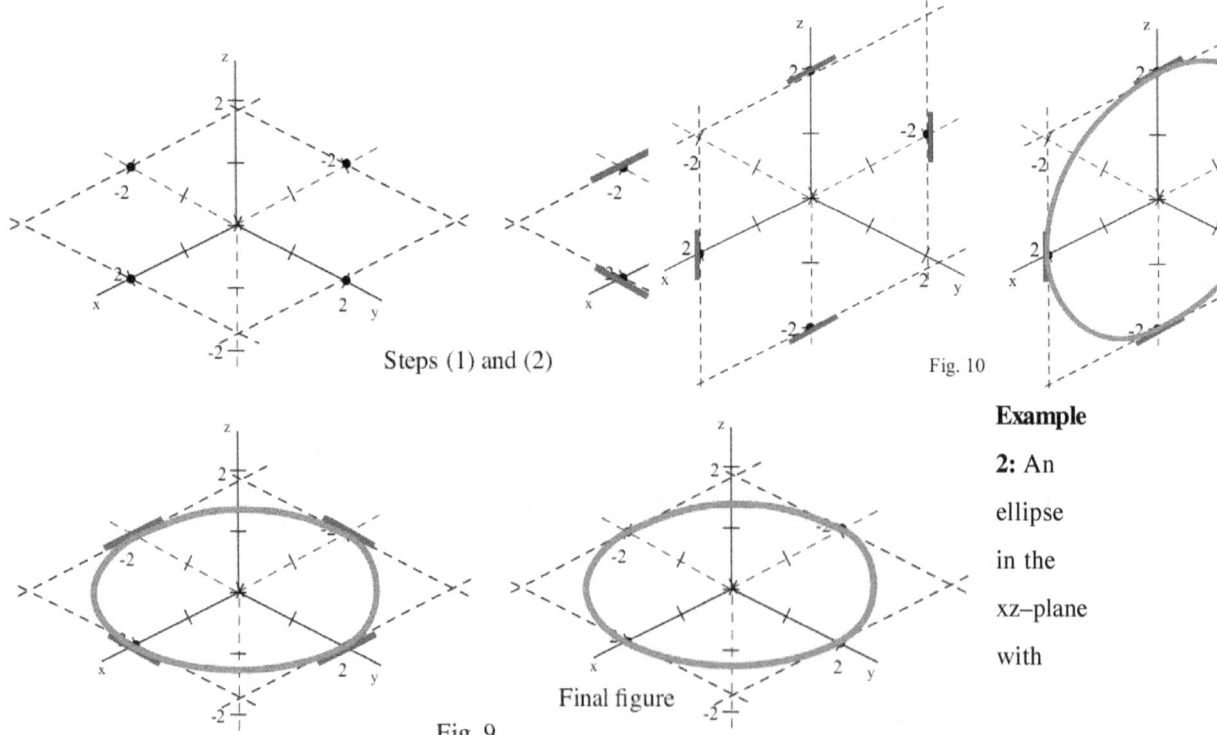

Steps (1) and (2)

Fig. 10

Example 2: An ellipse in the xz–plane with

Final figure

Fig. 9

vertices at (3,**0**,0),

(–3,**0**,0), (0,**0**,2), and (0,**0**,–2).

Fig. 10 shows the intermediate steps to construct this ellipse as well as the final result.

Practice 4: Sketch the ellipse with vertices at (1, **0**, 0), (0, **0**, 0), (0, **0**, 3), and (0, **0**, –3) on the XYZ

system in Fig. 11.

Practice 5: Sketch the ellipse with vertices at (3, 0, 2), (–3, 0, 2),

(0, 1, 2), and (0, –1, 2) .

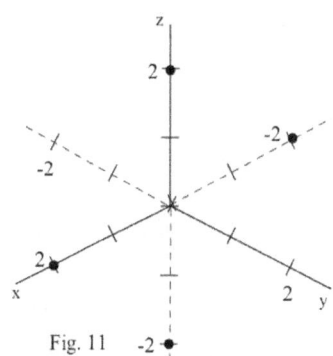

Practice 6: Sketch the elliptical cylinder $\dfrac{y^2}{9} + z^2 = 1$.

Challenge 2: Sketch the ellipse with vertices at (4,2,0), (2,0,3),

(0,2,6), and (2,4,3).

D . Parabolas

Fig. 11

Keys: Plot the vertex and one other point on the parabola. Make a parallelogram "frame" for the

parabola and plot a symmetric point. Sketch short "tangent segments" at the plotted points.

Example 3: Sketch the parabola lying in the xy-plane that satisfies $y = x^2 - 4x + 6$. (Fig. 12)

(1) Plot the vertex (2,1,0). When x = 1, then y = (1)2 – 4(1) + 6 = 3. Plot (1,3,0).

(2) Sketch a parallelogram "frame" by drawing lines through (2,1,0) parallel to the x–axis and the

  y–axis, and through (1,3,0) parallel to the x–axis and the y–axis.

(3) Plot the "symmetric point" (3,3,0) and draw a line through it parallel to the x–axis and the y–axis.

(4) Sketch a short "tangent segment" at the vertex (2,1,0) (this "tangent segment" lies on the side

  of the parallelogram). Add "tangent segments" at the other two plotted points (1,3,0) and (3,3,0)

  (these "tangent segments are **not** parallel to the sides of the parallelogram).

(5) Finish the sketch by smoothly connecting the "tangent segments."

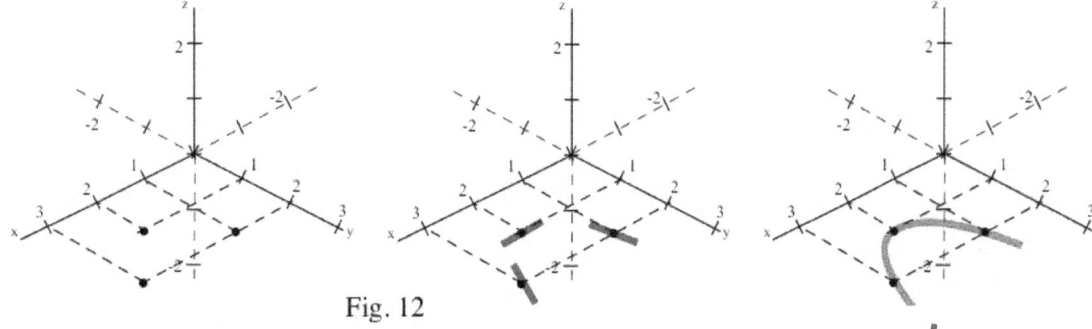

Fig. 12

Example 4: Sketch the parabola lying in the xz-plane that

satisfies $z = x^2 + 1$.

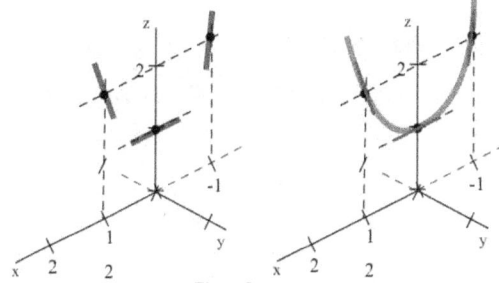

Fig. 13

Fig. 13 shows a beginning step and the final result.

Practice 7: Sketch the parabola with vertex at (0,2,0) that
contains the point (0,1,2) on the XYZ system in Fig. 14.

Practice 8: Sketch the parabola with vertex at (2,0,2) that contains
the point (0,0,0) .

Practice 9: Sketch the parabola $x = (y-1)^2$ on the plane z = 1.

Challenge 3: Sketch the parabola with vertex (4,2,0) that contains
the symmetric points (0,1,3) and (0,3,3).

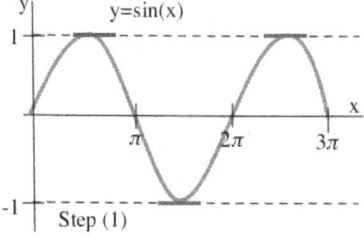

Fig. 14

E . General curves in coordinate planes

Keys: Start with a regular rectangular coordinate graph of the function and a few "tangent segments." Plot
a domain–range parallelogram in 3D along with the points and the "tangent segments."

Example 5: Sketch the curve y = sin(x), 0 ≤ x ≤ 3π, in the xy–plane. (Fig. 15)

(1) Graph y = sin(x), 0 ≤ x ≤ 3π, in the 2D rectangular coordinate system,

   label a few points and add "tangent segments" at those points.

(2) Sketch the domain–range parallelogram, 0 ≤ x ≤ 3π and –1 ≤ y ≤ 1.

(3) Sketch the points and "tangent segments" in the parallelogram in step (2).

(4) Finish the sketch by smoothly connecting the "tangent segments."

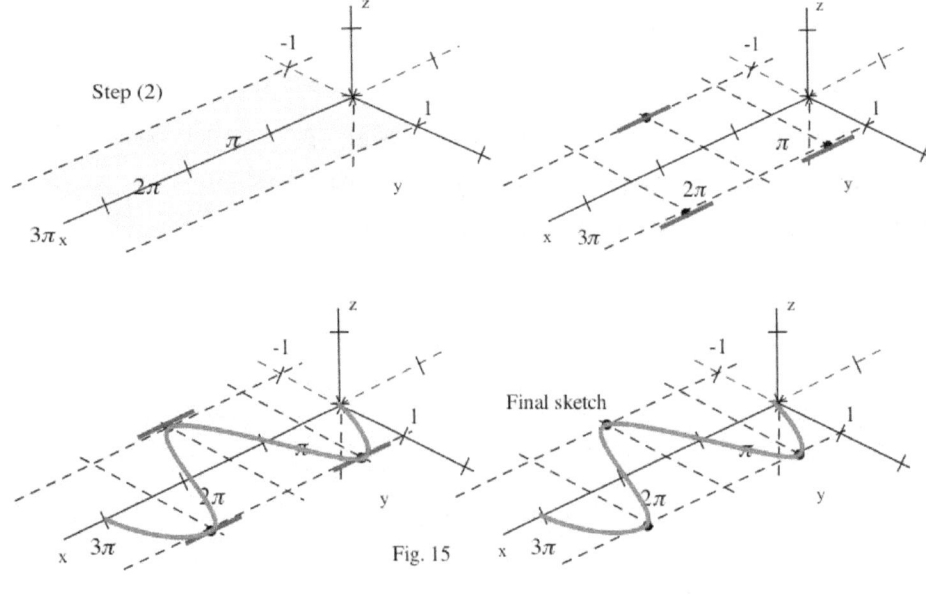

Fig. 15

Practice 10: On the XYZ system in Fig. 16 sketch a curve

(a) in the xz-plane (y=0) that satisfies z = 1 + cos(x), –π ≤ x ≤ 3π,

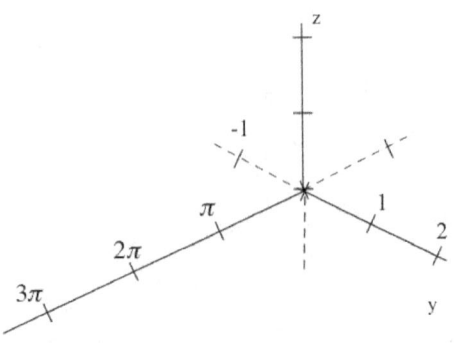

(b) in the y = 2 plane that satisfies z = 1 + cos(x), $-\pi \leq x \leq 3\pi$,.

Practice 11: On an XYZ system sketch the curve (a) in the

yz-plane (x=0) that satisfies z = y for $0 \leq y \leq 4$, (b) in the x = 2 plane that satisfies z = y for $0 \leq y \leq 4$.

Practice 12: Sketch the curve $y = 4 - x^2$, $-2 \leq x \leq 3$ (a) in the xy–plane and (b) in the z = 3 plane.

Challenge 4: Sketch the graph of z = | x | + | y | on the plane {y = –x, no restrictions on z} .

Final Note

Sketching 3D curves on a 2D piece of paper presents challenges for most of us, but with a bit of practice (and a systematic approach) most of us can learn to do decent sketches of a some simple 3D curves. It certainly helps if the curve lies in a plane.

11.1   Selected Answers

5.   (b) $|\mathbf{U}| = \sqrt{17}$, $|\mathbf{V}| = \sqrt{13}$

direction of $\mathbf{U}$ is $\langle 1/\sqrt{17}$, $4/\sqrt{17} \rangle$, direction of $\mathbf{V}$ is $\langle 3/\sqrt{13}$, $2/\sqrt{13} \rangle$

(c)   slope of $\mathbf{U}$ is $4/1 = 4$, slope of $\mathbf{V}$ is $2/3$

angle of $\mathbf{U}$ with x-axis is $\theta = \arctan(4) = 1.326 \ (\approx 76^\circ)$

angle of $\mathbf{V}$ with x-axis is $\theta = \arctan(2/3) = 0.588 \ (\approx 33.7^\circ)$

7.   (b) $|\mathbf{U}| = \sqrt{29}$, $|\mathbf{V}| = \sqrt{58}$

direction of $\mathbf{U}$ is $\langle -2/\sqrt{29}$, $5/\sqrt{29} \rangle$, direction of $\mathbf{V}$ is $\langle 3/\sqrt{58}$, $-7/\sqrt{58} \rangle$

(c)   slope of $\mathbf{U}$ is $-5/2$, slope of $\mathbf{V}$ is $-7/3$

angle of $\mathbf{U}$ with x-axis is $\theta = \arctan(-5/2) = -1.190 \ (\approx -68.8^\circ)$

angle of $\mathbf{V}$ with x-axis is $\theta = \arctan(-7/3) = -1.166 \ (\approx -66.8^\circ)$

9.   (b) $|\mathbf{U}| = \sqrt{25} = 5$, $|\mathbf{V}| = \sqrt{25} = 5$

direction of $\mathbf{U}$ is $\langle -4/5$, $-3/5 \rangle$, direction of $\mathbf{V}$ is $\langle 3/5$, $-4/5 \rangle$

(c)   slope of $\mathbf{U}$ is $3/4$, slope of $\mathbf{V}$ is $-4/3$

angle of $\mathbf{U}$ with x-axis is $\theta = \arctan(3/4) = 0.643 \ (\approx 36.9^\circ)$

angle of $\mathbf{V}$ with x-axis is $\theta = \arctan(-4/) = -0.927 \ (\approx -53.1^\circ)$

13. $\mathbf{U} = \mathbf{A} + \mathbf{B} - \mathbf{C} = \langle -1, 3 \rangle$, $\mathbf{V} = \mathbf{A} - \mathbf{B} + \mathbf{C} = \langle 3, 5 \rangle$

15. $\mathbf{V} = 3\langle 0.6, 0.8 \rangle = \langle 1.8, 2.4 \rangle$ 17. $\mathbf{V} = \langle 4.10, 2.87 \rangle$

19. $\mathbf{V} = \langle 7/\sqrt{10}, 21/\sqrt{10} \rangle$ or $\langle -7/\sqrt{10}, -21/\sqrt{10} \rangle$ 21. $\mathbf{V} = \langle 1/\sqrt{26}, 5/\sqrt{26} \rangle$ or $\langle -1/\sqrt{26}, -5/\sqrt{26} \rangle$

23. $\mathbf{V} = \langle 1, 0 \rangle$ or $\langle -1, 0 \rangle$ 25. $\mathbf{V} = \langle 1, 0 \rangle$ or $\langle -1, 0 \rangle$

31.   shadow on x-axis $1\mathbf{i} + 0\mathbf{j}$, on y-axis is $0\mathbf{i} + 4\mathbf{j}$

33.   shadow on x-axis $5\mathbf{i} + 0\mathbf{j}$, on y-axis is $0\mathbf{i} - 2\mathbf{j}$

37. $\mathbf{C} = \langle -4, -6 \rangle$ 38. $\mathbf{C} = \langle 3, 7 \rangle$

39. $\langle 25.36, 0 \rangle$ and $\langle 0, 54.37 \rangle$ 40. $\langle 96.59, 0 \rangle$ and $\langle 0, 25.88 \rangle$

41.   magnitude = 119.5 pounds, angle $\approx 39.2^o$ 42.   magnitude = 139 pounds, angle $\approx 21.4^o$

43.   magnitude = 268.45 pounds, angle $\approx 23.9^o$ 44.   (a) path = $\langle 230, 40 \rangle$ (b) aim 11.5^o south of east

46.   (a) You are 19.6 miles from home   (b) You should hike in the direction 34.6^o west of south

47.   (a)   The tension in each rope is 90.45 pounds.

(b)   The tension in the short rope is 97.1 pounds, The tension in the long rope is 59.2 pounds.

11.2   Selected Answers

13.   dist(A,B)= $\sqrt{5}$, dist(A,C)= $\sqrt{11}$, dist(A,d)= 5 , dist(B,C)= $\sqrt{6}$, dist(B,D)= $\sqrt{8}$, dist(C,D)= $\sqrt{18}$
      No three of these points are colinear.

15.   dist(A,B)= 6 , dist(A,C)= 3 , dist(A,d)= $\sqrt{6}$, dist(B,C)= 9 , dist(B,D)= $\sqrt{50}$, dist(C,D)= $\sqrt{11}$
      The points A, B, and C are colinear.

17.   corners: (1,2,3), (4,2,3), (4,4,3), (4,4,1), (1,4,1). volume = (3)(2)(2) = 12

19.   corners: (1,4,0), (1,5,0), (4,4,0), (4,4,3), (4,5,3). volume = (3)(1)(3) = 9

25. $(x-4)^2+(y-3)^2+(z-5)^2=9$ 26. $x^2+(y-3)^2+(z-6)^2=4$

27. $(x-5)^2+(y-1)^2+z^2=25$ 29.   center (3, -4, 1), radius = 4

30.   center (-2, 0, 4), radius = 5   31.   center (2, 3, 4), radius = 10

33.   empty set (no intersection), a point, a line   34.   empty set (no intersection), a line, a plane

35.   empty set (no intersection), a point, a circle   36.   empty set (no intersection), a point, a circle, a sphere

45.   (a) $\dfrac{16\pi}{3}$ (for half sphere)   (b) $\dfrac{8\pi}{3}$ (for quarter sphere)   (c) $\dfrac{4\pi}{3}$ (for 1/8 sphere)

46.   (a) 18π (b) 9π (c) $\dfrac{9\pi}{2}$

S1.   (1, 2, 0) on xy-plane, (1, 0, 3) on xz-plane, (0, 2, 3) on yz-plane

S2.   (4, 1, 0) on xy-plane, (4, 0, 2) on xz-plane, (0, 1, 2) on yz-plane

S3.   (a, b, 0) on xy-plane, (a, 0, c) on xz-plane, (0, b, c) on yz-plane

S4.   (4, 2, 0) to (1, 3, 0) on xy-plane, (4, 0, 1) to (1, 0, 3) on xz-plane, (0, 2, 1) to (0, 3, 3) on yz-plane

S7.   xy-plane: line segment from (0,0,0) to (4,0,0). yz-plane: line segment from (0,0,0) to (0,0,3),
      xz-plane: triangle with vertices (0,0,0), (4,0,3), and (4,0,2)

S8.   xy-plane: triangle with vertices (1,2,0), (4,3,0), and (2,3,0)
      xz-plane: triangle with vertices (1,0,3), (4,0,1), and (2,0,4)
      yz-plane: triangle with vertices (0,2,3), (0,3,1), and (0,3,4)

S10. (a) 0   (b) 10   S11. (a) 0   (b) 12

11.3   Selected Answers

5. $\mathbf{W} = \langle 6, -3, 18 \rangle$, $|\mathbf{U}| = 7$, $|\mathbf{V}| = 11$, $|\mathbf{W}| = \sqrt{369} \approx 19.21$

{ dir. of $\mathbf{U}$} $= \mathbf{U}/|\mathbf{U}| = \langle 2/7, 3/7, 6/7 \rangle$, { dir. of $\mathbf{V}$} $= \mathbf{V}/|\mathbf{V}| = \langle 2/11, -9/11, 6/11 \rangle$

{ dir. of $\mathbf{W}$} $= \mathbf{W}/|\mathbf{W}| = \langle 6/\sqrt{369}, -3/\sqrt{369}, 18/\sqrt{369} \rangle \approx \langle 0.31, -0.16, 0.94 \rangle$

7. $\mathbf{W} = \langle 14, -3, 32 \rangle$, $|\mathbf{U}| = 15$, $|\mathbf{V}| = 9$, $|\mathbf{W}| = \sqrt{1229} \approx 35.06$

{ dir. of $\mathbf{U}$} $= \mathbf{U}/|\mathbf{U}| = \langle 5/15, 2/15, 14/15 \rangle$, { dir. of $\mathbf{V}$} $= \mathbf{V}/|\mathbf{V}| = \langle 4/9, -7/9, 4/9 \rangle$

{ dir. of $\mathbf{W}$} $= \mathbf{W}/|\mathbf{W}| = \langle 14/\sqrt{1229}, -3/\sqrt{1229}, 32/\sqrt{1229} \rangle \approx \langle 0.40, -0.09, 0.91 \rangle$

9. $\mathbf{W} = \langle 21, 18, -2 \rangle$, $|\mathbf{U}| = 11$, $|\mathbf{V}| = 9$, $|\mathbf{W}| = \sqrt{769} \approx 27.73$

{ dir. of $\mathbf{U}$} $= \mathbf{U}/|\mathbf{U}| = \langle 9/11, 6/11, 2/11 \rangle$, { dir. of $\mathbf{V}$} $= \mathbf{V}/|\mathbf{V}| = \langle 1/3, 2/3, -2/3 \rangle$

{ dir. of $\mathbf{W}$} $= \mathbf{W}/|\mathbf{W}| = \langle 21/\sqrt{769}, 18/\sqrt{769}, -2/\sqrt{769} \rangle \approx \langle 0.76, 0.65, -0.07 \rangle$

11. $\mathbf{W} = \langle 26, 25, -2 \rangle$, $|\mathbf{U}| = 15$, $|\mathbf{V}| = 9$, $|\mathbf{W}| = \sqrt{1305} \approx 36.12$

{ dir. of $\mathbf{U}$} $= \mathbf{U}/|\mathbf{U}| = \langle 10/15, 11/15, 2/15 \rangle$, { dir. of $\mathbf{V}$} $= \mathbf{V}/|\mathbf{V}| = \langle 2/3, 1/3, -2/3 \rangle$

{ dir. of $\mathbf{W}$} $= \mathbf{W}/|\mathbf{W}| = \langle 26/\sqrt{1305}, 25/\sqrt{1305}, -2/\sqrt{1305} \rangle \approx \langle 0.72, 0.69, -0.06 \rangle$

13. $\mathbf{C} = \langle -8, -6, 6 \rangle$ 14. $\mathbf{C} = \langle 4, 1, -4 \rangle$ 15. $\mathbf{C} = \langle -3-e, -9-\pi, -1 \rangle$

17.   smallest magnitude is $\mathbf{C}$, largest magnitude is $\mathbf{D}$

18.   smallest magnitude is $\mathbf{D}$, largest magnitude is $\mathbf{C}$

23. $\langle 0, 1, 0 \rangle$, $\langle 0, 0, 1 \rangle$, $\langle 0, 3, -4 \rangle$ are all perpendicular to $\mathbf{A}$ as is ever non-zero vector with

x-coordinate equal to 0 (= vectors that lie in the yz-plane). There are an infinite number of

nonparallel vectors that are perpendicular to $\mathbf{A}$.

24. $\langle 1, 0, 0 \rangle$, $\langle 0, 3, 0 \rangle$, $\langle 5, -4, 0 \rangle$ are all perpendicular to $\mathbf{B}$ as is ever non-zero vector with

z-coordinate equal to 0 (= vectors that lie in the xy-plane). There are an infinite number of

nonparallel vectors that are perpendicular to $\mathbf{B}$.

25. $\langle 0, 0, 2 \rangle$, $\langle -2, 1, 0 \rangle$, $\langle 2, -1, 7 \rangle$ are all perpendicular to $\mathbf{C}$. There are an infinite number

of nonparallel vectors that are perpendicular to $\mathbf{C}$.

31. $x(t) = 3 + 4t$, $y(t) = 5 - t$, $z(t) = 1$

32. $x(t) = 1 + 4t$, $y(t) = 2 - 2t$, $z(t) = 3 + 2t$

33. $x(t) = 2 + 3t$, $y(t) = 3$, $z(t) = 6 - 5t$

11.4   Selected Answers

1. $\mathbf{A} \bullet \mathbf{B} = 3$, $\mathbf{B} \bullet \mathbf{A} = 3$, $\mathbf{A} \bullet \mathbf{A} = 14$, $\mathbf{A} \bullet (\mathbf{B}+\mathbf{A}) = 17$, and $(2\mathbf{A}+3\mathbf{B}) \bullet (\mathbf{A} - 2\mathbf{B}) = $ -101

2. $\mathbf{A} \bullet \mathbf{B} = 2$, $\mathbf{B} \bullet \mathbf{A} = 2$, $\mathbf{A} \bullet \mathbf{A} = 41$, $\mathbf{A} \bullet (\mathbf{B}+\mathbf{A}) = 43$, and $(2\mathbf{A}+3\mathbf{B}) \bullet (\mathbf{A} - 2\mathbf{B}) = $ -94

3. $\mathbf{U} \bullet \mathbf{V} = 2$, $\mathbf{U} \bullet \mathbf{U} = 41$, $\mathbf{U} \bullet \mathbf{i} = 6$, $\mathbf{U} \bullet \mathbf{j} = $ -1, $\mathbf{U} \bullet \mathbf{k} = 2$, and $(\mathbf{V} +\mathbf{k}) \bullet \mathbf{U} = 8$

4. $\mathbf{U} \bullet \mathbf{V} = 0$, $\mathbf{U} \bullet \mathbf{U} = 22$, $\mathbf{V} \bullet \mathbf{i} = 2$, $\mathbf{V} \bullet \mathbf{j} = 4$, $\mathbf{V} \bullet \mathbf{k} = $ -3, and $(\mathbf{V} +\mathbf{k}) \bullet \mathbf{U} = 2$

5. $\mathbf{S} \bullet \mathbf{T} = $ -3, $\mathbf{T} \bullet \mathbf{U} = $ -4, $\mathbf{T} \bullet \mathbf{T} = 35$, $(\mathbf{S} + \mathbf{T}) \bullet (\mathbf{S} - \mathbf{T}) = $ -14, and $(\mathbf{S} \bullet \mathbf{T})\mathbf{U} = \langle -3, -9, -6 \rangle$

7.   angle between $\mathbf{A}$ and $\mathbf{B}$ is 1.39 ($\approx 79.9^o$), angle between $\mathbf{A}$ and x-axis is 1.30 ($\approx 74.5^o$)

angle between $\mathbf{A}$ and y-axis is 1.01 ($\approx 57.7^o$), angle between $\mathbf{A}$ and z-axis is 0.641 ($\approx 36.7^o$)

9.   angle between $\mathbf{U}$ and $\mathbf{V}$ is 1.51 ($\approx 86.7^o$), angle between $\mathbf{U}$ and x-axis is 0.36 ($\approx 20.4^o$)

angle between $\mathbf{U}$ and y-axis is 1.73 ($\approx 98.98^o$), angle between $\mathbf{U}$ and z-axis is 1.25 ($\approx 71.8^o$)

11.   angle between $\mathbf{S}$ and $\mathbf{T}$ is 1.68 ($\approx 96.3^o$), angle between $\mathbf{S}$ and x-axis is 1.12 ($\approx 64.1^o$)

angle between $\mathbf{S}$ and y-axis is 2.63 ($\approx 150.5^o$), angle between $\mathbf{S}$ and z-axis is 1.35 ($\approx 77.4^o$)

13.   angle between $\mathbf{A}$ and $\mathbf{B}$ is 1.57 ($\approx 90^o$), angle between $\mathbf{A}$ and x-axis is 1.24 ($\approx 71.1^o$)

angle between $\mathbf{A}$ and y-axis is 2.08 ($\approx 119.1^o$), angle between $\mathbf{A}$ and z-axis is 0.624 ($\approx 35.8^o$)

15.   angle between $\mathbf{A}$ and $\mathbf{B}$ is 2.59 ($\approx 148.7^o$), angle between $\mathbf{A}$ and x-axis is 0.38 ($\approx 21.8^o$)

angle between $\mathbf{A}$ and y-axis is 1.95 ($\approx 111.8^o$), angle between $\mathbf{A}$ and z-axis is 1.57 ($\approx 90^o$)

17.   angle between $\mathbf{U}$ and $\mathbf{V}$ is 1.43 ($\approx 81.9^o$), angle between $\mathbf{U}$ and x-axis is 1.25 ($\approx 71.6^o$)

angle between $\mathbf{U}$ and y-axis is 1.57 ($\approx 90^o$), angle between $\mathbf{U}$ and z-axis is 0.322 ($\approx 18.4^o$)

19. 0.124 ($\approx 7.13^o$)   20. 2.48 ($\approx 142.1^o$)   21. 0.785 ($\approx 45^o$)

22. 1.23 ($\approx 70.5^o$)   23. 0.897 ($\approx 51.4^o$)   24. 0 ($= 0^o$)

25. 0 ($= 0^o$)   26. 1.57 ($\approx 90^o$)   27. 0.32 ($\approx 18.4^o$)

29. $\mathbf{N} = \langle -2, -1, 0 \rangle$ is one correct answer.   30. $\mathbf{N} = \langle 3, 0, 5 \rangle$ is one correct answer.

31. $\mathbf{N} = \langle 3, -7 \rangle$ is one correct answer.   32. $\mathbf{N} = \langle 3, 7 \rangle$ is one correct answer.

33. $\mathbf{N} = \langle -1, 1, 1 \rangle$ is one correct answer.   34. $\mathbf{N} = \langle -5, 0, 1 \rangle$ is one correct answer.

36. $\mathbf{N} = \langle 0, 2, -3 \rangle$ is one correct answer.   37. $\mathbf{N} = \langle 1, 1, 2 \rangle$ is one correct answer.

38. $\mathbf{N} = \langle 1, 0, 6 \rangle$ is one correct answer.   39. $\mathbf{N} = \langle 3, 2 \rangle$ is one correct answer.

40. $\mathbf{N} = \langle -2, 3 \rangle$ is one correct answer.   41. $\mathbf{N} = \langle 2, 3 \rangle$ is one correct answer.

43. $N = \langle 3, 2 \rangle$ is one correct answer.   45. $N = \langle 1, -4 \rangle$ is one correct answer.

46. $N = \langle 5, 1 \rangle$ is one correct answer.   47. $N = \langle 0, 3 \rangle$ is one correct answer.

57. $\text{Proj}_{\mathbf{B}}\, \mathbf{A} = \langle 25/34, 0, -15/34 \rangle$, $\text{Proj}_{\mathbf{A}}\, \mathbf{B} = \langle -1, 2, 0 \rangle$

59. $\text{Proj}_{\mathbf{B}}\, \mathbf{A} = \langle 0,0,0 \rangle$, $\text{Proj}_{\mathbf{A}}\, \mathbf{B} = \langle 0,0,0 \rangle$. $\mathbf{A}$ and $\mathbf{B}$ are perpendicular.

61. $\text{Proj}_{\mathbf{B}}\, \mathbf{A} = \langle 0,0,0 \rangle$, $\text{Proj}_{\mathbf{A}}\, \mathbf{B} = \langle 0,0,0 \rangle$. $\mathbf{A}$ and $\mathbf{B}$ are perpendicular.

63. $\text{Proj}_{\mathbf{B}}\, \mathbf{A} = \langle 0,-2,0 \rangle$, $\text{Proj}_{\mathbf{A}}\, \mathbf{B} = \langle -1/7, 2/7, -3/7 \rangle$

65. $\left| \text{Proj}_{\mathbf{B}}\, \mathbf{A} \right| = \left| \dfrac{\mathbf{A} \cdot \mathbf{B}}{|\mathbf{B}|} \right|$, $\text{Proj}_{\mathbf{A}}\, \mathbf{B} = \left| \dfrac{\mathbf{A} \cdot \mathbf{B}}{|\mathbf{A}|} \right| = \left| \dfrac{\mathbf{A} \cdot \mathbf{B}}{3|\mathbf{B}|} \right| = \dfrac{1}{3} \left| \dfrac{\mathbf{A} \cdot \mathbf{B}}{|\mathbf{B}|} \right|$. $\left| \text{Proj}_{\mathbf{B}}\, \mathbf{A} \right|$ is larger.

67.   distance is $3/\sqrt{2} \approx 2.12$ 68.   distance is 3.88   69.   distance is 3.06

70.   distance is 0.95   71.   distance is 1.40

73.   distance is 15.4 feet   74.   distance is 27.76 inches   75.   9.18 m

77.   work $= \dfrac{5000}{\sqrt{141}} \approx 421$ foot-pounds   78.   work $= 842$ foot-pounds

79.   work $= 480$ foot-pounds   80.   work $= 240$ foot-pounds

81.   work $= 1212.1$ foot-pounds   82.   A work $= 2349$ ft-lbs, B work $= 2441$ ft-lbs

83. $\mathbf{A} \bullet \mathbf{B} = 15$, angle $\approx 1.05 \ (= 60^{\circ})$ 84. $\mathbf{A} \bullet \mathbf{B} = 16$, angle $\approx 1.21 \ (\approx 69.3^{\circ})$

85.   angle $\approx 112^{\circ}$, "different"   86.   angle $\approx 19.9^{\circ}$, "very alike"

87.   angle $\approx 9.45^{\circ}$, "very alike"   88.   angle $= 90^{\circ}$, "different"

11.5   Selected Answers

1. 7   2. 7   3. $2x - 5y$ 4. $15 - ab$

5. 1   6. -1 7. -6 8. -42

9. $x + 7y - 5z$ 10. $4a + 10b - 6c$ 11. -77 12. $3x - 3x^2$

13. (a) $\langle -10, -5, 10 \rangle$ (b) 0   (c) 0   (d) 15

15. (a) $\langle -24, -8, 0 \rangle$ (b) 0   (c) 0   (d) $\sqrt{640}$

17. (a) $\langle 3, -11, -5 \rangle$ (b) 0   (c) 0   (d) ≈ 12.45

19. scalar   20. not defined   21. not defined   22. not defined

25. The angle between a vector **A** and itself is 0 so $\mathbf{A}\mathbf{x}\mathbf{A} = |\mathbf{A}|\,|\mathbf{A}|\sin(0) = 0$. Alternately,

   $|\mathbf{A}\mathbf{x}\mathbf{A}|$ = the area of the parallelogram determined by **A** and **A**, and that area is 0.

26. $|\mathbf{A}\mathbf{x}\mathbf{B}| = |\mathbf{A}|\,|\mathbf{B}|\,|\sin(\theta)|$ which is maximum when $|\sin(\theta)| = 1$, and $|\sin(\theta)| = 1$ when $\theta = \pm\pi/2$.

31. (a)   torque = $\mathbf{A}\mathbf{x}\mathbf{B}$ = $\langle 12\cos(-30^o),\ 12\sin(-30^o),\ 0 \rangle \mathbf{x} \langle 0,\ 70,\ 0 \rangle$

   $= \{840\cos(-30^o)\}\mathbf{k}\ \approx 727.46\mathbf{k}$ inch-pounds

   (b)   torque = $\mathbf{A}\mathbf{x}\mathbf{B}$ = $\langle 8\cos(-30^o),\ 8\sin(-30^o),\ 0 \rangle \mathbf{x} \langle 20\cos(40^o),\ 20\sin(40^o),\ 0 \rangle$

   $= \{160\cos(-30^o)\sin(40^o) - 160\sin(-30^o)\cos(40^o)\}\mathbf{k}\ \approx 150.35\mathbf{k}$ inch-pounds

33. Yes. {torque on **A** by **B**} + {torque on **A** by **C**} = $\mathbf{A}\mathbf{x}\mathbf{B} + \mathbf{A}\mathbf{x}\mathbf{C}$ = $\mathbf{A}\mathbf{x}(\mathbf{B}+\mathbf{C})$ = {torque on **A** by (**B**+**C**) }

34. parallelogram area = $|\mathbf{A}\mathbf{x}\mathbf{B}|$ = 18 , triangle area = 9

35. parallelogram area = $|\mathbf{A}\mathbf{x}\mathbf{B}|$ = 6 , triangle area = 3

37. triangle area = $\dfrac{\sqrt{180}}{2} \approx 6.71$ 38. triangle area = $\dfrac{\sqrt{(bc)^2 + (ac)^2 + (ab)^2}}{2}$

39. parallelpiped volume = $|(\mathbf{A}\mathbf{x}\mathbf{B})\bullet\mathbf{C}|$ = 17   40. parallelpiped volume = $|(\mathbf{A}\mathbf{x}\mathbf{B})\bullet\mathbf{C}|$ = 78

41. parallelpiped volume = $|(\mathbf{A}\mathbf{x}\mathbf{B})\bullet\mathbf{C}| = |a|\cdot|b|\cdot|c| = |abc|$ cubic units

42. tetrahedron volume = $\dfrac{1}{6}|36\mathbf{k}|$ = 6 cubic units   43. tetrahedron volume = $\dfrac{15}{6}$ cubic units

44. tetrahedron volume = 8 cubic units   45. tetrahedron volume = $\dfrac{|abc|}{6}$ cubic units

47. $A_{xy} = 4$, $A_{xz} = 4$, $A_{yz} = 8$, and $A_{xyz} = \sqrt{96}$ 48. $A_{xy} = 4$, $A_{xz} = 6$, $A_{yz} = 12$, and $A_{xyz} = \sqrt{196} = 14$

11.6   Selected Answers

1. $x(t) = 2 + 3t$, $y(t) = -3 + 4t$, $z(t) = 1 + 2t$ 3. $x(t) = -2 + 5t$, $y(t) = 1$, $z(t) = 4 - 3t$

5. $x(t) = 2 + t$, $y(t) = -1 + 5t$, $z(t) = 3 - 5t$ 7. $x(t) = 3$, $y(t) = -2 + 6t$, $z(t) = 1 - 2t$

9.   Lines intersect at the point $(2, -1, 3)$ when $t = 0$, $\theta = \arccos\left(\dfrac{9}{\sqrt{6}\sqrt{21}}\right) \approx \arccos(0.802) \approx 0.604 \ (\approx 36.7^{\circ})$

11. $L(0) = (1, 5, -2) = K(-2_ $. The lines intersect at the point $(1, 5, -2)$.
$\theta = \arccos\left(\dfrac{18}{5\sqrt{14}}\right) \approx \arccos(0.962) \approx 0.277 \ (\approx 15.8^{\circ})$

13. $5(x-2) + (-2)(y-3) + 4(z-1) = 0$ or $5x - 2y + 4z = 8$ 14. $3x + y - 5z = 22$

15. $0(x + 3) + 3(y - 5) + 0(z-6) = 0$ or $3y = 15$ 16. $2x - 2y + z = 0$

17. $(-6)(x-1) + (-6)(y-2) + (-12)(z-3) = 0$ or $x + y + 2z = 9$ 18. $y = 5$

19. $20x + 28y + 25z = 101$ 20. $-x + 2y - z = 0$

21. $z = 7$ 22. $x = 2$ 23. $3x - 2y + 5z = 23$

24. $2x + 3y - z = 0$ 25. $5x - 3y + 2z = 23$ 26. $y = 7$

27.   They intersect along the y-axis   29.   Plane intersects the x-axis at $x=10$

31.   They intersect at the point $(4, 2, 1)$ 32.   y-axis never intersects the plane $z = 3$

33. $x(t) = -26 + t$, $y(t) -57 + 3t$, $z(t) = t$. $\mathbf{N_1} = \langle 4, -2, 2 \rangle$, $\mathbf{N_2} = \langle 3, -2, 3 \rangle$
$\theta = \arccos\left(\dfrac{22}{\sqrt{24}\sqrt{22}}\right) \approx \arccos(0.957) \approx 0.294 \ (\approx 16.8^{\circ})$

34. $x(t) = 9 + 6t$, $y*(t) = -9t$, $z(t) = t$. $\theta \approx 1.066 \ (\approx 61.1^{\circ})$

35. $x(t) = 12 - \dfrac{22}{5}t$, $y(t) = 2 + \dfrac{1}{5}t$, $z(t) = t$. $\mathbf{N_1} = \langle 0, 5, -1 \rangle$, $\mathbf{N_2} = \langle 1, 2, 4 \rangle$
$\theta = \arccos\left(\dfrac{6}{\sqrt{26}\sqrt{21}}\right) \approx 1.311 \ (\approx 75.1^{\circ})$

37.   They intersect at the point $(-11/3, -1/3, -58/3)$. $\arccos\left(\dfrac{3}{\sqrt{30}\sqrt{21}}\right) \approx 1.451 \ (\approx 83.1^{\circ})$ so the
angle of intersection is $\theta = \pi/2 - 1.451 \approx 0.120 \ (\approx 6.9^{\circ})$

38.   They intersect at the point $(0, 6, -7)$. Angle of intersection is approximately $0.222 \ (\approx 12.7^{\circ})$

39.   They intersect at the point $(0, 8, 5)$. Angle of intersection is approximately $0.271 \ (\approx 15.5^{\circ})$

41.   Yes   42.   Yes   43. $x(t) = -7t$, $y(t) = 4t$, $z(t) = 3 + 9t$

44.   (a) $\approx 1.30 \ (\approx 74.5^{\circ})$ (b) $\approx 1.01 \ (\approx 57.7^{\circ})$ (c) $\approx 0.64 \ (\approx 36.7^{\circ})$

45. $\theta = \arccos\left(\dfrac{a}{\sqrt{a^2 + b^2 + c^2}}\right)$ with the xy-plane. $\theta = \arccos\left(\dfrac{b}{\sqrt{a^2 + b^2 + c^2}}\right)$ with the xz-plane.

$\theta = \arccos\left(\dfrac{c}{\sqrt{a^2 + b^2 + c^2}}\right)$ with the yz-plane.

47.   (a) $\theta = \arctan(481/378) \approx 0.905$ $(\approx 51.8^\circ)$ (b) $\cos(\varphi) \approx 0.382$ so $\varphi \approx 1.179$ $(\approx 84.8^\circ)$

(c) $\alpha = \arctan(481/534.6) \approx 0.733$ $(\approx 42.0^\circ)$

51.   distance ≈ 4.879 52.   distance ≈ 2.145 53.   distance ≈ 1.18 54.   distance ≈ 6.164

55.   distance ≈ 1.32 56.   distance ≈ 1.35

57.   (a)   The objects "crash" at the point (15, 24, 13) when t=6.   (b) Paths intersect fo {min. dist.} = 0

(c)   No, the objects crash, and their paths intersect.

58.   (a)   The objects "crash" at the point (2, 3, 4) when t=3.

59.   (a)   The objects do not crash. They are never at the same point at the same time.

(b)   The paths of the objects intersect: object A is at (0,1,5) when t=1 and B is at (0,1,5) when t=2.

(c)   {minimum distance between objects} ≈ 0.85 , {min. distance between paths} = 0 since paths intersect.

61.   (a)   The objects do not crash. They are never at the same point at the same time.

(b)   The paths of the objects do not intersect.

(c)   {minimum distance between objects} ≈ 2.79 , {min. distance between paths} = 1

63.   (a)   Shortest distance between the airplane and the car is $\sqrt{58} \approx 7.62$.

(b)   Shortest distance between paths is $13/\sqrt{5} \approx 5.81$.

Calculus Reference Facts

Derivatives

Notation: $D(f) = Df$ represents the derivative with respect to x of the function f(x)

$$D(f) = Df = \frac{d}{dx} f(x) = f'(x)$$

$D(k) = 0$ k a constant $D(f + g) = Df + Dg$ $D(f \cdot g) = f \cdot Dg + g \cdot Df$

$D(k \cdot f) = k \cdot Df$ k a constant $D(f - g) = Df - Dg$ $D(f/g) = \dfrac{g \cdot Df - f \cdot Dg}{g^2}$

$D(f^n) = n f^{n-1} \cdot Df$

$D(f(g(x))) = f'(g(x)) \cdot Dg$ (Chain Rule)

$D(\sin(f)) = \cos(f) \cdot Df$ $D(\tan(f)) = \sec^2(f) \cdot Df$ $D(\sec(f)) = \sec(f) \cdot \tan(f) \cdot Df$

$D(\cos(f)) = -\sin(f) \cdot Df$ $D(\cot(f)) = -\csc^2(f) \cdot Df$ $D(\csc(f)) = -\csc(f) \cdot \cot(f) \cdot Df$

$D(e^f) = e^f \cdot Df$ $D(a^f) = a^f \ln(a) \cdot Df$

$D(\ln|f|) = \dfrac{1}{f} \cdot Df$ Logrithmic Differentiation: $Df = f \cdot D(\ln|f|)$

$D(\log_a |f|) = \dfrac{1}{f \cdot \ln(a)} \cdot Df$

$D(\arcsin(f)) = \dfrac{1}{\sqrt{1 - f^2}} \cdot Df$ $D(\arctan(f)) = \dfrac{1}{1 + f^2} \cdot Df$ $D(\text{arcsec}(f)) = \dfrac{1}{|f|\sqrt{f^2 - 1}} \cdot Df$

$D(\arccos(f)) = \dfrac{-1}{\sqrt{1 - f^2}} \cdot Df$ $D(\text{arccot}(f)) = \dfrac{-1}{1 + f^2} \cdot Df$ $D(\text{arccsc}(f)) = \dfrac{-1}{|f|\sqrt{f^2 - 1}} \cdot Df$

Integrals — General Properties

$$\int k \cdot f \, dx = k \cdot \int f \, dx \qquad\qquad \int (f - g) \, dx = \int f \, dx - \int g \, dx$$

$$\int (f + g) \, dx = \int f \, dx + \int g \, dx \qquad\qquad \int f \cdot (g') \, dx = f \cdot g - \int g \cdot (f') \, dx$$

Integrals — Particular Functions

1. $\int k \, dx = k \cdot x + C$

2. $\int x^n \, dx = \frac{1}{n+1} \cdot x^{n+1} + C$ if $n \neq -1$ $\qquad$ $\int x^{-1} \, dx = \int \frac{1}{x} \, dx = \ln|x| + C$

3. $\int (ax+b)^n \, dx = \frac{1}{n+1} \cdot \frac{1}{a} \cdot (ax+b)^{n+1} + C \quad n \neq -1$

    (a) $\int (ax+b)^{-1} \, dx = \int \frac{1}{ax+b} \, dx = \frac{1}{a} \cdot \ln|ax+b| + C$

    (b) $\int \sqrt{ax+b} \, dx = \frac{2}{3a} \cdot (ax+b)^{3/2} + C$

    (c) $\int \frac{1}{\sqrt{ax+b}} \, dx = \frac{2}{a} \cdot (ax+b)^{1/2} + C$

4. $\int \frac{1}{x(ax+b)} \, dx = \frac{1}{b} \cdot \ln\left|\frac{x}{ax+b}\right| + C$

5. $\int \frac{1}{(x+a)(x+b)} \, dx = \frac{1}{b-a}\{\ln|x+a| - \ln|x+b|\} + C = \frac{1}{b-a} \ln\left|\frac{x+a}{x+b}\right| + C \quad a \neq b$

    (a) $\int \frac{1}{(x+a)(x+a)} \, dx = \int \frac{1}{(x+a)^2} \, dx = -\frac{1}{x+a} + C$

6. $\int x(ax+b)^n \, dx = \frac{(ax+b)^{n+1}}{a} \cdot \left\{\frac{ax+b}{n+2} - \frac{b}{n+1}\right\} + C \quad n \neq -1, -2$

    (a) $\int x(ax+b)^{-1} \, dx = \int \frac{x}{ax+b} \, dx = \frac{x}{a} - \frac{b}{a^2} \cdot \ln|ax+b| + C$

    (b) $\int x(ax+b)^{-2} \, dx = \int \frac{x}{(ax+b)^2} \, dx = \frac{1}{a^2} \cdot \left\{\ln|ax+b| + \frac{b}{ax+b}\right\} + C$

7. $\int \sin(ax) \, dx = -\frac{1}{a}\cos(ax) + C$ $\qquad\qquad$ 8. $\int \cos(ax) \, dx = \frac{1}{a}\sin(ax) + C$

9. $\int \tan(ax) \, dx = \int \frac{\sin(ax)}{\cos(ax)} \, dx = -\frac{1}{a}\ln|\cos(ax)| + C = \ln|\sec(ax)| + C$

10. $\int \cot(ax) \, dx = \int \frac{\cos(ax)}{\sin(ax)} \, dx = \frac{1}{a}\ln|\sin(x)| + C$

11. $\int \sec(ax) \, dx = \frac{1}{a}\ln|\sec(ax) + \tan(ax)| + C$ $\qquad$ 12. $\int \csc(ax) \, dx = \frac{1}{a}\ln|\csc(ax) - \cot(ax)| + C$

13. $\int \sin^2(ax) \, dx = \frac{x}{2} - \frac{\sin(2ax)}{4a} + C = \frac{x}{2} - \frac{\sin(ax)\cos(ax)}{2a} + C$

14. $\int \cos^2(ax)\, dx = \dfrac{x}{2} + \dfrac{\sin(2ax)}{4a} + C = \dfrac{x}{2} + \dfrac{\sin(ax)\cos(ax)}{2a} + C$

15. $\int \tan^2(ax)\, dx = \dfrac{1}{a}\tan(ax) - x + C$

16. $\int \cot^2(ax)\, dx = -\dfrac{1}{a}\cot(ax) - x + C$

17. $\int \sec^2(ax)\, dx = \dfrac{1}{a}\tan(ax) + C$

18. $\int \csc^2(ax)\, dx = -\dfrac{1}{a}\cot(ax) + C$

19. $\int \sin^n(ax)\, dx = \dfrac{-\sin^{n-1}(ax)\cos(ax)}{na} + \dfrac{n-1}{n}\int \sin^{n-2}(ax)\, dx$

    (a) $\int \sin^3(ax)\, dx = \dfrac{-\sin^2(ax)\cos(ax)}{3a} - \dfrac{2}{3a}\cos(ax) + C$

    (b) $\int \sin^4(ax)\, dx = \dfrac{-\sin^3(ax)\cos(ax)}{4a} + \dfrac{3x}{8} - \dfrac{3\sin(ax)\cos(ax)}{8a} + C$

20. $\int \cos^n(ax)\, dx = \dfrac{\cos^{n-1}(ax)\sin(ax)}{na} + \dfrac{n-1}{n}\int \cos^{n-2}(ax)\, dx$

21. $\int \tan^n(ax)\, dx = \dfrac{\tan^{n-1}(ax)}{(n-1)a} - \int \tan^{n-2}(ax)\, dx \qquad n \neq 1$

    (a) $\int \tan^3(ax)\, dx = \dfrac{1}{2a}\tan^2(ax) + \dfrac{1}{a}\ln|\cos(ax)| + C$

    (b) $\int \tan^4(ax)\, dx = \dfrac{1}{3a}\tan^3(ax) - \dfrac{1}{a}\tan(ax) + C$

22. $\int \cot^n(ax)\, dx = -\dfrac{\cot^{n-1}(ax)}{n-1} - \int \cot^{n-2}(ax)\, dx \qquad n \neq 1$

23. $\int \sec^n(ax)\, dx = \dfrac{\sec^{n-2}(ax)\tan(ax)}{(n-1)a} + \dfrac{n-2}{n-1}\int \sec^{n-2}(ax)\, dx \quad n \neq 1$

    (a) $\int \sec^3(ax)\, dx = \dfrac{1}{2a}\sec(ax)\tan(ax) + \dfrac{1}{2a}\ln|\sec(ax) + \tan(ax)| + C$

    (b) $\int \sec^4(ax)\, dx = \dfrac{1}{3a}\sec^2(ax)\tan(ax) + \dfrac{2}{3a}\tan(ax) + C$

24. $\int \csc^n(ax)\, dx = -\dfrac{\csc^{n-2}(ax)\cot(ax)}{(n-1)a} + \dfrac{n-2}{n-1}\int \csc^{n-2}(ax)\, dx \quad n \neq 1$

25. $\int \sin(ax)\sin(bx)\, dx = \dfrac{\sin((a-b)x)}{2(a-b)} - \dfrac{\sin((a+b)x)}{2(a+b)} + C \qquad a^2 \neq b^2$

26. $\int \cos(ax)\cos(bx)\, dx = \dfrac{\sin((a-b)x)}{2(a-b)} + \dfrac{\sin((a+b)x)}{2(a+b)} + C \qquad a^2 \neq b^2$

27. $\int \sin(ax)\cos(bx)\, dx = \dfrac{-\cos((a+b)x)}{2(a+b)} - \dfrac{\cos((a-b)x)}{2(a-b)} + C \qquad a^2 \neq b^2$

28. $\int x^n \sin(ax)\, dx = -\frac{1}{a} x^n \cos(ax) + \frac{n}{a^2} \int x^{n-1} \cos(ax)\, dx$

29. $\int x^n \cos(ax)\, dx = \frac{1}{a} x^n \sin(ax) - \frac{n}{a^2} \int x^{n-1} \sin(ax)\, dx$

30. $\int e^{ax}\, dx = \frac{1}{a} e^{ax} + C$ 31. $\int b^{ax}\, dx = \frac{b^{ax}}{a \cdot \ln(b)} + C$

32. $\int x^n e^{ax}\, dx = \frac{1}{a} x^n e^{ax} - \frac{n}{a} \int x^{n-1} e^{ax}\, dx + C$

    (a) $\int x e^{ax}\, dx = \frac{1}{a} x e^{ax} - \frac{1}{a^2} e^{ax} = \frac{e^{ax}}{a^2}(ax - 1) + C$

    (b) $\int x^2 e^{ax}\, dx = \frac{1}{a} x^2 e^{ax} - \frac{2}{a^2} x e^{ax} + \frac{2}{a^3} e^{ax} = \frac{e^{ax}}{a^3}(a^2 x^2 - 2ax + 2) + C$

33. $\int x^n b^{ax}\, dx = \frac{1}{\ln(b)}\left\{ \frac{1}{a} x^n b^{ax} - \frac{n}{a} \int x^{n-1} b^{ax}\, dx \right\} + C$

34. $\int \frac{1}{\sqrt{a^2 - x^2}}\, dx = \arcsin\left(\frac{x}{a}\right) + C$ 35. $\int \frac{1}{a^2 + x^2}\, dx = \frac{1}{a} \cdot \arctan\left(\frac{x}{a}\right) + C$

36. $\int \frac{1}{|x|\sqrt{x^2 - a^2}}\, dx = \frac{1}{a} \cdot \text{arcsec}\left(\frac{x}{a}\right) + C$ 37. $\int \frac{1}{a^2 - x^2}\, dx = \frac{1}{2a} \ln\left|\frac{x + a}{x - a}\right| + C$

38. $\int \ln(x)\, dx = x \cdot \ln(x) - x + C$ 39. $\int x^n \cdot \ln(x)\, dx = \frac{x^{n+1}}{(n + 1)^2}\left\{(n+1) \cdot \ln(x) - 1\right\} + C$

40. $\int \frac{1}{x \cdot \ln(x)}\, dx = \ln|\ln(x)| + C$

41. $\int e^{ax} \sin(nx)\, dx = \frac{e^{ax}}{a^2 + n^2}\left\{ a \cdot \sin(nx) - n \cdot \cos(nx) \right\} + C$

42. $\int e^{ax} \cos(nx)\, dx = \frac{e^{ax}}{a^2 + n^2}\left\{ a \cdot \cos(nx) + n \cdot \sin(nx) \right\} + C$

43. $\int \frac{1}{\sqrt{x^2 \pm a^2}}\, dx = \ln|x + \sqrt{x^2 \pm a^2}| + C$

44. $\int \sqrt{x^2 \pm a^2}\, dx = \frac{x}{2}\sqrt{x^2 \pm a^2} + \frac{1}{2} a^2 \ln|x + \sqrt{x^2 \pm a^2}| + C$

45. $\int x^2 \sqrt{x^2 \pm a^2}\, dx = \frac{1}{8} x \cdot (2x^2 \pm a^2)\sqrt{x^2 \pm a^2} - \frac{1}{8} a^4 \ln|x + \sqrt{x^2 \pm a^2}| + C$